Fruits and Vegetables Technologies

OP Chauhan
Editor

Fruits and Vegetables Technologies

Postharvest Processing and Packaging

Springer

Editor
OP Chauhan
Fruits and Vegetables Technology
Defence Food Research Laboratory, Ministry of Defence, DRDO
Mysore, Karnataka, India

ISBN 978-981-96-8432-8 ISBN 978-981-96-8433-5 (eBook)
https://doi.org/10.1007/978-981-96-8433-5

This Springer imprint is published by the registered company Springer Nature Singapore Pte Ltd.
The registered company address is: 152 Beach Road, #21-01/04 Gateway East, Singapore 189721, Singapore

Dedicated to My Loving Family

Preface

The global imperative to minimize food loss and waste has placed increasing emphasis on the critical role of postharvest technologies in ensuring food security and maximizing the nutritional and economic value of fresh produce. Fruits and vegetables, being highly perishable commodities, necessitate careful handling and appropriate preservation strategies from the moment of harvest until they reach the consumer. This book, *Fruits and Vegetables Technologies: Postharvest Processing and Packaging*, aims to provide a comprehensive and up-to-date resource on the multifaceted field of postharvest management.

Recognizing the diverse challenges and opportunities within this domain, this volume brings together the collective expertise of eminent personalities actively contributing to advancements in postharvest technology. The chapters within these pages delve into the intricacies of various postharvest operations, commencing with best practices for handling, harvesting, and transportation to minimize physical damage and maintain quality. It further explores a wide spectrum of storage technologies, ranging from traditional methods to sophisticated controlled and modified atmosphere environments designed to extend the shelf life of a diverse array of fruits and vegetables.

A significant portion of the book is dedicated to fruits and vegetables processing and preservation techniques. It meticulously covers fundamental unit operations alongside established methods such as dehydration, low-temperature preservation, freezing, canning, etc. Furthermore, it sheds light on the rapidly evolving landscape of non-thermal processing technologies and the growing importance of minimal processing techniques in delivering freshlike quality. Recognizing the crucial role of packaging in maintaining produce integrity, the book provides in-depth coverage of conventional packaging methods as well as the latest innovations in modified and controlled atmosphere packaging and storage. This book also addresses the processing of fruits and vegetables into value-added products. Dedicated chapters explore the technologies involved in the production of popular items such as jams, jellies, marmalade, juices, sauces, ketchups, etc., highlighting key processing parameters and quality considerations.

By encompassing both fundamental principles and recent breakthroughs, *Fruits and Vegetables Technologies: Postharvest Processing and Packaging* serves as a comprehensive resource for a broad audience. It is envisioned to be an invaluable guide for students seeking a thorough understanding of the subject, a rich source of information for researchers exploring new frontiers in postharvest science, and a practical reference for faculty involved in teaching and curriculum development. Moreover, food professionals across the supply chain, from producers to processors and retailers, will find this book a vital tool in navigating the complexities of preserving and delivering high-quality fruits and vegetables to consumers. We sincerely hope that this compilation will contribute significantly to the advancement and application of effective postharvest technologies, ultimately contributing to a more sustainable and efficient food system.

Mysore, Karnataka, India OP Chauhan

Contents

Contributors

Aashika Ashok Defence Food Research Laboratory, DRDO, Mysore, India

Mazia Ahmed Centre of Food Technology, University of Allahabad, Prayagraj, Allahabad, India

Bindvi Arora Division of Food Science and Postharvest Technology, ICAR-Indian Agricultural Research Institute, New Delhi, India

Mridusmita Barman Department of Food Engineering & Technology, Tezpur University, Tezpur, Assam, India

Vasundra Vijay Bhat Department of Applied Agriculture, Central University of Punjab, Bathinda, Punjab, India

Anu Chandel Defence Food Research Laboratory, Mysore, India

OP Chauhan Defence Food Research Laboratory, Defence Research and Development Organization, Mysore, Karnataka, India

Poonam Choudhary Agricultural Structures and Environmental Control Division, ICAR-Central Institute of Post-Harvest Engineering and Technology, Ludhiana, Punjab, India

Shreya Chaudhary Department of Food Science and Technology, National Institute of Food Technology Entrepreneurship and Management, Kundli, Sonipat, Haryana, India

Madhuresh Dwivedi Department of Food Process Engineering, National Institute of Technology, Rourkela, India

Prasoon Gunjan Division of Food Science and Postharvest Technology, ICAR-Indian Agricultural Research Institute, New Delhi, India

Unaiza Iqbal Centre of Food Technology, University of Allahabad, Prayagraj, India

S. J. Kale ICAR-National Institute of Secondary Agriculture, Ranchi, Jharkhand, India

Thejesh Kathirvelu Defence Institute of Bio-defence Technologies, Defence Research and Development Organization, Mysore, Karnataka, India

Devinder Kaur Centre of Food Technology, University of Allahabad, Allahabad, Uttar Pradesh, India

Manpreet Kaur Department of Applied Agriculture, Central University of Punjab, Bathinda, Punjab, India

Nita Khandekar ICAR-Central Institute of Agricultural Engineering, Bhopal, Madhya Pradesh, India

S. Ganga Kishore Department of Food Process Engineering, National Institute of Technology, Rourkela, India

Pankaj Kumar Transfer of Technology Division, ICAR-Central Institute of Post-Harvest Engineering and Technology, Ludhiana, Punjab, India

Radha Kushwaha Centre of Food Technology, University of Allahabad, Allahabad, Uttar Pradesh, India

S. G. Lekshmi Division of Food Science and Postharvest Technology, ICAR – Indian Agricultural Research Institute, New Delhi, India

Poonam Mishra Department of Food Engineering & Technology, Tezpur University, Tezpur, Assam, India

Shivangi Mishra National Institute of Food Technology Entrepreneurship and Management, Kundli, Sonipat, Haryana, India

M. K. Mrithula Mahalakshmi Defence Institute of Bio-defence Technologies, Defence Research and Development Organization, Mysore, Karnataka, India

Prerna Nath ICAR-Research Complex for Eastern Region, RS, FSRCHPR, Ranchi, Jharkhand, India

Swarajya Laxmi Nayak Medicinal and Aromatic Plants Research, Anand, Boriavi, Gujarat, India

Pradeep Singh Negi Fruit and Vegetable Technology Department, CSIR-Central Food Technological Research Institute, Mysuru, India

Singamayum Firdosh Nesha Department of Food Process Engineering, National Institute of Technology, Rourkela, India

Arun Kumar Pandey Amity Institute of Biotechnology, Amity University Rajasthan, Jaipur, India

Neha Pandey ICAR-National Soybean Research Institute, Indore, India

Vijay Paul Division of Plant Physiology, ICAR – Indian Agricultural Research Institute, New Delhi, India

C. P. Rakshith Patel Department of Food Process Engineering, National Institute of Technology, Rourkela, India

Rudragoud Policegoudra Defence Food Research Laboratory, DRDO, Mysore, India

Havalli Bommegowda Rashmi Department of Postharvest Management, College of Horticulture, University of Horticultural Sciences, Bagalkot, Mysuru, India

Pinki Saini Centre of Food Technology, University of Allahabad, Prayagraj, India

A. Sangeeta Department of Food Engineering & Technology, Tezpur University, Tezpur, Assam, India

Reshma Saroj Centre of Food Technology, University of Allahabad, Allahabad, Uttar Pradesh, India

Shruti Sethi Division of Food Science and Postharvest Technology, ICAR – Indian Agricultural Research Institute, New Delhi, India

Swati Sethi Food Grains & Oilseeds Processing Division, ICAR-Central Institute of Post-Harvest Engineering and Technology, Ludhiana, Punjab, India

Sakshi Sharma Defence Food Research Laboratory, Mysore, India

Arpit Shrivastava Amity Institute of Biotechnology, Amity University Rajasthan, Jaipur, India

Akash Deep Shukla Department of Applied Agriculture, Central University of Punjab, Bathinda, Punjab, India

P. M. Sindhu Division of Food Science and Postharvest Technology, ICAR – Indian Agricultural Research Institute, New Delhi, India

Heena Singh Department of Food Science and Technology, National Institute of Food Technology Entrepreneurship and Management, Kundli, Sonipat, Haryana, India

Rakhi Singh Department of Food Science and Technology, National Institute of Food Technology Entrepreneurship and Management, Kundli, Sonipat, Haryana, India

Yashi Srivastava Department of Applied Agriculture, Central University of Punjab, Bathinda, Punjab, India

Rahul Vashishth Department of Biosciences, School of Biosciences and Technology, Vellore Institute of Technology, Vellore, India

Janifer Raj Xavier Defence Institute of Bio-defence Technologies, Defence Research and Development Organization, Mysore, Karnataka, India

Sangeeta Yadav Centre of Food Technology, University of Allahabad, Prayagraj, India

About the Editor

OP Chauhan is currently working as Scientist "G" and Head of Fruits and Vegetables Technology Department at Defence Food Research Laboratory, Mysore, India. Dr. Chauhan holds M.Sc. and Ph.D. in Food Technology from G B Pant University of Agriculture and Technology, Pantnagar. His research interest includes postharvest handling of fruits and vegetables including high-pressure processing, pulsed electric field processing, microwave processing, thermal processing, modified atmosphere packaging, etc. His research findings have appeared in more than 150 international and national peer-reviewed journals. He has edited a book on "Non-thermal Processing of Foods" published by CRC Press, USA; a book on Advances in Food Chemistry—Food Components, Processing and Preservation published by Springer, Singapore; a book on "Food Science" published by NIPA, India; and a book on Fruits and Vegetables Technologies—Postharvest Processing and Packaging published by Springer, Singapore. He has published 38 book chapters as well. He also has 17 patents to his credit. He has presented more than 100 papers in various national and international conferences. He has served as the chief editor of *American Journal of Food Technology* and editorial board member of *Journal of Food Science and Technology*. Presently, he is serving as chief editor of *Journal of Food and Agriculture Research*, *Indian Food Industry*, *International Journal of Food and*

Fermentation Technology, *Pantnagar Journal of Research*, etc. Dr. Chauhan has supervised several Ph.D., M.Sc., M.Tech., and B.Tech. students. He is a recipient of INSA fellowship to work at the German Institute of Food Technologies (DIL), Quakenbruck, Germany, on Advanced Food Technologies besides being trained on high-pressure processing technology in UK. He has been instrumental in organizing several conferences, workshops, training programs, courses, etc. He has transferred 26 technologies to various firms for commercialization. Dr. Chauhan is the recipient of DRDO Young Scientist Award, AFSTI Young Scientist Award, Laboratory Scientist of the Year Award, Laljee Godhoo Smarak Nidhi Award (AFSTI), DRDO Technology Group Award, FICCI Best Postharvest Technology Innovation Award, DRDO Technology Spin-off Award, DRDO Technology Absorption Award, Bioved Fellowship, Dr JS Pruthi Award (AFSTI & AIFPA), National Science day medal, National Technology day medal, as well as several Best Paper and Poster Awards from different associations. He is Life Member of Association of Food Scientists and Technologists (India), Nutrition Society of India, and Indian Science Congress.

Chapter 1
Structure and Composition of Fruits and Vegetables

Arun Kumar Pandey, Arpit Shrivastava, Rahul Vashishth, and OP Chauhan

1.1 Introduction

Fruits and vegetables are essential parts of the human diet. The significance of studying the structure and composition of fruits and vegetables is to gain comprehensive insights into their physical characteristics, chemical composition, and nutritional value. This guides the development of postharvest techniques that preserve freshness and extend shelf-life (Rahman 2013). This examination provides a deeper understanding of the biological processes, functional properties, and potential applications of essential food sources. It also enables the refinement of food processing methods to retain or improve taste, texture, and nutritional content in food. Additionally, it informs farming practices to enhance crop yield, quality, and ensure food safety by managing microbial contamination to prevent spoilage and illness (Brasil and Siddiqui 2018).

In botanical terms, fruits are made of different combinations of tissues, such as an ovary that has grown larger, seeds, and other plant parts like receptacles (like those of apples, strawberries, and cashew apples), bracts, and peduncles (like those of pineapples). The word "fruit" derives from the Latin word "fructus," which means to enjoy and produce, whereas the dictionary definition of fruit is "the edible product of a plant or tree, consisting of the seed and its envelope, especially the latter when juicy and pulpy." The refined consumer-friendly definition of fruit is:

A. K. Pandey (✉) · A. Shrivastava
Amity Institute of Biotechnology, Amity University Rajasthan, Jaipur, India

R. Vashishth
Department of Biosciences, School of Biosciences and Technology, Vellore Institute of Technology, Vellore, India

OP Chauhan
Defence Food Research Laboratory, DRDO, Mysore, India

OP Chauhan (ed.), *Fruits and Vegetables Technologies*,
https://doi.org/10.1007/978-981-96-8433-5_1

"Plant products with aromatic flavors, which are either naturally sweet or normally sweetened before eating"; they are essentially dessert foods.

However, the concept of a vegetable can be understood in two distinct ways. One perspective views a vegetable as any plant that is grown for the parts of it that we eat; the other views a vegetable as those specific parts of a plant that we consume, such as celery stalks, carrot roots, potato tubers, onion bulbs, spinach leaves, artichoke flowers, and the fruits or seeds of various plants like apples, cucumbers, pumpkins, strawberries, tomatoes, beans, and peas. This broader interpretation views fruits as a type of vegetable. The term "vegetable" itself originates from a Latin word meaning "to grow" or "to thrive," reflecting the nature of plant life (Online Etymology Dictionary 2021).

The latter definition includes fruits as a subset of vegetables. The word "vegetable" comes from the Medieval Latin 'vegetabilis,' meaning "growing, or flourishing," which refers to plant growth (Ebabhi and Adebayo 2022).

The scope of the current chapter encompasses the understanding of the physical characteristics, internal structures, and external features of various fruits and vegetables across different species and cultivars, the macronutrient and micronutrient profiles, including carbohydrates, proteins, lipids, vitamins, minerals, antioxidants, and fiber content, metabolic processes, enzymatic reactions, and biochemical pathways involved in fruit ripening, postharvest physiology, and flavor development, and the exploration of innovative technologies, such as spectroscopy, chromatography, microscopy, and molecular biology techniques, for analyzing and characterizing fruits and vegetables. These insights can be implemented in the development of novel food products, functional ingredients, and value-added formulations. Therefore, by delineating the purpose and scope of studying the structure and composition of fruits and vegetables, researchers can focus their efforts effectively, generate meaningful insights, and contribute to advancements in agriculture, food science, nutrition, and public health.

1.2 Types and Classification of Fruits and Vegetables

1.2.1 Fruits

Fruits are traditionally classified into three main categories based on their developmental origin:

- Simple fruits
- Aggregate fruits (etaerio)
- Composite fruits (multiple)

1.2.1.1 Simple Fruits

Mature from monocarpellary ovary or multicarpellary syncarpous ovary.

(A) *Dry Fruits*

The pericarp remains dry and undifferentiated into three layers viz.:

(i) *Dehiscent or capsular*

(a) *Legume or pod:* the fruit emerges from both sutures and grows from a monocarpellary superior ovary with marginal placentation. The ovary is unilocular with numerous ovules, such as capsules (*poppy, okra*) or pepo (Cucumber, Pumpkin).

(b) *Follicle:* the fruit develops by a single suture and emerges from a superior, unilocular ovary, which is typically found in pairs or groups. Examples include milkweed, peonies *(Paeonia*), *calotropis*, and larkspur (*consolida*).

(c) *Siliqua:* the fruit is produced by a bicarpellary, syncarpous, superior ovary with parietal placentation. The ovary is initially unilocular but later becomes bilocular because of the production of a false septum known as a replum. Each locule contains numerous seeds, and the fruit develops from underneath to the top when mature. An illustration of a Cruciferae family would typically include mustard, radish, and turnip.

(d) *Silicula:* it is broad and flat and resembles siliqua, but its length and breadth are equal. Examples are shephard's purse (*Capsellabursa pastoris*) and candytuft (*Iberisamara*).

(e) *Capsule:* this gland arises from a superior, syncarpous, multicarpellary ovary with numerouslocules and axile placentation. These fruits are further categorized based on their mechanism of dehiscence as follows:

- Porous dehiscence of poppy (*Papaver*)
- Loculicidal dehiscence of cotton (*Gossypium*) and Okra (*Hibiscus esculentus*)
- *Septicidal dehiscence* in *Aristolochia*
- *Septifragal dehiscence* in *Datura*
- Transverse dehiscence of *Celosia*

(ii) *Indehiscent or achenial*: fruits do not rupture at maturity, and seeds remain enclosed within the pericarp.

(a) *Achene:* the fruit matures from a monocarpellary superior ovary; the pericarp does not fuse with the seed coat, and the seeds are unilocular and single-seeded. The examples are *Naravelia* spp. and *Clematis* spp.

(b) *Caryopsis:* the berry fruit has a fleshy pericarp and develops from a monocarpellary or multicarpellary syncarpous ovary. Examples of berry fruits include tomatoes, grapes, and bananas.

(c) *Cypsella:* the fruits are unilocular, single-seeded, and develop from a syncarpous, bicarpellary, inferior ovary with basal placentation. The fruit apex's is where the persistent, hairy calyx, or pappus, develops.

Examples include representatives of the Compositae family (*Taraxacum* and *Cosmos*).

(d) *Nut:* fruits grow from a superior ovary that is syncarpous, unilocular, multicarpellary, single-seeded, and has a rigid, stoney pericarp. Cashew nut (*Anacardium*), Water chestnut (*Trapa*), and Litchi (*Litchi chinensis*) are a few examples.

(e) *Samara:* the superior ovary, syncarpous, bicarpellary, and fruit develop from these organs and are single seeded. Like wings, the pericarp becomes horizontal. Example: Chilbil (*Holoptelea*).

(iii) *Schizocarpicoror splitting fruits*: fruits are multiseeded, dry fruits that split into single-seeded segments upon maturity called mericarps. However, the mericarps do not dehisce further.

(a) *Lomentum:* fruits grow from superior monocarpellary and unilocular ovaries. It is a legume that has been genetically modified. Fruits classified as bisuturals are single-seeded mericarps. Examples: *Tamarindus indica* (tamarind), *Mimosa* spp., and *Arachis hypogaea* (peanut).

(b) *Cremocarp:* fruits emerge from an inferior, syncarpous, and bicarpellary ovary, and split into two single-seeded mericarps upon ripening, connected by a carpophore (a slender stalk). The Umbelliferae (Apiaceae) family includes coriander (*Coriandrum sativum*), carrot (*Daucus carota*), and fennel (*Foeniculum vulgare*).

(c) *Regma:* fruits develop from the superior, syncarpous, multilocular, and tricarpellary (multicarpellary) ovary. They distinguish and then split into several carpels as they ripen. Each component is referred to as a coccus and contains one seed. *Geranium* spp. and castor (*Ricinus communis*) are two examples.

(d) *Carcerulus:* fruits mature from a superior, syncarpous, bicarpellary (or multicarpellary), multilocular ovary, where the false septation may increase the number of locules. A single-seeded mericarp divides into two parts upon maturity. Examples: *Althaea* spp., *Ocimum* spp. (basil), and *Malva* spp.

(e) *Double samara:* fruits develop from superior ovaries that are both syncarpous and bicarpellary. The pericarp ripens into two pieces and grows into two wings. Examples: *Acer* spp. (maple) (Sinha et al. 2012; Cerri and Reale 2020).

(B) *Succulent or Fleshy Fruits*

The pericarp is fleshy or fibrous and remains distinguishable into three layers: epicarp, mesocarp, and endocarp. These are indehiscent

(i) *Drupe*: larger syncarpous ovaries that can be monocarpellary or multicarpellary develop into fruits. Fruits usually contain one seed; numerous seeds are rare. There are three layers. The fruit's skin is shaped by the epicarp. The endocarp is hard and stoney, whereas the mesocarp is squishy or fibrous. Mango, coconut, peach, cherry, walnut, almond, etc., are a few examples.

(ii) *Berries*: using axile or parietal placentation, monocarpellary or multicarpellary, syncarpous, superior, or inferior ovaries develop into fruits. The fruit's rind is made by the epicarp, whereas the mesocarp gets meaty and the endocarp remains thin or membrane-like. Tomato, brinjal, guava, date, papaya, chiku, areca nut, etc., are few examples.

(iii) *Pepo*: the inferior ovary with parietal placentation is a tricarpellary, syncarpous, unilocular fruit development organ. Fruits have numerous seeds and a large placenta inside them. The mesocarp and endocarp are soft, but the epicarp forms the hard rind. The Cucurbitaceae family includes bottle gourds, cucumbers, and muskmelon.

(iv) *Hesperidium*: hesperidium is a multicarpellary, syncarpous, multilocular, superior ovary with axile placentation that emerges as fruits. Epicarp has several oil glands and is firm and leathery. The mesocarp is a fibrous, whitish portion connected to the epicarp. The membrane-covered endocarp extends inward to create separate chambers. Like in lemon and orange, the inner side of the endocarp contains a greater number of juicy unicellular hairs.

(v) *Balausta*: balausta fruits have numerous seeds and grow from a multicarpellary, syncarpous, multilocular, superior ovary. The endocarp is hard and creates irregular chambers to house seeds, whereas the mesocarp is thin and papery. The epicarp is irregular and leathery. The calyx on the fruit is persistent. While the tegmen hardens, the testa remains red and juicy. Example: pomegranate.

(vi) *Amphisarca*: the fruit matures from a superior ovary that is multicarpellary, syncarpous, and has many chambers. Several seeds are dispersed across the hard epicarp, fleshy mesocarp, and soft endocarp. Wood apple (*Aeglemarmelos*) is an example.

(vii) *Pome*: the inferior ovary, which is syncarpous and bi- or polycarpellary, develops into the fruit (Fig. 1.1). The fruit is surrounded by a bulging, meaty thalamus. Hence, it is a fake fruit. The pericarp is papery. The portion of the fruit that is edible is the fleshy, enlarged thalamus, as seen in examples like pear and apple (Sinha et al. 2012).

1.2.1.2 Aggregate Fruits (Etaerio)

These fruits are derivatives of a multicarpellary, apocarpous ovary of a single flower. Etaerio, the aggregate of fruitlets, is the name given to this type of fruit. There are four categories for them:

(i) *Etaerio of follicles*: every free carpel matures into a follicle, a fruitlet. A flower is etaerio when it has numerous follicles or etaerio of follicles. *Calotropis* and

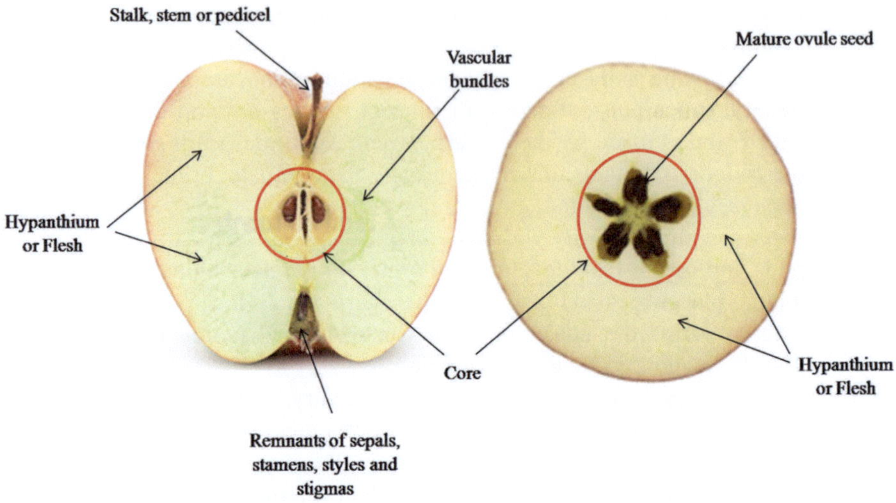

Fig. 1.1 Pome fruit of apple

Catharanthu, which are two follicular aggregates, and *Michelia*, which is an aggregate of many follicles, are some examples.

(ii) *Etaerio of achenes*: the achene etaerio is a collection of fruitlets that emerge from a single bloom. Roses, strawberries, and *clematis* are some of the examples.

(iii) *Etaerio of berries*: the aggregation of fruitlets produced from a single flower is referred to as an etaerio, with the custard apple (*Annona*) serving as a notable example.

(iv) *Etaerio of drupes*: the drupe etaerio is a collection of fruitlets that appear from the apocarpous ovaries of a single bloom. For example, raspberry (Sinha et al. 2012; Ramaswamy 2014).

1.2.1.3 Composite Fruits (Multiple)

These fruits, frequently referred to as infructescence, emerge from the entire inflorescence. Different types of composite fruits are:

(i) *Syconus*: a hypanthodium within the florescence gives rise to the fruit of syconus. The container becomes fleshy and hollow, and has a small opening at the top. It assumes the shape of a cup. On its inner surface, unisex flowers develop with male blooms on the upper side and female flowers on the lower side. The fruit is of the achene type, and its edible part develops through the fleshy receptacles, e.g., *Ficus*.

(ii) *Sorosis*: this type of fruit, including pineapple, mulberry, and jackfruit, emerges from a spike, spadix, or catkin inflorescence. The inflorescence appears as a single fruit because the fruits are tightly pressed together. The rind of jackfruit is spiky and rough due to the fusion of fruitlets, not the stigmas. In jackfruits, the seeds, perianth, and bracts become edible. The edible part of mulberries is the perianth surrounding dry achenes (Sinha et al. 2012; Ramaswamy 2014).

1.2.1.4 Fruit Structure

The exocarp, mesocarp, and endocarp are the three nearly distinct layers of the fruit wall that develop from the ovary wall. The ovary wall is also known as the pericarp. The ovary wall determines the fruit structure (Fig. 1.2), despite substantial variation along family lines. For instance, all fruit development is induced by fertilization, which induces growth and differentiation of the ovary wall into three layers. Pedicel, calyx, withered stamens, style, stigma, and even remnants of the corolla are present when the fruit is mature. Tissues other than the ovary wall that form part of a fruit are called accessory. A substantial amount of the fruit of strawberries (*Fragaria* sp.), apples (*Malus* sp.), and pineapples (*Ammas* sp.) is called accessory. The edible portion of the strawberry is a thick, pulpy central receptacle in which achenes (dry, one-seeded fruits) are encased.

The flower begins the process of reproduction when the seed completes its development and is ready for dispersal. During fertilization, the zygote forms an embryo, the ovule's integuments form the seed coat, and the primary endosperm nucleus forms an endosperm. The zygote nucleus splits quickly after fertilization, whereas the chief endosperm nucleus divides quickly to produce the endosperm, the nutrient-rich storage tissue that will nourish the seed upon germination. The zygote nucleus divides to generate a filament of numerous cells once the endosperm has emerged. The early-stage embryo (proembryo) is formed by a sequence of divisions initiated by the cell that is the most distant from the micropyle. The cell next to the micropyle splits and elongates at approximately the same time and forms the suspensor, which supports the embryo in the endosperm. While the two cotyledons have developed, more divisions produce a globular stage that eventually develops into a heart-shaped

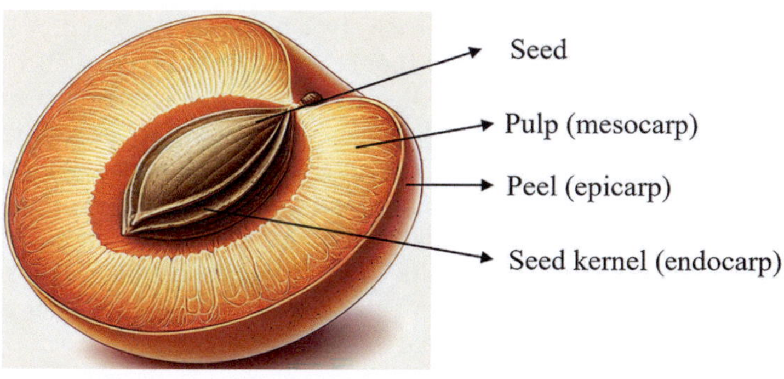

Fig. 1.2 Mature apricot (*Prunus armeniaca*) fruit

stage. Furthermore, the embryo develops a shoot tip at one side and a radicle (embryonic root), at the other.

The previously described phases are restricted to cotton (*Gossypium hirsutum*); the specifics of embryo development differ significantly among different plants. For example, the number of cotyledons two or one, respectively, distinguishes the embryos of dicotyledonous and monocotyledonous plants. The nucleus, endosperm, and integuments of the growing embryo are also changed in ways typical of the plant family to which the seed belongs.

The nucleus and endosperm are in the early stages of embryo development in most plants. This is particularly true for the nucleus, which usually serves as a source of nutrition in the early stages of embryogenesis. It endures in the perisperm, a food storage tissue found in sugar beets and many other species of seeds. Many monocot plants, such as onions (*Allium cepa*), contain seeds that retain endosperm as a food reserve. These include grasses like rice (*Oryza sativa*), which are of significant economic importance, and dangerous weed pests like yellow foxtail (*Setaria lutescens*). Apart from castor bean (*Ricinus communis*), very few dicot seeds maintain the endosperm as a food storage structure.

1.2.1.5 Chemical Composition of Fruits

(i) *Carbohydrates*: in addition to water, carbohydrates frequently comprise the most prevalent constituent group. From basic sugars to complex polymers formed of a multitude of sugar monomer units, they can be identified in a broad variety of molecular weights. 2–40 g/100 g of produce tissue can be made up of carbohydrates; small quantities can be found in some cucurbits, such as cucumbers, and high levels can be found in plants that store starch, like cassava. Sucrose, glucose, and fructose are the primary sugars found in fruits and vegetables; the quantity of each type of sugar differs depending on the type of product. While sucrose is found in just two-thirds of the produce, glucose and fructose are the major sugars found in all produce and frequently exist at equal levels. Fruit from tropical and subtropical regions has the highest sugar content. Among the vegetables, beetroot has the highest sugar concentration (8 g/100 g), with sucrose being the sole type of sugar found in it. A large portion of the carbohydrates found in fruit is in the form of dietary fiber, which is not digested in the human upper digestive tract but is instead metabolized in the lower intestine and excreted from the body as feces. This enhances the fruit's sensory appeal. In terms of carbohydrate polymers, cellulose, pectic compounds, and hemicelluloses comprise the majority of fiber. Lignin is a significant constituent of dietary fiber and is a complex polymer of aromatic chemicals connected by propyl units. Because humans are unable to release the enzymes necessary to break down polymers into simple monomer units that the intestinal tract can absorb, dietary fiber cannot be digested.

(ii) *Lipids*: lipids, present in less than 1 g/100 g of most fruits and vegetables, are linked to cell membranes and the protective cuticle layers that cover the produce's surface. The exceptions are the avocado and olive oil, which contain oil droplets comprising approximately 20 and 15 g of lipid per 100 g of cell, respectively. Health officials strongly advise people to consume more fruits and vegetables because of their relatively low lipid content, which is thought to be beneficial in reducing the incidence of heart disease. Monounsaturated fatty acids are present in even high-lipid-content products, such as avocados.

(iii) *Organic acids*: the majority of fruits and vegetables have higher amounts of organic acids than what is required for the Tricarboxylic acid (TCA) cycle or Krebs cycle and other metabolic reactions to function. Usually, the extra is kept isolated from other cellular constituents in the vacuoles. The organic acid content of lemon, lime, passion fruit, and black currant can vary by more than 3 g/100 g. Citric and malic acids are often the most common acids found in vegetables. Other important organic acids found in some commodities are isocitric acid in blackberries, oxalic acid in spinach, and tartaric acid in grapes. Organic acids have significant effects on flavor in addition to their metabolic significance, especially in fruit that has a good balance of sugar and acid.

(iv) *Minerals and vitamins*: vitamin C or ascorbic acid is a minor component of fruits and vegetables. It is essential in human nutrition for averting scurvy and possibly possesses protective antioxidant properties. Fruits and vegetables contain approximately 90% of the vitamin C that humans are required to consume. About 50 mg of vitamin C is required per day, and in many commodities, less than 100 g of tissue contains this amount of vitamin. Fruits and vegetables typically provide around 40% of the daily requirements for folic acid and vitamin A, making them significant nutritional sources of these nutrients. The body needs vitamin A to preserve the anatomy of the eye; blindness can result from a sustained shortage of vitamin A.

Produce excludes retinol, the active vitamin A molecule; nonetheless, humans can convert some carotenoids, such as β-carotene, to retinol. Vitamin A precursors comprise only 10% of the carotenoids found in fruits and vegetables. Although they lack vitamin A activities, all other carotenoids, including lycopene, a key pigment found in tomatoes, may nevertheless be significant as antioxidants. Anemia can be caused by a folic acid deficiency since it plays a role in RNA production. Since fetal spina bifida is linked to a folic acid shortage during early pregnancy, multiple countries have taken steps to reduce the risk of folic acid shortage by enriching specific foods with folic acid. It should be a top priority to maintain these vitamins during handling and storage, especially when the food will be consumed by people on marginally sufficient diets. Potassium is the primary mineral present in fruits and is found in most fruits at concentrations exceeding 200 mg/100 g.

To mitigate the consequences of diets that contain excessive sodium, health authorities in many countries recommend higher consumption of potassium, and the highest natural food supply of potassium is found in fruits and vegetables. While fruits and vegetables include several additional vitamins and critical minerals, their overall dietary requirements are often not significantly impacted by them. Although they frequently appear in forms indigestible to humans, iron and calcium can be found in foods at nutritionally relevant amounts. For example, most of the calcium in spinach exists as calcium oxalate, which is only weakly absorbed.

(v) *Volatiles*: at room temperature, many kinds of tiny molecules (less than 250) are formed by fruits and vegetables. These compounds exhibit a certain level of volatility. These substances are not important in terms of quantity (generally less than 10 mg/100 g), but they are significant in terms of giving fruit and to a lesser extent vegetables their distinct flavor and perfume. Many ornamental aromatic components are used as fumes for human consumption or as background odors in manufactured goods. More than 100 different volatiles are found in fruits, vegetables, and ornament. As analytical techniques for identifying these volatiles become more sensitive, the number of chemicals found in a given crop increases. Esters, alcohols, acids, and carbonyl compounds (aldehydes and ketones) comprise many substances. Whereas some of these substances, like ethanol, exist in all fruits and vegetables, others are distinctive to a person or species; esters, for instance, are found in most ripe fruits, whereas volatiles containing sulfur are found in tomatoes and Brassica crops.

(vi) *Water*: the majority of fruits comprise more than 80 g of water per 100 g, but certain tissues like melons, cucumbers, and lettuce have as much as 95 g of water. Because, individual cells vary significantly in their water content, significant variation in water content can occur within a species. The water content is determined by the water availability to the tissue at the harvest, which changes throughout the day in response to changes in relative humidity and temperature. In most vegetables, if you are looking for the crispier texture, then it is better to harvest when the maximum amount of water is readily available. Because leafy crops exhibit significant and quick variations in water content in response to environmental changes, harvest timing is a crucial factor to consider (Ramaswamy 2014; Wills 1987).

1.2.1.6 Nutritional Value of Fruits

Fruits are a rich source of carbohydrates, minerals, and dietary fiber, with high water content while being low in fat, protein, and energy. The nutritional composition of fruits depends on multiple factors, including variety, maturity, and growing conditions. Table 1.1 shows the range of nutritional content found in the edible part of some common fruits.

Apart from nutritional values, fruits are also considered as a good source of fat- and water-soluble vitamins, pigments, polyphenols, antioxidants, and other bioactive compounds which may offer significant health benefits to humans. Major nutrients of some common fruits and their specific health benefits are given in Table 1.2.

Table 1.1 Nutritional composition of edible part of some common fruits (*fresh weight basis*)

Fruit	Nutritional composition							References
	Water %	Carbohydrate %	Fat %	Protein %	Dietary fiber %	Ash %	Energy (kcal)	
Mango	78.90–82.80	16.20–17.18	0.30–0.53	0.36–0.40	0.85–1.06	0.34	62.10–190.0	Maldonado-Celis et al. (2019)
Orange	86.75–87.6	7.59–10.60	0.12–0.16	0.94–0.95	1.90–2.40	1.40	197.0	Etebu and Nwauzoma (2014), Gangakhedkar et al. (2021)
Banana	67.99–78.20	20.73–22.84	0.30–0.70	0.92–1.31	2.36–3.70	0.74–1.00	80.70–89.00	Ranjha et al. (2020)
Apple	84.50–85.60	12.10–14.80	0.10–0.20	0.26–0.35	2.00–2.40	0.19–0.30	52.0–57.0	Arnold and Gramza-Michalowska (2024)
Watermelon	91.00–95.00	2.12–6.32	0.03–0.10	0.39–0.65	0.12–0.21	1.72–1.98	11.71–28.63	Yimer and Tehulie (2020)
Guava	80.80–87.17	14.32–15.70	0.54–0.95	0.90–2.55	4.59–5.40	0.53–1.39	36–68	Chen et al. (2007), Medina and Herrero (2016)
Apricot	83.00–84.39	3.76–13.00	0.40–1.53	1.40–3.01	1.50–2.40	0.74–4.94	24.76–79.41	Sharif et al. (2015), Jaafar (2021)
Papaya	85.00–87.70	10.83–13.83	<0.10–0.171	0.702–0.843	0.490–0.632	0.364–0.460	200.5–245.3	Tripathi et al. (2011)

(continued)

Table 1.1 (continued)

Fruit	Nutritional composition							References
	Water %	Carbohydrate %	Fat %	Protein %	Dietary fiber %	Ash %	Energy (kcal)	
Kiwifruit	82.40–83.10	14.70–15.80	0.28–0.52	1.02–1.14	1.40–3.00	0.47–0.61	61.0–63.0	Richardson et al. (2018)
Plum	86.0–88.0	10.40–13.20	0.10–0.30	0.40–0.90	1.30–2.40	0.30–0.40	30.0–66.0	Walkowiak-Tomczak (2008)
Pineapple	81.20–87.24	11.82–13.70	0.11–0.13	0.53–0.55	1.40	1.80	45–52	Hossain et al. (2015), de Ancos et al. (2017)
Cherry	74.84–81.57	12.20–16.00	0.20–0.30	1.00–5.91	1.60–2.20	4.21–5.36	33.80–63.00	Mahmood et al. (2013), Yang et al. (2023) and McCune et al. (2011)
Strawberry	86.30–92.19	2.14–7.68	0.130–0.48	0.53–1.17	2.00–2.42	0.40–0.81	32.00–53.80	Giampieri et al. (2012), Hossain et al. (2016) and Ganhao et al. (2019)
Avocado	72.30–77.72	7.40–8.64	15.40–12.80	0.94–1.96	6.80–6.90	1.52–1.66	143.81–167.00	Dreher and Davenport (2013), Orhevba and Jinadu (2011)
Muskmelon	91.84–92.97	4.24–5.40	0.26–0.35	0.42–0.53	1.49–1.51	0.48–0.59	54.0–58.0	Manchali and Murthy (2020)

Table 1.2 Major nutrients of some common fruits and their specific health benefits

Fruit name	Major nutrients	Health benefits	References
Amla	Excellent source of vitamin C and rich composition of polyphenols	As an effective antioxidant Aids in the absorption of iron Antidiabetic, anti-hyperlipidemia, anticancer, anti-inflammatory, digestive tract and neurological protective activities	Mirunalini et al. (2012), Gul et al. (2022)
Apple	High levels of soluble fiber and pectin, vitamin A and C, and antioxidants	Lower levels of blood cholesterol (LDL) Combating diarrhea and constipation Historically believed that gout, rheumatism, and arthritis could be managed Enhances absorption and digestion It also boosts the body's defense against disease as well	Boyer and Liu (2004), Kyun and Jang (2016)
Apricot	Carotenoids and vitamin C suppliers	Reduced risk of heart disease, stroke, cancer, blood glucose stability, reduced iron deficiency, anemia, and cataracts	Al-Soufi et al. (2022), Fatima et al. (2018)
Banana	It is rich in carbohydrates, dietary fiber, potassium, magnesium, ascorbic acid, vitamin A, and B_6	It helps control blood pressure Gives energy Provides strong endurance Prevent constipation (ripe banana) Raw bananas prevent diarrhea Aiding in sleep A good source to treat anemia	Sidhu and Zafar (2018), Singh et al. (2016)
Blueberry	It is a good source of iron, fiber, and vitamin C	It helps reduce and control cholesterol level	Stull et al. (2024)
Cherry	Vitamins A, C, antioxidants and energy	Excellent source for hypertensive patients because it does not contain fat, sodium, or cholesterol Anti-inflammatory	Kelley et al. (2018)
Figs	Good source of vitamin, minerals, dietary fiber, and phenolic compounds	Oxidant, laxative, diuretic digestible, blood cleansing agent It is good for the liver Reduce body weight Beneficial in diabetes	Sandhu et al. (2023)
Gooseberry	Good source of provitamin A, vitamin C, potassium, and flavonoid	Function as an anticancer agent; beneficial for patients with diabetes and hypertension Anti-inflammatory, and anti-hepatotoxic activity	Ramadan (2011)

(continued)

Table 1.2 (continued)

Fruit name	Major nutrients	Health benefits	References
Grape fruit	Low in calories excellent source of flavonoids, potassium, and fiber an excellent means to get vitamin C Grapefruit that is red or pink consists of carotenoid and lycopene Also good in vitamin A and antioxidants	Effective in cholesterol reduction, normalization of hematocrit levels, and anticancer properties Helps prevent constipation, formation of stones, control blood pressure, and body weight Efficiently decreases cold-induced fever Conduct sound sleeps Assists the hardening of body tissue, including the liver and arteries	Dow et al. (2012)
Guava	Good sources of vitamin A, C and dietary fiber	Increases immune function Helps iron uptake Helps prevent infection Anti-diabetic, anti-cancer	Mathpal and Rathore (2022)
Kiwi-fruit	Good source of vitamin C	Activates immune function Aids in blood formation Lowering blood lipid levels, prevent CVD and cancer	Keith (2012)
Mango	Rich source of vitamins A and C rich in phytochemicals (antioxidant lutein) and dietary fiber	It improves the immune system Aiding in iron absorption Helps in the prevention of infection Help in digestive and gastrointestinal health	Swaroop et al. (2018)
Melons	Excellent source of pro-vitamin A and C Also contains vitamin E, B_6, fiber, and potassium	Provide the most effective and richest mineral elements Refreshing Alkalizing Mineralizing Oxidant Diuretic Anticlotting effect on the blood	Lester (1997)
Orange	Provide vitamin C, A, fiber, and potassium It also comprises folate, B complex vitamin and antioxidants	It offers the body powerful barriers, a capacity to defend from infectious disease, the ability to utilize iron, a potent antioxidant, has potential to prevent cancer, particularly stomach cancer, a drop in cholesterol, and the ability to strengthen and enhance blood vessels Retaining the strength of the cell walls; encourage capillary circulation Efficaciously treating overly acidic bodily conditions, such as constipation	Grosso et al. (2013), and Rawal et al. (2023)

(continued)

Table 1.2 (continued)

Fruit name	Major nutrients	Health benefits	References
Papaya	Rich in enzyme (pepsin), vitamins C, A, and E, calcium, fiber, and antioxidants	Diuretic, laxative, and refreshing. It improves the body's acid-alkaline balance. It is used to treat fevers, stomach ulcers, and other digestive disorders	Rahman (2013), Pandey and Chauhan (2018) and Pandey et al. (2020)
Peaches	Moderate concentration of vitamins C and fiber	Reduces the risk of heart disease; assists with weight loss because of its diuretic, cleansing, and purifying properties Prevents cancer, ulcers, rheumatic pain, and hypertension Prevents lung and liver infections	Bento et al. (2022)
Pears	Excellent source of vitamins C and fiber	Preventing heart disease, diabetes, ulcer, and cancer Laxative properties, lower plasma lipids, and improve gut health	Reiland and Slavin (2015)
Pineapple	Rich source of vitamin C moderate in vitamin A Good source of dietary fiber	Treats rheumatoid arthritis Treats diabetic ulcers Aids in general surgery Aids in digestion since it includes the enzyme "bromelain"	Ali et al. (2020)
Pomegranate	Poor source of calorie Good source of vitamin C, B_5, potassium and flavonoids	Treat diarrhea Treat cancer Treatment of the heart and liver infection	Vucic et al. (2019)
Strawberry	Rich in vitamin C, folic acid, and fiber without fat	Diuretic with great renal effects Used for treating gout and rheumatism	Afrin et al. (2016), Sravani et al. (2017)

1.2.2 Vegetable Types and Classification

Different classification systems are commonly used to categorize vegetables based on their specific characteristics. Though various classification systems exist, the widely accepted classification method for vegetables is mentioned below (Fig. 1.3).

1.2.2.1 Classification by Edible Part

(i) *Leafy greens*: this category includes vegetables such as spinach, lettuce, and kale. They are primarily used as leaves and are rich in vitamins and minerals.
(ii) *Root vegetables*: these are vegetables where the root parts are consumed, such as carrots, beets, and turnips. They are often high in fiber and starch.
(iii) *Cruciferous vegetables*: cruciferous vegetables, also known as brassicas, include broccoli, cauliflower, and cabbage. They are known for their respective anticancer properties.

Fig. 1.3 Vegetable classification based on different parts

(iv) *Fruit/vegetables*: vegetables like tomatoes, cucumbers, and peppers are botanically classified as fruits but are culinarily used as vegetables in cooking.

(v) *Pods and seeds*: vegetables including okra and snap peas fall under this category; they are harvested for their edible pods or seeds.

(vi) *Bulb vegetables*: onions, garlic, and leeks are examples of bulb vegetables, which are characterized by their growth as a bulb underground.

(vii) *Tubers*: potatoes and yams are fleshy underground storage organs that contain nutrients to support plant growth.

(viii) *Stems and shoots*: asparagus and bamboo shoots are examples of vegetables consumed for their stems or shoots (Sinha et al. 2012).

1.2.2.2 Structure of Vegetables

The root system consists of parts that take water and nutrients from the soil, whereas the shoot system comprises stems, leaves, flowers, and fruits. Stems provide support to plant leaves, where photosynthesis occurs, and flowers and fruits are essential for reproduction. Each of these parts of the plant serves a function throughout its life cycle (Figs. 1.3 and 1.4). For instance, roots anchor the plant to absorb water and nutrients, while the shoot system enables the plant to reach sunlight and exchange gases with its surrounding (Fig. 1.3). The structure of vegetables at the cellular level is quite complex and fascinating. Vegetables consist of various types of cells that come together to form the structure. The outer layer of cells is known as the tissue, which acts as a shield against elements. It contains a waxy layer called the cuticle, which helps in retaining moisture. The ground tissue forms the majority of the vegetable's interior and plays an important role in photosynthesis, storage, and providing support. It consists of parenchyma, collenchyma, and sclerenchyma cells. Within the plant, tissue comprising the xylem and phloem transports water, nutrients, and sugars throughout the plant system (Kole 2007; Nwosisi et al. 2019).

Each cell type has a different function within the plant, contributing to its overall form and function. Understanding this cellular structure helps in many areas, from agriculture to nutrition to food science (Dhenge et al. 2023; Wei et al. 2023). Vegetables are composed of various tissues and organs that contribute to their growth, development, and function. Here, we present a detailed look at the tissues and organs in vegetables:

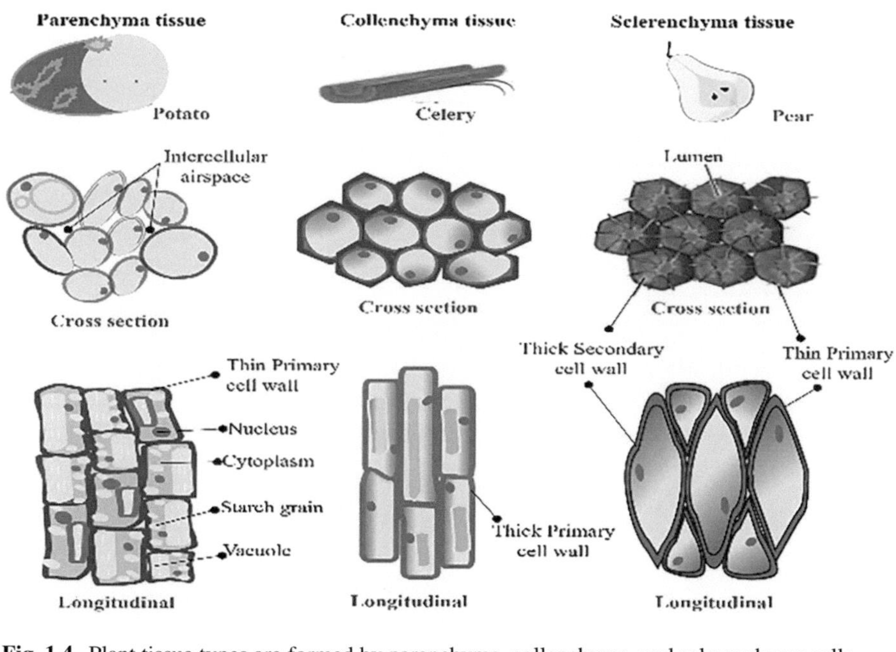

Fig. 1.4 Plant tissue types are formed by parenchyma, collenchyma, and sclerenchyma cells

Microscopic examination of vegetables shows a detailed analysis of the cellular and tissue organization of vegetables. This type of examination can reveal the intricate details of cell walls, vacuoles, nuclei, and other cellular components that make up vegetable tissue. Changes in morphology and composition can be evaluated by some of the microscopic techniques like light microscopy, as many food components' origins exhibit inherent fluorescence (autofluorescence). In plants, these include pigments (e.g., chlorophyll and carotenoids), and both high, and low molecular mass phenolic compounds (e.g. lignin and ferulic acid), electron microscopy, scanning electron microscopy, transmission electron microscopy, and conventional microscopy (LM) Light microscopy, (TEM) Transmission Electron Microscopy, and (SEM) Scanning Electron Microscopy have been used in food structure studies to obtain a qualitative description of the structure of samples. Localized ion techniques can be used in conjunction with conventional microscopies, to determine the distribution of structural elements, macromolecules, and elements of interest in the samples. Other techniques include the scanning probe microscopy technique, atomic force microscopy, magnetic resonance imaging, and acoustic microscopy.

A simplified overview of the cellular structure of vegetables (Fig. 1.5) is as follows:

Cell wall: composed mainly of cellulose, lignins, and proteins, cell walls provide rigidity and strength to plant cells.

Nucleus: contains the cell's DNA and acts as the control center.

Vacuoles: one or more vacuoles containing nutrients and waste products are present.

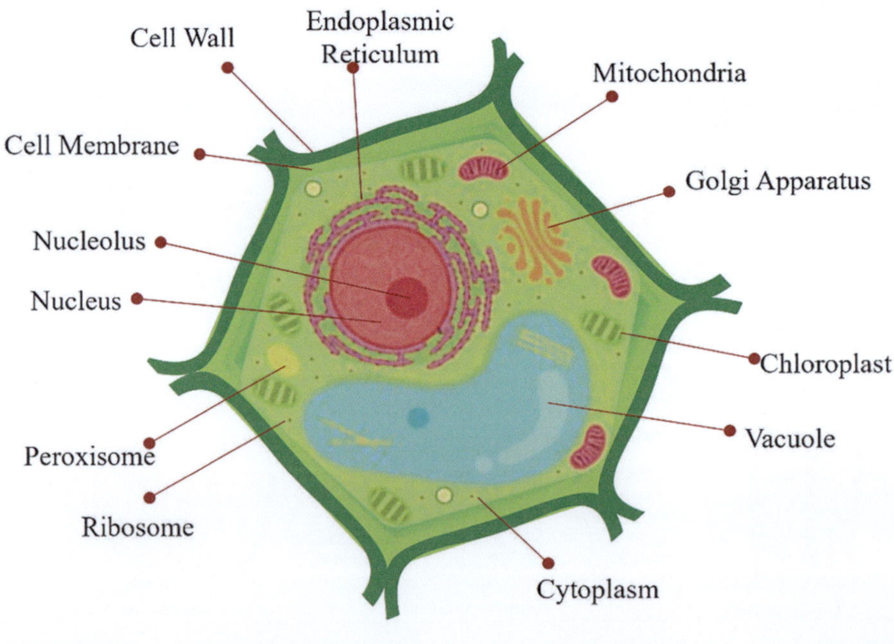

Fig. 1.5 Illustration of a plant cell

Mitochondria: mitochondria known as the powerhouses of the cell; they facilitate the conversion of sugars into adenosine triphosphate (ATP), the primary energy currency of the cell.

Chloroplasts: found in green parts of the plant and are responsible for photosynthesis.

Chromoplasts: develop from chloroplasts once chlorophyll has broken down and are responsible for the color of the vegetables.

Amyloplasts: contain grains of starch and are particularly prominent in starchy vegetables.

1.2.2.3 Composition of Vegetables

Understanding vegetable composition allows for informed dietary choices (Table 1.3 and 1.4). It helps to identify:

(i) *Specific nutrients*: vegetables include vitamins, minerals, and fiber. Knowing the content of each vegetable type allows for targeted selection to meet individual needs (e.g., focusing on iron-rich vegetables for those at risk of deficiency).

(ii) *Macronutrient balance*: vegetables vary in carbohydrate, protein, and fat content. This knowledge aids in creating meals with balanced macronutrients, promoting satiety, and maintaining stable blood sugar levels (Morrison and Laeger 2015).

(iii) *Phytochemicals*: vegetables are rich in phytochemicals, which have antioxidant and anti-inflammatory properties. Understanding their presence helps individuals select vegetables that meet their specific health goals.

1.2.2.4 Nutritional Differences Among Vegetables

The diverse vegetable kingdom offers a variety of nutritional profiles:

- *Color*: deeply colored vegetables (orange, red, green) are often rich in vitamin A precursors (carotenoids), vitamins C, vitamin K, and fiber.
- *Starchy* vs. *nonstarchy*: starchy vegetables provide sustained energy storage with their higher carbohydrate content, whereas nonstarchy options are lower in carbs and higher in water, vitamins, and minerals.
- *Cruciferous and allium vegetables*: cruciferous vegetables (broccoli, cauliflower, and kale) are known as glucosinolates with potential cancer-fighting properties. Allium vegetables (garlic, onions) contain organosulfur compounds that are linked to heart health benefits (Ismail et al. 2004).

Table 1.3 Macro- and micronutrients of some common vegetables

Vegetable	Water (%)	Carbohydrates (%)	Protein (%)	Fat (%)	Fiber (g/100 g)	Vitamin C (mg/100 g)	Potassium (mg/100 g)	Calcium (mg/100 g)	Iron (mg/100 g)	References
Broccoli	89	7	3	0.4	2.4	89.7	180	47	0.7	Kole (2007)
Carrot	88	10	0.9	0.2	2.5	5.4	190	40	0.4	Kole (2007)
Potato (white)	79	17	2	0.1	2.2	12.7	422	8	0.8	Dutt et al. (2020)
Spinach	90	3	3	0.3	2.5	28.1	558	60	2.8	Morelock and Correll (2008)
Tomato	94	4	1.2	0.2	1.2	23.7	237	14	0.4	de Bruijn et al. (2023)
Carrot	88.29	9.85	0.93	0.24	2.8	5.9	240	34	0.4	Nagraj et al. (2020)
Okra	88.6	8.2	2.1	0.2	1.70	16	90	84	1.2	Gemede et al. (2015)
Bottle gourd	94.5	0.339	0.062	0.002	–	10.1	150	26	0.20	Ahmad et al. (2022)

Table 1.4 Major phytochemicals in some vegetables and their potential health benefits

Vegetable	Major phytochemicals (estimated % range)	Potential health benefits	References
Broccoli	Sulforaphane (0.05–1.0%), glucosinolates (50–90% of total dry weight), carotenoids (0.1–1.0%)	May reduce cancer risk, support detoxification, and exert antioxidant effects	Kotecha et al. (2016)
Carrot	Beta-carotene (50–70% of total carotenoids), lutein (10–20% of total carotenoids)	Supports eye health and antioxidant effects	Office of Dietary Supplements (ODS) (National Institutes of Health 2020)
Garlic	Allicin (0.1–1.0% fresh weight), organosulfur compounds (varies depending on the type)	May improve heart health and, potential antimicrobial properties	Ismail et al. (2004); Bayan et al. (2014)
Onion	Quercetin (5–10% of dry weight), organosulfur compounds (varies depending on the type)	May reduce inflammation, antioxidant effects	Tsai et al. (2006)
Berries (blueberries, raspberries, etc.)	Anthocyanins (1–3% fresh weight), Ellagic acid (0.1–1.0% fresh weight)	May improve cognitive function, antioxidant effects	Kesse et al. (2017)
Tomatoes	Lycopene (0.5–8.0% of total carotenoids)	May reduce cancer risk and improve heart health	Ismail et al. (2004)
Leafy greens (spinach, kale, etc.)	Flavonoids (varies depending on type), carotenoids (varies depending on type)	May reduce inflammation and antioxidant effects	Ismail et al. (2004)

1.3 Factors Affecting Fruits and Vegetables Structure and Composition

There are many predisposing factors that influence the structure and composition of fruits and vegetables, including:

- *Climate*: the role of environmental factors, such as temperature, sunlight, rainfall, and humidity, is significant in fruits and vegetables composition. For instance, the warmer the temperature, the higher the sugar content; conversely, cooler temperatures may increase the acidity of fruits (Pereira 2000).
- *Light*: sunlight is essential for plant growth because it drives photosynthesis, which is crucial for the production of energy and nutrients. Different plants have different light requirements; furthermore, the factors such as elevation, latitude, season, and cropping systems can affect the amount of light received by crops.
- *Temperature*: temperature plays a vital role in various physiological processes in plants, such as photosynthesis, water absorption, transpiration, and regulation of

enzyme activity. Different crops have specific temperature requirements for optimal growth, and extreme temperatures can reduce growth, delay development, and impact yield and quality (Terefe and Versteeg 2019).

- *Genetics*: genetic makeup determines the structure and composition of fruits and vegetables. Sugar content, acidity, flavor profile, nutritional composition, etc., are the function of plant genetic makeup (Pereira 2000). Different varieties of the same fruit or vegetable can have different levels of nutrients, flavors, and textures. Breeding and genetic modification can also play a significant role in enhancing certain desirable traits in fruits and vegetables (Davis 2009; Nwosisi et al. 2019; Majhi et al. 2023).
- *Cultivar*: different cultivars of the same fruit or vegetable variety can have diverse compositions due to selective breeding for explicit traits like sweetness, size, color, disease resistance, etc. (Pereira 2000).
- *Soil*: the availability of essential nutrients to plants depends on the type and quality of soil. The mineral content, flavor, and overall nutritional value of fruits or vegetables depend on the soil composition (Pereira 2000). Soil pH affects the availability of nutrients to plants, with optimal levels typically between 6.5 and 7.5. Soil fertility, which includes nutrients such as nitrogen, phosphorus, and potassium, is crucial for providing adequate nutrition to plants and maintaining soil health.
- *Water availability*: Adequate water supply plays a crucial role in fruits and vegetables development. Soil moisture influences nutrient availability and root function, thereby affecting overall plant health and productivity. Fruits or vegetables size, juiciness, and overall quality also affected by the severe water stress or drought conditions (Pereira 2000).
- *Humidity*: humidity levels affect transpiration, nutrient uptake, and overall plant health. High humidity can slow down water usage by plants, leading to issues like rotting, whereas low humidity can cause wilting and reduce photosynthetic activity.
- Understanding and managing these environmental factors are essential for optimizing production, ensuring healthy plant growth, maximizing yields, and maintaining crop quality. Farmers need to consider these factors when planning and implementing agricultural practices to support sustainable and successful fruits or vegetables cultivation (Ziv and Fallik 2021).
- *Maturation and ripening*: the composition of fruit changes during the maturation and ripening process. Normally, ripening leads to an increase in sugar content, reduced acidity, and characteristic flavors and aromas development (Terry and Joyce 2004). In contrast, vegetables become more fibrous with reduced color intensity after maturation.
- *Harvesting practices*: the timing and method of harvesting influence the composition of fruits and vegetables. The flavor, nutritional quality, and size of fruits or vegetables can be ensured by harvesting at the optimal stage of ripeness or maturity (Terry and Joyce 2004).
- *Postharvest handling and storage*: proper handling, storage, and transportation prevent deteriorative changes in the structure and composition of fruits and

vegetables. Improper handling practices lead to changes such as loss of moisture, texture, and nutrient degradation (Terry and Joyce 2004).

- *Other factors*: these can include the method of cultivation (organic vs. conventional), use of fertilizers and pesticides, and maturity stage at harvesting. Each of these factors contributes to the final nutritional profile of fruits and vegetables.

1.4 The Importance of Classification

The facts required to differentiate between structural and compositional shifts and nutrient composition are provided by the classification of fruits and vegetables. The information can aid in designing the recommendation for eating to help people choose the right food to meet their dietary and fitness needs. Food guides are accessible in many countries and provide recommendations for intake in addition to graphical representations of food groups and subgroups. The focus of food recommendations is not only on the key nutritional components of fruits and vegetables but also on their availability (Sinha et al. 2012). In the case of vegetables, they can be grouped according to their use, botanical family, or both. Leafy vegetables, also known as greens, include plants such as spinach and kale. Vegetables used in salads, such as celery and lettuce, are another group. The *Brassica* family produces cole crops such as cabbage and cauliflower, except for Chinese cabbage. Root vegetables have a substantial root system and include carrot and beets. The Allium family includes bulb vegetables such as onions and garlic. The *Cucurbitaceae* family includes gourd such as cucumbers and pumpkins. Lastly, the *Solanaceae* family is known for solanaceous fruits such as tomatoes and peppers.

By categorizing fruits and vegetables based on their botanical characteristics or edible parts, we can better determine their vitamin and mineral content, flavor profile, and culinary use. This systematic approach also aids in agricultural practices from cultivation to harvest, ensuring that we can maximize the health benefits and culinary potential of each fruit or vegetable. Essentially, classification serves as a roadmap for navigating the diverse world of fruits and vegetables, allowing for more informed choices in both consumption and agriculture (Pennington and Fisher 2009).

1.5 Impact of Fruits and Vegetables Composition on Processing, Preservation, and Storage

The composition of fruits and vegetables has a significant effect on their processing, preservation, and storage.

1.5.1 Impact on Processing

(i) *Acidity*: special processing techniques are required for highly acidic fruits or vegetables like citrus species, to maintain quality and prevent spoilage.
(ii) *Sugar content*: fruits and vegetables, such as grape, mango, potato, etc., are highly susceptible to browning reactions during processing because they have high sugar content. The blanching techniques and antioxidant treatments exhibit a preventive effect on such fruits and vegetables.
(iii) *Enzyme activity*: enzymes in fruits and vegetables are responsible for undesirable changes in color, flavor, and texture during processing. Techniques such as blanching, heating, and cold treatments help in inactivating these enzymes.

1.5.2 Impact on Preservation

(i) *Water activity*: fruits and vegetables with high water activity are highly prone to microbial growth and spoilage. Drying, canning, and concentration can be effectively used to reduce water activity and extend the shelf-life.
(ii) *pH*: acidic fruits are less likely to support microbial growth and can be preserved using methods like pickling and canning. Vegetables are usually less acidic and more susceptible to microbial growth; hence, low-temperature preservation is required.
(iii) *Pectin content*: pectin reacts with water and acid to form a gel network in jams and preserves. This reduces the amount of available water thus regulating microbial growth and extending the shelf-life of the products.

1.5.3 Impact on Storage

(i) *Respiration rate*: some fruits and vegetables spoil faster due to their higher respiration rates, such as bananas, avocados, green leafy vegetables, etc. Temperature and atmosphere control can slow down the respiration process and extend the shelf-life.
(ii) *Ethylene production*: ethylene is a ripening hormone. Fruits that produce high levels of ethylene, such as apples and tomatoes, can cause other fruits to ripen faster when stored together.
(iii) *Moisture content*: fruits and vegetables with high moisture content are more susceptible to microbial growth and spoilage during storage. Proper humidity control and packaging can help maintain critical moisture levels (Salunkhe et al. 1991).

The composition of fruits and vegetables, including acidity, sugar content, enzyme activity, water activity, pH, pectin content, respiration rate, ethylene production, and moisture content, plays a crucial role in determining the most suitable processing, preservation, and storage methods to maintain quality and extend their shelf-life.

1.6 Conclusion

The highlight of this chapter is to explore the types, structures, chemical compositions, nutritional values, health benefits, factors influencing fruit composition, and advanced techniques used to analyze fruit composition. Studying the structure and composition of fruits and vegetables is the key to their application in nutrition, diet, and culinary sciences. Environmental factors, genetic variations, and postharvest handling influence fruit and vegetable composition. Understanding the composition of fruits and vegetables is vital for assessing their nutritional benefits, including variations among their different types during storage. Moreover, knowledge of fruit and vegetable structure and composition facilitates good and effective agricultural practices by guiding cultivation techniques to optimize fruit and vegetable quality and yield.

References

Afrin S, Gasparrini M, Forbes-Hernandez TY, Reboredo-Rodriguez P, Mezzetti B, Varela-Lopez A, Giampieri F, Battino M (2016) Promising health benefits of the strawberry: a focus on clinial studies. J Agric Food Chem 64:4435–4449

Ahmad MD, Ahmad I, El-Chaghaby G, Rashad S (2022) Nutritional and medicinal potential of bottle gourd (Lageneria siceraria): a mini review. Egypt J Bot 62:1–10

Ali MM, Hashim N, Aziz SA, Lasekan O (2020) Pineapple (*Ananas comosus*): a comprehensive review of nutritional values, volatile compounds, health benefits, and potential food products. Food Res Int 137:109675

Al-Soufi MH, Alshwyeh HA, Alqahtani H, Al-Zuwaid SK, Al-Ahmed FO, Al-Abdulaziz FT, Raed D, Hellal K, Nani NHM, Zubaidi SN, Asni NSM, Hamezah HS, Kmal N, Al-Muzafar H, Mediani A (2022) A review with updated perspectives on nutritional and therapeutic benefits of apricot and the industrial application of its underutilized parts. Molecules 27:5016

Arnold M, Gramza-Michalowska A (2024) Recent development on the chemical composition and phenolic extraction methods of apple (Malus domestics)—a review. Food Biopro Technol 17:2519–2560

Bayan L, Koulivand PH, Gorji A (2014) Garlic: a review of potential therapeutic effects. Avicenna J Phytomed 4(1):1

Bento C, Goncalves AC, Silva B, Silva LR (2022) Peach (Prunus Persica): phytochemicals and health benefits. Food Rev Int 38:1703–1734

Boyer J, Liu RH (2004) Apple phytochemicals and their health benefits. Nutri J 3:5

Brasil IM, Siddiqui MW (2018) Postharvest quality of fruits and vegetables: an overview. In: Siddiqui MW (ed) Preharvest modulation of postharvest fruit and vegetable quality. Academic, pp 1–40

Cerri M, Reale L (2020) Anatomical traits of the principal fruits: an overview. Sci Hortic 270:109390

Chen HC, Sheu MJ, Lin LY, Wu CM (2007) Nutritional composition and volatile compounds in guava. Fresh Produce 1:132–139

Davis DR (2009) Declining fruit and vegetable nutrient composition: what is the evidence? Hort Science 44(1):15–19

De Ancos B, Sanchez-Moreno C, Gonzalez-Aguilar GA (2017) Pineapple composition and nutrition. In: Lobo MG, Paull RE (eds) Handbook of pineapple technology: production, postharvest science, processing and nutrition, 1st edn. Wiley, p 221

de Bruijn J, Fuentes N, Solar V (2023) The effect of visible light on the postharvest life of tomatoes (*Solanum lycopersicum* L.). Horticulturae 9(1):94

Dhenge R, Rinaldi M, Rodolfi M, Barbanti D, Ganino T (2023) Modification of structural characteristics of vegetables by high-pressure processing: a review. Food Biosci 56:103407

Dow CA, Going SB, Show HHS, Patil BS, Thomson CA (2012) The effects of daily consumption of grapefruit on body weight, lipids, and blood pressure in healthy, overweight adults. Metabolism 61:1026–1035

Dreher ML, Davenport AJ (2013) Hass avocado composition and potential health effects. Crit Rev Food Sci Nutr 53:738–750

Dutt S, Raigond P, Singh B, Manjul AS, Chakrabarti SK (2020) Potato proteins. In: Raigond P, Singh B, Dutt S, Chakrabarti SK (eds) Potato. Springer, Singapore, pp 51–71. https://doi.org/10.1007/978-981-15-7662-1_4

Ebabhi A, Adebayo R (2022) Nutritional values of vegetables. In: Yildirim E, Ekinci M (eds) Vegetable crops: health benefits and cultivation. Intech Open, pp 3–14

Etebu E, Nwauzoma AB (2014) A review on sweet orange (Citrus Sinensis Osbeck): health, disease, and management. Am J Res Commun 2:33–70

Fatima T, Bashir O, Gani G, Bhat TA, Jan N (2018) Nutritional and health benefits of apricots. Int J Unani Integra Med 2:5–9

Gangakhedkar PS, Shinde ST, Mane RP, Gaikwad GP (2021) Studies on physico-chemical properties of sweet orange. Pharm Innov J 10:1524–1527

Ganhao R, Pinheiro J, Tino C, Faria H, Gil MM (2019) Characterization of nutritional, physicochemical, and phytochemical composition and antioxidant capacity of three strawberry "Fragaria × ananassa Duch." Cultivars ("Primoris", "Endurance", and "Portola") from western region of Portugal. Foods 8:0682

Gemede HF, Ratta N, Haki GD, Beyene AZWF (2015) Nutritional quality and health benefits of okra (Abelmoschus Esculentus): a review. Gob J Med Res: K Interdiscip 14:2

Giampieri F, Tulipani S, Alvarez-Suarez JM, Quiles JL, Mezzetti B, Battino M (2012) The strawberry: composition, nutritional quality, and impact on human health. Nutrition 28:9–19

Grosso G, Galvano F, Mistretta A, Marventano S, Nolfo F, Calabrese G, Buscemi S, Drago F, Veronesi U, Scuderi A (2013) Red orange: experimental models and epidemiological evidence of its benefits on human health. Oxidat Med Cell Long 2013:1. https://doi.org/10.1155/2013/157240

Gul M, Liu ZW, Haq IU, Rabail R, Faheem F, Walayat N, Nawaz A, Shabbir MA, Munekata PES, Lorenzo JM, Aasil RM (2022) Functional and nutraceutical significance of amla (Phyllanthus emblica L.): a review. Antioxidants 11:816

Hossain MF, Akhtar S, Anwar M (2015) Nutritional value and medicinal benefits of pineapple. Int J Nutr Food Sci 4:84–88

Hossain A, Begum P, Zannat MS, Rahman MH, Ahsan M, Islam SN (2016) Nutrient composition of strawberry genotypes cultivated in a horticulture farm. Food Chem 199:648–652

Ismail A, Marjan ZM, Foong CW (2004) Total antioxidant activity and phenolic content in selected vegetables. Food Chem 87:581–586

Jaafar HJ (2021) Effects of apricot and apricot kernels on human health and nutrition: a review of recent human research. Technium BioChemMed 2:139–162

Keith S (2012) Kiwifruit: overview of potential health beneftis. Nutr Today 47:133–147

Kelley DS, Adkins Y, Laugero KD (2018) A review of the health benefits of cherries. Nutrients 10:368

Kesse SH, Filippini T, Kromhout D et al (2017) Berry fruits intake and future cognitive decline in a middle-aged population. Neurology 89:765–771
Kole C (2007) Vegetables. In: Kole C (ed) Vol 5 of series genome mapping and molecular breeding in plants. Springer, Berlin/Heidelberg. https://doi.org/10.1007/978-3-540-34536-7
Kotecha R, Takami A, Espinoza JL (2016) Dietary phytochemicals and cancer chemoprevention: a review of the clinical evidence. Oncotarget 7(32):52517–52529
Kyun TK, Jang K (2016) Apple as a source of dietary phytonutrients: an update on the potential health benefits of apple. EXCLI J 15:565–569
Lester G (1997) Melon (Cucumis melo L.) fruit nutritional quality and health functionality. HortTechnology 7:222–227
Mahmood T, Anwar F, Bhatti IA, Iqbal T (2013) Effect of maturity on proximate composition, phenolics and antioxidant attributes of cherry fruit. Pak J Bot 45:909–914
Majhi PK, Bhoi TK, Sahoo KC, Mishra N, Tudu S, Das S, Ray M, Singh S, Samal I, Behera PP, Saini V (2023) Understanding the genetics and genomics of vegetable grafting to ensure yield stability. In: Singh S, Sharma D, Sharma SK, Singh R (eds) Smart plant breeding for vegetable crops in post-genomics era. Springer Nature Singapore, pp 69–98
Maldonado-Celis ME, Yahia EM, Bedoya R, Landazuri P, Loango N, Aguillon J, Restrepo B, Ospina JCG (2019) Chemical composition of mango (Mangifera indica L.) fruit: nutritional and phytochemical compounds. Frontiers. Plant Sci 10:1073
Manchali S, Murthy KNC (2020) Muskmelon. In: Jaiswal AK (ed) Nutritional composition and antioxidant properties of fruits and vegetables. Apple Academic Press, pp 533–546
Mathpal D, Rathore G (2022) An analysis of health benefits of guava. Int J Innov Res Eng Manag 9:239–242
McCune LM, Kubota C, Stendell-Hollis NR, Thomson CA (2011) Cherries and health: a review. Crit Rev Food Sci Nutr 51:1–12
Medina NNR, Herrero JVI (2016) Guava (Psidium guajava L.) cultivars: an important source of nutrients for human health. In: Simmonds MSJ, Preedy VR (eds) Nutritional composition of fruit cultivars. Apple Academic Press, pp 287–315
Mirunalini S, Vaithiyanathan V, Krishnaveni M (2012) Amla: a novel ayurvedic herb as a functional food for health benefits—a mini review. Int J Pharm Pharm Sci 5:1–4
Morelock TE, Correll JC (2008) Spinach. In: Prohens J, Nuez F (eds) Vegetables I: asteraceae, brassicaceae, chenopodicaceae, and cucurbitaceae. Springer Nature, New York, pp 189–218
Morrison CD, Laeger T (2015) Protein-dependent regulation of feeding and metabolism. Trends Endocrinol Metab 26:256–262
Nagraj GS, Jaiswal S, Harper N, Jaiswal AK (2020) Nutritional composition and antioxidant properties of fruits and vegetables. In: Carrot. Academic, pp 323–337
Nwosisi S, Dhakal K, Nandwani D, Raji JI, Krishnan S, Beovides-Garcia Y (2019) Genetic diversity in vegetable and fruit crops. In: Nandwani D (ed) Genetic diversity in horticultural plants. Springer, Cham, pp 87–125
Online Etymology Dictionary (2021) Vegetables—Origin and meaning of vegetable. www.etymonline.com
Orhevba BA, Jinadu AO (2011) Determination of physico-chemical propserties and nutritional content of avocado pear (Persea Americana M). Acad Res Int 1:372–380
Pandey AK, Chauhan OP (2018) Process optimization for development of vacuum fried papaya (Carica papaya) chips using response surface methodology. Agric Res 8:364. https://doi.org/10.1007/s40003-018-0375-x
Pandey AK, Kumar S, Ravi N, Chauhan OP, Patki PE (2020) Use of partial drying and freezing pre-treatments for development of vacuum fried papaya (Carica Papaya L.) chips. J Food Sci Technol 57:2310. https://doi.org/10.1007/s13197-020-04269-w
Pennington JAT, Fisher RA (2009) Classification of fruits and vegetables. J Food Compost Anal 22:S23–S31
Pereira A (2000) Plant genomics is revolutionizing agricultural research. Biotechnol Dev Monit 40:2–7

Rahman A (2013) Health benefits, chemistry and mechanism of Carica papaya a crowning glory. Adv Nat Sci 6:26–37
Ramadan MF (2011) Bioactive phytochemicals, nutritional value, and functional properties of cape gooseberry (Physalis peruviana): an overview. Food Res Int 44:1830–1836
Ramaswamy HS (2014) Post-harvest technologies of fruits & vegetables. DES tech Publications, Inc
Ranjha MMAN, Irfan S, Nadeem M, Mahmood S (2020) A comprehensive review on nutritional value, medicinal uses, and processing of banana. Food Rev Int. https://doi.org/10.1080/87559129.2020.1725890
Rawal P, Dudeja I, Pooja SA, Purewal SS, Kaur A, Swamy CT (2023) Orange. In: Purewal SS, Bangar SP, Kaur P (eds) Recent advances in citrus fruits. Springer, Cham. https://doi.org/10.1007/978-3-031-37534-7_4
Reiland H, Slavin BSJ (2015) Systematic reivew of pears and health. Nutr Today 50:301–305
Richardson DP, Ansell J, Drummond LN (2018) The nutritional and health attributes of kiwifruit: a review. Eur J Nutr 57:2659–2676
Salunkhe DK, Bolin HR, Reddy NR (1991) Storage, processing, and nutritional quality of fruits and vegetables. In: Fresh fruits and vegetables, vol I. CRC Press, p 323
Sandhu AK, Islam M, Edirisinghe I, Freeman BB (2023) Phytochemical composition and health benefits of figs (fresh and dried): a review of literature from 2000 to 2022. Nutrients 15:2623
Sharif MZ, Warriach AR, Ali MU, Akram MN, Ashfaq F, Raza A (2015) Proximate composition of apricot (Prunus armeniaca) fruit and kernel. Am Eurasian J Agric Envion Sci 15(10):2109–2112
Sidhu JS, Zafar TA (2018) Bioactive compounds in banana fruits and their health benefits. Food Qual Safe 2:183–188
Singh B, Singh JP, Kaur A, Singh N (2016) Bioactive compounds in banana and their associated health benefits—a review. Food Chem 206:1–11
Sinha NK, Sidhu J, Barta J, Wu J, Cano MP (eds) (2012) Handbook of fruits and fruit processing. Wiley
Sravani VJ, Ravi N, Roopa N, Kumar S, Pandey AK, Chauhan OP (2017) Use of high-pressure technology for the development of novel jam and its quality evaluation during storage. J Food Sci Technol 54:3562. https://doi.org/10.1007/s13197-017-2814-2
Stull AJ, Cassidy A, Djousse L, Johnson SA, Krikorian R, Lampe JW, Mukamal KJ, Nieman DC, Starr KNP, Rasmussen H, Rimm EB, Stote KS, Tangney C (2024) The state of the science on the health benefits of blueberries: a perspective. Front Nutr 11:1415737
Swaroop A, Bagchi M, Moriyama H, Bagchi D (2018) Health benefits of mango (Mangifera Indica L) and mangiferin. Jpn J Med 1:149–154
Terefe NS, Versteeg C (2019) Texture and microstructure. In: Fruits and vegetables, p 56
Terry LA, Joyce DC (2004) Elicitors of induced disease resistance in postharvest horticultural crops: a brief review. Postharvest Biol Technol 32(1):1–3
Tripathi S, Suzuki JY, Carr JB, McQuate GT, Ferreira SA, Manshardt RM, Pitz KY, Wall MM, Gonsalves D (2011) Nutritional composition of rainbow papaya, the first commercialized transgenic fruit crop. J Food Comput Anal 24:140–147
Tsai PW, Tsai JY, Lin YM (2006) Anti-inflammatory and reactive oxygen species scavenging effects of different fractions from water-extracted onion (*Allium cepa* L.) bulbs. J Agric Food Chem 54:7419–7426
Vucic V, Grabez M, Trchounian A, Arsic A (2019) Composition and potential health benefits of pomegranate: a review. Curr Pharm Des 25:1817–1827
Walkowiak-Tomczak D (2008) Characteristics of plums as a raw material with valuable nutritive and dietary properties – a review. Pol J Food Nutri Sci 58:401–405
Wei X, Fu T, He G, Zhong Z, Yang M, Lou F, He T (2023) Types of vegetables shape composition, diversity, and co-occurrence networks of soil bacteria and fungi in karst areas of Southwest China. BMC Microbiol 23(1):194
Wills RBH (1987) Composition of Australian fresh fruit and vegetables. Food Technol Aust 39:523–526

Yang Z, Li W, Li D, Chan ASC (2023) Evaluation of nutritional compositions, bioactive components, and antioxidant activity of three cherry tomato varieties. Agronomy 13:637
Yimer ZS, Tehulie NS (2020) Nutritional composition of different varieties of watermelon (Citrullus lanatus) fruit at gewane, Northeastern Ethiopia. J Curr Res Food Sci 1(1):16–22
Ziv C, Fallik E (2021) Postharvest storage techniques and quality evaluation of fruits and vegetables for reducing food loss. Agronomy 11:1133

Chapter 2
Physiology and Ripening of Fruits and Vegetables

S. G. Lekshmi, Shruti Sethi, Vijay Paul, P. M. Sindhu, and Swarajya Laxmi Nayak

2.1 Introduction

The physiology of fruits and vegetables plays a crucial role in their ripening process. Ripening, a complex biochemical phenomenon, is characterized by a series of physiological changes, including enzymatic activity, respiration rates, cell wall modification, and the synthesis of flavor compounds. Ethylene, a plant hormone, is a key regulator in the ripening process, triggering the expression of genes responsible for the conversion of starches to sugars, the degradation of chlorophyll, and the formation of color pigments such as carotenoids and anthocyanins. Furthermore, environmental factors such as temperature, humidity, and light significantly impact the rate and uniformity of ripening. This chapter explores the biochemical and physiological mechanisms underpinning the ripening of fruits and vegetables. Understanding the intricacies of ripening is critical for improving storage techniques, reducing post-harvest losses, and enhancing the nutritional and sensory qualities of fruits and vegetables. Additionally, it provides insight into the potential for genetic engineering and biotechnology, where advances could allow for the development of varieties with optimized ripening characteristics, improved nutritional content, and enhanced resistance to diseases and pests for better shelf life and marketability.

S. G. Lekshmi · S. Sethi (✉) · P. M. Sindhu
Division of Food Science and Postharvest Technology, ICAR – Indian Agricultural Research Institute, New Delhi, India

V. Paul
Division of Plant Physiology, ICAR – Indian Agricultural Research Institute, New Delhi, India

S. L. Nayåk
Medicinal and Aromatic Plants Research, Anand, Boriavi, Gujarat, India

OP Chauhan (ed.), *Fruits and Vegetables Technologies*,
https://doi.org/10.1007/978-981-96-8433-5_2

2.2 Growth and Development

Different types of flower parts are involved in the formation of diverse types of fruits, each containing its edible parts. Following fertilization, the ovary experiences multiple cycles of cell division and cell expansion, leading to a substantial increase in the fruit's volume and mass. During the phase of expansion, fruits undergo certain metabolic transformations which play a role in the production of diverse metabolites, such as phenolic compounds, starch, soluble sugars, anthocyanins, organic acids, and plant hormones etc. (Handa et al. 2012). Keeping in view the plant type and fruit type, there is noticeable diversity in metabolites finally present within various types of fruits.

The development of fleshy fruits can be categorized into four distinct stages, as outlined by Gillaspy et al. (1993). The initial stage is floral development, spanning from floral initiation to anthesis, during which the identity, number and shape of floral organs are established. The second stage involves cell division and development till the commencement of the postfertilization phase. The third stage involves cell enlargement and it is characterized by multiple rounds of endoreduplication and rapid cell expansion until the initiation of the ripening phase. The fourth and final stage is ripening itself, which begins once fruit growth concludes. Ripening involves rapid biochemical and structural transformations that finally determine aroma, texture, nutritional components (such as sugars, organic acids, amino acids) and color of the fruit (Ezura and Hiwasa-Tanase 2009). When a flower is successfully pollinated (except in cases of parthenocarpic development), the fruit transforms into an active carbohydrate sink due to higher meristematic activities. In certain fruits like currants and blackberries, cell division concludes by the time of pollination, although in majority of fruits, cell division persists during the postpollination phase. In avocado, cell division persists throughout the entire growth cycle of the fruit. For the majority of plant species, increase in cell size is the primary contributor to the expansion and the overall size of the fruit (Valero and Serrano 2010). For instance, kiwifruit takes more than 140 days post-full bloom to reach its maximum size, whereas apple and tomato achieve this in around 70 and 40 days, respectively. Many fruits follow a single sigmoid growth phase, while others display double or multiphasic growth patterns. Some fruits exhibit even more intricate growth patterns. The growth and development of fruits entail complex crosstalk among hormones and other regulatory networks (Anwar et al. 2015; Almansa et al. 2016).

The growth of fruit on a tree can be monitored through physical measurements like weight, length, width, and volume. The progression of these parameters typically exhibits a simple or double-sigmoid curve, and the specific curve shape depends on the type of fruit. Pear fruits exhibit a single sigmoid growth curve during development (Pei et al. 2020). Generally, the double-sigmoid pattern is distinctive in stone fruits such as cherry, peach, apricot, plum, and olive. Additionally, it is also observed in some nonstone fruits such as grapes and currants. In 1919, Connors divided the commonly recognized double-sigmoid growth curve into three stages: the initial phase marked by exponential growth (stage I), the lag phase (stage II), and the subsequent phase of second exponential growth (stage III). However, DeJong and Goudriaan (1989) later demonstrated that the double-sigmoid growth pattern in peach fruits essentially comprises only two distinct physiological phases of sink activity, challenging the conventional understanding of three stages. The traditional stage II in fruit growth appears to be a result of the temporal shift between these two physiological phases of sink activity. Fruit ripening typically begins after the fruit achieves physiological maturity, signifying its maximum size and the ability to ripen even post detachment from the mother plant. From an evolutionary standpoint, fruit ripening is a pivotal developmental stage that facilitates seed dispersal in flowering plants. During ripening, fruits acquire aromatic, flavorful and attractive qualities by accumulating volatile compounds, sugars and pigments that allure frugivores for effective seed dispersal (Giovannoni et al. 2017).

2.3 Components of Growth

The duration from flowering to the ripening of fruit exhibits considerable variation across different species, ranging from a brief 3 weeks for strawberries to an extended 60 weeks for Valencia oranges. For many fruits, this developmental span typically encompasses around 15 weeks. Within this timeframe, fruits undergo a substantial increase in both volume and weight, often scaling up several thousand times. The weight of the ripe fruit's flesh is intricately determined by factors such as cell number, cell volume, and cell density. The final count and size of cells at ripeness are influenced by the initial numbers and volumes at the time of flowering, as well as the pace and duration of both cell division and cell expansion in the subsequent phases of development (Seymour et al. 2013).

2.3.1 Cell Number

Cell division during anthesis is generally minimal. However, after anthesis, when the fruit changes into an active importer of reserves or photosynthate, numerous tissues undergo a meristematic transformation. This symbolizes the initiation of growth and fruit set. Fruits that are unable to trigger these vital processes often undergo abscission (Seymour et al. 2013). The phase of active cell division in the flesh is typically confined to the initial weeks of post anthesis. In apple, there are about two million cells during anthesis, which increases to about 40 million at harvest. Surprisingly, achieving these numbers requires 21 doublings before anthesis but only 4.5 doublings after that (Pearson and Robertson 1953). Grapes show a similar pattern, with the ovary cell count increasing from about 0.2 million to around 0.6 million after 40 days. Before anthesis, 17 doublings are required and after that 1.5 doublings complete the desired cell numbers (Coombe 1973). This basically highlights the significance of early ovary development for the growth of fruits and thereby the yield.

2.3.2 Cell Volume

While there is consensus regarding the significance of cell number in fruit development and its role in generating variability in fruit size, it is widely recognized that, for most species, the predominant factor contributing to total fruit expansion is the increase in cell volume. The expansion of cells in the flesh is intricately influenced by various factors, including the behavior of cell walls (such as plasticizing and deposition of wall material, including secondary wall development), turgor (involving water flux and differences in osmotic pressure inside and outside of cells) and constraints imposed by the extensibility of the skin or other surrounding layers. Each of these factors can also be influenced by growth regulators and environmental conditions.

2.3.3 Cell Density

The flesh of fruit undergoes the accumulation of various organic compounds, with their concentrations exhibiting a temporal increase. However, it is evident that the rise in concentration alone cannot explain the substantial amount of dry matter accumulated per fruit. Instead, this phenomenon is primarily attributed to the significant increases in cell volume (water), creating a matrix within which solutes can amass. The percentage of dry weight, mirroring cell density, undergoes changes throughout the ontogeny of a fruit. In the pericarp of grape, percentage of dry weight is high at anthesis due to a concentrated presence of tartrate, possibly exacerbated by the highly protoplasmic nature of the cells. As the grapes progress in development, this percentage decreases, reaching a minimum around the period of most rapid growth in the initial cycle. Subsequently, the values start to rise, initially due to the accumulation of malate and tartrate, followed by glucose and fructose (Possner and Kliewer 1985).

2.4 Phases of Growth Curve

Phase 1: Cell Division
In the initial phase of fruit development, known as cell division, the focus is on rapid multiplication of cells. This critical stage sets the foundation for the subsequent growth phases. During this phase, there are several rounds of cell division occurring in the fertilized ovule which leads to the development of a multicellular structure.

The newly formed cells play a vital role in determining the overall structure and potential size of the fruit. Polysomaty and somatic polyploidy are also observed in fruit flesh cells (Bourdon et al. 2009).

Phase 2: Determination of Fruit Size Through Cell Number
The second phase of fruit development is characterized by the determination of fruit size, primarily governed by the number of cells present. While cell division in the previous phase established the cellular framework, it is across this stage that the fruit decides its size depending on the final count of cells. The potential yield and size of the fruit are shaped by genetic influences, environmental factors, and nutrient availability, all of which play crucial roles. Coordination between different tissues and cell types becomes paramount in establishing a balance that contributes to the optimal growth and development of the fruit.

Phase 3: Cell Expansion
Following cell determination, the third phase, cell expansion, takes central stage. In this stage, the individual cells undergo substantial enlargement, contributing significantly to the final size and volume of the fruit. Cellular expansion involves the uptake of water and nutrients and the biosynthesis of various cellular components, such as carbohydrates, leading to an upsurge in cell turgor pressure. This causes the stretching and enlargement of cells, creating the characteristic size and shape of the mature fruit.

2.5 Types of Growth Curve

Single Sigmoid Growth Curve
The single sigmoid growth curve, often referred to as the 'S-shaped' curve, provides a visual representation of the growth trajectory of certain fruits over time, with time plotted on the x-axis and growth on the y-axis. This distinctive curve is emblematic of the four distinct phases of growth that these fruits undergo. In the initial phase, known as the lag phase, there is a period of slow growth as the fruit establishes itself. Subsequently, the log phase or exponential phase, ensues, characterized by rapid growth driven primarily by extensive cell division in the initial weeks following fertilization. Following this, the curve transitions into the diminishing phase, where the rate of growth gradually decreases. Finally, the curve reaches the stationary or steady phase, indicating a stabilization in growth as the fruit matures. The single sigmoid growth curve is particularly evident in fruits such as grapes, late-maturing peaches, and strawberries.

Double Sigmoid Growth Curve
Double sigmoid growth curve with the example of a grape berry is described as below

Stage 1: Rapid Cell Division and Initial Growth

The first stage of berry development in grape is marked by rapid cell division, during which the fruit achieves approximately half of its final weight and size. This crucial phase sets the foundation for subsequent growth processes, establishing the cellular structure that will support the berry's maturation.

Stage 2: Log Phase—Berry Growth Pauses

In the second stage, known as the log phase, berry growth temporarily pauses as the primary focus shifts to seed development and the initiation of chemical signaling processes. This phase is crucial in preparing the berry for the subsequent stages of softening and expansion. The duration of the log phase is variety-specific, longer log phase is often associated with late-ripening varieties. Conversely, varieties with relatively shorter log phase tend to ripen earlier.

Stage 3: Veraison to Harvest

The transition from veraison to harvest constitutes the third and final stage of grape berry development. This period is characterized by significant cell expansion within the berries. A notable transformation occurs as the berries shift from photosynthetic activity to heterotrophic metabolic activity. At the veraison stage, berries undergo a color change, marking the transition from a partial source to a prominent sink. This crucial phase sees the accumulation of key compounds such as sugars, proteins, tannins, and anthocyanins, which are accountable for color and aroma of berries. The berries become the focal point for nutrient accumulation, transforming them into fully ripened, flavorful grapes ready to harvest. Understanding these distinct developmental stages is essential for vineyard management and grape cultivation. It enables viticulturists to implement targeted practices at each phase, optimizing grape quality, flavor profiles and harvest timing. During the growth stages of grapes, distinct traits of each grape type and the overall prosperity of grape farming are determined by the intricate balance between cell division, chemical communication, and metabolic changes.

Triple Sigmoid Growth Curve

Triple sigmoid growth curve having five stages with the example of kiwi is described as below:

(i) Initial rapid growth, seeds reaching full size (0–9 weeks).
(ii) Slow growth, seeds harden and start to color, first very large respiratory response to ethylene (9–12 weeks).
(iii) Rapid growth, seeds become dark brown, response to ethylene increases (12–17 weeks).
(iv) Very little growth, seeds are dark brown, softening starts, soluble solids start to increase, respiratory response to ethylene rises to a maximum and then decreases (17–21 weeks).
(v) A smaller, but significant growth increase to approximately final size. The fruit matures, the seeds turn to very dark brown and free in the tissues.

The growth pattern of these fruit crops is aptly represented by a complex structure: a double sigmoid curve modified by a "shrinkage curve." This intricate model captures the late shrinkage phenomenon observed in fruits just before harvest, providing valuable insights into the factors influencing fruit size and maturation. It is possible to estimate a critical inflection point by analyzing the shape of the curve. This point marks the moment when the acceleration of fruit weight begins to decrease. The timing of the change point, where the transition from shrinkage to the second sigmoid growth phase occurs, is highly dependent on environmental factors. Specifically, the average temperature during the period of rapid initial growth plays a pivotal role. An earlier change point is affected by high temperatures which in turn affects the final size of the fruit during harvest. This dependency on temperature underscores the impact of climatic conditions on the late maturation dynamics of the fruit crop (Bebbington et al. 2009).

2.6 Ripening

Fruit ripening is an important stage of growth and development for any fruit wherein quality, metabolism, physical, and physiological changes occur. Knowledge about ripening behavior is essential as it determines the commodity's quality, shelf-life, and its postharvest management. Developmental regulation and genetics control the ripening of fruits and vegetables in a complex but in a very well-coordinated manner. Fruit characteristics, such as texture, color, flavor, and aroma, are significantly altered throughout the process of ripening. As fruits mature, they undergo intricate transformations that result in delightful sensory qualities such as taste, smell, appearance, and overall feel. Progress of ripening involves a series of genetic and biochemical events that contribute for the readiness of fruit towards consumption.

2.6.1 Changes During Ripening

2.6.1.1 Physical Changes

Color

The loss of green hue in fruits and vegetables becomes noticeable with ripening. With the progress of ripening, pigmentation alters due to the build-up of diverse pigments like β-carotene (mango, nectarine, peach, plum, citron, loquats apricot, sweet potato, broccoli, cantaloupe, winter squash, and carrot), anthocyanins due to which color ranges from light red to dark red or blue/purple (apple, blueberry, blackberry, raspberry, strawberry, black plum, blood orange, cherry, black and red grape, pomegranate, and onion), lycopene (tomato, watermelon, cranberry, persimmon, asparagus, papaya, pummelo, red carrots, pink-fleshed guava cv. Arka Kiran, and grapefruit cv. Ruby Red), lutein (carrot cv. Pusa Kulfi and marigold), xanthophylls (papaya), and violaxanthins (orange, kiwifruit, and pummelo cv. Chandler). This change in color serves as a crucial indicator for harvesting of fruits and vegetables. The loss of green color during ripening occurs primarily due to breakdown of chlorophyll molecules which is facilitated by the enzyme chlorophyllase. This breakdown is influenced by shift in pH (due to leakage of organic acids from vacuoles), oxidative stress, and production of secondary compounds like carotenoids, anthocyanins, and other pigments that can overshadow the presence of chlorophyll pigments (Wills et al. 1998; Sethi et al. 2011; Su-Mon et al. 2022). The chlorophyll molecule undergoes hydrolysis into chlorophyllide and phytol due to the action of chlorophyllase. Subsequently, a hydrogen atom replaces the central magnesium atom, resulting in the formation of a brownish pheophorbide (Kanayama and Kochetov 2015). Ultimately, pheophorbide is transformed into colorless compounds like chlorins and purpurins (Fig. 2.1). The prevalent maturity indicator in the majority of fruits and vegetables is the loss of green color, with a few exceptions such as avocado, kiwifruit, Granny Smith apple, pineapple guava (*Feijoa sellowiana*), custard apple, gooseberry, Anjou pear, and Thompson seedless grape.

Fruit Firmness

The primary constituents of fruit cell walls encompass pectin and cellulose, alongside limited quantities of hemicellulose and noncellulosic polysaccharides. The transformation and solubilization of pectic substances occur during the ripening phase of fruits. The softening of fruits and vegetables is attributed to the enzymatic breakdown, solubilization, and esterification of pectic substances as they offer structural support to cells and uphold their physical integrity.

Pectin within the middle lamella forms strong cross-links with calcium salts, established during the initial stages of cell development. The middle lamella is highly sensitive to heat, and its dissolution leads to the separation of plant cells. De-esterified pectins in the middle lamella associate with Ca^{2+} ions, and their removal causes cell detachment (Prasanna et al. 2007). Throughout ripening, the insoluble protopectin undergoes a conversion into highly soluble polyuronides, resulting in the softening of fruit texture. Consequently, the firmly bound, high-molecular-weight original pectin degrades, leading to fruit softening due to the transformation of protopectin (from insoluble and high-molecular-weight parent pectin into loosely bound soluble pectins) (Sulaimankhil et al. 2021).

During the softening process, neutral sugars like galactose and arabinose (also as components of neutral protein) and acidic pectin (rhamnogalacturonan) are lost from the cell wall. Protopectinase, pectin methylesterase, polygalacturonase, cellulase, and β-galactosidase are key enzymes involved in fruit softening. These enzymes act sequentially to degrade the pectic substance situated in the middle lamella of the cell wall (Fig. 2.2). Genes that are responsible for enzymes leading to degeneration of cell walls, namely, polygalacturonase, pectin methylesterase, expansin, galactosidase, and β-galactosidase have been found to be linked with the decrease in firmness of the fruit (Wang et al. 2021).

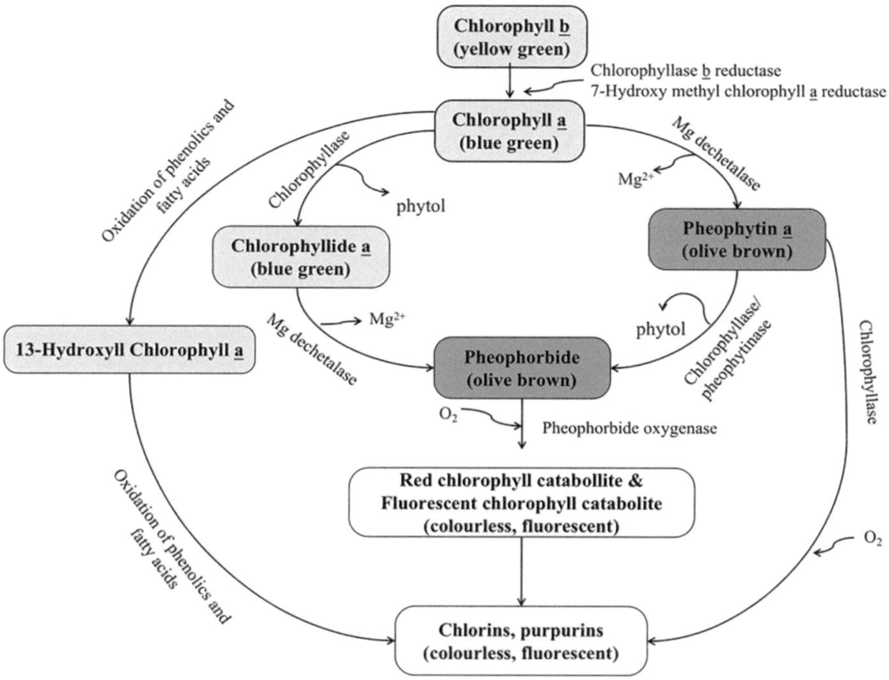

Fig. 2.1 Chlorophyll degradation pathway

2.6.1.2 Biochemical Changes

Starch

Depending on the fruit type and where it ripens (whether on or off the plant), sugar levels tend to rise either due to the import of sugars from the parent plant or the conversion of stored starch within the fruit. As the fruit undergoes maturation, the starch reserves it contains start to decompose into simpler sugars like glucose, fructose, or sucrose. This is a crucial stage in the ripening process. Enzyme invertase breaks down more complex sugars, such as sucrose, into simpler ones, such as glucose and fructose. For specific crops like potatoes and peas, the conversion of starch into sugars is a critical step. In peas, higher levels of sucrose are observed at immature stage, which then transform into starch as maturity is reached. Changes in carbohydrates involve conversions such as starch to sugar (unwanted in certain crops like potatoes and cassava but beneficial in other fruits), sugar to starch (unwanted in peas and sweet corn but advantageous in potatoes and cassava), and the catabolism of complex starch molecules into simple sugars, CO_2, and water through respiration (Kader 1985). Starch reserves within fruits significantly contribute to their sugar content. Bananas, for instance, experience a decrease in starch content throughout postclimacteric ripening stages, accompanied by a simultaneous increase in soluble sugars. The breakdown of starch into glucose, fructose, or sucrose is a defining

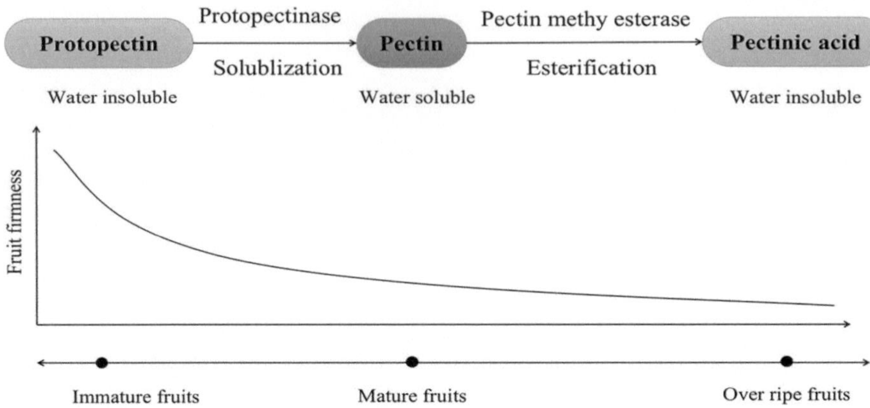

Fig. 2.2 Pectin degradation during fruit ripening

characteristic of ripening in many fruits, driven by a series of enzymes that degrade starch and produce energy.

Enzymes like α-amylase and β-amylase play crucial roles in fragmenting starch into glucose and maltose. Maltose can be converted to glucose via α-glucosidase or directly into glucose-1-phosphate with maltose phosphorylase, and further into glucose-6-phosphate through phosphoglucomutase. These reactions primarily occur in the chloroplasts when fruits are immature and rich in starch. The breakdown products, glucose or glucose-1-phosphate, are then utilized mainly in the cytoplasm. While the exact mechanism for the movement of breakdown products of starch within fruit cell membranes remains unclear, it appears that products are likely to move as six-carbon sugar phosphates or, more probably, as triose phosphates. Within the cytoplasm, the products of starch breakdown are either used in glycolytic respiration or they get converted back into glucose phosphate and fructose for sucrose synthesis. The enzyme sucrose synthase facilitates the formation of sucrose. In contrast, the breakdown of sucrose is likely catalyzed by invertase, which is commonly found in fruits and vegetables and becomes more active during ripening.

Organic Acids

Organic acids play a crucial role in fostering the growth and development of fruits while also bolstering defense mechanisms against various postharvest ailments. Typically, fruits lean towards acidity, while vegetables tend to be more alkaline, with some exceptions, like tomatoes and rhubarb. Fruits tend to accumulate more concentrations of organic acids. The overall acid levels in different fruits such as bananas and limes range approximately from 0.3% to 0.8%. The prominent organic acids present in fruits is citric acid (commonly found in mango, guava, pineapple, pear, tomato, and citrus fruits excluding sweet lime), malic acid (found in apple, banana, cherry, and watermelon), tartaric acid (abundant in grape and tamarind), quinic acid (present in kiwifruit and blueberry), and ellagic acid (notably in strawberry and jamun). In fruits, organic acid content usually declines gradually, especially as they get closer to the ripening phase. Nevertheless, in fruits such as bananas and pineapples, the acid content remains significantly high even after ripening. Generally, a decrease in acidity during ripening or storage of many fruits may be due to an increase in membrane permeability, which allows organic acids (e.g., malic acid in apples and pears) to be stored within respiring cells, as acid salts (malates). This decrease is often associated with a reduction in the amount of organic acids that are transported from the leaves. The reduction in organic acid content after fruit maturity could be due to a decrease in the ability to synthesize acids or due to translocation of sugar and also as a result of dilution with increase in the volume of fruit.

Flavor/Aroma Compounds

Flavor comprises three elements: aroma, taste, and mouthfeel. While the taste of fruits comes from a complex blend of sugars, organic acids, phenolics, and volatile compounds, the unique flavor profile of any particular produce is largely credited to specific volatile compounds that contribute to its aroma. The key compounds that contribute to the taste are esters, alcohols, aldehydes, acids, and ketones. These volatiles are natural, small molecule/compound (typically <250 Dalton). Usually, the range of flavor compounds found in fruits and vegetables is around 10 mg/100 g (Gundewadi et al. 2018). Important flavor imparting compounds in different fruits and vegetables are listed in Table 2.1.

Mineral and Vitamin Composition

The mineral levels, in most cases, remain relatively consistent as fruits ripen, meaning the amount of potassium in a green banana is likely to be similar to that in a ripe one. However, antioxidants generally tend to increase as fruits ripen. Vitamin C (L-ascorbic acid) serves crucial antioxidant and metabolic roles in both plant and animal systems. As the fruit grows, its vitamin C content tends to increase gradually. However, in fruits/vegetables undergoing ripening or postharvest handling, such as for pear, sweet potato, potato, asparagus, and okra, the levels of ascorbic acid decline. On the other hand, pome, stone, and berry fruits exhibit lower levels of ascorbic acid at the time of harvest. The rise in vitamin C content during ripening is associated with an enhancement in the synthesis of metabolic intermediates that facilitate the production of vitamin C precursors. Conversely, the decrease in vitamin C in later stages of ripening may result from its oxidation into dehydroascorbic acid. The ascorbic acid oxidase enzyme catalyzes this oxidation process. The total vitamin C content in vegetables is determined by measuring both types of vitamin C, namely, ascorbic acid and dehydroascorbic acid. Dehydroascorbic acid is the major initial product of ascorbic acid oxidation, formed through enzymatic and nonenzymatic oxidation processes. The surge in ascorbic acid content during the ripening in some fruits may be attributed to an increase in lipid peroxidation because fruit ripening is often considered as an oxidative process involving the turnover of active oxygen species. Notably, ripened bell pepper exhibit vitamin C content about 50% higher than that of its green stage while vitamin A content is almost ten times higher (Nerdy 2018). Differences were noticed in nutrient status during ripening between plant-attached and plant-detached tomato fruits. Plant-detached fruits had

Table 2.1 Flavor imparting compounds as present in different fruits and vegetables

Fruit/vegetable	Stage/flavor/odor	Compound
Fruits and vegetable	Grassy flavor Pungent, penetrating odor	Hexanal Acetaldehyde
Pineapple cv. Jaldoop	Alcohol flavor	Ethyl acetate
Apple	Green stage Ripe stage	Hexanal, 2-hexenal Ethyl 2- methylbutarate
Pear		Butyl ethanoate
Banana	Green stage Ripe stage Over ripe stage	2- hexenal Eugenol Isopentanol
Grapefruit		Nootakatone
Lemon		Citral
Mango		Ocimene, myrcene and dimethylstyrene
Papaya		Methyl butanoate
Strawberry		Furaneol
Raspberry		l-(π Hydroxyphenyl)-3-butanone
Cucumber		2, 6-nonadienal
Mushroom		l-Octen-3-ol, lenthionine
Potato		Ethyl pyrazine, dimethyl pyrazine
Grape	Muscat odor Foxy odor	Methyl salicylate Methyl anthranilate
Cabbage	Pungent odor	Allyl isothiocyanate
Cauliflower	Sulfur odor	Methanethiol
Radish		4-Methylthio-3-trans-butenyl isothiocyanate
Bell pepper	Green stage	3-Isopropyl-2-methoxypyrazine
Pumpkin		Hexanal, 1-hexanol, heneicosane
Muskmelon		Ethyl butyrate, ethyl 2-methylbutyrate, nonadienal, ethyl 3-(methylthio) propanoate

more of Mg, Ca, Mn, Zn and Cu in their outer pericarp tissue. Slow ripening of tomato was found to be associated with higher levels of total B and Cu and lower levels of total K and Ca and in outer pericarp of fruits (Paul and Srivastava 2007; Paul et al. 2012a).

Phenols and Flavonoids

Phenols and flavonoids are crucial bioactive compounds that are key to the quality and flavor profiles of fruits and vegetables. The ripening stage substantially impacts the overall content of phenolics and anthocyanins and their radical scavenging activities. As fruits and vegetables ripen, there is a gradual shift in specific secondary metabolites via initiation of phenolic pathways. Phenols play pivotal roles in both, fruit pigmentation and resistance against pathogens (Ghasemzadeh and Ghasemzadeh 2011). Various factors, including genotype, cultivation practices, preharvest environmental conditions, harvest maturity, storage, and processing techniques, profoundly affect the cumulation of phenolic content in fruits and vegetables. Kumar et al. (2018) recorded significant variation in phenolic compounds in 22 apple genotypes grown under the hilly region of India with chlorogenic acid and phloridzin dehydrate being the predominant ones. Likewise, Nayak et al. (2020, 2024) also reported significant variations in the phenolic content of citrus grown in the semiarid region as a function of genotype and stage of maturity. In most fruits, the phenolic content tends to decrease from the initial growth phases until reaching to the physiological maturity. Throughout growth and maturation, phenolic compounds undergo intricate metabolic processes, leading to significant compositional alterations in plants and plant-derived products (Prasanna et al. 2007). These compounds have gained attention due to their innate antioxidant capacities, which offer notable health benefits in preventing oxidative stress-related illnesses like cancer and cardiovascular diseases. Flavonoids, present in different horticultural products, exist in conjugated glycosylated or esterified forms (Liu 2004). Specific flavonoid compounds, namely, anthocyanins and quercetin are found in fruits and vegetables (El-Ramady et al. 2015).

2.6.1.3 Physiological Changes During Ripening

Respiration

Respiration involves breaking down of organic reserves like carbohydrates, proteins, and fats into simpler molecules, yielding energy as (ATP) adenosine triphosphate. This process catabolizes respiratory substrates such as carbohydrates, proteins, lipids, and organic acids. The type of substrate utilized in respiration is discernible by its respiratory quotient (RQ), which represents the ratio of produced CO_2 to consumed O_2. Different substrates possess varying RQ values. For carbohydrates, RQ will be 1; for organic acids, RQ is >1; while for lipids, RQ will be <1. Higher RQ values indicate the possibility of anaerobic respiration. Respiration, an exothermic reaction, contributes to elevating the temperature of commodities. This is usually done by alternate respiration or cyanide-resistant respiration. Climacteric fruits, in particular, show a higher fraction of alternate respiration for the fruits undergoing the ripening process and thereby enhancement in the temperature of fruits such as mango and banana. Its pace varies across commodities and their maturity stages, with higher rates usually inversely related to the shelf-life of fresh produce. Respiratory metabolism involves four primary pathways: glycolysis, citric acid cycle, electron transport and pentose-phosphate shunt. Unlike the varied and unique reactions in secondary metabolism, these pathways encompass sets of reactions common to all vascular plants. In postharvest research, the main aim has been to uphold quality while extending the shelf-life of harvested horticultural products and this primarily involves controlling the rate of respiratory metabolism (Fig. 2.3).

As fruits and vegetables undergo ageing, reserved food material is consumed through respiration to sustain energy levels and viability. This also results in loss of flavor, quality, and marketable weight, which is particularly significant for dehydrated products. The respiration rate primarily hinges on temperature, atmospheric composition, and mechanical damage. It serves as the primary driver for the deterioration of fresh produce as respiratory metabolic processes tend to double with every 10 °C rise in temperature (Fagundes et al. 2013). Altering the oxygen and carbon dioxide levels around the fruit significantly reduces the respiration, thereby extending their shelf-life. However, certain fruits like tomatoes, mangoes, bananas, and apples deviate from the standard respiratory pattern by exhibiting a marked increase in respiration during ripening, termed as a respiratory climacteric. In contrast, fruits like citrus, pineapples, and strawberries, categorized as nonclimacteric, do not undergo this significant rise in respiration during ripening. Despite that, nonclimacteric fruits undergo ripening-related changes, however at a slower pace compared to climacteric fruits. The respiratory rate of fruits and vegetables serves as an indicator to design storage conditions for maximizing their longevity.

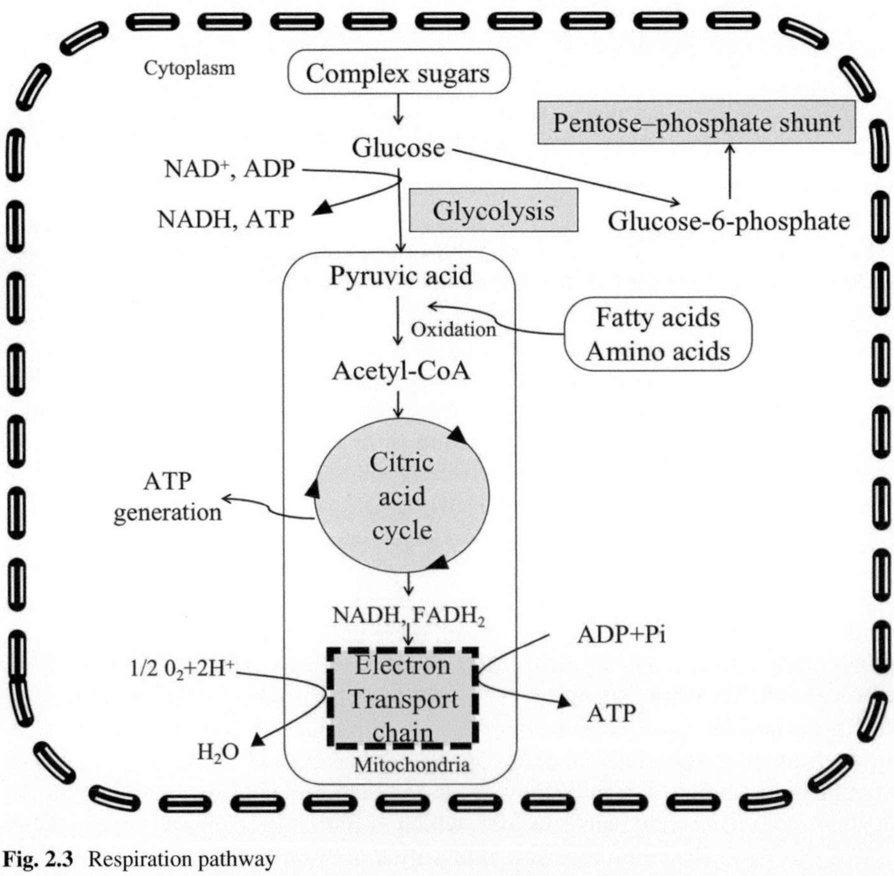

Fig. 2.3 Respiration pathway

2.7 Ethylene

Distinction between climacteric and nonclimacteric fruits can be done by observing their response to applied ethylene and the pattern of ethylene production during ripening. It has been established that all fruits produce small amounts of ethylene during their development. However, as they ripen, climacteric fruits produce considerably higher levels of ethylene compared to nonclimacteric ones. This disparity is evident in their internal ethylene concentrations at various development and ripening stages. Climacteric fruits exhibit varying internal ethylene concentrations, whereas nonclimacteric fruits show minimal changes during development and ripening. Ethylene can cause multiple increases in respiration in nonclimacteric fruits, which differs from the single but significant respiration boost that occurs in climacteric fruits.

A saying “a bad apple affects the whole bunch” holds literal significance. To understand how different produce ripens and how they should be stored together are crucial for prolonging their freshness and reducing postharvest losses. It is essential not to store high ethylene-producing produce with ethylene-sensitive ones. For instance, storing spinach, kale, or broccoli in the same bin as apples could lead to discoloration and rapid spoilage of these vegetables within a few days. Furthermore, avoid storing potatoes and onions together because potatoes emit ethylene and this causes premature sprouting in onions. Damaged, bruised, or punctured fruit also tend to produce more ethylene, hastening the ripening, softening, and spoilage of neighboring produce. It is advisable to bifurcate damaged fruits first and keep them separate from unblemished ones during storage. Some nonclimacteric fruits and vegetables are adversely affected by ethylene and show physiological disorders such as Russet spotting of lettuce, and bitterness in carrots (due to isocoumarin production).

Ethylene Biosynthesis

In 1980, Sgang Fa Yang discovered ethylene biosynthesis pathway, which is also referred as the Yang Cycle (Yang 1980). This process begins with methionine, a sulfur-containing amino acid, which is activated using ATP, leading to the formation of S-adenosyl methionine (SAM) and the release of pyrophosphate (Fig. 2.4). The enzyme SAM synthetase (ATP-methionine S-adenosyl transferase) catalyzes this reaction. In S-adenosyl methionine, the sulfur atom from methionine is bonded to the C-5 of the ribose group in adenosine. Interestingly, selenomethionine (Se-Met) has been discovered as a more effective substrate for ethylene biosynthesis than methionine. SAM synthetase shows a stronger affinity for selenomethionine compared to methionine. The subsequent stage in ethylene biosynthesis involves the conversion of SAM to 1-amino cyclopropane carboxylic acid (ACC) via the enzyme ACC synthase (ACS). This step is considered a bottleneck in ethylene production, and any factor influencing the activity of ACC synthase impacts ethylene synthesis. The final phase in ethylene biosynthesis is the oxidative breakdown of 1-amino cyclopropane carboxylic acid (ACC) by the enzyme ACC oxidase (ACO), previously known as ethylene-forming enzyme. This process results in the formation of ethylene, CO_2, and (HCN) hydrogen cyanide. Eventually, HCN is converted to β-cyanoalanine by the enzyme β-cyanoalanine synthase.

Ethylene Signaling

One of the initial hypotheses in ethylene signaling is known as the triple response. The exposure of etiolated seedlings to ethylene causes seedlings to exhibit an amplified curvature in the apical hook. This phenomenon, initially documented by Neljubow (1901), is identified as the triple response. This response has proven to be a valuable test for screenings, aiding in the discovery of numerous elements within the ethylene signaling pathway. A group of receptors that detect ethylene is present on endoplasmic reticulum (ER). There are several key components for ethylene signaling including a family of receptor viz. a protein kinase named constitutive triple response 1, an ER-localized transmembrane protein known as ethylene-insensitive 2 with unspecified biochemical activity, and transcription factors like EIN3, EIN3-like, and ethylene response factors (ERFs). These receptor proteins bind with ethylene and rely on the essential copper (Cu) cofactor, crucial for their functionality, which is provided by responsive to-antagonist-1 (RAN-1) located within the golgi complex. RAN-1 integrates copper into the hydrophobic regions of receptors, a process vital for the receptors to bind correctly. Ethylene, produced via the Yang Cycle, attaches to these receptors in the ER region, initiating a signaling

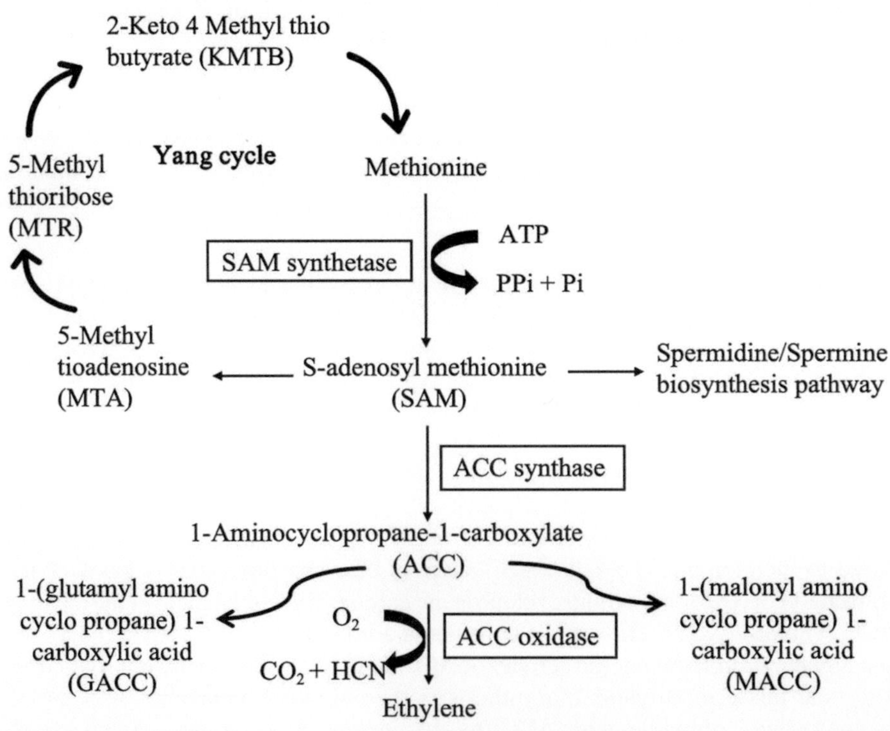

Fig. 2.4 Ethylene biosynthesis pathway

cascade to constitutive triple response-1 (CTR-1). The receiver domain of these receptors transfers phosphate to CTR-1, resembling rapidly accelerated fibrosarcoma kinase in the mitogen-activated protein kinase (MAPK) system. CTR-1 further activates EIN-2, which triggers the EIN-3 family of transcription factors, ultimately prompting the expression of ethylene responsive factor-1 (ERF-1). This activation leads to extensive alterations in gene expression, ultimately impacting the physiology of ripening and senescence. Kinase enzyme will be inactive, without phosphorylation and downstream responses are activated due to absence of mitogen-activated protein kinase (MAPKK) signals. In the absence of ethylene and in the case of mutant receptors, kinase activation occurs, along phosphorylation, which inhibits the downstream responses by producing MAPKK signals (Chen et al. 2005).

2.8 Classification of Fruits and Vegetables Based on Ripening Behavior

Fruits and vegetables are commonly categorized as either climacteric or nonclimacteric based on their respiration pattern and ethylene release rates. It is also true that there is not very clear and distinct boundary between these classifications as it is the sensitivity towards ethylene that plays a more important role (Paul et al. 2012b). Certain fruits such as guava, melon, Japanese plum, Asian pear, and pepper often challenge the traditional perception of fruit ripening as exclusively climacteric or nonclimacteric. These fruits display characteristics of both climacteric and nonclimacteric ripening, influenced by their specific cultivar or genetic makeup. For instance, while guava varieties Pedro Sato and Jen-Ju Bar are nonclimacteric, others such as Allahabad Safeda and most Indian varieties exhibit climacteric traits. Researchers have utilized tomatoes as a model for understanding the ripening mechanisms in climacteric fruits and strawberries as a model for nonclimacteric fruits.

Climacteric Fruits

Climacteric fruits and vegetables form a diverse group in the world of produce, distinguished by their unique ripening characteristics and behaviors. Understanding these distinctions is crucial for growers, suppliers, and consumers alike due to their impact on postharvest handling, storage, and overall quality. Climacteric fruits and vegetables undergo a specific ripening phase that sets them apart from the nonclimacteric group. Fruits such as bananas, tomatoes, apples, mangoes, and more, as well as certain vegetables like tomatoes and specific pepper varieties, exhibit a surge in respiration and ethylene production during their ripening process (Fig. 2.5). This phase, known as the climacteric phase, allows them to continue ripening even after they have been harvested.

During the climacteric phase, these fruits and vegetables undergo significant physiological and biochemical changes. For instance, cell wall degradation in fruits leads to a softening of texture while sugars increase, enhancing sweetness. There is also a decrease in acidity, balancing the overall taste profile. Simultaneously,

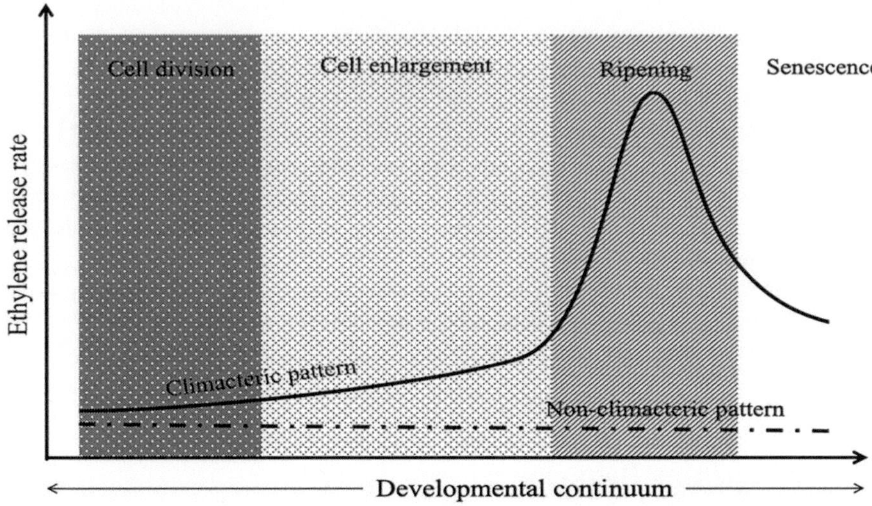

Fig. 2.5 Ethylene release pattern of climacteric and nonclimacteric commodities

aromatic compounds intensify, enhancing the flavor and fragrance, making these produce items more palatable and appealing (Prasad et al. 2022; Menaka et al. 2024; Jayarajan et al. 2024).

One of the primary challenges with climacteric fruits and vegetables lies in their postharvest management. Their ability to continue ripening after their harvest poses significant challenges in storage, transportation, and shelf-life maintenance. Ethylene, a gaseous plant hormone, plays a pivotal role in the continuation of the ripening process. Exposure to ethylene can hasten the ripening of neighboring produce, leading to accelerated softening and potential spoilage. To counter these challenges, various strategies are employed in the postharvest handling of climacteric fruits and vegetables. Controlled atmosphere storage, involving the adjustment of oxygen and carbon dioxide levels, is a popular method to slow down respiration and extend shelf-life. Additionally, the use of ethylene inhibitors or scavengers helps in mitigating the effects of ethylene exposure, and thereby prolonging the storage life of these produce items. Climacteric fruits and vegetables hold significant economic importance in global agriculture and the food industry. Their desirability, enhanced flavor, aroma, and nutritional value make them sought-after commodities in markets worldwide. However, their perishable nature requires careful handling, storage, and transportation practices to maintain quality and extended availability in the market. Ongoing research and advancements in technology are continuously refining postharvest handling and storage methods for climacteric produce. Presently, researchers aim to further understand the intricacies of climacteric fruit and vegetable biology, mainly focusing on ethylene regulation, genetic modifications, and innovative storage techniques. These endeavors aim to enhance fruit and vegetable quality, prolonged shelf-life, and reduction in postharvest losses and thereby ensuring consistent availability and high-quality produce for consumers globally.

Nonclimacteric Fruits

Nonclimacteric fruits do not undergo the autocatalytic ethylene production seen in climacteric fruits during ripening. However, some studies note a slight increase in ethylene in nonclimacteric fruits as well as in grapes and strawberries during ripening. In grapes, inhibiting ethylene perception affected various ripening processes, suggesting its involvement. Similarly, in citrus fruits, ethylene contributes to color changes. However, the timing of ethylene production in nonclimacteric fruits does not tightly align with the ripening phases seen in climacteric fruits. While ethylene might impact certain aspects of fruit ripening in nonclimacteric fruits, it is unlikely to control the entire process solely. Multiple factors, some unidentified so far, likely interact to drive ripening in these fruits. Common regulatory mechanisms shared between climacteric and nonclimacteric fruits have been suggested based on gene expression patterns in tomatoes and peppers. Studies comparing gene expression in different fruits, such as tomatoes, grapes, and peppers, have been conducted, but definitive conclusions still need to be made in view of various genetic differences among these species. Research exploring the genetic basis of climacteric and nonclimacteric fruit traits is going on including studies manipulating ethylene biosynthesis and ripening processes. Ripening determinants and their interactions in these fruits are more complex (Paul and Pandey 2016) and thereby need to be fully understood with further investigations.

Suppressed-Climacteric Fruits

Compared to their climacteric counterparts, these types of fruits produce only a small amount of ethylene, possibly due to their limited ability to convert ACC into ethylene. As a result of their inability to detect ethylene and/or develop fresh receptors, these fruits do not exhibit climacteric traits. During the final stages of ripening, fruits with this trait produce ethylene, but their ethylene levels are lower than those of typical climacteric fruits, and they exhibit a less pronounced respiratory climacteric. This phenomenon has been observed in certain cultivars of plum, apple, kiwifruit, and pear. Plum cultivars with suppressed climacteric properties, such as Late Santa Rosa, Casselman, Roysum Shiro, Ruby-Red, and Angeleno, produce detectable amounts of ethylene and undergo slow softening when exposed to exogenous ethylene or propylene.

Semiclimacteric Fruits

This demonstrates that the respiration rate peaks during the mature green stage, whereas ethylene production surges during the red-ripe stage, and conversely, they are absent during the opposite stages. There was no observable increase in the respiration rate during the transition in color. Commercially, peppers are harvested at the mature green stage. Pepper pods, due to their indeterminate growth habit, exhibit characteristics that align with both climacteric and nonclimacteric fruits, leading to their designation as semiclimacteric (Gross et al. 1986). Specifically, "Cherry Bomb" peppers at the mature-green stage tend to generate significant amounts of CO_2 compared to fruit at other maturity stages. This phenomenon is also observed in various other fruits such as cayenne pepper (*Capsicum annuum* L), guava, plum, ber, fig, olive, and muskmelon.

Pseudoclimacteric Fruits

Ethylene is a crucial regulator of ripening in climacteric fruits. However, it also plays a role in nonclimacteric fruits, probably by different metabolic networks in which changes in tissue sensitivity to the gas or interactions with other factors may be necessary. So, other ripening models (referred to as pseudoclimacteric), in addition to the classical climacteric and nonclimacteric, may be considered, comprising fruits with different ethylene requirements and sensitivity to even very small amounts of the gas (Grierson 2013). A pseudoclimacteric pattern of respiration was found in very young immature fruit of pomegranate cv. Wonderful (Ben-Arie et al. 1984) and also in oranges and grapefruit (Eaks 1970). Overall and generalized examples of climacteric and nonclimacteric fruits and vegetables are listed in Table 2.2.

In climacteric fruits and vegetables, the application of external ethylene accelerates their ripening process by triggering endogenous ethylene synthesis and activating genes associated with ripening. This leads to faster softening, color changes, conversion of starches to sugars, and the development of aroma and flavor. For commercial purposes, ethylene is often used to uniformly ripen climacteric fruits and vegetables after harvest. While in nonclimacteric fruits, the application of external ethylene has a limited impact on their ripening process. However, ethylene can influence certain aspects, such as degreening (removal of chlorophyll) in citrus fruits, making them visually appealing for the market. Despite this, the overall ripening mechanisms in nonclimacteric fruits are primarily regulated by other hormonal and environmental factors, and their response to ethylene remains minimal compared to climacteric commodities.

Alternate Respiration and Fruit Ripening in Climacteric and Nonclimacteric Fruits

It was found that alternate respiration contributes to the rise in the temperature of mango and banana fruits. During the ripening of mango, internal temperature increased from 29.0 to 38.9 °C. This rise was coupled with increase in total and alternate respiration of fruit (Kumar et al. 1990). Alternate respiration increased from 38% to 73% of total respiration during ripening of banana (Kumar and Sinha 1992). Later, similar findings were reported by Reddy and Srivastava (1998). A future line of work was thereby suggested wherein possibility of manipulating alternate respiration to influence the ripening and ripening-related changes can be probed (Paul et al. 2012b).

Other Factors Affecting Ripening and Intervarietal Differences in Ripening (Tomato and Mango)

Anatomical features determine the internal microenvironment and this regulates ripening rate and ripening-related changes in tomato fruit primarily by determining gaseous (O_2 and CO_2) and moisture exchange across the fruit's boundary (Paul and Srivastava 2006; Paul et al. 2010; Paul and Pandey 2016). Likewise, differences in surface morphology and internal anatomical features determine the ripening behavior of mango by regulating endogenous level of ethylene, moisture and O_2 to CO_2 ratio within the fruit (Paul et al. 2007, 2019).

Table 2.2 Examples of climacteric and nonclimacteric fruits and vegetables

Climacteric	
Apple	Apricot
Bael	Banana
Blackberry	Blueberry
Breadfruit	Cantaloupe
Cape gooseberry	Cherimoya
Datepalm	Durian
Fig	Guava
Jackfruit	Jujube
Mango	Mangosteen
Papaya	Passionfruit
Peach	Pear
Persimmon	Plum
Sapota	Soursop
Sugar apple	Tomato
Nonclimacteric	
Cacao	Cactus pear
Cashew	Carambola
Cucumber	Cherry
Grapefruit	Grape
Java plum	Indian gooseberry
Litchi	Lemon
Loquat	Longan
Orange	Mandarin
Pineapple	Pepper
Pomegranate	Dragon fruit
Rose apple	Pummelo
Tamarind	Strawberry
Olive	Watermelon

Systems of Ethylene Production

Discrimination between climacteric and nonclimacteric fruits has been made on the basis of the presence or absence of the climacteric rise in respiration and autocatalytic ethylene production. The main specificity of climacteric fruits is their capacity to produce ethylene through a so-called autocatalytic process in which ethylene stimulates its own biosynthesis. It has been demonstrated that two systems exist for the production/regulation of ethylene biosynthesis. System 1 refers to the low levels of ethylene production in immature developing fruit and vegetative tissues. The factors controlling system 1 are unknown, but different forms of ACC synthase (ACS) and ACC oxidase (ACO) are thought to be involved compared to system 2 (Paul et al. 2012b). Once climacteric fruit reaches a particular developmental stage, there is a transition from systems 1 to 2, which is responsible for the production of large amounts of ethylene necessary to initiate ripening-related actions. System 2 is autocatalytic in nature wherein the already produced ethylene triggers further production of ethylene. Many of nonclimacteric fruits and vegetative tissues do not show an active system 2 of ethylene production. However, exogenous treatment of developing fruits with ethylene can induce system 2 of ethylene production (at specified stages) and thereby even premature ripening, but such fruits may have poor eating quality.

2.9 Future Perspective

Understanding the physiology and ripening processes of fruits and vegetables presents numerous opportunities for future research and practical applications. Key areas for future exploration include genetic engineering and biotechnology, where advances could allow for the development of fruit varieties with optimized ripening characteristics, improved nutritional content, and enhanced resistance to diseases and pests. Metabolomics and systems biology can provide a deeper understanding of the metabolic pathways involved in fruit ripening, leading to the identification of key regulatory molecules and manipulation of metabolic pathways to enhance traits such as flavor, color, and nutritional value. Research into climate change adaptation is crucial to develop fruit varieties that can withstand extreme weather conditions and maintain quality and yield. Innovations in postharvest handling and storage technologies can extend shelf-life and maintain the quality of fruits and vegetables, including better packaging materials, advanced refrigeration techniques, and controlled atmosphere storage systems. Exploring sustainable agricultural practices that enhance soil health and reduce the environmental impact of fruit production is essential, involving organic farming methods, integrated pest management, and reducing the carbon footprint of fruit production and transportation. Research into nutritional enhancement through traditional breeding or biotechnological approaches can contribute to better public health outcomes by increasing levels of vitamins, antioxidants, and other beneficial compounds in fruits. Understanding consumer preferences and market trends can guide the development of fruit varieties that meet demands for taste, appearance, and convenience, with participatory breeding programs ensuring new varieties align with market needs. Developing new methods to manage ethylene production and sensitivity in both climacteric and nonclimacteric fruits can improve the control of ripening processes, involving ethylene inhibitors, genetic modifications, or novel storage technologies. Overall, the future perspective on the physiology and ripening of fruits and vegetables is promising, with potential benefits for producers, consumers, and the environment. Continued research and innovation will be key to addressing the challenges and opportunities in this field.

2.10 Summary

This chapter provides an in-depth explanation of the physiological and ripening processes in fruits and vegetables, highlighting the intricate cellular and metabolic transformations that contribute to fruit development and quality. Fruit growth involves significant cell division and cell expansion, leading to increased volume and mass, with metabolic activities producing vital compounds that vary across fruit types. The time from flowering to ripening varies by species, affecting the final size and weight of the fruit, and is marked by phases of cell division, expansion, and metabolic changes. Fruits exhibit different growth patterns, such as single, double, and triple sigmoid curves, describing the phases of development from rapid growth to maturation. Ripening is a crucial stage marked by changes in color, firmness, and biochemical composition, enhancing the sensory qualities of the fruit, making it palatable and nutritious. During ripening, starch converts to sugars, organic acids accumulate, and vitamin and antioxidant levels change, impacting the fruit's taste, nutritional value, and shelf-life. Ethylene plays a central role in ripening, especially in climacteric fruits, which exhibit a significant increase in ethylene production. The chapter outlines the biosynthesis, signaling, and effects of ethylene on fruit ripening. Fruits are categorized as climacteric or nonclimacteric based on their ethylene production and respiration patterns during ripening, with implications for postharvest handling and storage. Overall, understanding the physiology and ripening mechanisms of fruits and vegetables is essential for improving their quality, shelf-life, and postharvest management.

References

Almansa S, Hernandez F, Legua P (2016) Physico-chemical and physiological changes during fruit development and on-tree ripening of two Spanish jujube cultivars (*Ziziphus jujuba* Mill.). J Sci Food Agric 96:4098–4105

Anwar R, Mattoo AK, Handa AK (2015) Polyamine interactions with plant hormones: crosstalk at several levels. In: Kusano T, Suzuki H (eds) Polyamines: a universal molecular nexus for growth, survival and specialized metabolism. Springer, London, pp 267–302

Bebbington M, Hall AJ, Lai CD et al (2009) Dynamics and phases of kiwifruit (*Actinidia deliciosa*) growth curves. N Zeal J Crop Hortic Sci 37:179

Ben-Arie R, Segal N, Guelfat-Reich S (1984) The maturation and ripening of the 'Wonderful' pomegranate. J Am Soc Hortic Sci 109:898–902

Bourdon M, Frangne N, Mathieu-Rivet E et al (2009) Endoreduplication and growth of fleshy fruits. In: Progress in botany 71. Springer Berlin Heidelberg, Berlin/Heidelberg, pp 101–132

Chen YF, Etheridge N, Schaller GE (2005) Ethylene signal transduction. Ann Bot 95:901–915

Connors CH (1919) Growth of fruits of peach. New Jersey Agric Exp Station Ann Rep 40:82–88

Coombe BG (1973) The regulation of set and development of the grape berry. In: Symposium on growth regulators in fruit production, vol 34, pp 261–274

DeJong TM, Goudriaan J (1989) Modeling peach fruit growth and carbohydrate requirements: reevaluation of the double-sigmoid growth pattern. J Am Soc Hortic Sci 114:800–804

Eaks IL (1970) Respiratory response, ethylene production, and response to ethylene of citrus fruit during ontogeny. Plant Physiol 45:334–338

El-Ramady HR, Domokos-Szabolcsy E, Abdalla NA et al (2015) Postharvest management of fruits and vegetables storage. Sustain Agric Rev 15:65–152

Ezura H, Hiwasa-Tanase K (2009) Fruit development. In: Plant developmental biology-biotechnological perspectives, vol 1. Springer Berlin Heidelberg, Berlin/Heidelberg, pp 301–318

Fagundes C, Carciofi BAM, Monteiro AR (2013) Estimate of respiration rate and physicochemical changes of fresh-cut apples stored under different temperatures. Food Sci Technol 33:60–67

Ghasemzadeh A, Ghasemzadeh N (2011) Flavonoids and phenolic acids: role and biochemical activity in plants and human. J Med Plant Res 5:6697–6703

Gillaspy G, Ben-David H, Gruissem W (1993) Fruits: a developmental perspective. Plant Cell 5:1439

Giovannoni JJ, Nguyen C, Ampofo B (2017) The epigenome and transcriptional dynamics of fruit ripening. Ann Rev Plant Biol 68:61–84

Grierson D (2013) Ethylene and the control of fruit ripening. In: The molecular biology and biochemistry of fruit ripening, vol 36. Wiley, pp 43–73

Gross KC, Watada AE, Kang MS et al (1986) Biochemical changes associated with the ripening of hot pepper fruit. Physiol Plant 66:31–36

Gundewadi G, Reddy VR, Bhimappa BB (2018) Physiological and biochemical basis of fruit development and ripening-a review. J Hill Agric 9:7–21

Handa AK, Tiznado-Hernandez ME, Mattoo AK (2012) Fruit development and ripening: a molecular perspective. In: Altman A, Hasegawa PM (eds) Plant biotechnology and agriculture: prospects for 21st century. Elsevier, New York, pp 405–424

Jayarajan S, Sethi S, Awasthi OP et al (2024) Synergistic influence of melatonin- hydrocolloid coating on decay and senescence of nectarine (*Prunus persica* var. *nucipersica*) during supermarket storage conditions. Plan Theory 13:822–835

Kader AA (1985) Postharvest biology and technology an: overview. In: Postharvest technology of horticultural crops, Agriculture and natural resources pub (No. 3311). University of California, California, pp 1–10

Kanayama Y, Kochetov A (eds) (2015) Abiotic stress biology in horticultural plants. Springer, New York

Kumar S, Sinha SK (1992) Alternative respiration and heat production in ripening banana fruits (*Musa paradisiaca* var. Mysore Kadali). J Exp Bot 43:1639–1642

Kumar S, Patil BC, Sinha SK (1990) Cyanide resistant respiration is involved in temperature rise in ripening mangoes. Biochem Biophys Res Commun 168:818–822

Kumar P, Sethi S, Sharma RR et al (2018) Nutritional characterization of apple as a function of genotype. J Food Sci Technol 55:2729–2738

Liu RH (2004) Potential synergy of phytochemicals in cancer prevention: mechanism of action. J Nutr 134:3479S–3485S

Menaka M, Asrey R, Meena NK et al (2024) Effect of melatonin on biochemical changes, antioxidant system and oxidative membrane damage of Indian guava (cv. Barafkhana) during cold storage. S Afr J Bot 169:95–108

Nayak SL, Sethi S, Sharma RR et al (2020) Variation in fruit quality traits and health promoting compounds of citrus fruits grown in semi-arid region. Indian J Hortic 77:635–640

Nayak SL, Sethi S, Dubey AK et al (2024) Impact of on-tree storage and genotype on citrus granulation: Physico-chemical analysis. Erwerbs-obstbau. https://doi.org/10.1007/s10341-024-01049-0

Neljubow DJBCB (1901) Uber die horizontale Nutation der Stengel von *Pisum sativum* und einiger anderen Planzen. Beihefte zum Botanischen Centralblatt 10:128–139

Nerdy N (2018) Determination of vitamin C in various colours of bell pepper (*Capsicum annuum* L.) by titration method. ALCHEMY Jurnal Penelitian Kimia 14:164–177

Paul V, Pandey R (2016) Internal atmosphere of fruits: role and significance in ripening and storability. In: Pareek S (ed) Postharvest ripening physiology of crops. Series: innovation in postharvest technology. CRC Press/Taylor & Francis Group, Boca Raton, pp 359–412

Paul V, Srivastava GC (2006) Role of surface morphology in determining the ripening behaviour of tomato (*Lycopersicon esculentum* Mill.) fruits. Sci Hortic 110:84–92

Paul V, Srivastava GC (2007) Minerals in pericarp of tomato (*Solanum lycopersicum* L.) fruit and its ripening behaviour. J Appl Hortic 9:140–145

Paul V, Malik SK, Srivastava GC (2007) Intervarietal differences in the surface morphology and anatomy of mango (*Mangifera indica* L.) fruits. Phytomorphology 57:211–220

Paul V, Pandey R, Srivastava GC (2010) Ripening of tomato (*Solanum lycopersicum* L.). Part II: regulation by its stem scar region. J Food Sci Technol 47:527–533

Paul V, Pandey R, Ramesh KV et al (2012a) Role of mineral nutrients in physiology, ripening and storability of fruits. In: Nutriophysiological and molecular interventions for crop improvement under changing climate, vol 13. Scientific Publishers (India), pp 56–96

Paul V, Pandey R, Srivastava GC (2012b) The fading distinctions between classical patterns of ripening in climacteric and non-climacteric fruit and the ubiquity of ethylene—an overview. J Food Sci Technol 49:1–21

Paul V, Pandey R, Malik SK (2019) Varietal variations in rate of ripening and respiration of mango (*Mangifera indica* L.) fruits: anatomical substantiation. Plant Physiol Rep 24:340–350

Pearson JA, Robertson RN (1953) The physiology of growth in apple fruits IV. Seasonal variation in cell size, nitrogen metabolism, and respiration in developing Granny Smith apple fruits. Aust J Biol Sci 6:1–20

Pei MS, Cao SH, Wu L et al (2020) Comparative transcriptome analyses of fruit development among pears, peaches, and strawberries provide new insights into single sigmoid patterns. BMC Plant Biol 20:1–15

Possner DRE, Kliewer WM (1985) The localization of acids, sugars, potassium and calcium in developing grape berries. Vitis 24:229–240

Prasad K, Sharma RR, Asrey R et al (2022) Hydrocolloid edible coatings extend shelf life and reduce postharvest decay and maintain quality of mango fruits (*Mangifera indica* L.) under ambient storage. J Food Biochem. https://doi.org/10.1111/jfbc.14481

Prasanna V, Prabha TN, Tharanathan RN (2007) Fruit ripening phenomena–an overview. Crit Rev Food Sci Nutr 47:1–19

Reddy YV, Srivastava GC (1998) Malic enzyme and cyanide resistant respiration in mango during ripening. Indian J Plant Physiol 3:185–187

Sethi S, Srivastav M, Samuel DVK et al (2011) Evaluation of newly developed mango hybrids for their storage behaviour and peel colour. Indian J Agric Sci 81:58–61

Seymour GB, Ostergaard L, Chapman NH et al (2013) Fruit development and ripening. Ann Rev Plant Biol 64:219–241

Sulaimankhil Z, Sethi S, Sharma RR et al (2021) Influence of aqueous hexanal on quality of 'Royal Delicious' apple during cold storage. Acta Physiol Plant 43:134

Su-Mon M, Asrey R, Meena N et al (2022) Attenuating sugar spot and retaining quality of banana fruits by combined use of hot water and calcium lactate during storage. Food Sci Technol Int 30:49–60

Valero D, Serrano M (2010) Postharvest biology and technology for preserving fruit quality, 1st edn. CRC Press, Boca Raton, p 287

Wang Y, Deng L, Meng J et al (2021) Transcriptomic and metabolic analyses reveal the mechanism of ethylene production in stony hard peach fruit during cold storage. Int J Mol Sci 22:11308

Wills R, Mc Glasson B, Graham D et al (1998) Postharvest: an introduction to the physiology and handling of fruit, vegetables and ornamentals. CAB International, Oxford

Yang SF (1980) Regulation of ethylene biosynthesis. Hortic Sci 15(3):238–243

Chapter 3
Postharvest Handling, Packaging and Storage of Fruits and Vegetables

Anu Chandel and OP Chauhan

3.1 Introduction

Fruits and vegetables are crucial for a healthy diet, providing essential vitamins, minerals (mainly electrolytes), dietary fiber, and phytochemicals, which function as antioxidants, anti-inflammatory compounds and show protective mechanisms (Slavin and Lloyd 2012; Barman et al. 2018). However, these perishable commodities are susceptible to a multitude of postharvest challenges that significantly impact quality, safety, and economic viability throughout the supply chain. Effective postharvest handling practices are essential to reduce these losses and assure the delivery of fresh, high-quality produce to consumers. These challenges include physiological deterioration processes such as water loss and respiration, which can lead to wilting, weight loss, and reduced marketability (Brasil and Siddiqui 2018; Yahia and Carrillo-Lopez 2018). Additionally, fruits and vegetables are susceptible to mechanical damage during harvest, transportation, and processing, which can create entry points for microbial spoilage and decay organisms, compromising both safety and quality (Raju et al. 2018). Furthermore, certain fruits and vegetables are highly sensitive to ethylene gas, a natural plant hormone that promotes ripening. Exposure to ethylene can accelerate ripening and lead to overripe, unmarketable produce (Shewfelt and Prussia 2022; Florkowski and Banks 2021).

A. Chandel (✉) · OP Chauhan
Defence Food Research Laboratory, Mysore, India

OP Chauhan (ed.), *Fruits and Vegetables Technologies*,
https://doi.org/10.1007/978-981-96-8433-5_3

3.2 Postharvest Losses

Depending on the commodity and location, postharvest losses of fruits and vegetables worldwide are estimated to be between 20% and 50% (Elik et al. 2019). These losses translate into substantial economic repercussions for producers, distributors, and consumers. Furthermore, postharvest losses contribute to increased food insecurity, particularly in developing countries where access to fresh produce is often limited (Hasan et al. 2019). Implementing effective postharvest management practices can significantly reduce these losses, leading to improved food security, economic benefits for stakeholders, and a more sustainable food system (Gardas et al. 2017). With a substantial 9.3% share of global production, India is second in the production of fruits and vegetables worldwide (Singh and Sharma 2018). In 2020–2021, the nation's horticultural output reached a record high of 326.6 million tonnes, surpassing even food grain production (Ali et al. 2018). This impressive production translates to leading positions in various fruits and vegetables. India reigns supreme in mangoes (including mangosteens and guavas) with a staggering 43.80% of global production, while also holding the top spot for bananas (26.45% of global production) and ranking second globally for papayas (39.30% of global production). Similar success extends to vegetables, with India securing the number two position for potatoes, onions, cauliflowers, brinjal (eggplant), and cabbages (Food and Agriculture Organization Corporate Statistical Database (FAOSTAT)). As illustrated in Fig. 3.1, significant food wastage occurs throughout the entire food supply chain, from production to consumption.

3.3 Maximizing Quality and Shelf Life

Postharvest handling encompasses all practices employed from harvest to consumption, aiming to maintain the quality and safety of fruits and vegetables (Hussein et al. 2020). Effective postharvest practices extend the shelf life of produce by slowing down the natural deterioration processes that occur after harvest. The key steps involved in postharvest handling of fruits and vegetables are shown in Fig. 3.2. The key factors influencing postharvest quality are as follows:

- *Respiration*: fruits and vegetables are living tissues that continue to respire after harvest, consuming oxygen and releasing carbon dioxide. High respiration rates accelerate quality decline, leading to increased water loss, softening, and sugar degradation (Prasad et al. 2018). Proper postharvest practices such as precooling and controlled atmosphere (CA) storage can significantly reduce respiration rates, thereby extending shelf life.
- *Water loss*: fruits and vegetables are composed primarily of water, and excessive water loss leads to wilting, weight loss, and reduced marketability. Maintaining a high relative humidity during storage and transportation minimizes water loss and preserves the fresh appearance of produce (Anjali et al. 2024).

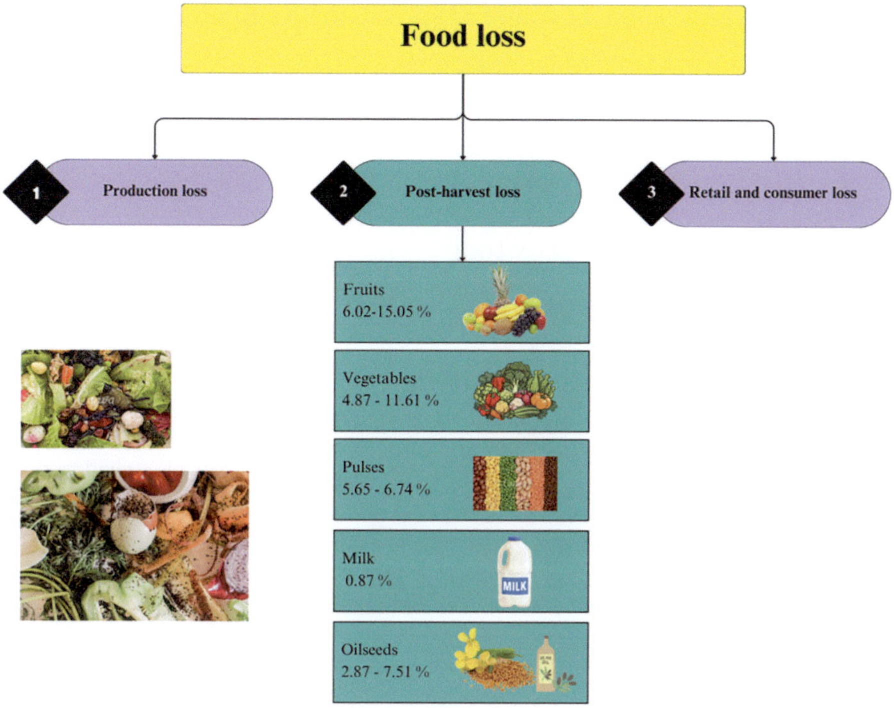

Fig. 3.1 Food wastage throughout the food supply chain

- *Ethylene sensitivity*: certain fruits and vegetables are highly sensitive to ethylene gas, a natural plant hormone that promotes ripening. Exposure to ethylene can accelerate ripening and lead to overripe, unmarketable produce (Lakshmi et al. 2017). Postharvest strategies such as application of ethylene-removal agents or storage in controlled atmosphere conditions with low ethylene levels can mitigate these detrimental effects (Blasco et al. 2017).
- *Mechanical damage*: rough handling during harvest, transportation, and processing can cause physical damage to fruits and vegetables, leading to bruising, cuts, and punctures. These injuries not only detract from the aesthetic appeal of the produce but also serve as entry points for microbial spoilage and decay organisms (Blasco et al. 2017). Implementing proper handling techniques and utilizing appropriate packaging materials can minimize mechanical damage and associated losses.
- *Temperature and humidity*: temperature and humidity are critical factors influencing postharvest quality. Exposure to high temperatures accelerates respiration, water loss, and enzymatic activity, leading to rapid deterioration. Conversely, low temperatures can induce chilling injury in certain fruits and vegetables, resulting in discoloration, pitting, and internal browning. Maintaining optimal

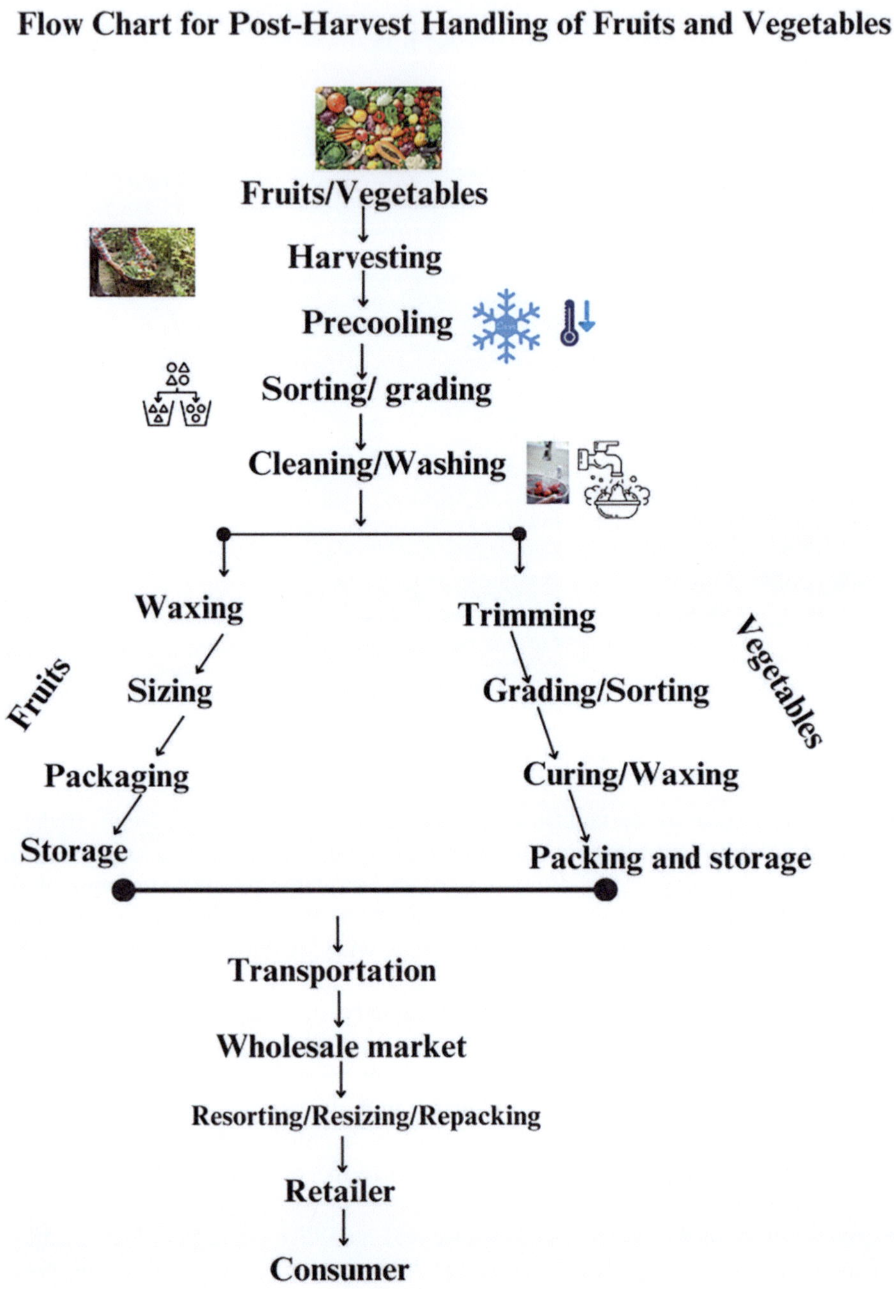

Fig. 3.2 Steps involved in postharvest handling of fruits and vegetables

temperature and humidity conditions during storage and transportation is essential for preserving quality and extending shelf life (Nema et al. 2023).

By understanding and managing these key factors through effective postharvest practices, significant improvements can be achieved in maintaining the quality, safety, and shelf life of fruits and vegetables, ultimately benefiting producers, consumers, and the environment.

3.4 Harvesting Practices

Harvesting marks the critical transition point between active growth and postharvest life for fruits and vegetables. Selecting the optimal harvest time is paramount for ensuring high-quality produce with extended shelf life (Xiao et al. 2023).

Making the right call on harvest timing is crucial for fruits and vegetables. This decision impacts not only the product's marketability but also ensures consumers receive a product with optimal eating quality. Two key concepts guide harvest timing: physiological maturity and horticultural maturity (Table 3.1) (Xiao et al. 2024). Physiological maturity refers to the stage when a fruit or vegetable reaches a sufficient level of development while still attached to the plant. This stage is often marked by biological processes that signal the completion of growth and the initiation of ripening. For many horticultural products, harvest coincides with this physiological maturity. Examples include fruits like apples or pears that reach a specific size, color, and firmness indicative of their readiness to detach from the tree. Similarly, vegetables like peas or corn are harvested when their pods or kernels have

Table 3.1 Comparison of horticultural and physiological maturity in fruits and vegetables

Feature	Horticultural maturity	Physiological maturity
Definition	The stage of growth at which a fruit or vegetable has the texture, flavor, and appearance that are desired for consumption	The stage of development where a fruit or vegetable has reached its full growth potential and is capable of ripening after harvest
Relationship to ripening	Physiological and horticultural maturity may or may not coexist	Fruits at physiological maturity are capable of ripening after harvest
Focus	Consumer preferences and market requirements	Internal development and physiological processes
Timing	Depending on the product and its intended application, this can happen at any point during the development process	Typically occurs before horticultural maturity
Factors considered	Appearance, flavor, texture, firmness, shelf life, and consumer preferences	Size, color, firmness, sugar content, starch-to-sugar ratio, and ethylene production

reached a specific size and developed their characteristic flavors and textures. Horticultural maturity, however, takes consumer preferences into account. It defines the stage at which a fruit or vegetable possesses the desired characteristics for consumption. This stage may or may not coincide with physiological maturity. For instance, tender, immature vegetables like baby carrots, asparagus spears, or green beans are harvested during their early growth phase to achieve a desired texture and flavor preferred by consumers (Li et al. 2011a, 2011b). Conversely, some vegetables like winter squash or pumpkins are left on the vine to mature fully and develop a harder, more durable exterior for long-term storage. Similarly, some fruits like tomatoes or melons are harvested at their peak ripeness for immediate consumption, while others like mangoes or pears may be picked slightly unripe to allow for ripening during transport and extend their shelf life.

To determine the ideal harvest time, growers rely on a combination of maturity indices. These indices can be visual, tactile, or involve measuring specific physiological parameters. Visual indicators include changes in skin color, texture, and development of characteristic blemishes or patterns (Erkan and Dogan 2019). Tactile cues like firmness or ease of detachment from the plant can also be used. Physiological indices might involve measuring sugar content in fruits using a refractometer or assessing starch-to-sugar ratios in certain vegetables. Each fruit or vegetable type typically has a combination of these indices that signal the optimal harvest window. The harvesting method itself is influenced by the type of produce and its stage of maturity (Ruiz-Altisent et al. 2004). For delicate fruits like berries or stone fruits, hand harvesting is often preferred to minimize bruising. Shears or clippers might be used to carefully remove fruits from the plant. Certain vegetables like leafy greens or herbs are also hand harvested. For firmer fruits and vegetables, mechanical harvesting techniques may be employed. Shaking machines can be used to dislodge fruits like apples or nuts from trees. For vegetables like potatoes or carrots, specialized machinery digs and lifts them from the ground.

3.5 Maturity Indices

A measurable point in the commodity's development is implied by the concept of maturity as the level of development that offers the final customer the minimum acceptable quality. It also suggests the necessity of methods to assess maturity. One metric that can be used to assess a commodity's maturity is its relative maturity index (Kumar et al. 2021). A variety of methods can be employed to assess fruit maturity, as shown in Fig. 3.3 and Table 3.2 represents maturity Indices for Common Horticultural Crops.

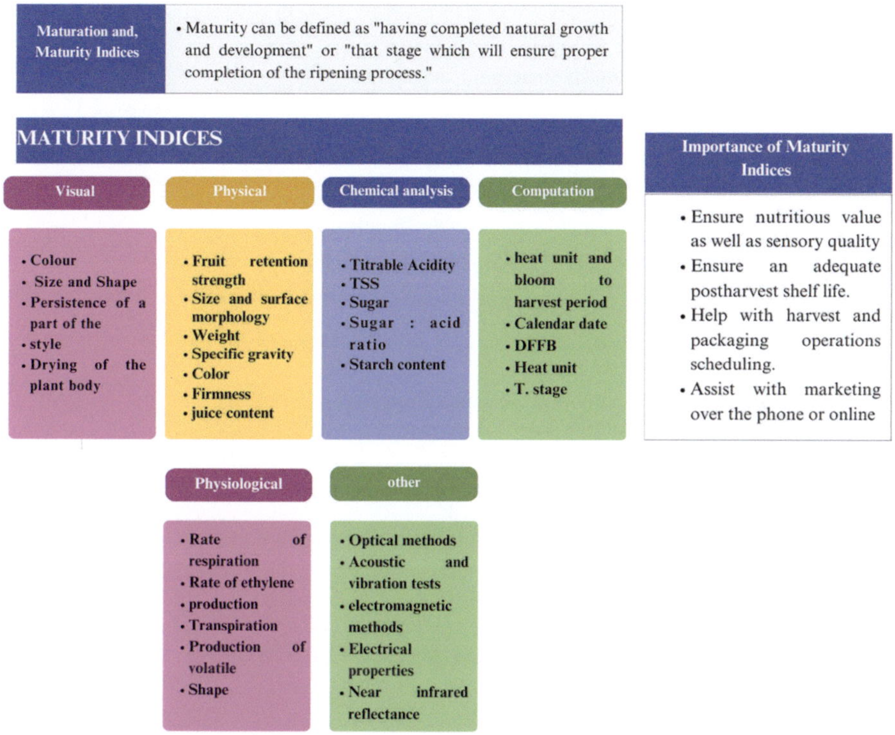

Fig. 3.3 Methods of estimating maturity of fruits and vegetables

Table 3.2 Maturity indices for common horticultural crops

Sl. no	Crop	Maturity indices
1.	Apples	Elapsed days from full bloom to harvest, development of abscission layer, size, mean heat units (during development), internal ethylene concentration, starch and sugar content
2.	Almond	Separation of hull from shell Abscission zone development
3.	Banana	Ridges on the surface of skin change from angularity to round, i.e., after the attainment of 3% full stages Peel color: shift from dark to light green Drying and brittleness of the leaves can be seen
4.	Ber	Color: yellow Specific gravity: less than 1
5.	Mango	TSS (%): 10–14 Specific gravity: 1.01–1.02
6.	Plums	TSS should be at least 17%, color, firmness
7.	Grapes (table)	TSS (%): 14–17.5 SSC/TA: 20 or higher

(continued)

Table 3.2 (continued)

Sl. no	Crop	Maturity indices
8.	Strawberry	Harvesting to be done when ½ to ¾ of the fruit has developed a red color
9.	Cherry	Color: light red TSS (%) = 14–15
10.	Watermelon	Dull hollow sound when thumped
11.	Tubers (onion, garlic and potato)	Top begins to dry and topple clown
12.	Okra	Tips of pods to be pliable Desirable fruit size (5–10 cm long) and tender Not fibrous
13.	Eggplant	Desirable-sized tender fruits having soft seeds Maturity days: 70–80
14.	Tomato	Color: Bright red Maturity days: 120–150 Fully developed seeds At least in one locule gel formation will start
15.	Mushroom	Veil completely intact Well-rounded caps

Physiological indices measures the ripeness. These include:

3.5.1 *Visual Indicators*

3.5.1.1 Fruit Color

Visual maturity indices are qualitative measurements that provide insights into the ripeness or maturity of fruits and vegetables by analyzing their color characteristics. These indices are widely employed in agriculture to facilitate grading, harvesting, and marketing activities. In many fruits, a shift in peel or skin color from green to yellow, red, or orange often indicates the onset of ripening. This visual cue is particularly valuable for fruits where skin color serves as a reliable indicator of maturity, allowing for efficient sorting and classification based on ripeness levels.

3.5.1.2 Shape and Size

The shape of a fruit can be a valuable indicator of its maturity stage. As fruits develop on the plant, their shape often undergoes changes that reflect internal physiological processes. For example, bananas become more rounded in cross-section as they mature, transitioning from a more angular shape. Similarly, mangoes exhibit changes in the relationship between their shoulders and the point of stalk attachment. In immature mangoes, the shoulders may slope away from the stalk, while in more mature mangoes, they become level or even raised above the attachment point. These visual indicators can reveal whether the fruit is ready to be picked. Additionally, harvest timing can be determined by changes in crop size. *Zea mays saccharata*, for example, sells smaller, less mature cobs as baby corn and partially ripe cobs as sweet corn. A banana's ripeness can be determined by measuring the width of each finger in a bunch. Growers can decide the best time to harvest fruit by using calipers to measure the caliper grade, or the maximum width of a finger halfway along the bunch.

3.5.2 *Physical Indices*

3.5.2.1 Fruit Retention Strength

The force necessary to detach a fruit from its parent plant serves as a valuable indicator of its maturity status. Immature fruits typically exhibit greater resistance to removal, requiring more force to detach compared to ripe fruits. This phenomenon is attributed to the presence of a stronger abscission layer, a specialized region at the base of the fruit stalk that facilitates detachment as the fruit matures. As the fruit ripens, the abscission layer weakens, reducing the force required for removal. This relationship between detachment force and maturity can be utilized as a practical method to assess the readiness of fruits for harvest, ensuring optimal quality and preventing premature or delayed picking.

3.5.2.2 Specific Gravity

Specific gravity, a measure of a substance's relative density compared to water, serves as a valuable tool in assessing fruit and vegetable maturity. It is determined by dividing the weight of a sample in air by its weight in water. As fruits mature, their specific gravity generally increases. This phenomenon can be attributed to the accumulation of sugars and other soluble compounds within the fruit tissue. While specific gravity is not routinely used to determine harvest time due to the challenges of obtaining accurate and representative samples, it can be employed in certain cases where suitable sampling techniques are developed. Post-harvest, specific gravity can be used to grade fruits based on maturity. By immersing fruits in a water tank, those that float are typically less mature than those that sink. This simple method provides a quick and practical way to categorize fruits according to their developmental stage.

3.5.2.3 Firmness

Firmness is a critical textural attribute of fruits and vegetables that undergo significant changes throughout the maturation and ripening process. As fruits ripen, they often become softer due to the breakdown of cell walls and the accumulation of soluble compounds. Excessive moisture loss can also contribute to textural changes, leading to a dry or leathery texture. Assessing firmness is essential for determining the optimal harvest time and ensuring that fruits are harvested at their peak quality. Traditional methods for assessing firmness rely on tactile evaluation, where harvesters gently squeeze the fruit to gauge its ripeness. However, this subjective approach can be influenced by individual variations in hand strength and pressure application. To obtain more objective measurements, pressure testers or penetrometers are employed. These devices measure the force required to insert a plunger of a specific size into the fruit's pulp. The force is typically measured in pounds or kilograms, providing a numerical value that indicates the firmness of the fruit. Firmness is particularly important for fruits like peaches and apricots, where softening is a key indicator of harvest ripeness. By monitoring firmness changes throughout the maturation process, growers can determine the optimal time to harvest these fruits and ensure they reach consumers at their peak quality.

3.5.3 Chemical Analysis

3.5.3.1 Total Soluble Solids (TSS)

TSS, primarily composed of sugars, play a pivotal role in assessing fruit maturity. As fruits ripen, starch is converted into sugars, leading to an increase in TSS. This phenomenon is particularly evident in climacteric fruits, where carbohydrates accumulate in the form of starch during maturation and are subsequently broken down into sugars upon ripening. In contrast, nonclimacteric fruits tend to accumulate sugars directly during maturation. A refractometer is a convenient and widely used instrument for measuring TSS in fruits. By placing a drop of fruit juice on the refractometer's prism and shining light through it, the amount of light refracted can be measured. This refraction is directly related to the concentration of soluble solids in the juice. The resulting reading, expressed in Brix units, provides a quick and reliable estimate of the fruit's sugar content. TSS measurements are valuable for determining fruit maturity in various regions worldwide. They offer a practical and objective method for assessing ripeness and ensuring that fruits are harvested at the optimal time. While a refractometer is commonly used for field measurements, laboratory techniques involving light transmission can also be employed for more precise TSS determinations. However, these laboratory methods may not be suitable for small-scale or village-level production due to their complexity and equipment requirements.

3.5.3.2 Acidity

Acidity, measured as titratable acidity (TA), is another important factor to consider when assessing fruit maturity. In many fruits, particularly citrus, acidity tends to decrease as the fruit matures on the tree. This reduction in acidity is often accompanied by an increase in sweetness. By measuring the TA of fruit juice and comparing it to the soluble solids content (SSC), a valuable ratio known as the brix: acid ratio can be calculated. This ratio provides a comprehensive assessment of fruit maturity, considering both sweetness and acidity. TA is determined by titrating a known volume of fruit juice with a standard alkaline solution, such as 0.1 N sodium hydroxide (NaOH). The endpoint of the titration is indicated by a color change using a suitable indicator like phenolphthalein. The volume of NaOH required to neutralize the acid in the juice sample corresponds to the TA value. By taking samples of fruits at different stages of maturity and measuring their TA, growers can establish a relationship between TA and optimal harvest time. While TA alone may not be sufficient to determine maturity in all cases, it serves as a valuable complementary indicator when combined with other factors like SSC and firmness.

3.5.3.3 Starch-to-Sugar Ratio

Starch content in fruits and vegetables decreases as they mature, converting into sugars. Measuring this ratio provides an indication of ripeness.

3.5.3.4 Ethylene Production

Certain fruits and vegetables produce ethylene gas as they ripen. Monitoring ethylene levels can help predict the onset of ripening and guide harvest timing.

3.5.4 Computation

3.5.4.1 Calendar Date

Calendar dates can serve as a reliable guide for harvesting perennial fruit crops grown in seasonal climates with relatively consistent weather patterns from year to year. This approach leverages the predictable timing of flowering and a relatively constant growth period between flowering and maturity. Temperature plays a crucial role in determining the timing of flowering, and the degree-day concept can be employed to estimate the number of days required for development from flowering to harvest for certain commodities. Farmers can use their accumulated experience to establish specific calendar dates for harvesting particular fruit crops, taking into account the unique characteristics of their local climate and the specific cultivars grown. While calendar dates provide a convenient and practical approach, it is important to supplement them with other maturity indices, such as visual observations, firmness measurements, and physiological tests, to ensure optimal harvest timing and maintain product quality.

3.5.4.2 Days from Full Bloom (DFFB)

Days from Full Bloom (DFFB) is a common maturity index used in horticulture to predict the optimal harvest time for various fruits and vegetables. By counting the number of days elapsed since full bloom, growers can estimate when the produce will reach its ideal stage of ripeness. However, DFFB can vary significantly depending on factors such as cultivar, location, and environmental conditions. Therefore, it is often used in conjunction with other maturity indices, such as visual observations, firmness measurements, and physiological tests, to ensure accurate harvest timing and maintain product quality.

3.5.4.3 Heat Units

Heat units, also known as growing degree days (GDDs), are a valuable tool for predicting crop maturity, particularly for crops that are sensitive to temperature fluctuations. Heat units accumulate over time based on the average daily temperature and a specific base temperature threshold. This threshold varies depending on the crop and is the minimum temperature required for growth and development. By calculating the accumulated heat units from the time of planting or flowering, farmers can estimate the expected harvest date with greater accuracy. Heat units can be calculated using the following formula:

$$GDD = \left(\text{max .daily temp.} + \text{min .daily temp.}\right) / 2 - \text{Base temp.}$$

The base temperature is subtracted from the average daily temperature to account for the fact that growth does not occur below this threshold. The resulting value is then summed over the growing season to determine the total heat units accumulated. By comparing the accumulated heat units for a particular crop to known maturity requirements, farmers can make more informed decisions about planting, harvesting, and marketing. This approach is particularly useful in regions with variable weather conditions, as it allows for adjustments based on the actual temperature experienced by the crop.

3.5.4.4 T-Stage

T-stage, a physiological maturity index, is a valuable tool for assessing the developmental stage of certain fruits and vegetables. It involves measuring the firmness of the fruit tissue using a penetrometer or other suitable device. The T-stage value represents the force required to penetrate the fruit's flesh to a specific depth. As fruits ripen, their firmness generally decreases, and the T-stage value increases. By monitoring T-stage changes throughout the growing season, growers can determine the optimal harvest time and ensure that fruits are picked at their peak quality. T-stage is particularly useful for fruits like apples and pears, where firmness is a critical indicator of maturity and consumer preference.

Several environmental and cultural factors can influence plant maturity. Temperature plays a crucial role in regulating plant growth and development, with warmer temperatures generally accelerating maturation. Soil conditions, including nutrient availability and water content, also impact plant growth and maturity. The size of planting material can affect the time it takes for a plant to reach maturity, with larger plants often maturing more quickly. Closer spacing between plants can increase competition for resources, potentially delaying maturity. Pruning intensity can also influence maturity, as it can affect the distribution of growth hormones and the allocation of resources within the plant. Girdling, a technique that involves partially cutting through the bark of a tree or vine, can stimulate fruit development and accelerate maturity.

Table 3.3 Difference between hand harvesting and mechanical harvesting

Feature	Hand harvesting	Mechanical harvesting
Method	Labor-intensive, individual fruit selection	Machine-operated, bulk harvesting
Efficiency	Slower, suitable for smaller-scale operations	Faster, suitable for large-scale production
Cost	Higher labor costs	Lower labor costs, but higher initial investment in equipment
Damage to produce	Minimal damage due to careful handling	Can cause bruising or damage if not handled properly
Suitability	Ideal for delicate fruits and vegetables	Suitable for firmer fruits and vegetables
Flexibility	Greater flexibility in selecting and sorting produce	Less flexibility, may require adjustments for different crop varieties
Environmental impact	Generally lower environmental impact due to less reliance on machinery	Higher environmental impact due to fuel consumption and emissions
Quality control	Higher level of quality control due to individual inspection	Lower level of quality control, may require additional sorting after harvesting
Labor requirements	Requires skilled labor for fruit selection and handling	Requires operators to operate and maintain machinery

3.6 Harvesting Methods

The choice of harvesting method for fruits and vegetables is a critical decision that significantly impacts product quality, efficiency, and economic viability. Table 3.3 shows the difference between hand harvesting and mechanical harvesting. The ideal method must be carefully selected to consider the delicacy of the crop, the need for speed, labor costs, and market requirements. Two main approaches prevail for harvesting of fruits and vegetables

Hand harvesting: this traditional method offers greater control and minimizes mechanical damage. It is often preferred for delicate fruits and vegetables like berries, leafy greens, and high-value crops.

Mechanical harvesting: this method is typically faster and more cost-effective for large-scale operations. Shaking machines, conveyor belts, and berry pickers are examples of mechanical harvesting technologies. However, these methods require careful adjustment and operator training to minimize damage.

Harvest timing also plays a vital role. Early morning hours, when temperatures are cooler, are generally preferred for harvesting most fruits and vegetables. Cooler temperatures reduce respiration rates and minimize heat stress on the produce (Nema et al. 2023). Using the right tools and equipment is essential. Hand-held shears, secateurs, and specialized knives are valuable for the precise harvesting of specific crops. Cushioning materials like mats and padded bins can help minimize bruising during mechanical harvesting (Fig. 3.4).

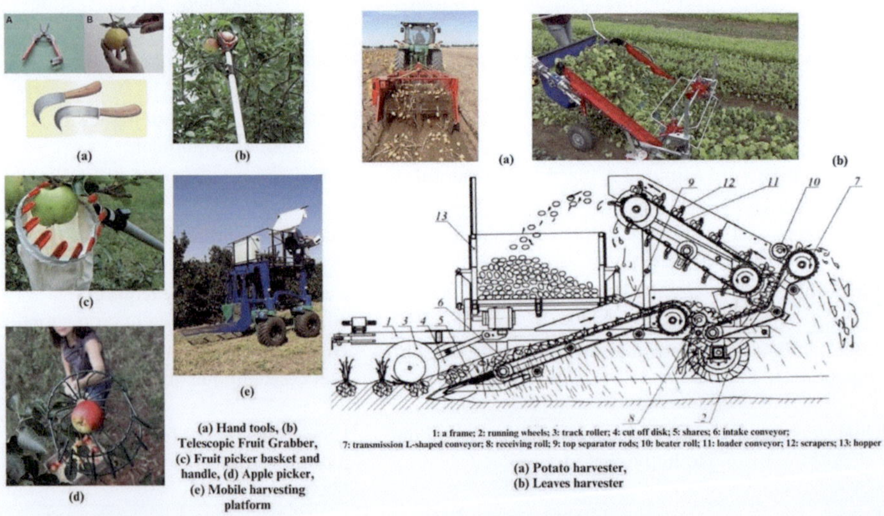

Fig. 3.4 Harvesting equipments

3.6.1 Harvesting of Fruits

Soft fruits like strawberries and raspberries, grown on low-growing plants, are typically harvested by carefully breaking the fruit stalk and placing them into containers. However, harvesting tree-borne fruits such as avocados, citrus fruits, mangoes, and apples presents unique challenges due to their height. Traditional methods involved using ladders to reach the fruit, which is a time-consuming and labor-intensive process. To improve efficiency, various harvesting techniques have been developed. One common approach involves using a long pole with a bag attached and a device for cutting or breaking the fruit stalk. This allows harvesters to reach higher branches without the need for a ladder. Additionally, picking platforms that can be raised and lowered are used to transport harvesters around orchards, enabling them to efficiently pick fruit from different heights. These innovations have significantly improved the efficiency and safety of harvesting tree-borne fruits (Li et al. 2011a, b).

The harvesting of fruits for the fresh market typically requires more delicate handling to minimize damage and preserve quality. Mechanical harvesting is less common for fresh market fruits due to the potential for bruising and other injuries that can accelerate deterioration. However, for fruits intended for processing, such as oranges for juice or grapes and soft fruits for preserves, mechanical harvesting methods are more widely used. Powerful wind machines can be employed to dislodge oranges from trees, followed by devices to collect the fallen fruit. Tractor-mounted machines equipped with combing fingers can efficiently harvest grapes and blackcurrants, removing both the fruit bunches and a significant portion of the leaves. Tree shakers, attached to the trunk, can also be used to dislodge fruit, with canvas sheets placed beneath the trees to break the fall and prevent bruising (Erkan and Dogan 2019). These mechanical harvesting methods are designed to maximize efficiency while minimizing damage to the fruit, making them suitable for processing applications where the highest level of fresh quality is not as critical.

3.6.2 Harvesting of Vegetables

Low-growing vegetables, like leafy greens or bush beans, are typically harvested using similar methods as soft fruits. They should be carefully cut or snapped off at the desired height. However, harvesting root crops requires a different approach. These crops, such as carrots, potatoes, or beets, must be dug from the soil. A garden fork or other suitable tool is inserted below the crop and used to pry it upwards. During this procedure, care must be taken to prevent harming the roots. Some root crops, like radishes, can be pulled from the soil by hand. Proper harvesting techniques are essential to preserve the quality and maximize the yield of both low-growing vegetables and root crops (Ruiz-Altisent et al. 2004).

Mechanical harvesting of vegetables like lettuce, cauliflowers, and cabbages often involves a combination of manual and mechanical processes. Vegetables are cut by hand and placed on conveyors on mobile packing stations that are slowly moved across the field. This allows for on-site packing and immediate transfer to vehicles for transportation to precoolers or directly to market. It is essential to process these vegetables promptly after harvest to minimize tissue damage and prevent enzymatic action that can lead to off-flavors. Mechanical harvesting of green beans can result in discoloration of the broken ends, necessitating careful handling and processing. For root crops, specialized tools like moldboard plows or subsoilers are used to dig them from the soil (Mathushika and Gomes 2021). These tools can be power-intensive, but using vibrating digger blades can reduce the required power by increasing the amplitude of vibration. Effective tuber separation requires sufficient tractor traction, which may be insufficient at low vibration amplitudes. By carefully selecting and optimizing mechanical harvesting methods, it is possible to balance efficiency with product quality and minimize postharvest losses (Belay 2021).

Once harvested, fruits and vegetables must be handled carefully to prevent bruising, cuts, or other injuries that can accelerate spoilage. Immediate precooling is crucial to slow down metabolic processes, reduce respiration, and prevent microbial growth. Various precooling methods, such as hydrocooling, forced-air cooling, and vacuum cooling, are employed to rapidly lower the temperature of the produce and maintain its freshness.

3.7 Precooling

A crucial postharvest procedure is precooling, which is quickly bringing harvested fruits and vegetables down to their ideal temperature following harvest and before delivery to a market or cold storage facility. It slows down the metabolic processes such as enzymatic activity, ethylene production and extends shelf life (Kattel et al. 2022; Dincer 1995). As temperature rises within the optimal range for plants, metabolic rates double for every 10 °C increase. By cooling the produce immediately after harvest, it is possible to minimize physiological deterioration, reduce respiration rates, and inhibit microbial growth (Zhang et al. 2023; El-Ramady et al. 2015). Technically, precooling also contributes to reducing the cold storage room's necessary refrigeration capacity.

3.7.1 Significance of Precooling

Extension of shelf life: by reducing metabolic rates, precooling can significantly extend the shelf life of the fresh produce, reduce spoilage and minimize losses during transportation and storage.

Prevention of chilling injury: certain fruits and vegetables are susceptible to chilling injury, a physiological disorder that occurs when they are exposed to temperatures below their tolerance limits. Precooling can help prevent chilling injury by ensuring that the produce is cooled gradually and evenly.

Improved market value: high-quality, fresh produce with a longer shelf life commands a higher market value. Precooling is essential for maintaining the marketability of fruits and vegetables.

Preservation of Quality by Lag Time Between Harvest and Cooling
It's very important to remove the field heat from fruits and vegetables to safe harvest or else it can cause deterioration of the fresh produce. Pre-cooling is considered the most effective, efficient, cost-effective preservation method (Duan et al. 2020; Raju et al. 2018). Precooling reduces wilting because transpiration and evaporation proceed more slowly at low temperatures, and it slows down the degradation and rotting process by delaying the proliferation of decay organisms. Precooling helps to maintain the freshness, flavor, texture, and color of fruits and vegetables by slowing down enzymatic activity and reducing water loss. Most commonly used precooling methods are: vacuum cooling, room cooling, hydro cooling, forced air cooling, etc. These techniques help to transfer heat at a faster rate from the produce to the cooling media such as water and air.

3.7.2 Precooling Methods

3.7.2.1 Room Cooling

It is a traditional, well-developed, low-cost and slow method of cooling. In this method, produce is simply loaded into a cool room and cool air is allowed to circulate among the cartons, sacks, bins or bulk load.

3.7.2.2 Forced-Air Cooling

Forced-air cooling is a highly efficient precooling method that involves passing cold air over the produce to rapidly reduce its temperature. This method is suitable for a wide range of fruits and vegetables and can be implemented on various scales of operations (Dincer 1995). By increasing the effective surface area of the produce through forced airflow, forced-air cooling can significantly reduce cooling times compared to natural cooling. This rapid cooling helps to minimize respiration, water loss, and the development of physiological disorders, thereby extending the shelf life of the produce.

3.7.2.3 Hydrocooling

This technique involves immersing produce in cold water, making it an effective and efficient method for rapidly reducing product temperature. This method is particularly well-suited for fruits and vegetables that can tolerate direct contact with water, such as leafy greens and root vegetables. Hydrocooling not only prevents water loss but can also potentially increase the moisture content of the produce. While hydrocooling is commonly used for unpackaged produce, its application for packed commodities is limited due to the challenges associated with water penetration through containers and the need for water-resistant packaging materials (El-Ramady et al. 2015). However, advancements in packaging technology and the development of water-permeable packaging materials are expanding the potential use of hydrocooling for a wider range of products. Hydrocoolers are typically classified into two main types: shower type and immersion type. Shower-type hydrocoolers spray cold water over the produce, while immersion-type hydro-coolers submerge the produce directly in cold water. The choice of hydrocooler depends on the type of produce, the desired cooling rate, and the specific requirements of the packaging.

3.7.2.4 Vacuum Cooling

It is a highly efficient precooling method that utilizes the principle of water evaporation to rapidly reduce product temperature. By creating a near-vacuum environment, the boiling point of water is significantly lowered, allowing water to evaporate at temperatures well below freezing. As water within the produce evaporates, it absorbs heat, cooling the product in the process (Kattel et al. 2022). The resulting water vapor is then condensed using refrigeration coils, recovering the heat and minimizing moisture loss. Vacuum cooling is particularly well-suited for products with high moisture content and a large surface area-to-volume ratio, such as leafy greens and certain vegetables. However, products with waxy skins or low surface area may not cool effectively using this method. To address moisture loss concerns, modern hydro-vacuum coolers incorporate water spraying systems that can minimize the loss of weight during the vacuum process. Vacuum cooling offers several advantages over other precooling methods. It is the fastest method available, capable of reducing product temperature by 11 °C in as little as 15 min. This rapid cooling helps to preserve quality and extend shelf life. Additionally, vacuum cooling is highly energy efficient, as nearly all the electricity consumed is used directly to reduce product temperature. The sealed environment of the vacuum chamber eliminates the need for additional cooling systems or energy-consuming equipment.

3.7.2.5 Ice Cooling

This involves packing produce with crushed or flaked ice to absorb heat and lower the temperature. However, the melting of ice can create a humid environment that may contribute to microbial growth and disease (Ruiz-Altisent et al. 2004). The temperature of the ice itself can be below the freezing point of certain products, even those considered tolerant to cold temperatures. This can result in freezing injury if the produce comes into direct contact with the ice. Additionally, the melting of ice can lead to increased moisture content in the produce, which can cause splitting, discoloration, and susceptibility to spoilage. The quality of the water used for making ice is also crucial, as it must meet high microbial standards to prevent contamination. The use of ice should be carefully considered, particularly for sensitive products and in situations where refrigerated transport is not readily available. While ice cooling can be a practical option for short-term storage or transportation, it is important to weigh the benefits against the potential drawbacks. The use of ice can add extra steps to the packing process, increase costs due to added weight and volume, and introduce the risk of freezing injury. For optimal preservation of fruits and vegetables, it is often recommended to combine ice cooling with other precooling methods or refrigerated transport to ensure consistent temperature control and minimize quality degradation.

3.7.2.6 Cryogenic Cooling

It utilizes the extreme cold of liquid nitrogen (−196 °C) or solid CO_2 (dry ice) to rapidly reduce product temperature. While this process can effectively cool produce, it poses the risk of freezing, which can damage the product's texture and quality. To prevent freezing, careful control of the evaporation rate and conveyor speed is essential (Zhang et al. 2023). Cryogenic cooling offers a relatively affordable installation cost compared to other precooling methods, making it suitable for seasonal crops where the investment can be justified over a limited usage period. However, the ongoing cost of liquid nitrogen or dry ice can be significant, making this method most practical for high-value products.

Following harvest and precooling, fruits and vegetables undergo sorting, grading, and washing to ensure consistent quality, minimize postharvest losses, and promote food safety throughout the supply chain. Sorting, grading, and washing are critical postharvest operations that ensure the quality, safety, and marketability of fruits and vegetables. These processes involve carefully inspecting and categorizing produce based on various factors, such as size, color, maturity, and the presence of defects. By removing damaged or substandard produce and ensuring uniformity within batches, sorting and grading contribute to the overall value and appeal of the product. Washing is another essential step to remove dirt, debris, and potentially harmful contaminants from the produce, ensuring food safety and consumer confidence (Shakeel et al. 2022; Vivek et al. 2019).

3.8 Sorting and Grading

Sorting and grading are crucial postharvest processes that categorize fruits and vegetables based on specific quality attributes. This ensures uniformity within a shipment, catering to consumer preferences and maximizing marketability.

3.8.1 Importance of Sorting and Grading

3.8.1.1 Maintaining Quality Standards

Maintaining Quality Standards: sorting removes fruits and vegetables with defects such as bruises, blemishes, diseases, or insect damage, ensuring that only high-quality produce reaches consumers.

3.8.1.2 Uniformity in Size and Ripeness

Uniformity in Size and Ripeness: grading categorizes produce based on size, maturity, and other quality parameters. This uniformity allows for efficient packing, transportation, and marketing strategies tailored to specific consumer preferences.

3.8.1.3 Market Differentiation

Market Differentiation: grading enables producers to cater to different market segments. For example, high-grade produce may be directed towards premium retailers, while lower grades may be suitable for processing or industrial use.

3.8.2 Sorting and Grading Methods

The choice between manual and mechanical sorting and grading depends on factors such as the type of produce, volume, and desired level of accuracy.

3.8.2.1 Manual Sorting and Grading

Manual Sorting and Grading: this traditional method utilizes human labor to visually inspect and categorize fruits and vegetables based on predetermined quality criteria. While offering greater flexibility and accuracy, particularly for delicate produce, manual sorting is labor-intensive and may not be feasible for large-scale operations.

3.8.2.2 Mechanical Sorting and Grading

Mechanical Sorting and Grading: technological advancements have led to the development of automated sorting and grading machines. These machines utilize various technologies such as weight sensors, color sorting based on machine vision systems, and firmness detectors to efficiently categorize produce based on pre-programmed parameters. While faster and more cost-effective for large volumes, mechanical sorting may not be as meticulous as manual methods and can struggle with certain defects or nonuniform produce shapes.

Several key factors are considered during this process, including size, maturity, color, and the presence of defects. Fruits and vegetables are categorized into different size ranges to facilitate efficient packing and handling. Maturity stage, assessed through visual cues or physiological indices, is crucial for determining shelf life and quality. Consistent and desirable color is another important attribute, especially for fruits and vegetables sold based on their appearance. Finally, removing produce with blemishes, bruises, or other defects helps to maintain overall quality and minimize the risk of spoilage. By carefully considering these factors, sorting and grading processes contribute to the efficient and safe handling of fruits and vegetables throughout the supply chain.

3.9 Washing and Phytosanitization Procedures

Washing and sanitation are essential steps in the postharvest handling process to ensure food safety and minimize postharvest decay. Fruits and vegetables can harbor various microorganisms, including bacteria, mold spores, and viruses, acquired from the soil, water, or during handling. Washing removes dirt, debris, and some surface contaminants, while sanitization treatments further reduce the microbial load on the produce surface. Common washing and phytosanitization techniques include:

3.9.1 Washing with Chlorinated Water

Washing with Chlorinated Water: chlorine, in various forms, has been a cornerstone of postharvest sanitation programs for decades. Its primary applications include inactivating pathogens on seeds, cuttings, irrigation water, equipment, and produce surfaces. Chlorine is commonly used to treat postharvest cooling water, enhance firmness with calcium treatments, and rehydrate produce during shipping. Sodium ($NaOCl$) and calcium hypochlorite ($CaCl_2O_2$) are the most widely used forms of chlorine for these purposes (Mishra et al. 2018). By effectively eliminating pathogens, chlorine helps to minimize their transmission and protect the quality and safety of fresh fruits and vegetables. However, excessive chlorine use can cause taste and quality issues (Suslow 2000).

3.9.2 Washing with Alternative Sanitizers

Washing with Alternative Sanitizers: beyond chlorine, several alternative sanitizers have gained prominence due to their effectiveness and reduced environmental impact. Ozone, a powerful biocide, surpasses chlorine, chlorine dioxide, hydrogen peroxide, and peracetic acid in antimicrobial potency (Tzortzakis and Chrysargyris 2017). Ethanol, traditionally used for medical disinfection, disrupts biological membranes and denatures proteins, demonstrating strong antimicrobial properties. Hydrogen peroxide, a potent oxidizing agent, generates reactive species that damage cellular components, leading to microbial inactivation. Organic acids, such as acetic acid and lactic acid, have antimicrobial properties and are approved for use in food processing (Meitha et al. 2020). Electrolyzed water is another emerging alternative to sodium hypochlorite, offering effective sanitization while minimizing environmental concerns. These alternative sanitizers provide a range of options for postharvest washing, ensuring the safety and quality of fruits and vegetables while considering environmental sustainability (Feliziani et al. 2016).

3.9.3 Ozonation

Ozone (O_3), a highly reactive allotrope of oxygen, has emerged as a potent and environmentally friendly disinfectant for various applications, including postharvest treatment of fruits and vegetables. Its strong oxidizing properties enable it to effectively inactivate a wide range of microorganisms, including bacteria, fungi, and viruses, without leaving harmful residues. Ozone's antimicrobial efficacy stems from its ability to oxidize cellular components, disrupting cell membranes and inactivating enzymes. It reacts with the cell walls of microorganisms, leading to cell death. Additionally, ozone can oxidize organic matter, reducing the load of contaminants on produce surfaces. However, the effectiveness of ozone treatment can be influenced by factors such as ozone stability, penetration limitations, and equipment costs.

3.9.4 Cold Plasma

Cold plasma technology, a nonthermal process involving the application of a high-voltage electric field to a gas, resulting in the generation of a highly reactive plasma state, has emerged as a promising phytosanitation technology. Unlike thermal plasma, cold plasma operates at low temperatures, allowing for the treatment of temperature-sensitive materials without causing damage (Sakudo et al. 2020; Bourke et al. 2017). To generate plasma, a suitable system must incorporate the following elements. Energy source can be electrical energy (either direct current or alternating current), thermal energy, optical energy (such as ultraviolet light), magnetic fields, or radio frequency (gamma or microwave) radiation. These energy sources increase the kinetic energy of electrons, leading to the formation of reactive components. Carrier gas, such as air, nitrogen, oxygen, helium, argon, or a combination of different gases, serves as the medium through which the plasma is generated. Specialized electrodes are required to create the electric field necessary for plasma formation. These electrodes can vary in shape and material depending on the specific application (Desai et al., 2024; Pandey and Chauhan, 2019; Mandal et al. 2018). By generating reactive oxygen and nitrogen species, cold plasma can effectively inactivate pathogens and reduce microbial populations on the surface of fruits and vegetables. This approach offers several advantages over conventional methods, including reduced chemical residues, enhanced safety for consumers, and potential for on-site application. Recent studies have demonstrated the efficacy of cold plasma in controlling various foodborne pathogens, such as *Escherichia coli, Salmonella, Listeria monocytogenes,* etc., on fruits and vegetables. Additionally, cold plasma treatment has been shown to improve the shelf life of produce by inhibiting postharvest decay and maintaining freshness (Pankaj and Keener 2017; Zainal et al. 2015).

The optimal washing and sanitizing method depends on the specific crop, its susceptibility to microbial contamination, and regulatory requirements. Implementing effective washing and sanitization procedures is crucial for ensuring the safety and marketability of fruits and vegetables throughout the postharvest chain. To effectively decontaminate surfaces and prevent the spread of pathogens, various technologies have been explored. Traditional methods like irradiation, UV light, ozone, and chlorine dioxide have been used for decades. More recent advancements include technologies based on advanced oxidative processes and gas plasmas. These diverse approaches offer a range of options for surface disinfection, ensuring the safety and quality of food products (Unni et al., 2011; Murray et al. 2017).

3.10 Packaging Materials and Techniques

Fruits and vegetables are inherently delicate and susceptible to damage throughout the postharvest journey. Packaging plays a crucial role in protecting agricultural produce from physical injury, minimizing moisture loss, and extending shelf life (Shinde et al. 2018). Agricultural produce packaging means placing the product within a container with packing materials such as cushioning pads, plastic trays, molded pulp trays, inserts, etc., to protect or to keep it from moving around. These materials are generally made of materials such as plastic films, waxed liners, etc.

Packaging needs to accomplish three main goals: (i) confine the product and make handling and marketing easier by standardizing the quantity or weight that fits inside the container. (ii) During transportation, storage, and marketing, shields the product from harm (impact, wounds, abrasion, and compression) and unfavorable environmental factors (relative humidity and temperature). (iii) Give purchasers details like kind, weight, quantity, choice or quality level, manufacturer's name, nation, and region of origin. Recipes, nutritional values, bar codes, and any other pertinent traceability information are also frequently provided. A well-designed packaging must be modified to fit the circumstances or particular care needed for the product (Elik et al. 2019; El-Ramady et al. 2015).

3.10.1 Functions of Packaging

Packaging protects produce from physical damage caused by crushing, impacts, vibrations, and compression during handling, transportation, and storage. This minimizes bruising, punctures, and other injuries that can lead to spoilage and quality loss.

3.10.1.1 Preservation

Preservation: packaging plays a very important role in preserving freshness and quality by influencing factors such as respiration rate, moisture loss, and ethylene gas exposure. Certain packaging materials can be designed to modify the surrounding atmosphere, extending shelf life and reducing spoilage.

3.10.1.2 Information and Marketing

Information and Marketing: packaging serves as a communication tool, providing consumers with information about the product, including variety, origin, grade, weight, and nutritional content. Packaging design can also be used for marketing purposes, influencing consumer purchasing decisions by incorporating attractive graphics and branding elements.

3.10.2 Types of Packaging Materials

A wide range of packaging materials is utilized for fruits and vegetables, each offering unique properties and functionalities:

Bulk bins: these are large, open-top containers used for transporting and storing bulk quantities of fruits and vegetables. They are often made of wood, plastic, or metal and can be stacked for efficient storage.

Wooden bulk bins: these traditional bins are durable and reusable, but they can be heavy and susceptible to damage and contamination.

Corrugated fiberboard bulk bins: these lightweight and recyclable bins offer good protection and are widely used for transporting fruits and vegetables.

Reusable plastic bulk bins: these bins are durable, hygienic, and can be reused multiple times, reducing environmental impact.

3.10.2.1 Bags and Sacks

The fibers used to make sacks are typically taken from plants, cleaned, dried, spun, and then weaved into sheets. Commonly used fibers include jute (*Corchorus capsularia*), which yields the highest grade, while *C. olitorius* and sisal (*Agave sisalana*) are also used. These sacks are known for their strength and durability, but they are susceptible to damage when exposed to water. Other natural fibers like kenaf, hemp, and abaca have also been used to produce bags and sacks for handling crops. While woven sacks provide good ventilation, they offer minimal protection against mechanical injuries.

3.10.2.2 Cardboard Boxes

Cardboard Boxes: corrugated cardboard boxes provide structural support and protection for heavier or more delicate fruits and vegetables. These boxes can be lined with various materials such as wax coatings or plastic films to enhance moisture control and protection.

3.10.2.3 Mesh Bags

Mesh Bags: open-mesh bags are commonly used for certain fruits and vegetables such as potatoes, onions, and citrus. These bags offer ventilation and allow for visual inspection of the produce.

3.10.2.4 Cushioning Materials

Cushioning Materials: padding materials such as wood wool, molded pulp inserts, or foam nets can be used within packaging to provide additional protection against impact and prevent bruising.

3.10.3 Film Types

3.10.3.1 Low-Density Polyethylene (LDPE)

Low-Density Polyethylene (LDPE) is a branched-chain polymer produced from the polymerization of ethylene monomer. It has a molecular weight ranging from 14,000 to 1,400,000 and a density of 0.910–0.935 g/cm^3. LDPE film is characterized by its permeability to gases, allowing for the passage of oxygen (O_2) and carbon dioxide (CO_2). At a thickness of 0.0254 mm and a temperature of 22–25 °C, LDPE film exhibits an oxygen permeability of 3900–13,000 cm^3/m^2/day and a carbon dioxide permeability of 7700–77,000 cm^3/m^2/day at 1 atmosphere. Additionally, the film's water vapor permeability is 6–23.2 g/m^2/day at 37.8 °C and 90% relative humidity. These permeability properties make LDPE suitable for various packaging applications, including food packaging.

3.10.3.2 HDPE

HDPE, characterized by its high crystallinity, linear molecular structure, and high molecular weight, exhibits superior tensile strength, stiffness, and hardness compared to low-density polyethylene (LDPE). With a typical density range of 0.995 to 0.970 g/cm^3, HDPE films offer excellent barrier properties against oxygen and carbon dioxide. Specifically, HDPE films with a thickness of 0.0254 mm have reported oxygen permeability rates ranging from 520 to 4000 cm^3/m^2/day at 1 atmosphere and 22–25 °C. Similarly, their carbon dioxide permeability ranges from 3900 to 10,000 cm^3/m^2/day under the same conditions. Additionally, HDPE demonstrates low water vapor permeability, typically ranging from 4 to 10 g/m^2/day at 37.8 °C and 90% relative humidity. These exceptional properties make HDPE a versatile and widely used packaging material for various applications, including food packaging, consumer goods, and industrial products.

3.10.3.3 Plastic Films: Polyethylene (PE)

Plastic Films: Polyethylene (PE) films are widely used due to their low cost, transparency, and water resistance. For certain applications, breathable films may be employed. These films incorporate microperforations that allow for controlled gas exchange, optimizing the atmosphere surrounding the produce.

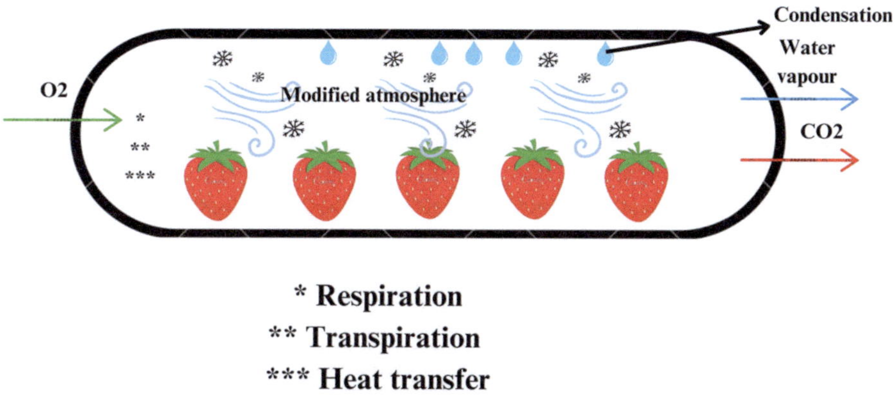

Fig. 3.5 Modified atmosphere packaging (MAP)

3.10.4 Modified Atmosphere Packaging (MAP)

Modified Atmosphere Packaging (MAP): modified atmosphere packaging (MAP) is a technique that involves altering the gas composition within a package to create an environment that is more conducive to preserving the quality of fruits and vegetables (Fig. 3.5) (Belay et al. 2016). The effectiveness of MAP depends on a delicate balance between the permeability of the packaging film to oxygen (O_2) and carbon dioxide (CO_2) and the respiration rate of the produce inside. Careful consideration of these factors is essential for designing optimal MAP systems that can maintain the freshness and extend the shelf life of perishable products. Understanding the dynamic interactions between the product, packaging, and atmosphere is crucial for developing effective MAP solutions (Soltani et al. 2015; Belay et al. 2016; Kader et al. 1989; Sandhya 2010; Mangaraj and Goswami 2009).

The selection of the most appropriate packaging material for a specific fruit or vegetable depends on several factors:

Product characteristics: physical attributes such as size, shape, ripeness, and susceptibility to mechanical damage or moisture loss all influence the choice of packaging material.

Transportation and storage conditions: factors such as transportation duration, storage temperature, and handling practices need to be considered when selecting packaging materials with adequate strength and protective properties.

Sustainability concerns: the environmental impact of packaging materials is a growing consideration. Biodegradable or compostable packaging materials are increasingly sought after to minimize environmental footprint.

Cost-effectiveness: balancing functionality and cost is crucial. Selecting packaging that provides adequate protection while remaining cost-effective is essential for commercial viability.

By carefully considering these factors, producers can select packaging materials that optimize the protection, preservation, and marketability of their fruits and vegetables throughout the postharvest supply chain.

Emerging intelligent packaging technologies: intelligent packaging is also well-known as smart packaging. It has the capability to recognize the properties of the food it surrounds or the atmosphere in which it is stored. It is furthermore able to warn the producer, retailer or customer regarding the condition of these food properties (Shinde et al. 2018).

3.11 Storage

Fruits and vegetables are stored for a variety of reasons. To enable a producer or group of growers to collect enough food to ship to market, it is a component of orderly marketing, where the storage period is often brief. While it is being sold, it might be kept in wholesale marketplaces for storage. To wait for a price increase, it could be preserved when the price is low at a given moment. To increase the length of their availability, some crops are kept in storage for extended periods of time. In this instance, the crop—and even particular cultivars—are cultivated specifically for long-term storage. Extended preservation of a crop can be costly and necessitate a high degree of crop expertise.

3.11.1 Store Design and Method

For as long as humans have cultivated crops, they have sought methods to preserve them beyond their immediate harvest. Early civilizations developed ingenious techniques to store fruits and vegetables for extended periods, often relying on natural processes and traditional wisdom.

3.11.1.1 In Situ/On Site/In Place

In situ storage basically refers to holding off on harvesting the crop until it is needed. With root crops, this is possible in some circumstances, but it does mean that the crop-growing area is still occupied and cannot be used for the planting of new crops. The crop can potentially be vulnerable to disease and pest attacks. It has been demonstrated that postponing harvesting in cassava might lower acceptance, starch content, and preharvest loss risk. The crop may be susceptible to the harmful effects of freezing or chilling injury in colder climates.

3.11.1.2 Burying

To retain and create a humid environment, root crops can be packed into materials like sawdust, coir dust, or even peat. In India, potatoes are typically kept underground in sand storage. When fruits, like plantains, are kept in a moist coir dust, they can stay green and preclimacteric for more than 20 days in the ambient circumstances of Jamaica by maintaining a high humidity level surrounding the fruit.

3.11.1.3 Pit or Trench Storage

Pit or Trench Storage: one of the oldest methods involved burying crops in pits or underground chambers. This technique was used to store root vegetables, grains, and fruits. The cool, moist environment of these pits helped to slow down spoilage. The trench or pit is excavated at the field's boundaries, where the crop has been planted. These should be positioned at a high point in the field if it slopes, particularly in areas that receive a lot of rainfall. Straw or other organic material can be used to line the trench or pit. The crop to be kept can then be placed within, topped with another layer of organic material, and finally covered with soil. Occasionally, wooden boards are laid out on the soil's surface before applying dirt. Venting trenches might be constructed all the way to the store's foundation in order to address the potential for rotting caused by inadequate ventilation. Straw can also be used to cover the ventilation openings at the top of the store to let air out but keep rain from getting inside. The climate will determine how deep the soil layer is. It may be necessary to apply up to 25 cm of soil in really cold weather to protect the crop from frost. In extremely hot weather, it can also be essential to cover the area with a thick layer of dirt or organic material to prevent significant daytime temperature increases. There is a considerable labor requirement for building, loading, and unloading, and storage times are often brief.

- *Drying*: dehydration was another common method used to preserve fruits and vegetables. This involved exposing the produce to sunlight or heat, removing moisture and inhibiting microbial growth.
- *Salting and curing*: salt was used to preserve meats and certain vegetables, such as olives and cabbage. The high salt concentration created a hostile environment for microorganisms.
- *Fermentation*: this process involves controlled microbial activity to transform and preserve food. Examples include pickling, sauerkraut, and yogurt.

3.11.2 Evolution of Storage Techniques/Modern Storage Practices

Over time, storage techniques have evolved in response to changing needs, technological advancements, and scientific understanding. Today, postharvest storage of fruits and vegetables is a highly sophisticated process that incorporates a combination of traditional methods and modern technologies. Cold storage, CA storage, MAP, and irradiation are widely used to preserve the quality and extend the shelf life of perishable produce. Additionally, advancements in packaging materials, transportation infrastructure, and supply chain management have contributed to the efficient and safe storage and distribution of fruits and vegetables.

3.11.2.1 Cold Storage

Cold Storage: refrigerated storage, or cold storage, is a critical component of post-harvest management for fruits and vegetables. It involves storing perishable produce in controlled temperature environments to extend their shelf life and maintain quality (Nummer and Brian 2002). This practice has been employed for centuries, with early civilizations utilizing natural cooling methods like ice pits or underground cellars to preserve food. The development of artificial refrigeration technology in the nineteenth century marked a significant milestone in the history of cold storage. Mechanical refrigeration systems enabled the creation of controlled environments where temperature and humidity could be precisely maintained (Watson et al. 2016). In general, the shelf life of fruit and vegetables increases with decreasing storage temperature until it reaches the freezing point. The crop's metabolic processes are connected to the rise in the pace of degradation. The respiration rate of crops grows exponentially with temperature within their physiological temperature range. This means that for every 10 °C increase in temperature, the crop's metabolism increases by two to three times. This is known as the Q10 value, and the van't Hoff rule predicts it. It is crucial to specify the precise temperature range over which the prediction is valid since many crops experience chilling harm at temperatures well above their freezing point. This technology revolutionized the storage and distribution of perishable goods, allowing for the preservation of fruits and vegetables far beyond their natural harvest seasons. Over time, cold storage facilities have become increasingly sophisticated. Advances in refrigeration technology, insulation materials, and temperature control systems have led to more efficient and effective cold storage solutions. Modern cold storage warehouses are equipped with precise temperature monitoring systems, humidity control mechanisms, and ventilation systems to ensure optimal storage conditions for various types of produce.

Refrigeration plays a vital role in preserving the quality and extending the shelf life of fruits and vegetables. By maintaining a cold environment, refrigeration helps to:

- Slow down metabolic processes: lower temperatures reduce the rate of respiration, enzymatic activity, and microbial growth, thereby preserving the freshness and nutritional value of the produce.
- Prevent spoilage: refrigeration can inhibit the growth of pathogens and fungi, reducing the risk of spoilage and ensuring food safety.
- Maintain quality attributes: by controlling temperature and humidity, refrigeration helps to preserve the color, texture, and flavor of fruits and vegetables.

- Extend shelf life: refrigeration can significantly extend the shelf life of perishable produce, allowing for better market distribution and reduced waste.

3.11.2.2 Controlled Atmosphere (CA) Storage

Controlled Atmosphere (CA) Storage: this technique involves creating a modified atmosphere within storage chambers by controlling the levels of oxygen, carbon dioxide, and nitrogen. CA storage is particularly effective for extending the shelf life of climacteric fruits (Fig. 3.6) (Falagán and Terry 2018).

3.11.2.2.1 Atmosphere Conditioning

Controlled atmosphere (CA) storage is a crucial technique for preserving the quality and extending the shelf life of fruits and vegetables. By reducing the oxygen level within the storage chamber, CA storage can inhibit respiration, delay ripening, and minimize microbial growth. Oxygen reduction can be achieved through various methods, including biological means such as fruit respiration or nonbiological methods like oxygen burning or nitrogen replacement. While biological methods may take several weeks to reach desired oxygen levels, nonbiological systems can rapidly reduce oxygen to 6–8% within 24 h. However, rapid cooling is essential to prevent asphyxia and ensure the fruits reach a temperature below 5 °C before

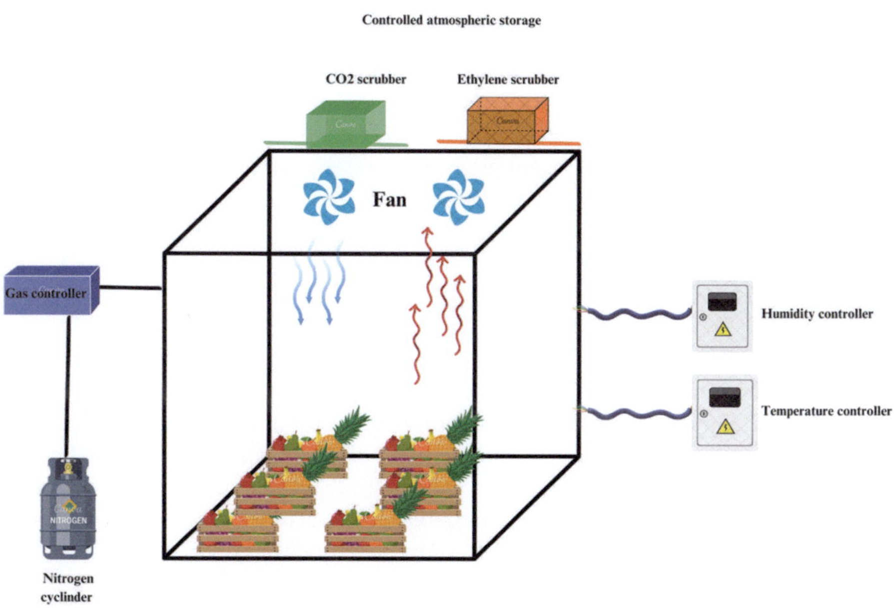

Fig. 3.6 Controlled atmosphere (CA) storage

lowering oxygen levels. Nonbiological methods, such as using nitrogen generators or liquid nitrogen, can quickly reduce oxygen levels but may be financially disadvantageous due to higher energy consumption. Nitrogen generators selectively separate air to produce an enriched nitrogen stream, providing a cost-effective option for CA storage (Thompson (2015); Bodbodak and Moshfeghifar 2016).

3.11.2.2.2 CA Storage Rooms

Creating a gas-tight environment is essential for successful controlled atmosphere (CA) storage. Traditional CA rooms often utilized steel walls lined with mastic to ensure gas tightness. Modern CA stores employ metal-faced insulated panels with gas-tight locking devices and sealed joints to achieve the same goal. Doors in CA rooms are typically equipped with rubber gaskets and sliding or vertical mechanisms to prevent air leakage. To accommodate pressure fluctuations during the storage season, pressure/vacuum relief valves or expansion bags can be installed. Pressure relief valves help maintain a safe pressure level within the room, while expansion bags can absorb pressure changes without affecting the oxygen content. The use of expansion bags is particularly valuable for ultra-low oxygen storage, ensuring optimal conditions for preserving the quality of fruits and vegetables (Guan et al. 2021).

3.11.2.2.3 Temperature Control

CA storage is most successful at low temperatures, making standard refrigeration units an essential component of CA stores. Temperature control is achieved using refrigerant pipes that circulate throughout the storage chamber, cooling the air. Common refrigerants include ammonia and chlorofluorocarbons. To ensure proper temperature management, CA stores are typically designed to be filled within a day, allowing for immediate cooling of the produce.

3.11.2.2.4 Humidity Control

Maintaining high relative humidity is crucial for preserving the quality of fruits and vegetables stored in controlled atmosphere (CA) environments. However, excessive humidity can lead to condensation on the produce, promoting spoilage. The refrigeration unit's cooling coils play a vital role in regulating humidity. If the refrigerant temperature is significantly lower than the store air temperature, water will condense on the coils, reducing the air's humidity and causing the stored crop to lose moisture through evapotranspiration. To prevent crop desiccation, it is essential to maintain a refrigerant temperature close to the store air temperature. Additionally, a variety of humidifying devices, such as spinning disc humidifiers, can be used to replenish moisture in the air. Secondary cooling techniques, like ice bank cooling, can also help maintain high humidity by indirectly cooling the store air, preventing direct contact between the cooling coils and the produce. By carefully managing humidity levels, it is possible to optimize storage conditions and minimize moisture loss in CA storage facilities (Yahia et al. 2019).

3.11.2.2.5 Gas Control

In modern CA storage facilities, the atmosphere is continuously monitored for carbon dioxide (CO_2) and oxygen (O_2) levels using infrared gas analyzers. These analyzers, calibrated with gas mixtures of known composition, provide real-time readings of gas concentrations. Ethylene analyzers are also employed to monitor ethylene levels, particularly in storage rooms where low ethylene is critical. To ensure proper ventilation and ethylene removal, regular checks are conducted on these systems. The high sensitivity of ethylene analyzers, with a minimum resolution of 0.2 ppm, makes them invaluable for detecting storage atmosphere issues, especially in ultra-ethylene-sensitive products like kiwi fruit. CA storage systems typically incorporate gas control systems with sensors located throughout the storage rooms. These sensors transmit data to a central controller, eliminating the need for sample tubing or pumps. This system enables precise control of gas levels and temperature in each individual storage room, ensuring optimal conditions for various products.

3.11.2.2.6 Scrubbers

Scrubbers are essential components of controlled atmosphere (CA) storage facilities for removing excess carbon dioxide (CO_2) produced by the stored produce. The choice of scrubber material, absorption mode, and air passage system depends on specific requirements and operating conditions. Activated charcoal scrubbers, widely used in CA storage, utilize the adsorption properties of activated carbon to capture and release CO_2. These scrubbers incorporate advanced electronic control systems and CO_2 analyzers to monitor carbon bed status and optimize scrubber cycles. By carefully managing CO_2 levels, scrubbers contribute to maintaining the desired atmosphere within CA storage rooms, ensuring optimal preservation of fruits and vegetables (Thompson et al. 2018).

3.11.2.3 Hypobaric Storage

A relatively little-known method of controlled atmospheric storage is called hypobaric storage or subatomic pressure. Pressure is kept lower than 101 kPa. It involves creating a low-oxygen environment at subatmospheric pressure ranges, which in turn lowers the rates of commodity respiration and metabolism kinetics and lengthens storage life (Kaur et al. 2021). There is a direct correlation between the decrease in oxygen partial pressure and the reduction in total pressure. The product is kept in a chamber with a partial vacuum. To keep the low-oxygen partial pressure ranges constant, saturated air is continually vented from the chamber. Typically, a 2% fall in oxygen concentration at standard atmospheric pressure is allowed by a 10 kPa drop in air pressure, which is equivalent to a 2.1 kPa partial pressure for oxygen (Vithu and Moses 2017).

3.11.2.3.1 Hypobaric Storage Unit and its Components

A humidifier, a refrigeration unit, a vacuum pump, and a product storage chamber make up a hypobaric system. Air passes through the airflow chamber and into the system. A vacuum gauge is used to measure the downstream pressure of the incoming low-pressure air, which is controlled by a needle valve. A humidifier allows air to enter to raise the relative humidity to between 80% and 100%. To achieve this, distilled water is routinely added to the humidifier using a valve to be removed from a water reservoir. By eliminating respiratory heat and any other extra heat from the surroundings, the humidification system stops water loss. A tube delivers saturated air to the vacuum storage chamber. In this chamber, food commodities are kept under vacuum (often between 4 and 400 mm Hg absolute). An essential factor is storage temperature, which is controlled between −2 and −15 °C with a refrigeration device that takes the shape of coolant tubes. The temperature variation during storage is displayed by a temperature gauge. A vacuum pump is included to guarantee the storage chamber's low pressure. The intake and output of air into and out of the system are managed by a pressure regulator/throttle valve.

Hypobaric treatments have been shown to effectively delay the senescence of various fruits and vegetables, including apples, asparagus, avocados, bananas, mangoes, cucumbers, parsley, mushrooms, tomatoes, and strawberries. By creating a modified atmosphere with low oxygen levels, hypobaric conditions can slow down the respiration of fresh produce and remove detrimental volatile metabolic products like ethylene and α-farnesene. This combination of factors helps to retard the ripening process and extend the shelf life of these perishable commodities (An et al. 2009).

3.12 Future Trends in Postharvest Management

Post-harvest science and technology facilitates the industries to deliver safe, nutritious, and fresh horticultural products to consumers at the end of the supply chain from farm to mouth (Palumbo et al. 2022; Al-Dairi et al. 2022).

3.12.1 Horticultural Genomics: Horticultural Genomics

To decrease postharvest losses and enhance the quality of a specific crop, genetic engineering techniques have been extensively employed over the past few decades to create genetically modified crops (Kumari et al. 2022). In several crops, postharvest losses have been reduced by using techniques like RNA interference and antisense RNA. Using antisense sequences that are complementary to the target sense RNA strand, antisense RNA technology functions as a regulatory molecule by attaching to the target sense strand through base pairing and preventing the production of certain genes. For instance, to extend the shelf life of potatoes and tomatoes, and to keep strawberries and tomatoes from softening (Rommens et al. 2006; Smith et al. 2002; Posé et al. 2013). Figure 3.7 shows the application of biotechnological techniques to enhance the postharvest shelf life and quality of horticultural crops. Table 3.4 shows the various genome sequencing projects for various horticultural crops.

Numerous studies have explored the possible roles in enhancing fruit quality, plant architecture, storage stability, and defense against biotic and abiotic challenges in plants (Kulus 2018). Wang et al. (2019) reported decrease in amylose content in sweet potato by using CRISPR/Cas technology by targeting GBSSI gene responsible for granule bound starch biosynthesis. In another study, Hunziker et al. (2022) targeted SIDDB1, SIDET1, SICYC-B genes responsible for the carotenoid metabolic pathway using Target AID technology, which resulted in an increase in lycopene content, beta carotene, and total carotenoid content. Through agrobacterium-mediated transformation, the N′ and C′ ends of eIF4E were targeted by Clustered Regularly Interspaced Short Palindromic Repeats(CRISPR/Cas)in cucumbers, conferring resistance against the papaya ringspot mosaic virus, cucumber vein yellow virus, and pumpkin mosaic virus (Chandrasekaran et al. 2016).

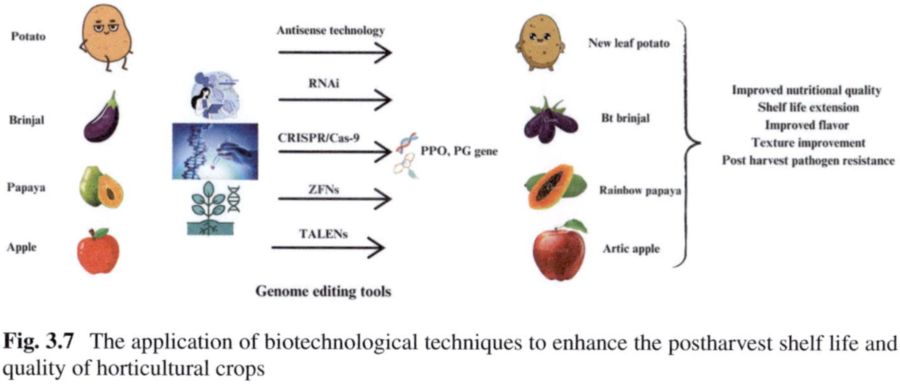

Fig. 3.7 The application of biotechnological techniques to enhance the postharvest shelf life and quality of horticultural crops

Table 3.4 Status of sequencing projects in horticultural crops

Crop	Scientific name	Genome size (Gb)	Chromosome number	Status	Sequencing center
Tomato	*Solanum lycopersicum*	950	12	Completed	International Tomato Genome Sequencing Consortium
Apple	*Malus domestica*	749	17	Completed	Genome 10 K Project
Strawberry	*Fragaria vesca*	247	7	Completed	The Genome 10 K Project
Grape	*Vitis vinifera*	486	19	Completed	International Grape Genome Consortium
Banana	*Musa acuminata*	525	11	Completed	International Banana Genome Sequencing Consortium
Peach	*Prunus persica*	239	16	Completed	Genome 10 K Project
Potato	*Solanum tuberosum*	840	12	Completed	Potato Genome Sequencing Consortium
Rice	*Oryza sativa*	370	12	Completed	International Rice GSP
Mouse-ear cress	*Arabidopsis thaliana*	125	5	Completed	Arabidopsis Genome Initiative
Citrus	*Citrus sinensis*	370	12	Completed	Citrus GSP
Cucumber	*Cucumis sativus*	367	7	Completed	Cucumber GSP
Lettuce	*Lactuca sativa*	1.7	9	Completed	Lettuce GSP
Onion	*Allium cep*	16	9	Completed	Onion GSP
Pepper	*Capsicum annuum*	3.5	12	Completed	Pepper GSP
Tomato	*Solanum lycopersicum*	950	12	Completed	International Tomato Genome Sequencing Consortium
Watermelon	*Citrullus lanatus*	400	11	Completed	Watermelon GSP
Kiwi	*Actinidia deliciosa*	1.1	29	Completed	International Kiwi Genome Sequencing Consortium
Mango	*Mangifera indica*	240	10	Completed	Mango GSP
Pineapple	*Ananas lucidus*	1.4	25	Completed	Pineapple GSP
Pomegranate	*Punica granatum*	240	16	Completed	Pomegranate GSP
Blueberry	*Vaccinium corymbosum*	1.8	12	Completed	Blueberry GSP
Cherry	*Prunus avium*	240	16	Completed	Cherry GSP

GSP Genome Sequencing Project

3.12.2 Nanotechnology

Nanotechnology, a rapidly evolving field, offers promising applications in various sectors, including agriculture, food, water, medicine, energy, and electronics. These materials exhibit exceptional properties compared to their bulk counterparts, such as increased surface area, enhanced reactivity, and unique optical and electrical characteristics. Nanomaterials are materials with at least one dimension in the nanometer range (1–100 nanometers). Nanotechnology-based strategies are being explored to improve the shelf life, safety, and sensory attributes of fresh produce (Neme et al. 2021). In the realm of food science, nanotechnology holds the potential to revolutionize food production, processing, and packaging. By utilizing nanomaterials, researchers are exploring innovative solutions to enhance the delivery of nutrients, improve food safety, and extend the shelf life of products (Fig. 3.8) (Sagar et al. 2022; Sohail et al. 2018; Chen et al. 2020). Nanomaterials can be incorporated into food packaging to enhance barrier properties and prevent spoilage. Additionally, nanotechnology-based sensors can monitor the quality and safety of food throughout the supply chain. These advancements contribute to the development of a more sustainable and efficient food industry, addressing challenges such as food security and waste reduction (Shan et al., 2023; Liu et al. 2020).

3.12.2.1 Nanotechnological Application

When used alone to package fresh fruit variants (FVs), traditional packaging materials including polyamide, polyethylene (PE), polypropylene, and polyvinyl chloride (PVC) typically have certain drawbacks. Over 350 million tons of plastic are

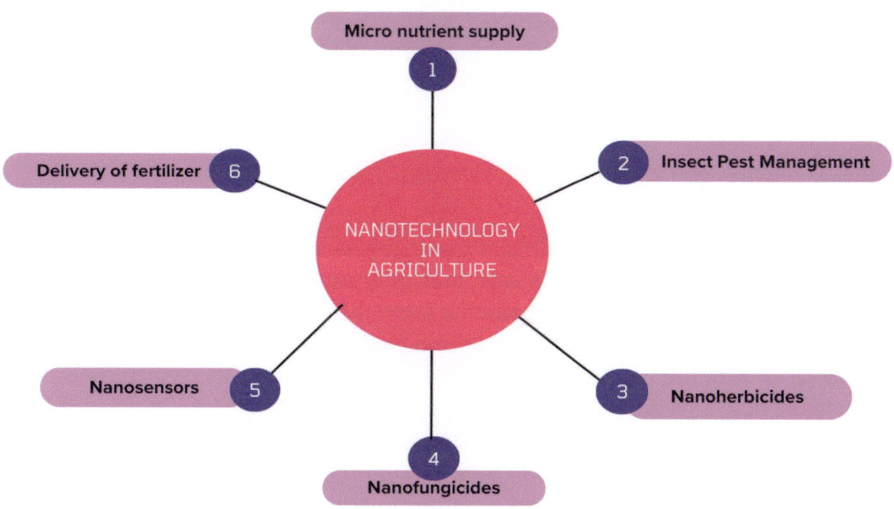

Fig. 3.8 Application of nanotechnology

Table 3.5 Packaging applications of nanomaterials—features and concerns

Application	Feature
Nanocomposite	Adding nanomaterials into the packaging films to enhance their barrier properties, durability, and physical characters
Nanocoatings	Incorporated either inside or outside Improves barrier properties
Surface biocides	Incorporated on the outer surface of the packaging material Have antimicrobial properties
Active packaging	Making packaging films with enhanced antimicrobial properties, having antioxidant characters
Intelligent packaging	Special sensors are incorporated to monitor and report the condition of the food

manufactured worldwide each year, with food packaging accounting for a significant portion of traditional plastic usage (Haghighi et al. 2020). This puts a significant strain on the environment. One approach that shows promise for addressing the drawbacks of conventional polymeric packaging made of plastic is the development of nanocomposites or nano coatings (Ghaani et al. 2016; Müller and Schmid 2019). They can be applied by coating, dipping, and spraying. Coating preservation technique creates a microatmosphere that prevents decay and postpones the senescence of fresh fruit variants. Because edible coating methods are eco-friendly, antiblastic, and help eliminate the usage of non-biodegradable disposable film, they are currently commonly utilized to preserve postharvest fruit variants (FVs). Because of its many uses and potential applications, nanopackaging has been extensively used in the postharvest preservation of fresh fruit varieties. Xu et al. (2017) reported ZnO nanoparticles to be effective in reducing the total colony forming units during the storage of carrots and extending the shelf life up to 40 days. Another study conducted for freshly cut Fuji apples where ZnO nanoparticles were used to pack the samples whereas control samples were packed in polyvinyl chloride film. Results showed that in nanomaterial packed samples (Table 3.5) (Li et al. 2011a, b).

3.12.3 Emerging Intelligent Packaging Technologies

Another prominent term for intelligent packaging is smart packaging. It can detect the characteristics of the food it is surrounded by or the environment in which it is kept. Additionally, it can alert the manufacturer, distributor, or buyer to the state of certain food qualities. While intelligent packaging differs from the concept of active packaging in a unique way, it may also be used to confirm the effectiveness and integrity of active package designs. Consequently, the packaging has sensors that alert consumers to the impacted goods and allow them to take action to undo the harmful changes made to the commodity. These packaging solutions come with instruments that can identify and provide information about the features and benefits of foods that are packaged, as well as provide an internal or external indicator for the history of the active ingredient and the assessment of the product's quality. When wireless communication, cloud data resources, data processing units, and supporting electronics and components such as indications, sensors and data carriers are integrated into the system, it becomes smarter.

3.12.4 *Time Temperature Integrator (TTI) Technology*

When put on a food product, a time temperature integrator (TTI) is a simple, inexpensive device that can provide a readily detectable change that indicates the full or partial temperature history of the product. The idea of TTI action is an irreversible transition that can be chemical, mechanical, microbiological, or enzymatic; it is usually expressed as a perceptible reaction in the process of mechanical deformation, color production, or color change. It is dependent on changes in temperature or time for biological response, such as microorganisms, spores, or enzymes, but it also depends on physical shifts or chemical reactions toward temperature and time for chemical or physical responses, such as acid–base reactions, melting, polymerization, etc.

3.12.5 *Biosensors*

Biosensors are highly useful in many areas. They are used in food processing, food safety, and food technology, and they have improved society's overall health by effectively detecting the presence and concentration of various microorganisms, such as *Vibrio cholera*, *Escheichia coli*, and *Clostridium* spp., as well as various toxins, such as nitrosamines, acrylamides, benzene, ethylbenzene, toluene, xylene, and benzo[a]pyrene, which are carcinogenic (Vallinayagam et al. 2022). A strong and cutting-edge analytical tool containing biological sensing elements, a "biosensor" has many uses in drug development, diagnosis, biomedicine, food safety and processing, environmental monitoring, defense, and security. Using an immobilized glucose oxidase electrode, the first biosensor developed by Clark Jr and Lyons (1962) measured glucose in biological samples by electrochemically detecting oxygen or hydrogen peroxide. A biosensor operates on the principle of selective recognition between a biological receptor and a target analyte. This interaction triggers a measurable signal that corresponds to the analyte's concentration or presence. The biological receptor, or recognition element, is a biomolecule or biological entity engineered to bind specifically with the target analyte. This binding event induces a biochemical or biophysical change, which is then converted into a detectable signal by a transducer. The amplified signal is processed and interpreted to provide a human-readable output. In essence, a biosensor integrates a bioreceptor, transducer, amplifier, processor, and display to enable accurate and efficient detection of target analytes (Nath 2024; Dash and Kaushik 2022; Shams et al. 2020). Biosensors, with their capacity for rapid, precise, and on-site contaminant detection, are poised to revolutionize food safety and quality monitoring. These innovative tools offer a significant advantage over conventional analytical methods (Meliana et al. 2024; Murugaboopathi et al. 2013).

3.12.6 *Information Technologies in Postharvest Management*

3.12.6.1 AI and IoT

Fruit industry sorting, grading, and packaging procedures are greatly aided by artificial intelligence (AI), a branch of computer science that focuses on smart and creative algorithms. Machine learning, robotics, natural language processing, computer vision, autonomous agents, and expert systems are just a few of the AI tools that improve marketing strategies, pricing decisions, production processes, predictive analytics, and customer experiences overall. The development of applications for smart agriculture is greatly aided by the Internet of Things (IoT) (Kasera et al. 2024). Cloud computing, edge computing, big data, artificial intelligence, and other cutting-edge technologies are also employed in smart agriculture. Farming operations for a particular crop can be broadly divided into three phases: preharvest, during harvest, and post harvest. Each step requires the completion of numerous tasks. Selection of seeds, preparation of the field, crop choice, etc., are all part of the preharvest stage. Irrigation, disease analysis, pathogen detection, etc., are part of the harvesting process. Storage, cooling, reaping, etc., are part of the postharvest stage (Pathmanaban et al. 2023).

3.12.6.2 RFID and QR Code

Radio-frequency identification (RFID) and quick response (QR) code technologies offer transformative solutions for postharvest management of fruits and vegetables. RFID tags, equipped with integrated circuits and antennas, wirelessly transmit unique identification information to RFID readers, enabling real-time tracking of product movement and location. QR codes, two-dimensional barcodes, store essential information like product origin, production date, batch number, and expiration date (Pavankumar et al. 2022). By scanning QR codes using smartphones or barcode scanners, consumers and supply chain stakeholders can instantly access comprehensive product details and traceability information. The integration of RFID and QR code technologies empowers stakeholders with enhanced visibility and control throughout the supply chain, facilitating efficient inventory management, quality control, and consumer engagement.

3.13 Conclusion

Ensuring quality and safety of fruits and vegetables from harvest to consumption necessitates a comprehensive understanding of postharvest challenges and the strategic application of effective handling practices. These perishable commodities are vital for human health due to their rich content of vitamins, minerals (mainly electrolytes), dietary fiber, and protective phytochemicals are vulnerable to physiological deterioration, mechanical damage, and ethylene-induced senescence. These factors contribute significantly to postharvest losses, impacting economic viability and the availability of nutritious food. Effective postharvest management, encompassing appropriate harvesting practices determined by maturity indices, careful harvesting methods for both fruits and vegetables, and timely precooling processes, forms the foundation for minimizing these losses. Furthermore, the strategic selection and application of packaging materials and techniques are crucial for providing physical protection, modifying the internal atmosphere, and extending shelf life. Understanding the diverse types of storage available, from ambient to controlled atmosphere, allows for tailored preservation strategies based on the specific requirements of different produce. Emerging technologies, such as nanotechnological applications and intelligent packaging incorporating biosensors, offer innovative solutions for enhanced monitoring and preservation. These advancements hold the potential to further minimize postharvest losses and ensure the delivery of fresh, high-quality fruits and vegetables to consumers. Ultimately, a holistic approach integrating best handling practices, appropriate packaging, optimized storage, and the adoption of novel technologies is essential for safeguarding the nutritional value and economic sustainability of the fruit and vegetable supply chain.

References

Al-Dairi M, Pathare PB, Al-Yahyai R et al (2022) Mechanical damage of fresh produce in postharvest transportation: current status and future prospects. Trends Food Sci Technol 124:195–207. https://doi.org/10.1016/j.tifs.2022.04.018

Ali A, Yeoh WK, Forney C, Siddiqui MW (2018) Advances in postharvest technologies to extend the storage life of minimally processed fruits and vegetables. Crit Rev Food Sci Nutr 58(15):2632–2649. https://doi.org/10.1080/10408398.2017.1339180

An DS, Park E, Lee DS (2009) Effect of hypobaric packaging on respiration and quality of strawberry and curled lettuce. Postharvest Biol Technol 52(1):78–83. https://doi.org/10.1016/j.postharvbio.2008.09.014

Anjali JA, Bamola A et al (2024) State-of-the-art non-destructive approaches for maturity index determination in fruits and vegetables: principles, applications, and future directions. FPPN 6(1):56. https://doi.org/10.1186/s43014-023-00205-5

Barman K, Sharma S, Siddiqui MW (eds) (2018) Emerging postharvest treatment of fruits and vegetables. CRC Press

Belay D (2021) Design, construction and performance evaluation of potato harvesters: a review. IRJET 8(4):2747–2771

Belay ZA, Caleb OJ, Opara UL (2016) Modelling approaches for designing and evaluating the performance of modified atmosphere packaging (MAP) systems for fresh produce: a review. Food Packag Shelf Life 10:1–15. https://doi.org/10.1016/j.fpsl.2016.08.001

Blasco J, Munera S, Aleixos N et al (2017) Machine vision-based measurement systems for fruit and vegetable quality control in postharvest. In: Measurement, modeling and automation in advanced food processing, pp 71–91. https://doi.org/10.1007/10_2016_51

Bodbodak S, Moshfeghifar M (2016) Advances in controlled atmosphere storage of fruits and vegetables. In: Eco-friendly technology for postharvest produce quality. Academic, pp 39–76. https://doi.org/10.1016/B978-0-12-804313-4.00002-5

Bourke P, Ziuzina D, Han L et al (2017) Microbiological interactions with cold plasma. J Appl Microbiol 12(2):308–324

Brasil IM, Siddiqui MW (2018) Postharvest quality of fruits and vegetables: an overview. In: Preharvest modulation of postharvest fruit and vegetable quality, pp 1–40. https://doi.org/10.1016/B978-0-12-809807-3.00001-9

Chandrasekaran J, Brumin M, Wolf D et al (2016) Development of broad virus resistance in non-transgenic cucumber using CRISPR/Cas9 technology. Mol Plant Pathol 17(7):1140–1153. https://doi.org/10.1111/mpp.12375

Chen S, Brahma S, Mackay J et al (2020) The role of smart packaging system in food supply chain. J Food Sci 85(3):517–525. https://doi.org/10.1111/1750-3841.15046

Clark LC Jr, Lyons C (1962) Electrode systems for continuous monitoring in cardiovascular surgery. Ann N Y Acad Sci 102(1):29–45. https://doi.org/10.1111/j.1749-6632.1962.tb13623.x

Dash S, Kaushik S (2022) Biosensors: current trends and future perspectives. In: Nanomaterials-based sensing platforms. Apple Academic Press, pp 1–35

Desai M, Chandel A, Chauhan OP et al (2024) Uses and future prospects of cold plasma in agriculture. Food Hum 2:100262. https://doi.org/10.1016/j.foohum.2024.100262

Dincer I (1995) Air flow precooling of individual grapes. J Food Eng 26(2):243–249. https://doi.org/10.1016/0260-8774(94)00049-F

Duan Y, Wang GB, Fawole OA et al (2020) Postharvest precooling of fruit and vegetables: a review. Trends Food Sci Technol 100:278–291. https://doi.org/10.1016/j.tifs.2020.04.027

Elik A, Yanik DK, Istanbullu Y et al (2019) Strategies to reduce post-harvest losses for fruits and vegetables. Strategies 5(3):29–39. https://doi.org/10.7176/JSTR/5-3-04

El-Ramady HR, Domokos-Szabolcsy É, Abdalla NA et al (2015) Postharvest management of fruits and vegetables storage. Sustain Agric Rev 15:65–152. https://doi.org/10.1007/978-3-319-09132-7_2

Erkan M, Dogan A (2019) Harvesting of horticultural commodities. In: Postharvest technology of perishable horticultural commodities. Woodhead Publishing, pp 129–159

Falagán N, Terry LA (2018) Recent advances in controlled and modified atmosphere of fresh produce. Johns Matthey Technol Rev 62(1):107–117. https://doi.org/10.1595/205651318X696684

Feliziani E, Lichter A, Smilanick JL (2016) Disinfecting agents for controlling fruit and vegetable diseases after harvest. Postharvest Biol Technol 122:53–69. https://doi.org/10.1016/j.postharvbio.2016.04.016

Florkowski WJ, Banks NH, Shewfelt RL et al (eds) (2021) Postharvest handling: a systems approach. Academic

Gardas BB, Raut RD, Narkhede B (2017) Modeling causal factors of post-harvesting losses in vegetable and fruit supply chain: an Indian perspective. Renew Sust Energ Rev 80:1355–1371. https://doi.org/10.1016/j.rser.2017.05.259

Ghaani M, Cozzolino CA, Castelli G et al (2016) An overview of the intelligent packaging technologies in the food sector. Trends Food Sci Technol 51:1–11. https://doi.org/10.1016/j.tifs.2016.02.008

Guan J, Lacombe A, Rane B et al (2021) A review: gaseous interventions for listeria monocytogenes control in fresh apple cold storage. Front Microbiol 12:782934. https://doi.org/10.3389/fmicb.2021.782934

Haghighi H, Licciardello F, Fava P et al (2020) Recent advances on chitosan-based films for sustainable food packaging applications. Food Packag Shelf Life 26:100551. https://doi.org/10.1016/j.fpsl.2020.100551

Hasan MU, Malik AU, Ali S, Imtiaz A et al (2019) Modern drying techniques in fruits and vegetables to overcome postharvest losses: a review. J Food Process Preserv 43(12):e14280. https://doi.org/10.1111/jfpp.14280

Hunziker J, Nishida K, Kondo A et al (2022) Phenotypic characterization of high carotenoid tomato mutants generated by the target-AID Base-editing technology. Front Plant Sci 13:848560. https://doi.org/10.1111/mpp.12375

Hussein Z, Fawole OA, Opara UL (2020) Harvest and postharvest factors affecting bruise damage of fresh fruits. HPJ 6(1):1–13. https://doi.org/10.1016/j.hpj.2019.07.006

Kader AA, Zagory D, Kerbel EL et al (1989) Modified atmosphere packaging of fruits and vegetables. Crit Rev Food Sci Nutr 28(1):1–30. https://doi.org/10.1016/j.lwt.2009.05.018

Kasera RK, Gour S, Acharjee T (2024) A comprehensive survey on IoT and AI based applications in different pre-harvest, during-harvest and post-harvest activities of smart agriculture. Comput Electron Agric 216:108522. https://doi.org/10.1016/j.compag.2023.108522

Kattel D, Thokar N, Subedi S (2022) A review on post-harvest precooling methods of fruits and vegetables. FAER 2(2):96–99. https://doi.org/10.26480/faer.02.2022.96.99

Kaur J, Aslam R, Saeed PA (2021) Storage structures for horticultural crops: a review. Environ Conserv J 22(SE):95–105. https://doi.org/10.36953/ECJ.2021.SE.2210

Kulus D (2018) Genetic resources and selected conservation methods of tomato. JABFQ 91. https://doi.org/10.5073/JABFQ.2018.091.019

Kumari C, Sharma M, Kumar V et al (2022) Genome editing technology for genetic amelioration of fruits and vegetables for alleviating post-harvest loss. Bioengineering 9(4):176. https://doi.org/10.3390/bioengineering9040176

Kumar M, Singh R, Sharma I (2021) Integrated management for post-harvest diseases of fruits and vegetables. Bio-Sci Res Bull:17–22. https://doi.org/10.5958/2320-3161.2021.00005.5

Lakshmi S, Pandey AK, Ravi N et al (2017) Non-destructive quality monitoring of fresh fruits and vegetables. DLSJ 2(2):103–110. https://doi.org/10.14429/dlsj.2.11379

Li P, Lee SH, Hsu HY (2011a) Review on fruit harvesting method for potential use of automatic fruit harvesting systems. Procedia Eng 23:351–366. https://doi.org/10.1016/j.proeng.2011.11.2514

Liu W, Zhang M, Bhandari B (2020) Nanotechnology–a shelf life extension strategy for fruits and vegetables. Crit Rev Food Sci Nutr 60(10):1706–1721. https://doi.org/10.1080/10408398.2019.1589415

Li X, Li W, Jiang Y et al (2011b) Effect of nano-ZnO-coated active packaging on quality of fresh-cut 'Fuji' apple. IJFST 46(9):1947–1955. https://doi.org/10.1111/j.1365-2621.2011.02706.x

Mandal R, Singh A, Singh AP (2018) Recent developments in cold plasma decontamination technology in the food industry. Trends Food Sci Technol 80:93–103. https://doi.org/10.1016/j.tifs.2018.07.014

Mangaraj S, Goswami TK (2009) Modified atmosphere packaging of fruits and vegetables for extending shelf-life-a review. Fresh Produce 3(1):1–31

Mathushika JM, Gomes C (2021) Emerging concepts and practices in post-harvest Management of Horticultural Crops Revisited. IJCSRR 4(8):859–876. https://doi.org/10.47191/ijcsrr/V4-i8-04

Meitha K, Pramesti Y, Suhandono S (2020) Reactive oxygen species and antioxidants in postharvest vegetables and fruits. Int J Food Sci 2020(1):8817778. https://doi.org/10.1155/2020/8817778

Meliana C, Liu J, Show PL et al (2024) Biosensor in smart food traceability system for food safety and security. Bioengineered 15(1):2310908. https://doi.org/10.1080/21655979.2024.2310908

Mishra V, Abrol GS, Dubey N (2018) Sodium and calcium hypochlorite as postharvest disinfectants for fruits and vegetables. In: Postharvest disinfection of fruits and vegetables. Academic, pp 253–272

Müller P, Schmid M (2019) Intelligent packaging in the food sector: a brief overview. Food Secur 8(1):16. https://doi.org/10.3390/foods8010016

Murray K, Wu F, Shi J (2017) Challenges in the microbiological food safety of fresh produce: limitations of post-harvest washing and the need for alternative interventions. Food Qual Saf 1(4):289–301. https://doi.org/10.1093/fqsafe/fyx027

Murugaboopathi G, Parthasarathy V, Chellaram C et al (2013) Applications of biosensors in food industry. Biosci Biotech Res Asia 10(2):711–714. https://doi.org/10.13005/bbra/1185

Nath S (2024) Advancements in food quality monitoring: integrating biosensors for precision detection. Sustain Food Technol 2:976–992. https://doi.org/10.1039/D4FB00094C

Nema P, Kheto A, Kumar P (2023) A review technological development in the grading of fruits and vegetables: a review. Agric Eng Int CIGR J 25(4)

Neme K, Nafady A, Uddin S et al (2021) Application of nanotechnology in agriculture, postharvest loss reduction and food processing: food security implication and challenges. Heliyon 7(12):e08539

Nummer BA, Brian A (2002) Historical origins of food preservation. National Center for Home Food Preservation. University of Illinois Extension

Palumbo M, Attolico G, Capozzi V et al (2022) Emerging postharvest technologies to enhance the shelf-life of fruit and vegetables: An overview. Food Secur 11(23):3925. https://doi.org/10.3390/foods11233925

Pandey AK, Chauhan OP (2019) Chapter 15: Use of plasma in food processing. In: Non-thermal processing of foods, p 283

Pankaj SK, Keener KM (2017) Cold plasma: background, applications and current trends. Curr Opin Food Sci 16:49–52. https://doi.org/10.1016/j.cofs.2017.07.008

Pathmanaban P, Gnanavel BK, Ananda SS et al (2023) Advancing post-harvest fruit handling through AI-based thermal imaging: applications, challenges, and future trends. Discov Food 3(1):27. https://doi.org/10.1007/s44187-023-00068-2

Pavankumar M, Megha M, Hegde SS et al (2022) Innovations in post-harvest techniques and food storage. In: Recent research innovations in agriculture

Posé S, Paniagua C, Cifuentes M et al (2013) Insights into the effects of polygalacturonase FaPG1 gene silencing on pectin matrix disassembly, enhanced tissue integrity, and firmness in ripe strawberry fruits. J Exp Bot 64(12):3803–3815. https://doi.org/10.1093/jxb/ert210

Prasad K, Jacob S, Siddiqui MW (2018) Fruit maturity, harvesting, and quality standards. In: Preharvest modulation of postharvest fruit and vegetable quality. Academic, pp 41–69. https://doi.org/10.1016/B978-0-12-809807-3.00002-0

Raju PS, Chauhan OP, Bawa AS (2018) Postharvest handling systems and storage of vegetables. In: Handbook of vegetables and vegetable processing, pp 247–264. https://doi.org/10.1002/9781119098935.ch10

Rommens CM, Ye J, Richael C et al (2006) Improving potato storage and processing characteristics through all-native DNA transformation. J Agric Food Chem 54(26):9882–9887. https://doi.org/10.1021/jf062477l

Ruiz-Altisent M, Ortiz-Cañavate J, Valero C (2004) Fruit and vegetables harvesting systems. In: Production practices and quality assessment of food crops volume 1: Preharvest practice, pp 261–285

Sagar NA, Kumar N, Choudhary R et al (2022) Prospecting the role of nanotechnology in extending the shelf-life of fresh produce and in developing advanced packaging. Food Packag Shelf Life 34:100955. https://doi.org/10.1016/j.fpsl.2022.100955

Sakudo A, Misawa T, Yagyu Y (2020) Equipment design for cold plasma disinfection of food products. In: Advances in cold plasma applications for food safety and preservation. Academic, pp 289–307

Sandhya (2010) Modified atmosphere packaging of fresh produce: current status and future needs. LWT 43(3):381–392. https://doi.org/10.1016/j.lwt.2009.05.018

Shakeel Q, Shaheen MR, Ali S et al (2022) Postharvest management of fruits and vegetables. In: Applications of biosurfactant in agriculture. Academic, pp 1–16

Shams R, Singh J, Ashraf S et al (2020) Application of biosensors in food quality control. J Postharvest Technol 8(1):53–74

Shan Y, Li T, Qu H et al (2023) Nano-preservation: an emerging postharvest technology for quality maintenance and shelf life extension of fresh fruit and vegetable. Food Front 4(1):100–130. https://doi.org/10.1002/fft2.201

Shewfelt RL, Prussia SE (2022) Challenges in handling fresh fruits and vegetables. In: Postharvest handling. Academic, pp 167–186. https://doi.org/10.1016/B978-0-12-822845-6.00006-3

Shinde R, Rodov V, Krishnakumar S et al (2018) Active and intelligent packaging for reducing postharvest losses of fruits and vegetables. Postharvest Biol Technol:171–189. https://doi.org/10.1002/9781119289470.ch7

Singh D, Sharma RR (2018) Postharvest diseases of fruits and vegetables and their management. In: Postharvest disinfection of fruits and vegetables. Academic, pp 1–52. https://doi.org/10.1016/B978-0-12-812698-1.00001-7

Slavin JL, Lloyd B (2012) Health benefits of fruits and vegetables. Adv Nutr 3(4):506–516. https://doi.org/10.3945/an.112.002154

Smith DL, Abbott JA, Gross KC (2002) Down-regulation of tomato β-galactosidase 4 results in decreased fruit softening. Plant Physiol 129(4):1755–1762. https://doi.org/10.1104/pp.011025

Sohail M, Sun DW, Zhu Z (2018) Recent developments in intelligent packaging for enhancing food quality and safety. Crit Rev Food Sci Nutr 58(15):2650–2662. https://doi.org/10.1080/10408398.2018.1449731

Soltani M, Alimardani R, Mobli H et al (2015) Modified atmosphere packaging: a progressive technology for shelf-life extension of fruits and vegetables. J Appl Packag Res 7(3):33–59

Suslow T (2000) Chlorination in the production and postharvest handling of fresh fruits and vegetables. In: Fruit and vegetable processing, pp 2–15

Thompson AK, Prange RK, Bancroft R et al (2018) Controlled atmosphere storage of fruit and vegetables. CABI

Thompson AK (2015) Fruit and vegetable storage: hypobaric, hyperbaric and controlled atmosphere. Springer

Tzortzakis N, Chrysargyris A (2017) Postharvest ozone application for the preservation of fruits and vegetables. Food Rev Int 33(3):270–315. https://doi.org/10.1080/87559129.2016.1175015

Unni LE, Chauhan OP, Raju PS et al (2011) High pressure processing of foods: present status and future strategies. Int J Food Ferment 1(1):49–62

Vallinayagam S, Paladhi AG, Pal K et al (2022) Multifunctional biosensor activities in food technology, microbes and toxins–a systematic mini review. Process Biochem 120:260–264. https://doi.org/10.1016/j.procbio.2022.06.019

Vithu P, Moses JA (2017) Hypobaric storage of horticultural products: a review. In: Engineering practices for agricultural production and water conservation, pp 155–170

Vivek K, Singh SS, Ritesh W et al (2019) A review on postharvest management and advances in the minimal processing of fresh-cut fruits and vegetables. JMBFS 8:1178–1187

Wang H, Wu Y, Zhang Y et al (2019) CRISPR/Cas9-based mutagenesis of starch biosynthetic genes in sweet potato (*Ipomoea Batatas*) for the improvement of starch quality. Int J Mol Sci 20(19):4702. https://doi.org/10.3390/ijms20194702

Watson JA, Treadwell D, Sargent SA et al (2016) Postharvest storage, packaging and handling of specialty crops: a guide for Florida small farm producers: HS1270/HS1270, 10/2015. EDIS 2016(1):18–18

Xiao F, Wang H, Li Y et al (2023) Object detection and recognition techniques based on digital image processing and traditional machine learning for fruit and vegetable harvesting robots: an overview and review. Agronomy 13(3):639. https://doi.org/10.3390/agronomy13030639

Xiao X, Wang Y, Jiang Y (2024) Review of research advances in fruit and vegetable harvesting robots. JEET 19(1):773–789. https://doi.org/10.1007/s42835-023-01596-8

Xu J, Zhang M, Bhandari B et al (2017) ZnO nanoparticles combined radio frequency heating: a novel method to control microorganism and improve product quality of prepared carrots. IFSET 44:46–53. https://doi.org/10.1016/j.ifset.2017.07.025

Yahia EM, Carrillo-Lopez A (eds) (2018) Postharvest physiology and biochemistry of fruits and vegetables. Woodhead Publishing

Yahia EM, Fadanelli L, Mattè P et al (2019) Controlled atmosphere storage. In: Postharvest technology of perishable horticultural commodities. Woodhead Publishing, pp 439–479. https://doi.org/10.1016/B978-0-12-813276-0.00013-4

Zainal MNF, Redzuan N, Misnal MF (2015) Brief review: cold plasma. J Teknol 74(10). https://doi.org/10.11113/jt.v74.4834

Zhang C, Zhou P, Mei J et al (2023) Effects of different pre-cooling methods on the shelf life and quality of sweet corn (*Zea mays L.*). Plan Theory 12(12):2370. https://doi.org/10.3390/plants12122370

Chapter 4
Postharvest Pathology

Sakshi Sharma and OP Chauhan

4.1 Introduction

Fruits and vegetables, being the important sources of vitamins, minerals, and fibers, play a significant role in the nutritional security of India. Globally, India is ranked second for fruits and vegetables production. In case of mango, bananas, guavas, and papaya fruits, India is the world's largest producer. For vegetables like tomatoes, potatoes, cabbage, cauliflower, and green peas, India ranked second in production globally. During 2021–2022, fruits are being cultivated in around 7.04 million ha, with a production level of 107.24 million MT while vegetables are being cultivated in 13.48 millionha with a production of 204.83 million MT in India. In horticultural commodities, occurrence of postharvest losses can be separated into five groups' viz., production/harvest, postharvest handling and storage, processing, distribution, and consumption. Postharvest losses include not only the losses in terms of produce but also the wastage of valuable resources like land, water, energy, and inputs used in the production of produce. According to Alao (2000), postharvest losses in case of fruits and vegetables are defined as the quantity of wholesome, edible produce (excluding moisture content) that is typically consumed by humans but, despite being intentionally harvested for this purpose, is not consumed for various reasons. Due to their composition of living tissues, fresh fruits and vegetables are highly perishable and particularly susceptible to such losses. Maintaining the vitality and health of these tissues, which consist of thousands of living cells, is essential throughout the marketing process. Therefore, reducing postharvest losses of fruits and vegetables also helped in increasing overall production.

Postharvest losses of fruits and vegetables are influenced by various factors, categorized as primary and secondary causes. Primary causes comprise mechanical, physiological, pathological, and environmental factors. Secondary causes involve

S. Sharma (✉) · OP Chauhan
Defence Food Research Laboratory, Mysore, India

OP Chauhan (ed.), *Fruits and Vegetables Technologies*,
https://doi.org/10.1007/978-981-96-8433-5_4

improper harvesting methods (due to inadequate equipment, rough handling, and harvesting delays), transportation issues (such as insufficient facilities and mechanical damage during transit), and inadequate storage, marketing, and legislation.

Reducing postharvest losses in fruits and vegetables can be realized using proper cultural practices, harvesting, transportation, storage, and pre- and postharvest treatments. Effective cultural practices, like fine soil tilling to prevent root forking in crops like carrots and radishes, adequate water supply to prevent cracking in carrots, radishes, tomatoes, and splitting of onion scales, and withdrawing irrigation before maturity to enhance the keeping quality of onions and garlic, are crucial. Additionally, the timely application of fertilizers and nutrients to maintain produce quality and staking vine crops like peas and tomatoes to prevent contact with the ground are essential measures (Danladi 2000). Excessive use of nitrogenous fertilizers can accelerate tissue deterioration, whereas proper potassium fertilization improves the shelf life of fruits and vegetables. Deficiencies in micronutrients such as molybdenum and boron can cause physiological disorders in crops such as cabbage and almonds, leading to poor shelf life and quality deterioration, while manganese deficiency in peas can result in marsh spots (Danladi 2000).

It is an established fact that consumers prefer freshly harvested, properly matured, insect- and disease-free fruits and vegetables with an attractive appearance. Therefore, vegetables should be harvested when they reach optimal size and are still tender. Most fruits are harvested at the proper maturity stage but before ripening and are later ripened artificially according to market demands (Opadokun 1987). The best time to harvest fruits and vegetables is during the cooler parts of the day, and the produce should be promptly moved to a packing area after harvest. Harvesting should be avoided during or immediately after rain, as this can generate situations suitable for microbial growth.

Fruits and vegetables require careful handling before being packed for transportation. Washing with clean water helps prevent wilting and enhances their appearance. Additionally, vegetables like cabbage, spinach, and lettuce should be trimmed before being harvested for the market (Hayatu 2000). Proper packaging significantly reduces wastage and protects against mechanical damage, undesirable physiological changes, and pathological deterioration during storage, transportation, and marketing. Effective packaging also preserves the freshness, succulence, and flavor of fruits and vegetables for an extended period (Yahaya 2005). All packages should include adequate ventilation to prevent physiological breakdown and the development of off-flavors. When packing for transportation, it is important to avoid bruising and damaging the fruits.

4.2 Pathogens Causing Postharvest Diseases

Various pathogens, including fungi and bacteria, cause diseases in fruits and vegetables. Among fungi (*Aspergillus, Botrytis, Colletotrichum, Diplodia, Monilinia, Penicillium, Phomopsis, Rhizopus, Mucor, Alternaria,* and *Sclerotinia*), along with bacteria like *Erwinia* and *Pseudomonas*, are mainly responsible for major postharvest losses (Barkai-Golan 2005). These pathogens are often weak, primarily invading damaged produce. Physical damage significantly contributes to postharvest deterioration and is the primary cause of these losses.

Injuries to produce can occur prior to harvest and/or after the harvest due to weather conditions, insects, birds, rodents, and farm equipment. Fruits may be damaged when dropped onto hard surfaces during handling, but this damage might not be immediately apparent. Bruising can manifest later, either externally (e.g., apples) or internally (e.g., potatoes) upon peeling. Overstocking of bulk produce in storage or overfilling packaging (e.g., grapes) can lead to compression bruising. Underfilled packaging can cause vibration damage especially during long-distance transportation. Various microorganisms can easily attack the damaged produce and result in progressive decay that can affect the entire batch (Snowdon 1990).

4.3 Infection Process

Microorganisms can lead to either preharvest or postharvest infection depending upon the occurrence of infection. Preharvest infection refers to the occurrence of infection on the produce which is still immature on the plant while postharvest infection refers to the occurrence of infection during harvesting, handling, and marketing operations. Mechanical injuries to the produce's peel, such as fingernail scratches, abrasions, rough handling, insect punctures, and cut stems, often help the postharvest infection. Infection can arise either through direct penetration of the cuticle or entry via stomata, lenticels, wounds, or abscission scar tissue. Additionally, the physiological condition of the produce, temperature, and periderm formation significantly influence the infection process, leading to two types of infections: preharvest and postharvest.

4.3.1 Preharvest Infection

The preharvest infection in various vegetables and fruits can arise by means of different pathways, line directly entry into peel, or an infection may arise via organic open areas, or through damaged areas. Pathogenic fungi often initiate infections on floral parts and developing fruits, where the infectivity stays dormant beyond harvesting. Once fruits start to ripen or its tissues begin to age, the host's resistance decreases, allowing the pathogen to grow (Barkai-Golan 2005). These latent infections significantly contribute to postharvest wastage, particularly in subtropical as well as tropical fruits, such as mango anthracnose, papaya anthracnose, banana crown rot, and citrus stem-end rot. An instance, the spores of the pathogen *Colletotrichum* can propogate in moistness over fruit exterior, with germ tube forming a structure called an "appressorium" that might or might not lead to penetration into fruit peel prior to occurring of dormancy of infectivity. Prasitic fungi as well as bacteria which are feeble can penetrate via organic open areas inside underdeveloped produce through natural openings like the stomata of the plant, or the lenticels' part, or the cracks that might occur during growing. These infections could remain dormant till fruit ripens and gets more susceptible. Healthy fresh produce is capable of inhibiting microorganisms for an extended duration (Barkai-Golan 2005). In an instance, spore of a pathogen *Phlyctaena vagabunda* can enter inside the lenticels part of apples prior to the harvesting stage, leading to rot nearby that part during storage. Entrance of the germ tubes of a pathogen *Colletotrichum gloeosporioides* spore in to the immature papaya, and entrance of germ tubes of *Monilinia fructicola* spores in to young stone fruits through stomata are examples of penetration through natural inlets (Barkai-Golan 2001). Several fruits and vegetables get infected by fungus through penetration through the lenticles, e.g., fungus *Dothiorella gregaria* in avocado, *Gloeosporium album* and *Gloeosporium perennans* in apple, *Penicillium expansum* which leads to the blue mold in the apple, *Erwinia cartovora* in potato causing soft-rot throughout storability, *Helminthosporium solani* in potato tubers causing the disease of silver scurf, etc. Another natural inlet from where penetration of fungi can occur is open calyx tube, e.g., *Alternaria alternata* in delicious apples and Na'ama tomato fruits; various pathogens can cause core rot in apples such as *Aspergillus niger*, *Botrytis cinerea*, *Mucor piriformis*, *Fusarium* sp., *Trichothecium roseum*, *Phomopsis mali*, *Penicillium funiculosum* and *Penicillium expansum* from the infection resulting through sinus between calyx and core cavity (Barkai-Golan 2001).

4.3.2 Postharvest Infection

Many fungi may lead to waste of produce cannot penetrate inside intact fruit peels, however may effortlessly invade via break or damage occurring to peels. Although damage may be microscopic, it allows pathogens existing on the host organism to proliferate. Additionally, the cuts stem often serves as a primary entry point for microorganisms, leading to stem-end rots, which are a major form of postharvest decay in various fresh produce (Barkai-Golan 2005). For instance, postharvest infections occurred because of *Sclerotinia* and *Colletotrichum* are frequent within number of produced fruit via straight peel permeation. These infections can occur at various florae portions, including flower infections, stem termination infections, as well as quiescent infections, as mentioned ahead:

4.3.2.1 Floral Infection

Many instances have been reported which are indicative of the infectivity occurring because of organisms happening via flower portions of several fruit crops. In an instance, the pathogen *Botrytis cinerea* growing upon blackcurrant fruit (McNicol and Williamson 1989) and raspberry (Dashwood and Fox 1988), and *Monilinia laxa* growing upon the fruit plum (Schlagbauer and Holz 1990), infect the produce at floral parts.

4.3.2.2 Stem-End Infection

Endophytic establishment of the inflorescence is a major infection route happening to mangoes that is because of the pathogen *Dothiorella dominicana* as stated by Johnson et al. 1992. Stem-end rot fungi like *Lasiodiplodia theobromae* and *Phomopsis citri* are restricted by the fruit's cuticle and wound periderm, entering only after abscission. To prevent this, postharvest 2,4-D treatment is used to retain the fruit buttons (Eckert and Eaks 1989). In tomatoes, gray mold at the stem-end can develop when *Botrytis* infects young fruit via flower portions.

4.3.2.3 Quiescent Infection

The interval separating infection at beginning and visual symptoms defines quiescent or the latent period (Swinburne 1983). "Quiescent" indicates an inactive association (parasitic) that might convert to effective association after some time (Barkai-Golan 2005). The inability of propagation/progress after a certain phase might be because of unfavorable circumstances provisionally enforced because of host. These conditions affect pathogens either directly or else via alteration of their infective capabilities. Gray mold in strawberries because of *Botrytis cinerea*, decay caused in apples by *Nectria galligena*, anthracnose in mango, avocado, papaya, and citrus family by *Colletotrichum gloeosporioides* or in banana by *Colletotrichum musae* are all examples of quiescent infection (Barkai-Golan 2001).

4.3.2.4 Penetration Through Wounds

Most organisms cannot penetrate the host's cuticle or the epidermis directly and typically need wounds or injuries to infect. Hence, they are commonly known as *wound pathogens*. Mechanical injury occurs after the harvesting process through which the pathogens can enter into host tissue and cause infection.

4.4 Key Influencers of Infection Onset

External conditions around plants play a critical role in the development of infections caused by pathogenic organisms, which can lead to successive postharvest losses. High temperatures and humidity favor postharvest decay, while chilling injuries make tropical and subtropical produce more vulnerable. Proper humidity levels can slow down postharvest spoilage by means of ripeness regulation and inhibiting pathogen propagation (Barkai-Golan 2005). Additionally, various other factors influence onset of any infection. Like, peels, acting like chosen mediums, are typically infected because of various fungus species. The vegetables with pH above 4.5 are more prone to bacterial rots. Matured fruit crops have been shown to be highly vulnerable to getting waste in comparison to the underdeveloped ones, so techniques that delay ripening (like low temperatures) hinder decay organism proliferation. Tubers, such as potatoes, cassava, yams, and sweet potatoes, can form specialized cell layers (wound periderm) at injury sites, limiting postharvest spoilage. Throughout marketable potato trading, periderm occurrence is encouraged due to storing them for around 10 days to 2 weeks at temperature above 7 °C and below 15 °C and around 95% relative humidity (RH), protocol called curing. Similar curing process, achieved through desiccation, reduces orange wastage by *Penicillium digitatum*. When oranges are stored at 30 °C and 90% humidity for several days, the peel becomes turgid, and lignin is synthesized in the injured tissue, which impedes microorganism entry and decay.

4.5 Postharvest Handling Under Extreme Weather Conditions

It is crucial to emphasize the significance of environmental factors that occur during the hours leading up to and immediately following harvest, especially under extreme conditions common in regions like low deserts and rainy tropics. Research by Fonseca et al. (2006) has demonstrated the direct impact of these environmental conditions on postharvest quality. A single stress event can cause significant physical changes in a plant's physiology and anatomy upon recovery. For instance, fever few plants developed trichomes after recovering from a wilting episode (Fonseca et al. 2006), and grafted watermelons formed adventitious roots and aerenchyma after a flooding event (Yetisir et al. 2006). Environmental factors during produce handling that can influence final quality include extreme temperatures, rain, and very low relative humidity (RH). As urban areas increasingly coexist with agricultural fields and orchards, other factors may become more critical in the future, such as high CO_2 emissions, low atmospheric pressure, and elevated ethylene levels (Fonseca 2009).

4.5.1 Postharvest Handling in the Tropics

Postharvest losses in tropical areas with high rainfall are the highest globally, with estimates around 50% in developing world along the tropical belt (FAO 1989). Notably, there is no significant relationship between solar radiation and mean temperature. However, fruits exposed to minimum temperatures above 22.4 °C in the 3 days before harvest show improved tolerance to quarantine heat treatments, a phenomenon linked to heat shock proteins. High relative humidity (RH) and temperatures around 30 °C enhance wound healing in citrus fruits by promoting lignin formation and antifungal compounds like scoparone and scopoletin, reducing pathogen penetration (Kinay et al. 2005).

Although high postharvest storage temperatures often degrade quality, certain herbs like marjoram benefit from temperatures up to 30 °C, which stimulate oil synthesis. Tropical climates exhibit less temperature fluctuation than other regions, and late-day harvesting (after 4 pm) has been found to improve the storage quality and rooting response of vegetative tissues due to higher carbohydrate levels (Rapaka et al. 2007). However, leafy vegetables such as Valeriana lettuce and fresh-cut rocket and spinach experience significant chlorophyll degradation with even a 6 °C increase in storage temperature (Ferrante et al. 2004).

Late-day harvests may extend shelf life due to diurnal metabolic changes, with proteins like xyloglucan endotransglucosylase-hydrolase more active later in the day, helping prevent cell degradation. Leaf-water relations also fluctuate daily, with stomata opening in response to light and temperature, affecting CO_2 uptake and productivity. Increased turgor pressure, enhanced by postharvest calcium treatments, can also prolong shelf life (Rico et al. 2007).

Extreme rainy conditions can exacerbate plant disease and food safety issues. While most pathogens thrive in high RH, some like *Escherichia coli* O157:H7 and hepatitis A virus have different survival patterns (Stine et al. 2005). To mitigate these risks, it is advisable to harvest when rain has stopped, allowing surface moisture to dry. Additionally, extreme weather can impact the efficiency of postharvest operations, as field workers are at higher risk of dehydration (Nag et al. 2007).

4.5.2 Postharvest Handling in the Desert

Deserts, ranging from sea level to 3500 m, are key agricultural zones despite their harsh conditions, such as cold nights and high daytime temperatures (Houston 2006). These extremes, often coinciding with harvest periods, can be particularly challenging. Desert crops are subject to intense sunlight, low RH, and high temperatures, which can inhibit even microbial life (McKay et al. 2003). Low RH is a significant concern, as freshly harvested produce rapidly loses weight, reducing both quality and marketability. A 5–10% weight loss can render fruits and vegetables unmarketable. The water loss rate, affecting the product's water status and quality, is influenced by temperature and RH (Fonseca 2009).

Water activity, or the differential between the vapor pressure of water in food and pure water, impacts the rate of water loss. Higher water activity means more water is released as vapor, reducing shelf life, especially at low RH (Fonseca 2006). For example, lettuce with high water activity has a shorter shelf life. Membrane integrity loss due to desiccation leads to tissue browning, while lower moisture content in beans enhances color retention and quality, extending shelf life from under 10 days to over 100 days with reduced initial moisture (Karathanos et al. 2006).

Herbs are particularly susceptible to weight loss at high temperatures, with their transpiration rates being double that of vegetables. However, drying herbs quickly does not typically affect their active compounds (Fonseca et al. 2006). Asparagus quality is highly dependent on postharvest conditions rather than growth cycle temperatures (Bhowmik et al. 2003). High temperatures at harvest can increase decay, as seen with Star Ruby grapefruit in France (Pailly et al. 2004). Overmature crops like broccoli and beans accumulate excessive fiber, exacerbated by high temperatures, and hormonal levels in produce are also impacted by extreme heat.

Water stress, common in deserts, can lead to significant anatomical changes in plants, such as trichome formation in feverfew after wilting (Fonseca et al. 2006). Severe stress at harvest negatively affects postharvest quality, as seen in the undesirable texture of stressed-grown potatoes. However, mild water stress at harvest can benefit produce quality, such as lettuce (Fonseca 2006). Mist treatments before harvest can prevent low-temperature breakdown in kiwifruit during storage (Sfakiotakis et al. 2005).

Seasonal timing of harvest is critical for crop shelf life, with stress conditions increasing beneficial compounds like polyphenols in cauliflower (Scalzo et al. 2007). Harvest time also affects pigment concentrations in apples, with notable differences in lutein and violaxanthin levels between June and November (Solovchenko et al. 2006). Postharvest light exposure is another factor; excessive sunlight can dramatically increase solanine content in potatoes to toxic levels, but using reflective tarpaulins can mitigate such effects, as seen with cherries (Schick and Toivonen 2002).

4.6 Postharvest Pathology of Different Fruits and Vegetables

4.6.1 Citrus Fruits

Citrus fruits (*Citrus* sp., Rutaceae) are ranked number one in value for global commerce. Citrus fruits include oranges, mandarins or tangerines, clementines, satsumas, lemons, limes, grapefruits, etc. Countries that grow *Citrus* include China, USA, Iran, Brazil, Turkey, Italy, India, Spain, Mexico, South Africa, Egypt, and Morocco.

Diseases from postharvest fruit infection: postharvest disorders of *Citrus* crops along with causative organisms have been mentioned in Table 4.1.

Table 4.1 Postharvest diseases and causative organisms of citrus fruits

Disease	Causative organism
Green and blue mold	*Penicillium digitatum*
	Penicillium italicum
Sour Rot	*Geotrichum citri-aurantii*
Brown rot or citrus gummosis	*Phytophthora* sp.
Greasy spot	*Mycosphaerella citri*
Alternaria brown spot or black rot	*Alternaria* sp.
Anthracnose	*Colletotrichum gloeosporiodes*
Phomopsis stem end rot or citrus melanose	*Diaporthecitri*
Diplodia stem end rot	*Lasiodiplodia theobromae*
	Botryodiplodia theobromae
Aspergillus rot	*Aspergillus niger*

4.6.1.1 Green and Blue Mold

Green mold, the causative organism of which is *Penicillium digitatum* Sacc., is known as a predominant postharvest disorder in *Citrus* fruits. Blue mold, the causative organism of which is *Penicillium italicum* Wehmer, holds less importance as its growth rate is slower than the former. Third *Penicillium* sp., *Penicillium ulaiense*, the cause of whisker mold has low virulence in comparison to *P. digitatum* and *P. italicum* that spoil the *Citrus* (quickly; 3–5 times) (Holmes et al. 1994). *P. digitatum* and *P. italicum* infect citrus fruits primarily through rind wounds, forming soft, water-soaked, circular lesions. These wounds typically occur during harvest, handling, or commercialization, though preharvest injuries from cracks or insect damage can also serve as entry points. Fruits infected shortly before harvest (within 3 days) may appear healthy and go undetected, while infections long before harvest often lead to premature fruit drop. Intact rinds prevent infection, as surface conidia cannot germinate without tissue damage. However, when conidia reach internal peel layers through injuries, infection becomes irreversible within 48 h at 20–25 °C. Both fungi are necrotrophs that secrete enzymes like polygalacturonases and cellulases to degrade peel tissue, with white aerial mycelium emerging at the lesion center and spreading outward. Sporulation begins after 3–5 days, depending upon the inoculum load, and develops a layer of conidia which is colored. After 7–8 days, in the green mold infections, inner lesion part becomes olive green in color. Thick, compact, nonspore-forming whitish colored mycelium surrounds the central lesion limited by a moderately steady spoilage peels. Regarding blue molds, a centermost spore-forming blue–blue greenish zone is encircled by a thin patch of nonspore-forming whitish colored mycelium bordered by soft as well as water-dipped peels.

The optimum temperature for germination and growth of both *Penicillium* sp. is 25 °C. However, green mold is more frequent on fruits at room temperature while blue mold can be more prevalent among fruits stored below 10 °C, particularly 3–5 °C. *P. italicum* is active at lower temperatures and water activities than *P.*

digitatum. On the other hand, green mold can overgrow blue mold in a mixed infection on fruit at room temperatures and thus is generally more common. Blue mold can spread by contact from fruit to fruit leading to pockets of decay involving various decayed fruits (Barmore and Brown 1982).

All marketable citrus species and cultivars are susceptible to green and blue molds, with rind condition—both physical and physiological—playing a greater role in susceptibility than genetic traits. Even so, diverse degrees of susceptibility are commonly observed among different cultivars. According to Eckert and Eaks (1989), general susceptibility to postharvest diseases typically decreases in this order: mandarins, lemons, oranges, and grapefruits.

4.6.1.2 Sour Rot

Sour Rot is known to infect all types of citrus fruits, including tangerine, orange, grapefruit, and lemon. In the Indian climate, sour rot is a major problem wherein the handling of citrus fruits is carried out on bare ground while packaging is carried out on paddy straw. Among all the citrus fruits, the Nagpur mandarin gets infected mainly during the months of harvest (March to April). However, the disease is not found on crops that are harvested in late winter months, i.e., November to December (Vinay et al. 2021).

The causative organism for sour rot disease is *Geotrichum citri-aurantii*. The shape of conidia of *G. citri-aurantii* is cylindrical to oval shape, white colored mycelium with prominent dichotomous branching. The ascospores of *G. citri-aurantii* occur in pairing; and the production of ascospores decay the lemon fruits in the field. *G. citri-aurantii* has been found to be widely present in the citrus grove soils. The arthroconidia spread to the fruit through dust or rain splashing or irrigation or insects on the fruit surface that causes infections at the site of injury on the fruits that are hanging lower on the trees. The citrus fruits get infected by insects through injuries caused by insects such as false coding moth, fruit-sucking moths and fruit flies (Vinay et al. 2021). The arthroconidia of *G. citri-aurantii* require high water activity to germinate. Therefore, warm and humid climate poses a serious issue to the citrus fruits industry.

Sour rot lesions are softer, gelatinous and can pierce deeper inside *Citrus*. This pathogen develops enzymes that can digest the tissue of citrus fruits into a liquid that drips onto the fruit surface and can be spread by insects, which can lead to quick dissemination of this disease (Smilanick et al. 2019). Initial infection of *G. citri-aurantii* is a water-soaked lesions on the fruit which develops into light brown to yellow. The mycelium develops, bearing a large amount of conidia. Subsequently, the fruits get soft, semisolid accumulated matter because of enzyme activity (Vinay et al. 2021).

At terminal phase of disorder, sour odor develops that entices *Dorsophilla* that again disseminates this disorder. This disorder generates quickly under temperatures around 28–30 °C, while temperatures below 10 °C are known to overpower the proliferation of *G. citri-aurantii* (Vinay et al. 2021).

4.6.1.3 Brown Rot

Brown rot or Citrus gummosis, the causative organism of which is *Phytophthora* group of spp.: *P. parasitica, P. citrophthora* and *P. palmivora*. Trunk gummosis and root rot are major damage-causing types in this disease. *P. parasitica* is generally linked to the root rot, though *P. citrophthora* has been shown to be associated with trunk gummosis.

In India, *P. parasitica* has been observed from orchards of Nagpur mandarin in Nagpur (Naqvi and Singh 2002), and *P. parasitica* Dastur has been reported to be linked with citrus disease in Assam and northwest areas (Vinay et al. 2021). *P. parasitica, P. citrophthora,* and *P. palmivora* have been known to severely affect 20–50% of Nagpur mandarin in Vidarbha region of Maharashtra, 20–100% of acid lime plantation in Andhra Pradesh, and 10–80% of *C. sinensis* and Kinnow mandarin in Punjab. Sporangia form in the top 0–30 cm of soil upon air contact and serve as the main inoculum source. Infections occur via zoospores or sporangia germination—either directly through a germ tube or indirectly by zoospore release. Temperature and soil moisture influence their production and germination. Spread occurs through water splash, wind-blown droplets, or soil water. The motile zoospores swim short distances using flagella or travel farther via water flow. *Phytophthora citrophthora* and *P. parasitica* are polyphagous, soil-borne pathogens that complete their life cycle in soil. In arid regions, infections are limited to roots and crown bark, while in humid areas, they can reach the upper trunk and fruit (Vinay et al. 2021).

The optimal growing conditions for *P. parasitica* are 85–90 °F while for *P. citrophthora* are 75–80 °F. The nutritional depletion and light are known to stimulate the production of sporangia from the mycelium. Sporangia might directly proliferate to zoospores during moist environment by the growth of germ tubes, or indirect germination requires free water stimulated by lower temperature. The chlamydospore generation of *P. parasitica* happens due to a hostile environment by growing fungus such as nutritional reduction, lower levels of O_2, lower temperatures around 59–64 °F. The chlamydospores of *P. parasitica* are dormant below the temperature of 59 °F, and thus, for stimulating germination temperature of 82–90 °F is preferred.

Citrus gummosis is illustrated with the moistening of nursery bed seedlings. Phytophthora infection has led to over 20% seedling mortality in central India. The disease is marked by firm bark with fine cracks that exude gum; this exudate washes away with rain but persists on the trunk in dry weather. Severe infections cause pale green leaves with yellow veins. Brown rot typically occurs in regions experiencing rainfall during early fruit maturity, beginning as light brown, leathery lesions. Under humid conditions, white mycelium may appear on the fruit surface. Fruits closer to the ground are more susceptible and often rot.

4.6.1.4 Greasy Spot

Significant leaf loss occurs due to this disease which decreases capability of photosynthesis in trees. This leads to weakening of trees and thus the tree loses the capability to hold a greater number of fruits. Mostly, this disease happens during late spring to early summer. Among all citrus fruits, greasy spot majorly affects fresh grapefruit.

Mycosphaerella citri is the causative organism for greasy spot disease in citrus fruits, and should be extracted from leaves and lesions of fruit quickly. This fungi develops greyish to greenish colony on regular culture media. The growth of this fungi is slow; colony grows to about 20 mm diameter beyond 3 weeks. The conidia can measure about 2–3.5 × 10–70 micrometer. Conidia are multiseptaed, insipid olive brownish in color and cylinder to little curved in shape. Ascospores of *M. citri* are marginally fusiform with one number septum containing fat droplets inside cell. These are about 2–3 × 6–12 micrometer.

The predominant source for inoculum for this disease is spores spread through air that are generated when conditions or leaves dropping on the ground putrefy. In areas where high rainfall and high temperature are common, the cycle of this pathogen lasts throughout the year; the occurrence of infectivity could be anytime. This organism generates ethylene that ultimately induces immature denudation (Vinay et al. 2021).

4.6.1.5 Alternaria Brown Spot or Black Rot

Four distinctive diseases are caused by *Alternaria* pathogen, viz., *Alternaria* brown spot of tangerines, *Alternaria* leaf spot of rough lemon, *Alternaria* black rot or fruit rot and Machan foliar on lime. *Alternaria* fruit rot is significant in areas where citrus production is being carried out for juice because of contamination of juice by black fungal mycelium present in the interior of the infected fruits.

The fruits infected with *Alternaria* may turn light in color few weeks before the color break in the healthy fruits. The infected fruits may appear normal from outside, and there may be little to no external evidence of infection. The diseased fruit has a brown to black discoloration at the "blossom" end. During storage, lemons are prone to this disease. Splitting may predispose oranges and large navel fruits to infection by the pathogen.

During flowering, the conidial spores become the primary cause of infection. Conidia grow alone or in chains, are dark black to brown with ellipsoidal to obclavate shape. Transverse septa range from 3 to 6 while longitudinal septa up to 3 in numbers. Conidia, if ellipsoidal, are beakless, or might have a cylindrical beak, 3–4 micrometer in diameter and up to 13 micrometers in length (Timmer et al. 2000).

Alternaria grows on dead citrus fruit tissues during wet weather. Airborne conidia are produced in this weather, the spores germinate and fungus grows on the button or stylar end of the fruit. Split or cracks in the fruit facilitate the penetration of the pathogen to the host fruit, which then grows into the central core causing a black decay. The conidia are released from sporulating brown spot lesions by rainfall or changes in relative humidity, and the spores spread through air and deposit on susceptible tissues. The fruit may dry and get black and mummy-like in appearance (Timmer et al. 2000).

4.6.1.6 Anthracnose

Anthracnose of sweet orange is produced by *Colletotrichum gloeosporioides*. The tissues that are injured by several environmental factors possess higher susceptibility to anthracnose infection. Anthracnose generally occurs on the fruits that have been injured by pest damage, sunburn, bruising, chemical burn or long storage durations. The lesions are usually 1.5 millimeters or higher in diameter with brown to blackish spots. Under dry conditions, the spores are brown to blackish in color, while under humid conditions, the spores are pinkish to salmon colored (Timmer et al. 2000).

C. gloeosporioides have septate mycelia, asexual fruiting body is the acervulus, and setae are found present. Spores germinate to form a resting structure which aids them in remaining dormant till an injury happens. Temperatures of 30–32 °C and high humidity (80–85%) favor the growth of this pathogen and increase the disease severity. Source of inoculum is primarily dormant mycelia, while secondary is conidia that are generated in acervulus (Timmer et al. 2000).

4.6.1.7 Citrus Black Spot

Citrus black spot disease is well known to affect all types of citrus fruits, all sweet oranges, lemon and grapefruits (Vinay et al. 2021). Das et al. (2018) have reported the occurrence of citrus black spot-like symptoms on Nagpur Mandarin, Mosambi sweet orange, and *C. sinensis* in Mohpa and Kachimet areas of Nagpur district in December 2016.

There are four types of symptoms of citrus black spot viz., hard spot, cracked spot, false melanose, and virulent spot. The hard spot is a sunken lesion having a black margin and a grey center, wherein the fruiting bodies containing the fungal spores are produced. The virulent spot is a sunken necrotic lesion with no definite borders, which varies from brown to black to brick red in color. False melanose involves small, black necrotic lesions of less than 1 mm diameter. These symptoms generally are found in young green fruits. Freckle spot is found on mature fruits after the postharvest, during storage periods. Freckle spots are round in shape, depressed, and reddish brown having red to brown border, with size varying from 1 to 3 mm.

Phyllosticta citricarpa is the causative organism for citrus black spot disease. Ascomata of this pathogen are formed on fallen and decomposing leaves. Asci are cylindrical in shape, clavate, each consisting of eight spores. Ascospores are 4.5 and 6.5 micrometer in width, and 12.5–16 micrometer in length, hyaline, nonseptate and swollen from the center. Conidia are 5.5–7 micrometer in width and 8–10.5 micrometer in length, obovate to elliptical shaped, hyaline, nonspetate and multiguttulate having a colorless appendage.

Primary source of inoculum is from leaf litter and ascospores which gets ejected in wet leaf litter. The spores move through air. Lesions on infected fruits, leaves, and branches act as additional inoculum source. The dark brown or black pycnidia (which produces conidia) are formed on the fruits, pedicles, and leaves. Conidia produced are no wind-borne but can spread to susceptible fruits by rain splash. Higher light intensity, high temperatures, drought, and lower tree vigor are favorable for this disease to occur.

4.6.1.8 Phomopsis Stem-End Rot or Citrus Melanose

This disease is known to affect all citrus varieties; however, it most severely affects lemons and grapefruits. The perfect stage of the pathogen causes melanose that are lesions on the fruit and foliage, while the imperfect stage causes Phomopsis stem-end rot.

The first symptom of Phomopsis stem-end rot is the development of softness toward the stem end of the fruit with no discoloration. Afterwards, the color starts to develop and turns to tan brown and ultimately black. The infected tissue is shriveled a little forming a ridge among the rotten and the healthy tissues. The pulp of the fruit becomes mushy with no effect to color. At later stages, off odors develop in the fruits. In case of melanose, smaller spots or scab-like lesions, such as tear-drop, mud-cake, and star melanose are developed.

This disease is caused by *Diaporthe citri* that produces hyaline two-celled ascospores with each cell, and two oil droplets or guttulae are present. Ascospores are found inside perithecia that are flask-shaped fruiting bodies. Perithecia are circular and flattened at the base with long black beaks tapering outwards. The beaks range from 200 to 800 micrometer in length, 40–60 micrometer in width, while the base of the perithecium varies from 125 to 160 micrometer. *Phomopsis citri* produces pycnidia that are ovoid shaped and dark colored with thick walls. The pycnidia are developed on deadwood and rotting fruits. Two types of conidia are produced by this pathogen: alpha and beta conidia. Alpha conidia are hyaline, one-celled containing two guttulae, while beta conidia are long slender rod-shaped spores, hooked at one end, and produced in older pycnidia. Alpha conidia are mainly accountable for spreading of spores at short distances, while the beta conidia are unable to germinate (Wolf 1926).

D. citri survives on dead wood and obtains energy from living as well as dead wood. The ascospores are discharged in a forcible manner, which is the primary reason for long distance spread of disease (Vinay et al. 2021).

The infection starts when ascospores (*Diaporthe* stage) or conidia (*Phomopsis* stage) land on the leaf tissue. During dry conditions (temperatures less than 17 and more than 35 °C), the spores die and infection does not happen. This pathogen requires 10–24 h of moisture for the germination of spores at a temperature of 25 °C, while 18–24 h of moisture at 15 °C (Vinay et al. 2021).

4.6.1.9 Diplodia Stem-End Rot

This disease is most prevalent in humid and warm regions, while not common in arid regions. In India, this disease is most prevalent in central region (Nagpur Mandarins), followed by the arid regions of Punjab (Kinnows). The pathogen infects fruits from the button at the stem end. Infection proceeds through the fruit core rapidly, resulting in the development of soft brown to black decay symptoms at both ends of the fruit. The pathogen develops unevenly on the rind and produces finger-like projections having black to brownish discolorations at the margin of the lesion.

The causative organism is *Diplodia natalensis* and its synonyms are *Lasiodiplodia theobromae* and *Botryodiplodia theobromae*. Pycnidia is subglobose to globose and single or grouped in a stroma and papillate. The pycnidia are produced on dead citrus twigs. Conidiophores are hyaline, sometimes septate and rarely branched, and cylindrical in shape. Conidia are developed in pycnidia. The young conidial cells are hyaline, granular without septum. The mature conidia are striated with one septum (two-celled). Perithecia are globose, papillate and produces ascospores. Ascospores are single-celled, hyaline, and elliptical to ellipsoidal shaped. Ascospores become airborne and are responsible for long-distance dispersal to new planted trees. The temperature of 30 °C is optimal for the growth of mycelia. The pathogen survives various seasons on dead wood. The rainfall water can carry spores from deadwood to the surfaces of immature fruits. Prior to harvest, pathogen is latent and does not cause any decay. The infections develop postharvest during high temperatures and humidity.

4.6.1.10 Aspergillus Rot

Aspergillus is a predominant pathogen linked with the spoilage of citrus fruits. *Aspergillus*-infected fruits are covered with black mycelial growth producing conidial heads, known as black mold. *Aspergillus niger* hyphae are branched and septate. Conidia are greenish, globose to subglobose having rough surface occurring in chains of two or higher. Conidiophores are short circular to semicircular in shape. The spores are dark colored and oval shaped. *Aspergillus niger* is a parasitic fungus which can penetrate fruit tissue via microwounds and bruises. Higher temperatures and humidity are encouraging for the development of this disease. The favorable temperature for the development of disease is 25–40 °C. Swift decay and spread of the pathogen occur at 30–35 °C. *Aspergillus* species are not grown at 15 °C and refrigerated temperatures.

4.6.2 Banana

Banana, available throughout the year, is an important fruit crop commercially. India is the leading country in terms of production globally. In India, major banana-producing states include Tamil Nadu, Maharashtra, Andhra Pradesh, Gujarat, and Karnataka (Snehalatharani et al. 2021). Banana is a highly perishable fruit and about 25–30% postharvest losses have been reported by Kachhwaha et al. (1992). The climacteric nature of banana is the major reason for attack of primary and secondary pathogens that deteriorate the quality of banana after harvest till they are consumed (Snehalatharani et al. 2021).

Postharvest Diseases of Banana: Postharvest diseases of banana along with their causative organisms have been mentioned in Table 4.2.

4.6.2.1 Crown Rot

Crown rot is the most general disease related with handling of bananas after harvest during storage. The incidences of occurrence of this disease globally vary from 4% to 98%.

The infection initiates from the crown as a dark brown or black rot, spreading to the crown and ultimately penetrating to the pedicels of distinct fingers. Lassois et al. (2010) have reported white or grey or pink colored mycelium with visible fruiting bodies on crowns or distinct fingers, in advanced stages. The symptoms of this disease develop after a few days in the storage or transport and not during handling after harvest. At earlier stages, the infection is superficial which enters the internal tissues progressively. This disease leads to quick degradation of fruit due to the production of ethylene from stressed tissues, necrotic fruits, and fungal mycelium, as reported by Ewané et al. (2012).

Table 4.2 Postharvest diseases and causative organisms of banana

Disease	Causative organism
Crown Rot	*Fusarium* sp.
	Colletotrichum musae
Cigar end rot	*Muscillium theobromae*
Anthracnose	*Colletotrichum musae*
Stem end rot	*Colletotrichum musae*
Fungal scald	*Colletotrichum musae*
	Fusarium pallidosporeum
Botrydiplodia finger rot	*Lasiodiplodia theobromae*
Main stalk rod	*Lasiodiplodia theobromae*
	Colletotrichum musae
Fuzzy pedicel	*Sporothrix*
	Fusarium
Squirter	*Nigrospora sphaerica*

The fungal pathogens linked with crown rot disease are *Musicillium theobromae*, *Colletotrichum musae*, *C. gloeosporiodes*, *Nigrospora sphaerica*, *Ceratocystis paradoxa*, *Lasidiplodia theobromae*, and species of *Cladosporium*, *Penicillium*, *Aspergillus*, *Fusarium*, and *Acremonium* (Snehalatharani et al. 2021). Crown rot fungi are saprophytic in nature. The pathogen inoculum settles in the field either by airborne fungal spores or waterborne fungal spores. *Fusarium* or *Colletotrichum* species initially enter the flower parts, which leads to transmission through air or water splash and thus sporulation.

Factors such as geography, season, source-sink ratio, age of bunch, etc., are known to affect the occurrence of this disease. Bananas from high-altitude areas are less prone to this disease in comparison to bananas grown in low-altitude areas. During rainy weather, bananas are more prone to crown rot disease, and in few countries yield losses up to 86% have been reported. Separation of banana hands from the bunch reduces total sink, and thus reduction in rate of infection of this disease. Ewané et al. (2012) reported a higher occurrence of crown rot disease in older fruits. Ewané et al. (2012) further reported the presence of a waxy layer on leaves and cell wall as well as its constituents (cuticle and suberin) containing phenolic compounds. These phenolic compounds, such as lignin, dopamine, ferulic acid, cyanidin-related compounds, and phytoalexins (irenolone, phenylphenalenones, and emenolone), have been reported to prevent the growth of crown rot fungi (Ewané et al. 2012; Snehalatharani et al. 2021). Cooking bananas possess lower susceptibility to this disease in comparison to dessert bananas (Snehalatharani et al. 2021).

4.6.2.2 Cigar End Rot

Cigar end rot disease holds economic importance in countries such as Africa (central, western, and southern regions), India, Canary Islands, Iran, South America, and Egypt. The diseased part of the fruit remained attached to the fingers and had an appearance of ash of cigar (Snehalatharani et al. 2021).

This disease is linked with damaging the banana fruit preharvest in the fields as well as postharvest during storage and transport. The field symptoms include localized necrosis at the tip of banana, with subsequent darkening and wrinkling of the skin. This disease causes early ripening of the fruit without maturity. On the black and wrinkly part of tip, under favorable conditions, powdery grey conidia may develop, which appear like a burnt tip, known as "cigar end." The localized darkening and wrinkling of the peel at the tip occur, which appear as a black band with a narrow chlorotic area between the healthy and the infected tissues (Gul et al. 2018).

This disease is caused by *Musicillium theobromae*, *Trachysphaera fructigena,* and *Verticillium theobromae* (Snehalatharani et al. 2021). The pathogens of cigar end rot may enter through the flower to the banana finger and lead to dry rot of the immature banana fingers. *M. theobromae* has been largely linked with cigar end rot in comparison to *T. fructigena*, as reported by Masudi and Bonjar 2012.

4.6.2.3 Anthracnose

Anthracnose is a major postharvest disease of banana, accounting for about 40% losses, as reported by Coelho et al. 2010. This disease normally starts in the field as a quiescent infection on the fruit. This disease reduces the quality and nutritional value of banana fruit, making the fruit unfit to consume.

As the fruit ripens, brown sunken spots on the peel emerge. These spots enhance in size, coalesce, and develop wide areas of sunken and brown blackish tissue. Under favorable conditions, spore masses of orange color may appear. Symptoms only arise on overripe fruits, and the pulp remains unaffected till that time.

The causative organism of Anthracnose disease in banana is *Colletotrichum musae*. The conidia and conidiophores are produced on the acervuli. The spores are produced on senescing banana tissue, bracts, leaves, discarded fruits, and fruit stems. Rain splash or irrigation water spreads the spores to the fruits in the field. The optimum temperature for the growth of *C. musae* is 27–30 °C, and the germination of conidia and formation of appressoria takes place within 1–2 days. The pathogen infects the fruit and stays in a latent phase till the onset of ripening. The conidia are hyaline, have no septum, and are ellipsoidal in shape. The size of the spores varies from 10 to 18 micrometer and 5–9 micrometer (Snehalatharani et al. 2021). The entrance of this pathogen is influenced by the deposition of phytoalexins during the ripening of the fruit.

4.6.2.4 Stem-End Rot

In this disease, decay originates from the cut end in the distinct fingers, when they are detached from the bunch. In this disease, the pedicel or the stem of the green banana is injured, and the rot develops. *Colletotrichum musae* and *Chalaraparadoxa* are the causal organisms for this disease. However, due to improvements in handling of fruits, this disease is of less economic importance nowadays.

4.6.2.5 Fungal Scald

This disease develops on fruits in transport kept under controlled atmosphere conditions for days higher than 2 weeks. The fungal scald occurs at the tip of the fruit which is in contact with the polythene liners when the moisture is present. At the fingertips of the green fruit, a reddish brown sunken spot is found at the bottom of the carton. During ripening, spots increase in size. *Colletotrichum musae* starts the scald, and subsequently secondary infections by *Fusarium pallidosporeum* occur.

4.6.2.6 Botrydiplodia Finger Rot

This disease is a soft rot that occurs at the tip below the flower remnants. The pulp transforms into a black mass causing the decay of complete fruit. The fruit skin gets wrinkled and blackened; fruit ripens immaturely and is encrusted with pycnidia. Under high humidity conditions, mycelium of grey to black color appears on the surface. The matured fruit possesses higher susceptibility to this disease. The causative organism for this disease is *Lasiodiplodia theobromae*. Pycnidia are black in color and flask shaped having short necks, and size is around 250–300 micrometer in diameter. Conidia are exuded as single hyaline cells which mature to two-celled, brown, unconstructed longitudinally striated spores. The conidia spread through wind and water. The finger rot starts at the distal end of the finger and through the wounds (Jones 2000).

4.6.2.7 Main Stalk Rod

In this disease, decay of the peduncle occurs when bananas are kept in a bunch. This rot spreads through the proximal end and engulfs the lower hands, leading to finger drop. This rot has a characteristic odor. The causative organisms associated with this disease are *Lasiodiplodia theobromae*, *Chalaraparadoxa* and *Colletotrichum musae* are causal organisms for this disease.

4.6.2.8 Fuzzy Pedicel

This disease has a characteristic fluffy grey to white colored mycelium that occupies the pedicel of the fruit during storage conditions. The pathogens associated with this disease, as isolated by Tarnowski et al. (2010) from the infected pedicels, are *Sporothrix* and *Fusarium*.

4.6.2.9 Squirter

Squirter disease is produced by *Nigrospora sphaerica*, and is the disease of the single fruit. The pathogen enters through cut pedicels and advances further as the fruit ripens. The fruit pulp gets infected and converts to a liquid mass, which may squirt if the fruit is squeezed. Bluish tan discoloration of the peel is visible as the fruit ripens. The conidia are black in color when mature having dimensions as 18×15 micrometer. These develop on dead plants, including grass and banana waste. This disease is of less economic importance commercially (Jones 2000; Snehalatharani et al. 2021).

4.6.3 Mango

Mango is an important fruit crop worldwide. India shares about 56% of global production in mango, and is the leading country for both area as well as production. More than 36 varieties are grown worldwide commercially, out of which, most popular in India are Dasheshari, Langra, Chausa, Alphonso, Totapuri, Fazli, Kesar, Neelum and Bagganpallu (Shukla et al. 2021). Haggag (2010) reported 17.36% postharvest losses in mango.

Postharvest Diseases of Mango: postharvest diseases of mango along with their causative organisms are mentioned in Table 4.3.

4.6.3.1 Anthracnose

This is the most serious postharvest disease in mango throughout the world. Fruits at all stages of growth are susceptible to this disease and produce the symptoms associated with the disease (Yenjit et al. 2004).

Table 4.3 Postharvest diseases and causative organisms of mango

Disease	Causative organism
Anthracnose	*Colletotrichum gloeosporioides*
Stem end rot	*Lasiodiplodia theobromae*
	Colletotrichum gloeosporioides
	Aspergillus niger
Black rot	*Aspergillus* sp.
Brown rot	*Lasiodiplodia theobromae*
Stem end soft rot	*Ceratocystis paradoxa*
Charcoal rot	*Macrophomina phaseolina*
Blue mold	*Penicillium expansum*
Grey mold decay	*Botrytis cinerea*
Alternaria rot	*Alternaria* sp.
Botryosphaeria rot	*Botryosphaeria ribis*
Rhizopus rot	*Rhizopus* sp.
Fusarium rot	*Fusarium oxysporum*
Rhizoctonia rot	*Rhizoctonia solani*
Cladosporium rot	*Cladosporium herbarum*
Phoma rot	*Phoma magniferae*
Hendersonia rot	*Hendersonia creberrima*
Macrophoma rot	*Macrophoma magniferae*
Yeasty rot	*Saccharomyces* spp.
Sclerotium fruit rot	*Sclerotium rolfsii*
Watery soft rot	*Sclerotinia sclerotiorum*
Mucor rot	*Mucor subtilissimus*
Pestalotiopsis rot	*Pestalotiopsis* sp.

The associated pathogen produces symptoms on every tender portion of the tree. The symptoms of disease involve blossom blight-shot hole spots on leaves and spots and lesions on fruits. The infection occurs on panicles as brown to black streaks on peduncles and spots on flowers, subsequently covering the entire panicle, which becomes blighted. Other typical symptom of disease is sunken necrotic lesions on the surface of the fruit and mummified fruits (Rivera-Vargas et al. 2006). The young fruits that get infected appear to have black spots and shriveled which ultimately fall down. The disease in infected fruits shows as dark lesions during the ripening stage. At earlier stages, the infection remains in epicarp; however, it moves towards the pulp in a very short time.

The disease is produced by *Colletotrichum gloeosporioides*. High humidity and rain during fruit development and maturity stages are favorable for this disease, and therefore a major constraint in many regions globally (Haggag 2010; Shukla et al. 2016; Shukla and Adak 2017). The infection is reported to be intensified at a temperature range of 20–30 °C. Necessary conditions for the development and germination of conidia are the availability of free water at the surface of the fruits or high humid conditions (for >10 h), which also acts as the source of inoculum. Prakash (2004) and Prakash et al. (2011) reported a storage temperature reaching to 40 °C during the ripening stage can cause serious rotting of the fruit.

4.6.3.2 Stem End Rot

Stem end rot is reported to be an economically significant disease of mango. The associated pathogen remains dormant on unripe fruits. As the fruit ripens, rotting at the stem end initiates and gets brown to grey and softens, covering the entire surface of the fruit (Prakash et al. 2011). Lesion formation can occur away from the stem end in the injured fruits. Physical contact between healthy and infected fruits can lead to decay of the former.

The causal organisms related with stem rot disease are *Lasiodiplodia theobromae*, *Aspergillus niger*, *Phomopsis magniferae*, and *Colletotrichum gloeosporioides* (Shukla et al. 2021). The symptoms vary according to the type of causative organism. Firmer lesions with distinctive margins are linked with *P. magniferae* infection. The fruiting bodies are dark and the size of a pinhead. The spore masses produced by *C. gloeosporioides* are pink colored (Prakash et al. 2011).

The pathogen may enter the tree through wounds or the natural openings, which leads to infection during later stages by penetration through stem ends. Free water or high humidity conditions spread the conidia, and act as a source of inoculum.

4.6.3.3 Bacterial Black Spot

This disease can be seen as light-colored water-soaked spots, which eventually turn to darkened star-shaped craters. The infection by the bacteria appears as oozing of gummy substance from cracks which arises due to the infection. The infected tissue of the bacteria might allow other saprophytic fungi to grow.

4.6.3.4 Other Fruit Rots

Several kinds of rots associated with mango are:

(i) Black rot: It is caused by *Aspergillus niger*, *A. flavus*, *A. nidulans*, and *A. chevalieri*. The fungi infect the fruit via wounds or cut ends. At the onset, infection is perceived as grey to pale brown colored, soft, and sunken spot. This later develops into dark brownish to blackish lesions spreading rapidly, with subsequent profuse growth over these lesions further. Temperature range of 30–36 °C is favorable for the development of infection.

(ii) Brown rot: It is caused by *Lasiodiplodia theobromae*. The characteristic symptom associated with this rot is necrotic patches of dark brownish to blackish color with no distinctive margins near to the stalk or the wounded fruit. At advanced stages, the fruit might turn dark and brittle having off-flavored pulp inside.

(iii) Stem end soft rot: It is caused by *Ceratocystis paradoxa*. Soft, watery, pale colored, and irregular lesions develop that eventually darkens because of formation of spores. Rancid odor develops after the rotting.

(iv) Charcoal rot: It is caused by *Macrophomina phaseolina*. Irregular leathery necrotic patches of black color at the distal end are a characteristic symptom of this disease. Rotting of the fruit and brown discoloration starts to occur, which advances and the complete fruit (surface and pulp) turns brown and rotten. This infection generally starts via cut ends.

(v) Blue mold: It is caused by *Penicillium expansum*. This is a soft brown decay, and the spore masses are white and powdery at earlier stages. At mature stages, spore mass turns blue. The infection generally develops at the stem end or wounds.

(vi) Grey mold decay: It is caused by *Botrytis cinerea*. This pathogen disperses through contact with adjoining fruits. It is a light brown firm decay, which at later stages is covered by grey or brown spore mass.

(vii) Alternaria rot: It is caused by *Alternaria tenussima* and *A. alternata*. At earlier stages, water-soaked circular spots can be seen. This later enlarges in size with black sunken center.

(viii) Botryosphaeria rot: It is caused by *Botryosphaeria ribis*. At the onset, softening of the spot occurs, which subsequently leads to brown discoloration and darkened flesh underneath. At advanced phases, darkened, sunken lesions of oval or round or elliptical shape having small black-colored bodies, with white or greyish mold growth over it develops.

(ix) Rhizophus rot: It is cause by *Rhizopus arrhizus* and *R. oryzae*. Soft decay followed by foul odor is the characteristic symptom of this rot.
(x) Fusarium rot: It is caused by *Fusarium oxysporum*. The infection starts at the stem end or through injuries. Initially, large dark brown water-soaked spots occur on fruits, which later turns to dull pink colonies. Putrescent odor is also present.
(xi) Rhizoctonia rot: It is caused by *Rhizoctonia solani*. Dark colored lesion formation followed by hard brown rotting occurs on the upper portion. Clear distinction between healthy and infected tissue is visible.
(xii) Cladosporium rot: Repace t is caused by *Cladosporium herbarum*. In this disease, lesions are covered with white spores.
(xiii) Phoma rot: It is caused by *Phoma magniferae*. The infection is brown to black spots on fruit epicarp. The peel is still hard but internal part is rotten.
(xiv) Hendersonia rot: It is caused by *Hendersonia creberrima*. In this disease, circular lesions having cracks are formed, which are brown in color. The damaged tissue is soft and moist.
(xv) Macrophoma rot: It is caused by *Macrophoma magniferae*. Circular lesion is found on the infected fruit, which are water-soaked and brown in color. At later stages, the deep brown mold covers the entire fruit surface.
(xvi) Yeasty rot: It is caused by *Saccharomyces* species. The characteristic symptom is oozing out of the juice and a fermented smell. This infection is known to take place through wounds.
(xvii) Sclerotium fruit rot: It is caused by *Sclerotium rolfsii*. It is frequent in mangoes that are harvested at later stages and subsequently heaped on the soil. The source of inoculum is sclerotia and the hyphal strands that survive in the soil.
(xviii) Watery soft rot: It is caused by *Sclerotinia sclerotiorum*. The tissues that are infected are soft and oozing occurs. A fungal growth of white color might appear on the infected fruit.
(xix) Mucor rot: It is caused by *Mucor subtilissimus*. White mold growth occurs at the injured fruit surface or at the stem. On the water-soaked lesions, growth of yellow- colored sporangia is often seen.
(xx) Pestalotiopsis rot: It is caused by *Pestalotiopsis versicolor* and *P. magniferae*. The smaller sized brown-colored spots appears initially, which enlarge, subsequently turning dark brownish spots having greyish white centers.

4.6.4 Tomato

Tomato is an important crop worldwide; however, it has higher susceptibility to postharvest losses that vary from 10% to 20%. Tomatoes are prone to both bacterial, and fungal diseases, prior to and postharvest (Devappa et al. 2021).

Postharvest diseases: Postharvest diseases of tomato along with their causative organism have been mentioned in Table 4.4.

Table 4.4 Postharvest diseases and causative organisms of tomato

Disease	Causative organism
Grey mold rot	*Botrytis cinerea*
Early blight	*Alternaria* sp.
Phoma rot	*Phoma destructive*
Bacterial soft rot and hollow stem	*Erwinia carotovora*
Southern blight	*Sclerotinia rolfsii*
Anthracnose	*Colletotrichum* sp.
Rhizopus rot	*Rhizopus stolonifera*

4.6.4.1 Grey Mold Rot

In this disease, localized abnormality in the somatic part such as tan to light brown watery region occurs. This then advances to a soft watery mass. Gray mold rot disease is produced by *Botrytis cinerea. B. cinerea is a* necrotrophic fungus which produces a grey to brown colored mass of spore on the infected part of the tissue. This fungus produces a huge number of asexual spores that germinate to produce appressorium, after coming in contact with the host. These fungal spores can survive for up to 3 months in the pruning leaf scars. The spores disseminate majorly through air or wind or water splashes. For growth of this pathogen, high humidity and temperature range of 18–24 °C is optimal. At this temperature, infection can occur within 5 h. Temperatures >28 °C suppress growth and spore generation by the pathogen. Stem lesions generally expand in concentric rings, which subsequently girdle the complete stem. Flower petals are prone, which then starts the infection to the pedicels and to the immature fruits. The lesions on the fruit are white colored, skin ruptures exposing the tissues. White rings are visible on the infected fruits, which resulted in the development of ghost spots (Devappa et al. 2021).

4.6.4.2 Early Blight

This disease caused by *Alternaria solani* and *A. alternata* is particularly observed during the storage of tomato. *A. solani* produces spots of "bulls' eye" patterns on the tomato. This is a necrotrophic fungus that reproduces asexually by producing conidia. The infection takes place by the conidia which may enter through direct penetration or wounds. The primary infection occurs on old leaves found close to ground level. The fungi grow and germinate producing spores, from which conidia develop and are released. These conidia infect the plants and plant tissues. Increase in inoculum from alternative hosts like infected tubers of Solanaceous family or through weeds aggravates the growth and development of this fungus. *A. alternata* is a weak pathogen and reproduces on organic matter and survives on dead and old leaves found in the field. This colonizes moribund tissue. The pathogen infects immature fruit and once the fruit ripens, the symptoms only become visible. The pathogen germinates to develop a black colored, velvet-like layer on the surface of the sunken lesions, during warm and humid climate. Spores are often formed on the shallow surface spots. The mycelium of *Alternaria alternata* is haploid, septate, and becomes dark due to pigmentation. The asexual spore is conidia which can be borne singly or in a chain of two on separate conidiophores. The spores require free water for germination, and can germinate within 120 min. The optimal temperature range for spore germination varies from 26.6 °C to 29.4 °C (best being 26.6 °C), and will germinate at this temperature in half an hour. The penetration inside the plant may require 3–12 h, and is influenced by the temperature. Lesions develop after 48–72 h after penetration.

4.6.4.3 Phoma Rot

The Phoma rot appears as various small-sized, dark brown to blackish spots forming concentric rings on leaves. On stems, these appear as dark brown lesions having concentric rings. The fruit lesions occur on the calyx end as small sunken lesions which later form black colored, leathery lesions with various numbers of pycnidia in the center. The disease is produced by *Phoma destructive* and the pathogen produces septate, branched and hyaline mycelium. The fungi survive in soil, infected plant debris and weeds. The plant becomes more susceptible to the infection by this pathogen because of insect feeding, mechanical damage, cracking, and pruning. The fungal conidia are produced at a temperature of 20 °C and optimum humidity. The conidia are released from the pycnidia and are spread via rains, overhead irrigation, etc. The pycnidia are subcutaneous and eventually become erumpent and dark subglobose, producing hyaline pycnospores (Devappa et al. 2021). Lower levels of nitrogen and phosphorus in the soil can make the plants susceptible for attack by the pathogen.

4.6.4.4 Bacterial Soft Rot and Hollow Stem

Initially, watery and soft spots appear on the fruit, which eventually lead to soft decay of the fruits. The infected area may be surrounded by yellow halo. Lesions appear on fruits may be raised, sunken and brown colored, which later turn to scab-like appearance. At advanced stages, complete fruit becomes a soft, watery mass. The pathogen liquefies the fruit tissue by breaking the pectate "glue" holding the plant cells together, and the infected fruit becomes slimy and collapses. *Erwinia carotovora* is the causal organism for this disease. This is a Gram-negative, rod-shaped, facultative anaerobic bacteria. It is nonspore forming and has peritrichous flagella (Devappa et al. 2021). The bacteria enter the tomato through wounds. Under humid conditions when free moisture is present, uninjured tissues can get infected. Humidity during storage and transport enhances the chances for pathogen development. Rains, ill-drained soils, and warmer temperatures favored the development of this pathogen and eventually developed disease.

4.6.4.5 Southern Blight

Dark brown lesions initially appear on the stem and subsequently girdle the stem and followed by yellowing and wilting of the leaves. After few days, round shaped, dark brown colored and structures of the size of a mustard seed, which are known as sclerotia, appear on the white fungal growth. These are round, soft, smaller in diameter, and light colored. The causal organism of this disease is *Sclerotinia rolfsii*. This is a necrotrophic, soilborne pathogen that produces white mycelium on the infected plants. The cells are hyaline, having thin cell walls and sparse cross-walls. The mature sclerotia are hard, having a distinctive rind. These sclerotia are mostly spherical, while a few are flattened or coalesce with others to form irregular sclerotia. This pathogen is known for not forming asexual fruiting structures or spores. For the spread of this pathogen, higher temperatures (around 29.4–35 °C) and high humidity are favorable. The pathogen infects plants directly or through wounds which may be caused by nematodes or insects. The germination of sclerotia is favored at the surface of the soil, while it slows down at a depth within the soil.

4.6.4.6 Anthracnose

Anthracnose is a general fungal disease and can cause significant loss in the yield and marketability of tomatoes. Small, sunken, watery, circular spots symptom appeared initially. Lesions appear as concentric rings, and acervuli which are fungal fruiting structures appear in the center of the lesions. Salmon colored spore mass is released from these fruiting bodies under favorable conditions. The lesions merge together to form large rotted regions. The disease is caused by *Coletotrichum coccoides*, *C. gloeosporioides*, and *C. dematium*. The colonies appear white with aerial mycelium having dark pigmentation. The colonies include various numbers of black colored, spherical-shaped sclerotia. The conidia are straight, fusiform, attenuated at the ends, and the appressoria appear clavate, tan with varying shapes. In late spring, the lower leaves and fruit became infected by germinating sclerotia and spores in the soil debris (Devappa et al. 2021). The fungal spores from the diseased plant tissues act as a source of inoculum. Water splash from rain and overhead irrigation usually disperses the fungal spores. Anthracnose can infect both green and ripe fruits; on infected green fruits, symptoms develop only after maturity, while, on ripe fruits, lesions become visible within 5–6 days after infection. The temperature range of 23.8–29.4 °C is found conducive for the growth and spread of this disease. Besides, extended periods of fruit and leaf wetness due to overhead irrigation and frequent rains are also conducive for the development of this disease (Devappa et al. 2021).

4.6.4.7 Rhizopus Rot

Rhizopus rot typically presents as water-soaked areas on the fruit, followed by puffy white fungal growth that collapses the outer tissue. Black, mustard-like spores on stalks are often visible, spreading across the fruit surface, often accompanied by a fermented smell.

The causative organism for this disease is *Rhizopus stolonifer* which is a mucoraceous fungi surviving in soil and debris. As soon as a spore comes in contact with the tissues that have been damaged, there is proliferation on the exterior which resulted into the production of thick mycelium. Cell-degrading enzymes, viz., pectinases and amylases, were produced by thick mycelium that denature the tissues resulting in contamination. Sporangiophores produce sporangia that contain various number of black sporangiospores. This pathogen might travel to the next fruit in contact and generally destroy the complete box material contents within some time. The damages occurred due to the action of insects, hail, and cracking at the time of harvesting are the primary entry sites for the pathogen (Bautista-Baños et al. 2014). This organism proliferated quickly at 27 °C with high relative humidity. Spore germination does not take place at 4 °C.

4.6.5 Onion and Garlic

Onion and garlic pertain to *Alliacease*, and *Allium* genus. *Allium cepa*, that is the bulb onion, and most popular spp. of *Allium*. The eatable part in both bulbs is the extended leaf scales. Regarding garlic, there is a modification of tissue to blade-less leaf, known as bulb scales. Regarding onion, outer swollen sheaths of the bulb develop from bladed leaves and the inner sheaths are bulb scales. Onions are seed-propagated and directly sown, while garlics are vegetatively propagated via cloves. The structure between bulb scales and leaves influences onion bacterial disease development (Hegde et al. 2021).

Pathogens infect onion and garlic crops in all development stages. Approximately 35–40% losses were observed because of several disorders in onion bulbs. In garlic, estimated postharvest losses have been found to be around 35–40%.

Postharvest Diseases: Postharvest diseases in onion and garlic along with the causative organisms have been mentioned in Table 4.5.

4.6.5.1 White Rot

Initially, yellowing of foliage and dieback symptoms appear in onion. When infected plants are uprooted, their roots appear rotten and the bulb base shows white or grey fungal growth, eventually forming small black spherical sclerotia. In garlic, white rot begins as fluffy white mycelium at the stem, spreading to the bulb and producing sclerotia on decayed tissue. *Sclerotium cepivorum* causes this disease; its sclerotia

Table 4.5 Postharvest diseases and causative organisms of onion and garlic

Disease	Causative organism
White rot	*Sclerotium cepivorum*
Basal rot	*Fusarium oxysporum*
Downy mildew	*Peronospora destructor*
Penicillium decay (blue mold rot)	*Penicillium* sp.
Onion smudge	*Colletotrichum circinans*
Purple blotch of onion	*Alternaria porri*
Southern blight or sclerotium bulb and stem rot	*Sclerotium rolfsii*
Bacterial brown rot	*Pseudomonas aeruginosa*
Bacterial soft rot	*Dickeya chrysanthemi*
	Pectobacterium caratovorum
Bacterial leaf streak and bulb rot	*Pseudomonas viridiflava*
Sour skin of onion	*Burkholderia cepacian*

remain dormant without a suitable host and spread via irrigation, equipment, or plant material. Sulfur compounds from onion and garlic roots trigger germination. Cool weather favors mycelial growth, which targets Allium roots, forming appressoria for host penetration. Infected plants lead to field outbreaks or postharvest rot. Primary inoculum comes from sclerotia in field debris, while secondary spread occurs through water movement. Proper drying of bulbs before storage prevents disease transmission.

4.6.5.2 Basal Rot

This disorder is additionally called "dieback" or "damping-off" disease of the seedlings (Le et al. 2021). Early indications are the yellowing in leaf, and subsequent formation of curls as well as withering. At advanced stages, rotting of bulbs and root abscission layer occurs, which leads to easy uprooting of plants and bulbs from roots. The causative pathogen associated is *Fusarium oxysporum*. Pathogen produces microconidia, macroconidia, and chlamydospores asexual spores. Microconidia is 5–12 microns in length and are without septa. Macroconidia are falcate shaped with 3–4 septa.

The pathogen is considered feeble due to plants that have been damaged by additional disease or bugs. The infestation begins at the base, moving towards the bulb. Organisms travel through infested seed, trash, or irrigated water. Chlamydospores are produced as inactive spores inside top soil which can stay up to numerous periods. The favorable temperature for disorder generation and optimal conditions favoring spread of disorder has been found to be 28–32 °C, and 27 °C, respectively. Hot conditions favor the disorder. The organism infests plants via root-hairs straightforwardly or indirectly via damaged areas (Hegde et al. 2021).

4.6.5.3 Downy Mildew

In onion, symptoms appear as brown to purplish colored fungus growing upon the leaves. Since disorder advances, lesions get insipid yellowish to brownish-colored necrotic regions that lead to girdling, with the downfall of leaves. The bulbs that are infected rot during storage. If an infected bulb is cultivated, the undergrowth converts to insipid greenish color.

In garlic bulbs, white colored furry matter appears upon leaf exterior having protruding yellowish staining. The decrease of young leaves and the stunting of old plants occur. Dehydration and collapse of leaf tissue occur. During storage, neck of the bulbs blackens and shrivels, while watery appearance of the scale occurs. Sprouting of premature bulbs may also be observed.

The causative pathogen is *Peronospora destructor*. Sporangia are pyriform to fusiform in shape and are associated with the sterigmata through the pointed end. Within the intercellular spaces, production of oogonia takes place.

Humid conditions favored the growth and proliferation of this pathogen. The dispersion of sporangiospores takes place via wind and rain. Chief source for sustenance of the disorder is planting bulbs and oospores in the infected trash. If these infected bulbs are planted, the pathogen proliferates with the onion plant, and production of sporangiospores takes place under favorable conditions. These spores are responsible for spreading the disease to the adjacent plants.

4.6.5.4 Penicillium Rot

Penicillium decay, additionally called as blue mold rot, causes damage to onions and other Alliums. The initial symptoms appear as water-soaked, pale yellow regions at exterior scale part. At advanced stages, green to bluish green, powdery mold appears on the lesion surface. At advanced stages, bulb might fragment to water-soaked rot. The disease spread in these crops via wounds, bruises, and uncured neck tissue.

Penicillium is the causal organism for this disease. Among those *P. digitatum*, *P. luteum*, *P. oxalicum*, and *P. expansum* are most dominant. This organism gerberates large quantities of conidia, the broom-shaped conidiophores. The branched conidiophores form a structure, known as the metulae. Conidia are oval to globuse shaped, single-celled and nonmotile in nature. The pathogen penetrates the host through wounds and uncured tissues in onion bulbs and garlic. Higher relative humidity and temperature range of around 21–25 °C are conducive conditions for propagation. Conidia can stay in air and transit to lengthy gaps through strong winds. The germination of conidia occurs immediately in the moist air (Sang et al. 2014; Hegde et al. 2021).

4.6.5.5 Onion Smudge

The dark greenish organism develops upon bulb initially. After maturity, they turn black having concentric rings upon scales. During high humidity conditions, disease disperses to the inner scales. The bulbs shrivel, sprout immaturely.

Colletotrichum circinan produce this disease. The conidia are fusiform in shape, hyaline to slight ochraceous in color with a projecting vacuole at the center. Thick-walled, dark brown chlamydospores are found which are in the shape of an egg. Acervulli develops upon stomata that develop below host cuticle. Secondary spreading agents of this disease are the rain splash and through wind. Warm and humid conditions and temperature varying from 20 °C to 26 °C are favorable for production of conidia.

4.6.5.6 Purple Blotch of Onion

The initial symptoms of this disease include watery lesions that develop and turn to prominent lesions that have a central white region and brown to purple border, and surrounding region is yellow in color. Later, lesions turn to dark brown and black colors having concentric rings, followed by sporulation. Finally, lesions girdle leading to the collapse and death of the leaves. Symptoms similar to the above mentioned appear on the seed stalks, which leads to the collapse of stalk and shriveling of seeds. The infection of the bulb starts from the neck region. At earlier stages, an infected region in the bulb appears bright yellowish which converts to a red wine color.

Casual organism is *Alternaria porri*. Mycelium colored, with branching, as well as with septa. Conidiophores develop as single or in groups, and are straight or geniculate or flexuous. The conidia develop on the tip of conidiophores and are obeclavate shaped, brown colored with transverse and longitudinal septa.

Conidia develops when prolonged humidity persists for greater than 12 h. After the dew dries out, the conidia become airborne and spreads through air. Beyond occurrence of infectivity, the indications become visible after 4 days.

4.6.5.7 Southern Blight or Sclerotium Bulb and Stem Rot

In this disease, outer scales of the bulb get infected, resulting in white-colored lesions. The tissues of bulb and neck soften and watery rot appears. On the surface of the bulbs, the appearance of mycelial growth of white color occurs. Sclerotia of pale brownish color grow over the infected bulbs or on debris and in soil.

Sclerotium rolfsii is the causative organism for this disease. The sexual stage of organism is *Athelia rolfsii*. The dispersion of the pathogen occurs through soil or irrigated water. Severity of disease increases in moist soil and temperature of around 25–30 °C. Below 15 °C temperature is unfavorable for growth, and multiplication (Hegde et al. 2021).

4.6.5.8 Bacterial Brown Rot

It is a destructive storage disease in onion in which rotting observed at the neck of the bulb. While squeezing the bulb, foul smell is also observed. The causative organism for this disease is *Pseudomonas aeruginosa*, which is a Gram-negative, rod-shaped bacterium with polar flagella. The temperature range conducive for the growth of this pathogen is 4–43 °C, with average growth observed at 37 °C. The pathogen secretes several lytic enzymes which degrades the polysaccharides present in the plant cell wall.

4.6.5.9 Bacterial Soft Rot

This disease is known to affect the onion bulbs during transport and storage and its symptoms are observed on mature bulbs only. At earlier stages, watery, pale yellow to light brown or greyish to white infected scales appear. As the rotting progresses, fleshy scales on bulb soften and become sticky, and breaking of inner portion of the bulb is noticed. Foul smell with thick watery liquid ooze out on the squeezing of the neck of infected bulb (Hegde et al. 2021). *Dickeya chrysanthemi* and *Pectobacterium carotovorum* are the causal organisms for the disease. The dispersion of this pathogen occurs through rain splash, irrigation water, and insects. Wounds occurred on the bulbs due to mechanical injuries and sunscald act as the entrance points of the injuries. Favorable conditions for spread of disease include warm and humid climate with temperature varying from 20 °C to 30 °C.

4.6.5.10 Bacterial Leaf Streak and Bulb Rot

Initial symptoms emerged on leaves as development of oval-shaped watery lesions and tip burn symptoms on leaves. Leaf streaking with variable streak length is the characteristics symptoms for this disease. The green leaf streaks slowly turn to black. As the disease advances, leaves dry out and collapse. With an increase in severity of the disease, leaf distortion and twisting symptoms appear. Characteristic dark spots on outer scales, while reddish brown discoloration on inner scales appears. A ring-like patterned rot develops because of the inhibition by the scales. *Pseudomonas viridiflava* produces the disease. The pathogen is also responsible for the leaf blight disease in other vegetables such as cabbage, cauliflower, pumpkin, brinjal, carrot, turnip, lettuce, and pea. Humid environmental conditions are conducive for the growth and development of the pathogen. Raindrops and splashes are responsible for the subsequent spread of this disease. High nitrogen content in the soil due to excessive use of fertilizers is favorable for the development of this disease.

4.6.5.11 Sour Skin of Onion

Initial symptoms are observed as light brown coloration in 1–2 leaves. Watery rot appears at the base of leaves and spreads to the neck region. The outermost scales and inner scales of the bulb remain not infected. Pale yellow to light brown, slimy decay is observed. Smell of vinegar from the infected bulb due to growth of secondary invaders, such as yeasts, is observed. The causal organism for this disease is *Burkholderia cepacian* which is a Gram-negative, obligately aerobic, rod-shaped and polar flagellated pathogen. This produces in fluorescent yellow to green pigment, which varies according to the type of strains. Heavy rains, overhead irrigation, and flooding are responsible for spreading of this pathogen. The infection begins through wounds. The optimal temperature condition for the growth is 30–35 °C. The disease is favored by rains and hot conditions over around 30 °C (Hegde et al. 2021).

4.6.6 Potato

Potato or *Solanum tuberosum* is an edible tuber and a perennial plant.

Postharvest Diseases of Potato These tubers are susceptible to various disorders after harvest. These postharvest diseases occur by several microorganisms. They can significantly impact potato eating quality, resulting in considerable economic losses.

Postharvest diseases of potato along with their causative organism have been mentioned in Table 4.6.

4.6.6.1 Blackleg

Blackleg is a bacterial disease affecting the underground stem and tubers of potatoes. It causes blackening at the stem base, decay of the seed piece, and stunted, discolored plants with a leathery, folded appearance. Infected stems are easily

Table 4.6 Postharvest diseases and causative organisms of potato

Disease	Causative organism
Blackleg	*Pectobacterium atrosepticum*
Dry rot	*Fusarium* sp.
Brown rot	*Ralstonia solanacearum*
Scab	*Streptomyces scabies*
	Streptomyces acidiscabies
	Spongospora subterranean
Sclerotium rot	*Sclerotium rolfsii*
Potato wart	*Synchytrium endobioticum*
Charcoal rot	*Macrophomina phaseolina*
Silver scurf	*Helminthosporium solani*
Late blight	*Phytophthora infestans*

pulled and show characteristic lesions, unlike leafroll which can appear similar. Symptoms vary in severity across sprouts, and infected plants may survive but remain dwarfed with spindly stalks. As the disease advances, complete water supply loss leads to plant browning and decay by secondary pathogens.

The xerophytic texture of the diseased plants is exhibited by stems and leaves alike. The foliage is discolored, usually light and of a metallic luster. The leaflets are folded, and the petiole and midrib are woody and lacking the elasticity and softness which characterize the normal organs. Not all the stems of a diseased plant are necessarily affected. Healthy sprouts may appear side by side with diseased ones, and the diseased sprouts may exhibit various degrees of injury. Plants which have been attacked rather early often continue to live for a considerable period. These plants remain naturally dwarfed; the stalks are spindling, the internodes shortened, and the leaves small. Plants attacked at a more mature age may attain full size, though they usually plants affected earlier in their life. In the last stages of the disease, when the rot has progressed far enough to cut off completely the water supply, the entire plant turns brown and sooner or later, depending on weather conditions, falls prey to the attack of saprophytic bacteria and fungi.

The causal organism for the disease is *Pectobacterium atrosepticum. P. atrosepticum* is a Gram-negative, nonspore-forming, facultative anaerobe with peritrichous flagella and rod-shaped bacteria. It is oxidase-negative, catalase-positive and has an optimal growth temperature of 27 °C. Molecular investigations have established that the main pathogenicity determinants of *P. atrosepticum* are the wide array of plant cell wall degrading enzymes, including pectinases, proteases and cellulases, secreted through the type I and II secretion systems. The enzymes break down pectinases and other molecules in the apoplast and plant cell walls, allowing the bacteria to move through the plant. Genomic studies have revealed that *P. atrosepticum* has all seven of the known Gram-negative bacterial secretion systems, which are used to export other virulence determinants. These include siderophores, putative effectors (although there are no known associated resistance genes in potato to this pathogen), nonribosomal peptide synthases, and the polyketide phytotoxin coronafacic acid. The expression of these and other determinants is controlled by an array of regulators, perhaps the most important of which is the N-acetyl-homoserine lactone-mediated quorum sensing, which interacts to fine tune the infection process (Van Gijsegem et al. 2021).

Pectobacterium atrosepticum thrives in cool, wet conditions. The disease often begins in the field but can continue to develop during storage, especially under poorly ventilated conditions. The infection occurs through wounds or natural openings in the tuber, with the bacteria spreading rapidly in moist environments. The infected tubers show black, wet, and soft decay, often beginning at the stem end and progressing inward. The decay is usually confined to the outer layers of the tuber, but can penetrate deeper in advanced stage of disease. The infected tubers may emit a strong, foul odor, similar to that of soft rot.

4.6.6.2 Dry Rot

Dry rot is one of the most significant postharvest diseases of potato, causing substantial economic losses worldwide. The disease primarily affects stored potatoes, leading to reduced market value and compromised seed quality (Tiwari et al. 2020).

The pathogens responsible for dry rot in potatoes belong to the genus *Fusarium*. The most common species are *Fusarium solani* var. *coeruleum*, *F. sambucinum* (also known as *F. sulphureum*), *F. avenaceum*, and *F. oxysporum*. These fungi are soil-borne and can survive in soil or on plant debris for long periods, making them persistent in agricultural environments. *Fusarium* sp. produce both asexual spores (conidia) and sexual spores (ascospores, in some species) that aid in their spread and survival. Conidia are the main inoculum for infection which is produced in abundance either on the surface of infected tubers or in soil. The pathogens can survive in the soil as chlamydospores (thick-walled resting spores) or in plant debris for several years. They can also survive on contaminated storage equipment, seed potatoes, and in soil adhering to tubers. The pathogenicity of *Fusarium* sp. varies among species and environmental conditions. Some strains are more aggressive, causing rapid and extensive decay, while others may result in slower disease progression (Tiwari et al. 2020).

The infection typically occurs through wounds or bruises sustained by the potatoes during harvesting, handling, or storage. The pathogens enter the tubers through these openings. Dry rot pathogens can infect potatoes at any time from harvest through to storage. However, the disease usually becomes apparent during storage, particularly when conditions are favorable for fungal growth. The optimal temperatures for *Fusarium* species growth range between 15 °C and 20 °C. The disease progresses more rapidly in warmer storage environments. High humidity levels (above 85%) in storage facilities can exacerbate the disease by promoting fungal growth and spore germination. Poorly ventilated storage conditions, combined with high humidity and suboptimal temperature management, significantly increase the risk of dry rot development.

Fusarium sp. is disseminated through contaminated soil, plant debris, or infected seed tubers. Once introduced into a storage facility, the pathogen can spread from infected to healthy tubers, especially if the conditions are conducive to fungal growth. Secondary spread can occur during handling or sorting of potatoes if proper sanitary measures are not observed (Tiwari et al. 2020).

Small, discolored, and sunken lesions on the surface of the potato tuber are the first visible sign of dry rot. These lesions are often brown or black and may be slightly depressed. These lesions expand, and become more sunken and wrinkled as disease progresses. The skin over these lesions may crack, revealing the dry, decayed tissue underneath. Beneath the surface lesions, the tuber flesh becomes dry, brown, and necrotic. The decayed tissue is typically dry, crumbly, and may have a corky texture. The rot often extends deep into the tuber, creating large, hollow cavities as the decay progresses. These cavities are usually lined with a dry, powdery material, which is the result of fungal mycelium and decayed tissue. Unlike some other postharvest diseases (e.g., soft rot), dry rot is not typically associated with a foul odor. However, if secondary bacterial infections occur, the tuber may develop an unpleasant smell.

4.6.6.3 Brown Rot

Brown rot disease, also known as bacterial wilt, is the most devastating disease affecting potato crops worldwide. Brown rot significantly impacts both the quality and yield of potato crops and poses a severe threat to the potato industry due to its rapid spread, persistence in the environment, and difficulty in control.

Wilting of the plant during the warmest period of the day is the earliest and most noticeable symptom of brown rot, followed by recovery at night. As the disease progresses, wilting becomes permanent. Infected plants often exhibit stunted growth, yellowing of leaves, and premature senescence. When stems are cut, a brown discoloration of the vascular tissue is visible. If squeezed, a milky white bacterial ooze may exude from the cut stem. Externally, infected tubers may not show any obvious signs of infection, making early detection difficult. However, in advanced stages, tubers may develop soft, sunken areas, especially around the stolon end. Internally, the tuber exhibits a brown, granular discoloration of the vascular ring. When the tuber is cut open, a slimy, brown rot may be seen, and squeezing the tuber often results in bacterial ooze emerging from the vascular tissue. Infected tubers may emit a characteristic foul odor because of the decay process, particularly when secondary bacterial infections occur.

Bacterium *Ralstonia solanacearum*, a highly destructive pathogen with a wide host range, is the causal organism of the disease. *R. solanacearum*, a Gram-negative, rod-shaped bacterium that belongs to a diverse species complex, which includes different strains with varying host ranges, pathogenicity, and geographic distributions. The pathogen is classified into five races and four biovars based on host range and biochemical characteristics. Race 3, biovar 2 is the most common strain affecting potatoes and is particularly adapted to cooler climates.

Ralstonia solanacearum produces several virulence factors, including extracellular polysaccharides (EPS), which help it to block xylem vessels and contribute to wilting symptoms. The bacterium can survive in a dormant state in soil and water, even without host plants. It forms biofilms that enhance its survival and facilitate infection of new plants. The bacterium's pathogenicity is largely determined by its ability to colonize and persist within the plant's vascular system. EPS production is critical, as it clogs the xylem and disrupts water flow. Other virulence factors include cell wall-degrading enzymes, which help the bacterium invade host tissues and facilitate systemic infection.

R. solanacearum primarily infects potato plants through the roots, often via natural openings, wounds, or cracks caused by nematodes or other mechanical injuries. Once inside the plant, the bacterium invades the vascular system, particularly the xylem vessels, where it multiplies rapidly, blocking water transport, and leading to the wilting and ultimate death of the plant. The pathogen thrives in warm, moist conditions, with 25–35 °C of optimum temperature. However, strains like race 3, biovar 2 can also survive and cause infections at cooler temperatures (15–25 °C), making it a threat in temperate regions. High soil moisture levels, especially waterlogged conditions, significantly enhance disease development by facilitating bacterial movement in the soil and increasing infection opportunities. Brown rot is more prevalent in acidic soils with a pH of 4.5–6.5, although the bacterium can survive in a wider range of soil types, including sandy, loamy, and clay soils.

The bacterium can survive in infested soil, contaminated water, infected plant debris, and host plants, including wild solanaceous weeds. It can persist in the soil for several years in the absence of its primary host. Spread of the pathogen occurs through contaminated irrigation water, movement of infested soil, farm machinery, and infected seed tubers. The disease can also spread during wet weather conditions when bacteria are washed into the soil from infected plants. The disease cycle begins with the bacterium surviving in the soil or plant debris. It is then transmitted to the potato plant through the roots. Once inside, the bacterium colonizes the vascular system, blocking the xylem vessels, causing wilting, and disrupting water flow, leading to characteristic symptoms of brown rot (Abd El-Rahman and Shaheen 2016).

4.6.6.4 Scab

Scab is a common and economically significant disease of potato that affects the appearance and marketability of tubers. This disease manifests primarily as surface blemishes, which, although not typically damaging to the potato's internal quality, severely reduce its commercial value. There are different types of scab diseases affecting potatoes, including common scab, acid scab, and powdery scab, each caused by different pathogens and influenced by specific environmental conditions.

Scab diseases of potatoes are complex, involving different pathogens and influenced by various environmental factors. Understanding the specific type of scab and tailoring management strategies accordingly can help reduce the occurrence and harshness of this problematic disease in potato production systems.

(i) Common scab: the causative organism for common scab is *Streptomyces scabies* and related *Streptomyces* species like *S. acidiscabies* and *S. turgidiscabies* can also cause similar symptoms. These pathogens are Gram-positive, filamentous, soil-borne actinomycetes that thrive in a wider range of soil forms and conditions.
(ii) Acid scab: the causative organism for acid scab is *Streptomyces acidiscabies*. Acid scab, also caused by a *Streptomyces* species, differs primarily in its ability to thrive in acidic soils (pH 4.5–5.2), which is uncommon for most scab pathogens.
(iii) Powdery scab: the causative organism for powdery scab is *Spongospora subterranean.* Unlike common and acid scab, powdery scab is caused by a

protozoan-like pathogen (plasmodiophorid), which is classified as a soil-borne slime mold. It not only affects the tubers but can also infect the roots, leading to stunted growth and yield loss.

Appearance of rough, raised, corky lesions on the potato surface is the most common symptom of this disease. These lesions can vary in size, shape, and depth, ranging from small, pitted spots to large, raised scabs. Lesions may be superficial or penetrate deep into the tuber, depending on the severity of the infection and environmental conditions. The lesions are often dark brown to tan and can coalesce, covering large portions of the tuber surface. Common scab generally does not affect the internal quality of the potato, although deep lesions can sometimes extend slightly below the skin (Braun et al. 2017).

Acid scab symptoms are similar to common scab, but lesions tend to be less raised and shallower. The disease thrives in more acidic conditions, unlike the common scab.

Powdery scab is characterized by small, raised, pimple-like pustules on the tuber surface. These pustules release a powdery mass of spores upon rupturing, leaving behind dark, and ragged pits. The pustules are usually more numerous and smaller than in comparison to common scab, giving the tuber a warty appearance. Infected roots develop galls, which can stunt plant growth and reduce tuber yield significantly. The pathogen can serve as a vector for the potato mop-top virus which further complicates the management.

Scab pathogens for common scab and acid scab infect potato tubers primarily through natural openings such as lenticels and wounds on young developing tubers. The bacteria invade the corky layer of the potato skin, producing lesions. These lesions released toxins that induced cell death and tissue collapse. Common scab is most severe in neutral to alkaline soils (pH 5.5–8.0). Acid scab is an exception, thriving in more acidic soils. Dry soil conditions favor the development of common scab, as high soil moisture inhibits the activity of *Streptomyces* species. The optimal temperatures for scab development are 15–25 °C. Warm and dry conditions during tuber formation are highly conducive to disease development. Potato varieties differ in susceptibility to scab, with some cultivars being more prone to infection due to genetic factors affecting skin type and structure (Braun et al. 2017).

In case of powdery scab, *Spongospora subterranea* infects tubers and roots via zoospores that are released from resting spores (cystosori) in the soil. The pathogen forms spore balls (sporosori) on the surface of tubers, leading to pustules that rupture and release powdery spores. Powdery scab thrives in cool, wet conditions and can tolerate a wider range of soil pH than common scab, although it prefers slightly acidic to neutral soils. The disease is favored by cool temperatures (10–18 °C), particularly in poorly drained or waterlogged soils (Merz and Falloon 2009).

Streptomyces species (common scab and acid scab pathogens) produce filamentous mycelium and chains of conidia (spores) that help them survive in soil for extended periods. These pathogens can survive as free-living saprophytes on organic matter in the soil, making them hardy and persistent even without a host crop. The production of thaxtomin A, a phytotoxin capable to inhibit cellulose synthesis in plant cells, is critical for the pathogenicity of *Streptomyces*. This toxin disrupts cell walls and leads to characteristic scab lesions.

Spongospora subterranea forms spore balls that can live in the soil for several years, withstanding harsh conditions and waiting for favorable conditions to germinate and infect new hosts. The pathogen's life cycle includes both a motile zoospore stage, which facilitates infection, and a resting spore stage that ensures long-term survival in the soil (Merz and Falloon 2009).

4.6.6.5 Sclerotium Rot

S. rot, also known as white mold or southern blight, is a severe fungal disease affecting a wide range of crops, including potatoes, and can result in significant losses, especially in warm, humid climates. Rapid decay of plant tissues, particularly near the soil line, and the formation of distinctive are small, hard, and brown resting structures known as sclerotia is characteristics feature of this disease. *Sclerotium* rot of potato is a destructive disease that thrives in warm, moist environments and poses a grave threat to potato production in susceptible regions.

Infected stems show water-soaked lesions at or near the soil line. These lesions quickly expand, girdling the stem and forcing wilting and collapse of the plant. White, cottony mycelial growth is often evident on the surface of infected stems, especially in warm, humid conditions. This white mold is a unique feature of *Sclerotium* rot. As the disease advances, numerous small, round, tan to dark brown sclerotia form on the infected plant tissue and surrounding soil surface. Tuber infection usually begins at points of contact with the soil or where the tubers have wounds or bruises. Symptoms include soft, watery rot that spreads rapidly through the tuber flesh. Infected tubers may also show white mycelial growth on their surface, particularly at the stolon end, where infection often begins. The occurrence of sclerotia on the tuber surface is a key diagnostic feature; these sclerotia resemble mustard seeds and may be embedded in the decaying tissue. Although primarily a stem and tuber disease, in some cases, the pathogen can cause leaf blight if the fungus spreads upward. This is not common but can occur under conditions of high humidity conditions and dense canopy. Plants infected with *Sclerotium rolfsii* often exhibit sudden wilting and yellowing of foliage. The wilting is usually irreversible, and the plant finally collapses as the stem tissue decays.

The soil-borne pathogen *Sclerotium rolfsii* (also known as *Athelia rolfsii* in its sexual stage) belonging to the Basidiomycota is the casual organism of the disease. It has broad host range, attacking more than 500 plant species, including vegetables, legumes, and ornamental plants. The pathogen produces sclerotia, which are the primary survival structures. The small, round, mustard seed-like structures allow the fungus to continue in the soil for extended periods, even in adverse conditions.

The infection begins when sclerotia present in the soil germinate in the existence of susceptible plant tissue, high temperatures, and moisture. Germination produces mycelium that invades the plant tissue, secreting enzymes and toxins that break down cell walls, leading to tissue rot. The pathogen can infect potato plants through direct contact with sclerotia, mycelium, or hyphae in the soil. It usually starts at the stem base near the soil line, but it can also infect tubers, particularly those close to or on the soil surface. The optimal temperatures for *Sclerotium rolfsii* range from 25 °C to 35 °C. The disease is most severe in warm climates and can be particularly problematic in regions with high summer temperatures. High soil moisture and humidity condition found to be helpful for the germination of sclerotia and the fungus growth. Prolonged duration of wet soil or frequent irrigation increased the risk of infection. The fungus can survive well in a wide range of soil types and pH levels but prefers acidic to neutral soils (pH 5–7). Heavier, poorly drained soils exacerbate the disease's severity due to increased moisture retention. The sclerotia serve as the primary means of survival for *Sclerotium rolfsii*, allowing the pathogen to persist in the soil for several years in the absence of a host. These structures are highly resistant to environmental extremes, including desiccation and high temperatures. The spread occurs primarily through the movement of contaminated soil, water, plant debris, and infected seed tubers. The fungus can also be spread by farm equipment, animals, and irrigation water.

The disease cycle begins with the sclerotia germinating in the soil under favorable conditions. The emerging mycelium colonizes nearby plant tissues, causing rapid rot. The fungus then produces new sclerotia on infected tissue, completing the cycle and ensuring future disease outbreaks.

S. rolfsii produces sclerotia, which are the key to its survival in adverse conditions. These structures can remain viable in the soil for numerous years, and germinate when conditions become favorable. The pathogen also produces a dense, white mycelium that can quickly colonize plant tissues. Mycelial growth is rapid, particularly under warm, moist conditions, contributing to the aggressive nature of the disease. The fungus secretes a range of enzymes, including cellulases, pectinases, and oxalic acid, which degrade plant cell walls and facilitate tissue colonization. Oxalic acid is particularly important in pathogenesis as it lowers the pH of the infected tissues, enhancing enzyme activity and suppressing host defense responses.

4.6.6.6 Potato Wart

Potato wart disease is the most severe disease of potatoes. It poses a significant threat because of its high persistence in the soil and the potential for severe yield loss. The disease primarily affects potato tubers and underground stems, leading to unsightly growths or "warts" that render the tubers unmarketable. Due to its economic impact and the difficulty of eradication, potato wart is considered a quarantine disease in many countries, strictly regulated to prevent its spread. Potato wart disease is a major threat to potato production because of its impact on tuber quality, yield, and marketability. The disease's persistence in the soil, along with its survival ability for decades, makes it one of the most difficult potato diseases to manage (Franc 2007).

The most distinctive symptom of potato wart disease is the formation of irregular, warty growths or galls on tubers, stolons, and occasionally underground stems.

These growths can range in size from small, cauliflower-like protrusions to large, grotesque masses that engulf the entire tuber. Initially, the warts are light green or whitish, blending with the color of healthy tissue. As the warts mature, they turn brown or black due to secondary infection and necrosis. The surface of the warts is rough, irregular, and spongy, often appearing as malformed clusters of tissue with a distinct, scaly appearance. Unlike other potato diseases, potato wart does not typically affect the internal flesh of the tuber directly. The symptoms are mostly external, though severe infections can lead to overall tuber rot and decay due to secondary pathogens. The disease rarely affects above-ground parts of the plant. In severely infected cases, it may cause mild stunting or chlorosis, but however, these secondary symptoms are less common. Infected plants often suffer from reduced vigor and stunted growth due to the energy diverted to wart formation. The galls disrupt nutrient and water flow, compromising the plant's overall health and productivity (Franc 2007).

The causal pathogen for this disease is the obligate parasitic fungus *Synchytrium endobioticum. S. endobioticum* is a soil-borne, obligate biotrophic chytrid fungus and belongs to *Chytridiomycota*, a group of motile spores (zoospores) producing fungi. The pathogen is unique because it lacks hyphae; instead, it infects plant cells directly, leading to the formation of characteristic wart-like growths on infected tissues. The fungus survives in the soil as highly resistant sporangia (resting spores), which can survive for decades, making eradication extremely challenging (van de Vossenberg et al. 2022).

The disease cycle begins when resting sporangia in the soil germinate under favorable conditions, releasing motile zoospores. These zoospores swim through soil water films and infect susceptible plant tissues. Zoospores penetrate the epidermal cells of the potato's growing points, such as eyes, stolons, and tubers, initiating the formation of warty growths. The infection disrupts normal cell division, causing hypertrophy (enlarged cells) and hyperplasia (increased cell numbers). The pathogen completes its life cycle within the host cells, forming new sporangia which are released subsequently back into the soil as the infected tissue decays, perpetuating the disease.

The optimal conditions for *S. endobioticum* infection and development occur at temperatures between 10 °C and 20 °C. This disease is most dominant in temperate regions with cool, moist growing seasons. High soil moisture is crucial for the germination of resting spores and the movement of zoospores. Prolonged periods of wet conditions, poor drainage, and high soil humidity favor disease development. Heavy, clayey, and soils with poor drainage are mainly conducive to potato wart disease due to their ability to retain moisture, which facilitates the pathogen's life cycle. Susceptibility varies among potato cultivars, with some varieties exhibiting high resistance due to genetic factors. The disease is somewhat host-specific to potatoes and as such affected all members of *Solanaceae* family.

The sporangia of *S. endobioticum* can live in the soil for 30–50 years without a host, making fields infested with the pathogen unusable for potato cultivation for extended periods. The disease spreads through the movement of contaminated soil, infected seed tubers, water runoff, and farm machinery. Contaminated clothing, animals, and wind can also disseminate the pathogen over short distances. International trade and transport of contaminated seed potatoes and soil are major pathways for the long-distance spread of the disease, necessitating strict quarantine measures.

The resting sporangia are the thick-walled spores, that are the primary survival structure of the fungus. They remain dormant in the soil until favorable conditions trigger germination. Upon germination, each sporangium releases numerous motile zoospores that swim through soil water to infect new hosts. Zoospores penetrate host cells, forming plasmodia that develop into new sporangia within the infected tissue. This process induces abnormal cell growth, resulting in the characteristic warts. As infected tissues decompose, new resting sporangia are released back into the soil, where they can remain viable for decades, awaiting suitable conditions to restart the cycle (van de Vossenberg et al. 2022).

4.6.6.7 Charcoal Rot

Charcoal rot is a severe disease of potatoes. The name charcoal coat comes from condition of infected tissue as dark, charcoal-like appearance of infected tissues, characterized by the presence of numerous microsclerotia within the plant tissue. The disease primarily affects the vascular system, leading to severe wilting, stunted growth, and tuber rot. Charcoal rot of potato is a challenging disease to manage due to its adaptability to warm, dry conditions and its ability to persist in the soil for years. The most severe pathogen's impact occurs when plants are under stress, particularly during hot and dry periods.

Initial symptoms include wilting of the lower leaves, especially during hot days, with partial recovery at night. As the infection progresses, the entire plant may become stunted, wilted, and eventually die. Infected plants may exhibit yellowing of the leaves, followed by premature leaf drop. The affected plants often appear drought-stressed due to impaired water uptake caused by the blocked vascular system. Affected roots and lower stems exhibit dark, necrotic lesions. The internal vascular tissues are often discolored, turning dark brown to black due to the proliferation of fungal hyphae and microsclerotia. As the infection advances, the roots, lower stem, and sometimes the tubers develop a characteristic charcoal-gray appearance due to the build-up of microsclerotia. This symptom is diagnostic of the disease and is visible when the bark is scraped away. Tuber rot usually starts at the point of contact with the infected soil and appears as dry, black, shrunken lesions. These lesions are often sunken and covered with microsclerotia, giving the tubers a blackened, ashy look. The affected tubers may become hard and shriveled, with black internal specks, especially around the vascular ring.

Infected plants suffered from reduced growth due to disrupted nutrient and water transport. This leads to lower tuber quality and quantity. Severe infections can cause significant yield losses, especially in stressed plants. Charcoal rot can result in up to 60% yield reduction in heavily infested fields.

The causal organism for this disease is the soil-borne fungus *Macrophomina phaseolina* belonging to the Ascomycota division. It is characterized by its production of microsclerotia, which are tiny, black, hard structures that serve as the primary survival mechanism of the fungus. The pathogen has a broad host range, affecting over 500 plant species, and thrives in warm temperatures (28–35 °C) and dry conditions, making it particularly problematic in arid and semi-arid regions. This pathogen is highly destructive, especially in warm, dry conditions, and affects a wide range of crops beyond potatoes, including beans, soybeans, and maize.

The primary inoculum for charcoal rot is microsclerotia, which persist in the soil and plant debris for several years. These structures can survive in the soil without a host and are resistant to extreme environmental conditions, including heat and drought. Microsclerotia germinate under favorable conditions of high temperatures and low soil moisture, producing hyphae that invade plant roots and vascular tissues. The fungus colonizes the xylem, obstruct movement of water and nutrient, which leads to wilting and decay. Infection usually begins in the roots and lower stems, spreading upward through the vascular system. The pathogen enters through natural openings, such as root hairs and wounds, or directly penetrates the plant epidermis.

M. phaseolina thrives in high temperatures, with optimal growth between 28 °C and 35 °C. The disease is most severe during hot, dry weather when plant stress is elevated. Dry soil conditions are conducive for the development of charcoal rot. Water stress in plants makes them increasingly prone to infection, as the pathogen exploits weakened tissue. Wet conditions, conversely, are less conducive to the pathogen's activity. Sandy, well-drained soils are particularly conducive to charcoal rot, as these soils retain less moisture and can reach high temperatures. Poorly drained or heavily compacted soils are less favorable due to higher moisture retention.

The fungus spreads primarily through infected soil, contaminated plant debris, and movement of infested soil on farm equipment. Microsclerotia can also be spread through irrigation water and infected seed tubers. Microsclerotia can survive in the soil for 2–15 years, dependent on environmental circumstances. They remain dormant until favorable conditions prompt germination, allowing the pathogen to persist in fields even without a host.

The primary survival structures, microsclerotia, are small, compact masses of fungal cells that form within the infected plant tissue. These structures are resistant to desiccation and high temperatures, allowing the fungus to survive adverse conditions. Once germinated, the hyphae rapidly invade the host plant's vascular system, blocking water transport and leading to wilting and tissue death. The fungus produces a variety of enzymes that degrade plant cell walls, facilitating colonization. *M. phaseolina* produces toxins, enzymes, and secondary metabolites that degrade plant tissue, enhance its spread, and suppress host defenses. The pathogen's ability to thrive under water stress conditions gives it a competitive advantage over the plant's defense mechanisms (Abbas et al. 2014).

4.6.6.8 Silver Scurf

Silver scurf is a common fungal disease of potatoes. This disease affects the tuber skin, causing unsightly blemishes that reduce marketability and tuber quality. Although it primarily affects the skin, silver scurf can also lead to increased water loss from the tubers, resulting in shriveling and weight loss during storage. The disease is mainly problematic in storage facilities, where it can spread rapidly under favorable conditions. Silver scurf is a significant postharvest disease of potatoes, primarily affecting the tuber skin and reducing market quality. The disease thrives in warm, humid conditions, making proper storage management essential for control (Errampalli et al. 2001; Tiwari et al. 2021).

The most characteristic symptom of silver scurf is the appearance of circular, silver-gray patches on the tuber skin. These lesions are often irregular in shape and size, and they give the skin a shiny, metallic appearance when viewed in light. As the disease progresses, the affected areas may darken, turning from silver to gray or brown. The discoloration remains superficial, affecting only the skin layer without penetrating the flesh of the tuber. Infected tubers lose moisture more rapidly than healthy ones, leading to shriveling and wrinkling, particularly in prolonged storage. This results in a decrease in tuber weight and quality. Conidia form on the surface of lesions and appear as fine, dark, powdery growths under magnification. These spores are the primary agents of disease spread, especially in humid conditions.

Silver scurf affects the aesthetic quality of potatoes, making them less appealing to consumers and reducing their market value. Tubers with extensive silver scurf lesions are often rejected for fresh markets. The disease leads to increased water loss, causing tubers to become light, wrinkled, and less firm. In storage, this can significantly reduce the overall crop yield and quality.

The causal pathogen for this disease is *Helminthosporium solani* (syn. *Spondylocladium atroroseum*). *H. solani* is a soil-borne and seed-borne fungal pathogen belonging to the phylum Ascomycota. It primarily infects the outer layers of the potato tuber, causing cosmetic damage that significantly affects market value. The fungus produces conidia (asexual spores) on the surface of infected tubers that play a critical role in the disease cycle by facilitating spread and reinfection in storage environments.

Infected seed tubers and soil contaminated with conidia are the major source of inoculum. The pathogen can survive on the surface of seed potatoes and in soil residues, ready to infect new crops during planting. *H. solani* produces conidia on the surface of infected tubers. These spores are easily dispersed by water splash, wind, and physical contact, particularly in humid storage environments. The fungus infects the periderm (outer skin) of the potato tuber, often entering through natural openings, wounds, or lenticels. Infection is superficial, primarily affecting the skin without penetrating deeply into the flesh.

Warm temperatures, with optimal growth occurring between 15 °C and 25 °C, are favorable conditions for the development of disease. However, the pathogen can survive and remain active over a wide temperature range, from cool field conditions to warm storage environments. High relative humidity, especially in storage (over

90%), promotes spore germination and the spread of *H. solani*. Moisture on tuber surfaces creates ideal conditions for the fungus to thrive, making humidity control in storage critical. Poorly managed storage conditions with fluctuating temperatures and inadequate ventilation significantly increase the risk of silver scurf development and spread. In the field, damp soils with prolonged wet conditions can also contribute to disease establishment.

Infected seed potatoes are the main source of introducing silver scurf into new fields. Using infected seed can lead to early infection of daughter tubers during the growing season. The fungus survives between growing seasons on infected seed tubers, in plant debris, and in the soil. In storage, spores on infected tubers can remain viable and infect healthy potatoes, especially when stored in close proximity. Once in storage, the disease easily spread to healthy tubers from infected tubers through direct contact, handling, and spore movement within the storage environment.

H. solani produces conidia, which are easily dispersed within the storage environment. The conidia are formed on dark, thread-like structures called conidiophores that develop on the surface of infected tubers. Under suitable conditions, conidia germinate and produce germ tubes that penetrate the potato skin. The infection is primarily limited to the outer layers, where the pathogen multiplies and produces new spores. The pathogen's ability to produce large numbers of conidia, combined with its persistence on seed tubers and in the soil, allows it to effectively survive between growing seasons and spread within storage facilities (Errampalli et al. 2001; Tiwari et al. 2021).

4.6.6.9 Late Blight

Late blight is one of the most devastating diseases of potatoes and tomatoes worldwide. It is notorious for causing the Irish Potato Famine in the 1840s and continues to be a significant challenge for potato production. The disease affects both foliage and tubers, leading to significant yield losses if not properly managed. Late blight can develop rapidly under favorable environmental conditions, making it crucial for farmers to adopt timely and effective control measures. Late blight remains one of the most challenging diseases for potato growers worldwide, requiring constant vigilance and an integrated approach to management. Late blight can spread quickly under favorable conditions, leading to severe crop losses if not managed effectively.

The earliest symptoms on leaves appear as small, irregularly shaped, water-soaked spots. These lesions typically begin at the leaf margins and can expand rapidly under moist conditions. As the disease progresses, the lesions turn dark brown or black and may have a chlorotic (yellow) halo around them. In humid conditions, a white, fuzzy growth of sporangia may be visible on the underside of infected leaves. Severely affected leaves may curl, collapse, and die, leading to defoliation. In advanced cases, the entire plant can be blighted, with blackened stems and collapsed foliage. Infected stems develop dark, elongated lesions that can girdle the stem, leading to the collapse of the above-ground parts of the plant. Stem infections are particularly dangerous as they allow the pathogen to move rapidly from foliage to tubers. Late blight tuber lesions appear as firm, brownish-purple, dry rots on the surface of the tuber. These lesions often extend into the flesh, creating brown, granular, dry rot that can lead to significant tuber loss. Infected tubers are prone to secondary infections by other pathogens, such as soft rot bacteria, which can lead to complete breakdown of the tuber in storage. Tubers infected with late blight can continue to rot in storage, especially if stored under warm, humid conditions. The disease can spread from tuber to tuber, leading to significant postharvest losses.

The causal pathogen for this disease is the oomycete *Phytophthora infestans*. *P. infestans* is an oomycete, often mistakenly referred to as a fungus, but it belongs to the kingdom Stramenopila, which is more closely related to algae. This pathogen is notorious for its ability to spread rapidly under cool, wet conditions. The pathogen produces sporangia, which release zoospores in the presence of water. These motile spores are responsible for initiating new infections on potato plants.

The primary sources of inoculum for late blight are infected seed tubers, volunteer plants (potato plants that grow from tubers left in the field from the previous season), and cull piles (discarded potatoes). The pathogen overwinters in infected tubers and plant debris.

Under favorable conditions, *P. infestans* produces sporangia on the surface of infected plant tissues. These sporangia can be dispersed by wind, rain splash, or contaminated tools and equipment, spreading the disease rapidly across fields. In the presence of free water, sporangia release zoospores, which swim towards plant tissues. These zoospores are highly infectious and can quickly penetrate plant surfaces, particularly through stomata or wounds. Late blight thrives in cool, wet conditions. The optimal temperature for disease development is between 10 °C and 20 °C. High humidity (above 90%) or prolonged leaf wetness (more than 12 h) is critical for sporangia production, zoospore release, and infection. Frequent rainfall or irrigation increases the likelihood of late blight outbreaks by providing the necessary moisture for spore germination and dispersal. Overhead irrigation can exacerbate the problem by increasing leaf wetness. Dense canopies that retain moisture, poorly drained soils, and frequent fog or dew contribute to a microclimate conducive to late blight. Fields with limited airflow are more prone to rapid disease spread.

Sporangia can be carried by the wind over long distances, spreading the disease rapidly from field to field. This airborne spread makes late blight particularly difficult to contain once an outbreak occurs. Zoospores can move through soil water, infecting tubers underground. Contaminated irrigation water or surface runoff can also spread the pathogen within and between fields. The pathogen survives between growing seasons in infected tubers left in the field or in storage. These infected tubers can sprout and produce inoculum, initiating new infections in the following season.

P. infestans produces sporangia, which are spore-bearing structures that form on the surface of infected plant tissue. Sporangia are responsible for spreading the disease to new hosts. Sporangia release zoospores, which are motile spores that swim in water and can infect plant tissues directly. Zoospores are produced in cool, moist conditions and are crucial for initiating infections in new plants. The pathogen survives between seasons in infected tubers, plant debris, and cull piles. In cooler climates, it may survive in soil or on volunteer plants that grow from tubers left in the field (Henfling 1987).

4.7 Strategies for Mitigating Postharvest Pathogen in Fresh Produce

Mitigating diseases after harvesting in fresh produce aims to avoid disease till their consumption, hence, inhibition, suppression, and suspending disease indications at the time of transport and storing need due considerations (Barkai-Golan 2005).

Physical, chemical as well as biological treatments are used to regulate postharvest diseases among fresh produce. Effectiveness of these methods largely depends upon how fast it reaches to organism, the degree of infectivity, as well as host's sensitivity level. These treatments have been described briefly hereunder:

4.7.1 Physical Treatments

Physical treatments for postharvest disease control include temperature management (cooling and heating), magnetic fields, and ionizing radiation. These may involve ultrasound, microwave, infrared, UV, X-rays, gamma rays, and cathode rays—some of which are strongly fungicidal, while others offer limited efficacy (Eckert and Sommer 1967).

4.7.1.1 Use of Ionizing Radiation

The main advantage of ionizing radiation is its ability to kill pathogens without significantly raising the produce's temperature, though it involves high initial capital. Gamma irradiation has been successful by penetrating deeply and inactivating pathogens (Yahaya 2005). Ionizing radiation, or radurization, is widely accepted for its ease of use, lack of residue, and minimal energy requirements. It extends shelf life by pasteurizing surfaces, inhibiting sprouts, and slowing senescence. Examples of its use are provided below.

Irradiation at 25–35 krad effectively delays the natural ripening of bananas without compromising fruit quality. Ionizing radiations prevent sprout growth in potatoes, onions, and other bulbs (10–15 krad).

Jeong and Jeong et al. (2017) reported that combining irradiation with other treatments has been developed. The synergistic effects of gamma irradiation and thermal techniques (38 °C, 96 h) resulted in a five to about ten times increment in spore deactivation of postharvest organisms (Temur and Tiryaki 2013). Thermal technique enhance resistance by pathogens to the host because of prompting the production of repressive substances. The combined use of both significantly decreased fungus growth in various crops.

Combined method of hot water and gamma irradiation in tomatoes resulted in infection rates of 1.7% for *Botrytis cinerea* and 10% for *Rhizopus stolonifer*. Additionally, dipping mangoes in hot water followed by 0.5 kGy irradiation completely prevented decay from *Alternaria alternata* for 8 days at 23 °C (Spalding and Reeder 1986). Similar studies have been earlier reported (El-Samahy et al. 2000; Barkai-Golan et al. 1993).

Another study suggested that combining irradiation with cold storage is successful to control of diseases after harvesting, for instance, in apples (Mostafavi et al. 2012).

However, combining hot water treatment with irradiation is rarely applied to fruits, as it can negatively affect the quality of clementines and has produced inconsistent results across studies (Palou et al. 2007). Palou et al. (2007) reported that sodium carbonate, a cost-effective alternative to synthetic fungicides for controlling citrus postharvest disease, is more effective when combined with an X-ray dose of 0.875 kGy against *Penicillium digitatum* and *Penicillium italicum* than when used alone. Several studies have also demonstrated the inhibitory effect of nanosilver particles (NA) in sterilization (Jung et al. 2014). Although gamma irradiation alone showed no antifungal activity at 1 kGy, it exhibited strong antifungal effects when combined with NA or nanosized silica silver at concentrations above 1 μg/L.

4.7.1.2 Use of Low Temperature or Cooling

Low temperature not only inhibits pathogen growth but also delays the ripening process, making it a vital strategy for minimizing decay in many fresh produces. Controlling or managing the temperature played an important role in reducing physiological decline, avoiding moisture loss and shriveling, and reducing disease incidence. It was well established that the metabolic activities increase by two- or threefolds with every increase in 10 °C temperature in between 0 °C and 30 °C.

Generally, storage at just above the freezing temperature does not harm or cause the chilling injury to the host. The tissues turn brown as the effect of the chilling injury and sometimes disease resistance is reduced. At near 0 °C temperature, generally no pathogenic development has been found in most of the fungi.

However, chilling injury is common in many fruits and vegetables, so storage and transport should generally occur above 0 °C. Chilling injury causes tissues to turn brown and can reduce disease resistance.

There are several factors which affect the efficiency of refrigeration. These factors include—postponing of thermal elimination, nonexistence of satisfactory low temperature resources, taking out the produce from refrigeration to complete the ripening process, unable to utilize the desired level of low temperatures due to susceptibility of the produce to chilling injury, and the capacity of few postharvest organisms in fruits and vegetables to develop at below the freezing point temperature.

4.7.1.3 Thermal Treatments

Killing of pathogenic fungi using heat treatment is old, and first stated by Fawcett (1922) prevention of spoilage in *Citrus*. Heat may be applied to fruits and vegetables in several ways such as hot water dipping (HWD), vapor heat, or hot dry air or by hot water rinsing and brushing (Lurie 1998; Fallik et al. 1996, 2001).

Hot water dips (HWDs) were proven successful to control fungal pathogens because fungal spores and latent infections are typically located on the surface or within the first few cell layers under the peel of fruits and vegetables (Kalamaki et al. 2009). Postharvest dips to control decay are usually applied for just a few minutes at higher temperatures than those used for disinfestation, as only the surface of the produce needs to be heated (Lara et al. 2009).

4.7.2 Chemical Treatments

Chemical control of postharvest diseases of fruits and vegetables has become an integral part of the handling and successful marketing process.

Controlling pathogenic microorganisms that cause losses in fresh produce often require the use of fungicides. In addition to fungicides, various chemicals can extend the storage life of fruits and vegetables by delaying ripening and senescence. Postharvest treatment with gibberellic acid has been shown to significantly delay banana ripening (Hayatu 2000). Potassium permanganate is used as an ethylene absorbent to maintain ethylene levels below the threshold. Maleic hydrazide (MH) has proven highly effective in slowing the ripening process of various fruits.

The effectiveness of chemical treatments for disease control in perishable crops depends on several factors, including the initial spore load, the depth of infection within the host tissues, the infection's growth rate, temperature, humidity, and the chemical's ability to penetrate the host tissues. Additionally, the chemical must be nonphytotoxic (i.e., it should not damage the host tissue) and must comply with local food additive regulations.

Around 20 organic compounds have been extensively studied as postharvest treatments for controlling diseases in perishable crops. First-generation postharvest fungicides, such as sodium o-phenylphenate, dichloran, and sec-butylamine, are effective in preventing decay caused by wound-invading pathogens like *Penicillium* and *Rhizopus*. However, they are less effective against latent and deep-seated infections. Since 1965, newer fungicides have been developed that are more effective against latent infections and offer protective and antisporulant properties for controlling certain postharvest diseases.

Fungicides for controlling postharvest diseases of fruits and vegetables can be applied in various ways, which will be discussed further.

4.7.3 Preharvest Chemical Treatments

Preharvest chemical treatment is useful in those cases where fungicides are unable to penetrate deeply into the tissues. In this method, broad-spectrum protective fungicide was applied to the fruits on the plant in the field itself. Preharvest spraying of fungicides can result in a significant reduction of the pathogens in the harvested produce.

Preventive sprays control the penetration of the fungal zoospores into the fruits, which are on the tree. Additionally, the time of spray is of considerable importance. It should be applied before the rain as germination of the zoospores depends on water (Timmer and Fucik 1975). To prevent anthracnose (*Colletotrichum gloeosporioides*) infection in various tropical and subtropical fruits, protective sprays in the plantation have also been widely used. The casual organisms of anthracnose penetrate the young fruits on the tree and establish a dormant infection therefore spray on the developing fruits checks spore germination and subsequently formation of aspersoria and infective hyphae, which are the inactive stages of the fungus.

4.7.4 Sanitation

Fresh produce usually gets wounded in the process of handling and can come into contact with pathogens. As far as disease proliferation is concerned, the pathogen availability of wound in the host for penetration is essentially requited. Therefore, the absence of either of pathogen or wound will definitely to prevent the disease development (Barkai-Golan 2005). Reduction in wound intensity could be achieved by cautious handling procedures during fruit drop-down phases. Besides, physiological damage because of temperature regulation, lack of O_2, and environment factors favors the attack of organisms. Similarly, careful handling during transit, storage, packaging or storing the fruits or vegetables will play a significant role in reducing magnitude of wounds which ultimately reduces the chances of infection (Barkai-Golan 2005).

4.7.5 Postharvest Chemical Treatments

Injury occurring to fresh fruits and vegetables produce formed at the time of handling cause primary invasion of organisms, and hence, the chemical treatment might noticeably reduce spoilage during storability. Besides, lenticels and stomata, the natural opening in the host surface, are the other potential sensitive points for infection.

Use of aluminum-containing salts having a minimal concentration of 1–10 mM exhibited sturdy inhibition against *Fusarium sambucinum*, *Alternaria solani*,

Pythium sulcatum, *R. stolonifera* and *B. cinerea*. Use of 1 mM of aluminum chloride and aluminum sulfate effectively inhibits the mycelial growth of pathogens by 47 and 100%, respectively. Application of aluminum sulphate having concentration of 5 mM effectively inhibits dry rot and cavity spot. Based on these findings, Kolaei et al. (2013) opined that various aluminum-containing salts can be used as an alternative to synthetic fungicides. Youssef et al. (2014) proved antifungal effect of both sodium carbonate and bicarbonate on *P. digitatum* in citrus fruits.

4.7.6 Growth Regulators

Plant growth regulators are well known to postpone senescence and the onset of fruit rots. Indole acetic acid and MH against *Aspergillus* rot and *Rhizopus* rots of papaya fruits, and planofix (NAA), used at 0.01%, were found effective in all rots except *Fusarium* rot as post inoculation treatment. For effective control of the postharvest fungal diseases caused by *Rhizopus* spp., growth regulators (2,4-D, 2,4,5-T, NAA, ascorbic acid, and gibberellic acid) were found suitable as reported by Thakur et al. (1974). Gibberellic acid, as a postharvest treatment, significantly delayed the ripening of banana (Hayatu 2000). Likewise, Tak et al. (1985) described the effectiveness of rovral (500 ppm), MH (100 ppm), and hydrogenated peanut oil to prevent fruit roting in apples. Potassium permanganate for absorption of ethylene kept ethylene level beneath limit.

The postharvest plant hormones played a major role in shelf-life extension of vegetables. Cytokinins are particularly effective in preserving leafy vegetables, while gibberellins have reported to slow down the ripening process of tomatoes (Hayatu 2000).

4.7.7 Methods of Using Fungicides

For regulating postharvest diseases dipping, electrostatic sprays, dusting, fumigation, and use of chemical pads are various methodologies for application of fungicides. These methods are described hereunder:

4.7.7.1 Dip Method

In this method, the produce or its part is immersed or dipped in water containing an appropriate concentration of a chemical, which is toxic to the disease-causing fungi (Daines 1970). The produce may be passed below a shower of the diluted chemical, which is called as "cascade application." For improving the effectiveness of dips, use of additives or warm waters can be applied. The additives include wetting agents or acids. The wetting agents, e.g., Teepol or Triton X-100 reduce the surface tension and allow a better coating of the chemical on the produce. The acids such as citric acid lowered the pH of the fungicide and can make it more effective. The effectiveness of the fungicidal treatment may be enhanced by heating the water into which fruits are kept. It was found that to the control of mango anthracnose, postharvest dips in fungicides in cold water rarely reduced the level of infection, but when fungicide is applied in hot water, complete control could be achieved. Benomyl (500 ppm) applied in water at 55 °C for 5 min reported to be effective with no damage to the fruit. In this case, the fruits are passed over rollers, which constantly revolve the fruits to confirm even exposure of fruits from all sides to the spray. The cascade applicator is a modification of this method whereby the fungicide is applied as a liquid under the cut crowns, which are placed upwards, because this is the area where the fungi are likely to attack the fruit.

4.7.7.2 Electrostatic Spray

In this method, pesticides broken to small drops, thus assisting in effectual electric charge has been utilized for spraying in fields. Increased uniformity application of spraying solutions as well as loss of their biological activity are key advantages. Electrical sprayed particles exhibit similar electrical charge, thereby causing repulsion among each other. They are attracted towards crop and then form a thin even layer/coat. This technique is being used in managing decay in several fruits in many advanced countries.

4.7.7.3 Dusting

Several types of dust have been shown for prevention of postharvest disorders after application to fresh produce. Fungicide exhibiting certain chemical agents formulated with inert carriers like talc are used for postharvest disease control. For example, these preparations are being loaded in stores in case of potatoes.

4.7.7.4 Fumigation

Fumigation involves filling a sealed space with gaseous pesticides to eliminate pests and pathogens. It is commonly used after harvest to protect fruits during storage, transport, or prior to processing. Sulphur dioxide (SO_2) is widely applied in grapes to control *Botrytis cinerea*, with effective treatments ranging from 5 to 18 ppm or 1% SO_2 for 20 min (Ryall and Harvey 1959). However, SO_2 can be corrosive and may bleach dark grapes at high concentrations. To reduce these risks, sodium metabisulfite pads are used for slow SO_2 release.

Other fumigants include acetaldehyde (5000 ppm for 24 h), which reduced grape decay by 92% without leaving residues (Avissar et al. 1989), and diphenyl wraps in citrus to prevent fungal growth. Compounds like 2-aminobutane and ecnazene have also shown potential in fumigating potatoes and citrus, though their application is technically challenging and often limited to specialized contractors (Eckert 1969; Burden and Wills 1989).

4.7.7.5 Chemical Pads

For the first time, paper pads infused with fungicidal agents were developed specifically for bananas to prevent infections during transportation. Known as crown pads, these innovations are designed to protect the cut crowns of fruits from fungal infections. Each pad is crafted from multiple layers of soft paper, which are soaked in a fungicide—commonly thiabendazole—and then dried.

The efficient approach in using fungicides and antibiotics is to mix them into wax emulsions (Alao 2000; Hayatu 2000). It is crucial to avoid using harmful chemicals whenever possible. If use of such chemicals cannot be avoided, then the consumption of these treated fruits and vegetables will essentially be delayed by at least 1–2 months. This precaution helps mitigate potential health risks, particularly respiratory and digestive disorders, which may arise from consuming contaminated produce (Alao 2000; Hayatu 2000).

4.7.8 Biological Control

Biological control refers to controlling of diseases by applying biological agents to the host, thereby preventing disease progression by spoilage organisms. It can be achieved by (i) use of resistant varieties, (ii) organic product, and (iii) antagonistic organisms, for preventing postharvest diseases in the produce.

4.7.9 Developing and Utilizing Resistant Cultivars

Developing resistant cultivars in contradiction to spoilage organisms is found to be highly dependable, cost efficient as well as ecological method of disease management. However, it is clear from available information that it has received less attention in developing and subsequently using resistant cultivars or the hybrids against postharvest spoilage causing organisms. Using the biotechnological tool or conventional methods, genetically resistant cultivars can be developed which can subsequently be used for production. Incorporation of desirable traits into the susceptible varieties using the varieties resistant to postharvest pathogens in fruit and vegetable varieties is another approach. Usually cultivars having very thin peels, lesser tannins, more sugar, however, but regrettably, associated factor help in vulnerability to the postharvest organisms. It has to be recognized by breeder that the resistance to postharvest disorders is quite different than resistance under field conditions. In this case, resistance is required after harvesting of the produce. Therefore, specific breeding strategies must be developed for utilization in that resistance forms.

4.7.10 Manipulation of Resistance Responses in Harvested Commodities

Many factors accountable to develop resistance within the harvested produce, which are mentioned ahead:

4.7.10.1 Wound Healing

Wound healing in fruits involves cell-mediated deposition of lignin-like substances that enhance resistance to pathogens. In lemons, increased resistance to *G. candidum* was correlated with higher lignin-like deposit at wound sites (Baudoin and Eckert 1985). Apples wounded immediately after harvesting showed faster healing and greater resistance compared to those wounded during storage (Skene 1981).

4.7.10.2 Enzyme Inhibitors

Many researchers like Fielding (1981) and Abu-Goukh and Labavitch (1983) reported that manipulation of inhibitors of pectinolytic enzymes in harvested fruits of peaches, plums, and pears led to effective disease control in fruits.

4.7.10.3 Induction of Resistance

Host defenses in fruits can be activated through chemical, biological, and physical means. Ethylene stimulates phenylalanine ammonia-lyase, a key enzyme in the shikimate pathway that drives the synthesis of defense-related secondary metabolites like phenols, phytoalexins, and lignins (Kuc 1982). Low doses of UV-C light (<280 nm) have been shown to induce resistance against storage rots in citrus, apples, peaches, sweet potatoes, and onions by triggering host defense responses (Chalutz et al. 1992; Lu et al. 1991).

Natural compounds like chitosan, found in fruit waxes, inhibit pathogens such as *R. stolonifer*, *C. gloeosporioides*, *B. cinerea*, and *A. alternata*, while also eliciting phytoalexin production and delaying ripening (El-Ghaouth et al. 1992; Wilson et al. 1994). Biocontrol agents, including *Pichia guilliermondii* and *Candida* spp., induce PAL activity, chitinase expression, and structural defenses such as papillae formation in fruit tissues (Wisniewski and Wilson 1992; Wilson et al. 1994).

These findings highlight the potential of combining physical treatments, natural elicitors, and microbial antagonists as sustainable alternatives for managing postharvest diseases. Further research into such elicitors could enhance biological control strategies in fruits and vegetables.

4.7.10.4 Polyamine Biosynthesis Inhibitors

Polyamines, such as putrescine, spermidine, spermine, and cadaverine, are low molecular weight nitrogenous compounds essential for nucleic acid function and are present across both prokaryotes and eukaryotes. In fungi, their biosynthesis can be disrupted by targeting the enzyme ornithine decarboxylase, without harming plant tissues. Difluoromethylornithine is a selective inhibitor of this enzyme, offering potential to control postharvest diseases without affecting arginine decarboxylase activity (Metcalf et al. 1978; Galston and Kaur Sawhney 1982).

4.7.11 Natural Plant Products for Biological Control

Numbers of constitutive and inducible compounds with antimicrobial properties were present in fruits and vegetables (Sharma and Pongener 2010). These compounds can be used for biological control in or on the plants where they are produced or extracted and applied to other harvested commodities. Natural additives, like essential oils and extracts from fruit, have shown good antimicrobial and antifungal activities with biopolymeric compatibility, thus have potential to be used as edible coating formulations (Kanmani and Rhim 2014). Low production cost, low toxicity, and good biodegradability (Maswada and Abdallah 2013) improved antioxidant properties due to abundance of polyphenols (Silva-Weiss et al. 2014), environmentally friendly and possible to find a semipermeable barrier against gases and water vapor, thereby lowering chances of microbial attack (Dhall 2013) are the several advantages offered by the natural products. Besides, edible coatings can be mix well with natural or synthetic active values to effectively avoid microbial decay (Falguera et al. 2011).

Dubey and Kishore (1988) reported that application of essential oils at concentrations between 500 and 2000 ppm obtained from leaves of *Melanleuca leucadendron*, *Ociumum canum*, and *Citrus medica* were found to save various stored commodities from biodegradation by *Aspergilus flavus* and *Avicularia versicolor*. Many fruit volatiles produced by ripening peaches were found as highly fungicidal (Wilson et al. 1987). Potential of cinnamon oil as natural antifungal was established by (Xing et al. 2010). They found cinnamon oil having concentrations of 2% (v/v) and 3% (v/v) to effectively prevent the growth of fungi viz., *Rhizopus nigricans, A. flavus*, and *P. expansum* in wound-inoculated Lingwu Long Jujube and Sand Sugar Orange fruits under in-vivo conditions. Similarly, lemongrass oil vapor at the concentration of 28 μL L1 was found utmost effective against anthracnose caused by *C. gloeosporioides* in artificially inoculated papaya fruit, with no significant change in quality parameters of papaya (Ali et al. 2015).

Karim et al. (2016) demonstrated the usefulness of methanol and chloroform extracts of eight *Cistaceae* species to prevent citrus sour rot decay, which was caused by *Geotrichum citri-aurantii*. Of these, methanol extracts of these plant species found more effective under both in vitro and in vivo conditions. The methanol *Cistus* extract also found to have antifungal activities. Various compounds like benzaldehyde, methyl salicylate, and ethyl benzoate completely inhibited the growth of *Monilinia fructicola* and *B. cinerea*. Prasad and Stadelbacher (1973) reported that acetaldehyde vapor effectively controls the *B. cinerea* and *R. stolonifer* rots of strawberry and raspberry. Immersion of lemons in an ethanol at 10% (w/v) solution for short duration or more at 44 °C or higher found to lower the green mold incidence by nearly 80% with no significant injury to the fruits (Smilanick et al. 1995). However, the ethanol treatment was found less effective due to brief immersion in a heated sodium carbonate (3%, w/v) solution, particularly when the fruits were held in storage for long periods. Hence, this practice has not been implemented commercially. Immersion of lemons in hydrogen peroxide solutions (5–15%, w/v) reduce in green mold incidence moderately, coupled with more risk of injury to the fruits (Smilanick et al. 1999).

Eight Moroccan medicinal and aromatic plants like *Halimium umbellatum, Hammada scoparia, Ighermia pinifolia, Inula viscosa, Asteriscus graveolens, Bubonium odorum, Thymus leptobotrys, and Anvillea radiata* were extracted with petroleum ether, chloroform, ethyl acetate, and methanol. These plant extracts exhibited potential to control blue mold of citrus, suggesting that more research work is needed to find new compounds with better activity than the chemical fungicides that are currently available (Askarne et al. 2012). The organic solvent extracts of several plants have been tested against citrus pathogens. Treatment of mandarin fruit with methanol extracts of *Cistus villosus*, *Halimium umbellatum*, and *Ceratonia siliqua* effectively control the citrus sour rot (Askarne et al. 2012). Similarly, Gatto et al. (2011) reported effective control of green mold using methanol extracts from *Sanguisorba minor.*

Yang et al. (2013) reported that combination of tea polyphenol with tea saponin lowered lesion diameter of gray mold disease caused *B. cinerea* in grapes. TP and TS have exhibited synergistic antifungal interactions. The inhibitory effect of acetic acid on microorganisms has been found to control postharvest diseases in several fruits and vegetables. Sholberg and Gaunce (1995) have shown that low concentrations of acetic acid in air were extremely effective in controlling of *B. cinerea* conidia on apple fruits with no phytotoxic effect. Fumigation with acetic acid protected grapes from in modified atmosphere packaging at 0 °C for up to 2 months (Moyls et al. 1996). It provides an alternative for extending shelf-life of grapes spoilage. Being natural and widely available compound use of acetic acid in fumigation had advantages as little or no residual hazard at the low levels required to kill fungal spores. Commercial use of acetic acid as effective fumigant was reported on apricots, plums (Liu et al. 2002), grapes (Sholberg et al. 2000), and cherries (Chu et al. 2002), which regulates various aspects of plant development and response to environmental stresses. Jasmonates (jasmonic acid and methyl jasmonates combined together) are naturally occurring plant growth regulators which belongs to class olypines derived from oxygenase-dependent oxidation of fatty acids. Droby et al. (1999) found that postharvest application of jasmonates reduced green mold decay (*P. digitatum*) of Marsh Seedless grapefruit. Tripathi and Dubey (2004) also found jasmonates applied at low concentration, as potential postharvest treatment to enhance natural resistance and to reduce decay in fruits. Hydrolysis of glucosinolates, a natural product of the members of *Crucifereae* family, produce isothiocyanate, thiocyanate, and nitrile, which have been employed successfully in modified atmosphere packaging or in gaseous treatment before storage. Exposure of pear fruit to allyl-isothiocyanate (AITC), a natural occurring flavor compound in mustard, resulted in good control of blue mold, including a TBZ-resistant strain on pears (Mari et al. 2002). The use of AITC, produced from purified sinigrin or from *Brassica juncea*, against *P. expansum* appeared very promising as an economically viable alternative with moderately low impact on environment (Mari et al. 2003).

Tosi et al. (1996) reported that propolis, a natural resinous material obtained from leaf buds and bark of poplar and conifer trees, possess antibacterial and antifungal property. It inhibits the postharvest pathogens, such as *B. cinerea* and *P. expansum* (Lima et al. 1999). Babu and Reddy (1986) have reported that extracts of *Eucalyptus globulua*, *Punica granatum*, *Lawsonia inermis*, and *Datura stramonium* were effectively control the fruit rot of lemon. Similarly, Tulsi leaf has shown capacity to check spore germination, growth, total proteins, and pectinolytic and cellulolytic enzymes of various rot of pathogens.

Mohapotra et al. (2000) found that while low concentrations of phenols (3–5 μg/mL) are utilized by fungi for normal metabolic processes, higher concentrations (20 μg/mL) inhibit fungal growth. Chitosan and its derivatives exhibit notable plant protective and antifungal properties (Allan and Hadwiger 1979). They can be applied in various forms, including solutions, powders, or as wettable coatings for seeds and fruits. Chitosan has also proven effective as an alternative control method against blue mold in harvested red delicious apples (de Capdeville et al. 2002). In vitro antimicrobial tests demonstrated that chitosan hydrochloride (CH) completely inhibited both the mycelial growth of *Botrytis cinerea* and the bacterial growth of *Pectobacterium carotovorum* subsp. *carotovorum*. In vivo studies showed antimicrobial activity lasting 5–7 days on lettuce and up to 20–25 days on kiwi fruits. This coating effectively combats gray mold caused by these pathogens during postharvest phases of fruit and vegetable products. The research highlighted the film-forming ability of CH, which maintains its physicochemical properties even after dissolution in water (Fortunati et al. 2017).

The potential of various concentrations of 24-epibrassinolide (EBR) (EBR1, 0.1 mg L^{-1} EBR; EBR2, 0.4 mg L^{-1} EBR; EBR3, 0.8 mg L^{-1} EBR)-treated table grapes exhibited a higher berry firmness and a lower percentage of weight loss, berry drop, and decayed berries. EBR-treated berries showed an increase in activities of defense-related enzymes including superoxide dismutase, POD, catalase, and PAL compared with control berries inoculated by *B. Cinerea* causing gray mold rot. It was reported that exogenous application of EBR could play a protective role in controlling the gray mold of postharvest grapes and maintaining the postharvest quality during cold storage (Liu et al. 2016).

Some essential oils were found to enhance the storage life of fruits and vegetables by controlling postharvest diseases (Dixit et al. 1995). For example, thymol is an essential oil obtained from thyme (*Thymus capitatus*) and its fumigation at a low concentration (2–4 mg L^{-1}) has reduced the incidence of gray mold rot and brown rot in apricot (Liu et al. 2002) with no phytotoxicity. Perumal et al. (2016) have found thyme oil as an effective fumigant at a concentration of 66.7 μL L^{-1} against postharvest pathogens of mango fruits stored at 25 °C for 6 days. It has potential to be used as an alternate natural fungicide to manage postharvest diseases in mango. Treatment of oranges with the essential oils of *Mentha arvensis*, *Ocimum canum*, and *Zingiber officinale* has been found to control blue mold, thereby enhancing their shelf-life (Tripathi 2001). Similarly, mustard, mobil, castor, and paraffin oils have also been found effective against *Rhizopus* rot of mango. Gupta and Pathak (1990) found groundnut oil to be quite effective against *Alternaria*, *Colletotrichum*, and *Aspergillus* rots of papaya.

Sugar analogs have long been known to effect metabolic processes of fungi and plant cells (Moore 1981). El-Ghaouth et al. (1995) discovered the efficacy of 2-deoxy-D-glucose against apple mold and brown rot of peach. The compound also enhances the biocontrol activity of some antagonists. Homeopathic drugs of plant origin can prevent infection when applied on harvested fruits (Khanna and Chandra 1976). They were found effective against various rots of guavas, mangoes, and citrus.

4.7.12 Use of Microbial Antagonists

Microbial antagonists (natural and artificially induced) have also been reported to controlling postharvest diseases of fruits and vegetables successfully. Both types of antagonists are described below:

4.7.12.1 Natural Antagonists

Antagonists that naturally occur on the surfaces of fruits or vegetable peels are termed natural antagonists. Natural antagonists can be isolated from fruit surfaces such as apples and citrus for use as biological control agents. Chalutz and Wilson (1990) observed that when concentrated washings from citrus fruit surfaces were plated on agar medium, only bacteria and yeast were present. However, after diluting these washings, rot fungi appeared on the agar, suggesting that the yeast and bacteria might be inhibiting fungal growth. Additionally, it has been observed that fruits and vegetables that are washed become more prone to decay compared to those that remain unwashed.

4.7.12.2 Artificially Introduced Antagonists

Those microbial antagonists which are artificially induced postharvest pathogens are known as artificially induced antagonists. They were found more effective in controlling postharvest diseases than other means of biological control.

4.7.12.3 Mode of Action of Microbial Antagonists

Understanding the mode of action of microbial antagonists is crucial for two in developing effective procedures and selection of effective antagonists for specific applications. Current information highlights that antibiotic production is a major mechanism through which antagonists control postharvest diseases. Additionally, antagonists may suppress pathogen activity through nutrient competition, direct parasitism, and potentially by inducing resistance in fruits during postharvest handling.

4.7.12.4 Enhancing the Bio-efficacy of Microbial Antagonists

Efforts to develop nonfungicidal methods for controlling postharvest decay in fruits and vegetables have included various approaches. These methods encompass environmental modifications, such as storing produce at temperatures that inhibit pathogen development, adjusting relative humidity and atmospheric conditions, and using treatments like hot air or water (Smoot and Melvin 1965; Fallik et al. 1995). Other strategies include enhancing resistance through elicitors or irradiation (Stevens et al. 1996), applying generally regarded as safe (GRAS) substances (Smilanick et al. 1997, 1999), using wax formulations (De Lima et al. 2014), and sterilizing fruits and handling water with UV irradiation or ozone (Tukey 1993). Despite these efforts, no single method has proven satisfactory in controlling decay when used alone. Effective commercial control of postharvest diseases requires a higher level of efficiency than managing tree-, field-, or soil-borne diseases. Currently, bio-control agents alone do not achieve this level of control. Various strategies have been recommended to increase the usefulness of bio-control agents.

4.7.12.5 Apply Mixed Cultures

Various researchers (Fukui et al. 1999; Gutesky et al. 2001) advocated the use of antagonistic mixtures to improve the biocontrol effectiveness in fruits and vegetables. Using of mixture of antagonists in lower quantity in apples found to control decay caused by broader spectrum of pathogens (Janisiewicz 1996). The advantages of using an antagonist mixture are clearly established; however, approval from the industry involved in producing biocontrol agents is needed.

4.7.12.6 Addition of Low Level of Fungicides

Adding a low concentration of fungicide to bio-control agents has been advised to increase their effectiveness. Droby et al. (1998) reported that biocontrol products such as ASPIRE and BIOSAVE-110 when combined with small amounts of synthetic fungicides exhibited the complete control. For instance, citrus fruits treated with *Candida oleophila* and thiabendazole at a concentration of 200 mg/mL showed decay control comparable to that achieved with commercial fungicides.

4.7.12.7 Addition of GRAS Chemicals

Those chemicals which are not injurious to human health are known as GRAS (generally regarded as safe) chemicals. These chemicals found effective in controlling postharvest diseases of fruits and vegetables when used with antagonists. Calcium chloride ($CaCl_2$), calcium carbonate ($CaCO_3$), calcium bicarbonate, calcium propionate, and ethylenediaminetetraacetic acid are some examples of effective GRAS chemicals. Use of $CaCl_2$ (2%) has found to enhance the defensive effect of some antagonistic yeasts in pome fruits. *Rhizopus* rot of peach has been controlled effectively by applying an aqueous suspension (108 cfu mL^{-1}) of *Kloeckera apiculata* in 2% $CaCl_2$. Wisniewski et al. (1995) and Zhang et al. (2005) reported that although $CaCl_2$ solution alone did not reduce decays but when its 2% solution applied along with the yeast antagonist *Candida* sp., it increased biocontrol of both gray and blue molds on apples and pears. Droby et al. (1997) reported that use of 68 mM $CaCl_2$ alone and in combination of an antagonist, *P. guilliermondii* strain US-7, reduced the green mold decay in grapefruit by 43% and 97%, respectively.

Janisiewicz and Conway (1998) found that pressure infiltration of 0.27 M $CaCl_2$ into apples improved calcium content of the mesocarp by threefold and also lowered the blue mold decay by 50%. Further, they found that combining $CaCl_2$ pressure infiltration with an antagonist, *P. syringae* (isolate ESC-11 used in Biosave 110), resulted relatively higher reduction in blue mold decay. Smilanick et al. (1995, 1997) reported that application of GRAS chemicals like $NaCO_3$, ethanol, and sodium bicarbonate have been found to reduce conidial germination of *P. digitatum* in many fruits. Combining treatments of 3% $NaCO_3$ with antagonist *P. syringae* (ESC-10) was found relatively better in controlling green mold on citrus in comparison to when applied alone (Smilanick et al. 1999). The efficacy of *Cryptococcus laurentii* for control of green mold was improved when combined with 2% sodium carbonate (Zhang et al. 2004).

Use of ethanol (10%) and ethanol-resistant *S. cerevisiae* strains 1440 or 1749 together were found to reduce the gray mold decay in apples by more than 90% whereas either treatment alone did reduce the decay to lower magnitude (Mari and Guizzardi 1998). Use of ethanol (10%) was also reported to reduce green mold of lemons by less than 5% (Smilanick et al. 1995). Pretreatment of lemons with sodium bicarbonate further reduced green mold decay in light green and yellow lemons (El-Ghaouth et al. 2000). In case of cherry, pre- and postharvest application of

Aureobasidium pulluans in combination with $CaCl_2$ or sodium bicarbonate were found to effectively reduce the spoilage (Ippolito et al. 2005). Wang et al. (2010) reported that in cherry tomatoes, application of $CaCl_2$ further enhanced the effectiveness of the marine antagonist *Rhodosporidium paludigenum* in checking postharvest diseases. They further reported that $CaCl_2$ alone did not have any direct influence on the population growth of *R. paludigenum* in nutrient yeast dextrose broth (NYDB) cultures or in cherry tomato wounds. However, the combined treatment with 1×10^8 cells mL^{-1} of *R. paludigenum* and $CaCl_2$ having concentration 0.5–2% concentration found to have more reduction to black rot caused by *A. alternata* in cherry tomato wounds, significantly higher than those of *R. paludigenum* or $CaCl_2$ alone.

4.7.12.8 Addition of Nutrients and Plant Products

Significant increase in effectiveness of microbial antagonists can be achieved by incorporating nitrogenous compounds such as L-asparagine and L-proline (Janisiewicz 1996; El-Ghaouth et al. 2000), as well as the sugar analog 2-deoxy-D-glucose. For instance, when applied to fruit wounds, the combination of *Candida saitoana* with a low concentration (0.2% w/v) of 2-deoxy-D-glucose effectively controls decay in apples, oranges, and lemons caused by *Botrytis cinerea*, *Penicillium expansum*, and *Penicillium digitatum* (El-Ghaouth et al. 2000). Additionally, Liu et al. (2010) demonstrated that *Hanseniaspora uvarum* combined with TP (Talc powder) or used alone effectively controlled gray mold in grapes, with TP alone proving effective at all concentrations. The combination of TP (at 0.5and 1.0%) with *H. uvarum* (1×10^6 cfu/mL) lead to a lower incidence of gray mold, and TP did not prevent the growth of *H. uvarum in vitro* but significantly increased its population in vivo. Similarly, the combination of TP with *Candida ernobii* was effective against the postharvest disease *Diplodia natalensis* in citrus fruits (Liu et al. 2010). TP's role was dual: it directly inhibited spore germination and mycelial growth of *D. natalensis* while indirectly enhancing *C. ernobii* populations in vivo.

The combination of *Candida saitoana* with 2-deoxy-D-glucose also demonstrated improved curative activity compared to existing microbial biocontrol products. *C. laurentii* applied at 1×10^8 cfu/mL, alone or in combination with methyl jasmonate (200 μmol/L), reduced lesion size caused by *Monilinia fructicola* and *Penicillium expansum* on peach fruits (Yao and Tian 2005). In another study, treatments using *Meyerozyma guilliermondii* strain 443 with 2% gelatin or 2% liquid carnauba wax reduced anthracnose in postharvest papayas by 50%. Electron micrographs of treated fruits showed excellent yeast adhesion to the surface. Formulations based on 2% starch, 2% gelatin, and 2% carnauba wax were found to be the most effective in preventing fungal diseases in postharvest papayas (De Lima et al. 2014).

4.7.12.9 Microbial-Physical Integrated Approaches

Fallik et al. (1995) demonstrated that prestorage hot air treatment at 38 °C for duration of 4 days significantly reduced or eliminated blue and grey molds decay in apples caused by *Penicillium expansum*. This heat treatment effectively eradicated decay in apples that had been inoculated with *P. expansum* for up to 12 h prior to heating. Additionally, applying heat enhanced the efficacy of bio-control using heat-tolerant yeasts when treated apples were inoculated with the pathogen up to 24 h after treatment (Leverentz et al. 2001). Efficiency of *Pseudomonas glathei* in controlling green mold by inhibiting the germination of *Penicillium digitatum* conidia was found to be more by use of heat (Huang et al. 1995).

The use of the antagonistic yeast *C. laurentii*, alone or combined with short hot water dips (HWDs at 55 °C for 30 s), significantly reduced *Rhizopus* rot in strawberries. Individually, HWDs and *C. laurentii* lowered infection rates from 96.7% to 65% and 63.3%, respectively, while their combination further reduced infection to 43.3%, indicating strong synergistic effects (Zhang et al. 2007). Similarly, *Pseudozyma fusiformata* AP6, *Metschnikowia fructicola* AP47, and *Aureobasidium pullulans* PL5, when combined with HWDs (55 °C, 50 s), showed promise in controlling postharvest brown rot in peaches (Zhang et al. 2010). However, environmental variability poses challenges to biocontrol efficacy. Mostafavi et al. (2011) demonstrated that combining gamma irradiation (0.3 kGy) with *P. fluorescens* inhibited *P. expansum* spore germination more effectively than irradiation alone (0.6 kGy), highlighting the potential of integrated treatments for enhanced control.

Irradiating root vegetables like carrots (Mercier et al. 1993) and sweet potatoes (Stevens et al. 1990) with UV-C light (less than wavelength of 280 nm) has been shown to enhance resistance and reduce postharvest decay. However, applying UV-C irradiation to pome, stone, and citrus fruits, especially when combined with bio-control agents, has generally been ineffective (Wilson et al. 1994). This lack of success is primarily due to the narrow range of optimal UV-C doses required for different fruits and variations in resistance based on fruit maturity and storage conditions. While UV-C irradiation combined with biocontrol treatments may help control postharvest decay, its primary benefit is in reducing pathogen propagules through its phytosanitary effects. Lurie et al. (1995) found that the effectiveness of *Pseudomonas guilliermondii* against *Penicillium digitatum* improved when oranges were stored at an optimal low temperature under controlled atmospheric conditions.

4.7.13 Alternate pH Treatments

Postharvest treatments are crucial for ensuring economic sustainability and reducing the risk of foodborne illnesses. However, conventional methods and chemical agents currently employed in the industry often pose significant ecological, well-being, and safety concerns, thereby complicating efforts toward sustainable food systems. There is an urgent need to develop and adopt alternative approaches that are both environmentally friendly and economically viable, particularly those capable of effectively eliminating microbial contamination while minimizing harm to human health and the ecosystem. Hypochlorous acid (HOCl) has emerged as a promising alternative for the postharvest treatment of fruits and vegetables, owing to its strong antimicrobial properties. Moreover, the ability to generate HOCl on-site reduces dependency on single-use plastic packaging and aligns with multiple United Nations Sustainable Development Goals (SDGs), including SDG 3 (Good Health and Well-being), SDG 6 (Clean Water and Sanitation), SDG 7 (Affordable and Clean Energy), SDG 9 (Industry, Innovation and Infrastructure), SDG 11 (Sustainable Cities and Communities), and SDG 13 (Climate Action).

4.8 Conclusions

Managing postharvest diseases is a complex challenge that requires a collaborative approach involving scientists, administrators, and policymakers. Future research should focus on several key areas to improve disease management. Advanced diagnostics using cutting-edge biotechnological tools are essentially required for precisely identifying postharvest diseases. Precise assessment of the economic impact of major fruit rots is crucial for understanding the scale of losses. Developing reliable methods for forecasting disease outbreaks will help in proactive management. Research should explore the potential use of nonpathogenic or attenuated pathogen strains for controlling diseases and identify compounds that can be applied before harvest to enhance fruit resistance to pathogens. Understanding the nature of both constitutive and induced resistance in harvested fruits is important for developing effective treatments. Additionally, improving the efficacy of physical, chemical, biological, and cultural methods, and integrating these methods, will provide a more comprehensive approach to disease management. Biotechnological tools should be utilized to develop fruit varieties with built-in resistance and high-quality characteristics. Strategies to enhance wound defense mechanisms in harvested fruits are also needed. Finally, identifying genes that promote the growth and effectiveness of epiphytic antagonists will contribute to more effective postharvest disease control.

References

Abbas MF, Naz F, Irshad G (2014) Important fungal diseases of potato and their management–a brief review. Mycopathologia 11(1):45–50

Abd El-Rahman AF, Shaheen HA (2016) Biological control of the brown rot of potato, Ralstonia solanacearum and effect of bacterization with antagonists on promotion of potato growth. Egypt J Biol Pest Control 26(4):733–739

Abu-Goukh AA, Labavitch HM (1983) The in vivo role of 'Bartlett' pear fruit polygalacturonase inhibitors. Physiol Plant Pathol 23:123–135

Alao SEL (2000) The importance of post-harvest loss prevention. Paper presented at graduation ceremony of school of food storage technology. Nigerian Stored Products Research Institute, Kano, pp 1–10

Ali A, Wee Pheng T, Mustafa MA (2015) Application of lemongrass oil in vapour phase for the effective control of anthracnose of 'Sekaki' papaya. J Appl Microbiol 118(6):1456–1464

Allan CR, Hadwiger LA (1979) The fungicidal effect of chitosan on fungi of varying cell wall composition. Exp Mycolog 3(3):285–287

Askarne L, Talibi I, Boubaker H, Boudyach EH, Msanda F, Saadi B, Serghini MA, Aoumar AAB (2012) In vitro and in vivo antifungal activity of several Moroccan plants against Penicillium italicum, the causal agent of citrus blue mold. Crop Prot 40:53–58

Avissar I, Marinansky R, Pesis E (1989) Postharvest decay control of grape by acetaldehyde vapour. Acta Hortic 258:655–660

Babu KJ, Reddy SM (1986) Efficacy of some indigenous plant extracts in the control of lemon-rot by two pathogenic fungi. Natl Acad Sci Lett 9(5):133–134

Barkai-Golan R (2001) Postharvest diseases of fruits and vegetables: development and control. Elsevier

Barkai-Golan R (2005) Postharvest diseases of fruits and vegetables. A Division of Reed Elsevier India, Pvt. Limited, New Delhi

Barkai-Golan R, Padova R, Ross I, Lapidot M, Davidson H, Copel A (1993) Combined hot water and radiation treatments to control decay of tomato fruits. Sci Hortic 56(2):101–105

Barmore CR, Brown GE (1982) Spread of Penicillium digitatum and Penicillium italicum during contact between citrus fruits. Phytopathology 72:116–120

Baudoin ABAM, Eckert JW (1985) Development of resistance against Geotrichumcandidum in lemon peel injuries. Phytopathology 74:174–179

Bautista-Baños S, Bosquez-Molina E, Barrera-Necha LL (2014) Rhizopus stolonifer (soft rot). In: Postharvest decay. Academic Presjeons, pp 1–44

Bhowmik PK, Matsui T, Suzuki H, Kosugi Y (2003) Glutamine synthetase in harvested asparagus spears: changes in activity and gene expression during storage. Asian J Plant Sci 2(1):102–107

Braun S, Gevens A, Charkowski A, Allen C, Jansky S (2017) Potato common scab: a review of the causal pathogens, management practices, varietal resistance screening methods, and host resistance. Am J Potato Res 94:283–296

Burden OJ, Wills RBH (1989) Prevention of postharvest food losses: fruits, vegetables and root crops, Training series 17/2. Food and Agriculture Organization of the United Nations, Rome, p 157

Chalutz E, Wilson CL (1990) Postharvest biocontrol of green and blue mold and sour rot of citrus fruit by Debaryomyceshansenii. Plant Dis 74(2):134–137

Chalutz E, Droby S, Wilson CL, Wisniewski ME (1992) UV-induced resistance to postharvest diseases of citrus fruit. J Photochem Photobiol B Biol 15(4):367–371

Chu YF, Sun JIE, Wu X, Liu RH (2002) Antioxidant and antiproliferative activities of common vegetables. J Agric Food Chem 50(23):6910–6916

Coelho AFS, Dias MSDC, Rodrigues MLM, Leal PAM (2010) Controlepós-colheita da antracnose da banana-prata anãtratada com fungicidas e mantida sob refrigeração. Ciência e Agrotecnologia 34:1004–1008

Daines RH (1970) Effects of fungicide dip treatments and dip temperatures on postharvest decay of peaches. Plant Dis Rep 54:764–767

Danladi DK (2000) Nigeria agriculture: The efficiency factor announce this article to your friends, 13, pp 1–3

Das AK, Nerkar S, Kumar A (2018) First report of Phyllostictacitricarpa causing citrus black spot on Citrus sinensis and C. reticulata in India. Plant Dis 102(8):1661

Dashwood EP, Fox RA (1988) Infection of flowers and fruits of red raspberry by Botrytis cinerea. Plant Pathol 37(3):423–430

de Capdeville G, Wilson CL, Beer SV, Aist JR (2002) Alternative disease control agents induce resistance to blue mold in harvested 'Red Delicious' apple fruit. Phytopathology 92(8):900–908

De Lima JR, Marto F, Viana P, Lima FA, Pieniz V, Rocha L (2014) Efficiency of a yeast-based formulation for the biocontrol of postharvest anthracnose of papayas. Summa Phytopathol 40:203–211

Devappa V, Sangeetha CG, Vinay MR, Snehalatharani A, Jhansirani N, Srinivas P (2021) Postharvest diseases of tomato and their management. In: Postharvest handling and diseases of horticultural produce. CRC Press, pp 327–336

Dhall RK (2013) Advances in edible coatings for fresh fruits and vegetables: a review. Crit Rev Food Sci Nutr 53(5):435–450

Dixit SN, Chandra H, Tiwari R, Dixit V (1995) Development of a botanical fungicide against blue mould of mandarins. J Stored Prod Res 31(2):165–172

Droby S, Wisniewski ME, Cohen L, Weiss B, Touitou D, Eilam Y, Chalutz E (1997) Influence of CaCl2 on Penicillium digitatum, grapefruit peel tissue, and biocontrol activity of Pichia guilliermondii. Phytopathology 87(3):310–315

Droby S, Cohen L, Daus A, Weiss B, Horev B, Chalutz E, Katz H, Keren-Tzur M, Shachnai A (1998) Commercial testing of Aspire: a yeast preparation for the biological control of postharvest decay of citrus. BiolControl 12(2):97–101

Droby S, Porat R, Cohen L, Weiss B, Shapiro B, Philosoph-Hadas S, Meir S (1999) Suppressing green mold decay in grapefruit with postharvest jasmonate application. J Am Soc for Hortic Sci 124(2):184–188

Dubey NK, Kishore N (1988) Book of abstracts, exploitation of higher plant products as natural fumigants. In: Proceedings of fifth international congress on plant pathology, Kyoto, Japan, p 423

Eckert JW (1969) Chemical treatments for control of postharvest diseases. World Rev Pest Control 8:116–137

Eckert JW, Eaks IL (1989) Postharvest disorders and diseases of citrus fruit. In: Reuther W, Calavan EC, Carman GE (eds) The citrus industry, vol 5. The University of California, Division of Agricultural Science, Oakland, pp 179–260

Eckert JW, Ratnayake M (1994) Role of volatile compounds from wounded oranges in induction of germination of Penicillium digitatum conidia. Phytopathology 84(7):746–750

Eckert JW, Sommer AF (1967) Control of diseases of fruits and vegetables by postharvest treatment. Annu Rev Phytopathol 5(1):391–428

El-Ghaouth A, Arul J, Asselin A, Benhamou N (1992) Antifungal activity of chitosan on postharvest pathogens: induction of morphological and cytological alterations in Rhizopus stolonifer. Mycol Res 96:769–779

El-Ghaouth A, Wilson CL, Wisniewski ME (1995) Sugar analogs as potential fungicides for postharvest pathogens of apple and peach. Plant Dis 79(3):254–258

El-Ghaouth A, Smilanick JL, Wisniewski M, Wilson CL (2000) Improved control of apple and citrus fruit decay with a combination of Candida saitoana and 2-deoxy-D-glucose. Plant Dis 84(3):249–253

El-Samahy SK, Youssef BM, Askar AA, Swailam HMM (2000) Microbiological and chemical properties of irradiated mango. J Food Saf 20(3):139–156

Errampalli D, Saunders JM, Holley JD (2001) Emergence of silver scurf (Helminthosporiumsolani) as an economically important disease of potato. Plant Pathol 50(2):141–153

Ewané CA, Lepoivre P, de Lapeyre L, de Bellaire LL, Lassois L (2012) Involvement of phenolic compounds in the susceptibility of bananas to crown rot. A review. Biotechnol Agron Soc Environ 16(3):393–404

Falguera V, Quintero JP, Jiménez A, Muñoz JA, Ibarz A (2011) Edible films and coatings: structures, active functions and trends in their use. Trends Food Sci Technol 22(6):292–303

Fallik E, Temkin-Gorodeiski N, Grinberg S, Davidson H (1995) Prolonged low-temperature storage of eggplants in polyethylene bags. Postharvest Biol Technol 5(1–2):83–89

Fallik E, Grinberg S, Alkalai S, Lurie S (1996) The effectiveness of postharvest hot water dipping on the control of grey and black moulds in sweet red pepper (Capsicum annuum). Plant Pathol 45(4):644–649

Fallik E, Tuvia-Alkalai S, Feng X, Lurie S (2001) Ripening characterisation and decay development of stored apples after a short pre-storage hot water rinsing and brushing. Innov Food Sci Emerg Technol 2(2):127–132

FAO (1989) Prevention of postharvest food losses fruits, vegetables and root crops, a training manual, FAO training series no. 17/2. FAO, Rome

Fawcett HS (1922) Packing house control of brown rot. Citrograph 7:232–234

Ferrante A, Incrocci L, Maggini R, Serra G, Tognoni F (2004) Colour changes of fresh-cut leafy vegetables during storage. J Food Agric Environ 2(34):40–44

Fielding A (1981) Natural inhibitors of fungal polygalacturonases in infected fruit tissues. Microbiol 123:377–381

Fonseca JM (2006) Postharvest quality and microbial population of head lettuce as affected by moisture at harvest. J Food Sci 71(2):M45–M49

Fonseca JM (2009) Postharvest handling under extreme weather conditions. In: Florkowski WJ, Prussia SE, Shewfelt RL, Brueckner B (eds) Postharvest handling: a systems approach, Food science and technology series. Academic/Elsevier Inc., New York, pp 539–559

Fonseca JM, Rushing JW, Rajapakse NC, Thomas RL, Riley MB (2006) Potential implications of medicinal plant production in controlled environments: the case of feverfew (Tanacetum parthenium). HortScience 41(3):531–535

Fortunati E, Giovanale G, Luzi F, Mazzaglia A, Kenny JM, Torre L, Balestra GM (2017) Effective postharvest preservation of kiwifruit and romaine lettuce with a chitosan hydrochloride coating. Coat 7(11):196

Franc GD (2007) Potato wart, the American phytopathological society education contributed content. https://www.apsnet.org/edcenter/apsnetfeatures/Pages/PotatoWart.aspx

Fukui R, Fukui H, Alverez AM (1999) Comparisons of single versus multiple bacterial species on biological control of anthurium blight. Phytopathology 89:366–373

Galston AW, Kaur Sawhney R (1982) Polyamines: are they a new class of plant growth regulators? In: Wareing PF (ed) Plant growth substances. Academic, London/New York, pp 451–564

Gatto MA, Ippolito A, Linsalata V, Cascarano NA, Nigro F, Vanadia S, Di Venere D (2011) Activity of extracts from wild edible herbs against postharvest fungal diseases of fruit and vegetables. Postharvest Biol Technol 61(1):72–82

Gul N, Khanzada AM, Khanzada MA, Rajput AQ, Shah GS, Sahito OM (2018) Responses of banana cigar end rot pathogen to chemical fungicides. Plant Prot 2(2):35–44

Gupta AK, Pathak VN (1990) Epidemiology and management of papaya fruit rots. Summa Phytopathol 16:92–105

Gutesky R, Shtienberg D, Elad Y, Dinoor A (2001) Combining biocontrol against to reduce the variability of biological control. Phytopathology 91:621–627

Haggag WM (2010) Mango diseases in Egypt. Agric Biol J North Am 1(3):285–289

Hayatu M (2000) Post-Harvest physiological studies of some selected members of family Solanaceae. Kano, pp 2–25

Hegde GM, Vinay JU, Shamarao J (2021) Recent advances in the management of postharvest diseases of onion and garlic. In: Postharvest handling and diseases of horticultural produce, pp 349–362

Henfling JW (1987) Late blight of potato, vol 4. International Potato Center

Holmes GJ, Eckert JW, Pitt JI (1994) Revised description of Penicillium ulaiense and its role as a pathogen of citrus fruits. Phytopathology 84(7):719–727

Houston J (2006) Evaporation in the Atacama Desert: an empirical study of spatio-temporal variations and their causes. J Hydrol 330(3–4):402–412

Huang Y, Deverall BJ, Morris SC (1995) Postharvest control of green mould on oranges by a strain of pseudomonas glathei and enhancement of its biocontrol by heat treatment. Postharvest Biol Technol 5(1–2):129–137

Ippolito A, Schena L, Pentimone I, Nigro F (2005) Control of postharvest rots of sweet cherries by pre-and postharvest applications of Aureobasidium pullulans in combination with calcium chloride or sodium bicarbonate. Postharvest Biol Technol 36(3):245–252

Janisiewicz W (1996) Ecological diversity, niche overlap, and coexistence of antagonists used in developing mixtures for biocontrol of postharvest diseases of apples. Phytopathology 86(5):473–479

Janisiewicz WJ, Conway WS (1998) Integrating biological control and calcium treatment for controlling. HortScience 33(1):105–109

Jeong RD, Jeong MA, Park MR (2017) Gamma irradiation-induced disease resistance of pear (Pyrus pyrifolia "Niitaka") against Penicillium expansum. J Phytopathol 165(9):626–633

Johnson GI, Mead AJ, Cooke AW, Dean JR (1992) Mango stem end rot pathogens-fruit infection by endophytic colonisation of the inflorescence and pedicel. Ann Appl Biol 120(2):225–234

Jones DR (ed) (2000) Diseases of banana, Abacá and Enset. CABI Publishing

Jung YJ, Kim KT, Kim JY, Yang SY, Lee BG, Kim SD (2014) Bioconcentration and distribution of silver nanoparticles in Japanese medaka (Oryziaslatipes). J Hazard Mater 267:206–213

Kachhwaha M, Chile A, Khare V, Mehta A, Mehta P (1992) A new fruit rot disease of banana. Indian J Mycol Plant Pathol 22(2):211

Kalamaki MS, Alexandrou D, Lazari D, Merkouropoulos G, Fotopoulos V, Pateraki I, Aggelis A, Carrillo-Lopez A, Rubio-Cabetas MJ, Kanellis AK (2009) Over-expression of a tomato N-acetyl-L-glutamate synthase gene (SlNAGS1) in Arabidopsis thaliana results in high ornithine levels and increased tolerance in salt and drought stresses. J Exp Bot 60(6):1859–1871

Kanmani P, Rhim JW (2014) Antimicrobial and physical-mechanical properties of agar-based films incorporated with grapefruit seed extract. Carbohydr Polym 102:708–716

Karathanos VT, Bakalis S, Kyritsi A, Rodis PS (2006) Color degradation of beans during storage. Int J Food Prop 9(1):61–71

Karim H, Boubaker H, Askarne L, Talibi I, Msanda F, Boudyach EH, Saadi B, Ait Ben Aoumar A (2016) Antifungal properties of organic extracts of eight Cistus L. species against postharvest citrus sour rot. Lett Appl Microbiol 62(1):16–22

Khanna KK, Chandra S (1976) Control of tomato fruit rot caused by Fusarium roseum with homoeaopathic drugs. Indian Phytopathol 29:269–272

Kinay P, Yildiz F, Sen F, Yildiz M, Karacali I (2005) Integration of pre-and postharvest treatments to minimize Penicillium decay of Satsuma mandarins. Postharvest Biol Technol 37(1):31–36

Kolaei EA, Cenatus C, Tweddell RJ, Avis TJ (2013) Antifungal activity of aluminium-containing salts against the development of carrot cavity spot and potato dry rot. Ann Appl Biol 163(2):311–317

Kuc J (1982) Induce immunity to plant disease. Bios 32:854–860

Lara MV, Borsani J, Budde CO, Lauxmann MA, Lombardo VA, Murray R, Andreo CS, Drincovich MF (2009) Biochemical and proteomic analysis of 'Dixiland' peach fruit (Prunus persica) upon heat treatment. J Exp Bot 60(15):4315–4333

Lassois L, Jijakli MH, Chillet M, Lapeyre D, de Bellaire L (2010) Crown rot of bananas: preharvest factors involved in postharvest disease development and integrated control methods. Plant Dis 94(6):648–658

Le D, Ameye M, De Boevre M, De Saeger S, Audenaert K, Haesaert G (2021) Population, virulence, and mycotoxin profile of *Fusarium* sp. Associated with basal rot of *Allium* sp. In Vietnam. Plant Dis 105(7):1942–1950

Leverentz B, Conway WS, Alavidze Z, Janisiewicz WJ, Fuchs Y, Camp MJ, Chighladze E, Sulakvelidze A (2001) Examination of bacteriophage as a biocontrol method for Salmonella on fresh-cut fruit: a model study. J Food Prot 64(8):1116–1121

Lima G, Arru S, De Curtis F, Arras G (1999) Influence of antagonist, host fruit and pathogen on the biological control of postharvest fungal diseases by yeasts. J Ind Microbiol Biotechnol 23(3):223–229

Liu WT, Chu CL, Zhou T (2002) Thymol and acetic acid vapors reduce postharvest brown rot of apricots and plums. HortScience 37(1):151–156

Liu HM, Guo JH, Cheng YJ, Luo L, Liu P, Wang BQ, Deng BX, Long CAA (2010) Control of gray mold of grape by Hanseniasporauvarum and its effects on postharvest quality parameters. Ann Microbiol 60:31–35

Liu Q, Xi Z, Gao J, Meng Y, Lin S, Zhang Z (2016) Effects of exogenous 24-epibrassinolide to control grey mould and maintain postharvest quality of table grapes. Int J Food Sci Technol 51(5):1236–1243

Lu JY, Stevens C, Khan VA, Kabwe MK, Wilson CL (1991) The effect of ultraviolet irradiation on shelf life and ripening of peaches and apples. J Food Qual 14:299–305

Lurie S (1998) Postharvest heat treatments. Postharvest Biol Technol 14(3):257–269

Lurie S, Droby S, Chalupowicz L, Chalutz E (1995) Efficacy of Candida oleophila strain 182 in preventing Penicillium expansum infection of nectarine fruits. Phytoparasitica 23:231–234

Mari M, Guizzardi M (1998) The postharvest phase: emerging technologies for the control of fungal diseases. Phytoparasitica 26:59–66

Mari M, Leoni O, Iori R, Cembali T (2002) Antifungal vapour-phase activity of allyl-isothiocyanate against Penicillium expansum on pears. Plant Pathol 51(2):231–236

Mari M, Bertolini P, Pratella GC (2003) Non-conventional methods for the control of post-harvest pear diseases. J Appl Microbiol 94(5):761–766

Masudi S, Bonjar GHS (2012) Fulfillment of Koch's postulates for in vitro pathogenicity of Musicillium theobromae (Turconi) Zare and W. Gams as the cause of banana cigar end rot disease. J Plant Prot Res 52(4):410–414

Maswada HF, Abdallah SA (2013) In vitro antifungal activity of three geophytic plant extracts against three post-harvest pathogenic fungi. Pak J Biol Sci 16(23):1698–1705

McKay CP, Friedmann EI, Gómez-Silva B, Cáceres-Villanueva L, Andersen DT, Landheim R (2003) Temperature and moisture conditions for life in the extreme arid region of the Atacama Desert: four years of observations including the El Nino of 1997–1998. Astrobiology 3(2):393–406

McNicol RJ, Williamson B (1989) Systemic infection of black currant flowers by Botrytis cinerea and its possible involvement in premature abscission of fruits. Ann Appl Biol 114(2):243–254

Mercier J, Arul J, Ponnampalam R, Boulet M (1993) Induction of 6-methoxymellein and resistance to storage pathogens in carrot slices by UV-C. J Phytopathol 137(1):44–54

Merz U, Falloon RE (2009) Powdery scab of potato—increased knowledge of pathogen biology and disease epidemiology for effective disease management. Potato Res 52:17–37

Metcalf BW, Bey P, Danzin C, Jung MJ, Casara P, Vevert JP (1978) Catalytic irreversible inhibition of mammalian ornithine decarboxylase (EC 4.1. 1.17) by substrate and product analogs. J Am Chem Soc 100(8):2551–2553

Mohapotra NP, Pati SP, Ray RC (2000) In vitro inhibition of Botryodiplodia theobromae (Pat.) causing Java black rot in sweet potato by phenolic compounds. Ann Plant. Prot Sci 8(1):106–109

Moore D (1981) Effects of hexose analogues on fungi: mechanisms of inhibition and of resistance. New Phytol 87(3):487–515

Mostafavi HA, Mirmajlessi SM, Fathollahi H, Minassyan V, Mirjalili SM (2011) Evaluation of gamma irradiation effect and pseudomonas flourescens against Penicillium expansum. Afr J Biotechnol 10(54):11290–11293

Mostafavi HA, Mirmajlessi SM, Mirjalili SM, Fathollahi H, Askari H (2012) Gamma radiation effects on physico-chemical parameters of apple fruit during commercial post-harvest preservation. Radiat Phys Chem 81(6):666–671

Moyls AL, Sholberg PL, Gaunce AP (1996) Modified-atmosphere packaging of grapes and strawberries fumigated with acetic acid. HortScience 31(3):414–416

Nag PK, Nag A, Ashtekar SP (2007) Thermal limits of men in moderate to heavy work in tropical farming. Ind Health 45(1):107–117

Naqvi SAMH, Singh S (2002) Fungal diseases of citrus: diagnosis and management. Technical Bulletin 5, 61

Opadokun JS (1987) Reduction of post-harvest losses in fruits and vegetables. In: Joint national crop protection workshop. IAR, Zaria, pp 3–26

Pailly O, Tison G, Amouroux A (2004) Harvest time and storage conditions of 'Star Ruby' grapefruit (Citrus paradisi Macf.) for short distance summer consumption. Postharvest Biol Technol 34(1):65–73

Palou L, Marcilla A, Rojas-Argudo C, Alonso M, Jacas JA, del Río MÁ (2007) Effects of X-ray irradiation and sodium carbonate treatments on postharvest Penicillium decay and quality attributes of clementine mandarins. Postharvest Biol Technol 46(3):252–261

Perumal AB, Sellamuthu PS, Nambiar RB, Sadiku ER (2016) Antifungal activity of five different essential oils in vapour phase for the control of Colletotrichum gloeosporioides and Lasiodiplodia theobromae in vitro and on mango. Int J Food Sci Technol 51(2):411–418

Prakash OM (2004) Diseases and disorders of mango and their management. In: Diseases of fruits and vegetables volume I: diagnosis and management. Springer Netherlands, Dordrecht, pp 511–619

Prakash K, Vijayakumar RM, Balamohan TN, Sundhar Singh SD (2011) Effect of drip irrigation regimes and fertigation levels on yield and quality of mango cultivar 'alphonso' under ultra high density planting. In: Global conference on augmenting production and utilization of mango: biotic and abiotic stresses. 1066p. pp 147–150

Prasad K, Stadelbacher GJ (1973) Control of postharvest decay of fresh raspberries by acetaldehyde vapor. Plant Dis Rep 57(9):795–797

Rapaka VK, Faust JE, Dole JM, Runkle ES (2007) Effect of time of harvest on postharvest leaf abscission in lantana (Lantana camara L.'Dallas Red') unrooted cuttings. HortScience 42(2):304–308

Rico D, Martin-Diana AB, Frias JM, Barat JM, Henehan GTM, Barry-Ryan C (2007) Improvement in texture using calcium lactate and heat-shock treatments for stored ready-to-eat carrots. J Food Eng 79(4):1196–1206

Rivera-Vargas LI, Lugo-Noel Y, McGovern RJ, Seijo T, Davis MJ (2006) Occurrence and distribution of Colletotrichum spp. on mango (Mangifera indica L.) in Puerto Rico and Florida, USA. Plant Pathol J 5

Ryall AL, Harvey JM (1959) The cold storage of Vinifera grapes. United States Department of Agriculture Handbook, p 159

Sang MK, Han GD, Oh JY, Chun SC, Kim KD (2014) Penicillium brasilianum as a novel pathogen of onion (Allium cepa L.) and other fungi predominant on market onion in Korea. Crop Prot 65:138–142

Scalzo RL, Bianchi G, Genna A, Summa C (2007) Antioxidant properties and lipidic profile as quality indexes of cauliflower (Brassica oleracea L. var. botrytis) in relation to harvest time. Food Chem 100(3):1019–1025

Schick JL, Toivonen PM (2002) Reflective tarps at harvest reduce stem browning and improve fruit quality of cherries during subsequent storage. Postharvest Biol Technol 25(1):117–121

Schlagbauer HE, Holz G (1990) Infection and colonization of plum blossoms by Monilinialaxa. Phytophylactica 22(4):419–422

Sfakiotakis E, Chlioumis G, Gerasopoulos D (2005) Preharvest chilling reduces low temperature breakdown incidence of kiwifruit. Postharvest Biol Technol 38(2):169–174

Sharma RR, Pongener A (2010) Natural products for postharvest decay control in horticultural produce: a review. Stewart Postharvest Rev 6(4):1–9

Sholberg PL, Gaunce AP (1995) Fumigation of fruit with acetic acid to prevent postharvest decay. HortScience 30(6):1271–1275

Sholberg P, Haag P, Hocking R, Bedford K (2000) The use of vinegar vapor to reduce postharvest decay of harvested fruit. HortScience 35(5):898–903

Shukla PK, Adak T (2017) Anthracnose disease dynamics of mango orchards in relation to humid thermal ratio under subtropical climatic condition. J Agrometeorol 19(1):56–61

Shukla PK, Adak T, Misra AK (2016) Weather based models in predicting powdery mildew of mango for Malihabad mango belt. J Eco Friendly Agric 11(2):142–148

Shukla PK, Tahseen F, Gundappa B, Nidhi K (2021) Postharvest diseases of mango and their management. In: Postharvest handling and diseases of horticultural produce. CRC Press, pp 189–200

Silva-Weiss A, Bifani V, Ihl M, Sobral PJDA, Gómez-Guillén MC (2014) Polyphenol-rich extract from murta leaves on rheological properties of film-forming solutions based on different hydrocolloid blends. J Food Eng 140:28–38

Skene DS (1981) Wound healing in apple fruits; the anatomical response of Cox's Orange Pippin at different stages of development. J Hortic Sc 56:145–153

Smilanick JL, Margosan DA, Henson DJ (1995) Evaluation of heated solution of sulfur dioxide, ethanol, and hydrogen peroxide to control postharvest green mould of lemons. Plant Dis 79:742–747

Smilanick JL, Michael IF, Mansour MF, Mackey BE, Margosan DA, Flores D, Weist CF (1997) Improved control of green mold of citrus with imazalil in warm water compared with its use in wax. Plant Dis 81(11):1299–1304

Smilanick JL, Margosan DA, Mlikota F, Usall J, Michael IF (1999) Control of citrus green mold by carbonate and bicarbonate salts and the influence of commercial postharvest practices on their efficacy. Plant Dis 83(2):139–145

Smilanick JL, Erasmus A, Palou L (2019) Citrus fruits. CRC Press, Postharvest pathology of fresh horticultural produce, pp 3–54

Smoot JJ, Melvin CF (1965) Reduction of citrus decay by hot-water treatment. Plant Dis Rep 49:463–467

Snehalatharani A, Devappa V, Sangeetha CG (2021) Postharvest diseases of banana and their management. In: Postharvest handling and diseases of horticultural produce. CRC Press, pp 201–210

Snowdon AL (1990) A colour atlas of post-harvest diseases and disorders of fruits and vegetables Volume 1: General introduction and fruits. Wolfe Publishing, p 302

Solovchenko AE, Avertcheva OV, Merzlyak MN (2006) Elevated sunlight promotes ripening-associated pigment changes in apple fruit. Postharvest Biol Technol 40(2):183–189

Spalding DH, Reeder WF (1986) Decay and acceptability of mangos treated with combinations of hot water, imazalil, and gamma radiation. Plant Dis 70

Stevens C, Khan VA, Tang AY, Lu JY (1990) The effect of ultraviolet radiation on mold rots and nutrients of stored sweet potatoes. J Food Prot 53(3):223–226

Stevens C, Wilson CL, Lu JY, Khan VA, Chalutz E (1996) Plant hormones induced by ultraviolet light-C for control ling postharvest disease of tree fruits. Crop Prot 15:129–134

Stine SW, Song I, Choi CY, Gerba CP (2005) Application of microbial risk assessment to the development of standards for enteric pathogens in water used to irrigate fresh produce. J Food Prot 68(5):913–918

Swinburne TR (1973) The resistance of immature Bramley's seedling apples to rotting by Nectrixgalligena Bres. In: Byrde RJW, Cutting CV (eds) Fungal pathogenicity and the plant's response. Academic, London/New York, pp 365–382

Swinburne TR (1983) Quiescent infections in post-harvest diseases. In: Dennis C (ed) Post-harvest pathology of fruits and vegetables. Academic, London, pp 1–21

Tak SK, Verma OP, Pathak VN (1985) Control of Alternaria rot of apple fruits by postharvest application of chemicals. Indian Phytopathol 38(3):471–474

Tarnowski TL, Pérez-Martínez JM, Ploetz RC (2010) Fuzzy pedicel: a new postharvest disease of banana. Plant Dis 94(5):621–627

Temur C, Tiryaki O (2013) Effect of Gamma Irradiation on Penicillium expansum Isolated from "Golden Delicious" Apples. The J Turk Phytopathol 40(1-2-3):1–12

Thakur DP, Chennulu VV, Kanwar ZS, Kherata RBS (1974) Growth regulators in control of post-harvest fungal disease of fruits and vegetables. Indian Phytopathol 27:532–536

Timmer LW, Fucik JE (1975) The effect of rainfall, drainage, tree spacing and fungicide application on the incidence of citrus brown rot. Phytopathology 65(3):241–242

Timmer LW, Zitko SE, Gottwald TR, Graham JH (2000) Phytophthora brown rot of citrus: temperature and moisture effects on infection, sporangium production, and dispersal. Plant Dis 84(2):157–163

Tiwari RK, Kumar R, Sharma S, Sagar V, Aggarwal R, Naga KC, Lal MK, Chourasia KN, Kumar D, Kumar M (2020) Potato dry rot disease: current status, pathogenomics and management. 3 Biotech 10:1–18

Tiwari RK, Kumar R, Sharma S, Naga KC, Subhash S, Sagar V (2021) Continuous and emerging challenges of silver scurf disease in potato. Int J Pest Manag 68(1):89–101

Tosi B, Donini A, Romagnoli C, Bruni A (1996) Antimicrobial activity of some commercial extracts of propolis prepared with different solvents. Phytother Res 10(4):335–336

Tripathi P (2001) Evaluation of some plant products against fungi causing postharvest diseases of some fruits. (Ph.D. thesis), Department of Botany, Banaras Hindu University, Banaras, India

Tripathi P, Dubey NK (2004) Exploitation of natural products as an alternative strategy to control postharvest fungal rotting of fruit and vegetables. Postharvest Biol Technol 32(3):235–245

Tukey B (1993) Overview of ozone use at Snokist Growers. Tree Fruit Postharvest J 4:14–15

van de Vossenberg BT, Prodhomme C, Vossen JH, van der Lee TA (2022) Synchytriumendobioticum, the potato wart disease pathogen. Mol Plant Pathol 23(4):461–474

Van Gijsegem F, Hugouvieux-Cotte-Pattat N, Kraepiel Y, Lojkowska E, Moleleki LN, Gorshkov V, Yedidia I (2021) Molecular interactions of Pectobacterium and Dickeya with plants. In: Plant diseases caused by dickeya and pectobacterium species, pp 85–147

Vinay JU, Jahagirdar S, Devappa V, Ashtaputre SA, Singh D (2021) Recent advances in postharvest handling and management of citrus diseases. In: Postharvest handling and diseases of horticultural produce, pp 175–188

Wang Y, Yu T, Xia J, Yu D, Wang J, Zheng X (2010) Biocontrol of postharvest gray mold of cherry tomatoes with the marine yeast Rhodosporidium paludigenum. Biol Control 53(2):178–182

Wilson CL, Franklin JD, Pusey PL (1987) Biological control of Rhizopus rot of peach with Enterobacter cloacae. Phytopathology 77(2):303–305

Wilson CL, El-Ghaouth A, Chalutz E, Droby S, Stevens C, Lu JY, Kahn VA, Arul J (1994) Biological control of postharvest diseases of fruits and vegetables: an emerging technology. Annu Rev Phytopathol 27:425–441

Wisniewski ME, Wilson CL (1992) Biological control of postharvest diseases of fruits and vegetables: recent advances. HortScience 27(2):94–98

Wisniewski M, Droby S, Chalutz E, Eilam Y (1995) Effects of Ca2+ and Mg2+ on Botrytis cinerea and Penicillium expansum in vitro and on the biocontrol activity of Candida oleophila. Plant Pathol 44(6):1016–1024

Wolf FA (1926) The perfect stage of the fungus which causes melanose of citrus. J Agric Res 33(7–12):621

Xing Y, Li X, Xu Q, Yun J, Lu Y (2010) Antifungal activities of cinnamon oil against Rhizopus nigricans, Aspergillus flavus and Penicillium expansum in vitro and in vivo fruit test. Int J Food Sci Technol 45(9):1837–1842

Yahaya SM (2005) Contribution of harvest to pathogenic and non-pathogenic losses of vegetables grown in Kano State-Nigeria. Bayero University, Kano, p 12

Yang XP, Jiang XD, Chen JJ, Zhang SS (2013) Control of postharvest grey mould decay of nectarine by tea polyphenol combined with tea saponin. Lett Appl Microbiol 57(6):502–509

Yao HJ, Tian SP (2005) Effects of a biocontrol agent and methyl jasmonate on postharvest diseases of peach fruit and the possible mechanisms involved. J Appl Microbiol 98(4):941–950

Yenjit P, Intanoo W, Chamswarng C, Siripanich J, Intana W (2004) Use of promising bacterial strains for controlling anthracnose on leaf and fruit of mango caused by Colletotrichum gloeosporioides. Walailak J Sci Technol 1(2):56–69

Yetisir H, Caliskan ME, Soylu S, Sakar M (2006) Some physiological and growth responses of watermelon [Citrullus lanatus (Thunb.) Matsum. and Nakai] grafted onto Lagenaria siceraria to flooding. Environ Exp Bot 58(1–3):1–8

Youssef K, Sanzani SM, Ligorio A, Ippolito A, Terry LA (2014) Sodium carbonate and bicarbonate treatments induce resistance to postharvest green mould on citrus fruit. Postharvest Biol Technol 87:61–69

Zhang HY, Fu CX, Zheng XD, He D, Shan LJ, Zhan X (2004) Effects of Cryptococcus laurentii (Kufferath) Skinner in combination with sodium bicarbonate on biocontrol of postharvest green mold decay of citrus fruit. Bot Bull Acad Sin 45:159
Zhang H, Zheng X, Fu C, Xi Y (2005) Postharvest biological control of gray mold rot of pear with Cryptococcus laurentii. Postharvest Biol Technol 35(1):79–86
Zhang H, Zheng X, Wang L, Li S, Liu R (2007) Effect of yeast antagonist in combination with hot water dips on postharvest Rhizopus rot of strawberries. J Food Eng 78(1):281–287
Zhang D, Spadaro D, Garibaldi A, Gullino ML (2010) Selection and evaluation of new antagonists for their efficacy against postharvest brown rot of peaches. Postharvest Biol Technol 55(3):174–181

Chapter 5
Non-destructive Quality Monitoring of Fruits and Vegetables

Havalli Bommegowda Rashmi and Pradeep Singh Negi

5.1 Introduction

Fruits and vegetables, being nutritious are mostly preferred for consumption in fresh form, however, they are perishable in nature and slightest deviation in postharvest management will impact their marketability. Consumers are not only concerned on the quality parameters of fruits and vegetables but also on their safety, pesticide residues and toxin retentions. Quality monitoring of fruits and vegetables is the most crucial step before they reach to local or export market or processing industries and finally to consumers. The assessment of maturity indices based on end use becomes the prominent stage for the fruit and vegetable value chain management, influencing quality, palate, and salability of the produce (Naik et al. 2017).

Monitoring of quality is crucial for maintaining freshness, assurance of nutritive value retention, ensuring safety, and optimizing processing and storage capabilities of fruits and vegetables. Various non-destructive techniques are employed in the horticultural value chain to assess the quality of produce without causing damage. Firmness is a key indicator of freshness and shelf life. Firmness testing methods, such as acoustic firmness measurement, assess the mechanical properties of fruits and vegetables without causing damage. These non-destructive methodologies have a decisive role in fruit physiology research, post-harvest management, and quality control, enabling scientists, growers, and distributors to monitor fruit health, development, and ripening processes accurately.

H. B. Rashmi
Department of Postharvest Management, College of Horticulture, University of Horticultural Sciences, Bagalkot, Mysuru, India

P. S. Negi (✉)
Fruit and Vegetable Technology Department, CSIR-Central Food Technological Research Institute, Mysuru, India
e-mail: psnegi@cftri.res.in

OP Chauhan (ed.), *Fruits and Vegetables Technologies*,
https://doi.org/10.1007/978-981-96-8433-5_5

5.2 Fruit and Vegetable Quality

Quality can be defined as the degree of excellence. Quality parameters of fruits and vegetables include external and internal quality factors as depicted in Fig. 5.1. Overall quality parameters include size, shape, colour, freshness, appearance, texture, flavour (taste and aroma), nutrition, dry matter, oil, and freeness from defects. Post-harvest biochemical and physiological deterioration of crops affects their garden-fresh form, taste, aroma, mass, turgidity, and nutritive significance (Elik et al. 2019), which need to be monitored both in the standing plant in the orchards and after harvest at the warehouse.

5.3 Need and Assessment of Quality Parameters by Various Non-destructive Methods (NDM)

Non-destructive methods provide continuous gauging of biochemical and physical information of fruits and vegetables devoid of any damage to the internal or external content of crops. It compiles both quantitative and qualitative data of the final quality assessment of crops at one instance without separate analysis (El-Mesery et al. 2019).

Non-destructive analysis is done on the external skin layer of the fruits and vegetables, devoid of any invasive procedure impacting nutritive value and quality.

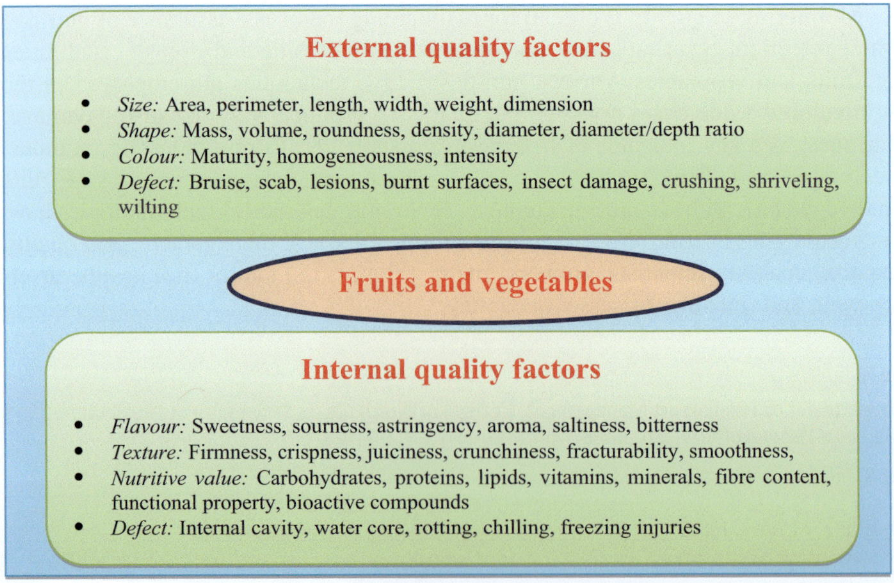

Fig. 5.1 Quality parameters of fruits and vegetables

These methods provide information on fruit and vegetable physiognomies like structural, mechanical, and bio-physico-chemical attributes, which are specific to their variety and species (El-Mesery et al. 2019).

The application of conventional destructive quality monitoring techniques is more labour intensive, slow and time consuming, destructive to the fruit and vegetable tissues, and requisite of pre-analysis material preparation (Magwaza et al. 2013; Kyriacou and Rouphael 2018). Owing to the above reasons, there is a great demand to design non-invasive, dynamic, fast, easy, and accurate analysis procedures for regulating quality and safety (Liu et al. 2015), with nominal test sample preparation (Cozzolino 2022). NDM for quality monitoring examines external and internal properties of fruits and vegetables without inducing destruction to their inherent nature (El-Mesery et al. 2019). The application of spectral quantification methodologies in the quality evaluation is emerging immensely in the current period (Escárate et al. 2022). Adoption of bio-sensing approaches as a favourable replacement for conventional quality assessment is the need of the hour (Okere et al. 2022). Non-destructive instrumental sensory analysis consists of various techniques to measure colour, appearance, flavour, texture, and nutritive value. NDM provides accurate readings, which can be correlated to the physico-chemical parameters of fruits and vegetables, so that researchers can gain a mechanistic understanding of observed differences. These methods are more sensitive to minute variations between samples, hence can detect any inclinations in quality deterioration in fruits and vegetables before human detection (Brosnan and Sun 2004). NDM produces huge data on continuous mode; hence, making them excellent monitors in quality control operations.

To overcome the demerits of destructive quality assessment methods, scientists are seeking the assistance of high-tech-non-destructive methods such as computer vision methods, leveraging image processing, biosensors, nano-sensors, and AI models to automated optimization of maturity indices of crops (Cardenas-Perez et al. 2017; Dhanya et al. 2022; Xiao et al. 2023).

5.4 Non-destructive Techniques

In the field of optics, techniques such as image analysis measure quality parameters like size, shape, colour, and exterior defects. Reflectance, transmittance, and absorbance spectroscopy cover colour, phyto compounds, and interior defects, whereas laser spectroscopy techniques analyse texture, visco-elasticity, defects, and shape. In the field of dynamics, techniques such as vibrated excitation (texture, visco-elasticity, ripening index), sonic (texture, visco-elasticity, interior cavity, density), ultrasonic (interior cavity, structure, texture, tenderness), X-ray, and CT (interior cavity, structure, ripening index) are utilized to assess the quality of fruits and vegetables. The electro-magnetic field utilizes techniques such as impedance (water, density, sugar, interior cavity) and MRI (sugar, lipids, water, interior defects, and structure) for non-destructive quality assessment. Some commonly used NDMs (Fig. 5.2) are summarized below.

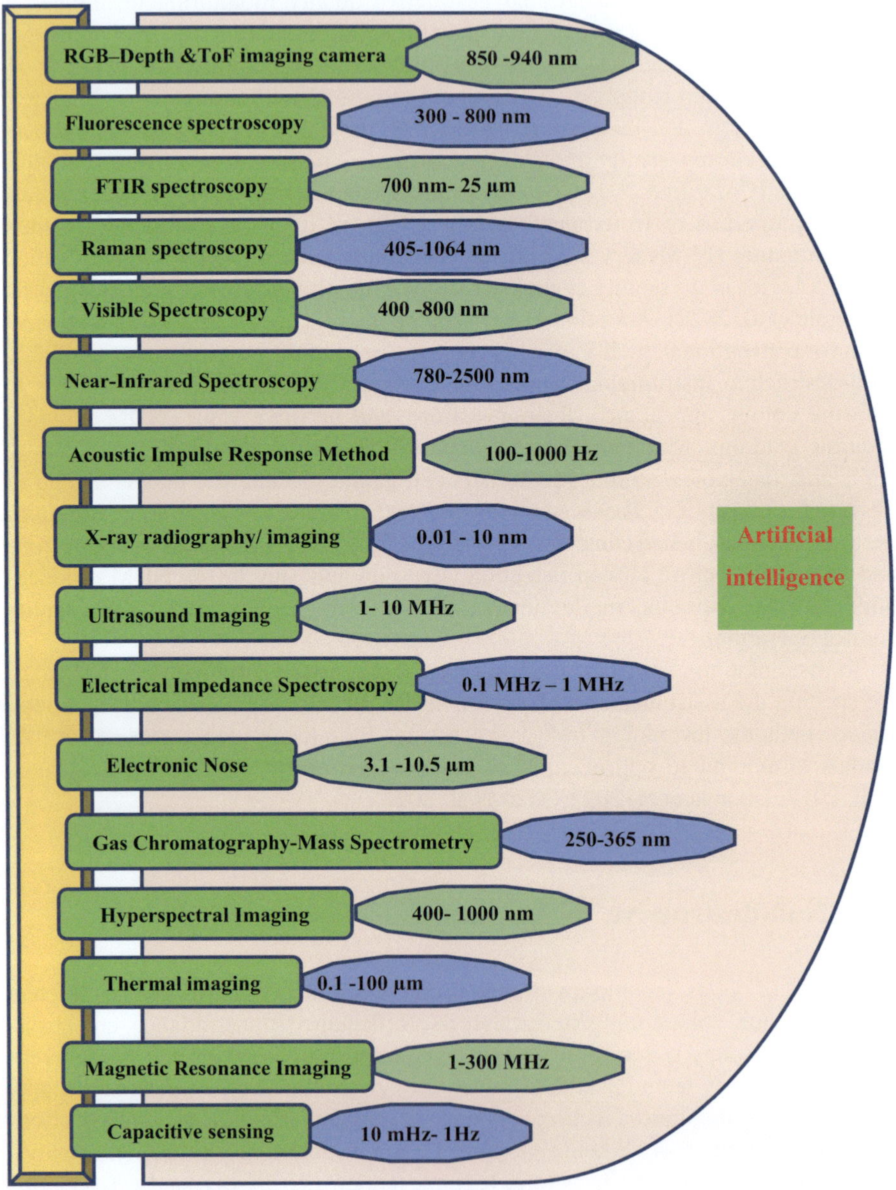

Fig. 5.2 Different non-destructive techniques based on spectrum

5.4.1 Visual Inspection

Visual inspection of fruits and vegetables is done for physical attributes such as colour, glossiness, size, shape, and surface defects. Being a subjective method, it is an important initial step in quality assessment. Colour is perceived by the human eye's retina when light is reflected from the fruits and vegetables, and there is no colour without light. Colour perception depends upon the type and intensity of light, physical and chemical attributes of fruits and vegetables, and a person's ability to categorize colour range. Colour evaluation can be either a subjective or an objective method of inspection. In the subjective method, humans can perceive the colours based on their perception or with the aid of colour charts. The advantage of this method is it is faster and easier than the objective method. It requires no sophisticated equipment to visualize. Colour charts will gives exact references for matching the colours as in crops like banana, tomatoes, and nectarines. Colour perception entirely depends on the available light quality and quantity when perceiving the colour. Glossiness is the visual feature of quality that depends on the ability of a surface to reflect light. Freshly harvested crops exhibit a bright and glossy surface, which will later disappear due to physiological loss of weight and other post-harvest handling measures.

Uniform and characteristic shapes are the most reliable quality characteristics. Misshapen and malformed products might be susceptible to mechanical injury, hence avoided by customers. The size of the commodity is important based on its intended utility. Surface defects such as cuts, bruises, disease and pest injuries, low temperature injury symptoms, and physiological disorders need to be evaluated by a scoring system with 1 to 5 or 10 scores rating from no defect to severely defective. Consumers assume larger size as good quality and large fruits are at the appropriate stage of maturity, hence, crops need to be thoroughly visually examined and assessed as non-destructive quality monitoring (Bhargava and Bansal 2021).

5.4.2 Colorimetric Method

Colorimetric method measures the colour of fruits and vegetables using a colorimeter or spectrophotometer. Colorimeters are typical instruments used for regular quality assessments, whereas spectrophotometers record wavelength-by-wavelength spectra of the reflecting and/or transmitting properties of samples (Hunter Lab 1995). They provide quantitative data on parameters such as skin colour, which can indicate the degree of ripeness, maturity, and potential defects. It assesses colour by gauging the external appearance of the sample and determines colour scores (L^*, a^*, and b^*). The major advantages of the colorimetric method are less variability in colour measurement; it can accurately assess the slightest differences in colour, and its efficient automated system and even portable handheld machines are user friendly (Aguilar-Hernández et al. 2021).

5.4.3 Red, Green, Blue-Depth, and Time-of-Flight Imaging

Time-of-flight (ToF) is an active depth sensing camera that records the time taken by light to travel from the camera to the sample and backwards. Time delay needs to be recorded, so that the time-of-flight camera can estimate the distance of the sample from the camera, and generate a depth map, each pixel encodes the distance to the analogous point in the scene. An IR structured-light pattern can be projected onto the scene enabling optical triangulation in ToF. It estimates the 3D structure directly, without the need for conventional computer-vision algorithms. Real-time utility of ToF are robot navigation, 3D reconstruction, and human–machine interaction (May et al. 2009). The time-of-flight camera consists of an irradiation section, an optical lens, an imaging sensor, a control component, and a computing system. The significance of ToF is that it works in multi-dimensions, such as detection range, angle, frame rate, light resistance, and economic. Time-of-flight cameras are advantageous in situations, where the precision of mm level is essential (Hansard et al. 2012). These systems are economically viable and easily usable; but applicable to only non-occluded fruit and not suitable under direct daylight (Wang et al. 2017). Spectroscopy and depth imaging systems, alongside machine vision, determine size parameters such as length, breadth, thickness, and volume of crops up to 90% accuracy (Sanchez et al. 2020).

5.4.4 Fluorescence Spectroscopy

Here the object is rendered to a definite wavelength of light from an exterior source of illumination, such as a laser or lamp. It is a technique that analyses a sample's fluorescence properties by measuring the amount of fluorescence light it emits. The instrument is greatly sensitive and distinguishes even minute quantities of fluorescent composites in the testing sample. Fluorescence investigations are typically conducted with radiation in the ultraviolet to visible regions of the electromagnetic spectrum, which ranges from 250 to 700 nm. Fluorescence spectroscopy is a technique to study a sample's structure, function, and reaction dynamics, such as disclosing light-induced conformational variations in proteins, geneses of charge transfer reactions, solvent relaxation occurrences, and local conformational alterations. Fluorescence decay happens in the time frame of ~100 f sec to nanoseconds, hence analysis necessitates short light pulses and high temporal resolution instrumentation (Arnaut 2021). Chlorophyll fluorescence imaging measures the efficiency of photosynthesis and the state of chlorophyll in fruit tissues. Changes in chlorophyll fluorescence can indicate stress, physiological disorders, or growth, development, and ripening processes in fruits (Dhiedt et al. 2021; Gupta et al. 2022).

5.4.5 *Fourier Transform Infrared (FTIR)*

Spectral measurements of samples are done in FTIR. It is based on vibrational spectroscopy, which utilizes illumination interference to categorize the biochemical constitution of fruit and vegetable specimens by generating an IR absorption spectrum (Larkin 2017). FTIR functions at the mid-infrared region range 2500–25,000 nm and creates chemical functional configuration by acquiring standard vibrational and rotational stretching modes of molecules (Egidio et al. 2009; Ali et al. 2023). FTIR is fast and easy to perform with the ability to measure many quality parameters at a time, preferably with a good signal-to-noise ratio. Simple sample preparation protocols are followed (Lan et al. 2020).

5.4.6 *Raman Spectroscopy*

Raman spectroscopy works on high spatial resolution, vibrational, and complementary, which is a quick, easy, sensitive, and selective technique. Raman spectroscopy identifies molecules as well focuses on chemical bonding and intramolecular bonds in the sample. Vibrational frequencies are precise to chemical bonds and symmetry of molecules; hence the fingerprint region of organic molecules ranges from 500 to 1500 cm^{-1} wave number. It has the ability to detect water-rich molecules and the spatial distribution of the molecules. Raman spectroscopy is of different types such as surface-enhanced Raman, resonance Raman, tip-enhanced Raman, polarized Raman, stimulated Raman, transmission Raman, spatially offset Raman, and hyper ramanspectroscopy (Li et al. 2016; Munera et al. 2017; Wang et al. 2021; Pandiselvam et al. 2022). It offers rapid and on-the-spot assessment and provides evidential information on chemical and structural molecules, thus facilitating manifold evaluation by nominal test specimen preparation. Raman spectroscopy has many advantages, such as being quicker and cheaper than methods such as FTIR and TED-GC-MS, non-invasive technique ensuring real-time monitoring, and creating 2D spectral maps. It analyses the samples in all the states of matter (solids, liquids, or gases), whereas sample preparation is minimal. Vacuum is not a pre-requisite for Raman spectroscopy, hence making it an economical equipment (Boyaci et al. 2015).

5.4.7 *Visible/Near-Infrared Spectroscopy (VNIR/NIRS)*

The near-infrared spectroscopy technique measures the absorption of near-infrared light by organic molecules in the produce in the electro-magnetic spectral range of 780–2500 nm (Gupta et al. 2022). NIR is a real-time analysis, which measures multiple components concurrently with fast acquisition of spectra. It requires nominal test sample preparation for analysis (Srivichien et al. 2015; Arendse et al. 2021). It can provide information about internal quality attributes such as content of starch, sugar, moisture, acidity, firmness, and fruit quality (Pandiselvam et al. 2022). VNIR spectroscopy measures the reflectance of light in the visible and near-infrared spectra. It is utilized for assessing external quality attributes like colour and skin defects (Gabriels et al. 2020).

NIRS measures the peel colour in bananas, tomatoes, and mangoes, as changes in chlorophyll pigment indicate the maturity of fruits (Chauhan et al. 2017). NIRS measures fruit firmness, SSC, dry matter, hygroscopicity, toughness, sugar, and acid (Ravindran et al. 2015). NIRS is useful in apple fruit quality prediction, such as thickness, cell wall composition, dry substance, soluble solids, and sourness index (Lan et al. 2020), and glucose, malic acid, and dry substance in onion and apple puree (Lan et al. 2020). NIRS analyses morphological disparities in pomegranate varieties through transmittance and reflectance modes in the range of 400–1100 nm along with standard normal and multiplicative scatter correction (Khodabakhshian et al. 2019).

5.4.8 Acoustic Impulse Response Method (AIRM)

Acoustic technology utilizes sound waves from the object to communicate, collect, or process the information. This system is most important in assessing maturity, firmness, density, moisture content, texture, internal defects, and cavity detection in fruits and vegetables. Acoustic technology remains functional in several media, such as air, liquid, and solids; hence, it can be applied to various fields and environments. AIR frequency exploits the inherent frequencies of the testing sample that is acquired by measuring the sound produced by hitting the sample. Later, the Fourier transformations are utilized for recording the signal. It creates a stiffness factor, depending on the fruit's mass and first resonance frequency (Muramatsu et al. 2015). Advantages of acoustic vibration techniques are being fast, economical, and suitable for online operation in the production chain. It predicts optimal ripeness of fruits, and variations in texture characteristics are understandable by analysing frequency bands. Frequency range varies from 5 MPa for green mature fruit to 0.5 MPa for over-ripe fruit stage (Studman 1999). AIRM generates auditory velocity of the vibration using many instruments such as a plastic plunger, a laser Doppler vibrometer, an acceleration pickup, and an wooden ball pendulum for hitting the fruits such as bananas, mangoes, peaches, pears, melons, persimmons, apples, and watermelons (Nishani et al. 2017). Acoustic impulse-response techniques are used to analyse internal damage and biochemical changes in the cell walls of apples (Landahl et al. 2005). Acoustic impulse response techniques exhibited sensitivity to disparities in fruit texture and internal characters in apple and watermelon (Yamamoto et al. 1980).

5.4.9 X-Ray Radiography/Imaging

X-ray radiography is used to identify internal defects such as bruises, internal decay, and insect damage without damaging the produce. It provides detailed information about the internal structure and properties by displaying density differences using gray scale levels. It helps in sorting fruits and vegetables based on quality, the identification of healthy fruits and those with physiological disorders, disease detection, or pest damages (Abd Shaib et al. 2017; Diels et al. 2017). Advantages of X-ray radiography are the detection of contaminants present in fruit and vegetables with different densities than the testing sample being inspected, as well as being non-invasive. Dose of X-rays used for inspection is less than that used for irradiation, hence does not affect the safety, quality, or nutritional value of foods. Differential densities inside the tissues of fruits are affected by moisture, rigid tissues, infested holes, void spaces, disease, and physiological disorder symptoms and are denoted by diverse gray scale stages (Chauhan et al. 2017). X-ray radiography is most widely utilized to segregate citrus, lettuce, and tomato by identifying exterior defects (Barcelon et al. 2000; Hsiao et al. 2021). It is also utilized for the fruit quarantine by fluoroscopy images to detect problematic parts (Iqbal et al. 2018).

5.4.10 Ultrasound Imaging

Ultrasound imaging is used to assess the internal quality parameters such as maturity defects, sugar, firmness, moisture, and lipids of fruits and vegetables by measuring acoustic properties such as density and sound speed. It can detect internal defects and assess ripeness. Low power or high-frequency ultrasound monitors the composition and physico-chemical properties of fruits and vegetables and processed products during handling, processing, and storage, which is decisive in controlling the food properties and improving its quality. Low energy or low-intensity ultrasound has frequencies higher than 100 kHz at intensities below 1 W·cm^2, which can be utilized for non-invasive analysis and monitoring of various fruits and vegetables during processing and storage to ensure high quality and safety. Ultrasonic imaging has advantages of being low-priced, simple and energy saving, hence it has become an emergent technology for examining and altering food by utilizing non-ionizing radiation for functioning. They can be portable, facilitated by laptop-sized ultrasound machines. This system is less costly than other sophisticated imaging systems like MRI. Ultrasonic waves have their capability to penetrate the produce and non-destructively extract the material characteristics that might be correlated with quality indices of the fruits and vegetables (Awad et al. 2012; Ikeda et al. 2015). In the ultrasonic testing method, fruit tissue attributes and maturity stage of peaches were correlated through the attenuation coefficient and ultrasonic wave velocity for pH and acidity (Abolghasemi et al. 2009). The amplitude of the ultrasound waves conveyed through fruit skins is augmented when melon fruit colour changes from green to yellow, demonstrating a strong association between the maturity index and acoustic attenuation (Mizrach et al. 1991).

5.4.11 Electrical Impedance Spectroscopy (EIS)

The working principle of EIS is based on the ratio of the voltage applied and its analogous current. In this system, there is an alteration in magnitude and a phase shift due to frequency-dependent response. Electrical impedance is analysed by the 4 probe procedure through utilization of an impedance analyser, in which alternating voltage is applied at a specific frequency array. EIS is the most promising system for probing the charge transmission and storing (Pathak et al. 2023). EIS measures the electrical properties of produce, such as conductivity and capacitance. It can provide information about freshness, maturity, and internal defects. EIS measurement of fresh fruit tissues assesses physiological parameters such as tissue integrity, water content, and cell membrane permeability. They are also used for the detection of pesticide residues in fresh fruits and vegetables. An effective analytical technique for electrochemical systems has shown a wide application for food quality and safety assessment. Advantages are fast in action, non-destructive, good adaptability, and potential on-line detecting system to replace traditional methods to realize time, cost, and quality grading (Islam 2018; Bakirhan et al. 2018; Simić et al. 2023). Electrical Impedance Spectroscopy recorded impedance changes of Kiwi fruits at ripening in the frequency (50 Hz to 1 MHz) (Bauchot et al. 2000). Bruise detection by impedance measurements was carried out in apple varieties using Ag/AgCl electrodes implanted at different places with wide signal frequencies (50 Hz to 1 MHz) (Jackson and Harker 2000). Impedance changes during ripening in mangoes at the frequency of 1–200 kHz were recorded in an electrical model. Effective resistance exhibited greater disparities at low frequency (1 kHz) at the initial phase of the ripening process, reaching a highest at the fifth day and then reduced with further ripening and senescence (Rehman et al. 2011).

5.4.12 Gas Chromatography-Mass Spectrometry (GC-MS)

It analyses the volatile compounds emitted by fruits and vegetables, which provides information about aroma, flavour, and ripeness, which are commonly used in automated sorting and grading systems. Gas exchange measurements, including techniques like gas chromatography or portable gas analysers, monitor respiratory activity in fruits. This helps in understanding respiration and ethylene evolution, which influence fruit ripening and quality. Aroma volatiles are precisely quantified using GC-MS, but it is a laborious procedure and the system is costlier. In GC-MS, the testing solution injected into the equipment passes into a gas stream, which carries the testing solution into a separation tube or column. Helium or nitrogen gas is utilized as a carrier gas. Numerous volatile constituents are segregated within the column segment in the equipment. GC-MS provides multidimensional data, contributing to both qualitative and quantitative understanding of the information. It detects compounds at very low concentrations, facilitating regulatory assessments and quality control in fruits and vegetables. GC-MS is a very competent method and precise, capable of separating and analysing complex food mixtures quickly, and provides a high peak capacity. But it necessitates samples to be volatile and thermally constant, which confines its applicability. Off-flavours produced in fruit and vegetables during storage can be estimated by quantifying volatile acids or amines by GC-MS analysis (Su et al. 2017; Lebrun et al. 2008).

5.4.13 Electronic Nose (E-Nose)

E-nose systems impersonate the human olfactory system and can detect volatile compounds emitted by the fruits and vegetables to assess quality attributes such as freshness, ripeness, and spoilage (Srivastava and Sadistap 2018). E-nose detects volatile compounds in the headspace of a sample bottled in the package. E-nose recognizes the overall health condition of the crop, growing situations, and recognizes biotic or abiotic stresses. E nose has many advantages of speedy analysis of bulk samples, early detection of pests and pathogens, hence ensuring timely intervention. E-nose records variations in volatile aroma compounds in the ripening of both climacteric and non-climacteric crops (Brezmes et al. 2000; Saevels et al. 2003; Pathange et al. 2006). Electronic sensors were used to measure the taste and odour in broccoli (Hong et al. 2022). E-tongue identifies the taste of tartness, umami tastes, salty, sweetness, astringency, and bitterness of the fruits or vegetables.

5.4.14 Hyperspectral Imaging (HSI)

It is a scanning spectroscopy based on local chemical composition, which creates a 2D anatomic oxygenation map of the imaged tissue. It is utilized for the identification and quantification of chemical constituents, besides determining their position or spatial dispersal in the sample. HSI captures a pile of images, developing 3D structure of multivariate image data (hypercube), consisting of a spectrum for each pixel in the image. It combines both the benefits of image analysis and spectrometry to investigate varied structures and acquire chemical information in a single image (Jiménez-Carvelo et al. 2021). HSI determines both spectral and spatial information, along with the concurrent assessment of many parameters in fruits and vegetables. HSI has the advantages of high resolution and quick analysis in NDM. It provides an extra dimension to the traditional imaging method, hence making it more accurate in analysis. It detects impurities, minerals, toxic metal depositions, and foul odours emitted from fruits and vegetables. Samples for HSI analysis can range from a minimum to a large group. HSI machines are compact, cost-effective, and convenient to use. It helps in detecting the defects, assess ripeness, and classify produce based on quality attributes as well as on their spectral signatures, which is accessible in different algorithms and utilized in assessing internal mechanical damage of blueberry (Wang and Zhai 2018; Metlenkin et al. 2022). Hyperspectral imaging is utilized in assessing banana fruit quality and maturity stages (Rajkumar et al. 2012; Chandrasekaran et al. 2019).

5.4.15 *Thermal Imaging*

It measures the surface temperature of fruits, which can reflect physiological processes such as respiration, transpiration, and metabolic activity. Thermal imaging measures temperature by assessing IR beams emitted by the external layer of the sample. Thermal imaging takes up this discharged radiation and frames pseudoimages of the thermal emission on the external layer known as thermograms. The IR band correlates to the electromagnetic spectrum range of 0.75–100 μm (Gowen et al. 2010). It provides insights into fruit ripening dynamics and quality changes. Thermal imaging has many applications in safety and quality assessment of fruits and vegetables, such as temperature validation, bruise and foreign body detection, and fruit quality evaluation. It is a passive and non-invasive approach, which can capture through all the visual barriers and in any diverse weather vagaries. The durability of thermal imaging cameras is more, thus making it cost-effective in the long-term usage (Vadivambal and Jayas 2011; Pathmanaban et al. 2019). IR ranges from 700 to 1000 nm, and the methodology of generating the infrared range is known as thermographic photography (Bhargava and Bansal 2021). Reliability of thermal imaging is based on its sensitivity, speed of detection, and spatial and intensity resolution of the image (Prakash 2000). Infrared thermal imaging approaches and optical fibre sensors attached to RGB LEDs are utilized to assess the colour in fruit crops such as pineapple and mango (Ali et al. 2022; Yahaya et al. 2011).

5.4.16 *Magnetic Resonance Imaging (MRI)*

MRI offers the non-invasive imaging of interior structures and physiological processes of fruit, such as water distribution, tissue density, and cell structure. It helps in understanding fruit development and ripening dynamics. MRI detects and tracks the process of physiological development of storage disorders over time. MRI scans are applicable for a wide range of tropical and temperate fruits and vegetables, hence being versatile in nature and work mode. NMR spectroscopy, NMR relaxometry, and MRI come under nuclear magnetic resonance imaging (MRI) techniques.

NMR spectroscopy works by nuclear spin and magnetic resonance. Relaxation times and signal intensities of NMR spectroscopy depict water, mobility, and molecular interactions in the fruits and vegetables (Ebrahimnejad et al. 2018). This system determines constituents of samples such as water, lipids, sugar, soluble solids, and acids, which usually change during fruit maturation (Newton et al. 2017; Yodh and Chance 2017). Magnetic resonance imaging (MRI) has been used to scan the surface morphology of fruits such as grapes, especially cuticle formation, banana, and litchi (Prasad et al. 2018). MRI can be used to determine the composition of fruit, especially moisture and lipids in the fruit. Ice crystals formed in the fruits and vegetables restrict NMR signals, thus helping in detecting the freezing injury symptoms induced at storage (Newton et al. 2017; Yodh and Chance 2017).

5.4.17 *Capacitive Sensing*

It evaluates output voltage and capacitance of fruits, which measures quality parameters such as TSS and moisture (Aziz et al. 2021). The sine wave frequency generator and its analogue plates capacitor samples are utilized to bring out the variations in capacitance in the sample. A capacitive sensor has a high-frequency oscillator, in which the sensing surface has two metal electrodes. When the sample is brought near the sensing surface of the sensor, it gets into the electrostatic field of the sensor's electrodes, which changes the oscillator's capacitance. Detecting range of a capacitive sensor is around 25 mm or 1 inch. Relative permittivity correlation with quality parameters of fruits will be further studied in capacitive sensing. The capacitive sensor is simple in fabrication and changeable according to the sample, it detects through packaged containers, cheaper equipment with high sensitivity, and operates with a minimum magnitude of force. It works with a high resolution of 0.003 mm and good frequency response (Soltani et al. 2010).

5.5 Artificial Intelligence (AI) Models

AI helps in atomizing and augmenting the maturity indices evaluation, analysing quality attributes, and identifying microbial contamination and physiological disorders (Cardenas-Perez et al. 2017; Dhanya et al. 2022; Xiao et al. 2023). AI systems are utilized in grading fruits and vegetables based on parameters such as size, shape, and colour that contribute to the consumer's acceptability. Pre-processing methods involved in the AI add to their precise accuracy and competence. Computer vision and CNN models are recognition systems to carry out sorting and grading horticultural crops with specificity to their variety and species, particularly in export or domestic markets (Dubey and Jalal 2012). Machine vision systems use cameras and image processing algorithms to analyze external quality attributes such as size, shape, colour, and defects. They are commonly used in automated sorting and grading systems (Hitchman et al. 2016). Machine vision system takes numerous sample images for the sake of volume estimation, which demonstrates probable high accuracy in quality prediction (Yildiz et al. 2019). Successive projection algorithms were used for selecting the ideal wavelengths which are correlated to anthocyanin content in lychee fruit. Radial basis function for support vector regression and neural network models pooled as a lone prototype proved to be a good performer in calculating and envisioning anthocyanin variations in lychee fruit during storage (Huang et al. 2017).

Fruits and vegetables quality monitoring using various NDMs is listed in Table 5.1.

5.6 Constraints in Non-destructive Analysis

With all the advantages of non-destructive techniques discussed above, there are a few constraints and major challenges in adopting these techniques as described here.

Major disadvantage in visual inspection method is that it might lead to a lot of variation due to differential perception of colours by human eyes, which might result in higher chances of human error. The colorimetric method has a lacuna over the subjective method as it incurs significant initial investment for equipment (Aguilar-Hernández et al. 2021). The ToF camera exhibits limited field of view and impacts the quality of range measurements by more dynamic in-object reflection, as well as presently available models of ToF exhibit poor data quality (May et al. 2009).

FTIR has the disadvantage of getting a representative background, as well difficult to read the interferogram (Lan et al. 2020). Raman spectroscopy has the demerits of showing weak Raman scattering, fluorescence interference, less replicative, redundant data set, high cost, and lower operational speed. Raman spectroscopy gives inadequate depth of penetration. Fluorescence from the specimen or impurities will give unintelligible spectral information. Cosmic rays affect detector during data collection, hence producing high-intensity spikes in the spectra (Li et al. 2016; Wang et al. 2021).

X-ray equipment are expensive; they require high-voltage power supplies, pose potential health risk to operators working with X-ray inspection systems, snd show difficulty in detecting low-density contaminants like hair, paper, and insect fragments.

NIR spectroscopy has limited penetration depth, too many time-consuming calibration procedures, and complex signal interpretation (Srivichien et al. 2015; Arendse et al. 2021). NIR spectroscopy methods are surface-sensitive and analyse

Table 5.1 Non-destructive methods, their components, and principles for quality assessment in fruits and vegetables

Sl. No.	Non-destructive methods	Principle of non-destructive measurements	Components of the non-destructive measurements	Crop/sample	Parameters assessed	References
1.	Attenuated total reflectance-Fourier transform infrared (*ATR-FTIR*) spectroscopy	Total internal reflectance when infrared light collides food sample (high refractive index) at the junction, where infrared light reflected	Infrared light source, interferometer (beam splitter), collimator, 2 mirrors, detector	Peel and fresh banana pulp	Ripening stage	Sinanoglou et al. (2023)
2.				Mango juice	Added sugar content, TSS, real juice content	Jha and Gunasekaran (2010)
3.				Freeze-dried fruits and vegetables	Cell wall compositions	Canteri et al. (2019)
4.				Apple	Quality	Lan et al. (2020)
5.	Fourier transform infrared spectroscopy (FTIR)	Vibrational spectroscopy—Light interference to categorize parameters by creating IR absorption or emission spectra	Infrared light source, interferometer, beam splitter, collimator, 2 mirrors, detector	Banana	Maturity	Zhang et al. (2021)

(continued)

Table 5.1 (continued)

Sl. No.	Non-destructive methods	Principle of non-destructive measurements	Components of the non-destructive measurements	Crop/sample	Parameters assessed	References
6.	Hyperspectral imaging (HSI)	Combination of imaging and spectroscopy	Imaging unit, illumination (light source), sample stage, computer	Banana	Soluble solid content, acidity	Chu et al. (2022)
7.				Potato	Bruises	Ye et al. (2018)
8.				Banana	Grading	Mesa and Chiang (2021)
9.				Eggplant	Surface stiffness, peel gloss	Martínez (2013)
10.				Potato	Soluble solid content	Su and Sun (2019)
11.				Kiwi	For chlorfenuron content	Dong et al. (2017)
12.				Jujube	Volatile insecticides (chlorpyrifos and imidacloprid) content	Su et al. (2017)
13.				Avocado	Defects	Metlenkin et al. (2022)
14.				Pineapple-dehydrated	Water activity	Aozora et al. (2022)
15.				Apple	Ethylene concentration	Bratu et al. (2021)
16.	Thermal imaging	Infrared radiation is emitted from the sample to generate images	Optic system, detector, amplifier, display	Apple	Maturity	Prasad et al. (2018)
17.				Tomato	Maturity	Prasad et al. (2018)
18.				Citrus	Maturity	Gan et al. (2020)
19.				Apple	Bruise detection	Varith et al. (2003)
20.				Pineapple	Quality prediction	Ali et al. (2022)
21.				Mango	Colour	Yahaya et al. (2011)

Sl. No.	Non-destructive methods	Principle of non-destructive measurements	Components of the non-destructive measurements	Crop/sample	Parameters assessed	References
22.	Visible/Near-infrared spectroscopy (V/NIR)	Spectrum of absorption and reflection, in which NIR light interacts with the chemical composition of fruits. Vibrational bands are identified by functional groups perceived by NIR in the wavelength (780–2500 nm)	Illumination source, sample accessor, monochromator, detector, optical components (lenses, optical fibres)	Mango	Firmness	Flores et al. (2008)
23.				Papaya	Starch, soluble solid content	Purwanto et al. (2015)
24.				Avocado	Moisture, dry matter, maturity	Olarewaju et al. (2016)
25.				Avocado	Texture	Uarrota and Pedreschi (2022)
26.				Mango	Maturity	Jha et al. (2014)
27.				Banana	TSS, pH	Ali et al. (2018)
28.				Mango	Firmness	Mishra et al. (2020)
29.				Banana	Firmness	Zude (2003)
30.				Mango	TSS, TA, firmness, ripeness index	Rungpichayapichet et al. (2016)
31.				Mango	TSS, Vitamin C	Kusumiyati and Suhandy (2021)
32.				Chillies	Vitamin C	Wang et al. (2011)
33.				Apple	Chlorophyll	Merzlyak et al. (2003)
34.				Watermelon	Quality	Tian et al. (2007)
35.				Navel orange	Soluble solids	Liu et al. (2008)
36.				Macadamia nut	Internal nut defects and oil content	Rahman et al. (2021)
37.				Pineapple	Nitrate	Srivichien et al. (2015)

(continued)

Table 5.1 (continued)

Sl. No.	Non-destructive methods	Principle of non-destructive measurements	Components of the non-destructive measurements	Crop/sample	Parameters assessed	References
38.	Raman spectroscopy	Vibrational spectroscopy uses laser beams in the infrared region (2500 to 25,000 nm)	Monochromatic laser, wavelength separator, detector	Sweet potato	Moisture content, Carotenoids	Sebben et al. (2018)
39.				Potato	Grading	Morey et al. (2020)
40.				Mango	Carotenoids	Bicanic et al. (2010)
41.				Cassava	Starch	Cardoso and Jesus Poppi (2021)
42.				Carrot	Carotenoids	Lawaetz et al. (2016)
43.				Tomato	Carotenoids	Hara et al. (2018)
44.				Pomegranate	Tannins	Vargas-Murga et al. (2016)
45.	Acoustic impulse response method	Sound waves produced by the food sample to evaluate inherent features	Impacting balls, which generate sound from the sample.Microphone, Fourier transformation to capture sound data to identify largest resonant frequency	Kiwifruit	Firmness	Muramatsu et al. (2015)
46.				Apricot	Firmness	Petrisor et al. (2010)
47.				Papaya	Firmness	Chauhan et al. (2015)
48.				Persimmon	Elasticity index, texture index	Taniwaki et al. (2009)
49.				Apple	Sensory crispness, crunchiness, hardness	Zdunek et al. (2011)
50.				Apple	Firmness	Zude et al. (2006)

Sl. No.	Non-destructive methods	Principle of non-destructive measurements	Components of the non-destructive measurements	Crop/sample	Parameters assessed	References
51.	Ultrasound imaging	Sound waves with frequencies exceeding the hearing limit of the human ear (~20 kHz) utilized. Sound propagates through fruits and vegetables as mechanical waves induce alternative compressions and decompressions	Pulse generator, oscilloscope, ultrasound probes, transducer, pulse generator, function generator	Potato	Defects	Cheng and Haugh (1994)
52.				Tomato	Firmness	Mizrach (2007)
53.				Orange juice	Sugar, viscosity	Kuo et al. (2008)
54.				Plum	Maturity, sugar	Mizrach (2004)
55.	Electrical impedance spectroscopy (EIS)	Voltage or current is applied on the food sample to gauge its impedance at an appropriate frequency array	Microcontroller-based portable systems; electrical circuit computer-based software platforms	Avocado	Ripening process	Islam (2018)
56.				Onion	Dehydration process	Simić et al. (2023)

(continued)

Table 5.1 (continued)

Sl. No.	Non-destructive methods	Principle of non-destructive measurements	Components of the non-destructive measurements	Crop/sample	Parameters assessed	References
57.	E-nose	Chemical vapour sensor arrays detect stereo-chemical features of volatile molecules responsible for odour or flavour. Sensor array data investigate by statistical or structural algorithms to discriminate and identify volatile patterns	Sample delivery system, detection system (chemical sensor array), computing system	Apple	Aroma components	Brezmes et al. (2000), Saevels et al. (2003) and Pathange et al. (2006)
58.				Peach	Aroma components	Brezmes et al. (2000); Di Natale et al. (2001a, 2001b, 2002) and Infante et al. (2011)
59.				Mango	Aroma components	Lebrun et al. (2008), Zakaria et al. (2011)
60.				Apricot	Aroma components	Parpinello et al. (2007)
61.				Pear	Aroma components	Brezmes et al. (2000)
62.				Mango	Volatiles in maturity	Nouri et al. (2014)
63.				Tomato	Aroma components	Gomez et al. (2006a)
64.				Orange	Aroma components	Qiu and Wang (2015)
65.				Mandarin	Aroma components	Gomez et al. (2006b), Qiu and Wang (2015)
66.				Cherry	Aroma components	Benedetti et al. (2010)

Sl. No.	Non-destructive methods	Principle of non-destructive measurements	Components of the non-destructive measurements	Crop/sample	Parameters assessed	References
67.	Machine vision technique	Digital cameras and image analyser for the automation of visual reviews	Light, imaging unit, image capture board, computer hardware and software	Apples	Surface bruise	Gao et al. (2014)
68.				Strawberry	Sorting	Nagata et al. (1997)
69.				Raspberry fruit	Small insects in holes	Okamoto et al. (2013)
70.	Capacitive sensing	It produces an electrical field to detect the sample from the sensor's detection end	Sensor, oscillator circuit, non-metallic containers	Papaya	TSS moisture	Aziz et al. (2021)
71.				Banana	Maturity	Soltani et al. (2010)
72.				Mango	Ripening	Ariffin and Rosman (2024)
73.				Banana	Maturity	Masoumi et al. (2024)
74.				Papaya	Ripening	Bhosale and Sundaram (2015)
75.				Citrus	Sugar	Bhosale (2017)
76.				Highbush blueberry	Anthocyanin	Ribera-Fonseca et al. (2016)
77.				Apple	Ethylene, firmness, starch	Doerflinger et al. (2016)
78.				Banana	Ripening	Sankhe (2015)
II	*Quality assessment in combination with two or more non-destructive methods*					
79.	FTIR, FTNIR		Pineapple	Soluble solid content, pH, acidity	Egidio et al. (2009)	
80.	FTIR, NIR, machine vision		Pineapple	Quality	Ali et al. (2023)	

(continued)

Table 5.1 (continued)

Sl. No.	Non-destructive methods	Principle of non-destructive measurements	Components of the non-destructive measurements	Crop/sample	Parameters assessed	References
81.	NIR, HSI		Sweet potato	Varietal recognition	Su et al. (2019)	
82.	NIR, HIS		Pineapple	Water activity	Aozora et al. (2022)	
83.	NIR, HIS		Mango	Defect	Rivera et al. (2014)	
84.	HSI, Machine Learning, Deep learning		Papaya	Maturity	Garillos-Manliguez and Chiang (2021)	
85.	Vis–NIR, Machine Learning		Mango	Colour	Gabriels et al. (2020)	
86.	Vis-NIR, HIS		Avocado	Fatty acid	Kämper et al. (2020)	
87.	SWIR, HIS		Potato	Black spot	López-Maestresalas et al. (2016)	
88.	Raman spectroscopy, nanotechnology		Mango	Pesticide residue	Pham et al. (2022)	
89.	Machine learning-based analysis		Apple, lemon, mango	Damage	Bobde et al. (2021)	

the external layer of fruits and vegetables, thus limiting their utility for certain applications (Dhiedt et al. 2021; Gupta et al. 2022). NIR needs previous technical know-how for a precise quality parameter that necessitates an assessment utilizing a reference standard methodology.

It is doubtful whether HSI analysis has practical applicability for fruits and vegetables in large quantities and greater challenge of enhancing data handling (Cortes et al. 2019; Si et al. 2022). HSI analysis requires sophisticated hardware and software for rapid investigation of vast hyperspectral data (Xu et al. 2023) and machine learning algorithms, particularly for deep learning methods (Caceres-Hernandez et al. 2023). A lengthy procedure and chemometrics techniques are essential to derive relevant information for HSI, which are its major drawbacks (Rajkumar et al. 2012; Chandrasekaran et al. 2019).

The major disadvantage of E-nose technology is that it is not efficient in non-climacteric fruits volatile assessment. E-nose technology shows difficulty in differentiating among specific target molecules and decreasing cross-sensitivity. The technical challenge lies in the intricacy of fabricating sensor arrays with manifold sensing elements, like nano-wires prepared from various composite ingredients, which are fabrication and upkeep issues. E-nose is less sensitive to variations and is impacted by environmental parameters such as temperature and humidity, causing drift in sensors of E-nose, which makes the quantification of individual bioactive compounds very challenging (Hu et al. 2013).

The major drawback of the acoustic method is the requirement for specialized apparatus and standardization protocols, which restrict its practical utility. Ultrasound is not as sophisticated as MRI or CT, thus unsuitable for thorough imaging (Awad et al. 2012). MRI machines are costly with heavy maintenance charges, thus making the procedure comparatively costlier than other methods. However, MRI is known to give resolution better than acoustic imaging systems (Dhiedt et al. 2021; Gupta et al. 2022).

Relatively long measurement time is the major disadvantage of electrical impedance spectroscopy in comparison to other methods (Nováková et al. 2024) and difficulty in analysing EIS data in terms of simple equivalent electrical circuits (Macdonald 1990). Thermal imaging techniques are expensive and challenging to detect the temperature if the sample is separated or covered by a non-transparent material such as glass. It requires good analytical expertise and an extensive knowledge base to evaluate the results of imaging (Prakash 2000).

5.7 Conclusion

Non-destructive quality monitoring methods are crucial in the present world to face the challenge of feeding a growing population and the necessity of a healthy supply chain of perishable produce. They are most important for the maintenance of freshness, assurance of nutritive value retention, ensuring safety, and optimizing the processing and storage capabilities of fruits and vegetables. Various non-destructive techniques adopted in the quality management are red, green, blue-depth and TOF camera imaging; FTIR, fluorescence, Raman, and visible/near-infrared spectroscopy; acoustic impulse response method, X-ray radiography, ultrasound imaging, electrical impedance spectroscopy, machine vision systems, electronic nose, hyperspectral imaging, thermal imaging, magnetic resonance imaging, capacitive sensing techniques, along with advanced artificial intelligence models such as computer vision and convolutional neural network models. They are employed in the horticultural value chain to assess the quality of produce rapidly without causing damage to internal or external tissues. Further, there are many technical glitches to be addressed to adopt all these non-destructive methods for the bulk assessment of perishables. They are the most promising technologies, as well a combination of one or more non-destructive methods leads to efficient quality control and management in enhancing the postharvest shelf life of fruits and vegetables.

References

Abd Shaib MF, Rahim RA, Muji SZM et al (2017) Investigating maturity state and internal properties of fruits using non-destructive techniques – a review. Telkomnika 15:1574–1584

Abolghasemi R, Emadi B, Aghkhani MH et al (2009) Determination of peach maturity using ultrasonic waves. Iran Food Sci Tech Res J 5(1):63–74

Aguilar-Hernández MG, Nunez-Gomez D et al (2021) Quality parameters of spanish lemons with commercial interest. Food Secur 10(1):1–135. https://doi.org/10.3390/foods1001006

Ali MM, Janius RB, Nawi NM et al (2018) Prediction of total soluble solids and PH in banana using near infrared spectroscopy. J Eng Sci Technol 13(1):254–645

Ali MM, Hashim N, Aziz SA et al (2022) Quality prediction of different pineapple (*Ananas comosus*) varieties during storage using infrared thermal imaging technique. Food Control 138:1–9. https://doi.org/10.1016/j.foodcont.2022.108988

Ali MM, Hashim N, Bejo SK et al (2023) Innovative non-destructive technologies for quality monitoring of pineapples: recent advances and applications. Trends Food Sci Technol 133:176–188. https://doi.org/10.1016/j.tifs.2023.02.005

Aozora DW, Tantinantrakun A, Thompson AK et al (2022) Near infrared hyperspectral imaging for predicting water activity of dehydrated pineapples. Res Militaris 12:11–33. https://doi.org/10.3390/foods12142793

Arendse E, Nieuwoudt H, Magwaza LS et al (2021) Recent advancements on vibrational spectroscopic techniques for the detection of authenticity and adulteration in horticultural products with a specific focus on oils juices and powders. Food Bioprocess Technol 14(1):1–225. https://doi.org/10.1007/s11947-020-02505-x

Ariffin MA, Rosman R (2024) Non-destructive measurement on Kwini mango fruit using capacitive sensing technique at 250kHz to 2MHz for ripeness determination. J Adv Res Appl Sci 47(1):123–139

Arnaut L (2021) Experimental methods. In: Arnaut L (ed) Chemical kinetics, 2nd edn. Elsevier, pp 37–91. https://doi.org/10.1016/B978-0-444-64039-0.00004-2. ISBN 9780444640390

Awad TS, Moharram HA, Shaltout OE et al (2012) Applications of ultrasound in analysis processing and quality control of food a review. Food Res Int 48(2):410–427

Aziz NAA, Jusoh MZ, Rosman R (2021) Relationship of total soluble solid (TSS) and capacitance value of papaya fruit using capacitive sensing technique in ISCI 2021-2021. In: IEEE symposium on computers and informatics, Kuala Lumpur, Malaysia, pp 51–57. https://doi.org/10.1109/ISCI51925.2021.9633402

Bakirhan NK, Uslu B, Ozkan SA (2018) The detection of pesticide in foods using electrochemical sensors. In: Food safety and preservation. Academic, pp 91–141

Barcelon EG, Tojo S, Watanabe K (2000) Nondestructive ripening assessment of mango using an X-ray computed tomography. Int Agric Eng J 9:73–80

Bauchot AD, Harker FR, Arnold WM (2000) The use of electrical impedance spectroscopy to assess the physiological condition of kiwifruit. Postharvest Biol Technol 18:9–18

Benedetti S, Spinardi A, Mignani I et al (2010) Non-destructive evaluation of sweet cherry (*Prunus avium* L.) ripeness using an electronic nose. Ital J Food Sci 22(3):298

Bhargava A, Bansal A (2021) Fruits and vegetables quality evaluation using computer vision: a review. J King Saud Univ Comput Inf Sci 33(3):243–575. https://doi.org/10.1016/j.jksuci.2018.06.002

Bhosale AA (2017) Detection of sugar content in citrus fruits by capacitance method. Procedia Eng 181:466–471

Bhosale AA, Sundaram KK (2015) Non-destructive method for ripening prediction of papaya. Procedia Technol 19:623–630

Bicanic D, Dimitrovski D, Luterotti S et al (2010) Estimating rapidly and precisely the concentration of beta carotene in mango homogenates by measuring the amplitude of optothermal signals, chromaticity indices and the intensities of raman peaks. Food Chem 121(3):832–385. https://doi.org/10.1016/j.foodchem.2009.12.093

Bobde S, Jaiswal S, Kulkarni P et al (2021) "Fruit quality recognition using deep learning algorithm," in 2021 International Conference on Smart Generation Computing, Communication and Networking (SMART GENCON), Pune, India, 1–5.Boyaci IH, Temiz HT, Genis HE et al (2015) Dispersive and FT-Raman spectroscopic methods in food analysis. RSC Adv 5(70):56606–56624

Bratu AM, Popa C, Bojan M et al (2021) Non-destructive methods for fruit quality evaluation. Sci Rep 11(1):7782

Brezmes J, Llobet E, Vilanova X et al (2000) Fruit ripeness monitoring using an electronic nose. Sens Actuators B Chem 69(3):223–229

Brosnan T, Sun D-W (2004) Improving quality inspection of fruit products by computer vision – a review. J Food Eng 61:3–16

Caceres-Hernandez D, Gutierrez R, Kung K et al (2023) Recent advances in automatic feature detection and classification of fruits including with a special emphasis on watermelon (*Citrilluslanatus*): a review. Neurocomputing 526:62–79. https://doi.org/10.1016/j.neucom.2023.01.005

Canteri MHG, Renard CMGC, Le Bourvellec C et al (2019) ATR-FTIR spectroscopy to determine cell wall composition: application on a large diversity of fruits and vegetables. Carbohydr Polym 212:186–196. https://doi.org/10.1016/j.carbpol.2019.02.021

Cardenas-Perez S, Chanona-Perez J, Mendez-Mendez JV et al (2017) Evaluation of the ripening stages of apple by means of computer vision system. Biosyst Eng 159:46–58

Cardoso KVG, Jesus Poppi R (2021) Cleaner and faster method to detect adulteration in cassava starch using raman spectroscopy and one-class support vector machine. Food Control 125:107917. https://doi.org/10.1016/j.foodcont.2021.107917

Chandrasekaran I, Panigrahi SS, Ravikanth L et al (2019) Potential of near-Infrared (NIR) spectroscopy and hyperspectral imaging for quality and safety assessment of fruits: an overview. Food Anal Methods 12(11):2438–2585. https://doi.org/10.1007/s12161-019-01609-1

Chauhan OP, Chitravati K, Raju PS (2015) New innovations in postharvest handling of fruits and vegetables. In: Paper presented at 2nd national seminar on hitech horticulture: challenges and opportunities, Babasaheb Bhimrao Ambedkar University Lucknow India, pp 88–192

Chauhan OP, Lakshmi S, Pandey AK et al (2017) Non-destructive quality monitoring of fresh fruits and vegetables. Def Life Sci J 20(2):103

Cheng Y, Haugh CG (1994) Detecting hollow heart in potatoes using ultrasound. Trans ASABE 37(1):217–222

Chu X, Miao P, Zhang K, Wei H, Fu H, Liu H, Jiang H, Ma Z (2022) Green banana maturity classification and quality evaluation using hyperspectral imaging. Agriculture (Switzerland) 12(4):1–185. https://doi.org/10.3390/agriculture12040530

Cortes V, Blasco J, Aleixos N et al (2019) Monitoring strategies for quality control of agricultural products using visible and near-infrared spectroscopy: a review. Trends Food Sci Technol 85:138–148. https://doi.org/10.1016/j.tifs.2019.01.015

Cozzolino D (2022) An overview of the successful application of vibrational spectroscopy techniques to quantify nutraceuticals in fruits and plants. Foods 11(3):1–11. https://doi.org/10.3390/foods11030315

Dhanya VG, Subeesh A, Kushwaha NL et al (2022) Deep learning based computer vision approaches for smart agricultural applications. Artif Intell Agric 6:211–229

Dhiedt E, Verheyen K, De Smedt P et al (2021) Early tree diversity and composition effects on topsoil chemistry in young forest plantations depend on site context. Ecosystems 24(7):1638–1653

Di Natale C, Macagnano A, Martinelli E et al (2001a) The evaluation of quality of post-harvest oranges and apples by means of an electronic nose. Sens Actuators B Chem 78:26–31

Di Natale C, Macagnano A, Martinelli E et al (2001b) Electronic nose based investigation of the sensorial properties of peaches and nectarines. Sens Actuators B Chem 77(1–2):561–566

Di Natale C, Zude-Sasse M, Macagnano A et al (2002) Outer product analysis of electronic nose and visible spectra: application to the measurement of peach fruit characteristics. Anal Chim Acta 459(1):107–117

Diels E, van Dael M, Keresztes J et al (2017) Assessment of bruise volumes in apples using X-ray computed tomography. Postharvest Biol Technol 128:24–32

Doerflinger FC, Nock JF, Shoffe YA et al (2016) Non-destructive maturity assessment of 'Empire' apples treated with pre-harvest inhibitors of ethylene perception and production with a delta absorbance (DA) meter. Acta Hortic. https://doi.org/10.17660/ActaHortic.2016.1119.32

Dong J, Guo W, Zhao F et al (2017) Discrimination of Hayward kiwifruits treated with Forchlorfenuron at different concentrations using hyperspectral imaging technology. Food Anal Methods 10:477–486

Dubey SR, Jalal AS (2012) Robust approach for fruit and vegetable classification. Procedia Eng 38:3449–3453

Ebrahimnejad H, Ebrahimnejad H, Salajegheh A et al (2018) Use of magnetic resonance imaging in food quality control: a review. J Biomed Phys Eng 8(1):127

Egidio VD, Sinelli N, Limbo S et al (2009) Evaluation of shelf-life of fresh-cut pineapple using FT-NIR and FT-IR spectroscopy. Postharvest Biol Technol 54(2):87–925. https://doi.org/10.1016/j.postharvbio.2009.06.006

Elik A, Yanik DK, Istanbullu Y et al (2019) Strategies to reduce post-harvest losses for fruits and vegetables. Int J Sci Technol Res 5(3):29–395. https://doi.org/10.7176/jstr/5-3-04

El-Mesery HS, Mao H, Abomohra AE (2019) Applications of non-destructive technologies for agricultural and food products quality inspection. Sensors (Basel, Switzerland) 19(4):846. https://doi.org/10.3390/s19040846

Escárate P, Farias G, Naranjo P et al (2022) Estimation of soluble solids for stone fruit varieties based on near-infrared spectra using machine learning techniques. Sensors 22(16):1–115. https://doi.org/10.3390/s22166081

Flores K, Sanchez MT, Perez-Marin DC et al (2008) Prediction of total soluble solid content in intact and cut melons and watermelons using near infrared spectroscopy. J Near Infrared Spectrosc 16(2):91–98. https://doi.org/10.1255/jnirs.771

Gabriels SH, Mishra P, Mensink MG et al (2020) Non-destructive measurement of internal browning in mangoes using visible and near-infrared spectroscopy supported by artificial neural network analysis. Postharvest Biol Technol 166:111206. https://doi.org/10.1016/j.postharvbio.2020.111206

Gan H, Lee WS, Alchanatis V et al (2020) Active thermal imaging for immature citrus fruit detection. Biosyst Eng 198:291–303

Gao H, Zhu F, Cai J (2014) A review of non-destructive detection for fruit quality. In: IFIP advances in information and communication technology. Springer, Berlin/Heidelberg. https://doi.org/10.1007/978-3-642-12220-0_21

Garillos-Manliguez CA, Chiang JY (2021) Multimodal deep learning and visible-light and hyperspectral imaging for fruit maturity estimation. Sensors (Switzerland) 21(4):1–185. https://doi.org/10.3390/s21041288

Gomez AH, Hu GX, Wang J et al (2006a) Evaluation of tomato maturity by electronic nose. Comput Electron Agric 54:44–52

Gomez AH, Wang J, Hu GX et al (2006b) Electronic nose technique potential monitoring mandarin maturity. Sensors Actuators B Chem 113(1):347–353

Gowen AA, Tiwari BK, Cullen PJ et al (2010) Applications of thermal imaging in food quality and safety assessment. Trends Food Sci Technol 21(4):190–200

Gupta AK, Koch P, Yumnam M et al (2022) Biosensors involved in fruit and vegetable processing industries. In: Biosensors in food safety and quality: fundamentals and applications, pp 111–134

Hansard M, Lee S, Choi O et al (2012) Time-of-flight cameras: principles, methods and applications. Springer Science & Business Media

Hara R, Ishigaki M, Kitahama Y et al (2018) Excitation wavelength selection for quantitative analysis of carotenoids in tomatoes using raman spectroscopy. Food Chem 258:308–313. https://doi.org/10.1016/j.foodchem.2018.03.089

Hitchman S, Wijk K, Davidson Z (2016) Monitoring attenuation and the elastic properties of an apple with laser ultrasound. Postharvest Bioltechnol 121:71–77

Hong SJ, Yoon S, Lee J et al (2022) A comprehensive study for taste and odor characteristics using electronic sensors in broccoli floret with different methods of thermal processing. J. Food Process Preserv 46(4):e16435

Hsiao WT, Kuo WC, Lin HH et al (2021) Assessment and feasibility study of lemon ripening using x-ray image of information visualization. Appl Sci 11(7):3261

Hu Y, Lee H, Kim S et al (2013) A highly selective chemical sensor array based on nanowire/nanostructure for gas identification. Sens Actuators B Chem 181:424–431

Huang L, Zhou Y, Meng L et al (2017) Comparison of different CCD detectors and chemometrics for predicting total anthocyanin content and antioxidant activity of mulberry fruit using visible and near infrared hyperspectral imaging technique. Food Chem 224:1–10. https://doi.org/10.1016/j.foodchem.2016.12.037

HunterLab (1995) Colorimeters vs. spectrophotometers. Applications note. Insight Color 5(6):2

Ikeda T, Choi PK, Ishii T et al (2015) Firmness evaluation of watermelon flesh by using surface elastic waves. J Food Eng 160:28–33

Infante R, Rubio P, Meneses C et al (2011) Ripe nectarines segregated through sensory quality evaluation and electronic nose assessment. Fruits 66(2):109–119

Iqbal Z, Khan MA, Sharif M et al (2018) An automated detection and classification of citrus plant diseases using image processing techniques: a review. Comput Electron Agric 153:12–32

Islam M (2018) Electrical impedance spectroscopy and tomography: applications on plant characterization. Doctoral dissertation, University of Saskatchewan

Jackson PJ, Harker FR (2000) Apple bruise detection by electrical impedance measurement. HortScience 35:104–107

Jha SN, Gunasekaran S (2010) Authentication of sweetness of mango juice using fourier transform infrared-attenuated total reflection spectroscopy. J Food Eng 101(3):337–342. https://doi.org/10.1016/j.jfoodeng.2010.07.019

Jha SN, Narsaiah K, Jaiswal P et al (2014) Nondestructive prediction of maturity of mango using near infrared spectroscopy. J Food Eng 124:152–157. https://doi.org/10.1016/j.jfoodeng.2013.10.012

Jiménez-Carvelo AM, Martin-Torres S, Cuadros-Rodríguez L et al (2021) Non targeted fingerprinting approaches. In: Food authentication and traceability. Academic, pp 163–193

Kämper W, Trueman SJ, Tahmasbian I et al (2020) Rapid determination of nutrient concentrations in hass avocado fruit by vis/NIR hyperspectral imaging of flesh or skin. Remote Sens 12(20):1–195. https://doi.org/10.3390/rs12203409

Khodabakhshian R, Emadi B, Khojastehpour M et al (2019) A comparative study of reflectance and transmittance modes of Vis/NIR spectroscopy used in determining internal quality attributes in pomegranate fruits. J Food Meas Charact 13: 3130–3139. https://doi.org/10.1007/s11694-019-00235-z

Kuo FJ, Sheng CT, Ting CH (2008) Evaluation of ultrasonic propagation to measure sugar content and viscosity of reconstituted orange juice. J Food Eng 86(1):84–90

Kusumiyati AAM, Suhandy D (2021) Fast simultaneous and contactless assessment of intact mango fruit by means of near infrared spectroscopy. AIMS Agric Food 6(1):172–845. https://doi.org/10.3934/AGRFOOD.2021011

Kyriacou MC, Rouphael Y (2018) Towards a new definition of quality for fresh fruits and vegetables. Sci Hortic 234:463–469. https://doi.org/10.1016/j.scienta.2017.09.046

Lan W, Renard CMGC, Jaillais B et al (2020) Fresh, freeze-dried or cell wall samples: which is the most appropriate to determine chemical, structural and rheological variations during apple processing using ATR-FTIR spectroscopy? Food Chem 330:127357. https://doi.org/10.1016/j.foodchem.2020.127357

Landahl S, Gonzalez J, Van Linden V et al (2005) The acoustic impulse-response of apples in relation to internal damage and in relation to apoplasticp H. Acta Hortic 682:1403–1410

Larkin P (2017) Infrared and raman spectroscopy: principles and spectral interpretation. Elsevier, Amsterdam

Lawaetz AJ, Christensen SMU, Clausen SK et al (2016) Fast cross cultivar determination of total carotenoids in intact carrot tissue by raman spectroscopy and partial least squares calibration. Food Chem 204:7–13. https://doi.org/10.1016/j.foodchem.2016.02.107

Lebrun M, Plotto A, Goodner K et al (2008) Discrimination of mango fruit maturity by volatiles using the electronic nose and gas chromatography. Postharvest Biol Technol 48(1):122–131

Li JL, Sun D-W, Cheng J-H (2016) Recent advances in non-destructive analytical techniques for determining the total soluble solids in fruits: a review. Compr Rev Food Sci Food Saf 15(5):897–9115. https://doi.org/10.1111/1541-4337.12217

Liu Y, Sun X, Ouyang A (2008) Nondestructive measurement of soluble solid content of navel orange fruit by visible-NIR spectrometric technique with PLSR and PCA- BPNN. J Food Eng 85:652–631

Liu D, Zeng XA, Sun DW (2015) Recent developments and applications of hyperspectral imaging for quality evaluation of agricultural products: a review. Crit Rev Food Sci Nutr 55:1744–1757. https://doi.org/10.1080/10408398.2013.777020

López-Maestresalas A, Keresztes JC, Goodarzi M et al (2016) Non-destructive detection of blackspot in potatoes by vis-NIR and SWIR hyperspectral imaging. Food Control 70:229–241. https://doi.org/10.1016/j.foodcont.2016.06.001

Macdonald DD (1990) Some advantages and pitfalls of electrochemical impedance spectroscopy. Corrosion 46(3):229–242

Magwaza LS, Ford HD, Cronje PJ et al (2013) Application of optical coherence tomography to non-destructively characterise rind breakdown disorder of 'NulesClementine' mandarins. Postharvest Biol Technol 84:16–21. https://doi.org/10.1016/j.postharvbio.2013.03.019

Martínez AAM (2013) Non-destructive approaches for quality evaluation of eggplant (*Solanum melongena* L. cv. Traviata). Department of Bioresource Engineering Macdonald campus, McGill University Montreal, Quebec, Canada

Masoumi F, Gottardo A, Ibba P et al (2024) Capacitive impedance analysis for noncontact assessment of fruit quality and ripening. In: IEEE transactions on AgriFood electronics

May S, Droeschel D, Holz D et al (2009) Three-dimensional mapping with time-of-flight cameras. J Field Robot 26(11–12):934–965

Merzlyak MN, Solovchenko AE, Gitelson AA (2003) Reflectance spectral features and non-destructive estimation of chlorophyll, carotenoid and anthocyanin content in apple fruit. Postharvest Biol Technol 27:197–211

Mesa AR, Chiang JY (2021) Multi-input deep learning model with RGB and hyperspectral imaging for banana grading. Agriculture (Switzerland) 11(8):1–18. https://doi.org/10.3390/agriculture11080687

Metlenkin DA, Platov YT, Platova RA et al (2022) Non-destructive identification of defects and classification of hass avocado fruits with the use of a hyperspectral image. Agron Res 20(2):326–340. https://doi.org/10.15159/AR.22.027

Mishra P, Woltering E, Harchioui N (2020) Improved prediction of 'Kent' mango firmness during ripening by near-infrared spectroscopy supported by interval partial least square regression. Infrared Phys Technol 110:103459. https://doi.org/10.1016/j.infrared.2020.103459

Mizrach A (2004) Assessing plum fruit quality attributes with an ultrasonic method. Food Res Int 37(6):627–631

Mizrach A (2007) Nondestructive ultrasonic monitoring of tomato quality during shelf-life storage. Postharvest Biol Technol 46(3):271–274

Mizrach A, Galili N, Rosenhouse G et al (1991) Acoustical, mechanical and quality parameters of winter grown melon tissue. Trans ASABE 34(5):2135–2138

Morey R, Ermolenkov A, Payne WZ et al (2020) Non-invasive identification of potato varieties and prediction of the origin of tuber cultivation using spatially offset raman spectroscopy. Anal Bioanal Chem 412(19):4585–4945. https://doi.org/10.1007/s00216-020-02706-5

Munera S, Besada C, Blasco J et al (2017) Astringency assessment of persimmon by hyperspectral imaging. Postharvest Biol Technol 125:35–41

Muramatsu N, Sakurai N, Yamamoto R et al (2015) Comparison of a non-destructive acoustic method for firmness measurement of kiwifruit. Postharv Biol Technol 12(3):221–228. https://doi.org/10.1016/S0925-5214(97)00054-9

Naik S, Patel B (2017) Machine vision based fruit classification and grading-a review. Int J Comput Appl 170(9): 22–34.

Nagata M, Cao Q, Bato P et al (1997) Basic study on strawberry sorting system in Japan. In: 1997 ASAE annual international meeting technical papers. ASAE, St. Joseph

Newton M, Breeds E, Morris R (2017) Advances in electronics prompt a fresh look at continuous wave (CW) nuclear magnetic resonance (NMR). Electronics 6(4):89. https://doi.org/10.3390/electronics6040089

Nishani S, Deshpande S, Gundewadi G (2017) Use of acoustics as non-destructive techniques: a review. Int J Curr Microbiol Appl Sci 6:2468–2476

Nouri FG, Chen Z, Maqbool M (2014) Monitoring mango fruit ripening after harvest using electronic nose (zNoseTM) technique. In: 5th international conference on food engineering and biotechnology IPCBEE 65. IACSIT Press Singapore

Nováková K, Papež V, Sadil J et al (2024) Review of electrochemical impedance spectroscopy methods for lithium-ion battery diagnostics and their limitations. Monatshefte für Chemie-Chem Mon 155(3):227–232

Okamoto H, Murakami M, Kataoka T et al (2013) Machine vision for detecting insects in hole of raspberry fruit. IFAC Proc 46:350–354

Okere EE, Arendse E, Tsige AA et al (2022) Pomegranate quality evaluation using non-destructive approaches: a review. Agriculture (Switzerland) 12(12):1–255. https://doi.org/10.3390/agriculture12122034

Olarewaju OO, Bertling I, Magwaza LS (2016) Non-destructive evaluation of avocado fruit maturity using near infrared spectroscopy and PLS regression models. Sci Hortic 199:229–236. https://doi.org/10.1016/j.scienta.2015.12.047

Pandiselvam R, Kaavya R, Monteagudo SI et al (2022) Contemporary developments and emerging trends in the application of spectroscopy techniques: a particular reference to coconut (*Cocos nucifera* L.). Molecules 27(10):1–22. https://doi.org/10.3390/molecules27103250

Parpinello GP, Fabbri A, Domenichelli S et al (2007) Discrimination of apricot cultivars by gas multisensory array using an artificial neural network. Biosyst Eng 97(3):371–378

Pathak AK, Garg H, Yokozeki T et al (2023) Monitoring techniques to measure the rapid curing of polymer composites. In: Rapid cure composites. Woodhead Publishing, pp 71–84

Pathange LP, Mallikarjunan P, Marini RP et al (2006) Non-destructive evaluation of apple maturity using an electronic nose system. J Food Eng 77(4):1018–1023

Pathmanaban P, Gnanavel BK, Anandan SS (2019) Recent application of imaging techniques for fruit quality assessment. Trends Food Sci Technol 94:32–42

Petrisor C, Lucian Radu G, Balan V et al (2010) Rapid and non-destructive analytical techniques for measurement of apricot quality. Rom Biotechnol Lett 15:5213–5216

Pham UT, Phan QHT, Nguyen LP et al (2022) Rapid quantitative determination of multiple pesticide residues in mango fruits by surface-enhanced raman spectroscopy. Processes 10(3):1–14. https://doi.org/10.3390/pr10030442

Prakash A (2000) Thermal remote sensing: Concepts, issues and applications. Int Arch photogramm remote Sens 33(B1; PART 1):239–243

Prasad K, Jacob S, Siddiqui MW (2018) Fruit maturity harvesting and quality standards. In: Pre harvest modulation of postharvest fruit and vegetable quality. Academic, pp 41–69. https://doi.org/10.1016/B978-0-12-809807-3.00002-0

Purwanto YA, Budiastra IW, Darmawati E et al (2015) Measurement of starch and soluble solid content in papaya using near infrared spectroscopy. J Chem Pharm Res 7(6):112–165

Qiu S, Wang J (2015) Application of sensory evaluation, HS- SPME GC-MS, E-nose, and E-tongue for quality detection in citrus fruits. J Food Sci 80(10):S2296–S2304

Rahman A, Wang S, Yan J et al (2021) Intact macadamia nut quality assessment using near-infrared spectroscopy and multivariate analysis. J Food Compos Anal 102:104033

Rajkumar P, Wang N, Eimasry G et al (2012) Studies on banana fruit quality and maturity stages using hyperspectral imaging. J Food Eng 108(1):194–200. https://doi.org/10.1016/j.jfoodeng.2011.05.002

Ravindran A, Ravindran A (2015) A review on non-destructive techniques for evaluating quality of fruits. Int J Eng Res Technol 2278–0181

Rehman M, Abu Izneid AJA, Abdullah MZ et al (2011) Assessment of quality of fruits using impedance spectroscopy. Int J Food Sci Technol 46:1303–1309

Ribera-Fonseca A, Noferini M, Rombolá AD (2016) Non-destructive assessment of highbush blueberry fruit maturity parameters and anthocyanins by using a visible/near infrared (vis/NIR) spectroscopy device: a preliminary approach. J Soil Sci Plant Nutr 16(1):174–186

Rivera N, Ve'lez Gómez-Sanchis J et al (2014) Early detection of mechanical damage in mango using NIR hyperspectral images and machine learning. Biosyst Eng 122:91–98. https://doi.org/10.1016/j.biosystemseng.2014.03.009

Rungpichayapichet P, Mahayothee B, Nagle M et al (2016) Robust NIRS models for non-destructive prediction of postharvest fruit ripeness and quality in mango. Postharvest Biol Technol 111:31–40. https://doi.org/10.1016/j.postharvbio.2015.07.006

Saevels S, Lammertyn J, Berna AZ et al (2003) Electronic nose as a non-destructive tool to evaluate the optimal harvest date of apples. Postharvest Biol Technol 30(1):3–14

Sanchez PDC, Hashim N, Shamsudin R et al (2020) Applications of imaging and spectroscopy techniques for non-destructive quality evaluation of potatoes and sweet potatoes: a review. Trends Food Sci Technol 96:208–221. https://doi.org/10.1016/j.tifs.2019.12.027

Sankhe D (2015) Ripeness inspection system for banana. In: CAE Proceedings on international conference on communication technology. International Journal of Computer Applications (0975-8887)

Sebben J, Antônio ES et al (2018) Development of a quantitative approach using raman spectroscopy for carotenoids determination in processed sweet potato. Food Chem 245:12–31. https://doi.org/10.1016/j.foodchem.2017.11.086

Si W, Xiong J, Huang Y et al (2022) Quality assessment of fruits and vegetables based on spatially resolved spectroscopy: a review. Food Secur 11(9):1–215. https://doi.org/10.3390/foods11091198

Simić M, Stavrakis AK, Kojić T (2023) Parameter estimation of the Randles equivalent electrical circuit using only real Part of the impedance. IEEE Sensors J 23(5):4922–4929. https://doi.org/10.1109/JSEN.2023.3238074

Sinanoglou VJ, Tsiaka T, Aouant K et al (2023) Quality assessment of banana ripening stages by combining analytical methods and image analysis. Appl Sci (Switzerland) 13(6):3533

Soltani M, Alimardani R, Omid M (2010) Prediction of banana quality during ripening stage using capacitance sensing system. Aust J Crop Sci 4(6):443–447

Srivastava S, Sadistap S (2018) Non-destructive sensing methods for quality assessment of on-tree fruits: a review. J Food Meas Charact 12(1):497–526

Srivichien S, Terdwongworakul A, Teerachaichayut S (2015) Quantitative prediction of nitrate level in intact pineapple using vis-NIRS. J Food Eng 150:29–34. https://doi.org/10.1016/j.jfoodeng.2014.11.004

Studman CJ (1999) Fruit and vegetables: fruit and vegetable quality. In: CIGR (ed) CIGR handbook of agriculture engineering, vol 4, pp 243–272

Su WH, Sun DW (2019) Rapid determination of starch content of potato and sweet potato by using NIR hyperspectral imaging. HortScience 54:S38

Su WH, Sun DW, He JG et al (2017) Variation analysis in spectral indices of volatile chlorpyrifos and non-volatile imidacloprid in jujube (*Ziziphus jujuba* Mill.) using near-infrared hyperspectral imaging (NIR-HSI) and gas chromatograph-mass spectrometry (GC–MS). Comput Electron Agric 139:41–55

Su WH, Bakalis S, Sun Da W (2019) Chemometrics in tandem with near infrared (NIR) hyperspectral imaging and fourier transform mid infrared (FT-MIR) microspectroscopy for variety identification and cooking loss determination of sweet potato. Biosyst Eng 180:70–86. https://doi.org/10.1016/j.biosystemseng.2019.01.005

Taniwaki M, Hanada T, Sakurai N (2009) Postharvest quality evaluation of Fuyu and Taishuu persimmons using a non-destructive vibrational method and an acoustic vibration technique. Postharvest Biol Technol 51:80–85

Tian HQ, Ying YB, Lu HS et al (2007) Measurement of soluble solids content in watermelon by Vis/NIR diffuse transmittance technique. J Zhejiang Univ Sci B 8:105–110

Uarrota VG, Pedreschi R (2022) Mathematical modelling of hass avocado firmness by using destructive and non-destructive devices at different maturity stages and under two storage conditions. Folia Hortic 34(2):139–150. https://doi.org/10.2478/fhort-2022-0011

Vadivambal R, Jayas DS (2011) Applications of thermal imaging in agriculture and food industry – a review. Food Bioprocess Technol 4:186–199

Vargas-Murga L, de Rosso VV, Mercadante AZ et al (2016) Fruits and vegetables in the Brazilian Household Budget Survey (2008–2009): carotenoid content and assessment of individual carotenoid intake. J Food Compos Anal 50:88–96

Varith J, Hyde GM, Baritelle AL et al (2003) Non-contact bruise detection in apples by thermal imaging. Innov Food Sci Emerg Technol 4(2):211–218

Wang M, Zhai G (2018) Application of deep learning architectures for accurate and rapid detection of internal mechanical damage of blueberry using hyperspectral transmittance data. Sensors (Switzerland) 18(4):1–145. https://doi.org/10.3390/s18041126

Wang X, Xue L, He X et al (2011) Vitamin C content estimation of chilies using Vis/NIR spectroscopy. In: Proceedings of the 2011 international conference on electric information and control engineering (ICEICE), Wuhan, China. IEEE, New York, pp 1894–1897

Wang Z, Walsh KB, Verma B (2017) On-tree mango fruit size estimation using RGB-D images. Sensors (Switzerland) 17(12):1–155. https://doi.org/10.3390/s17122738

Wang K, Li Z, Li J et al (2021) Raman spectroscopic techniques for non-destructive analysis of agri-foods: a state-of-the-art review. Trends Food Sci Techno1 18:490–504. https://doi.org/10.1016/j.tifs.2021.10.010

Xiao F, Wang H, Xu Y et al (2023) Fruit detection and recognition based on deep learning for automatic harvesting: an overview and review. Agronomy 13(6):1625

Yahaya OKM, Matjafri MZ, Aziz AA (2011) Non-destructive quality evaluation of fruit by color based on RGB LEDs system. In: 2nd international conference on electronic design ICED 1001, pp 230–233. https://doi.org/10.1109/ICED.2014.7015804

Yamamoto H, Iwamoto M, Haginuma S (1980) Acoustic impulse response method for measuring natural frequency of intact fruits and preliminary applications to internal quality evaluation of apples and watermelons. J Texture Stud 11(2):117–136

Ye D, Sun L, Tan W et al (2018) Detecting and classifying minor bruised potato based on hyperspectral imaging. Chemometr Intell Lab Syst 177:129–139. https://doi.org/10.1016/j.chemolab.2018.04.002

Yildiz F, Özdemir AT, Uluışık S (2019) Evaluation performance of ultrasonic testing on fruit quality determination. J Food Qual 1:6810865

Yodh A, Chance B (2017) Spectroscopy, and imaging with diffusing light. Phys Today 48:34–40

Zakaria A, Shakaff AYM, Adom AH et al (2011) *Magnifera indica* cv. Harumanis classification using E-Nose. Sensor Lett 9(1):359–363

Zdunek A, Cybulska J, Konopacka D et al (2011) Evaluation of apple texture with contact acoustic emission detector: a study on performance of calibration models. J Food Eng 106:80–87

Zhang L, Huang Y, Sun F et al (2021) The effect of maturity and tissue on the ability of mid infrared spectroscopy to predict the geographical origin of banana (*Musa cavendish*). Int J Food Sci Technol 56(6):2621–2275. https://doi.org/10.1111/ijfs.14960

Zude M (2003) Non-destructive prediction of banana fruit quality using VIS/NIR spectroscopy. Fruits 58:135–142. https://doi.org/10.1051/fruits:2003001

Zude M, Herold B, Roger J et al (2006) Non-destructive tests on the prediction of apple fruit flesh firmness and soluble solids content on tree and in shelf life. J Food Eng 77:254–260. https://doi.org/10.1016/j.jfoodeng.200

Chapter 6
Minimal Processing of Fruits and Vegetables

Reshma Saroj, Radha Kushwaha, and Devinder Kaur

6.1 Introduction

Growing urbanisation, broken lifestyles, nutritional consciousness, and dual income have all contributed to the increasingly demanding consumer. Convenience and quality are the food market's new goals. Fresh fruit and vegetable consumption has increased due to consumers' growing interest in these healthful foods that are low in cholesterol, little to no fat and sodium, high quantities of vitamins and minerals, fibre, and, as of late, possible antioxidants. In this sense, the foods of the future would be minimally processed frozen fruits and vegetables. Fast-food chains and the lodging business can benefit greatly from less processing. Airlines have consistently expressed a need for pre-cut salads, shredded cabbage, and onions. Strict hygienic practices in processing, advanced packaging technology, and maintaining the proper temperature during storage and transit are all necessary for this area to succeed. It is evident that for a considerable amount of time, researchers in the domains of food science, nutrition, technology, associated disciplines, and engineering have examined and investigated the many aspects of minimum processing (MP) food, particularly in the areas of uncooked, fresh, and cold-preserved foods. The practice of preserving trimmed and chopped fruits and vegetables without significantly altering their fresh qualities is known as minimal processing. These goods work well in salads, curries, oatmeal, and custard-like dishes. Necessary conditions for minimal fruit and vegetable processing include high-quality raw resources, such as selecting suitable cultivars or varieties, applying appropriate cultivation techniques, and ensuring optimal harvesting and storage conditions to maintain the freshness, safety, and overall quality of the final product. Sorting, sizing, and grading the raw materials ensure that the final goods are uniform and standardised for purchase and sale. Size, form, colour, hardness, flavour, friability, bruising, illness,

R. Saroj · R. Kushwaha · D. Kaur (✉)
Centre of Food Technology, University of Allahabad, Allahabad, Uttar Pradesh, India

OP Chauhan (ed.), *Fruits and Vegetables Technologies*,
https://doi.org/10.1007/978-981-96-8433-5_6

soundness, cut surfaces, and chemical composition are the most crucial grade factors. Products that are undersized, overripe, and flawed are segregated from those that are of a respectable calibre. Thorough cleaning and washing both before and after peeling: for the majority with fresh produce and minimally processed fruits and vegetables, cleaning and washing may be the only preservation procedures. Cleaning is the process of getting rid of foreign objects. Cleaning is a sort of separating action that involves cleaning fruits and vegetables, containers, and equipment of debris, twigs, stalks, dirt, sand, mud, insects, and pesticide and fertiliser residues. Along with these methods, the cleaning process also includes picking, screening, destoning, floating, and gravity-based separation of light from heavy components. In a minimally processed fruit and vegetable processing line, human contact with the product should be minimised, and washing should take place in a separate chamber with few openings. High-grade water is suitable for washing: the essential component of the minimally processed goods is water. Water quality and source must be taken into account. It is imperative that the water utilised is potable, devoid of any unwanted flavour or odour, and colourless.

6.2 Minimal Processing

Fruits and vegetables that have had minimal processing are typically sorted, cleaned, washed, peeled, trimmed, and deseeded before being cut to a precise size (Fig. 6.1). Fruits and vegetables with minimal processing (MPFV) must have the scent, taste, texture, and aesthetic appeal of freshly harvested food in addition to appearing fresh. Therefore, only high-quality, fresh vegetables should be utilised as the initial ingredient in minimal processing. Additionally, MPFV needs to be healthy, safe, and nourishing.

A more elusive, ambiguous topic could be the categorization of cucumbers that have been cleaned and waxed. These goods would likely not be classified as MP foods since, despite being packed (in wax), they are undamaged too. Pre-cut or sliced cucumbers, on the other hand, are classified as MP foods and have a shorter shelf-life than waxed intact cucumbers. Fresh produce that is unspoiled and may go through standard postharvest handling procedures, including size, grading, cleaning, waxing, and storing in a modified atmosphere (MA), should not be included in MP fruits and vegetables. A subset of goods typically categorised as chilled meals that have been kept include frozen foods. However, MP foods differ from frozen foods in that, based on the product or mixes of commodities, MP foods require cold-chain refrigeration or somewhat higher temperatures, while foods that are frozen need to be transported and kept at −10 °C or below (Wiley 1994). Table 6.1 deliberates the range of minimally processed (MP) chilled fruits and vegetables food storage systems.

Food that is frozen has a different overall quality than food that is similar to fresh food from MP. The consumer can frequently tell the textural changes between MP

Fig. 6.1 Standard production flow diagram for preparing minimally processed fruits and vegetables

Harvesting

↓

Receiving

↓

Pre-cooling

↓

Washing & Disinfection

↓

Peeling, Trimming, De-seeding

↓

Cutting into specific sizes

↓

Sorting for defects

↓

Dipping ← Antimicrobial Agent, Anti-browning Agent

↓

Drying change to surface moisture

↓

Packaging & Labeling

↓

Storage at 3-6°C

↓

Distribution

and frozen food. For example, some sous vide meal producers freeze their goods, which essentially turns the food into a frozen dish.

Food with minimal processing (MP) does not display "commercial sterility", which is a characteristic of thermally processed food. This is a significant

Table 6.1 Range of minimally processed (MP) chilled fruits and vegetables food storage systems ([*]Product freshness)

Food preservation and processing categories	Not preserved, maybe home cooked	Minimally processed, chilled, and possibly prepared at home	Processed, frozen, and kept in the cold; it can be made at home	Irradiation, electronic processing methods	Dehydrated	Canning, sterilisation, pasteurisation, ultra-high temperature processing, and frying are examples of heat processing	[*]Non-thermal and electronic food processing and preservation methods
Food quality colour, flavour, texture	Fresh	Like-fresh	Slightly to fully modified[**]	Slightly modified[**]	Slightly to fully modified[**]	Fully modified[**], but safe, no microbial contamination	Slightly modified[**]
Storage, shelf-life	May not be refrigerated	Requires transportation, refrigeration, and storage	Necessitates market display, transportation, and frozen or chilled storage	Necessitates cold storage	Shelf stable at ambient temperatures	Extended shelf-life under ambient conditions	Shelf stable, requires refrigerated transportation, storage
Packaging	Could not be packaged	Needs to be packaged	Needs to be packaged	Needs to be packaged	Needs to be packaged	Requires hermetic, aseptic packaging	Needs to be packaged
Cell integrity	No change	No change	Changes	Changes	Changes	Total destruction	Changes
Enzymes (activity and structure)	No change	No change	Changes	Changes	Changes	Total destruction	Changes
Phytochemicals, vitamins, proteins	No change	No change	Changes	Changes	Changes	Highly changed	Changes
Free radical formation	None	None	A few radical free	Free radicals formation	A few radical free	Free radicals formation	Free radicals formation

[*]Non-thermal and electro-technologies of the preservation and processing of food consists of:
[a]High pressure, ultra-high pressure (HPP), [b]Plasma, [c]Pulsed light and pulsed electric fields, [d]Ohmic heating, [e]Ionizing radiation, f Pulsed X-ray, [g]Ultrasound, [h]High-voltage arc discharge, [i]Magnetic fields, [j]Dense phase carbon dioxide, [k]Microwaves, [l]Electron beam processing
[**]Product freshness

distinction between MP and thermally processed foods. According to this definition, no live microorganism (including spores) of public health concern can reproduce, and foods that have undergone thermal processing are free of microorganisms that can multiply in an unrefrigerated environment. Applying heat, lowering pH, or controlling water activity can all be used to achieve this "sterility". Figure 6.2 shows the product both in its original packaging and after surface modifications like peeling or slicing. There is a chance that minimally prepared vegetables and fruits that

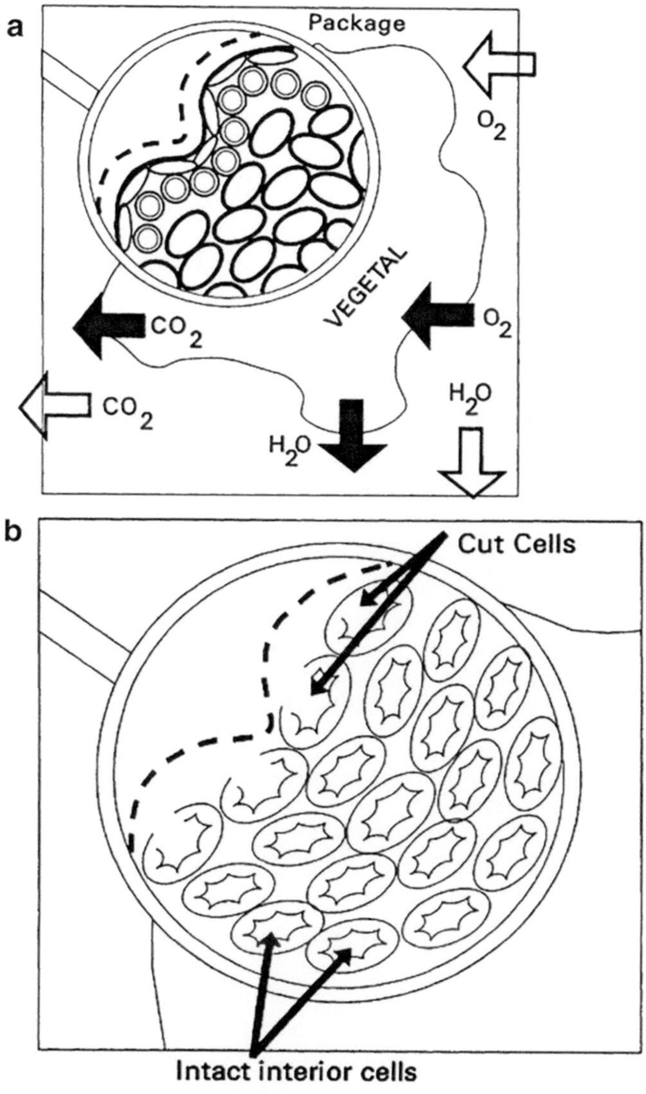

Fig. 6.2 Diagram showing the product both in its original packaging and after surface modifications like peeling or slicing (CTIFL, France)

are judged safe will be included in the definition of "commercially sterile". If the term "commercial sterility" or "microbial and enzymatic safety" is indeed applicable to MP fruits and vegetables, then the limits of microbes and enzymes for these commodities have not yet been established. Food that has been thermally treated materials cannot be compared to fresh. However, the relatively new notion of minimal processing foods can benefit from the vast quantity of numerous studies on how fruits and vegetables that are fresh, frozen, irradiated, dried, and thermally treated are handled, processed, and preserved (Rolle and Chism III 1987).

6.3 Standard of Produce that is Fresh and other Plant Foods with Minimal Processing

As shown in Table 6.2, a number of factors influence the calibre and duration of minimally processed produce from the farm to the fork. The quality attributes of MP fruits and vegetables have been investigated and recorded by numerous researchers (Diaz-Mendoza et al. 2016; Kårlund et al. 2014). One physiological process that is essential to a plant's life is leaf senescence. Cellular structures are broken down, large amounts of macromolecules are broken down, and nutrients are efficiently transferred from ageing foliage to developing organs or tissues that sink. According to Diaz-Mendoza et al. (2016) and Yildiz (2007), the extension of shelf-life in MP fruits and vegetables will be directly impacted by a reduction of senescence and cell membrane peroxidation. The health advantages of minimally processed fruits and vegetables have been highlighted by numerous studies conducted in numerous countries during the past 20 years (Crowe and Sheppard 2011; Carter et al. 2010).

A growing body of epidemiological and experimental research suggests that eating fruits and vegetables has numerous health advantages (Hughes 2011; Cho et al. 2010). Plant diets contain around 10,000 phytochemicals, which include anti-inflammatory and antioxidant compounds, among many others (World Cancer

Table 6.2 An overview of the variables influencing the general standard of the fruits and vegetables with the least amount of processing

Parameters	Affects
Agricultural activities used before harvest	Constituency and phytonutrients
Crop-growing techniques	Dimensions, form, scars, and pesticide residues
Harvest procedures	Colour, maturity, and brix/acid
After-harvest management	Biochemical alterations and mechanical harm
Processing	Browning, microbial development, and weight loss
Storage, handling, and packaging	Loss, softness, deterioration, weight, and surface drying
Changes in fruit and vegetable respiration rates (O_2/CO_2) from harvest to fork	Senescence, respiration rate variations, ethylene generation, yellowing, water loss, size changes, phytonutrient loss, and shelf-life

Research Fund 2007). For instance, in humans, polyphenols have cardio-protective and antioxidant properties (Anna et al. 2014).

6.4 Treatment Methods that Reduced the Quantity of Fruit and Vegetable Processing

6.4.1 Harvesting and Receiving

Just as crucial as understanding how to cultivate vegetables is knowing when to pick them. Every vegetable has a best time of year to be harvested. Certain veggies can tolerate being left in the garden for a little while, while others can quickly transform from being nice and soft to tough and bitter. Prior to harvesting, the following elements should be taken into account since they could be a crucial first step in the fruit and vegetable processing system.

To prevent contamination and damage during handling, fresh product of good quality must be chosen by a well-trained individual. During fresh-cut processing, the handler is responsible for making sure that sanitary tools such as padded cartons, sacks, and sharp harvesting knives are used. Large enterprises employ mechanical harvesters to handle produce closer to the field, extending its shelf-life and preserving its quality. Employees are required to follow hygienic procedures and wear clean clothes.

6.4.2 Pre-cooling

To increase the shelf-life of fresh-cut food, pre-cooling should ideally try to remove field heat prior to processing or storage. Numerous approaches, including forced air cooling, vacuum cooling, chilled water baths, and ice-water packed item mixes, can be used to accomplish it. It is important to maintain a temperature that will prevent fruits and vegetables from becoming too cold. To prevent cross-contamination, fresh produce that is still intact should be kept apart from produce that has been prepared beforehand and cleaned within a tiny cold storage space. Employees should also manage stock rotation using the First-in, First-out (FIFO) policy.

When harvesting leafy vegetables that do not need to be harvested mechanically, they are usually cooled right away to eliminate field heat, sorted to get rid of debris, cleaned to get rid of dirt, and then wrapped or packed for retail and shipping. Vegetables are typically wrapped as entire plants since chopping them may damage the cells and release ethylene, which accelerates senescence and reduces shelf-life. Storing at low temperatures is crucial for maintaining the quality of leafy greens. However, storing vegetables below refrigerator temperature might cause chilling harm and quick deterioration of quality. Fresh vegetable postharvest losses in underdeveloped nations without access to refrigeration can reach up to half of the harvest. For roots and legumes, whose harvesting is frequently carried out by machines, some sorting and grading are done in the field or at collection stations. These vegetables are often processed in large quantities, and further preparation steps are taken prior to delivery. Vegetables that must be stored for extended periods of time may need to be treated to stave off insects, tiny animals, and microbial degradation.

Cucumbers are one crop that may be made more durable and aesthetically pleasing by washing and waxing them.

6.4.3 Washing

According to Seymour et al. (2002) and Soliva-Fortuny et al. (2002), the first step in minimal fresh fruit processing usually involves cleaning the entire fruit to remove undesirable detritus, pesticide traces, plant remnants, soil, insects, and foreign objects, as well as to decrease the rate of the enzymatic discoloration processes. MPFV has been cleaned using a variety of techniques. Washing can be done on a small scale by running water through a filter in a washbasin or under a tap continually. Generally, two washing tanks are used: one for removing severe soiling and one for final washing using cleaner tanks that contain chlorinated water. Chlorination is not very successful after pathogens have invaded their host, although several scholars have utilised sodium or calcium hypochlorite and other salts for fruit surface cleanliness and sterilisation to avoid microbial inoculation (Hong and Gross 1998). For instance, the duration of fresh-cut tomatoes can be extended by surface sterilising whole tomatoes with sodium hypochlorite, potassium bicarbonate, calcium chloride, and calcium lactate (Hong and Gross 1998; Artes et al. 1999; Gil and Allende 2012). Furthermore, Aguayo et al. (2003) disinfected whole melons prior to their minimal fresh processing using a pH 6.4 water solution containing 1.4 mM of acidified hypochlorite. For optimal results while washing, it is important to keep an eye on the chlorine content, pH, temperature, and contact time. Wash water disinfection levels can also be tracked via ORP (oxidation reduction potential) measurements.

6.4.4 Peeling, Trimming, and Deseeding

Abrasive peelers, large-scale automatic trimmers, and small-scale knives are some of the equipment types utilised for peeling, trimming, and deseeding. Some large processing companies utilise automated peelers with abrasive rollers to peel carrots and potatoes. Certain high-volume crops have prompted the invention of machinery that makes use of steam, lye, and air under high pressure to facilitate the peeling process. However, hand peeling works well for modest jobs.

6.4.5 Cutting Operations/and Sorting for Defects

A variety of processes, including chopping, slicing, shredding, peeling, dicing, and sectioning, may be included in the cutting procedure. Many small-scale companies primarily need human labour to complete these tasks. It is necessary to take the cutting dimensions into account as well. Strict adherence to Good Hygienic Practice (GHP) is necessary to prevent contamination when cutting. Employees should receive the attention they deserve. They need to be well trained and use safety gear, including aprons, gloves, and hair nets. Sorting is important because it separates defective items from the others, improving shelf-life and homogeneity of the completed product, fresh-cut.

6.4.6 Dipping

It is an alternative to dip produce in a mixture of an acidulant and an antioxidant or in an anti-softening agent such as calcium chloride, like a mixture of ascorbic acid and citric acid. Pre-treatments based on a combination of several chemicals, such as 1% ascorbic acid and 0.2% citric acid, have also been used to control enzymatic browning in fresh-cut apples. It has been found that dipping apple slices with additional browning inhibitors, like cysteine or 4-hexylresorcinol, can effectively delay the browning of the cut surface. Additionally, some edible coatings can extend the shelf-life of products by lowering respiration. Edible coatings lower respiration, water loss, and oxidation reaction rates by acting as a semipermeable barrier against the migration of oxygen, carbon dioxide, moisture, and solvents.

6.4.7 Drying

Produce dipping necessitates removing the moisture from the surface of the product before packaging to prevent the formation of mould and other microorganisms and texture degradation. A few mechanical and manual techniques used to remove water from produce are solar drying, convective drying, freeze drying, osmotic drying, and radiation drying:-

6.4.7.1 Solar Drying

Since the dawn of time, the sun, a free and limitless source of energy, has been used to dry agricultural products. It is as old as the universe itself. There are two types of solar drying: direct and indirect. This is a conventional technique that dries food items by using the sun. Before being packed, produce is dried off by exposing it to the sun for a few days. This occurs frequently in poor nations where fuel is expensive and in short supply. The chamber, chimney, and wind-ventilated dryer types are examples of solar dryers. The indirect technique of solar drying uses a solar system to produce heat, which is then transferred into the food to be dried through air flow. To get rid of the evaporating moisture, the top of the drying chamber is vented.

6.4.7.2 Convective Drying

Modern heating equipment uses the convective drying method to remove water from agricultural produce by heat transfer. This involves applying hot air to effectively remove moisture from food products and deliver heat to them.

6.4.7.3 Freeze Drying

Freeze drying is a method that crystallises the solvent (often water) and/or suspension medium at low temperatures and then sublimates them out. Water can sublimate, or change directly from a solid to a gaseous state without melting. Thus, freeze drying is the technique of freezing a food product and using direct sublimation to remove the food's related solvents.

6.4.7.4 Osmotic Drying

The process of using a hypertonic solution to dry food is known as osmotic drying. Plant tissues are dried by immersing them in a hypertonic solution, which causes a concentration difference between the moisture of the food to be dried and the solution.

The high osmotic pressure of the hypertonic solution provides the force needed for the diffusion of water from the tissue into the solution. The osmotic solution's solutes counter-diffusionally diffuse into the tissue at the same time that water does, and food products' solutes do the same.

6.4.7.5 Radiation Drying

Radiation drying is an alternate technique that might be utilised to get around the issues with hot air drying. Food goods are dried using microwave radiation, which uses electromagnetic energy. This is the process of spreading microwave heat over space using an electric and magnetic field. It has been discovered that using microwave heating to dry food items results in higher-quality dried goods that require less time and heat to eliminate moisture.

6.5 Factors and Processing Operations that Affect the Quality of Minimally Fresh Processed Plant Foods

The shelf-life and microbial quality of raw prepared fruit and vegetables can be affected by a number of factors, including good agricultural practices in crop cultivation, good hygienic practices during harvest and handling, the quality of washing water, processing technologies, packaging methods and materials, transportation, and processing and storage temperatures (Artés and Artés-Hernández 2000; Artés 2004). Figure 6.3 lists the most common parameters influencing the shelf-life of minimally fresh processed fruits, vegetables, and drinks. Traceability is a key concept in today's food supply chain, which often consists of multiple steps from storage and transportation to consumption (Allende et al. 2004).

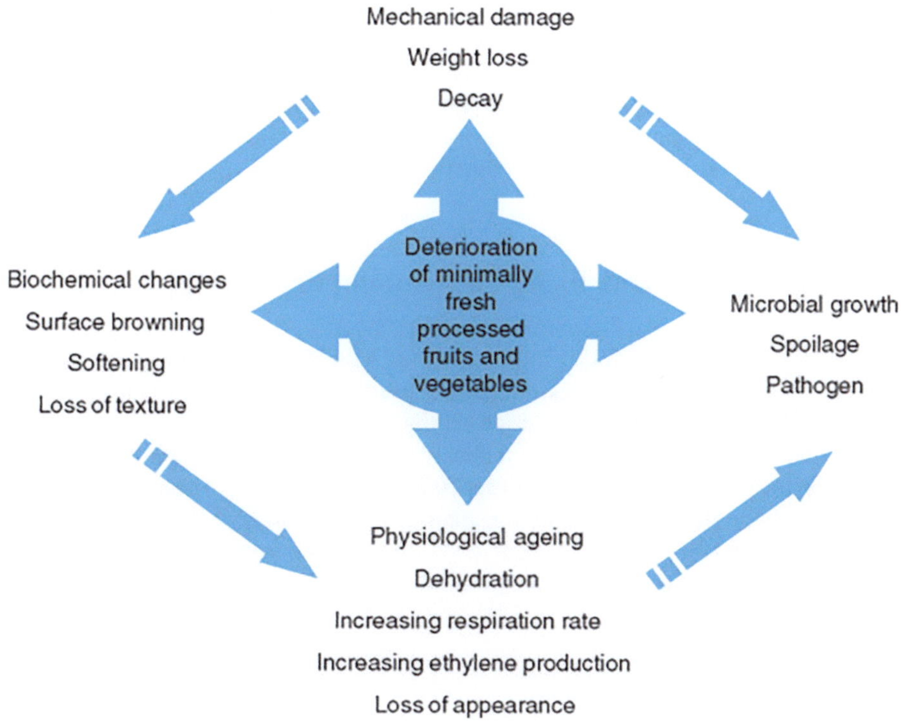

Fig. 6.3 Factors influencing the shelf-life and deterioration of slightly fresh processed fruits and vegetables

6.6 Packaging

The market has seen a rise in the popularity of minimally processed fresh produce, which is a response to consumer desires for convenient foods. Additional preparation procedures for these veggies include washing, sorting, grading, chopping, and packing into containers suitable for retail sales. Vacuum packing and modified-atmosphere (MA) packaging are used to increase the items' shelf-life. Usually, an atmosphere with low oxygen content and high carbon dioxide content replaces the air. The respiration rate and, consequently, the senescence of chopped vegetables can be slowed by this altered atmosphere. The most popular goods in the marketplaces of America and Europe are different kinds of chopped lettuces mixed with shredded cabbage, carrots, and other vegetables. With the use of "clean room" concepts in modern packaging processes, vegetable products like stir-fry mix and salad mix can have shelf-lives that are almost identical to those of the whole plants. The items have a long enough shelf-life to be delivered to customers even when they are sent in refrigerated containers to foreign countries. Vegetables that have undergone minimal processing typically do not have preservatives and have not been heated or chemically treated. The drawback of these items is that they require refrigeration for storage, which restricts their use to industrialised nations.

6.7 Emerging Technologies for Minimal Processing

To increase the shelf-life following the various MPFV manufacturing preparation phases, preservation procedures should be used. These strategies can lessen or improve MPFV's vulnerability to microbiological or physiological spoiling, which is why the food industry must continuously conduct research to create new techniques or improve current ones in order to increase shelf-life. These technologies were initially produced in research on small-scale labs, where it is inexpensive and easy to regulate the parameters and conditions of the equipment. This section describes novel developing technologies and discusses how they affect quality.

6.7.1 Processed Sous Vide

The process of preparing uncooked food in vacuum-sealed packaging, often known as "sous vide" cooking, originated in France and is now widely used to produce superior prepared meals with a short shelf-life in the fridge. Given the name suggests, the unprocessed substance is vacuum-packed in multilayer plastic packaging, boiled in water, or removed from oxygen by air/steam mixes or microwaves operating at a temperature lower than 100 °C. It is quickly cooled in an ice bath or blast chiller to +3 °C. Nutrient loss is reduced, and outstanding quality is produced.

Additionally, there is less chance of post-processing contamination, and catering has more freedom and a wider selection of products.

This technique is being used to treat a large assortment of fish, meat, and vegetable items, mostly for the institutional market but also finding its way into the Benelux and French consumer markets (Clausen 2010; Schellekens 1996). Good hygiene control and monitoring are essential to the sous vide's quality and safety goods throughout the whole preparation, packing, processing, and distribution chain. It is highly advised to use a Hazard Analysis Critical Control Point (HACCP) method. The possibility of pathogens that thrive and survive at low temperatures, such as *Listeria monocytogenes*, is especially high for sous-vide products. According to guidelines, a product should be heated to 70 °C for at least 2 min before being distributed and stored at 3 °C or lower, and for a maximum of 7 days.

6.7.2 Bio-control

While the primary method of controlling postharvest diseases is still the use of synthetic fungicides, a number of factors, including the requirement that minimise after harvest, the application of chemicals, have accelerated the creation of substitute strategies, like finding microbial antagonists to regulate postharvest decay (Nguyen and Carlin 1994; Ippolito and Nigro 2003). Holzapfel (1998) recommended against using protective microorganisms as a replacement for appropriate production and packing of marginally recently prepared food items; rather, they should be used in conjunction with excellent production procedures. Therefore, it may be thought of using biocontrol cultures to improve current hurdle technology in order to stop the spread of infections (Breidt and Fleming 1997). Many foods have been fermentatively preserved for generations using lactic acid bacteria (LAB), and attempts have even been made to preserve fresh fruit and vegetables (Vescovo et al. 1996; Carlin et al. 1995). Since LAB may be capable of acidifying or fermenting vegetable substrates, the only consideration that needs to be made is that the product's quality needs to be assessed after application (Carlin et al. 1995). Many different H_2O_2 enzymes (lysozyme and the lactoperoxidase system with H_2O_2), organic acids (lactic and acetic acids), and low-molecular-weight compounds that promote bacterial growth are examples of the inhibitory metabolites of LAB. Thus, to reduce spoiling and harmful bacteria in fresh-cut vegetables, introduce strains that produce antibacterial agents or utilise culture seeps through in conjunction with chilled storage and sanitary handling procedures (Torriani and Massa 1994). Bacteriocins (nisin and others) and reuterin, diacetyl, and fatty acids are examples of metabolites (Breidt and Fleming 1997). Additionally, more bacteriocins and organic acids, which prevent the spread of pathogens in food.

6.7.3 High Hydrostatic Pressure

Processing under high hydrostatic pressure involves applying a pressure treatment between 4000 and 9000 atm to food products. In order to extend the shelf-life of food goods, high hydrostatic pressure is employed in order to stop microbial growth and some enzymes. The degree of microbial inactivation varies depending on the food's composition, pH, and osmotic pressure, and ambient temperature. Gram-negative bacteria are known to be suppressed at a reduced pressure compared to their Gram-positive counterparts. It is possible to control the suppression of microbial spores by utilising frigid temperatures in conjunction with the high-pressure treatment.

Citrus juices that have not been pasteurised and have a shelf-life of approximately 17 months can be produced with high hydrostatic pressure while maintaining a flavour that is comparable to fresh juice and preserving vitamin C (Farr 2018). One of the most significant advantages of this new method is that fruit juice is placed inside the product's plastic container, which is then compressed for 1–30 min at a pressure of between 4000 and 6000 atm. As a result, it is simple to prevent product contamination following packaging.

6.7.4 Pulsed Electric Fields

Pulsed electric fields (PEF) are a newly developed technology that aims to stop microbes from surviving without causing major changes to organoleptic qualities. Its foundation is the application of brief electric pulses, either in batch or continuous flow, to samples positioned in a chamber between two electrodes, with field intensities between 0.1 and 80 kV/cm. Bipolar, square wave, oscillatory, exponentially decaying, or oscillatory pulses with different temperatures can all be used to apply an electric field. It is being done at a lower or moderate intensity (0.1–5 kV/cm). Research is a possible instrument to boost secondary metabolism in plants and trigger the synthesis of chemicals of interest. In addition, the objective determines the intensity used: high intensity (15–40 kV/cm) enables the inactivation of microbes.

Electropermeabilization, which happens when a field of electricity is introduced, is the basis for PEF's mode of action. It can be applied to encourage the contents inside the cells of any microbes present in the item to seep out, thereby rendering the bacterium inactive. Theories that complement one another have been put out to explain the effects of an electric field on living cells. Sale and Hamilton presented the dielectric rupture theory in 1968. It illustrates how applying an electric field causes cells' inherent trans membrane potential to rise, resulting in localised membrane thinning. Consequently, reversible pores form; nevertheless, irreversible pores will form if the intensity of the electric field increases.

The electrical conductivity of the medium and the cell radius both affect the trans-membrane potential. According to Schoenbach et al. (2000), pores can only enlarge to a certain point before they become irreversible. It takes a sufficiently long pulse duration to reach critical size. These ideas propose that separate and/or complementary mechanisms cause bacterial membranes to break. Optimising treatment conditions for specific products, equipment, and microorganisms is still crucial in order to render them inactive. In general, enzymes are more resilient to electric fields that pulse. Certain enzymes, such as lipoxygenase, polyphenol oxidase, pectin methyl esterase, and peroxidase (all of vegetable origin), may continue to function after treatment and have an unfavourable effect on the calibre of the product. Because of this, maintaining quality requires the application of extra supplementary treatments that guarantee enzymatic inactivation.

6.7.5 Ultrasound

Treatments with ultrasound (US) are renowned for their high-purity product attainment, simplicity, low cost, and repeatable outcomes. A longitudinal wave that exceeds the top frequency threshold (20,000 Hz) perceptible to humans is referred to as US. Applying this non-thermal, non-destructive, non-invasive, and nonintrusive technology to MPFP can stop microbiological and biochemical spoiling. Little, fleeting air bubbles are created by the high energy used in ultrasonic treatments, and these bubbles both create and intensify cavitation effects. Applying a lot of energy can prevent the previously listed enzymes that cause quality deterioration, including browning.

6.7.6 High-Intensity Pulsed Light

A new non-thermal method called high-intensity pulsed light (PL) works by applying brief light pulses at specific frequencies to food surfaces. When compared to other forms of light, UV radiation, which can be controlled by filters, has demonstrated a more potent antibacterial impact. Furthermore, by strengthening biosynthetic pathways, the strain that UV radiation causes in the live cells of vegetables and fruits encourages the gathering of specific bioactive substances. The primary theory is that PL serves to improve plant tissues' metabolic stress, leading to oxidative damage that sets off secondary metabolism and, depending on the product, induces the creation of phytochemicals like carotenoids or phenolic compounds. The exact mechanism underlying this kind of reaction is not known.

6.7.7 Cold Plasma

This is a non-thermal, chemical- and water-free device that operates at air pressure. It is produced by adding energy to an ionised gas mixture, which encourages the build-up of active substances such as UV rays, charged particles, and radicals, free radicals being the most potent against microbes. This method is being researched further and is applied as an antibacterial therapy. The variability of the produced plasmas, which varies under numerous conditions, is one of the cited drawbacks. It is therefore difficult to compare the efficacy of particular circumstances with that of other investigations.

6.7.8 Plasma Sterilization at Atmospheric Pressure

A unique non-thermal sterilisation technique using atmospheric pressure-produced OH-radicals from pulsed discharge plasma has been presented by Mizuno (2007). Nebulised hydrogen peroxide is exposed to pulsating discharge plasma within a cylindrical concentric electrode at 15–22 kV and 230–240 Hz, using argon as the carrier gas. The technique, according to the scientists, can be effectively used to sterilise dry food products and surfaces at room temperature quickly and continuously. Research on using plasma for sterilisation and decontamination is expanding, and it has shown that comparatively cool ionised gases can kill bacteria cells very quickly.

6.7.9 Air Ion Bombardment

Bacteria-contaminated food mounted on agar can be killed by a straight jet of negatively charged air, as shown by UK researchers (Sutherland and Brew 2018). Availability duration, distance to the source, ion charge, and ion flow rate were the determining factors. The essential parts are a conducting surface that is grounded which holds the food that has to be cleaned, a high-voltage generator, and a negative ion emitter with a pointed tip. It is thought that the microbial cell membrane-destroying superoxide radical anion is the primary source of the microbiocidal action. Initial research has shown that treatment for a few minutes was efficient in killing both yeast and Gram- and Gram+ bacteria, including spore formers. It seems like an easy and reasonably priced solution. Minimal food processing using non-thermal techniques involves treatments that preserve food quality while ensuring safety; for example, 55 treatment may have comparable applications to intense light pulses, as both methods can effectively inactivate microorganisms without significantly affecting the sensory or nutritional properties of the product.

6.8 Packaging and Storing of Minimal Processing of Fruits and Vegetables

Packaging plays a crucial part in the production of minimally prepared fruits and vegetables. Modified atmospheric packaging (MAP) is the greatest researched packaging technique for processed uncooked fruits and vegetables. The fundamental idea behind MAP refers to an altered environment able to produce actively by combining materials for permeabilised wrapping with a certain gas combination or passively by using appropriate permeable packaging materials. Both seek to achieve the ideal gas balance inside the package, whereby a product's respiration activity is minimised while assuring that levels of carbon dioxide (CO_2) and oxygen (O_2) do not pose a risk to the product. The overall goal is to have a gas composition with the remaining nitrogen, 2–3% O_2, and 2–2% CO_2 (Gil and Allende 2012). Coatings that are edible are one potential "packaging" technique for prolonging the preservation of minimally treated fruits and vegetables after harvest. These are very thin material layers that are edible as a component of the entire food product. According to Baldwin et al. (1995), coatings can seal in flavour volatiles, minimise respiration, prevent moisture loss, limit the entry of oxygen, delay the generation of ethylene, and contain additives such as antioxidants that prevent microbiological growth and discolouration. Chilling and humidity management are two crucial preservation obstacles. Most bacterial infections can proliferate quickly on fresh-cut vegetables when stored at 10 °C or higher. The temperature at which something is stored matters when using vacuum or MAP packaging. Temperature fluctuations ought to be prevented. Increased warmth promotes the growth of pathogens and accelerates spoilage. Temperature fluctuations lead to in-pack condensation, which hastens the spoiling process. Abuse of temperature occurs in many areas of the supply chain, including handling of consumers, retail displays, storage, and transportation. If this is a serious issue, it could be required to limit the shelf-life, say to 5–7 days at 5–7 °C, so that psychrotrophic microorganisms do not have enough time to grow and make toxins.

Products should satisfy one or more of the subsequent criteria if their shelf-life is more than 10 days, there is a possibility the storage temperature may rise over 3 °C:

- A minimal application of heat, like 90 °C for 10 min
- A food that has a pH of 5 or lower and a 3.5% (aqueous) salt content overall
- Water activity values of 0.97 or lower during the entire meal, "aw"
- Any mix of heat and preservatives that has been demonstrated to stop *C. botulinum* from growing and producing toxins

6.8.1 Edible Coatings

The main purposes of edible coatings are to control gas exchanges between the surrounding environment and the tissue, as well as to minimise water loss. Furthermore, antimicrobials, antioxidants, or bioactive chemicals can be used in edible coatings to stop microbiological spoilage, firmness, colour, and flavour changes, as well as the development of off-odours. Edible coating formulations can contain a variety of substances, such as lipids, proteins (from either plant or animal origin), and polysaccharides (such as cellulose, pectin, starch, alginates, chitosan, and mucilages). Since each of them has benefits and downsides, formulations frequently combine many distinct ingredients to increase the efficacy of the coating. Anaerobic fermentation or enzymatic degradation can occasionally lead to the development of unfavourable firmness or colour changes, or alcoholic off-flavours.

Coating formulas and other features need to be optimised based on the commodity to be able to prevent these negative impacts on the quality. On the other hand, fresh-cut goods that retain their carotenoids, vitamin C, anthocyanins, and total phenolic content may have higher nutritional quality. By altering the rate at which fruit tissues respire and postponing the metabolic processes that cause deterioration, modified atmosphere packaging, or MAP, is used to increase the durability and calibre of goods. Low O_2 and a moderate CO_2 level concentration are advised in order to accomplish this goal; however, the best circumstances will vary depending on the species and handling parameters. MAP hinders the absorption of foreign odours or the loss of naturally occurring flavour components in food coatings. The makeup of the atmosphere, the particular product, and the storage circumstances all affect texture, colour, and health-related components.

6.9 Safety of Minimally Processed Fruits and Vegetables

In today's global food market, consumers have greater demands for safe, high-quality, and affordable products, making food safety crucial for public health (van Schothorst 2002; Lang 1998).

A worldwide, more thorough strategy has been implemented in conjunction with the need to address significant occurrences such as Bovine Spongiform Encephalopathy, commonly referred to as "mad cow disease", and others.

This strategy mostly concentrates on risk assessment and the introduction of new rules regarding hygiene, hazard analysis critical control points (HACCP), traceability, and the elimination of hazardous items from the market (van Schothorst 2002; Mathews Jr et al. 2003).

However, in the era of globalization, food supply chains exceed national boundaries, leading to an increase in the prevalence and internationalization of health issues (Mathews Jr et al. 2003; Ohlsson 1994).

Despite sincere efforts, the problem of food safety has never been more severe, as over 200 diseases, from cancer to diarrhoea, are caused by pathogen contamination (Uyttendaele et al. 2016). Therefore, it is increasingly evident that traditional methods of ensuring food safety are insufficient, particularly as our food systems grow more intricate (Boqvist et al. 2018). The complexity of food supply chains brought about by globalization emphasises the necessity for an all-encompassing approach to prevent the development of microbial infections and antibiotic resistance in processed food consumption. Table 6.3 discusses some specifications and quality control, and safety measures for minimally processed fruits and vegetables. From agriculture production to consumption, this shared responsibility recognises the interconnectedness of the entire food chain. Due to the complexity of the global supply chain it operates in, the food business faces challenges like longer supply

Table 6.3 Some specifications and quality control and safety measures for minimally processed fruits and vegetables

Raw material	Use premium raw materials and the appropriate cultivar variety under the right growing, harvesting, and storage circumstances
Hygiene	Maintain stringent hygienic standards and ethical production methods
Processing	Utilise critical control point principles and hazard analysis during processing at low temperatures
Washing	Use high-quality water with the proper pH and bacteria, and wash thoroughly both before and after peeling
Additives	To avoid disinfection or browning, use gentle chemicals in your washing water
Storage	Raw fruits and vegetables should be kept at the right temperature and humidity level, shielded from insects and rodents
Ingredients	Make certain that ingredients are at least "food grade" quality and are purchased from reliable vendors
Sanitation	Using ultraviolet (UV) light to render bacteria inactive and reduce product contamination
Blanching	Instead of using boiling water, use steam to preserve soluble substances, such as water-soluble vitamins
Firmness	Measure the firmness of generally uniform fruit, including apples and stone fruits, using a penetrometer

chains that lead to longer transit times and quality hazards. It is increasingly evident that in order to tackle these issues more thoroughly, foodborne outbreaks should be collaboratively examined by the human, animal, and environmental health sectors, with an emphasis on One Health concepts. This method stresses cross-sectoral cooperation while acknowledging the interdependence of health systems. The identification of sources of contamination and crucial phases in the food production chain has been delayed by the absence of cooperation throughout the entire chain. It should be sought to close the "farm to fork" gap by integrating animal, food, environment, and human elements in the food supply chain using an approach known as One Health, particularly using genomics.

6.10 Assuring Safety in the Fresh-Cut Processing Chain

During processing, fresh-cut produce is extremely susceptible to contamination. Therefore, every effort must be taken to reduce the possibility of contamination in the chain from the manufacturer to the customer. The following is a summary of factors that help reduce risk in the production-consumption chain.

6.10.1 Harvesting and Post-harvest Handling

- Cleaning and disinfecting harvesting machinery
- Wild animals and birds should not be allowed in packing houses
- Preventing cross-contamination while being delivered to the process
- Reducing bruising and cutting

6.10.2 Fresh-Cut Processing

- A program for cleaning tools and surfaces
- Good initial inspection and disinfection

Avoiding excessive peeling or cutting

- Reducing or eliminating human interaction with manufactured food
- Antimicrobial dipping and efficient washing
- Preventing contamination after dipping

6.10.3 Packaging/Distribution/Retail

- Picking the right packing materials
- Monitoring the microbiological quality of products that are packaged
- Labeling with a modest shelf-life
- A well-made automobile
- Appropriate vehicle loading procedures
- The loading of chilled cabinets
- Retailer and customer education

6.11 HACCP for the Effective Running of a Fresh-Cut Plant

Since 1988, HACCP ideas have been worked upon and integrated into food safety laws in other nations worldwide. The method was initially taken from "Failure Mode and Effect Analysis" in the Engineering System. Pillsbury/NASA refined it further for the American Space Program. HACCP decreased the possibility that astronauts would experience negative consequences from eating tainted food while in orbit. Finding Critical Control Points (CCPs) in the processes of food production and preparation is the foundation of this food safety methodology. Food safety for human consumption will be guaranteed by closely watched CCPs.

The conventional approach of testing the final product without understanding the risks associated with preparation has been eliminated by using HACCP.

The HACCP system consists of seven principles:

Principle 1: Analyse the hazards. Products may be affected by three primary types of hazards: biological, chemical, and physical. Management should identify the risks to food safety and the preventive steps the plant can use to mitigate these risks.

Principle 2: Determine the steps or processes in a food process where control can be implemented, or CCPs. As a consequence, a food safety concern is either eliminated or reduced to a level that is acceptable.

Principle 3: Set important limits for every CCP. The maximum or lowest value that must be maintained at a critical control point in order to prevent, eliminate, or minimise a physical, biological, or chemical hazard to a level that is tolerable is known as a critical limit.

Principle 4: Specify the criteria for CCP monitoring. At every critical control point, monitoring is required to make sure the process is under control.

Principle 5: Take corrective action. These are steps to take in the event that monitoring shows a departure from a predetermined critical limit. A plant's HACCP plan must specify the corrective measures to be performed in the event that a critical limit is not reached, per the final rule. The purpose of corrective actions is to ensure that any product affected by a deviation—whether harmful to health or otherwise compromised—is prevented from reaching consumers.

Principle 6: Document critical control points and critical limit monitoring, verification operations, and processing deviation handling in a written HACCP plan and other record-keeping procedures.

Principle 7: Make sure the HACCP system is operating as planned by establishing processes. Validation makes sure that the plans accomplish their intended goals, which is to guarantee the manufacturing of safe goods. The plants will have to verify their own HACCP plans. The process of verification guarantees that the HACCP plan is sufficient and operating as planned. Microbial sampling and analysis, critical limits, CCP records, and an examination of HACCP plans are some examples of verification techniques.

The entire process must be assessed for potential risks when creating a HACCP strategy. A decision-making tree would be used to identify which hazards could be managed and measured, and a HACCP flow chart would display the CCPs. If the final rinse water and metal detection CCPs are not up to par, they can be tracked and managed. This is what a danger must have in order to be considered a CCP.

6.12 Advantages of Minimal Processing of Fruits and Vegetables

One significant benefit of the creation and meticulous categorization of MP meals is that they offer a foundation for the division of the conventional field of cold preservation, as defined by Potter (1986), into two discrete domains: frozen foods and refrigerated foods. By grouping every frozen item combined and every chilled food in a single category, the classification of frozen and refrigerated foods would be significantly simplified if these two sections were separated. Without a doubt, the categories of prepared foods and commodities should be subdivided into refrigerator and other preservation categories. When handling some varieties of MP fruits and vegetables, it is important to take into account the "living tissue" notion, which sets them apart from many other chilled items. Because MP fruits and vegetables contain living tissue and active enzymes, their shelf-life should likewise be restricted to 3–6 months from harvest to consumption. Current knowledge gathered from many years of technological advancements in food science research that may be applicable to MP foods has to be updated and arranged. Furthermore, there is a need for research to bring MP foods up to level with the more widely utilised preservation techniques that are now in use and for which a large number of food processing enterprises are called. The food business appears to be interested in intensifying competition among the vast family of MP foods.

Effectively addressing the variety of food product categories that use this technology, along with the packaging systems necessary for producing minimally processed (MP) foods, would require extensive coverage across many volumes. Although it can be difficult to differentiate MP fruits and vegetables from other MP foods—especially in the context of regulations and quality control—this volume will focus specifically on MP fruits and vegetables. While it can be challenging to distinguish MP fruits and vegetables from other MP foods in some of the book's chapters, particularly those dealing with regulations and quality control, MP fruits and vegetables will receive special attention in this volume. Specialists in plant physiology, pathology, biochemistry, biotechnology, microbiology, food science, food engineering, and packaging are among the professionals in this field. To mention a few specialties, they are needed in this rapidly expanding field of food science. To provide the markets with MP fruits and vegetables with a high-quality shelf-life that is prolonged beyond 8 and 10 days, typically sufficient regarding the European retail market, a really interdisciplinary effort will be required. According to Lioutas (1988), for MP fruits and vegetables to be competitive in US markets, they probably must possess a minimum shelf-life of 21 days. Fruits and vegetables that have been trimmed, peeled, cut, and shredded undergo minimal processing by being submerged in a solution that typically contains potassium meta-bisulphite (0.01–0.09%), ascorbic acid (200–300 ppm), citric acid (0.5–1.2%), sodium benzoate (0.03–0.08%), and salt (3–6%). This process is carried out within a few hours at 20–25 °C or at 5 ± 10 °C for multiple hours.

Non-acidic products are occasionally subjected to extremely moderate in-pack heat treatments, lasting 5 min at 70–80 °C in order to reduce their microbial burdens. However, to prevent heat-induced deterioration of the chlorophyll, the treatment's intensity is continuously regulated and kept to a minimum. Pre-treatments that involve minimal processing, such as ionising radiation, soaking in solutions, and other non-thermal methods, present a promising approach for enhancing the nutritional value of sliced fruits and vegetables that are ready to consume. It has also been discovered that food-borne pathogens in juices, fruits, and vegetables that have been pre-cut can be checked using ionising radiations at 1–5 kGy. To render microorganisms that have gotten into the food inactive, radiation is applied throughout the food stuff. After washing, items are removed, and wetness is extracted using a basket centrifuge running at 1000 rpm. This process is known as gentle spin-drying. Peeling, chopping, slicing, and/or shredding gently, any method used to cut or break fruits and vegetables into more manageable, consistent pieces with a defined shape and size is referred to as size reduction. Reducing size could be a crucial step in improving taste, digestibility, and handling convenience. Unit activities such as dicing, slicing, peeling, and cutting harm perishable fruits and vegetables, whose quality deteriorates quickly at room temperature. Near the sliced surface, loss of cellular integrity disrupts the segregation of enzymes and substrates and provides pathways for pathogen entry and growth substrate, which raises respiratory activity and ethylene generation, further leading to accelerated senescence. Furthermore, it might also cause the oxidative processes and browning,causing enzymes to become active. After cutting, the item needs to be used right away or stored at 5 ± 10 °C to slow down the rate of these reactions.

6.13 Future Trends

Much more research is required to produce minimally processed produce items, including fruits and vegetables with superior nutritional value, microbiological security, as well as sensory excellence. At 5 °C, it is possible to achieve a 7–8-day shelf-life; however, certain items may require a 2–3-week shelf-life. It is necessary to learn more about the development of harmful bacteria or dietary alterations in very little processing, long-lasting fruits and vegetables. The requirement for a combined strategy, where uncooked materials, distribution, packing, handling, and processing all need to be correctly controlled to enable prolongation of shelf-life, is a defining element of minimum processing. The minimal processing of fresh food requires additional advancements in hurdle technology, which makes use of naturally occurring preservatives such as lactic acid bacteria inhibitors, as well as the matching of suitable processing techniques and ingredient pairings. Plant geneticists are likely to produce carefully chosen and developed hybrids or cultivars tailored to the specific needs of minimal processing in the future. Produce meant to be processed as little as possible will probably be developed in carefully regulated environments (Martinez and Whitaker 1995). To make unit processes like peeling and shredding more delicate, more work needs to be put into them. It makes no sense to process food roughly, so lowering its quality, and then attempt to mitigate the damage by adding preservatives later on. Particularly active areas for development include edible films and active packaging methods, as well as more porous plastic films that are more in line with how fruits and vegetables breathe.

6.14 Conclusion

The last 10 years have seen a significant growth in customer concerns about convenience and health, which has led to a desire for fresh-like MP fruits and vegetables. Today's food market is more consumer-driven, technologically sophisticated, and competitive. The market now offers more processed fruit and vegetable products and more creative packaging than it did 25 years ago. Some significant variables influencing the food industry and new product development are technological advancements in food processing, consumer needs, taste preferences, and new developments in scientific study. Scientific research and food biotechnology advancements have also had a significant influence. One such technology that has emerged as a result of shifting consumer demands and the demand for high-quality, fresh goods is minimal processing. In layman's terms, minimal processing (MP) refers to pre-cut fruits and vegetables that are packed in appropriate materials to provide for a sufficient shelf-life and fresh appearance. Let us first review the history of minimal processing and look at some of the issues that have affected customers and the food business as a whole.

References

Aguayo E, Allende A, Artés F (2003) Keeping quality and safety of minimally fresh processed melon. Eur Food Res Technol 216:494–499. https://doi.org/10.1007/s00217-003-0682-7

Allende A, Aguayo E, Artés F (2004) Quality of commercial fresh processed red lettuce throughout the production chain and shelf life. Int J Food Microbiol 91:109–117. https://doi.org/10.1016/S0168-1605(03)00373-8

Anna PD, Jimenez-Martinez J, Tabuteau H et al (2014) Mixing and reaction kinetics in porous media: an experimental pore scale quantification. Environ Sci Technol 48(1):508–516. https://doi.org/10.1021/es403105b

Artés F (2004) Refrigeration for preserving the quality and enhancing the safety of plant foods. J Int Acad Refrig LXXXIV(1):5–25

Artés F, Artés-Hernández F (2000) Innovacionesindustriales en el procesadomínimo de frutas y hortalizas. CTC, Revista Agroalimentación e IndustriasAfines 7:29–33

Artes F, Conesa MA, Hernandez S, Gil MI (1999) Keeping quality of fresh-cut tomato. Postharvest Biol Technol 17(3):153–162. https://doi.org/10.1016/S0925-5214(99)00044-7

Baldwin EA, Nisperos-Carriedo MO, Baker RA (1995) Use of edible coatings to preserve quality of lightly (and slightly) processed products. Crit Rev Food Sci Nutr 35(6):509–524. https://doi.org/10.21273/HORTSCI.30.1.35

Barbosa-Cánovas GV, Rodriguez JJ (2005) Thermodynamic aspects of high hydrostatic pressure food processing. In: Novel food processing technologies, pp 183–205. https://doi.org/10.1111/1541-4337.12763

Boqvist S, Söderqvist K, Vågsholm I (2018) Food safety challenges and one health within Europe. Acta Vet Scand 60:1. https://doi.org/10.1186/s13028-017-0355-3

Breidt F, Fleming HP (1997) Using lactic acid bacteria to improve the safety of minimally processed fruits and vegetables. Food Technol Chica 51(9):44–51. https://doi.org/10.3390/microorganisms8060952

Carlin F, Nguyen-The C, da Silva AA (1995) Factors affecting the growth of listeria monocytogenes on minimally processed fresh endive. J Appl Microbiol 78(6):636–646. https://doi.org/10.1111/j.1365-2672.1995.tb03110.x

Carter DA, D'Souza F, Simkins BJ, Simpson WG (2010) The gender and ethnic diversity of US boards and board committees and firm financial performance. Corp Govern Int Rev 18(5):396–414. https://doi.org/10.1111/j.1467-8683.2010.00809.x

Cho CH, Roberts RW, Patten DM (2010) The language of US corporate environmental disclosure. Acc Organ Soc 35(4):431–443. https://doi.org/10.1016/j.aos.2009.10.002

Clausen DR (2010) Processing factors influencing acceptability in extractions from complex subjects. Unpublished ms, Stanford University

Crowe M, Sheppard L (2011) A general critical appraisal tool: an evaluation of construct validity. Int J Nurs Stud 48(12):1505–1516. https://doi.org/10.1016/j.ijnurstu.2011.06.004

Diaz-Mendoza M, Velasco-Arroyo B, Santamaria ME, González-Melendi P, Martinez M, Diaz I (2016) Plant senescence and proteolysis: two processes with one destiny. Int J Genet Mol Biol 39:329–338. https://doi.org/10.1590/1678-4685-GMB-2016-0015

Farr JE (2018) Evaluating the mechanism of ascorbic acid bleaching of Anthocyanins and proposed ways to mitigate the interaction. Doctoral dissertation, The Ohio State University

Gil MI, Allende A (2012) Minimal processing. In: Decontamination of fresh and minimally processed produce, pp 105–120. https://doi.org/10.1002/9781118229187

Holzapfel WH (1998) The gram-positive bacteria associated with meat and meat products. In: The microbiology of meat and poultry, vol 31, pp 35–74

Hong JH, Gross KC (1998) Surface sterilization of whole tomato fruit with sodium hypochlorite influences subsequent postharvest behavior of fresh-cut slices. Post Harvest Biol Technol 13(1):51–58. https://doi.org/10.1016/S0925-5214(97)00072-0

Hughes C (2011) Changes and challenges in 20 years of research into the development of executive functions. Infant Child Dev 20(3):251–271. https://doi.org/10.1002/icd.736

Ippolito A, Nigro F (2003) Natural antimicrobials in postharvest storage of fresh fruits and vegetables. In: Natural antimicrobials for the minimal processing of foods, pp 201–234. https://doi.org/10.1533/9781855737037.201

Kårlund A, Moor U, Sandell M, Karjalainen RO (2014) The impact of harvesting, storage and processing factors on health-promoting phytochemicals in berries and fruits. PRO 2(3):596–624. https://doi.org/10.3390/pr2030596

Lang T (1998) The new globalisation, food and health: is public health receiving its due emphasis? J Epidemiol Community Health 52:538–539. https://doi.org/10.1136/jech.52.9.538

Lioutas TS (1988) Challenges of controlled and modified atmosphere packaging: a food companýs perspective. Food Technol (Chicago) 42(9):78–86

Martinez MV, Whitaker JR (1995) The biochemistry and control of enzymatic browning. Trends Food Sci Technol 6(6):195–200. https://doi.org/10.1016/S0924-2244(00)89054-8

Mathews KH Jr, Bernstein J, Buzby JC (2003) International trade of meat/poultry products and food safety issues. In: Buzby JC (ed) International trade and food safety: Economic theory and case studies, Agricultural Economic Report No 828. Economic Research Service, Washington, DC, pp 48–73

Mizuno A (2007) Industrial applications of atmospheric non-thermal plasma in environmental remediation. Plasma Phys Control Fusion 49:5A. https://doi.org/10.1088/0741-3335/49/5A/S01

Nguyen-the C, Carlin F (1994) The microbiology of minimally processed fresh fruits and vegetables. Crit. Rev. Food Sci. Nutr, 34: 371–401

Ohlsson T (1994) Minimal processing-preservation methods of the future: an overview. Trends Food Sci Technol 5(11):341–344. https://doi.org/10.1016/0924-2244(94)90210-0

Potter NN (1986) Cold preservation and processing. Food Sci:201–245

Rolle RS, Chism GW III (1987) Physiological consequences of minimally processed fruits and vegetables. J Food Qual 10(3):157–177. https://doi.org/10.1111/j.1745-4557.1987.tb00856.x

Schellekens M (1996) New research issues in sous-vide cooking. Trends Food Sci Technol 7(8):256–262. https://doi.org/10.1016/0924-2244(96)10027-3

Schoenbach KH, Joshi RP, Stark RH, Dobbs FC, Beebe SJ (2000) Bacterial decontamination of liquids with pulsed electric fields. IEEE Trans Dielectr Electr Insul 7(5):637–645. https://doi.org/10.1109/94.879359

Seymour IJ, Burfoot D, Smith RL, Cox LA, Lockwood A (2002) Ultrasound decontamination of minimally processed fruits and vegetables. Int J Food Sci Technol 37(5):547–557. https://doi.org/10.1046/j.1365-2621.2002.00613.x

Soliva-Fortuny RC, Grigelmo-Miguel N, Hernando I et al (2002) Effect of minimal processing on the textural and structural properties of fresh-cut pears. J Sci Food Agric 82(14):1682–1688

Sutherland EJ, Brew BJ (2018) Human immunodeficiency virus and the nervous system. Neurol Clin 36(4):751–765. https://doi.org/10.1002/jsfa.1248

Torriani S, Massa S (1994) Bacteriological survey on ready-to-use sliced carrots. LWT-Food Sci Technol 27(5):487–490. https://doi.org/10.1006/fstl.1994.1097

Uyttendaele M, Franz E, Schlüter O (2016) Food safety, a global challenge. Int J Environ Res Public Health 13:67. https://doi.org/10.3390/ijerph13010067

van Schothorst M (2002) Microbiological risk assessment of foods in international trade. Safety Science, 40(4): 359–382. https://doi.org/10.1016/S0925-7535(01)00055-8

Vescovo M, Torriani S, Orsi C et al (1996) Application of antimicrobial-producing lactic acid bacteria to control pathogens in ready-to-use vegetables. J Appl Microbiol 81(2):113–119. https://doi.org/10.1111/j.1365-2672.1996.tb04487

Wiley RC (1994) Preservation methods for minimally processed refrigerated fruits and vegetables. In: Minimally processed refrigerated fruits & vegetables, pp 66–134. https://doi.org/10.1007/978-1-4939-7018-6_6

World Cancer Research Fund/American Institute for Cancer Research (2007) Food, nutrition, physical activity, and the prevention of cancer: a global perspective. WHO, Washington DC. wcrf.org/about-the-report

Yildiz M (2007) E-government research: reviewing the literature, limitations, and ways forward. Gov Inf Q 24(3):646–665. https://doi.org/10.1016/j.giq.2007.01.002

Chapter 7
Freezing Preservation of Fruits and Vegetables

Rakhi Singh, Shivangi Mishra, Shreya Chaudhary, and Heena Singh

7.1 Introduction

Fruits and vegetables are rich in essential nutrients and antioxidants, offering numerous health benefits. Despite being harvested, they continue to respire at the cellular level, making them living entities prone to spoilage (Grover and Negi 2023). This chemical reaction and higher water activity make them prone to spoilage by microbes; it also leads to the sensory and physicochemical loss of these commodities. In this regard, freezing emerges as the most economical, convenient, and effective method to preserve their nutritional value and sensory attributes by inhibiting microbial growth and enzyme activity (James et al. 2015). The frozen food market has grown significantly due to advancements in cold chain markets, the retail landscape, and increasing demand for convenience food. Specifically, the market for frozen fruits and vegetables, valued at 25.4 billion dollars in 2021, is expected to reach 30.1 billion dollars by 2027 (Grover and Negi 2023). Preservation is a centuries-old technique that has been employed by human beings to reduce the spoilage of food items, ensuring their availability beyond their natural seasonality. Freezing helps maintain the nutritional integrity of these perishable produce, extends their availability, reduces food waste, and enhances convenience for consumers. This preservation method is a potent tool that helps retain the product's nutritional and sensory attributes compared to other techniques, such as canning or drying. At lower temperatures, chemical, enzymatic, and microbial activities reduce and delay, preserving the natural goodness of fruits and vegetables. During freezing, the product is subjected to a temperature of −18 °C or below, resulting in a decreased rate of deterioration in fruits and vegetables (Jha et al. 2024).

R. Singh (✉) · S. Mishra · S. Chaudhary · H. Singh
Department of Food Science and Technology, National Institute of Food Technology Entrepreneurship and Management, Kundli, Sonipat, Haryana, India

OP Chauhan (ed.), *Fruits and Vegetables Technologies*,
https://doi.org/10.1007/978-981-96-8433-5_7

Freezing has a rich history spanning thousands of years, showcasing humanity's resourcefulness in reducing spoilage rates and extending shelf life. From early societies preserving food in naturally cold settings to the invention of motorized refrigeration in the 1800s, the advancement of freezing methods has been remarkable. The beginning of motorized refrigeration systems enabled the widespread distribution of frozen foods, laying the foundation for the frozen food industry (Evans 2008). In the early twentieth century, Clarence Birdseye's pioneering quick-freezing technique revolutionized the industry, making frozen fruits, vegetables, and prepared meals more accessible and appealing to consumers by preserving their quality and texture. Throughout the twentieth century, freezing techniques continued to evolve with improvements in equipment, packaging, and freezing technologies. It was observed that the value of frozen commodities lies in the size and distribution of ice crystals, affecting their physicochemical and textural properties. Rapid freezing produces tiny, evenly distributed ice crystals, preserving texture and quality. In contrast, slow freezing leads to large, unevenly distributed ice crystals, causing damage to texture and quality. This damage includes dislocation, perforation, and deformation of cell walls in fruits and vegetables, resulting in high drip loss upon thawing. Ideal freezing conditions involve the formation of fine ice crystals uniformly distributed within cells and tissues to minimize damage and maintain overall product quality (James et al. 2015). However, despite efforts to optimize freezing techniques, issues related to freezing and thawing persist, posing challenges to the quality of frozen products. Incidentally, various new and innovative methods of freezing have been introduced, including blast freezing, cryogenic freezing using liquid nitrogen or carbon dioxide, and individually quick-frozen (IQF) methods, which are being used further to enhance the quality and efficiency of frozen food preservation. These advancements enabled manufacturers to develop various frozen products with minimal flavor, texture, and nutritional value loss. In recent decades, freezing techniques have become more sophisticated and tailored to specific food products. Advancements in freezing technology, such as cryogenic freezing tunnels and spiral freezers, allow for precise control over freezing parameters, ensuring optimal preservation while minimizing energy consumption and processing time. Additionally, innovations in packaging materials and techniques, such as vacuum sealing and modified atmosphere packaging, help maintain frozen foods' quality and freshness during storage and transportation (Muthukumarappan et al. 2018). The benefits of freezing techniques in preserving fruits and vegetables are illustrated in Fig. 7.1. This chapter explores traditional freezing preservation methods and their associated challenges. Furthermore, it examines advanced freezing techniques, including isochoric freezing, ultrasound-assisted freezing, and impingement freezing, and their impacts on fruits and vegetables.

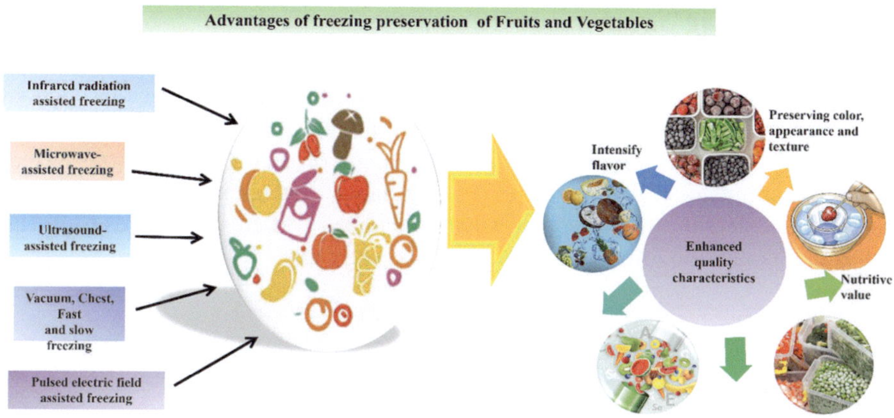

Fig. 7.1 Advantages of freezing preservation of fruits and vegetables

7.2 Principle of Freezing Process

Freezing involves carefully extracting heat from the product in a controlled and consistent manner. The process continues until the product reaches thermal equilibrium, ensuring stability and preservation (De Ancos et al. 2006). The freezing process involves several distinct stages, each contributing to the transformation of food from its initial state to a frozen product suitable for storage.

7.2.1 Pre-freezing Stage

Product temperature is lowered gradually by removing the heat from the product, bringing its temperature down to freezing. The initial freezing temperature varies depending on the product and its moisture content.

7.2.2 Super-Cooling

The stage where the temperature is already below freezing but the water in the food matrix remains in the liquid state despite being below the freezing temperature, due to the lack of a nucleation point needed for crystal formation, it is often temporary and is not always observed.

7.2.3 Freezing Stage

During this phase, the water content of food begins to transform into ice throughout the entire mass. In an ideal system, the temperature remains constant, but in real situations, it gradually decreases with the extraction of latent heat. The transformation of water into ice crystals involves two key stages: nucleation and crystal growth. Nucleation marks the starting point, where organized clusters of particles form and act as the foundation for subsequent crystal development.

7.2.4 The Sub-Freezing Stage

It is the final stage in the freezing process where the product temperature is lowered to the desired storage level, with most of the sensible heat being removed during this phase. Various factors influence the time and the freezing process, including the commodity size (particularly thickness) and shape of the product, food matrix composition, heat transfer properties, and the temperature of the cooling medium (Liu et al. 2020). From a physical perspective, fruits and vegetables are primarily composed of dilute aqueous solutions and have freezing points below 0 °C. The extent of depression is influenced by the concentration of dissolved compounds in the water phase, not by the total amount of water present. It is crucial to rapidly pass through the temperature range of 1–5 °C during freezing, as this range is where most irreversible changes occur. Additionally, foods should traverse the solidification phase—the zone of maximum ice crystal formation, which is typically in the range of 0 °C and – 3.9 °C—within 30 min to minimize quality degradation. Frozen foods are characterized by the formation of a high proportion of ice from their water content. The speed of the cooling process significantly impacts ice crystal size, with faster freezing producing smaller, more uniform crystals. During freezing, ice crystals initially form in the extracellular liquid due to its higher freezing point compared to the intracellular fluid. In slow-freezing conditions, these extracellular crystals grow larger, leading to an increase in the solute concentration of the remaining liquid. This creates a gradient that drives water out of the cells via osmosis, causing significant dehydration and shrinkage. As a result, slow freezing typically produces fewer but larger ice crystals, situated between the contracted cells, which can compromise the texture and quality of the product (Bonat Celli et al. 2016). Alternatively, rapid freezing removes the heat so rapidly that there is minimal time available for the dehydration of the cells, and ice crystals will also form in the cells (Kobayashi and Suzuki 2019).

7.3 Traditional Freezing Preservation Method of Fruits and Vegetables

Traditional freezing preservation methods have been pivotal in increasing the shelf stability and preserving quality and sensory attributes of F&V in the food processing sector. These methods encompass a range of techniques, each with its unique approach to freezing produce. From the widely used air blast freezing method to the cutting-edge cryogenic freezing technology, these traditional methods have ensured the availability of high-quality frozen fruits and vegetables year round. However, a comprehensive examination of their mechanisms, advantages, and limitations is imperative to optimize their efficacy and sustainability in food preservation practices (Kobayashi and Suzuki 2019). The freezing process could be crucial in freezing perishable foods like fruits and vegetables. Frequently used freezing techniques encompass air-blast freezing, which involves subjecting food items to high-velocity air streams at approximately −30 °C; contact freezing, whereby food, either packaged or unpackaged, is situated on or between chilled metal surfaces and cooled via conduction; immersion freezing, where food is submerged in frigid brine solutions; cryogenic freezing, entailing the exposure of food to ultra-low temperatures facilitated by cryogenic fluids like liquid nitrogen or solid carbon dioxide; and hybrid approaches that integrate various combinations of these methods (Grover and Negi 2023). The most significant reduction in bacterial viability occurs within the temperature range of −4 °C to −10 °C, primarily attributable to cold shock, which disrupts their metabolic activity and may even lead to cessation. Slower freezing rates afford bacteria the opportunity to acclimate to the altered environmental conditions, underscoring the necessity for rapid freezing of food products. These freezing techniques incude air-blast freezers, fluidized bed freezers, impingement freezers, liquid immersion freezers, plate freezers, liquid nitrogen freezers, and carbon dioxide freezers. The air blast freezer is particularly noted for its versatility, consistent and uniform freezing rate (Chevalier et al. 2000). Other freezing methods, such as plate freezing is a type of contact freezing known for a faster cooling rate. The following are some traditional freezing preservation methods that have been used for years.

7.3.1 *Air Blast Freezing*

This method of freezing is widely employed across the globe for freezing a diverse range of food products, ranging from raw to processed items. They are highly versatile in nature due to their effectiveness and usability with a wide range of products. These freezers are utilized to transition food products from a chilled state to their intended storage temperature (depending on the product) using air temperatures typically ranging from −35 °C to −45 °C. Freezing times can vary from 12 to 48 h based on the specific requirements of the product (Dempsey and Bansal 2010). Yu and Liao (2016) investigated the three-step processing treatment combined with the air blast freezing method on strawberries and observed the effect of different variables such as air velocity, air temperature, and cryoprotectant. The result of this study revealed that the quality of the commodity is mainly affected by super-cooling capacity and cryoprotectant. Similarly, Da Silva et al. (2022) investigated the effect of freezing rate in control air and velocity atmosphere on the quality parameters of frozen strawberries. Strawberries' firmness was enhanced by 72%, with a 5% reduction in drip loss of its original value at freezing conditions from −20 °C /0.4 $m{\cdot}s^{-1}$ to −53 °C/6.5 $m{\cdot}s^{-1}$. A study on the freezing of tomatoes by air blast freezing (−30 °C and 3 ms^{-1} air velocity) revealed that physicochemical properties such as pH, ascorbic acid content, and sensory attributes of tomato were better retained in air blast freezing than in normal −18-degree refrigeration condition (Rikasa et al. 2022).

7.3.2 *Cryogenic Freezing*

Cryogenic freezing operates on the principle of heat absorption from the product, achieved through the phase change of the refrigerant (or cryogen) within the freezing system, which utilizes extremely low temperatures to rapidly freeze fruits and vegetables. This technique exposes products to cryogenic gases (liquid nitrogen or CO_2), with temperatures below −100 °C (−148 °F). The intense cold cryogenic gases enable rapid heat transfer, resulting in the quick freezing of the product (Loayza-Salazar et al. 2024). The process begins with the liquefied cryogenic gas spraying or circulating the fruits and vegetables. These gases absorb heat rapidly from the product, causing the temperature of the food to drop rapidly.

This rapid cooling forms ice crystals within the product, effectively preserving its texture, flavor, and nutritional content. Various studies have been done on the cryogenic freezing preservation method of F&V. A study (2018) on the cryogenic preservation of dates revealed that the produce's texture, color, and flavor were better preserved by cryogenic freezing than by conventional freezing. Quick cryogenics freezing of cabbage slices prevents the texture parameters, such as hardness and springiness, from deterioration during freezing. Additionally, microbial spoilage is also prevented (Wang et al. 2019). Cryofreezing of raspberries enhances their total solid content, acid value, and color parameters during storage (Gales et al. 2022). The rapid freezing process helps to minimize damage to cellular structures and preserve the physicochemical attributes of the produce, making them available to consumers year-round while minimizing food waste.

7.3.3 *Immersion Freezing (IF)*

For rapid freezing, liquid nitrogen and carbon dioxide are widely used refrigerants, significantly accelerating freezing rates due to their high thermal coefficients. In the immersion method, food is directly exposed to a chilled aqueous solution, often utilizing sodium or calcium chloride in binary brine solutions. Studies have explored this freezing technique for various F&V, including carrots, peas, potatoes, and fruit juices. Immersion freezing offers a dual advantage over air blast freezing, potentially reducing energy consumption by at least 25%. For instance, while an air blast tunnel takes 15–20 min to freeze peas, immersion freezing achieves the same in less than 2 min using a ternary solution. Moreover, immersion freezing produces less product dehydration, yielding higher-quality end products. Recent applications of IF have showcased numerous benefits in enhancing the quality and storage stability of foods such as apples, strawberries, potatoes, cheese, and broccoli. IF diminishes drip loss, enhances color retention, and preserves the veracity of cell structure.

7.4 Drawbacks of Traditional Freezing Methods

The freezing preservation of fruits and vegetables is crucial due to higher water activity and sensitivity to phase changes, which can lead to quality deterioration. While traditional freezing methods like air blast, cryogenic, immersion, and plate freezers have been widely used for preserving fruits and vegetables, they come with drawbacks, such as equipment corrosion, safety concerns, off-flavor development, bigger ice crystal formation, and high operating costs. Freeze damage in fruits and vegetables can occur due to ice formation, resulting in textural losses, tissue damage, and the release of deteriorative enzymes. Freeze cracking, particularly prevalent in cryogenic freezing due to high freezing rates, is another critical issue. Factors influencing freeze damage include freezing method, product characteristics, and storage conditions. Air fluidization freezing is a promising method with high freezing rates and attractive product quality. However, it has drawbacks, such as lower surface heat transfer coefficients and high fan energy consumption. Immersion freezing offers advantages like high heat transfer rates and fine ice crystal formation, but faces challenges like uncontrolled solute uptake and operational issues with immersion liquids. Recent advancements in heat transfer, fluid dynamics, and process control have led to the development of innovative freezing systems. However, selecting the appropriate freezing method involves balancing costs, quality, and feasibility considerations.

7.5 Innovative Freezing Methods

The freezing process comprises three main steps: cooling, freezing, and further cooling. Initially, in the cooling phase, the food is exposed to the freezing medium, extracting sensible heat until the temperature reaches the freezing point. Similarly, in freezing fruits and vegetables, heat and mass transfer occur, and it induces (Bonat Celli et al. 2016) a rapid temperature rise that releases latent heat and forms ice crystals, a process involving nucleation and crystal growth (Muthukumarappan et al. 2018). The effectiveness of freezing lies in ice crystal characteristics, which depend on freezing kinetics, ice nucleation rate, and crystal growth. A higher freezing rate yields more uniformly distributed, smaller, and rounder ice crystals. Therefore, optimizing and controlling freezing methods are required to minimize the structural damage by promoting smaller, more uniform ice crystal formation. Various technologies, including high-pressure freezing, anti-freeze proteins, ultrasound-assisted freezing, and supercooling technology, aim to enhance freezing rates, control ice nucleation, and decrease freezing periods, thereby maintaining temperature stability (Liu et al. 2020). Recent advancements in freezing technology focus on process control enhancements to elevate freezing rates and reduce costs. Ice crystal nucleation and growth, significant factors affecting fruit and vegetable quality during freezing, can be manipulated by temperature-dependent rate processes. High-pressure freezing ensures instantaneous and homogeneous ice formation, while impingement jet freezing offers a rapid and cost-effective freezing method (Muthukumarappan et al. 2018). Novel freezing systems have emerged to improve organoleptic quality and production economics by enhancing heat removal rates or modifying product structure.

7.5.1 Pressure-Assisted Freezing

Freezing of fruits and vegetables while applying pressures ranging from 200 to 400 MPa has attracted significant research interest in recent years (Cheng et al. 2021). High-pressure freezing involves high-pressure shift-assisted freezing (PSF) and high-pressure-induced crystallization (HPIC) (Evans 2008). All these techniques aid conventional techniques and enhance the overall quality attributes of the product.

High-pressure freezing is a process of cooling a sample under constant pressure up to its phase transition temperature (James et al. 2015). At atmospheric temperature, water increases its volume upon freezing and results in more cellular damage to the product. However, during high-pressure-assisted freezing, various ice forms with numerous shapes and sizes are formed with a higher volume-to-mass ratio than water. These ice crystals do not expand their volume and exist in a non-crystalline state, causing less damage to the cellular part of the product. Under a pressure of 200 MPa, the freezing point is significantly lowered to approximately −22 °C. This reduction facilitates a vitrification depth of around 200 μm, enabling the effective freezing of objects with thicknesses ranging from 0.4 to 0.6 mm (Fikiin 2008). In PSF, food undergoes cooling under high pressure to subzero temperatures without experiencing a phase transition. Instead, it freezes while under pressure and remains in this state until the pressure is relieved. Upon pressure release, rapid nucleation happens across the food, yielding small, uniform, evenly dispersed ice crystals. Consequently, this process leads to shorter phase transition periods, minimizes mechanical stress, and produces smaller ice crystals uniformly spread throughout the product (James et al. 2015). An innovative and new version of PSF is known as HPCI (high-pressure carbonic impression), as documented and demonstrated by Xu et al. (2014), which involves subjecting slices of carrot to pressurization within a chamber infused with high-pressure carbon dioxide. This methodology proves efficacious in rapidly freezing slices to temperatures below −1 at 6 MPa. Comparative analysis revealed that higher retention of nutrients reduces drip loss and texture loss in the subjected samples under HPCI than in the conventional mode. Another method of pressure freezing is HPIC, which is applied after traditional freezing. It works under the principle of phase transition of ice (I) to ice (III) and back to ice (I), useful for reducing microbial load in the sample (Van Buggenhout et al. 2007). Pressure-assisted freezing is also beneficial to maintain the microstructure of fruits and vegetables. It was observed that HPSF preserves the microstructure of mangoes and peaches when subjected to 200 MPa pressure (Otero et al. 2000).

7.5.2 *Impingement Freezing*

Utilizing impingement technology to enhance surface heat transfer in air and cryogenic freezing systems stands out as a recent innovation that has seen commercial application (Marazani et al. 2017). Impingement jet freezing utilizes high-velocity air or gas streams to rapidly freeze products. Impingement jets with velocities reaching up to 50 ms^{-1} disrupt the static surface boundary layer of gas enveloping a food product (Salvadori and Mascheroni 2002). This disruption creates a more turbulent medium surrounding the product, significantly enhancing heat exchange within this zone. Consequently, freezing occurs at a substantially faster rate compared to conventional freezing methods. This method facilitates rapid heat transfer and uniform freezing, thereby reducing processing time and maintaining the quality of fruits and vegetables (Sriwimon and Boonsupthip 2011). This freezing method is best suited to food that has a high surface area with respect to weight. Studies have shown that 20 mm thick product slices are efficiently frozen by the impingement freezing method (Newman 2001). The techniques are also very effective for products requiring rapid surface freezing. Various studies have been done on the effect of impingement freezing on the quality parameters of fruits and vegetables. Impingement jet freezing of partially ripe mango reduces texture and drip loss during freezing (Sriwimon and Boonsupthip 2011). Immersion freezing (IF) systems are readily accessible through various equipment suppliers, offering complete commercial solutions. Typically, these systems feature a modular design, enabling them to handle diverse throughput requirements effectively. However, it is worth noting that the initial investment for an impingement freezer tends to be higher compared to a traditional air blast freezer. Additionally, operating costs may be elevated due to increased fan power consumption (Table 7.1).

7.5.3 Ultrasound-Assisted Freezing

The expansion of the frozen food sector is spurring research into enhancing current freezing methodologies and innovating new technologies. These research efforts are primarily directed toward maintaining the freshness and inherent qualities of frozen foods, including texture, color, flavor, and nutritional content. The increasing demand for frozen fruits and vegetables is significantly fueling the development of advanced freezing technologies. An emerging area of interest is the use of power ultrasound for freezing the food commodities, which has demonstrated efficacy in both initiating instant nucleation and facilitating subsequent crystallization processes (Xin et al. 2014b). The application of high-power ultrasound technology has shown promising results by increasing the heat transfer rate through mechanisms such as acoustic cavitation (Grover and Negi 2023). The principle behind this phenomenon is the minimization of resistance to heat transfer due to the thinning of the boundary layer at the solid and water interface (Barbosa-Cánovas et al. 2017). This mechanism helps in the reduction of freezing time and crystal size during ultrasound-assisted freezing.

Numerous studies have been conducted to investigate the effect of UAF on, potatoes, apples, and broccoli (Delgado et al. 2009; Xin et al. 2014c, b; Kiani et al. 2015). The findings conclude that UAF can be effectively utilized for the produce with minimal effect on the texture and color of the samples. UAF (0.250–0.412 W/cm^2) treatment of broccoli significantly improved physical, chemical, and microstructure attributes than traditional methods (Xin et al. 2014c). Similarly, a combination of ultrasound with osmotic dehydration also enhances the freezing rate, so the smaller ice crystals result in better texture and color retention in tomatoes (Zhu et al. 2018). Other studies also supported that the retention of microstructure and controlled cooling before nucleation were better with ultrasound application (Comandini et al. 2013; Cheng et al. 2014). Studies on potatoes by Comandini et al. (2013) and strawberries by Cheng et al. (2014) revealed that ultrasound accelerates the rate of freezing and reduces the freezing duration. This effect was attributed to the reduction in supercooling, consequently accelerating the onset of nucleation. Interestingly, both investigations noted an extension in the phase transition times with ultrasound treatment. Experimental investigations conducted at the laboratory scale by Li and Sun (2002b) demonstrated that UAF immersion, at ~16 W, elicited a notable 14% reduction in freezing duration for potato sticks from −1 to −7 °C. UAF of potatoes shows lesser tissue damage in microstructure analysis in comparison to

Table 7.1 Impact of various freezing techniques on numerous quality characteristics of fruits and vegetables

Produce	Freezing method	Treatment condition	Key findings	References
Whole strawberry	Fast freezing	Air speed 1.2 ms^{-1} at −30 °C and storing in a freezer at −25 °C for 4 months	Better color guard Increased antioxidant capacity	Rayman Ergün et al. (2021)
Whole raspberry	Liquid nitrogen freezing	Freezing in −80 °C/15 min stored at −24 °C for 12 months	No significant effects on texture and color Slight reduction in antiradical efficiency	Grover and Negi (2023)
Apple tissue	PEF-assisted vacuum freezing	Pulse from 1 to 10 kV/cm at 40 °C	PEF-treated samples hold more water during the rehydration process	Lammerskitten et al. (2019)
Frozen green pepper and carrot cantaloupe	Microwave- and infrared-assisted freeze-thawing	Freeze-thawing using infra-red for 20 s, 10 s rest, freeze-thawing for 10 s via the microwave-assisted method, and standing for 30 s	Infrared heating before microwave gave better quality, but loose ascorbic acid	Chen et al. (2022)
Whole aronia berry	Frost-free freezing	−20 °C for 8 months	Nonsignificant effects on TPC (Total Phenolic Content) Increase in the polyphenol oxidase enzyme	King et al. (2022)
Blanched broccoli	Fluidized freezer	−40 °C ± 1 °C/10 min and frozen storage at −20 °C for 12 months	Vitamin C content in ascorbic acid increased	Gonzalez-Hidalgo et al. (2019)
Potato and apple pieces	Microwave-assisted freezing	Constant microwave power (MW) was applied at 167 W/kg and pulsed MW power at 500 and 667 W/kg with a pulse width of 10 s and 20	No significant effects on texture, color, and drip loss, but lesser freezing damage	Jha et al. (2020)
Blueberry pulp	Chest freezing	−18 °C for 6 months of storage	No significant effects on phenolic content Decline in anthocyanin content	Orqueda et al. (2021)

(continued)

Table 7.1 (continued)

Produce	Freezing method	Treatment condition	Key findings	References
Whole balanced kale	Blast freezing	−20 °C/90 min and – 30 °C/120 min and storage for 12 months	9% reduction in tocopherol content Reduction in micronutrient	Grover and Negi (2023)
Cherries	Antifreeze peptides (AFPs)	Immersion of the sample in a 0.1 mg/mL concentration of AFPs in 9 g/L of saline before freezing	Reduced drip loss, improved textural properties, higher anthocyanin content, and total phenolic content	Kong et al. (2017)
Strawberry	Antifreeze protein treatment	Use of 0.02 g/L AFPs during vacuum impregnation	Lower drip loss Less damage to the microstructure	da Silva et al. (2019)
Star fruit	*Drimys angustifolia* (AFPs-I) anti-freeze protein	Immersion and vacuum infiltration, concentration of 0.1 mg/ml	Lower drip loss Firmness higher	Provesi et al. (2019)
Cucumber, onion, carrot, and zucchini	*Tenebrio molitor* Antifreeze protein	1 mg/mL	Cell protection Texture preservation	Song et al. (2015)
Fresh-cut eggplants	Dehydrofreezing	By vacuum dryer at 30 °C, 2.5 kPa	Improved surface, color, and texture preservation	Wu et al. (2009)
Carrot, bell pepper, and cucumber	Dehydrofreezing	Convective dehydrofreezing Drying at 0 °C and 70% RH for 15 h Followed by wrapped storage for 12 h at 0 °C and 70% RH	Lower drip loss Firmer product	Schudel et al. (2021)
Peach and mango	High-pressure freezing	Supercooling at 200 MPa to −20 °C, Whole fruits	Reduced freeze burn with smaller ice-crystal size	Otero et al. (2000)

(continued)

Table 7.1 (continued)

Produce	Freezing method	Treatment condition	Key findings	References
Carrot	High-pressure freezing	100, 200, 300, 400, and 500 MPa for −20 °C for 25 min Carrot disks of 15 mm diameter and 5 mm thickness	At 200, 340, and 400 MPa resulted in improved texture and histological structure At 100 MPa, firmness decreased At 500 MPa, high strain resulted in pectin release and structural damage	Fuchigami et al. (1997)
Cabbage	High-pressure freezing	100–700 MPa for 45mins at −20 °C	200 and 340 MPa have the least pectin loss and structural damage	Fuchigami et al. (1998)
Mushroom	Ultrasound-assisted freezing	300 W, 20 kHz Mushroom cubes (17 mm*17 mm*17 mm)	Increased ice nucleation, reduced nuclear size,and improved quality	Islam et al. (2015)
Broccoli	Ultrasound-assisted freezing	20 or 30 kHz frequency, 60 s on/60 s off duty cycle, and 0, 125, 150, 175, or 190 W power Blanched small florets of 5 ± 0.5 cm diameter, 2 ± 0.5 cm length	Freezing time reduced, lower drip loss, preserved microstructure, and firmness	Xin et al. (2014a)
Radish	Ultrasound-assisted freezing	28 kHz, 420 W, duty cycle of 50%, and working time of 120 s at −18 °C Predegassed samples, cylindrical pieces 1.7 cm* 2.0 cm (diameter *height)	Improved freezing efficiency and improved quality attributes	Tian et al. (2020)
Fruit puree/ juice	Ultrasound-assisted freezing	59 kHz and 300–840 W at −20 °C	Improve the freezing efficiency of viscous food	Tian et al. (2021)

(continued)

Table 7.1 (continued)

Produce	Freezing method	Treatment condition	Key findings	References
Whole potato	Ultrasound-assisted freezing	Frequencies of 20, 28, and 40 kHz at −18 °C	Enhanced freezing rate Better preservation of nutrients under multi-frequency ultrasound	Zhu et al. (2020)
Guava cubes	Magnetic field-assisted freezing	Field intensity: 2.4, 5.6, and 8.8 mT Cube sizes: 2, 3, and 4 cm	Lower phase transition time with increasing field intensity, decreasing cube size Drip loss higher for smaller size and higher field intensity Color change increased with field intensity and size	Panayampadan et al. (2022)
Peach, Apple, and Indian jujube	Magnetic field-assisted freezing	0-45mT field intensity with 20 mm × 15 mm × 10 mm block size	Reduction in drip loss and hardness	Leng et al. (2023)
Mango slices	Magnetic field-assisted freezing	10, 50, and 100 mT field intensity Slices 2 × 8 × 2 cm (W × L × D)	50 mT OMF intensities provided supercooling and preserved the quality with extended shelf life	Kang et al. (2021)

conventional methods (Sun and Li 2003). The accelerated freezing rate achieved at elevated ultrasonic levels was identified as the primary factor contributing to this outcome. Consistent findings were observed in experiments involving apple cylinders, where the freezing duration decreased by up to 8% compared to conventional immersion freezing. Additionally, the analysis of freezing curves suggests that ultrasound has the potential to induce primary nucleation (Delgado et al. 2009). The application of ultrasound during immersion freezing is not limited to the initiation of the nucleation, but it also contributes to enzyme inactivation, such as polyphenol oxidase inactivation in mushrooms (Islam et al. 2014). The advantages of ultrasound-assisted freezing are given in Fig. 7.2.

Ultrasound waves are applied to fruits and vegetables during freezing to enhance the rate of heat transfer and ice nucleation, resulting in tiny crystals. Ultrasound-assisted freezing is particularly effective for fruits that have more water, such as

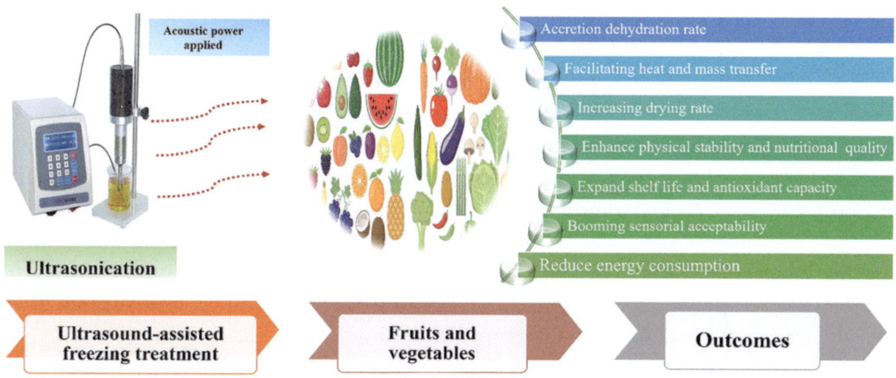

Fig. 7.2 Advantages of using ultrasound-assisted freezing for fruits and vegetables

apples and strawberries. UAF also significantly reduces drip losses and improves the texture of the product 0.

7.5.4 Magnetic Resonance-Assisted Freezing

Freezing with an induced magnetic field has been recognized as a technique capable of augmenting the freezing rate and enhancing the overall quality of frozen food products. Unlike conventional refrigeration systems, this method is a supplementary approach, aiding existing freezing processes and methodologies to achieve improved product quality and faster freezing rates. The significant influence of magnetic fields on the properties of water, analogous to the phenomena observed in the movement of tides known as geomagnetism, underscores its potential for enhancing freezing processes (Muthukumarappan et al. 2018).

The rapid transmission of cold air into the core of the product facilitates uniform and swift cooling. This prevents freezing at the outer surface, which could otherwise facilitate easy transmission of cold air to the inner sections, accelerating the cooling rate and potentially causing supercooling of free water within the product. Additionally, the spinning nuclei exhibit processional motion around the field's direction, thereby augmenting the thermal vibration of hydrogen nuclei, supercooling effects, and facilitating heat transfer throughout the product (Kaur and Kumar 2020). The process removes latent heat from water without freezing it, causing a temperature shift that is proportional to the square of the applied magnetic field. It disrupts hydrogen bonds, fragments water clusters, and forms small ice crystals, preserving the quality of frozen products in appearance, flavor, and fragrance. Reducing water cluster size increases nonfreezable bound water, enhancing product freshness (Owada and Saito 2010).

The utilization of electromagnetic fields started in the food industry with the effect of inhibiting bacteria (Boe et al. 1968), on the ripening behavior of tomatoes (Yoshimura 1989), on sterilizing of Japanese Sake (Bai et al. 2011), and on thawing frozen food. Recognizing the advantages of magnetic field application in freezing, numerous authors and research institutions have undertaken experimental studies to assess its impact on freezing various products, including water, aqueous solutions, tissues, foods, nanofluids, cells, organisms, and organs (Kaur and Kumar 2020). The concept was introduced in Japan, originating from experiments involving the freezing of water droplets exposed to magnetic fields ranging from 0 to 0.5 T. These experiments demonstrated notable supercooling of water molecules, which enhanced the stability of hydrogen bonding. Magnetic-assisted freezing has recently been utilized to preserve fruits and vegetables. A study by Qiao et al. (2023) on the effect of static magnetic field strength (0–10 mT) assisted freezing on pear fruit revealed that this technique reduces the juice and total solid loss. However, no significant effect is observed on the texture parameters of the fruit. This research suggested that magnetic field-assisted freezing positively affected the quality of frozen pears. Further efforts are necessary to enhance product quality. To investigate this, a study was carried out to examine the effects of electromagnetic field systems with intensities of 2.5 mT (0.025 kGauss) and 3.0 mT (0.030 kGauss) on the freezing behavior and quality of whole frozen tomatoes. The results revealed a significant reduction in freezing time, with samples fully frozen within 220–260 min compared to the standard freezing time of 1460 min. This was accompanied by a notable decrease in freezing point and an extension in the degree of supercooling. The thawing loss was significantly lower for samples frozen under electromagnetic influence (0.02%) compared to the control group (1.73%). Additionally, electromagnetically frozen tomato samples displayed freshness and textural attributes closely resembling those of fresh samples even after thawing, suggesting a potential method for preserving perishable goods for extended periods (Kaur et al. 2022).

The study examined the impact of static magnetic field (SMF) intensities ranging from 0 to 45 mT on apple, peach, cucumber, and Indian jujube at a macro scale. Various parameters such as ice crystal size, duration between −1 °C and −5 °C, drip loss, and texture were analyzed and compared. The findings revealed a tendency of static magnetic fields to decrease ice crystal size across the four types of fruits and vegetables. However, the optimal SMF intensity varied depending on the specific food type. Notably, the beneficial effect of SMF in reducing ice crystals was observed exclusively in Indian jujube (Leng et al. 2023). All the available studies on SMF show effectiveness in retaining nutrients and preventing loss during freezing. However, more studies are needed to understand better the effect of magnetic fields on highly perishable fruits and vegetable commodities. Pilot-scale studies can bring more information on the effect of these techniques at a large scale.

7.5.5 Sub-Zero Temperature

The distinct sub-zero temperature range, including ice temperature (ranging from 0 °C to the freezing point), supercooling (from the freezing temperature (FT) to the nucleation temperature (NT) without ice crystal formation), and super chilling (below the freezing temperature (FT) with partial freezing), has been identified as a promising approach for preserving fresh produce (Stonehouse and Evans 2015). Storing fruits and vegetables within the sub-zero temperature range, avoiding freezing but maintaining temperatures between ice temperature and supercooling, can sustain the vitality of tissue cells. This approach prevents post-harvest chilling damage and preserves cell structures, thus upholding the quality of the products (Fukuma et al. 2012). Numerous studies have explored the supercooling properties of specific vegetables, namely garlic, shallots, peppers, and potatoes, which involve temperatures below their freezing points without the formation of ice crystals (James et al. 2011), and studies on supercooling properties for fruits like strawberries have also shown promising results (Kobayashi and Suzuki 2019).

7.5.6 Ice Temperature

Ice-temperature storage has been shown to notably influence various fruits and vegetables by reducing ethylene production, slowing respiration rates, minimizing nutritional degradation, preserving texture and sensory quality, inhibiting membrane lipid peroxidation, and inhibiting microbial growth (Liu et al. 2020). Studies on fruits and vegetables, like swedes, and turnips, (Helland et al. 2016), *Toona sinensis* (Zhang et al. 2011), hawthorn (Wang et al. 2018), and tomatoes (Wang et al. 2018) have reported their suitability for ice-temperature preservation. Additionally, several innovative and hurdle ice-temperature technologies have been explored and successfully applied for preserving fruits and vegetables. Low-temperature acclimation with ice temperature (−0.7 °C) improves the antioxidant capacity, extends shelf life, and reduces postharvest loss in peaches (Wang and Xu 2018). Similarly, a combined treatment of 1.0 μL/L1- methyl cyclopropane with ice temperature (−1.0 °C) significantly inhibited the ethylene production and the respiration rate; it also reduced the pulp loss, total soluble solids, and titratable acid in hawthorn fruits (Wang et al. 2018). A combined treatment of ice temperature (−2.0 °C) with short-wave UV effectively reduced respiration intensity and weight loss, and improved the ascorbate peroxidase and phenylalanine ammonialyase of blueberries (Liu et al. 2020).

7.5.7 Super-Cooling

Super-cooling, also known as freezing point depression, undercooling, or sub-cooling, is a phenomenon where the temperature of a solution or food material drops below its freezing point (FT) without the formation of ice crystals. The effective use of supercooling technology depends on three critical aspects: the ability to achieve supercooling, the extent to which the freezing point can be lowered before nucleation occurs, and the stability of the supercooled state within the specific food matrix (James et al. 2009). Successful application of supercooling technology hinges on three key factors: the feasibility of supercooling, the maximum extent of freezing point depression before nucleation, and the stability of the supercooled state in the particular food matrix. Investigations into the super-cooling phenomenon in diverse vegetables, including garlic cloves and bulbs, broccoli, cauliflower, carrots, leeks, parsnips, and shallots, have found significant variations in the occurrence of super-cooling before nucleation, freezing temperature (FT), and the maximum degree of super-cooling among different species (James et al. 2011). In the study, it was observed that garlic and shallot samples demonstrated the capability of undergoing super-cooling, while only 40% of parsnip samples exhibited this phenomenon. Additionally, certain vegetables like garlic and shallots could maintain a super-cooled state for an extended period, reaching temperatures as low as −10 °C for some samples, without freezing. These vegetables were more stable than other species (James et al. 2011).

Workmaster et al. (1999) conducted a study to investigate the supercooling capability at sub-zero temperatures of cranberry plants and fruits. It was observed that cranberry fruit could maintain a supercooled temperature of up to −8 °C for 1 h following nucleation in solutions when inoculated with ice-nucleating bacteria. In contrast, for the stem, the ice crystal formed rapidly upon solution introduction, propagating into the leaves. Additionally, it was noted that at the immature stage, the berries exhibited shorter super-cooling durations and were unable to reach temperatures as low as mature fruits, which could be attributed to the changes in structural and nutritional properties. Another study on the super-cooling capacity of frozen strawberries revealed that the mean freezing temperature (FT) and nucleation temperature (NT) differed between the surface and center positions of strawberries. Specifically, the mean NT and FT for surface positions were recorded as −4.1 °C and −1.3 °C, respectively, while for center positions, they were −1.3 °C and −0.6 °C, respectively (Martins and Lopes 2007).

7.5.8 Super-Chilling

Super-chilling is increasingly recognized as a vital method for prolonging the shelf life of the harvest, followed by proper refrigerated storage and freezing, due to the remarkable preservation benefits. Within this temperature range, the product experiences partial or slight freezing, yet it demonstrates a degree of tolerance to this state (Liu et al. 2020). This technique has shown several beneficial effects on preserving the quality of specific fruits and vegetables. A study to investigate the effect of supercooling on the physiological and biochemical properties of strawberries revealed significant inhibition of respiratory intensity, as well as degradation of nutrients such as vitamin C and pectin with extension in the storage duration, by more than 2 months (Lin and Jing 2004).

In several studies, super-chilling has been observed to cause freezing injury to the tissues of fruits and vegetables. For example, Zhang et al. (2012) investigated the impact of various low temperatures (−0.5 °C, −1.0 °C, −1.5 °C, and 2.0 °C) on different cherry varieties. They found that temperatures below the freezing point caused chilling damage to all analyzed tissues, with the severity increasing as the temperature decreased. This finding suggests that the impact of super-chilling varies depending on factors such as tissue structure, chemical composition, and physicochemical reactions between species. Consequently, a deeper understanding of freezing injury mechanisms and tissue tolerance is essential for successfully applying super-chilling to fruits and vegetables.

7.5.9 Isochoric Freezing

Isochoric freezing, pioneered by Rubinsky et al. (2005), presents a novel approach to food preservation at temperatures below freezing, safeguarding it against detrimental ice-related damage. Within an isochoric setup, variations in pressure and temperature occur concurrently while maintaining a constant volume throughout the process (Powell-Palm et al. 2019). This technique operates on the principles of Le Chatelier, which posits that in a closed environment, as water freezes and expands, it generates elevated pressure, impeding further ice formation. By imposing volume constraints within an isochoric system, ice expansion is restricted, thereby creating hydrostatic pressure within the system (Grover and Negi 2023).

During isochoric freezing, the phase diagram of pure water illustrates a trajectory traversing liquid water, ice III, and ice I. The pressure buildup within the system during this process lowers the solution's freezing point. Consequently, ice crystal growth persists until equilibrium is reached between the system's freezing point and the ambient temperature. At this juncture, a dual-phase system emerges, comprising a frozen solid segment and a liquid portion, coexisting under pressure with a temperature lower than the atmospheric freezing point of the solution. This scenario enables the storage of food materials within the liquid volume of the system, shielding them from the adverse effects of freezing while prolonging their shelf life due to a reduced metabolic rate. The extent of ice formation in an isochoric system depends on the temperature differential between the system's actual temperature and the freezing point of the external solution at atmospheric pressure (Nida et al. 2021).

Several investigations have examined the effects of isochoric freezing on fruits and vegetables. A significant decrease in potato browning was observed in samples subjected to isochoric freezing compared to isobaric freezing at the same temperature of −5 °C (Lyu et al. 2017). Isochoric freezing at −2.5 °C effectively preserved lycopene in grape tomatoes, with a retention rate of 90% after 4 weeks, in contrast to IQF and isobaric freezing at the same temperature, which resulted in losses of 67% and 52%, respectively (Bilbao-Sainz et al. 2021). Spinach leaf turgidity and integrity were well maintained under isochoric conditions (29.7 ± 0.2 MPa, −4 ± 0.9 °C), with distinct cell walls observed around the cells. In contrast, isobaric freezing under the same temperature conditions led to cellular decompartmentalization, accompanied by reduced weight and thickness of the leaves (Bilbao-Sainz et al. 2020).

A review of studies by Zhu et al. (2021) highlights the potential drawbacks of increased pressure during isochoric freezing, including the disruption of pectin structure in peels, as well as the potential loss of juiciness and firmness in porous or soft fruits and vegetables such as eggplant, strawberry, and pepper. Similarly, research on pomegranate arils subjected to constant volume conditions at 12 MPa/−2.5 °C revealed a 3% higher mass loss than samples frozen under isobaric conditions (Bilbao-Sainz et al. 2021). Micrographic analysis of the frozen arils further demonstrated broken and irregularly shaped cells, indicating significant changes in tissue structure due to the hydrostatic pressure generated during freezing. These findings underscore the need to establish optimized pressure and temperature protocols for freezing various commodities. Exploring these aspects is essential to harness the full potential of isochoric freezing technology for widespread industrial applications.

In the field of food preservation, both traditional freezing methods and innovative techniques hold significant importance, particularly concerning their effects on the quality parameters of fruits and vegetables. Traditional methods like air blast and plate freezing have long been employed for their simplicity and reliability. However, these methods may have drawbacks, such as slower freezing rates and potential texture degradation due to the formation of large ice crystals. On the other hand, innovative freezing techniques, such as ultrasound-assisted and high-pressure processing, offer promising alternatives with the potential to address these challenges. These innovative methods often leverage advanced technology to achieve rapid freezing rates and minimize texture degradation, thus preserving the quality parameters of fruits and vegetables more effectively. By comparing the effects of traditional and innovative freezing methods on quality parameters such as texture, color, flavor, and nutrient retention, a comprehensive understanding can be gained to inform the selection of the most suitable preservation technique for different fruits and vegetables. The comparison of various techniques is further elucidated in Table 7.2.

Table 7.2 Comparative study of various freezing preservation technologies of fruits and vegetables

Novel techniques	Principle	Advantages	Disadvantages	References
High-pressure freezing	Suppression of the freezing point under high pressure inhibits the formation of the crystalline structure required for ice formation. High pressure restricts the molecular motion of water molecules, making it difficult to arrange in the lattice structure of ice	Supercooling results in high nucleation rates Reduced cell damage due to smaller nuclei formation	High cost Complex operation Limited capacity Sample alteration in case of high-pressure	Grover and Negi (2023)
Ultrasound-assisted freezing	The cavitation phenomenon induced by ultrasound waves plays a pivotal role in creating microstreaming of the fluid surrounding the food material	Enhances heat and mass transfer rates, reducing freezing time Provide a site for ice nucleation, resulting in more and smaller ice crystals. Therefore, reduced cell damage	High cost Complex operations Cavitation damage in the case of high-intensity ultrasound waves	Xin et al. (2015)
Impingement	Convective heat transfer using high-velocity air jets	High product quality due to uniform cooling Minimal damage to product due to short exposure time High throughput	Freeze burn in case of nonoptimized operations Not suitable for thick or irregularly shaped foods, due to limited penetration High velocity may result in damage due to high force	Sutariya and Sunkesula (2020)
Magnetic field assisted freezing	Magnetic fields exert force on charged particles like water molecules, which try to align with the changing magnetic field. This affects the formation of ice crystals during freezing	Enhanced supercooling, resulting in smaller crystal size and high nucleation rates Improved product quality due to minimal cell damage	High cost High energy consumption Nonuniform freezing in the case of varied field strength	Grover and Negi (2023)

(continued)

Table 7.2 (continued)

Novel techniques	Principle	Advantages	Disadvantages	References
Isochoric freezing	Freezing below freezing temperature by external pressure is applied that counteracts the volume change of the substance, keeping it constant throughout the process	Less energy consumption Minimal damage due to limited volume expansion and homogeneous nucleation Improved product safety due to high-pressure application	Tissue damage due to high-pressure development Limitation of size and quantity Longer processing time	Bilbao-Sainz et al. (2021)
Nucleation proteins	They facilitate the nucleation by providing a scaffold for the molecules to assemble and lower the energy barrier for nucleation	Shorter freezing time Increased nucleation with smaller nuclei formation, minimal tissue damage	Tissue damage due to ice crystal aggregation Neurogenerative disorders in the case of abnormal protein aggregation	Sun et al. (2023)
Antifreeze proteins	Lower the freezing point of the water without significantly affecting its melting point	Minimal recrystallization during storage Suppress nuclei growth Minimal tissue damage	Limited effectiveness in complex systems Health implications due to interaction in the biological system	Baskaran et al. (2021)

7.6 Pre-freezing Treatment

Pre-freezing treatments play a vital role in the overall effectiveness of freezing as a preservation method for various food products, including fruits, vegetables, meats, and seafood. These treatments are designed to prepare food items for freezing, optimize their quality, and enhance their preservation attributes. The importance of pre-freezing treatments lies in their ability to improve freezing efficiency, maintain product quality, and prolong shelf life. For instance, blanching or par-cooking can reduce the time required to freeze completely. Enzymes are deactivated by partially cooking or blanching foods before freezing, the microbial load is reduced, and water content is partially removed. This accelerates the freezing process and minimizes the formation of ice crystals, which can negatively impact texture and flavor. Similarly, syruping or bringing before freezing can help retain the texture and flavor of the fruits and vegetables and protect them from chilling injury during freezing. Coating a layer of sugar before freezing also prevents the produce from chilling injury or freezer burn, as sugar acts as a cryoprotectant (Li et al. 2017).

7.6.1 Dehydro-Freezing

Dehydro-freezing has recently been used by various authors and explored its potential to reduce the loss due to the traditional freezing method. Dehydro-freezing is a freezing process in which the product dehydrates to a desirable moisture content before freezing. Where moisture removal is involved through osmosis, this process is known as osmotic dehydro-freezing. The primary purpose behind this technique is to remove extra water from fruits and vegetables that are rich in water content so that the development of excess crystals and damage from them can be prevented. This treatment works on the reduction of water content; hence, the other thermal properties such as heat capacity, enthalpy, and freezing point are reduced (Tocci and Mascheroni 2008). It was also observed that dehydro-freezing retains and improves color, texture, and nutrients, decreasing enzymatic reactions in products (James et al. 2014). In some studies, commodities underwent osmodehydration either by sugar (fruits) or sodium chloride (vegetables), and the results of the findings suggested that this method significantly improves the texture and sensory parameters. Strawberries are dipped in 50% sugar (glucose and oligofructose) and 3.5 and 1.5% NaCl and $CaCl_2$ solution, respectively. The addition of sugar and reduced moisture content were observed to enhance firmness, vitamin C, and sensory characteristics for prolonged storage life and less water loss during thawing, maintaining tissue integrity. Similarly, osmodehydro-freezing of mango by mixing a solution of sucrose, glucose, and fructose improved the texture and reduced drip loss (Li et al. 2023). Freezing of banana slices has also become possible because of osmotic dehydration. It was observed that dipping the banana slices in 65% sucrose solution for 4 h before air blast freezing improves the color and texture of the banana slices (Atares et al. 2011).

A study assessed the effects of integrating osmotic dehydration and cryogenic freezing processes on the quality of fruits and vegetables, using tomato pieces as a model. The osmotic dehydration step involved immersing tomatoes in a 55% (w/w) maltodextrin solution at 35 °C for 24 h, resulting in a 15.74% reduction in water content and a 0.22% increase in solid content. This process enhanced the tomatoes' visual appeal, evidenced by increased redness and a lycopene concentration of 1244.78 μg/g on a dry mass basis.

When subjected to cryogenic freezing at −40 °C using liquid nitrogen, the osmotic pretreatment significantly influenced the freezing dynamics. Osmotically treated samples exhibited a faster freezing time of 3.13 min and a higher freezing rate of 16.53 °C/min compared to untreated samples, which froze in 4.48 min with a freezing rate of 11.58 °C/min. The benefits of osmotic pretreatment were also observed after thawing. Pretreated samples retained more lycopene and color while exhibiting reduced drip loss compared to their untreated counterparts. Furthermore, the structural integrity of cells in the pretreated samples was better preserved, with less cell collapse.

This research highlights the potential of osmotic dehydration as a preparatory step to enhance the quality of tomatoes during cryogenic freezing. The approach not only preserves color and nutritional content but also minimizes texture degradation, offering a promising technique for maintaining the quality of frozen fruits and vegetables (Li et al. 2017).

7.6.2 *Cryoprotective Agent*

Cryoprotectants serve as vital agents in the freezing process of fruits and vegetables by preserving their quality and nutritional integrity. As these perishable goods undergo freezing, the formation of ice crystals poses a significant threat to their cellular structure. Cryoprotectants, such as sugars and polyols, mitigate this risk by lowering the freezing point of water, thereby reducing the formation of large ice crystals that can rupture cell walls. By safeguarding against such damage, cryoprotectants help maintain the produce's texture, flavor, and appearance, ensuring they retain their texture and color after thawing. Moreover, these compounds mitigate osmotic shock during freezing, minimizing cellular dehydration and preserving the natural moisture content of fruits and vegetables (Bhattacharya 2018). This preservation of moisture not only contributes to the produce's juiciness but also helps retain its vibrant color and firm texture. Furthermore, incorporating cryoprotectants extends the shelf life of frozen fruits and vegetables by preventing deterioration during storage, thereby reducing food waste and enhancing food security (Neri et al. 2020). Cryoprotectants serve as invaluable allies in freezing, facilitating the retention of quality attributes and nutritional value in frozen fruits and vegetables. The studies found that sweet cherries subjected to freezing by osmoblanching in a mixture of glucose:fructose (2:1 ratio) showed significant improvement in antioxidant activity compared to control samples, which were not treated with cryoprotectant compounds (Chaovanalikit and Wrolstad 2004). Some other studies also explored the new cryoprotective agents other than sugars, such as xanthan gum, carboxymethyl cellulase, and alginate. It was observed that 0.4% coating of xanthan on sliced carrots prevented texture loss during freezing (Maity et al. 2011). Similarly, applying xanthan gum to potatoes, turnips, and carrots prevents drip loss during thawing (Downey 2002). Using pectin as a pretreatment potentially preserves texture during thawing (Reno et al. 2011).

7.6.3 *Blanching*

Blanching is a well-known treatment that retains the texture, flavor, and color of the products under freezing. In addition, it is also useful in the inactivation of enzymes and the reduction of water from fruits and vegetables. It was observed that a blanched sample, when placed for quick freezing form tiny ice crystals, thereby preserving the texture of fruits and vegetables (Xin et al. 2015). Blanching is a crucial step in the processing of mushrooms before freezing. This involves treating mushrooms in water or aqueous solutions containing combinations such as citric acid with L-ascorbic acid or lactic acid with L-ascorbic acid. Such treatments effectively mitigate undesirable textural changes that may occur during freezing and extended frozen storage (Jaworska and Bernaś 2010). In addition to texture preservation, blanching plays a significant role in minimizing the degradation of pigments, including chlorophylls and carotenoids (Lisiewska et al. 2004). The process reduces tissue deterioration and lowers oxygen levels in plant cells, which helps to stabilize and retain the natural color of mushrooms. Consequently, blanching not only enhances the visual appeal but also supports the overall quality of frozen mushrooms during storage. Various new blanching techniques, such as ad infra-red blanching, high-pressure blanching, and ultrasound blanching, have been introduced recently. These blanching treatments provide rapid and uniform heating to the subjected food product; hence, maximum nutrients retained in the sample are lost during long-term heating treatment. Manzocco et al. (2008) blanched whole apples using a radio frequency of 6.2 kV for 3 min. They found that it takes a shorter time to inactivate the enzymes than the conventional blanching method, saves energy, and reduces waste. These novel techniques, such as microwave blanching, require a lesser amount of water or no water at all to blanch the fruits and vegetables. The loss of nutrients from leaching in water during conventional blanching is also prevented. Overall, conventional or novel blanching treatment plays a significant role in preserving the fruits and vegetables' nutrients, color, and texture.

7.7 Quality Assessment and Monitoring

There is a growing consumption of frozen fruits and vegetables driven primarily by factors like the necessity to minimize post-harvest deteriorative losses in the produce, ensuring year-round availability, extension of shelf life, and preservation of nutritional quality. To ensure that the objectives of processing are met, it is necessary to assess the quality and monitor the storage conditions of the frozen fruits and vegetables. The quality assessment of frozen fruits and vegetables is a complex interplay between their physicochemical attributes specific to horticultural commodities (like fruits and vegetables) and consumer perception. The challenge in establishing a universally applicable definition of quality for frozen produce arises from the involvement of various stakeholders across the supply chain, each functioning essentially as a consumer concerning the preceding chain member (Kyriacou and Rouphael 2018). Over the years, many methodologies with advanced instrumentation have emerged, designed to mimic human sensory perceptions to assess the quality and its associated attributes. These developments have been accompanied by a diverse range of instrumental sensors developed for real-time and nondestructive testing applications to minimize wastage (Schmilovitch and Mizrach 2013).

7.8 Methods to Assess Quality Characteristics of Frozen Fruits and Vegetables

Assessment methods can be classified as nondestructive and destructive based on the nature of testing. The significant difference lies in their effect on the product. Nondestructive tests are more recent and sustainable options that do not damage the produce, which allows consumption after the assessment. On the other hand, destructive tests are cost-effective as they require more straightforward instruments and are more accurate as they directly measure the product's properties. However, these tests render the product unusable for further consumption. A flow chart for the quality assessment of fruit and vegetables is given in Fig. 7.3.

7.8.1 *Physical Evaluation*

7.8.1.1 Visual Inspection

Food regulatory bodies have set standards for the maximum allowable defects, like poor trimming, extraneous matter, discoloration, freeze burn, and bruises, depending on the type of produce or the assigned grade. Codex Alimentarius Commission and the Food & Agriculture Organisation of the United States have jointly set the standards for quick frozen fruits (excluding berries) and vegetables under codex stan 320–2015 and 319–2015, respectively. To inspect these defects, the internal quality control of the industry employs panellists to visually inspect and record the defects in the frozen fruit and vegetable quality. Adherence to the law is necessary

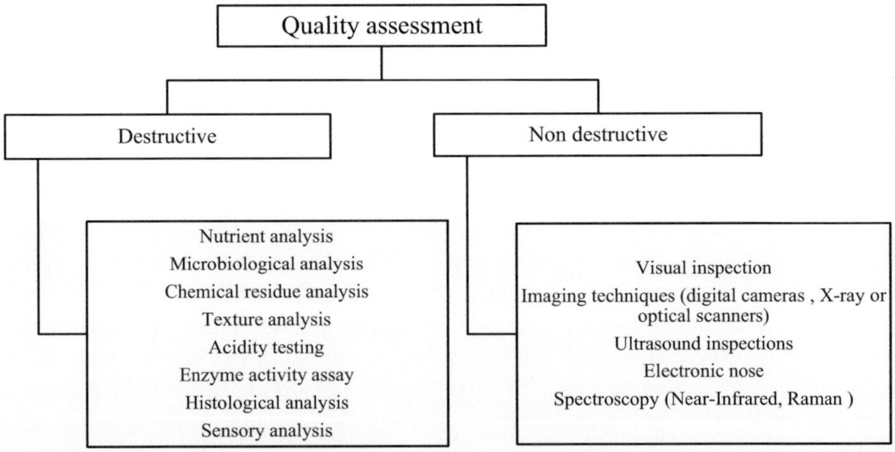

Fig. 7.3 Flowchart of quality assessment

to gain consumer trust, maintain product integrity, and contribute to the overall safety and quality of frozen products.

7.8.1.2 Color Grading

Color is a critical sensory factor in the food industry, which predominantly impacts the consumer decision of acceptance or rejection of the produce (fruits and vegetables); it defines the perceived freshness of the commodity. Various color spaces are utilized for assessing the freshness of commodities, which includes a variety of options, like the HIS space (H = hue, I = intensity and S = saturation), the Hunter Color Lab space, the Commission Internationale d'Eclairage (CIE) Lab* space, where L* at 100 represents lightness, +a* denotes red, -a* denotes green,+b* denotes yellow and -b* denotes blue—the CIE XYZ space, and RGB color spaces. Standard formulas are used to calculate color values like the Whiteness Index (WI), Browning Index (BI), and Total Color Differences (ΔE) (Akter et al. 2024). Uniformity of color holds significant importance for the frozen produce industry, ensuring consistency in appearance is important for visual appeal and marketability. The browning index is an important grading factor for cut frozen fruits like apples, pears, bananas, avocadoes, and apricots and vegetables like potatoes, mushrooms, eggplants, celery, and lettuce which are prone to enzymatic browning (Parreño and Torres 2005).

7.8.1.3 Textural Properties

Texture, in scientific terms, is used to describe the combined structural and mechanical attributes of a food item. This includes its physical composition, arrangement, and how it responds to external forces. Additionally, texture encompasses the sensory experience when the food is touched or consumed, such as how it feels in the hand or oral cavity. Texture evaluation involves measuring various parameters like hardness, chewiness, cohesiveness, adhesiveness, and resilience, which collectively contribute to the overall perception of the food's texture (Javadikasgari et al. 2018). Textural properties of frozen fruits and vegetables are largely influenced by freezing method, temperature, freezing rate, and thawing parameters (time, temperature). In frozen fruits and vegetables, loss of turgor occurs due to a decrease or absence of pressure within the cells of the plant which alters the textural properties with limped or wilted appearance. Texture profile Analysis (TPA), compression test, puncture test, and tensile strength test are used for measuring the mechanical properties of fruit tissue before and after freezing (Schmilovitch and Mizrach 2013).

7.8.1.4 Juiciness

Juiciness is an important factor in fruits and vegetables. The frozen product must meet the attributes of fresh fruits, like juiciness in order to have customer satisfaction and market acceptability. Frozen fruits are consumed after thawing, while vegetables can be consumed directly after thawing or as thaw-cooked meal. After thawing, it is required that the frozen produce mimics the attributes of the fresh produce. In case of improper freezing and storage, the drip losses during thawing and cooking can be high, leading to low juiciness and therefore poor acceptability. The common technique to assess the juiciness is by measuring the amount of expressible juice of thawed produce and comparing it with fresh produce (Neuman 2006).

$$\text{Expressible Juice}(\%) = \left(\frac{\text{volume of juice extracted}}{\text{weight of sample}}\right) * 100$$

7.8.2 *Chemical Assessment*

7.8.2.1 Vitamins

Water-soluble vitamins are key markers to measure the losses during freezing. Ascorbic acid losses are measured to assess the losses in frozen products due to their high-water solubility and reduction during freeze–thaw cycles. Therefore, ascorbic acid degradation serves as a stability index during freezing and frozen storage. Folate, a vitamin B complex and water-soluble compound, is another key indicator that is lost during freezing or during frozen storage. HPLC and UV detection are commonly used assessment techniques for these vitamins (Erickson 2005).

7.8.2.2 Glucosinolates

These are sulfur and nitrogen-containing glycosides, serving as secondary metabolites in the plant kingdom. Cruciferous plants, primarily belonging to the Brassicaceae family, such as cabbage, broccoli, Brussels sprouts, and collard greens, are the main sources of these phytochemicals. Being water-soluble and susceptible to enzymatic degradation, these can be utilized as a quality indicator in frozen produce. Enzyme activity, as well as factors like pre-treatment methods such as blanching, can lead to a reduction in glucosinolate levels during freezing or inadequate storage of frozen products.

7.8.2.3 Microbial Assessment

It is important to evaluate the microbiological safety of fruits and vegetables because freezing does not destroy the microorganisms, and frozen produce does not necessarily receive a pretreatment that can reduce the microbial count (Tanner 1934). TPC (total plate count) and enzyme activity are essential to ensure that the product does not undergo spoilage during storage.

7.8.3 Analytical Assessment

7.8.3.1 Microstructure Investigation

Examination of the microstructure of frozen fruits and vegetables serves various purposes, it helps to monitor the formation of ice crystals during freezing, assess the cell damage endured during freezing, storage, and thawing. Additionally, it helps to establish the correlations between microstructure and mouthfeel. Understanding the relationship between structure and function is essential for controlling and enhancing the quality of frozen foods. The methods used for microstructure investigation include light microscopy, cold-stage light microscopy, confocal laser scanning microscopy, electron microscopy (SEM: scanning electron microscopy, TEM: transmission microscopy), X-ray tomography, and MRI (magnetic resonance imaging).

7.8.3.2 Volatile/Flavors

Flavor is a fusion of taste and aroma. In addition to visual sensory evaluations, analytical techniques such as gas chromatography-mass spectrometry (GC-MS) are also used for the profiling of volatile compounds, which are responsible for the aroma of frozen products. This helps in understanding the chemical composition of aroma profiles and detecting any off-flavors or undesirable odors in frozen products. Novel sensory systems, like electronic noses or chemosensory systems, are also used for odor detection. The electronic nose sensory systems change their properties, like conductivity, oxidation/reduction, or oscillating frequency, when dealing with different odor-volatile molecules. These sensors respond to the volatiles present in the environment, generating “smell prints” which are specific for the different aroma characteristics (Neuman 2006).

7.8.3.3 Sensory Assessment

Sensory is one of the key assessments that evaluate the consumer expectations and acceptance of a product. For the sensory assessment, a panel is selected, products are given a code as their true nature cannot be revealed to the panellists, and assessment is done on an individual basis. This assessment can be descriptive, discriminative (triangle test, ranking test), and affective (hedonic scale testing). To minimize the bias in the sensory assessment, it is important to randomize the sample presentation and randomly select the population (panellists) to ensure inclusion of all age, demographic, and socio-economic groups (Chambers et al. 2005).

7.9 Monitoring Techniques for Evaluating Changes During Freezing and Storage

During freezing, fruits and vegetables undergo changes like alteration in their nutritional profile, texture, flavor, moisture value, and color values. Alteration of nutrients is due to loss of water-soluble components like vitamins C and B, which are lost at high rates due to drip loss. Additionally, prolonged inadequate storage conditions like temperature abuse can accelerate nutrient degradation. Texture alterations are inevitable during freezing, but controlling the size of nuclei, nucleation rates, and storage can minimize the texture deterioration. Improper storage conditions cause stress and break down of the cellular structure of produce, resulting in a mushy texture with a crimped appearance. Maintaining the temperature during storage is also important to prevent any enzymatic degradation, as temperature abuse can activate enzymes like lipoxygenase and peroxidase, which are responsible for the off flavors and enzymatic browning reactions. Temperature abuse during storage results in the melting of smaller ice crystals, releasing the water molecules. Upon subsequent temperature drop, these water molecules refreeze and deposit on the surface of larger ice crystals, resulting in a further increase in the size of ice crystals. The enlarged crystal size results in secondary damage to cells and tissues of fruits and vegetables (Wu et al. 2021). Thus, structural damage results in loss of tissue moisture and drip loss, which diminishes consumer appeal with textural and visual alterations. Therefore, it is essential to maintain and monitor the optimum freezing process and storage conditions depending on the cultivar in order to prevent freeze burn, textural, nutritional losses, and enzymatic degradation (Table 7.3).

Table 7.3 Monitoring parameters for fruits and vegetables before and after freezing

Monitoring parameters	
Monitoring before processing	Product type, initial quality of raw product, required pretreatment, and effectiveness of the pretreatment
Monitoring during the process	The freezing conditions (freezing methods and rate)
Monitoring during storage	The storage time, stacking arrangement, type of freezer, temperature of the base of the freezer, any temperature fluctuations, and integrity of the packaging material

7.10 Challenges and Solutions

Owing to the ease of processing and cost-effectiveness, low-temperature processing and storage are widely used in the food industry to increase the product's shelf life. However, like other processing techniques, freezing and frozen storage also present some demerits like freeze damage/freeze burn (cellular frost dehydration, membrane leakage) to the fruits and vegetables. It is important to understand the underlying damage mechanism to find an effective solution. The major reason for freeze injuries is the formation of ice crystals. The plasma membrane is responsible for maintaining cell integrity by maintaining the turgor pressure. During freezing, the ice crystal grows and punctures the plasma membrane, and it loses its turgidity. Further, with an increase in the ice crystal size, it punctures the cell wall, causing tissue damage and the release of degenerative enzymes. The lost turgidity contributes to the textural loss (lost firmness), giving the frozen fruits and vegetables a limp appearance. This effect is more pronounced in fruits as they have softer structures, which make them more prone to freeze damage. The breakdown of the cell wall releases enzymes like cellulase, pectinase, and protease, which break pectins and hemicellulose, leading to cell wall collapse and drip loss. Further, the release of the enzymes by cell wall collapse contributes to the change in color, odor, and flavor of the fruits and vegetables.

To prevent the occurrence of freeze burn in fruits and vegetables in order to maintain their integrity, underlying methods can be used to safeguard the produce and maintain optimalfreshness (Fig. 7.4).

7.10.1 Quality Produce

Fresh produce with optimal quality attributes is the foremost requirement to minimize losses. Selecting the fruit of optimum maturity is necessary, as over-ripe produce are more prone to freeze damage. Overripe fruits and vegetables have softer tissue and excess water content as well as have already undergone enzymatic changes; therefore, during freezing, their cell wall are easily damaged, leading to freeze damage.

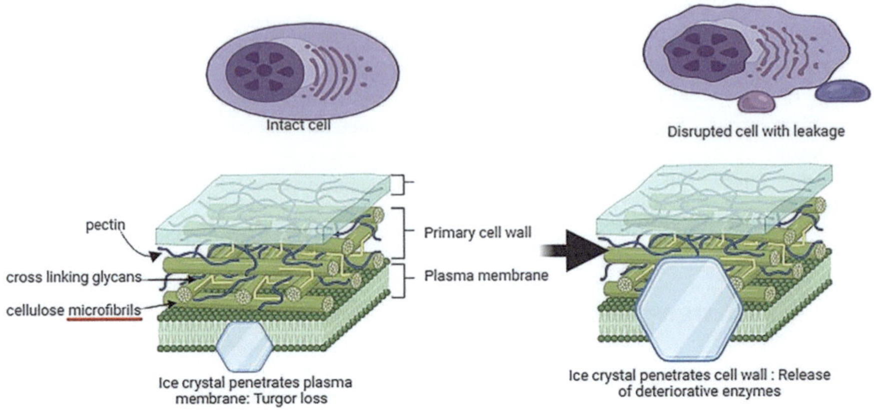

Fig. 7.4 Graphical representation of slow freezing impact on ice crystallization and cell structure of fruits and vegetables

7.10.2 *Quick Freezing*

7.10.2.1 High-Pressure Freezing

When water is frozen at atmospheric pressure (1 atm), ice I, having a density lower than water, is formed with the volume expansion of 9% and 13% at 0 °C and −20 °C, respectively. At ambient freezing conditions, the high-volume expansion is responsible for tissue damage and poor product quality (Li and Sun 2002a). Figure 7.5, water phase diagram (Chaplin 2015), depicts that temperature and pressure manipulation can help to produce complex ice types which have a minimal effect on the membrane and provide improved quality characteristics of the produce. At high-pressure conditions like 207.5 MPa, the freezing point of water is reduced from 0 °C to −22 °C; this phenomenon of suppression in the freezing temperature of liquid by the manipulation of freezing conditions is called supercooling. During supercooling, the rate of ice nucleation is increased, complex high-density ice types are formed with minimum volume expansion, resulting in minimal damage to the fruits and vegetables during freezing. For every 1 °C decrease in the supercooling temperature, the ice nucleation rate increases by ten-fold (Xin et al. 2015).

7.10.2.2 Dehydro-Freezing

Fruits and vegetables have a high water content of up to 95%, which makes them more prone to large crystal formation and freeze damage. Dehydro-freezing is a widely adopted technology where partial water removal from fruits and vegetables is done before subjecting them to freeze treatment. It has multiple advantages like reduced freezing time, weight, and reduced number and size of crystal formation

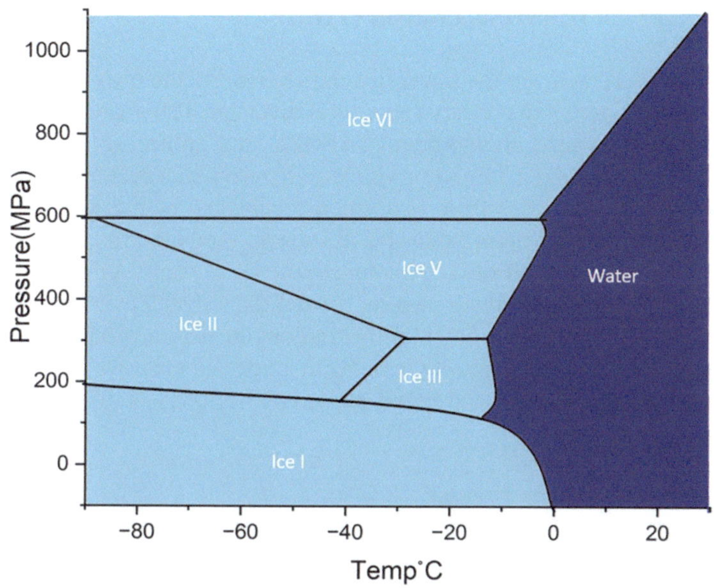

Fig. 7.5 Schematic illustration of water phase. (Source: Salzmann et al. 2009)

due to a lower volume of water. Hot-air drying, vacuum drying, microwave drying, and osmotic dehydration are the commonly utilized dehydration techniques for the partial reduction of the water content in fruits and vegetables (James et al. 2014). The degree of dehydration varies according to the produce and freezing conditions.

7.10.2.3 Ultrasound-Assisted Freezing

Ultrasound is a novel technology used in various food applications. Hurdle technology sono-crystallization utilizes ultrasound waves in the range of 20 kHz to 100 kHz along with freezing at subzero temperatures. Ultrasound induces the nucleation process, increases the freezing rate, heat, and mass transfer, which leads to controlled crystal formation with smaller crystal size. The cavitation phenomenon of ultrasound waves is responsible for the microstreaming of the fluid, which enhances the heat and mass transfer rates along with the initiation of nucleation. The effectiveness of ultrasound assistance in freezing depends on various factors like the duration, intensity, and frequency of ultrasonic waves. The optimal conditions vary according to the produce, stage of maturity, and freezing temperatures. Ice-nucleating proteins (INPs): INPs are ice-regulating materials that induce heterogenous nucleation at high nucleation rates and are extracted from various Gram-negative bacteria.

7.10.2.4 Thermal Hysteresis Proteins (THPs)

THPs are proteins that lower the freezing point of water without significantly affecting its melting point. On the basis of structural diversity, THPs have been classified (Fig. 7.6) into antifreeze glycoproteins (AFGPs) and antifreeze proteins (AFPs) type I, II, III, and IV, which help to survive in sub-zero temperatures (Barrett 2001). These proteins are extracted from fish, microorganisms, plants, and insects from polar regions, which help them to survive in subzero temperatures. The mechanism of action of THPs is through surface binding to the ice crystals. As THPs bind to the ice crystals, they alter the growth pattern, restrict their size expansion, and prevent recrystallization due to environmental fluctuations in frozen produce and prevent freeze burn. This helps to minimize the tissue damage and preserve freshness during freezing and storage (Griffith and Ewart 1995).

7.10.2.5 Monitoring the Freezer

It is important to have optimum storage conditions, such as the temperature of the freezer, to maintain the quality of the produce; installing temperature sensors and setting threshold temperatures help to set alerts in case of any temperature change.

7.10.2.6 Optimal Thawing

Thawing conditions that leave minimal damage are usually mentioned on the produce package. The optimum thawing conditions vary from produce to produce, but it is necessary to gradually thaw the produce under refrigerator conditions to prevent the texture, flavor, and nutritional value from being affected.

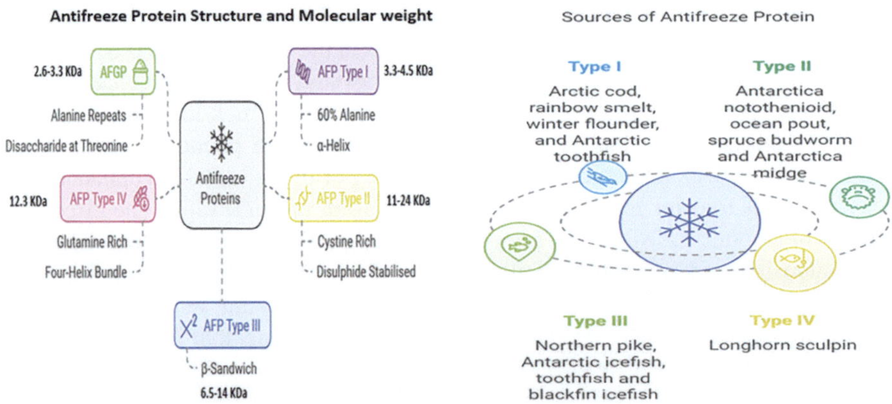

Fig. 7.6 Thermal hysteresis proteins used in freezing preservation. (Source: Barrett 2001)

7.11 Application and Industrial Practices

The busy lifestyle of modern society, with an increasing awareness of nutrition-rich food, has led to a significant surge in demand for frozen foods, mainly fruits and vegetables, owing to their associated benefits. The market for frozen fruits and vegetables has expanded significantly, with government agencies and researchers affirming their safety and nutritional value. Li et al. (2017) compared the nutritional profiles of various vegetables and fruits by classifying them into three categories: fresh, stored (5 days), and frozen produce. Based on the findings, it was concluded that frozen fruits and vegetables had a better nutritional profile when compared to fresh-stored. These key factors are the driving force in the frozen produce industry (Table 7.5).

Table 7.5 Industrial application of freezing preservation for fruits and vegetables

Product	Product type	Form	Application
Frozen fruits	Chunks, slices, diced cubes	Frozen individually or in mixes	Off-season fruit availability For baking, smoothies, and desserts
Vegetables	Pods, diced	Individual or in mixes	Off-season vegetable availability Soups, stews, casseroles, and side dishes
Root and leafy vegetables	Diced cubes	Blanched before freezing	Off-season vegetable availability Soups, stews, casseroles, and side dishes
Fruit pulp and puree	Fruits are processed into pulp and puree for ease of operation	–	Used in baking smoothies, desserts As baby foods
Frozen vegetable medley	Mix of pods, diced vegetables, and blanched-diced vegetables	Mix of different vegetables	Convenient nutritious blend that can be stir-fried, added to soups, pasta, and casseroles
Frozen ready-to-cook meals	Diced, sliced, and blanched	Mix of vegetables according to the meal	Vegetables are a crucial part of ready-to-cook meals
Frozen fruit concentrates	Concentration of extracted juices by removing a significant amount of water	–	Diluted with water for making juices, added to smoothies, desserts, sauces, marinades, salad dressing, cakes, and muffins for frosting, garnishing, or soaking syrups

7.12 Conclusion

Preserving the quality of perishable fruits and vegetables requires precise freezing methods that maintain consistency from storage to transportation. Traditional techniques like impingement freezing, air blast freezing, and cryogenic freezing have long been employed in the industry, albeit with drawbacks concerning quality retention. However, recent years have witnessed the emergence of novel freezing methods such as pressure-assisted, ultrasound-assisted, and isochoric freezing, each offering distinct advantages and limitations. Efficiency and quality preservation are key considerations in selecting a freezing method. Immersion freezing, for instance, stands out for its rapid freezing rate facilitated by high heat transfer coefficients in liquid media, leading to reduced energy consumption compared to air blast freezing. Conversely, cryogenic freezing boasts ultra-rapid freezing rates but may incur higher operating costs and safety concerns. The preservation of quality encompasses various aspects like texture, color, flavor, and nutrient retention. Immersion freezing, with its minimal dehydration, excels in maintaining texture and color, while other methods may prioritize nutrient or flavor retention. Thus, a comprehensive comparative analysis is crucial for choosing the most suitable freezing technique, weighing factors such as freezing rate, energy consumption, cost-effectiveness, and quality preservation. In addition to the freezing methods, pretreatment techniques play a crucial role in enhancing the effectiveness of preservation. Methods such as blanching, osmotic dehydration, and pretreatments with antioxidant solutions can mitigate enzymatic browning, preserve color, and extend shelf life. Moreover, quality monitoring throughout the freezing process is essential to maintain the desired attributes. Techniques such as sensory evaluation, instrumental analysis, and microbial testing provide valuable insights into the quality of frozen fruits and vegetables.

In conclusion, while conventional freezing methods have paved the way for fruit and vegetable preservation, novel techniques offer promising alternatives with improved efficiency and quality retention. By considering factors like freezing rate, energy consumption, and quality preservation, along with employing appropriate pretreatment and quality monitoring measures, producers can optimize the freezing process to meet specific requirements and priorities, ensuring the delivery of high-quality frozen produce to consumers.

References

Akter T, Bhattacharya T, Kim J et al (2024) A comprehensive review of external quality measurements of fruits and vegetables using nondestructive sensing technologies. J Agric Food Res 15:101068. https://doi.org/10.1016/j.jafr.2024.101068

Atares L, Sousa Gallagher MJ, Oliveira FAR (2011) Process conditions effect on the quality of banana osmotically dehydrated. J Food Eng 103:401–408. https://doi.org/10.1016/j.jfoodeng.2010.11.010

Bai X, Shirakashi R, Nishio S (2011) The dielectric loss characteristic of ice by dielectric heating method for the thawing of foods or biomaterials. Trans Jpn Soc Refrigerating Air Condition Eng 16:263–271

Barbosa-Cánovas GV, Buckle K, Caetano da Silva Lannes S et al (eds) (2017) Global food security and wellness, 1st edn. Springer, New York

Barrett J (2001) Thermal hysteresis proteins. Int J Biochem Cell Biol 33:105–117

Baskaran A, Kaari M, Venugopal G et al (2021) Anti freeze proteins (Afp): properties, sources and applications—a review. Int J Biol Macromol 189:292–305. https://doi.org/10.1016/j.ijbiomac.2021.08.105

Bhattacharya S (2018) Cryoprotectants and their usage in cryopreservation process. In: Cryopreservation biotechnology in biomedical and biological sciences

Bilbao-Sainz C, Sinrod AGJ, Dao L et al (2020) Preservation of spinach by isochoric (constant volume) freezing. Int J Food Sci Technol 55:2141–2151. https://doi.org/10.1111/ijfs.14463

Bilbao-Sainz C, Sinrod AJG, Dao L et al (2021) Preservation of grape tomato by isochoric freezing. Food Res Int 143:110228. https://doi.org/10.1016/j.foodres.2021.110228

Boe A, Do J, Salunkhe D (1968) Tomato ripening: effects of light frequency, magnetic field, and chemical treatments. Econ Bot 22:124–134

Bonat Celli G, Ghanem A, Su-Ling Brooks M (2016) Influence of freezing process and frozen storage on the quality of fruits and fruit products. Food Rev Intl 32:280–304. https://doi.org/10.1080/87559129.2015.1075212

Chambers E, McGraw S, Smiley K (2005) Sensory analysis of frozen foods. In: Handbook of frozen food processing and packaging, pp 561–576. https://doi.org/10.1201/b11204-30

Chaovanalikit A, Wrolstad R (2004) Total anthocyanins and total phenolics of fresh and processed cherries and their antioxidant properties. J Food Sci 69:FCT67–FCT72

Chaplin M (2015) Water phase diagram. pp 1–5

Chen B, Zhang M, Wang Y et al (2022) Comparative study of conventional and novel combined modes of microwave- and infrared-assisted thawing on quality of frozen green pepper, carrot and cantaloupe. LWT 154:112842. https://doi.org/10.1016/j.lwt.2021.112842

Cheng X, Zhang M, Adhikari B et al (2014) Effect of ultrasound irradiation on some freezing parameters of ultrasound-assisted immersion freezing of strawberries. Int J Refrig 44:49–55. https://doi.org/10.1016/j.ijrefrig.2014.04.017

Cheng L, Zhu Z, Sun D-W (2021) Impacts of high pressure assisted freezing on the denaturation of polyphenol oxidase. Food Chem 335:127485. https://doi.org/10.1016/j.foodchem.2020.127485

Chevalier D, Sequeira-Munoz A, Le Bail A et al (2000) Effect of freezing conditions and storage on ice crystal and drip volume in turbot (Scophthalmus maximus). Innovative Food Sci Emerg Technol 1:193–201. https://doi.org/10.1016/S1466-8564(00)00024-2

Comandini P, Blanda G, Soto-Caballero MC et al (2013) Effects of power ultrasound on immersion freezing parameters of potatoes. Innovative Food Sci Emerg Technol 18:120–125. https://doi.org/10.1016/j.ifset.2013.01.009

da Silva RM, Ferreira C, Provesi JG, Amante ER (2019) Effect of the antifreeze protein on the microstructure of strawberries (Fragaria ananassa Duch). Br J Food Technol 22:1–12. https://doi.org/10.1590/1981-6723.21818

Da Silva DL, Silveira AS, Ronzoni AF, Hermes CJL (2022) Effect of freezing rate on the quality of frozen strawberries (Fragaria x ananassa). Int J Refrig 144:46–54. https://doi.org/10.1016/j.ijrefrig.2022.07.006

De Ancos B, Sánchez-Moreno C, De Pascual-Teresa S, Cano M (2006) Fruit freezing principles. In: Handbook of fruits and fruit processing, pp 59–79

Delgado AE, Zheng L, Sun D-W (2009) Influence of ultrasound on freezing rate of immersion-frozen apples. Food Bioprocess Technol 2:263–270. https://doi.org/10.1007/s11947-008-0111-9

Dempsey P, Bansal P (2010) Air blast freezers and their significance to food freezing: a review. Proc ENCIT 2010:05–10

Downey G (2002) Quality changes in frozen and thawed, cooked purėed vegetables containing hydrocolloids, gums and dairy powders. Int J Food Sci Technol 37:869–877
Erickson MC (2005) Chemical measurements. In: Handbook of frozen food processing and packaging, pp 535–560. https://doi.org/10.1201/noe0849396274.ch23
Evans JA (ed) (2008) Frozen food science and technology. Blackwell Pub, Oxford/Ames
Fikiin K (2008) Emerging and novel freezing processes. In: Frozen food science and technology, pp 101–123
Fuchigami M, Kato N, Teramoto A (1997) High-pressure-freezing effects on textural quality of carrots. J Food Sci 62:804–808. https://doi.org/10.1111/j.1365-2621.1997.tb15459.x
Fuchigami M, Kato N, Teramoto AI (1998) High-pressure-freezing effects on extural quality of Chinese Cabbage. Food Sci 63:122–125
Fukuma Y, Yamane A, Itoh T et al (2012) Application of supercooling to long-term storage of fish meat. Fish Sci 78:451–461. https://doi.org/10.1007/s12562-011-0460-6
Gales O, Jones J, Swarts N (2022) An analysis on the impacts of cryogenic freezing on raspberry quality. Adv Hortic Sci 36(4):293–301
Gonzalez-Hidalgo I, Moreno DA, García-Viguera C, Ros-García JM (2019) Effect of industrial freezing on the physical and nutritional quality traits in broccoli. Food Sci Technol Int 25:56–65
Griffith M, Ewart KV (1995) Antifreeze proteins and their potential use in frozen foods. Biotechnol Adv 13:375–402. https://doi.org/10.1016/0734-9750(95)02001-J
Grover Y, Negi PS (2023) Recent developments in freezing of fruits and vegetables: striving for controlled ice nucleation and crystallization with enhanced freezing rates. J Food Sci 88:4799–4826. https://doi.org/10.1111/1750-3841.16810
Helland HS, Leufvén A, Bengtsson GB et al (2016) Storage of fresh-cut swede and turnip: effect of temperature, including sub-zero temperature, and packaging material on sensory attributes, sugars and glucosinolates. Postharvest Biol Technol 111:370–379. https://doi.org/10.1016/j.postharvbio.2015.09.011
Islam MN, Zhang M, Adhikari B et al (2014) The effect of ultrasound-assisted immersion freezing on selected physicochemical properties of mushrooms. Int J Refrig 42:121–133. https://doi.org/10.1016/j.ijrefrig.2014.02.012
Islam MN, Zhang M, Fang Z, Sun J (2015) Direct contact ultrasound assisted freezing of mushroom (Agaricus bisporus): growth and size distribution of ice crystals. Int J Refrig 57:46–53. https://doi.org/10.1016/j.ijrefrig.2015.04.021
James C, Seignemartin V, James SJ (2009) The freezing and supercooling of garlic (Allium sativum L.). Int J Refrig 32:253–260. https://doi.org/10.1016/j.ijrefrig.2008.05.012
James C, Hanser P, James SJ (2011) Super-cooling phenomena in fruits, vegetables and seafoods. In: 11th International Congress on Engineering and Food (ICEF 2011), pp 22–26
James C, Purnell G, James SJ (2014) A critical review of Dehydrofreezing of fruits and vegetables. Food Bioprocess Technol 7:1219–1234. https://doi.org/10.1007/s11947-014-1293-y
James C, Purnell G, James SJ (2015) A review of novel and innovative food freezing technologies. Food Bioprocess Technol 8:1616–1634. https://doi.org/10.1007/s11947-015-1542-8
Javadikasgari H, Soltesz EG, Gillinov AM (2018) Surgery for atrial fibrillation. In: Atlas of cardiac surgical techniques, pp 479–488
Jaworska G, Bernaś E (2010) Effects of pre-treatment, freezing and frozen storage on the texture of Boletus edulis (Bull: Fr.) mushrooms. Int J Refrig 33:877–885
Jha PK, Chevallier S, Xanthakis E et al (2020) Effect of innovative microwave assisted freezing (MAF) on the quality attributes of apples and potatoes. Food Chem 309:125594. https://doi.org/10.1016/j.foodchem.2019.125594
Jha PK, Chapleau N, Meyers P-E et al (2024) Can cryogenic freezing preserve the quality of fruit matrices during long-term storage compared to the mechanical method? Appl Food Res 4:100374. https://doi.org/10.1016/j.afres.2023.100374
Kang T, You Y, Hoptowit R et al (2021) Effect of an oscillating magnetic field on the inhibition of ice nucleation and its application for supercooling preservation of fresh-cut mango slices. J Food Eng 300:110541. https://doi.org/10.1016/j.jfoodeng.2021.110541

Kaur M, Kumar M (2020) An innovation in magnetic field assisted freezing of perishable fruits and vegetables: a review. Food Rev Intl 36:761–780. https://doi.org/10.1080/87559129.2019.1683746

Kaur M, Kumar M, Sethi V (2022) Maintaining the freeze thawing characteristics of tomato through development and evaluation of magnetic field-assisted freezing system. Food Process Preserv 46. https://doi.org/10.1111/jfpp.16900

Kiani H, Zhang Z, Sun D-W (2015) Experimental analysis and modeling of ultrasound assisted freezing of potato spheres. Ultrason Sonochem 26:321–331. https://doi.org/10.1016/j.ultsonch.2015.02.015

King ES, Noll A, Glenn S, Bolling BW (2022) Refrigerated and frozen storage impact aronia berry quality. Food Prod Process and Nutr 4:3. https://doi.org/10.1186/s43014-021-00080-y

Kobayashi R, Suzuki T (2019) Effect of supercooling accompanying the freezing process on ice crystals and the quality of frozen strawberry tissue. Int J Refrig 99:94–100. https://doi.org/10.1016/j.ijrefrig.2018.11.045

Kong CHZ, Hamid N, Ma Q et al (2017) Antifreeze peptide pretreatment minimizes freeze-thaw damage to cherries: an in-depth investigation. LWT 84:441–448. https://doi.org/10.1016/j.lwt.2017.06.002

Kyriacou MC, Rouphael Y (2018) Towards a new definition of quality for fresh fruits and vegetables. Sci Hortic 234:463–469. https://doi.org/10.1016/j.scienta.2017.09.046

Lammerskitten A, Mykhailyk V, Wiktor A et al (2019) Impact of pulsed electric fields on physical properties of freeze-dried apple tissue. Innovative Food Sci Emerg Technol 57:102211. https://doi.org/10.1016/j.ifset.2019.102211

Leng D, Zhang H, Tian C et al (2023) Static magnetic field assisted freezing of four kinds of fruits and vegetables: micro and macro effects. Int J Refrig 146:118–125. https://doi.org/10.1016/j.ijrefrig.2022.10.018

Li B, Sun DW (2002a) Novel methods for rapid freezing and thawing of foods – a review. J Food Eng 54:175–182. https://doi.org/10.1016/S0260-8774(01)00209-6

Li B, Sun D-W (2002b) Effect of power ultrasound on freezing rate during immersion freezing of potatoes. J Food Eng 55:277–282. https://doi.org/10.1016/S0260-8774(02)00102-4

Li J, Chotiko A, Kyereh E et al (2017) Development of a combined osmotic dehydration and cryogenic freezing process for minimizing quality changes during freezing with application to fruits and vegetables: a combined osmotic dehydration and cryogenic freezing process. J Food Process Preserv 41:e12926. https://doi.org/10.1111/jfpp.12926

Li J, Zhao J, Dong X et al (2023) Comparison of immersion freezing, osmo-dehydrofreezing and air freezing on freezing parameters and physicochemical properties of mango (*Mangifera indica* L.). Int J Food Sci Technol 58:4344–4353. https://doi.org/10.1111/ijfs.16536

Lin X, Jing S (2004) Effects of subfreezing storage on the freshness of strawberry. Beverage Fast Frozen Food Ind 1:11–13

Lisiewska Z, Kmiecik W, Słupski J (2004) Contents of chlorophylls and carotenoids in frozen dill: effect of usable part and pre-treatment on the content of chlorophylls and carotenoids in frozen dill (Anethum graveolens L.), depending on the time and temperature of storage. Food Chem 84:511–518

Liu D-K, Xu C-C, Guo C-X, Zhang X-X (2020) Sub-zero temperature preservation of fruits and vegetables: a review. J Food Eng 275:109881. https://doi.org/10.1016/j.jfoodeng.2019.109881

Loayza-Salazar S, Siche R, Vegas C et al (2024) Novel Technologies in the Freezing Process and Their Impact on the quality of fruits and vegetables. Food Eng Rev 16:371. https://doi.org/10.1007/s12393-024-09371-9

Lyu C, Nastase G, Ukpai G et al (2017) A comparison of freezing-damage during isochoric and isobaric freezing of the potato. PeerJ 5:e3322. https://doi.org/10.7717/peerj.3322

Maity T, Chauhan O, Shah A et al (2011) Quality characteristics and glass transition temperature of hydrocolloid pre-treated frozen pre-cut carrot. Int J Food Prop 14:17–28

Manzocco L, Anese M, Nicoli MC (2008) Radiofrequency inactivation of oxidative food enzymes in model systems and apple derivatives. Food Res Int 41:1044–1049

Marazani T, Madyira DM, Akinlabi ET (2017) Investigation of the parameters governing the performance of jet impingement quick food freezing and cooling systems—a review. Procedia Manuf 8:754–760. https://doi.org/10.1016/j.promfg.2017.02.097

Martins RC, Lopes VV (2007) Modelling supercooling in frozen strawberries: experimental analysis, cellular automation and inverse problem methodology. J Food Eng 80:126–141. https://doi.org/10.1016/j.jfoodeng.2006.05.009

Muthukumarappan K, Tiwari B, Swamy GJ (2018) Refrigeration and freezing preservation of vegetables. In: Siddiq M, Uebersax MA (eds) Handbook of vegetables and vegetable processing, 1st edn. Wiley, pp 341–363

Neri L, Faieta M, Di Mattia C et al (2020) Antioxidant activity in frozen plant foods: effect of Cryoprotectants. Freezing Process Frozen Storage Foods 9:1886. https://doi.org/10.3390/foods9121886

Neuman MR (2006) Physical measurements. Medical devices and systems, pp 46-1–46–18. https://doi.org/10.4135/9781452225814.n4

Newman M (2001) Cryogenic impingement freezing utilizing atomized liquid nitrogen for the rapid freezing of food products. pp 145–151

Nida S, Moses JA, Anandharamakrishnan C (2021) Isochoric freezing and its emerging applications in food preservation. Food Eng Rev 13:812–821. https://doi.org/10.1007/s12393-021-09284-x

Orqueda ME, Torres S, Verón H et al (2021) Physicochemical, microbiological, functional and sensory properties of frozen pulp of orange and orange-red chilto (Solanum betaceum Cav.) fruits. Sci Hortic 276:109736

Otero L, Martino M, Zaritzky N et al (2000) Preservation of microstructure in peach and mango during high-pressure-shift freezing. J Food Sci 65:466–470. https://doi.org/10.1111/j.1365-2621.2000.tb16029.x

Owada N, Saito S (2010) Quick freezing apparatus and quick freezing method. U.S. Patent 7,810,340

Panayampadan AS, Shafiq Alam M, Aslam R et al (2022) Effects of alternating magnetic field on freezing of minimally processed guava. LWT 163:113544. https://doi.org/10.1016/j.lwt.2022.113544

Parreño WC, Torres MDA (2005) Quality and safety of frozen vegetables. In: Handbook of frozen food processing and packaging, pp 377–416. https://doi.org/10.1201/b11204-22

Powell-Palm MJ, Aruda J, Rubinsky B (2019) Thermodynamic theory and experimental validation of a multiphase isochoric freezing process. J Biomech Eng 141:081011. https://doi.org/10.1115/1.4043521

Provesi JG, Valentim Neto PA, Arisi ACM, Amante ER (2019) Extraction of antifreeze proteins from cold acclimated leaves of Drimys angustifolia and their application to star fruit (Averrhoa carambola) freezing. Food Chem 289:65–73. https://doi.org/10.1016/j.foodchem.2019.03.055

Qiao J, Zhang M, Fang Z, Fu J (2023) Effect of static magnetic field assisted freezing on the product quality of Korla fragrant pear. J Food Process Eng 46(10):e14404.https://doi.org/10.1111/jfpe.14404

Rayman Ergün A, Yanat M, Baysal T (2021) The effects of the novel home freezing system on microstructure, color, antioxidant activity, and microbiological properties of strawberries. Int J Refrig 121:228–234. https://doi.org/10.1016/j.ijrefrig.2020.10.013

Reno MJ, Prado MET, de Resende JV (2011) Microstructural changes of frozen strawberries submitted to pre-treatments with additives and vacuum impregnation. Food Sci Technol 31:247–256

Rikasa A, Senarathne S, Wimalasiri K (2022) Determination of the effect of freezing methods on the quality of frozen tomato pieces. Asian J Dairy Food Res 41:317–322

Rubinsky B, Perez PA, Carlson ME (2005) The thermodynamic principles of isochoric cryopreservation. Cryobiology 50:121–138. https://doi.org/10.1016/j.cryobiol.2004.12.002

Salvadori VO, Mascheroni RH (2002) Analysis of impingement freezers performance. J Food Eng 54:133–140. https://doi.org/10.1016/S0260-8774(01)00198-4

Salzmann CG, Radaelli PG, Mayer E, Finney JL (2009) Ice XV: A new thermodynamically stable phase of ice. Phys Rev Lett 103(10):1–4. https://doi.org/10.1103/PhysRevLett.103.105701

Schmilovitch Z, Mizrach A (2013) Instrumental assessment of the sensory quality of fruits and vegetables. Woodhead Publishing Limited

Schudel S, Prawiranto K, Defraeye T (2021) Comparison of freezing and convective dehydrofreezing of vegetables for reducing cell damage. J Food Eng 293:110376. https://doi.org/10.1016/j.jfoodeng.2020.110376

Song AY, Oh YJ, Kim JE et al (2015) Cold plasma treatment for microbial safety and preservation of fresh lettuce. Food Sci Biotechnol 24:1717–1724. https://doi.org/10.1007/s10068-015-0223-8

Sriwimon W, Boonsupthip W (2011) Utilization of partially ripe mangoes for freezing preservation by impregnation of mango juice and sugars. LWT Food Sci Technol 44:375–383. https://doi.org/10.1016/j.lwt.2010.08.012

Stonehouse GG, Evans JA (2015) The use of supercooling for fresh foods: a review. J Food Eng 148:74–79. https://doi.org/10.1016/j.jfoodeng.2014.08.007

Sun D-W, Li B (2003) Microstructural change of potato tissues frozen by ultrasound-assisted immersion freezing. J Food Eng 57:337–345. https://doi.org/10.1016/S0260-8774(02)00354-0

Sun L, Zhu Z, Sun DW (2023) Regulating ice formation for enhancing frozen food quality: materials, mechanisms and challenges. Trends Food Sci Technol 139:104116. https://doi.org/10.1016/j.tifs.2023.07.013

Sutariya SG, Sunkesula V (2020) Food freezing: emerging techniques for improving quality and process efficiency a comprehensive review. Elsevier

Tanner FW (1934) Microbiological examination of fresh and frozen fruits and vegetables. Am J Public Health Nations Health 24:485–492. https://doi.org/10.2105/ajph.24.5.485

Tian Y, Chen Z, Zhu Z, Sun DW (2020) Effects of tissue pre-degassing followed by ultrasound-assisted freezing on freezing efficiency and quality attributes of radishes. Ultrason Sonochem 67:105162. https://doi.org/10.1016/j.ultsonch.2020.105162

Tian Y, Zhang Z, Zhu Z, Sun DW (2021) Effects of nano-bubbles and constant/variable-frequency ultrasoundassisted freezing on freezing behaviour of viscous food model systems. J Food Eng 292:110284. https://doi.org/10.1016/j.jfoodeng.2020.110284

Tocci AM, Mascheroni RH (2008) Some thermal properties of fresh and osmotically dehydrated kiwifruit above and below the initial freezing temperature. J Food Eng 88:20–27. https://doi.org/10.1016/j.jfoodeng.2007.10.014

Van Buggenhout S, Grauwet T, Van Loey A, Hendrickx M (2007) Effect of high-pressure induced ice I/ice III-transition on the texture and microstructure of fresh and pretreated carrots and strawberries. Food Res Int 40:1276–1285. https://doi.org/10.1016/j.foodres.2007.08.008

Wang D, Xu RNL (2018) Effects of low temperature acclimation on the quality of peaches stored at ice temperature. J Refrigeration Technol 41:17–20

Wang L, Zhang X, Zhao Y et al (2018) Effect of treatment with 1-methylcyclopropene or ethylene absorbent on the quality of hawthorn fruits during controlled freezing-point storage. Shipin Kexue/Food Sci 39:243–249

Wang R, Xing Y, Li X et al (2019) Microstructure and quality of cabbage slices (*Brassica oleracea* L. var. *capitata* L.) as affected by cryogenic quick-freezing treatment. Int J Food Prop 22:1815–1833. https://doi.org/10.1080/10942912.2019.1681449

Workmaster BAA, Palta JP, Wisniewski M (1999) Ice nucleation and propagation in cranberry uprights and fruit using infrared video thermography. J Am Soc Hortic Sci 124:619–625

Wu L, Orikasa T, Tokuyasu K et al (2009) Applicability of vacuum-dehydrofreezing technique for the long-term preservation of fresh-cut eggplant: effects of process conditions on the quality attributes of the samples. J Food Eng 91:560–565. https://doi.org/10.1016/j.jfoodeng.2008.10.021

Wu J, Zhang M, Bhandari B, Yang CH (2021) Drip loss control technology of frozen fruit and vegetables during thawing: a review**. Int Agrophys 35:235–250. https://doi.org/10.31545/intagr/142289

Xin Y, Zhang M, Adhikari B (2014a) The effects of ultrasound-assisted freezing on the freezing time and quality of broccoli (Brassicaoleracea L. var. botrytis L.) during immersion freezing. Int J Refrig 41:82–91. https://doi.org/10.1016/j.ijrefrig.2013.12.016

Xin Y, Zhang M, Adhikari B (2014b) Ultrasound assisted immersion freezing of broccoli (Brassica oleracea L. var. botrytis L.). Ultrason Sonochem 21:1728–1735. https://doi.org/10.1016/j.ultsonch.2014.03.017

Xin Y, Zhang M, Adhikari B (2014c) Freezing characteristics and storage stability of broccoli (Brassica oleracea L. var. botrytis L.) under Osmodehydrofreezing and ultrasound-assisted Osmodehydrofreezing treatments. Food Bioprocess Technol 7:1736–1744. https://doi.org/10.1007/s11947-013-1231-4

Xin Y, Zhang M, Xu B et al (2015) Research trends in selected blanching pretreatments and quick freezing technologies as applied in fruits and vegetables: a review. Int J Refrig 57:11–25. https://doi.org/10.1016/j.ijrefrig.2015.04.015

Xu Z, Guo Y, Ding S et al (2014) Freezing by immersion in liquid CO2 at variable pressure. Innovative Food Sci Emerg Technol 22:167–174. https://doi.org/10.1016/j.ifset.2013.06.005

Yoshimura N (1989) Jikisakkingijutsu no Kaihatsu to Riyou. Bio Industry 6:435–441

Yu A-C, Liao H-J (2016) Evaluation of three-step pretreatment combined with air blast or Cryomechanical freezing in improving the quality of frozen strawberries (Fragaria×ananassa Duch. cv. Harunoka). Int J Food Eng 12:153–163. https://doi.org/10.1515/ijfe-2015-0160

Zhang K, Zhu Z, Nong S et al (2011) Effects of controlled freezing point technology combined with modified atmosphere package on preservation for grape. Food Res Dev 32:126–130

Zhang P, Liu H, Chen S et al (2012) Effect of low-temperature treatment on chilling injury in cherry fruits. Food Sci (NY) 33:303–308

Zhu Z, Chen Z, Zhou Q et al (2018) Freezing efficiency and quality attributes as affected by voids in plant tissues during ultrasound-assisted immersion freezing. Food Bioprocess Technol 11:1615–1626. https://doi.org/10.1007/s11947-018-2103-8

Zhu Z, Zhang P, Sun DW (2020) Effects of multi-frequency ultrasound on freezing rates and quality attributes of potatoes. Ultrason Sonochem 60:104733.https://doi.org/10.1016/j.ultsonch.2019.104733

Zhu Z, Li T, Sun D-W (2021) Pressure-related cooling and freezing techniques for the food industry: fundamentals and applications. Crit Rev Food Sci Nutr 61:2793–2808. https://doi.org/10.1080/10408398.2020.1841729

Chapter 8
Canning of Fruits and Vegetables

Shruti Sethi, Bindvi Arora, S. G. Lekshmi, and Prasoon Gunjan

8.1 History of Canning

Food was preserved using a variety of techniques, such as drying, smoking, salting, and fermenting, before canning was invented. Although these methods increased the meal's shelf life, they frequently altered the texture and flavor of the food. As the population increased and worldwide travel extended, the need for better food preservation became more pressing, necessitating a longer-lasting food supply. The modern canning process was invented in the early nineteenth century by French chef and confectioner Nicolas Appert. In 1809, Appert published a book titled "L'Art de Conserver Les Substances Alimentaires" (The Art of Preserving Animal and Vegetable Substances), in which he described a method of preserving food by sealing it in glass jars and heating it. His method was based on the principle that heating food in a sealed container would prevent spoilage by killing microorganisms. Appert's work was spurred by the French government's call for innovative methods to preserve food for military and naval purposes during the Napoleonic Wars. His successful experiments with glass jars eventually led to the widespread adoption of his technique, laying the foundation for modern canning. While Appert's glass jars were a significant advancement, they were not ideal for all types of food and were prone to breakage. The next major innovation in canning came with the development of metal cans. In 1810, Englishman Peter Durand patented the use of tin cans for food preservation. Durand's invention was further refined by other inventors, including John Hall and Bryan Donkin, who improved the canning process and the design of the cans (Bell 1965). Durand's metal cans were initially used primarily by the British military and navy. They proved to be more durable and practical for transport compared to glass jars. However, the early cans were difficult

S. Sethi (✉) · B. Arora · S. G. Lekshmi · P. Gunjan
Division of Food Science and Postharvest Technology, ICAR-Indian Agricultural Research Institute, New Delhi, India

OP Chauhan (ed.), *Fruits and Vegetables Technologies*,
https://doi.org/10.1007/978-981-96-8433-5_8

to open and required a can opener, which was not widely available until the late nineteenth century. In 1813, Doukin, Hall, and Gamble introduced the practice of postprocessing incubation for canned foods. In 1825, Kensett and Duggett were granted a U.S. patent for preserving food in cans. The same year, Thomas Kenett developed the first kettle pan. In 1837, Winshow became the first to can corn directly from the cob. By 1839, tin cans were widely used in a limited number of states. In 1845, Elliott introduced canning to Australia.

One of the significant challenges with early metal cans was opening them. The solution came with the invention of the can opener in 1858 by Ezra Warner. Warner's design, which used a blade to cut through the can's lid, greatly improved the convenience of canning and made canned food more accessible to the general public. Later, in 1864, Louis Pasteur discovered that food spoilage was caused by microorganisms that were destroyed at elevated temperatures, and the technique was known as "Pasteurization." This understanding helped to form a scientific basis for establishing and revolutionizing the canning industry (Fitzgerald 1923).

Another critical advancement in canning technology was the introduction of pressure canning. Andrew Shriver developed the first retort pressure cooker in 1873, and in the early twentieth century, researchers discovered that high temperatures achieved through pressure cooking could effectively kill the bacteria responsible for botulism in low-acid foods such as vegetables and meats. This led to the development of the pressure canner, a sealed pot that uses steam pressure to reach higher temperatures than boiling water alone. This innovation significantly expanded the range of foods that could be safely preserved using canning methods. In 1890, Max Ann developed the first double-seaming machine for canning. Five years later, in 1895, Russell conducted the first bacteriological study of canned foods. Bitting, in 1916, introduced an index that outlined the relationship between processing time and temperature for various food products. By 1920, mathematician Ball developed mathematical calculations for heat sterilization in the canning process. The very next year, in 1921, horticulturists Magoon and Culpepper proposed methods for exhausting and vacuuming processed products. Finally, in 1928, the canning industry completed its calculations for heat processing.

The twentieth century saw the rise of home canning as a popular method for preserving fruits, vegetables, and meats. With advancements in technology and the availability of canning jars and equipment, more households began to engage in canning as a way to save money and enjoy homegrown produce. Companies like Ball and Kerr became household names, providing reliable canned products and promoting canning as a practical and enjoyable activity.

In recent decades, canning technology has continued to evolve, with improvements in jar design, sealing methods, and processing guidelines. Modern canning practices emphasize food safety and include rigorous standards to prevent contamination and ensure the long-term preservation of canned goods. After 2000, many innovative interventions came in the canning industry, including smart canning technology that monitors temperature and pressure during the canning process, ensuring optimal safety and quality. BPA (Bisphenol A)-free can linings, improving safety and addressing consumer health concerns regarding chemicals in food packaging, were also considered. Automation and robotics in canning were introduced to improve efficiency, reduce labor costs, and enhance food safety through precise monitoring. The development of research-based guidelines by organizations such as the US Department of Agriculture and the National Center for Home Food Preservation has further enhanced the safety and effectiveness of home canning.

8.2 Can Manufacture

Cans used for food packaging are made of metal. Metals are a group of elements where the bonding electrons are free to move large distances and are not confined to one location. This chemical property renders many desirable attributes to metals, such as thermal and electrical conductivity, ductility, and malleability. This makes metals suitable to be molded, cut, and designed into various forms, making them highly suitable to be transformed into cans. Metals have high mechanical strength, toughness, excellent barrier and sealing properties, and have high temperature tolerance to be suitable for high temperature processing operations.

Conversely, some metals are highly reactive, making them unsuitable for use as food packaging materials. As inertness is one of the key requirements of cans, they must be composed of materials that are safe for contact with food. Thus, typically inert food-grade metals such as aluminum or tin-coated stainless steel are used for can manufacture. Metal food containers must comply with strict safety regulations set by government bodies and international organizations. Aluminum, tinplate, tin-free steel, polymer-coated steel, and stainless steel are the primary types of metals used in food packaging. The metals commonly used for can manufacturing are discussed below.

Aluminum

Aluminum is a highly sought-after material for can manufacturing due to its exceptional properties. Its lightweight nature significantly reduces transportation costs, making it an environmentally friendly option. Moreover, aluminum is incredibly recyclable, allowing it to be reprocessed indefinitely without compromising its quality. This circular economy aspect enhances its sustainability appeal. Corrosion resistance is a key advantage of aluminum. It is naturally resistant to rust and corrosion, ensuring the safety and integrity of the can's contents. This is particularly important for products that require long-term storage or transportation. Furthermore, aluminum provides an excellent barrier against moisture, oxygen, and light. This barrier function is crucial for preserving the freshness and quality of the product inside the can. For instance, in beverage cans, aluminum helps prevent the loss of carbonation and ensures that the product retains its original flavor. The versatility of aluminum is another factor contributing to its popularity. It can be easily shaped and formed into various can sizes and shapes, accommodating a wide range of products. This flexibility allows manufacturers to meet diverse consumer preferences and market demands. In some cases, aluminum can interact with certain food products, leading to changes in taste, color, or even metal contamination. This is particularly a concern with acidic or highly reactive foods.

Stainless Steel

Stainless steel is a durable, corrosion-resistant metal that offers a long-lasting packaging solution. It is nonreactive with most foods, making it ideal for reusable containers, trays, and water bottles. Stainless steel cans are known for their strength and longevity, making them suitable for products that require extended shelf life or harsh storage conditions. Stainless steel is nonreactive with most foods and beverages, preventing any unwanted chemical interactions that could affect taste or quality. Being highly malleable and ductile, stainless steel can often be made into cans that have a sleek and modern appearance, which can enhance the perceived value of the product. A major limitation of stainless steel is that if not handled or stored properly, stainless steel cans can potentially lead to metal contamination of the food product. Steel can be reactive to high acid and high protein food materials and can cause food quality and safety issues. Acidic foods can be particularly problematic. The acidity can cause a chemical reaction between the stainless steel and the food, leading to the release of metal ions into the product. This can result in a metallic taste, discoloration, or even health concerns if the metal contamination is significant. Foods with high salt content or those with sulfur-containing amino acids can accelerate the corrosion process of stainless steel, increasing the likelihood of metal contamination. To counteract this limitation of stainless steel, tin plate, i.e., tin-coated stainless steel plate, is a modification used for manufacturing cans to package high-acid and high-protein foods.

Tinplate

Tinplate, a steel sheet covered in tin, is a durable and corrosion-resistant material that shields food from outside influences. Tin plate is a low-carbon mild steel sheet varying in thickness from 0.15 to 0.5 mm with a coating of tin between 2.8 and 17 gsm on both sides. Tinplate offers many advantages over other types of metals because it has excellent mechanical strength and good fabrication qualities such as drawability, ductility, malleability, weldability, nontoxicity, and lubricity. By preventing direct contact of food with steel, it helps preserve the color and flavor of the packaged food. Tinplate is also recyclable and widely used in canned goods. To prepare a tin plate, a hot dip tinning method has been used traditionally. In the process, the stainless steel sheets are dipped into a molten tin bath at a temperature of around 230–250 °C. The tin adheres to the steel surface during cooling. In recent times, electrolytic tin plating is used, which is a multistep process. The steel sheets are first cleaned to prepare them for tinning. During the tinning process, stainless steel sheets are immersed in an electrolytic bath containing tin ions. A direct current is applied between the sheets (cathode) and a tin anode. The tin ions from the anode are deposited onto the stainless steel surface, forming a thin layer of tin. Electrolytic tinning offers several advantages over other methods of tinning stainless steel, such as hot-dip tinning. It allows for precise control of the coating thickness and can be used to produce coatings with a more uniform appearance. Additionally, electrolytic tinning is generally a more energy-efficient and environmentally friendly process than the conventional hot-dip tinning. The tinned sheets may undergo additional treatments like annealing or chromium plating to enhance their properties.

8.3 Can Manufacturing Process

Metal cans come in two types: two-piece and three-piece. Three-piece cans have a separate body and two lids, while two-piece cans are made from a single piece of metal that is formed into a base and sides, with one lid. Two-piece cans are often used for beverages such as juices and sodas and are mainly made of aluminum, while three-piece cans are commonly used for fruits and vegetables and are mainly made of stainless steel.

Two-Piece Can Manufacture

A two-piece can is manufactured by precisely cutting, drawing, and trimming steel coils into cylindrical bodies. Separate bottoms are formed through cupping and flanging. The body and bottom are then joined together using specialized seaming machines. The process of forming a two-piece can begin with selecting high-quality metal plates that meet specific standards for strength, ductility, and corrosion resistance. The plates are often coated with tin or chromium to enhance corrosion protection, improve printability, and provide a smooth surface for food or beverage contact. The steel plates are cut into strips of appropriate width using laser cutting technology. Circular blanks are cut from the strips. The blanks are drawn into cylindrical shapes using a drawing press. The blanks are cupped into a shallow dish shape using a cupping press. Excess material is trimmed from the drawn body to form it into the shape of the can. The edges of the cupped bottoms are flanged to create a sealing surface. To make the top lid circular, blanks are cut from the steel coil. The lid is attached to the can body using a seaming machine.

The two processes by which two-piece cans are made are as follows.

1. Drawn and ironed: In this process, the steel coil is cut into strips and drawn into cylindrical shapes using a drawing press. The cylindrical iron strip is then ironed over the drawn steel body. This process involves passing the steel body through a series of ironing dies, gradually reducing its diameter and increasing its wall thickness (Fig. 8.1).

2. Drawn and redrawn: In this process, the steel coil is cut into strips and drawn into a cylindrical shape using a drawing press. This initial drawing operation forms the basic can body. The drawn body is then subjected to one or more redrawing operations. In each redrawing step, the can is passed through a smaller die, reducing its diameter and increasing its wall thickness. This process is repeated until the desired dimensions are achieved (Fig. 8.2).

Three-Piece Can Manufacture

Three-piece cans are metal containers consisting of a cylindrical body, a bottom, and a separate top. This design offers flexibility in terms of can size, shape, and

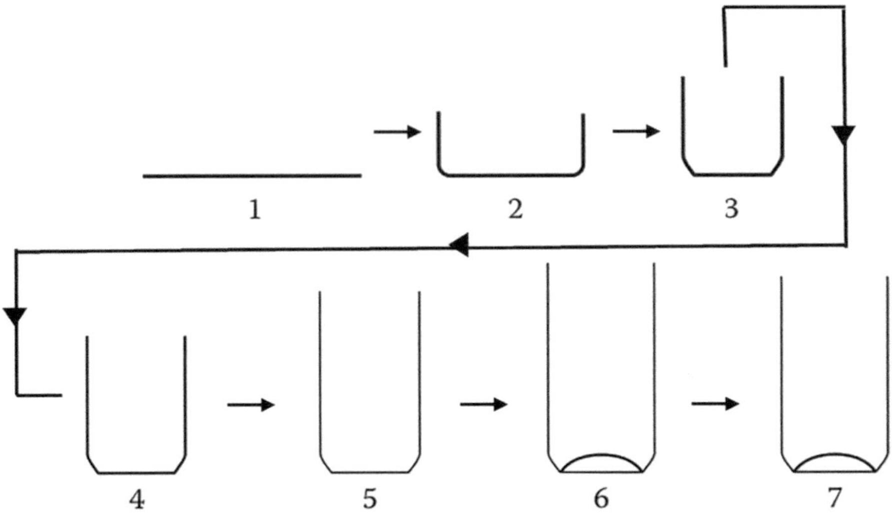

Fig. 8.1 Drawn and ironed can manufacture: 1. Circular disc cut from metal sheet; 2, 3. Drawn into a cup; 4, 5, 6. walls ironed to elongate; and 7. Cut from top to uniform height

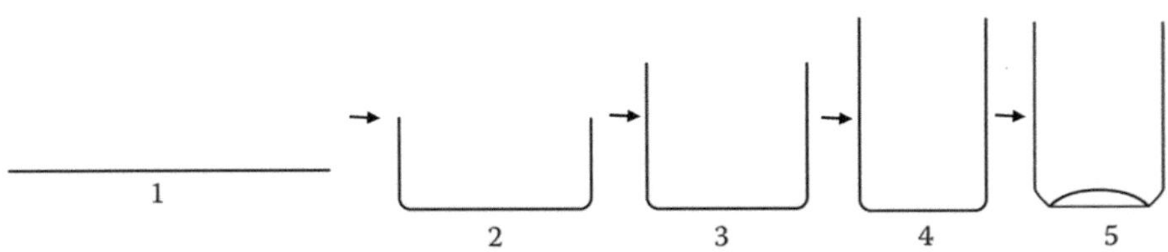

Fig. 8.2 Draw and redrawn can manufacture: 1. Circular disc cut from metal sheet; 2. Drawn into a cup; 3, 4. redrawn to increase wall height and decrease diameter; and 5. Cut from top to uniform height

closure options, making it suitable for a wide range of products. To manufacture three-piece cans, the steel coil is cut into strips of appropriate width to make the central body. The strip is made into a hollow cylinder and sealed from the edge. For making lids for the two sides of the can, circular blanks are cut from the steel coil. The blanks are cupped into a shallow dish shape using a cupping press. The edges of the cupped bottoms are flanged to create a sealing surface. The bottom is joined to the body using a seam-forming machine to form a complete can body.

Different sizes of cans are available in the market for canning of fruits and vegetables, as given in Table 8.1.

Lacquering

Lacquering is a process used to apply a protective coating to the inside or outside of metal cans. This coating helps to prevent corrosion, improve the can's appearance, and enhance its functionality. When applied to the inside of the can, it protects the product from contact with the metal surface. They can be food-safe or nonfood-safe,

Table 8.1 Can sizes and their capacity

S. no.	Trade name of can	Size (diameter × height) (mm)	Capacity (approx. in cubic cm)
1.	No. I (Picnic)	211 × 400	323.5
2	No. 211 Cylinder	211 × 414	401
3.	No. 300	300 × 407	450
4.	No. I Tall	301 × 411	494
5.	No. 2	307 × 409	608
6.	Jumbo	307 × 510	763
7.	No. 3 Cylinder	404 × 700	1529
8.	No. 5	502 × 510	1748

depending on the intended use of the can. Lacquer can be applied to the can surface by the following methods.

(i) Dip coating: The cans are dipped into a tank of lacquer, ensuring even coverage.
(ii) Spray coating: The lacquer is sprayed onto the cans using specialized equipment.
(iii) Roller coating: The cans are rolled through a bath of lacquer to uniformly coat the surface.

Acid-Resistant Cans

Acid-resistant food cans, also known as C-enamel cans, are specifically designed to withstand the corrosive effects of acidic food products. These cans are coated with a special lacquer that prevents the acidic foods from reacting with the metal of the can, ensuring the safety and integrity of the contents. Common products packed in acid-resistant food cans include raspberry, strawberry, red plum, colored grapes, tomato sauce, ketchup, paste, etc. The pH level of the food product is a crucial factor in determining whether an acid-resistant can is necessary. Foods with a pH below 4.6 are generally considered acidic and require lacquered cans.

Sulfur-Resistant Cans

Sulfur-resistant (SR) cans are specifically designed to withstand the corrosive effects of sulfur-containing foods and beverages. These cans often feature a special coating or lining that provides a barrier between the sulfur-rich content and the metal can. Sulfur compounds can react with the metal of a standard can, leading to discoloration, a metallic taste, and even the release of harmful substances into the food or beverage. Sulfur-resistant cans help prevent these issues, ensuring the safety and quality of the product. Some fruits and vegetables, such as beets, onions, and garlic, contain sulfur compounds that can react with metal cans. Certain beverages, like dark beer and some wines, may contain sulfur compounds that require the use of sulfur-resistant cans. Some cans may feature an epoxy lining that provides a protective barrier against sulfur corrosion.

8.4 Effect of Canning on Physical and Nutritional Quality Attributes

Canning is a method of preserving food using heat treatment and sealing it in containers to prevent spoilage. For optimal results, fruits and vegetables should be as fresh as possible. Fruits should be mature, firm, and free from defects, while vegetables are best when tender. Although canning significantly extends shelf life, it can also affect the physical and nutritional quality of the preserved foods in several ways. The canning process can have some serious effects on the organoleptic and nutritional properties during storage due to chemical reactions. The changes in the physical and nutritional properties can take place through the following:

(a) Sterilization process causing destruction of heat-labile nutrients.
(b) Leaching of water-soluble nutrients from the fruits and vegetables into the syrup or brine solution, which is usually discarded during the consumption
(c) Chemical destruction during storage, depending on a variety of factors such as temperature, residual oxygen, and the metallic surface of the container.

8.5 Canning Process

The canning process of fruits and vegetables includes the following general steps or unit operations.

Selection of Fruits and Vegetables
For canning, it is essential to use fresh, high-quality fruits and vegetables, as the quality of the raw materials directly impacts the final product. Fruits should be evenly ripe, firm, and free from defects. Any diseased, overripe, or insect-infested fruits and vegetables should be discarded. Immature fruits should also be avoided, as they tend to shrink and toughen during the canning process. Vegetables should be tender, and all produce should be clean and uncontaminated. The flowcharts for the canning of fruits/vegetables and fruit juices are depicted in Fig. 8.3.

Sorting
It is essential to carefully discard any rotten or damaged fruits and vegetables. Raw materials should be sorted based on their maturity level, size, and color to ensure uniform quality in canned products. Sorting can be done either manually or mechanically by screw and roller graders. Whole fruits like white peaches, pears, apricots, mangoes, pineapples, cherries, grapes, and plums are sorted as a whole, while smaller fruits such as berries, cherries, grapes, and plums are typically sorted after being cut into pieces.

Washing
It is essential to thoroughly wash fruits and vegetables with water to remove dust, debris, and any chemical residues. This process also helps eliminate bacteria from

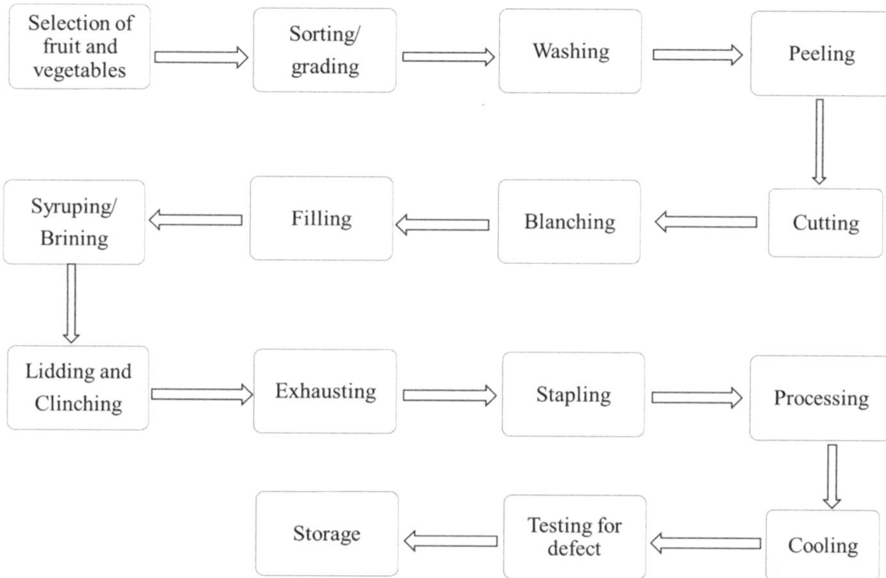

Fig. 8.3 Flowchart for the canning process of fruits and vegetables

their surfaces. Both hot and cold water can be used for washing. To sterilize the produce, 150 ppm of chlorine or a diluted potassium permanganate solution can be added to the water. Typically, manual washing involves soaking the produce in a water tank, though spraying water is considered the most effective method for cleaning.

Peeling

After being cleaned, fruits and vegetables are prepared for canning. The peeling process can be done manually with a knife, by machine, or through techniques like heat treatment or using a lye solution, which is made from caustic soda (Table 8.2). To soften and loosen the skin, produce such as peaches and potatoes are first treated with steam or boiling water and then quickly cooled in cold water. This makes it easy to remove the skin by hand or with the help of water pressure. For lye peeling, fruits and vegetables like peaches, apricots, oranges, and sweet potatoes are immersed in a boiling lye solution (1–2% caustic soda) for 30 s to 2 min. Afterward, they are thoroughly rinsed in cold running water to remove any residual alkali, and in some cases, they are also washed in a solution containing 0.5% citric or hydrochloric acid for further cleansing.

Cutting

Fruits and vegetables should be chopped into the desired shapes, such as fingers, cubes, or slices, using either a knife or a machine. A specialized coring knife can be used to simultaneously remove the seeds, pith, and core.

Table 8.2 Methods of peeling fruits and vegetables

S. no.	Methods of peeling	Specification	Application
1.	Knife peeling (manual)	The peeling knife with a curved blade and a special guard to regulate the depth of peeling can be used for uniform peeling in case of irregular fruit shapes	All fruits and vegetables
2.	Mechanical knife peeling	Mechanical peeling, coring, and cubing machines are used	Apple, pear, pineapple
3	Hot water peeling	Commodities are scalded in steam or boiling water to soften and loosen skin, which is subsequently removed manually	Tomato
4	High-pressure steam/water peeling	In flash-steam peeling, fruits and vegetables are placed into a slow-rotating pressure vessel (rotating at 4–6 rpm). High-pressure steam, typically at 1500 kPa, is introduced into the vessel to ensure all surfaces of the food are exposed for a specific time, depending on the type of fruit or vegetable. When the pressure is rapidly released, steam trapped beneath the skin causes the surface to peel away. Most of the removed skin is carried away with the steam, and any remaining skin traces are washed off with additional water sprays for a clean finish	Potato, tomato
5	Abrasive peeling	The food is placed onto carborundum rollers or into a rotating bowl lined with carborundum crystals, which act as an abrasive surface. As the food rotates against this abrasive surface, and with a continuous flow of water, the skin is effectively removed from the surface of the food	Potato, ginger, carrots
6	Flame peeling	The peeler uses a conveyor belt to carry and rotate the food through a furnace heated to over 1000 °C. In this process, the outer layer and root hairs of onions are burned off, and the charred skin is then manually removed	Brinjal, onion, garlic
7	Lye peeling	Items are peeled by dipping them in boiling lye (1–2%) for 1–2 min, depending on the lye strength, temperature, and maturity of the produce. Afterward, they are dipped in cold water. The hot lye loosens the skin, which is then gently rubbed off by hand. To neutralize the alkali, the produce can be briefly dipped in a dilute solution of hydrochloric or citric acid. This method is quick and efficient, minimizing waste and reducing peeling costs. The effectiveness of lye peeling depends on the concentration and temperature of the lye, the duration of exposure, and the level of agitation during the process	Orange segments, peach, apricot, nectarines, pears

Blanching

During blanching, cut fruits and vegetables are either immersed in boiling water or steamed for two to five minutes, followed by rapid cooling under cold running water. The blanching time and temperature vary depending on the type of produce. The process is considered complete when the peroxidase enzyme is inactivated, which indicates sufficient blanching.

The purposes of blanching include the following.

- Deactivating enzymes that cause discoloration and off-flavors.
- Reducing volume and mass to facilitate easier packing.
- Lowering the microbial load of raw materials.
- Enhancing the green color of vegetables like spinach and peas.
- Removing unwanted acids and the astringent taste of peels, thereby improving flavor.
- Releasing trapped gases to reduce pressure on the can seals during processing.

Filling

Tin cans are commonly used as containers for canning. These open-top sanitary cans can be opened from either end and are washed with hot water before use. Prepared fruits and vegetables are filled into the cans either manually or mechanically. While plain tins are typically used, lacquered tins are preferred for colored foods such as strawberries, red plums, and black grapes. For canned fruits, the drained weight should be at least 50%, and for berries, at least 40%. Similarly, the drained weight of canned vegetables must be no less than 55%, though this requirement primarily applies to tomatoes. As a result, the fruit and vegetable content in cans generally occupies around 60% of the can's capacity.

Syruping and Brining

Syrup is a mixture of sugar and water, typically poured over fruits during canning. Hot sugar syrup (79–82 °C) is poured into the cans, leaving a headspace of 0.3–0.5 cm. Syrup concentrations usually range from 10°B to 55°B (percent sugar). To make 20°B syrup, 250 g of sugar is dissolved in one liter of water, while 50°B syrup requires one kilogram of sugar dissolved in one liter of water. Sometimes, ascorbic or citric acids are added to the syrup to enhance flavor and nutritional value. Brine is a solution of table salt and water, used in the canning of vegetables. A saline solution with 1–3% salt is applied, heated to 79–82 °C, leaving 0.3–0.5 cm of headspace. The purpose of syruping/brining fruit is to:

- Enhance flavor
- Fill the void between the fruit/vegetable pieces in the can
- Facilitate heat transfer

Lid or Clinch

Before exhausting, the filled cans are loosely sealed with a lid. However, this method has some drawbacks, such as the risk of contents spilling and lids flipping. In modern canning, the clinching process has replaced this method. During clinching, the lid is partially secured or hemmed, leaving it slightly loose to allow steam, air, and gases produced during the exhausting process to escape.

Exhaustion

Before sealing, nearly all air must be removed from the can's contents. This process, called exhausting, expels gases from the container, reducing the risk of product discoloration from oxidation and preventing tinplate corrosion and pinholes during storage. Additionally, proper exhaustion helps preserve nutrients, particularly vitamin C. Exhausting can be done mechanically or by heating. In the heat method, cans pass through an exhaust box or hot water tank, where steam is generated. For fruit preserves, exhausting is done at 82–100 °C for 7–10 min, or until the temperature at the can's center reaches 79 °C. For vegetables, it is carried out at 90–100 °C for 7–10 min, or until the center reaches 79 °C. Proper exhaustion reduces stress on the can seam. The time and temperature of the process depend on the size and contents of the can, but should be sufficient to create a vacuum of 12–15 inches of mercury (Hg) in processed and cooled cans. The remaining space in the can is called headspace.

Sealing or Stapling

The process of seaming is used to seal cans after exhausting, utilizing a double-seaming machine. This method, also known as stapling, involves securing the can lids through a sealing or closing technique that creates a double seam to ensure a tight and secure closure.

Processing

Commercial sterilization refers to the process of heating and cooling canned foods to destroy microorganisms and preserve the food. Many bacterial spores are heat-resistant and can only be eliminated through prolonged cooking at extremely high or low temperatures. However, such intense treatment can negatively affect the food's quality. Therefore, the right combination of processing time and temperature is essential to fully eradicate bacterial growth without compromising the texture, flavor, or appearance of the food. Fruits and acidic vegetables can be processed effectively at 100 °C (boiling water temperature) due to the natural acidity in these foods, which inhibits the growth of bacteria and spores. These microorganisms also do not thrive in the high sugar content typically used in fruit canning. Nonacidic vegetables, except tomatoes and rhubarb, require higher processing temperatures, usually between 115 and 121 °C.

The size of the can affects both the temperature and the processing time—larger cans require longer processing times. For fruits and acidic vegetables, open cookers, continuous stirrer-less cookers, and continuous stirrer cookers are commonly used. Nonacidic vegetables are processed in closed retorts using steam pressure. When placing sealed cans in a retort, the water level is maintained 2.5–5.0 cm above the cans. The retort's lid is securely fastened, and steam is used to bring the appliance to the required temperature. The sterilization process begins when the water reaches a boil, and the time is monitored from this point. Once the processing time is completed, the heater is turned off, and the vent is opened. The cans are removed after the pressure returns to zero, and the lid is carefully taken off.

Cooling

After processing, the cans are rapidly cooled to approximately 39 °C to halt the cooking process. Several methods can be employed for cooling, including immersing the hot cans in a tank of cold water, spraying them with cold water, using a pressure cooker to circulate cold water, or exposing the cans to air. The immersion method is the most commonly used. To maintain the sterility of the cooling water, a chlorine solution at a concentration of one to two percent can be added. If the cans are not cooled promptly after processing, the quality of the preserves can deteriorate. For example, peaches and pears may darken, tomatoes can become brown and bitter, and peas may develop a mushy and overcooked texture.

Defect Testing

Before products are released for sale, they must be inspected for any defects. Finished cans are checked for leaks or faulty seals. This can be done by tapping the top of each can with a short steel rod. A dull, hollow sound indicates a leaky or defective seal, while a clear, ringing sound signifies a proper seal.

Storage, Labeling, and Packaging

Before storage, cans must be thoroughly dried, as even small amounts of moisture can lead to corrosion. They should be kept in a cool, dry place, as high temperatures can diminish the product's shelf life. Prolonged exposure to elevated temperatures can cause hydrogen expansion and lead to perforation. Basements are an ideal storage option during summer, providing a temperature difference of approximately 6–8 °C compared to the outside air. Once the cans are labeled, packed, and shipped in either cardboard or wooden boxes, they are ready for sale. Depending on the type of raw materials used and the product's shelf life, canned goods can typically be stored for 1–2 years (Fig 8.4).

The process of canning a few fruits and vegetables is discussed below.

Canning of Pineapple

According to Codex Alimentarius, canned pineapple is defined as a product made from fresh, frozen, or previously canned mature pineapple that meets the specifications of *Ananas comosus* (L.) Merr. The majority of pineapples used for canning come from four main cultivars: Cayenne, Singapore, Queen, and Red Spanish. High-quality canned pineapple can only be achieved when the fruit is harvested at peak ripeness and processed with minimal delay. In pineapple canneries, freshly picked pineapples are transported from the field to the cannery and processed on the same day, making this fruit one of the fastest-handled in the industry. The pineapple is processed by removing the peel and core and is then packed in water or another

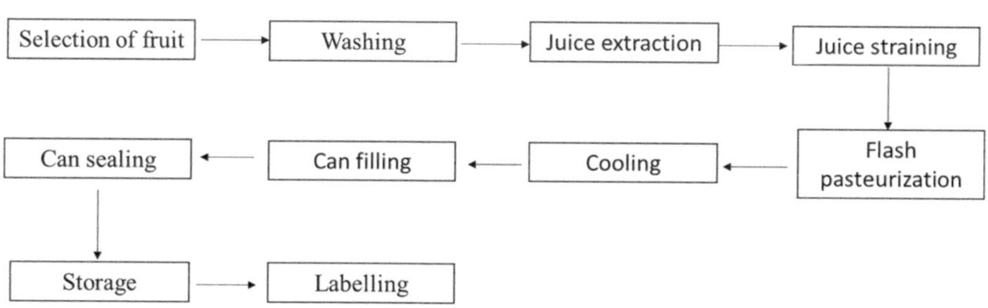

Fig. 8.4 Flowchart for canning of fruit juices

suitable liquid medium. It may also include nutritive sweeteners, seasonings, or other ingredients appropriate for the product. The canned pineapple is heat-processed either before or after sealing in a container to prevent spoilage.

Canned pineapple can be packed in the following styles.

- Whole: A cylindrical unit with the core removed.
- Slices, spiral slices, or rings: Uniformly cut circular slices or rings made from peeled and cored pineapple cylinders.
- Half slices: Uniformly cut semicircular halves of slices.
- Quarter slices: Uniformly cut one-fourth portions of slices.
- Broken slices: Arc-shaped portions that may vary in size and shape.
- Spear or fingers: Long, slender pieces cut radially and lengthwise from the cored pineapple cylinder, typically measuring 65 mm or longer.
- Tidbits: Relatively uniform, wedge-shaped sections cut from slices or portions, predominantly ranging from 8 to 13 mm thick.
- Chunks: Short, thick pieces cut from thick slices or from peeled, cored pineapple, generally measuring more than 12 mm in both thickness and width and less than 38 mm in length.
- Diced or cubes: Reasonably uniform, cube-shaped pieces, typically with edges measuring 14 mm or less.
- Pieces: Irregular shapes and sizes that do not fall into a specific category and exclude "chunks" or "chips."
- Chips: Small, irregularly shaped pieces of pineapple, similar to the remnants left after dicing, which may be included in the crushed style.
- Crushed or crisp cut: Finely cut, shredded, grated, or diced pieces of pineapple, which may include chips within the crushed mixture.

Cans made from internally plain (unlacquered) differentially coated electrolytic tinplate are preferred. The syrup used ranges from 20 to 40° B, with the specific Brix level depending on the product quality. In several Hawaiian plants, waste pineapple products, such as cores and residue from sizing (most of which is utilized for crushed pineapple), are processed through presses to extract juice. This juice undergoes some purification and is incorporated into the syrup formulation. This method reduces the amount of added sugar, thus lowering the syrup's cost. Additionally, during syrup production from the juice, citric acid is precipitated as calcium citrate, resulting in a valuable by-product.

Canning of Pears

Not all pear varieties are ideal for canning. The Bartlett pear is considered the best for this purpose due to its consistent size, regular shape, fine texture, vibrant color, and superior flavor—qualities that set it apart from other varieties. Harder pears are unsuitable for canning because they contain clusters of stone cells. There are established standards for canned pears, such as those outlined in the Codex Alimentarius (CODEX STAN 61-1981, Rev. 1-2001), along with additional regulations specific to different regions.

Canned pears can be packed in the following styles.

- Whole: Can be peeled or unpeeled, with cores either removed or left intact.
- Halves: Peeled or unpeeled, with stems and cores removed, cut into two roughly equal parts.
- Quarters: Peeled and cut into four approximately equal pieces.
- Sliced: Peeled and cut into wedge-shaped slices.
- Diced: Peeled and cut into cube-like pieces.
- Pieces or irregular pieces: Peeled, with irregular shapes and sizes.

Pears are peeled and cored either mechanically or by hand. Harder pears are typically peeled by hand, as their tougher skin can be difficult for mechanical peelers to handle. For hand-peeled pears, it is common to give them a brief steam or water blanch (20–30 seconds) before peeling. Once peeled, the pears are halved, and the core, stem, blossom ends, as well as any bruised or discolored areas, are carefully removed. To prevent browning after peeling, pears should be kept submerged in water until canned. If not processed immediately, they should be placed in a weak brine solution (1%) or a 1% citric acid solution. Because pears have moderate acidity, they do not necessarily require heavy syrup, though some processors prefer a sweeter syrup. The syrup should be added while hot, typically between 71 and 82 °C (160–180 °F). Syruping is performed using rotary, straight-line machines, or PVS (Proportionate Volumetric Syrup) systems. The PVS machine is advantageous because it conserves syrup by ensuring a consistent fill. When combined with a steam-flow sealer, it eliminates the need for thermal exhausting. In cases where cans are exhausted, this is best done in a steam exhaust box until the temperature at the center of the cans reaches 77–82 °C (171–180 °F). Pears are processed in boiling water at 100 °C (212 °F), with sterilization time ranging from 15 to 35 min, depending on the can size, syrup concentration, and closing temperature. Pears packed in heavy syrup require a longer processing time than those in light syrup. It is recommended to process pears until the center of the can reaches a temperature of 96 °C (205 °F). For harder pear varieties, a longer time may be needed to properly soften the fruit, even though the sterilization process may already be complete. Excessive heating or delayed cooling should be avoided, as this can cause the product to develop pink discoloration.

Canned vegetables can be just as nutritious as fresh vegetables. Studies show that after cooking, most frozen, fresh, and canned vegetables retain similar levels of most nutrients. Since canned vegetables are processed when very fresh, they are at their peak nutrition. While some nutrients, such as vitamin C and B vitamins, are lost during the canning process because they are water-soluble and heat-sensitive, the heat treatment also increases the availability of beneficial compounds like lycopene and beta-carotene. A common concern with canned vegetables is their sodium content, as salt (sodium chloride) is often added to preserve flavor. However, the amount of salt can be reduced without compromising taste and quality. Canned vegetables generally fall in the low-acid pH range, meaning that unless they are acidified, they must be sterilized to a minimum F_o level to ensure commercial sterility and shelf stability at room temperature. Process times and temperatures for low-acid products are recommended by the National Food Processors Association in Bulletin 26-L and Bulletin 30-L. While these guidelines provide good benchmarks for processing, they should be verified by a qualified thermal process authority for each specific product and recipe. It is essential to conduct heat penetration tests to confirm the safety of the processing parameters. Exceeding the critical control factor of fill weight constitutes a processing deviation and requires assessment by a thermal processing expert. The minimum initial temperature refers to the average temperature of the contents in the coldest can in the retort when the steam is turned on to start the process. The Codex Alimentarius International Food Standards (CODEX STAN 297–2009) offers guidelines for packing styles and quality criteria for canned vegetables. Some vegetables are packed in cans made from differentially coated tinplate with a high tin coating on the inside to preserve the product's bright color. For other products that may react with the tinplate and cause discoloration or staining, fully lacquered cans are used. Certain aggressive products may require both high tin coatings and internal lacquer to prevent the product from corroding the can over its shelf life. Blanching is a critical operation for vegetables, where air, water, or steam is used to heat the product to approximately 88–99 °C. When water or steam is employed, the process is referred to as wet blanching (Ranjan et al. 2017). The benefits of blanching have been discussed in the previous section. During blanching, the product can either be submerged in hot water or conveyed through a belt exposed to hot water or steam. For example, blanching vegetable soybeans at 100 °C for 90 s reduces initial lipoxygenase activity by 99% (Mozzoni et al. 2009). However, the blanching operation generates a significant portion of effluents in a canning plant and can lead to nutrient leaching. In recent years, innovative dry blanching techniques have been explored, such as microwaving, infrared radiation, and high-pressure treatment (Kalathur et al. 2013; Ranjan et al. 2017). Ancos et al. (1999) reported that dry blanching has several advantages over wet or conventional blanching, including reduced time needed to inactivate enzyme complexes and minimal to no leaching of vitamins, volatiles, pigments, carbohydrates, and other water-soluble components.

Low-acid products (pH > 4.6) such as vegetables are processed in retorts, where the temperature typically ranges from 110 to 135 °C. Depending on the type of retort, the heating medium may consist of water, steam, or a combination of steam and air. After thermal treatment is completed, the cans are cooled to a temperature of 30–40 °C. Once cooled, the cans are removed from the retort, labeled, and stored at room temperature, which should not exceed 30 °C to reduce the risk of thermophilic bacterial growth. For instance, vegetable soybeans (pH 6.5) can be processed with a brine solution containing 150 g/L NaCl and 2.9 g/L $CaCl_2$ in a can measuring 6.83 cm in diameter and 8.26 cm in height for 9 min at 121 °C to achieve a lethality equivalent to 4 min. This treatment is sufficient for achieving a 12-log reduction in microbial load (Mozzoni et al. 2009).

Canning of Asparagus

Canned asparagus consists of the tender and edible portions of either peeled or unpeeled stems from various asparagus varieties. There are two main types of asparagus that are canned: white and green. White asparagus is produced by covering the shoots with soil during growth (a process known as earthed up). This method prevents exposure to sunlight, inhibiting photosynthesis and resulting in white shoots. The stalks are harvested at least 15 cm below the ground's surface, before chlorophyll can develop. In contrast, green asparagus stalks are allowed to grow about 10–15 cm above the ground before being cut just below the surface. The butt ends of the stalks are cut off first, either mechanically or by hand, discarding the tough ends. Quality criteria ensure uniformity in the length and diameter of asparagus spears within each pack.

- Length: The lengths of the spears should be reasonably uniform, meaning at least 75% of the units should not deviate by more than 1 cm from the most common length, while at least 90% should not deviate by more than 2 cm.
- Diameter: Each product must adhere to the specified diameters for its individual grade. No more than 25% of the units in any container should belong to adjacent size groups.

Grading is performed based on color, quality, and size.

- No. 1 Grade: This grade comprises perfect spears with round, well-shaped heads. These are further regraded by size and color. Some packers may categorize asparagus into as many as seven sizes and three colors, although five sizes and two colors are typically sufficient for the No. 1 grade.
- No. 2 Grade: This grade includes spears that are somewhat flowered or have rough stalks with poorly shaped heads. These are regraded into two sizes: small and large. Different grades are kept in separate boxes and undergo careful inspection as they proceed to the cutting stage.
- No. 3 Soup Grade: This grade consists of broken, crooked, and otherwise deformed spears.

Green asparagus is generally graded into two to four categories, primarily based on the size of the stalks. Crooked or broken stalks that are still tender may be packed as fancy "cuts" for soup stock, while tougher stalks are typically packaged as standard "cuts" for soup stock. The various grades of asparagus are blanched separately to ensure uniformity and quality. The stalks are arranged in specialized blanching baskets, which can be sourced from canning machinery manufacturers, with the tips positioned upward. These baskets are then lowered into water heated to a temperature range of 77–93 °C (170–200 °F). In some instances, blanching is performed in boiling water, necessitating a reduction in processing time. After blanching, the asparagus is washed and thoroughly cooled using sprays of cold water. The cooled asparagus is then removed from the blanching baskets and placed onto filling tables. An alternative blanching method involves passing the filled baskets through a hot water bath on a continuous conveyor. For cut asparagus, blanching is done in a continuous blancher. This continuous process allows for the tougher parts of the stems to receive a longer blanching time compared to the tenderer blossom ends. Once filled, a 2–3% salt solution at a temperature of 88–93 °C (190–200 °F) is introduced into the cans. The hot brine should be sufficient to cover the tips when the can is closed and inverted. Discoloration can occasionally occur when packing green asparagus; adding 0.1% citric acid to the brine can help mitigate this issue. Cans with internal lacquer are recommended for this type of packaging, as the corrosion rate may be higher in acidified packs compared to nonacidified ones.

Canning of Green Beans

Beans intended for canning are typically purchased based on size, with smaller beans fetching higher prices. If beans need to be stored for any period before canning, they should be spread out in a layer no deeper than one foot to allow for ventilation, preventing overheating that could compromise their quality. Mechanically harvested beans often contain dirt, leaves, and pieces of vines. These contaminants are removed using mechanical shakers and air cleaners. The washing process may involve an agitating immersion washer, followed by a spray washer or pumping the beans through water, followed by transferring them into a squirrel-cage washer equipped with sprays. It is essential to ensure thorough washing to eliminate all foreign matter clinging to the beans. After washing, beans are conveyed to the preparation area, where snippers remove the stem ends. A pregrader sorts the beans into small, medium, and large sizes, allowing for optimal operation of the snippers tailored to each size.

Beans are passed through specially designed size graders to separate them into the desired sizes. Small sizes are typically packed whole, while larger sizes are cut either transversely into lengths of 2.5–3.8 cm (1–1.5 inches) for regular "cut beans," or into longer segments of 5.1 cm (2 inches) for "long cuts," or slit lengthwise for packs labeled as "French style," "julienne," "shoestring," and "cross cut." The cutting occurs postblanching to prevent the accumulation of small cut pieces and dislodged bean seeds that would otherwise result from slitting pods prior to blanching. Blanching is a crucial step in successfully canning beans; it helps remove occluded or dissolved air and gases, slightly shrinks the beans, and makes them pliable for proper canning. However, overblanching can lead to a slimy softness on the surface of the beans and increase the amount of fine particles in the liquid. Therefore, blanching should be kept to a minimum-just long enough to allow the beans to bend without breaking. When filling whole beans, a semiautomatic "hand pack" filler is commonly employed. The cans are placed in a tray covered with a metal plate that has apertures fitting the open tops of the cans. The tray is filled with beans and vigorously agitated until most beans drop into the cans, after which hand operators complete the filling. Automatic fillers for whole beans are also widely used today. When beans are filled by hand, brine should be added at a temperature of at least 93 °C (200 °F). For mechanical filling, brine is typically added simultaneously by the filler. The brine strength ranges from 1.5% to 5.2% (w/v) salt. Some canners also add up to 1.5% (w/v) sugar to the brine.

Beans are processed not only to sterilize but also to cook them. The following processes focus solely on sterilization. For more mature beans, longer processing times than those listed may be beneficial to enhance flavor. While these processes provide good guidelines for processing, they should always be verified by a competent thermal process authority for specific products and recipes. Conducting heat penetration tests is recommended to ensure the safety of the processing parameters. If the fill weight exceeds the critical control factor, it constitutes a processing deviation and must be evaluated by a competent thermal processing authority.

8.6 Effect of Canning on the Quality of Fruits and Vegetables

Texture

The high temperatures used in canning can cause breakdown of intercellular structures, which leads to cell separation, and damage or destruction of semipermeable cell membranes (Lyu et al. 2024). This breakdown results in a softer, sometimes mushy texture compared to fresh produce. For instance, canned tomatoes often have a more delicate, less firm texture than fresh ones. Blanching before canning helps preserve color and texture. However, it can also contribute to some textural changes by softening the vegetables before canning, since blanching is also a heat treatment. Increasing cooking time results in a softening of texture that can be attributed to the gelatinization and hydrolysis of starch in vegetables such as cowpea and beans (Rosello et al. 2024). Also, the denaturation of proteins as a result of heat treatment can cause textural changes. This is because heat causes the proteins' hydrogen bonds and other noncovalent bonds to break, which changes their conformation (Amaya-Farfan 2021). Calcium treatment before heat processing can prevent the textural degradation. Pectin molecules have carboxyl groups that can form bonds with calcium ions. This cross-linking creates a more stable gel-like structure, which helps maintain the firmness of the produce during processing and storage. Calcium contributes to improving the stiffness of cell walls and also reduces the activity of the pectinase enzyme, which is a major cause of cell degeneration (Cash and Sinha 2023).

Color

Canning can lead to a loss of vibrant colors due to the breakdown of pigments and the impact of heat. For example, the bright green of fresh spinach may dull to a more olive hue after canning. Heat treatment can result in the formation of pheophytin due to the displacement of magnesium in chlorophyll by two hydrogen atoms. With more intense heat, the carbomethoxy group in chlorophyll may be eliminated, producing pirochlorophyll. Additional heating can either remove a magnesium atom from pirochlorophyll or the carbomethoxy group from pheophytin, leading to the creation of piropheophytin (Putriani et al. 2020). Acidic foods (like tomatoes) tend to retain their color better during canning, while less acidic foods like beans and peas might experience more significant color changes. Maintaining an appropriate pH for canned foods is crucial not only for food safety but also for preserving the color of the food product (USDA 2015). Salt can also influence color, as it can cause some foods to become discolored. Before canning, many fruits and vegetables undergo enzymatic browning, which can darken their color. Blanching before canning can help inactivate these enzymes, preserving the original color better. Different pigments respond differently to heat and processing. For example, anthocyanins (found in berries) can change color under varying pH levels, while carotenoids (found in carrots) are relatively stable. Total carotenoid concentrations show no significant change after heat processing. In fact, there may be a slight increase in total carotenoids following (Le Bourvellec et al. 2018). Among the four carotenoids examined by the researchers, only cis-β-carotene significantly increased after heat processing. Several studies have noted an increase in cis-isomers in vegetables subjected to thermal treatment. Specifically, thermal processing is known to lead to the formation of the 13-cis-isomer (Schieber and Carle 2005).

Flavor

The flavor of canned fruits and vegetables may differ from fresh due to the heat used during processing. Some delicate volatiles may be lost, which affects the flavor. Canning can sometimes enhance flavors in certain cases, such as with tomatoes, where the cooking process can concentrate and deepen the flavor. The Maillard reaction, which occurs between amino acids and sugars during heating, leads to the development of new flavor compounds. It can enhance the flavor but may also create unexpected tastes in some cases (Van Boekel 2006). Caramelization is another chemical reaction that can alter the flavor significantly. When cans of apple sauce are stacked at temperatures exceeding 41 °C (105 °F), caramelization occurs, leading to an undesirable flavor that may be slightly burnt or unappealing (Lopez 1987). The flavor may also be affected by lipid oxidation, causing rancid odor (Secer et al. 2020). Certain species of spore-forming psychrophilic bacteria can spoil refrigerated canned foods, leading to the production of gas, off-flavors, and unpleasant odors (Erkmen and Bozoglu 2016). Positively, the canning process helps to remove the astringent flavor of some fruits like dates due to degradation of tannins during heat processing (Homayouni et al. 2015; Watson 1973).

Vitamins and Minerals

The heat used in canning can cause a reduction in the levels of certain vitamins, especially the thermolabile ones. Vitamin C and some B vitamins (such as folate) are particularly susceptible to degradation during the canning process. For example, canned produce may have lower vitamin C levels compared to fresh fruits and vegetables. Canning can result in significant losses of water-soluble vitamins, including ascorbic acid (vitamin C) and various B vitamins (thiamin, B1; riboflavin, B2; niacin, B3; folate, B9; and vitamin B6). Among these, ascorbic acid and thiamin are particularly sensitive to degradation (Lucci et al. 2016). Minerals (such as potassium, calcium, and magnesium) are generally stable during the canning process. They are less affected by heat and remain largely intact in canned fruits and vegetables (Rickman et al. 2007). Ash content remains intact during thermal processing; however, losses can occur due to the physical separation of vegetable components or leaching into water during cooking or processing. Raw vegetables typically contain less than 3% sodium (dry weight). In contrast, adding common salt to the canning brine raises the sodium levels in canned vegetables. Canned foods can also absorb iron from uncoated (nonenameled) cans (Martin-Belloso and Llanos-Barriobero 2001).

Antioxidants

A variety of bioactive compounds, including carotenoids, polyphenols, tocopherols, flavonoids, and trace elements, enhance the overall antioxidant capacity through their synergistic interactions. Antioxidant compounds are known to leach into brine or sugar solutions, and they may also degrade during thermal processing (Arampath and Dekker 2020). While some antioxidants may degrade, others can become more bioavailable. For example, the carotenoids in canned tomatoes are often more readily absorbed by the body than those in raw tomatoes. Heat processing and homogenization of the tomato increase the levels of cis-lycopene, which is the most bioavailable form for the human body. Cis-lycopene is believed to be more easily absorbed by micelles and chylomicrons for transport to tissues, while all-trans lycopene tends to aggregate in the intestine, forming crystals that hinder micellar absorption (Shi et al. 2004).

Carbohydrates and Fiber Content

Carbohydrates are less susceptible to chemical changes during heat preservation compared to many other food components. Levels of total and available carbohydrates in vegetables remain stable throughout the canning process and during subsequent storage of the canned products. This stability contributes to the nutritional value of canned fruits and vegetables, making them a viable option for maintaining dietary fiber intake (Ramesh 2007).

The fiber content in fruits and vegetables typically remains stable during the canning process. Although the breakdown of cell walls may compromise the structural integrity of the fiber, the overall fiber content is largely preserved. Key components of dietary fiber, such as cellulose, hemicelluloses, and pectins, play a crucial role in providing structure and texture to plant foods. Canning mainly alters the physical structure of fiber rather than its total quantity. While soluble fiber may dissolve into the canning liquid, reducing its visibility, insoluble fiber generally retains its form (Rani et al. 2021).

Phenolic Compounds

Phenolic compounds are secondary metabolites found in plants that play a crucial role in reducing reactive oxygen species. They hold significant pharmaceutical value and exhibit various biological activities, making them key contributors to the overall antioxidant capacity of fruits and vegetables. The primary types of phenolic substances include monophenols, phenolic acids, hydroxycinnamic acid derivatives, and flavonoids (Hu et al. 2022). Canning sometimes will increase the total phenol content, such as in canned raspberries (100 °C, 28 min) and blueberries (100 °C, 22 min) (Sablani et al. 2010). While some phenolic compounds, such as hydroxycinnamic acids, 3-O-caffeoylquinic acid, and 5-O-caffeoylquinic acid, present in apricots get degraded due to thermal processing or leaching out to the solution (Le Bourvellec et al. 2018). In plants, phenolic acids may be found both free and bound; the latter is conjugated to different elements in either glycosylated or ester-bound forms. As the temperature rises, the phenolic acids may be liberated from their bound state. On the other hand, some of the phenolic acids could be eliminated when the temperature rises (Maghsoudlou et al. 2019).

Antinutritional Compounds

Plants often produce antinutrients as a defense mechanism against herbivores and to adapt to challenging growing conditions. These antinutrients include compounds such as goitrogens, phytic acid, saponins, tannins, amylase inhibitors, lectins, anti-vitamin factors, and protease inhibitors. Proteinase inhibitors in seeds can hinder the action of digestive enzymes like trypsin and pepsin, thereby limiting the breakdown and absorption of proteins in the gut. Similarly, tannins can interfere with digestive enzymes, leading to decreased absorption of essential nutrients, including proteins and carbohydrates (Salim et al. 2023). During canning, these antinutritional factors are mainly inhibited. Moreover, phytates can be degraded by the phytase enzyme during canning, as the high temperatures involved in the process enhance phytase activity. This increased enzyme activity facilitates the breakdown of phytates, thereby improving mineral bioavailability (Antoine et al. 2022).

Similarly, saponins and tannins are also reduced during canning due to the activation of specific enzymes. For instance, the activity of polyphenol oxidase (PPO) increases with heat, leading to the hydrolysis of tannins. This enzymatic action not only decreases the levels of tannins but also mitigates their potential to interfere with nutrient absorption. Overall, the heat treatment during canning effectively reduces various antinutritional factors, improving the nutritional profile of the final product (Antoine et al. 2022).

8.7 Spoilage of Canned Fruits and Vegetables

Canned foods are preserved through aseptic processing in hermetically sealed containers. Although they are thermally processed under sterile conditions, they are prone to spoilage owing to microbial or nonmicrobial factors (Fig. 8.5). This results from faulty processing operations or faulty containers.

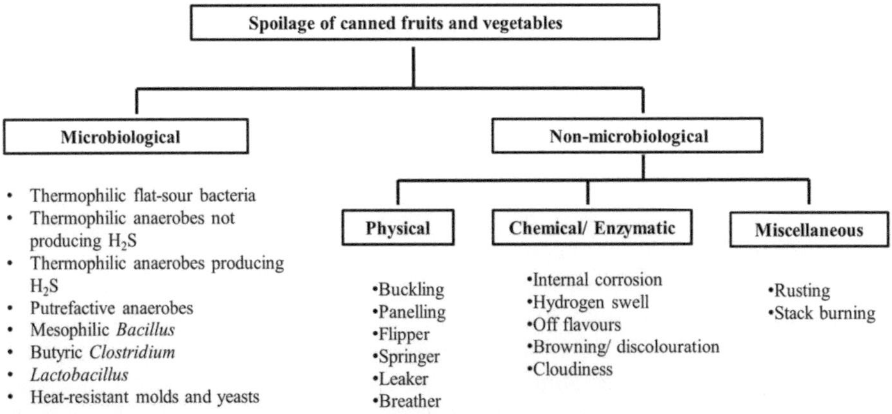

Fig. 8.5 Causative factors for spoilage of canned foods

Microbiological Spoilage

Canning of fruits and vegetables is carried out to ensure commercial sterility of the product. The microbial contamination may occur with or without the production of gas inside the can. Gaseous spoilage is manifested by bulging or swelling of cans due to the production of gases such as hydrogen or carbon dioxide. Nongaseous spoilage cannot be detected by the external appearance of the can and is detectable only after opening of the can, as observed by visible microbial growth, off flavor, and sour taste accompanied by discoloration of contents (Table 8.3).

Further, the pH level of the contents in the food governs the type of microorganisms that may cause spoilage.

Low-acid foods (pH > 5.20) such as certain vegetables, corn, and pears are spoiled by thermophilic flat-sour spoilage bacteria, sulfide-producing anaerobic bacteria, and mesophilic putrefactive anaerobic bacteria.

Table 8.3 Spoilage-causing microorganisms and their defects

Microorganisms	Source of contamination	Manifestation of spoilage
Flat-sour causing (*Geobacillus stearothermophilus*, *B. coagulans*)	Plant equipment, sugar, starch, and soil	Flat can (no gas production), gradual loss of vacuum, souring of product, abnormal odor, cloudy appearance
Thermophilic anaerobes (*T. thermosaccharolyticum*)	Sugar, dried milk, starch, flour, cereals, soiled vegetables	Can swell on one or both ends, can may burst due to gas production (CO_2, H_2), souring of product, off odor.
Sulfide spoilage (*D. nigrificans*, *C. bifermentans*)		Flat can, H_2S production with "rotten egg" smell, iron sulfide precipitate causing blackening
Putrefactive anaerobes (*C. sporogenes*, *C. putrefaciens*, *C. botulinum*)		Can swell that may burst due to gas production, increase in pH, putrid odor, production of ammonia, indoles, CO_2, H_2, and H_2S
Aerobic spore formers (*B. subtilis*, *B. mesentericus*, *B. polymyxa*, *B. macerans*)	Leak in the can	Flat or light swelling of can, acid and gas production, coagulation.
Butyric anaerobic spoilage (*C. butyricum*, *C. pasteurianum*)		Can swell, which may burst due to fermentation and gas (CO_2, H_2, not H_2S), butyric odor
Acid spoilage (*B. thermoacidurans*, *A. acidoterrestris*)		Flat can or swelling with gas (butyric anaerobes), off-flavors.
LAB spoilage (*L. brevis*, *L. mesentericus*, *L. dextranicum*, *L. mobilis*)		Can swell that usually burst by gas (CO_2), ropiness in product, acid production, off-odor.
Yeasts (*Saccharomyces*, *Zygosaccharomyces*)		Can swell that may burst, leakage of cans, fermentation (CO_2 production), yeasty odor.
Molds (*B. fulva*, *Aspergillus* *Penicillium*, *Citromyces*)	High sugar	Flat or slight swelling can, moldy taste, odor, color change

Medium-acid canned foods (pH 5.20–4.60). Vegetable mixtures can be spoiled by thermophilic anaerobic bacteria and putrefactive anaerobic bacteria.

Acid canned foods (pH 4.6–4.0). Tomatoes, pears, figs, pineapples, olives, cucumbers, and other fruits are spoiled by mesophilic spore-forming bacteria, *Lactobacillus*, butyric anaerobic bacteria, and aciduric bacteria.

High-acid canned foods (pH < 4.0). Pickles, grapefruit, citrus juices, ketchup, acid vegetables, pineapples, peaches, and sauerkraut are generally spoiled by yeasts, molds, and lactic acid bacteria (LAB).

Microbiological contamination in canned products may result from the following reasons:

1. Preprocess spoilage
2. Leakage through seams
3. Under processing (allows survival and growth of mesophilic microorganisms)
4. Inadequate cooling (allows germination and growth of thermophilic spore formers)

Chemical Spoilage

Chemical interaction between the steel base of the can and the food ingredients results in the spoilage of the canned product. The types of chemical spoilage are given in Fig. 8.5. Packing high acid foods (strawberries, apples, cider) in unlacquered or improperly lacquered cans results in hydrogen swell due to the action of the acid of food on the iron base of the can. The cans swell due to the production of hydrogen gas but should not be confused with microbiologically spoiled cans, as the food is safe for consumption. Poor exhausting, imperfections in tinning/lacquering of the inside of the cans, and high acidity of foods are the causative factors for the chemical reactions in canned products. Other manifestations of chemical spoilage of canned foods include brown discoloration, cloudiness of the contents, and off-flavor development. Contact of food with minerals such as iron, zinc, or copper in cooking utensils or water may result in discoloration. Cloudiness may result from the leaching of starch from certain vegetables in the brine. Selection of immature, overripe, or inferior fruit for canning can result in poor flavor of the canned food/juice.

Physical Spoilage

Any change in the unopened cans showing external faults is included in the physical spoilage of cans. Faulty retort operation, under exhausting, overfilling, too high internal vacuum, and rough handling are the major causes of external distortion of the cans. As shown in Fig. 8.5, the major physical defects of cans are as follows.

1. Buckling: This is the permanent distortion of the can ends caused by excessive internal pressure.
2. Paneling: Flat sides on the can body and/or permanent distortion (collapsing) of the can body caused by internal vacuum and excessive external pressure developed during processing or cooling.

3. Flipper: This results from a lack of vacuum. When only one end of the can is slack or slightly bulged. That end will remain flat if pressed in. Cans that bulge when sharply and squarely struck end-down on a flat surface are flippers, provided that the bulged end remains flat when pressed.
4. Springer: One end of a can bulges. Manual pressure on the bulged end forces the opposite end out, or the same end will spring out with the release of pressure. If both ends bulge but only one will remain flat when pressed, the can is a Springer. Springers result from moderate positive pressure in the can.
5. Leaker: Containers that have leaked.

Miscellaneous Spoilage

Rusting of cans is a major spoilage of canned foods. Storage of cans in humid atmospheric conditions, exposure to corrosive fumes, and fluctuations in storage temperatures lead to rusting of cans. Also, careless handling of cans and use of rusted trolleys, heavy embossing of can ends for coding, corrosive label plates/adhesives, and overcooling of cans are some other reasons for rusting.

8.8 Disadvantages of Cans

The disadvantages associated with metal packaging materials are that they are heavy compared to plastic packaging material, metals are opaque, and that there is no see through possibility. Metals also have a higher chance of reaction with the packaged food product as compared to glass and plastic materials, and the lacquering of metal cans is important.

Canning offers a reliable method for preserving fruits and vegetables for a long duration and ensures their availability to far-off locations and beyond their natural season. By effectively eliminating microorganisms and inhibiting enzymatic activity, canning extends the shelf life of these perishable foods. Historically, canned foods have been essential components of emergency food supplies, ensuring food security during crisis. The canning process not only ensures food safety but also maintains nutritional value and flavor, making it a valuable tool for home cooks and commercial food producers alike. While canning provides a convenient way to store food, it is essential to adhere to strict guidelines to prevent foodborne illnesses. Proper sterilization, sealing, and storage are crucial steps in the canning process. By understanding the principles and techniques of canning, the food industry can contribute to sustainability through reduction of food waste as well as through recyclability and reusability of cans. Future research should focus on advancements in sterilization techniques and packaging materials to further enhance the quality and sustainability of canned foods. Challenges such as rising costs of raw materials and energy, as well as evolving consumer preferences, need to be addressed to further the advancement in canning technology. However, by embracing innovation, focusing on sustainability, and prioritizing food safety, the canning industry can continue to thrive and meet the evolving needs of consumers worldwide.

References

Amaya-Farfan J (2021) Denaturation of proteins, generation of bioactive peptides, and alterations of amino acids. In: Chemical changes during processing and storage of foods. Academic Press Inc., Elsevier pp 21–84

Ancos DB, Cano MP, Hernandez A et al (1999) Effects of microwave heating on pigment composition and colour of fruit purees. J Sci Food Agric 79:663–670

Antoine T, Georgé S, Leca A et al (2022) Reduction of pulse "antinutritional" content by optimizing pulse canning process is insufficient to improve fat-soluble vitamin bioavailability. Food Chem 370:e131021

Arampath PC, Dekker M (2020) Thermal effect, diffusion, and leaching of health-promoting phytochemicals in commercial canning process of mango (*Mangifera indica* L.) and pineapple (*Ananas comosus* L.). Food Secur 10:46

Bell RA (1965) Origins of the canning industry. Trans Newcomen Soc 38:145–151. https://doi.org/10.1179/tns.1965.011

Cash JN, Sinha NK (2023) Canning of vegetables. In: Processing vegetables. Routledge, New York, pp 49–64

Codex Alimentarius, Codex Standard for Canned Mushrooms (CODEX STAN 55–1981)

Codex Alimentarius, Codex Standard for Certain Canned Vegetables (CODEX STAN297–2009)

Erkmen O, Bozoglu TF (2016) Food microbiology, principles into practice, vol 2. John Wiley & Sons, West Sussex.

Fitzgerald JG (1923) Pasteur and the science of bacteriology. Sci Mon 17:23–29

Homayouni A, Azizi A, Keshtiban AK et al (2015) Date canning: a new approach for the long time preservation of date. J Food Sci Technol 52:1872–1880

Hu W, Sarengaowa Guan Y et al (2022) Biosynthesis of phenolic compounds and antioxidant activity in fresh-cut fruits and vegetables. Front Microbiol 13:e906069

Kalathur HV, Girish KG, Hebbar UH (2013) Infrared assisted dry-blanching and hybrid drying of carrot. Food Bioprod Process 91:89–94

Le Bourvellec C, Gouble B, Bureau S et al (2018) Impact of canning and storage on apricot carotenoids and polyphenols. Food Chem 240:615–625

Lopez A (1987) A complete course in canning and related processes. Canning Trade. Inc., Baltimore, pp 21218–24576

Lucci P, Pacetti D, Loizzo MR et al (2016) Canning: impact on food products quality attributes. In: Food processing technologies. CRC Press, Boca Raton, pp 41–60

Lyu J, Yu X, Wang F et al (2024) Cell wall modifications during storage: changes in the texture behavior of the canned yellow peach. Innov Food Sci Emerg Technol 91:e103549

Maghsoudlou Y, Asghari Ghajari M, Tavasoli S (2019) Effects of heat treatment on the phenolic compounds and antioxidant capacity of quince fruit and its tisane's sensory properties. J Food Sci Technol 56:2365–2372

Martin-Belloso O, Llanos-Barriobero E (2001) Proximate composition, minerals and vitamins in selected canned vegetables. Eur Food Res Technol 212:182–187

Mozzoni LA, Morawicki RO, Chen PY (2009) Canning of vegetable soybean: procedures and quality evaluations. Int J Food Sci Technol 44:1125–1130

Putriani N, Perdana J, Meiliana et al (2020) Effect of thermal processing on key phytochemical compounds in green leafy vegetables: a review. Food Rev Intl 38:783–811

Ramesh MN (2007) Canning and sterilization of foods. In: Handbook of food preservation. CRC Press, Boca Raton, pp 603–642

Rani J, Shelke AR, Bansal B et al (2021) Effect of processing and storage on the physicochemical composition and quality of fresh, canned and dehydrated okra fruits. Mater Today Proc 45:4410–4416

Ranjan S, Dasgupta N, Walia N et al (2017) Microwave blanching: an emerging trend in food engineering and its effects on *Capsicum annuum* L. J Food Process Eng 40:e12411

Rickman JC, Bruhn CM, Barrett DM (2007) Nutritional comparison of fresh, frozen, and canned fruits and vegetables II. Vitamin A and carotenoids, vitamin E, minerals and fiber. J Sci Food Agric 87:1185–1196

Rosello S, Beltran J, Rivera A et al (2024) Are the sensory qualities of prestigious traditional varieties of beans preserved in commercial canned products? A comprehensive sensorial and chemical study on Ganxet beans. LWT 203:116413

Sablani SS, Andrews PK, Davies NM et al (2010) Effect of thermal treatments on phytochemicals in conventionally and organically grown berries. J Sci Food Agric 90:769–778

Salim R, Nehvi IB, Mir RA et al (2023) A review on anti-nutritional factors: unraveling the natural gateways to human health. Front Nutr 10:1215873

Schieber A, Carle R (2005) Occurrence of carotenoid cis-isomers in food: technological, analytical, and nutritional implications. Trends Food Sci Technol 16:416–422

Secer OM, Guneser BA, Guneser O (2020) Prediction of shelf-life and kinetics of quality changes in canned stuffed grape leaves. LWT-Food Sci Technol 132:e109850

Shi J, Naughton LM, Kakuda Y et al (2004) Bioavailability of lycopene from tomato products. Prevent Nutr Food Sci 9:98–106

USDA (2015) Complete guide to home canning. Agric Inf Bull 539

Van Boekel MAJS (2006) Formation of flavour compounds in the Maillard reaction. Biotechnol Adv 24:230–233

Watson EK (1973) Tannins in fruit extracts as affected by heat treatments. Family Consumer Sci Res J 2:112–118

Chapter 9
Beverages and Concentrates

A. Sangeeta, Mridusmita Barman, and Poonam Mishra

9.1 Introduction

Fruits and vegetables play an essential role in nutritional well-being of humans as they are a rich reservoir of several bioactive components. Phytochemical components (tannins, terpenoids, alkaloids, etc.) present in fruits exhibit several health benefits including antiallergic, anti-inflammatory and anticarcinogenic.

Fruit beverages are generally prepared from fresh fruits or preserved fruits. Fruit beverages, juices and concentrates represent a diverse array of products that play significant roles in human diets, nutrition and food industries worldwide. These products are derived from fruits, offering abundant flavours, nutrients and bioactive compounds, making them not only delicious but also beneficial for health. Awareness of the ingredients included, manufacturing processes and potential health risks of fruit-based beverages has drawn more attention in recent years. Fruit juices mostly consist of 80–89% water, along with several other vital nutrients. Water is a source of mineral salts, it plays a significant role as a diluent of orally absorbed substances and thus facilitates digestive processes (Daud et al. 2023). Fruit juice concentrate, with its high soluble solids content of 65–70%, is an effective way to reduce packing and transportation costs while guaranteeing consistent product quality. Preservation is enhanced by the increased sugar and acid content.

This chapter provides a comprehensive overview of fruit beverages, juices and concentrates, encompassing their production processes, nutritional profiles and physiological effects on human health.

A. Sangeeta · M. Barman · P. Mishra (✉)
Department of Food Engineering & Technology, Tezpur University, Tezpur, Assam, India
e-mail: poonam@tezu.ernet.in

OP Chauhan (ed.), *Fruits and Vegetables Technologies*,
https://doi.org/10.1007/978-981-96-8433-5_9

9.2 Fruit Beverages

Fruit beverages are manufactured using fresh unprocessed fruits or semi-processed products, e.g., fruit concentrates and fruit puree. Fruit beverages are broadly classified into two categories, i.e., fruit juice containing 100% pure juice without added ingredients and fruit drinks containing only 10% fruit juice with added ingredients. (Mani-López et al. 2019).

Juice is the liquid fraction expelled from cells or tissues (Mihalev et al. 2018). Fruits are vital part of human diet as they are rich source of several nutrients including vitamins, minerals and dietary fibre (Slavin and Lloyd Beate 2012). "Fruit juice is defined as the fluid portion extracted from the cells or tissues of ripe fruits by the application of mechanical pressing or squeezing an extract or an extractable fluid content of cells or tissues made by mechanically squeezing or pressing out the natural liquid contained in ripe fruits without using any heat or solvent" (Aykas and Rodriguez-Saona 2024).

The fruit juice market has grown significantly over the past 20–30 years, as evidenced by an impressive rise in consumption. This expansion has progressed beyond simply offering different flavours and types of juices to include juices enriched with vitamins and/or minerals and several other 'functional' ingredients (Caswell 2009). The global juice market is growing and it is mainly influenced by the increasing health awareness amongst consumers.

9.2.1 Classification of Fruit Juices

According to the Fruit Juice Directive, fruit juices are classified into six products (Fig. 9.1):

- *Fruit juice or "direct juice" or "not from concentrate (NFC) juice":* Fruit juice is described as "the product that is fermentable but an unfermented product produced from the edible portion of fruit which is sound and ripe, either fresh or preserved giving the characteristic flavour, colour and taste of the original fruit."
- *Fruit juice from concentrate*: These are the fruit juices that have been previously concentrated and then subsequently restored to their original form by the addition of water.
- *Concentrated fruit juice*: The product obtained after removing a certain proportion of water content from it by means of concentrating techniques like evaporation under vacuum.
- *Water-extracted fruit juice*: It is the juice obtained through diffusion with water from fruits such as dehydrated whole fruit (dried plum) or pulpy fruit (acai berry), where extraction of juice by mechanical means is difficult.
- *Dehydrated/powdered fruit juice*: "Fruit powder" is produced by complete removal of water from fruit juice or concentrated fruit juice.

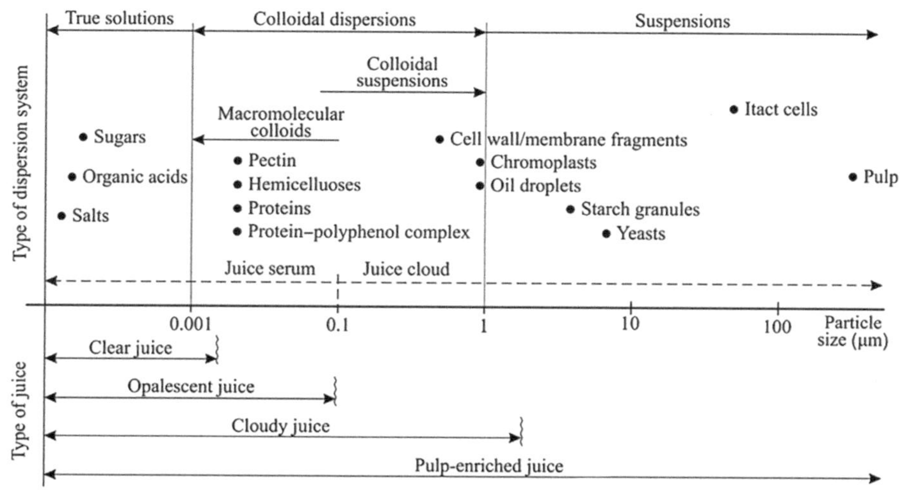

Fig. 9.1 Different types of fruit juice and their dispersion system composition. (Adapted from Mihalev et al. 2018)

- *Fruit nectar:* Fruit nectar is a product containing 20 percent of fruit juice (or any of the above-mentioned categories) and 15 percent of sugar and 0.3 percent of acidity (Mihalev et al. 2018; Chandel et al. 2018).

9.2.1.1 Based on Post-extraction Processing

- *Single-strength direct juice* is the simplest type where juice doesn't undergo any concentration process. These types can be produced for immediate consumption without heat treatment on a small-scale level or full commercial level with appropriate preservation approach including pasteurization or UHT sterilization with aseptic packaging.
- *Concentrated juice:* concentration of fruit juice is done using the thermal evaporation method, followed by preservation using high temperature or low temperature (Varnam and Sutherland 1994).

9.2.2 Fruit Juice Processing

Certain fruits like citrus/orange, lime, passion fruit, mango, apple, berries, etc., are some of the most common fruits used in juice production (Mushtaq 2018). Several factors including size, nature, shape, harvesting characteristics severely impacts the processing. Processing involves a number of stages including procurement of fruits at their optimum stage of maturity, expressing juice, clarification (physical/

chemical/enzymatic), filtration, concentration and then subsequent packaging and storage (Taylor 2016). Fruit maturity is considered as a vital quality factor in selection of fruit for juice industry, it is a stage when fruits attains its full natural development whereas another term called ripening refers to the stage where the fruit attained the optimal growth for its utilization, followed by senescence where respiratory pathways progressively decline and is preceded by several complex changes in structure and composition (Anwar et al. 2019; Mushtaq 2018). Determining the ratio of sugar and organic acids is a traditional approach to estimate fruit maturity. Along with this several other approaches including visual method, which largely depends on the external appearance of the fruit (skin colour, size, etc.), Physical method involving specific gravity, firmness, etc., Chemical method that determines titratable acidity (TA), total soluble solids (TSS), TSS:TA ratio, starch content, etc., Computation method covers the entire lifetime of the fruit from bud initiation stage until optimum fruit development and physiological method estimates the aroma development, respiration, etc. Maturity indices of fruits and vegetables are evaluated using two common approaches: destructive and non-destructive.

Non-destructive technique refers to assessing fruit quality without causing harm to their structure, like skin colour by visual assessment or using colour charts; Optical methods which uses light transmission properties to evaluate the fruit maturity, as internal optical density changes with the fruit maturity (due to the conversion of free water to bound water and vice versa depending on the maturity stage); Firmness is related to the fruit texture, is analysed using instruments, such as tenderometer, penetrometer, pressure testers; aroma, is also an indicator of fruit maturity as fruits synthesize volatile chemicals which gives them their characteristic odour used as a maturity index; leaf changes, this characteristic is used to determine when fruits & vegetables should be harvested for example, in root crops the condition of the leaves indicates the crop condition growing below ground (Prasad et al. 2018). Sensors and non-destructive techniques (NDT) are the new innovative approaches being used for assessing the fruit quality, for example, electric nose, electric tongue, magnetic resonance imaging (MRI), machine vision technique and X-ray computed tomography (CT) (Table 9.1) (Anwar et al. 2019).

Approaches to evaluate fruit ripeness or ripening by destroying them are referred to as destructive methods, for example, penetration, peeling fruit flesh or mesocarp, conducting a chemical reaction and applying an external force to a fruit (Fig. 9.2) (Islam et al. 2024).

9.2.2.1 Washing

Washing aids in reducing the physical and chemical contaminants from the incoming fruits prior to entering the juice extraction unit (Mushtaq 2018). The surface of raw materials is usually contaminated with a high microbial load (up to 10^5–10^9 microorganisms per gram).

The initial load only drops by three to five orders of magnitude after effective washing (Horvath-Kerkai 2012). It is an important practice to clean the fruits

Table 9.1 Quality determination of fruits using non-destructive methods

Method	Technique	Parameters
Optics	Image analysis, reflectance, transmittance and absorbance spectroscopy, laser spectroscopy	Size, shape, colour, external defect, internal component, visco-elasticity, firmness
X-ray	X-ray image and CT	Internal cavity and structure, ripeness
UV-ray	UV-ray and spectroscopy	Carotenoids and anthocyanin contents
NIR	Reflectance, transmittance, interactance and transflectance	Internal component
Mechanics	Ultrasonic sonic Vibrated excitation Ultrasound	Internal cavity and structure, ripeness. Internal cavity and structure, firmness, ripeness, Visco-elasticity, firmness, ripeness Maturity, oil and sugar content
Electromagnetic	Impedance MR/MRI	Moisture content, density, sugar content, internal cavity, density, sugar content, moisture content, oil content, internal defect and structure
Machine vision technique	Image processing technology, computer software and hardware	Surface colour, ratio of total soluble solid to titratable acidity

Adapted from Kumari et al. (2018)

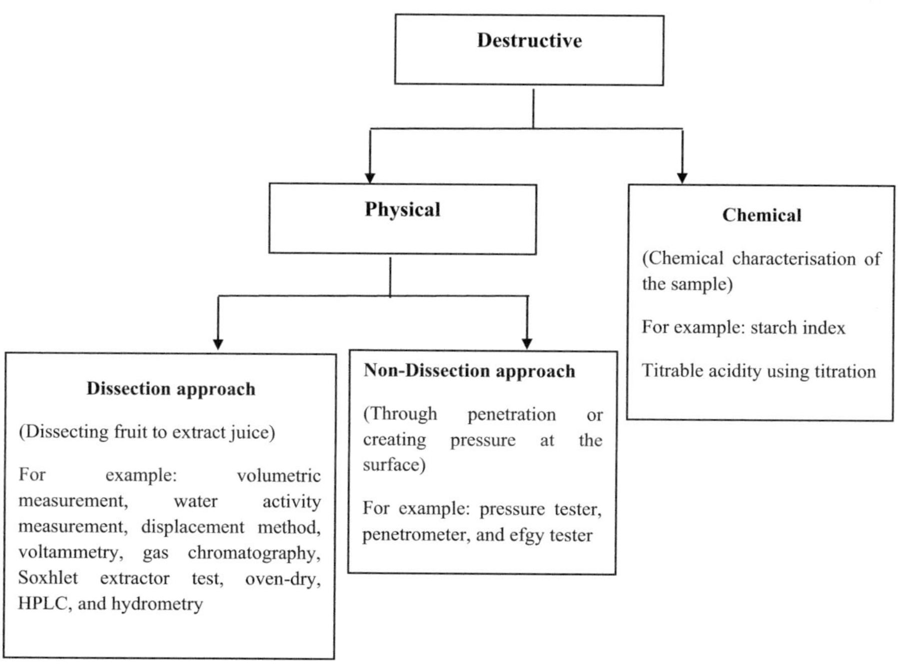

Fig. 9.2 Destructive methods to measure fruit maturity

immediately after harvesting to improve product appearance and edibility as well as to remove residues of field-applied chemicals and harmful pathogens that would shorten their shelf life (Oyeleke et al. 2014).

Bacteria including pathogens such as *Escherichia coli* O157:H7, *Salmonella* and *Listeria monocytogenes* are naturally present in the environment where fruits are grown. Fruit juices exhibit antimicrobial properties because of their acidic nature (pH 3.2 to 4.0); however, it has been reported that *Salmonella* and *E. coli* O157:H7 can survive in orange juice and apple cider, respectively.

Many techniques have been investigated for decontaminating orange surfaces, including hot water, lactic acid with sodium hypochlorite, hydrogen peroxide, chlorine dioxide (gas or solution), etc. (Martínez-Gonzáles et al. 2011). Numerous studies have been conducted on the use of mechanical surface cleaning systems, such as spinning brushes and/or high-pressure water sprays, to remove mites and other surface arthropods. Pressurized sprays (2240–5516 kPa) help in successful elimination of surface arthropods (Oyeleke et al. 2014). Conventional approaches like water showers or flowing water or pressurized air were predominant for decontaminating the fruits, which possess severe limitations of large quantities of water requirement, mechanical damage to fruits and are ineffective in removing oleoresins, waxes and other persistent materials from the fruit surface. In recent times, juice-manufacturing plants comprises additional components (e.g. fibrous brushes) that helps in effectively eliminating waxes, physical debris, pesticide residues, etc. (Mushtaq 2018). Another mechanical system called a rotary fruit washing machine has an elementary design and a robust construction. It is a continuous system containing fruits being continuously fed into the hopper, where they are cleaned by the rotary drum's gentle tumbling action and then removed through the other end, it is suitable for mango, apple, carrot and other root crops but unsuitable for fragile crops (such as tomato) (Oyeleke et al. 2014).

In order to prevent microbial inoculation, a chemical method of washing fruits and vegetables has been employed, employing salts such as calcium or sodium hypochlorite to sterilize the fruit's surface.

Sodium hypochlorite (NaClO) is one of the most widely favoured salts for sanitizing the surface of fruits, which works best at low pH levels. However, in general, pH values between 6 and 7.5 should be chosen to reduce the danger of corrosion on metallic processing equipment (Artés and Allende 2014).

9.2.2.2 Preparation of Fruits

Preparation of fruits is another essential stage in fruit juice processing. Preparation of fruits entails a number of steps, such as removing the stem, which is necessary for some fruit species (such as plums, cherries and sour cherries) in order to preserve the juice's colour and other desirable qualities. This process can be done either by hand or by mechanical means. However, a mechanized stem removal approach works best for raw materials with rigid structures and uniform sizes (Horvath-Kerkai 2012). Machines consisting of different types of metallic cylinders, cutting machines, blades, crushers, or hammers are used for carrying out different preparation operations of peeling, trimming and removal of skin and other unwanted parts. Impingement method is commonly used for extracting pomegranate seeds, and it operates with pressurized air-jets which open the fruit, separate it into two halves, and extract the seeds from the pomegranate.

In recent times, a number of innovative processes have been developed that facilitate the opening of fruit without cutting, extraction of the arils with minimal damage, separation of the arils from extraneous materials, and the delivery of clean arils to a packing machine (Mushtaq 2018).

The four main categories of peeling techniques are enzymatic, chemical, thermal and mechanical abrasion (Caceres et al. 2012).

Chemical peeling is an intricate process that involves mass and heat transmission as well as a simultaneous chemical reaction. In this approach, the product is immersed in a hot solution of caustic soda (lye solution) for a specific duration of time.

Skin separation is caused by a reaction between the lye solution and the cell walls of the epidermis and hypodermis (Barreiro et al. 2006).

The fruit's ripeness and cultivar selection determine the effectiveness of fruit peeling with caustic soda; overmatured fruit loses a significant amount of its appearance and tissue during treatment. The concentration, temperature and immersion time of the lye solution determine the effectiveness of the peeling process (Guldas 2003).

9.2.2.3 Fruit Juice Extraction

Traditional methods are commonly used for obtaining juice from the fruit, which includes maceration of a fruit by hand or peeling, chopping, breakage and pressing of the fruit. The traditional approach has certain drawbacks including high energy and labour consumption, minimum juice yield as well as an unsanitary approach. Thus, this traditional method of juicing can be used neither in small nor in medium productions to satisfy local commercial needs (Medvedkov et al. 2021). The process of juice extraction from selected, sorted, washed and trimmed fruits involves various stages including crushing, juice separation from solid fruit flesh particles or juice extraction, juice treatment and preparation for filling. Crushing comprises chopping and preparation for pressing, where the former involves smashing and cutting the fruit, which increases the surface area and aids in cell fluid elimination. Fruits like apples have a firm texture and are chopped by crushing, whereas fruits like red currants having a soft texture are only cracked (Horvath-Kerkai 2012). Chopped fruit preparation involves procedures to increase juice yield with better aroma, flavour and colour properties and prevent undesirable changes (mechanical, chemical, biological, etc.). This operation can be carried out using a variety of methods, including enzymatic, heat treatment, freezing, vibration, electroplasmolytic procedures, etc. The prepared fruits are crushed or milled using an array of machines that provide channels which enable the juice extracted from the fruit to be transferred into a mesh of an appropriate particle size (Mushtaq 2018). There are three common methods for the extraction of fruit juice. The first method involves cleaning, followed by crushing and cutting into pieces, then heating and finally pouring and sieving without pressing onto a wet muslin cloth. The second method is done using a fruit press or a fruit mill, followed by heating the juice at 60 °C and then straining through a muslin or cheese cloth. The third approach utilizes a juice steamer, where the steam and heat extract the juice from the fruits (Oyeleke and Olaniyan 2007).

9.2.2.3.1 Fruit-Grinding Mills

Fruit-grinding mills consist of a rotating shaft with attached sharp knives/blades. The fruits entering the grinding mills (Fig. 9.3) get crushed by the rotating shaft with blades which are fed into the pressing machine. These types of machines are mainly utilized for cutting large-sized fruits with soft flesh (e.g. apple, pear and mango).

For the citrus fruit juices, Finger cup extractors consist of two half cups enclosed by finger-like projections along with a strainer which aids in separating the juice from the strongly flavoured peel oils in the exterior of the fruit (Mushtaq 2018).

9.2.2.3.2 Fruit Machinery Corporation (FMC)

It is generally employed for citrus fruits. The extractor is made of stainless steel and consists of two cups, one inverted above the other. The feed is fed into the lower cup and then the upper cup descends for extraction. Concurrently, a perforated stainless-steel tube cuts a plug in the fruit's bottom by forcing itself up through a channel in the lower cup. The juice is forced out through the tube's perforations as long as

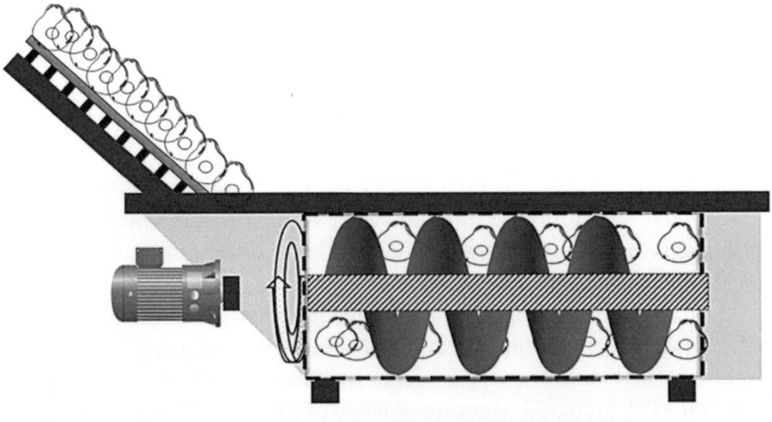

Fig. 9.3 Cylindrical fruit crushing machine. (Adapted from Mushtaq 2018)

pressure is applied; solids are removed at the end of the process cycle. When the fruit is compressed between the two cups, oil glands (present on the outer peel)get ruptured, releasing oil and juice gets expelled from the juice sacs. The expressed juice passes through a series of extractors, followed by the ‘finishing stage’ where juice passes through a series of screens for removing excess pulp from the juice. The recovered pulp from the above process is used for generating soluble solids with the help of counter-current extraction with water. The recovered and washed pulp is a rich source of carbohydrates and hence is utilized as ‘filler’ in livestock feed blends (Taylor 2016).

9.2.2.3.3 Novel Extraction Techniques

9.2.2.3.3.1 Gamma Radiation

In recent times, gamma irradiation has gained significant attention (Fig. 9.4).

When food comes into contact with high-energy photons from gamma irradiation, it energises electrons within the atoms. This excitation or ionization produces highly reactive free radicals, which contain unpaired electrons that may pair up with the outer shell electrons of the atoms that constitute cellular components.

When ionizing radiation is applied to food, it produces hydrogen and hydroxyl radicals from water molecules. Several reactions including hydrolysis of pectin followed by depolymerization of carbohydrate polymers producing sugars, take place in fruits and vegetables, which increases the sweetness and softness of the product. During processing, the yield of fruit juice rises because of the softening of the internal tissues (Kalaiselvan et al. 2018).

9.2.2.3.3.2 Enzymes

The plant cell wall is made of pectin substances, cellulose and hemicellulose, which act as its structural constituents. The cell wall is strengthened by cellulose and hemicellulose, and the cellulose network is cemented together by pectic substances. In fruit juice processing, the incorporation of commercial enzyme preparations is intended to break down the pectic materials found in the central lamella of the plant cells, minimizing their effect on final product’s quality.

Commonly used commercial enzymes in the fruit juice industry are pectinases, cellulases and tannases, also known as macerating enzymes, which are used in the extraction and clarification of fruit juice. Degradation of pectic compounds is catalysed by pectinase via depolymerization (hydrolases and lyases) and de-esterification (esterases). Another class of enzymes called cellulases possesses a significant capacity to process and convert cellulosic biomasses into beneficial products. Tannin acylhydrolases, or tannases, are used in the production of fruit juice and tea, among other industrial uses. Microorganisms such as *Aspergillus*, *Paecilomyces*, *Lactobacillus* and *Bacillus* are produced by tannases, which function across a broad temperature and pH range. Several factors affect enzyme-assisted extraction

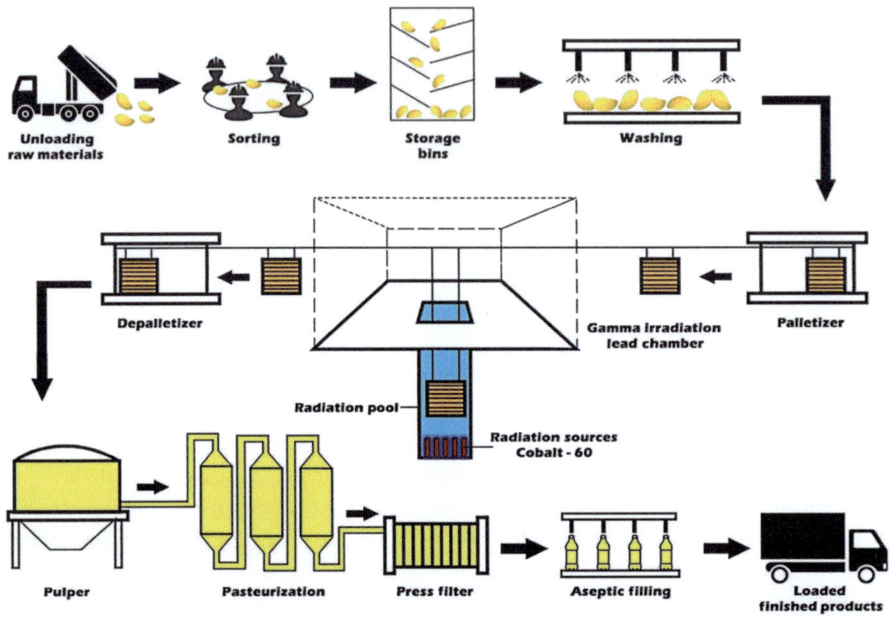

Fig. 9.4 Flow chart for the gamma-radiation aided extraction technique used in the processing of fruit juices. (Adapted from Kalaiselvan et al. 2018)

techniques, including enzyme types, combinations, incubation times, pH levels, and temperatures. It is evident that using a variety of pectinolytic enzymes together yields higher yield values than using just one type of enzyme. Sustainable production is positively impacted by the application of enzyme-assisted extraction. By optimizing the extraction from raw materials in the fruit juice industry, this method reduces waste flow, uses less energy, and depends less on chemicals to clean the equipment (Ramadan 2019; Danalache et al. 2018).

9.2.2.3.3.3 Ultrasound Extraction of Fruit Juices

For controlled cell disintegration, a novel nonthermal approach called ultrasonication (20 kHz to 1 MHz) is employed. Cavitation bubbles are generated as a result of acoustic cavitation spurred by ultrasonic waves propagating through the pulp. Combination of high and low pressure, i.e., rarefaction and compression, generates mechanical waves within the pulp. Ultrasound treatment is reported to show minimal effect on the functional, nutritional, and organoleptic characteristics of heat-sensitive foods (Vivek et al. 2018). According to a number of research findings, the extraction yield for pineapple mash juice production increased by 10.8% when ultrasonic treatment was used alone. Ultrasonic treatment increases the shelf life and microbial safety, it also helps in enhancing the content of total acids, sugars, vitamin C and phenolics (Başlar et al. 2016).

In the food industry, ultrasonic frequencies have been employed as an alternative to thermal processing. These waves can be classified as high-energy (frequency between 20 and 500 kHz and intensities greater than 1 W/cm^2) and low energy (frequency >100 kHz and intensity <1 W/cm^2), where the former type of waves are used in solid foods to produce acoustic and hydrodynamic cavities causing several physical and chemical changes in solid food; while the latter is used for fluidic foods, where they induce surface erosion, particle size reduction, and collapse between bubbles. The most commonly reported applications of ultrasonic frequencies in the juice industry are the following: the liberation or extraction of target analytes from the complex biological mixture; functional group alterations; microbial decay or inactivation; and the dissolution or clarification of coagulated moieties (Mushtaq et al. 2018).

9.2.2.3.4 Separation of Juice

Depending on the kind and quality of the fruit, several theories and techniques are used to separate juice from prepared or crushed fruit. A simple pressing technique is generally used for pulpy citrus fruits, whereas in the case of pineapple, apple, etc., disintegrators or blades are commonly used for the liberation of juices. The separation machines mainly works on the principle of exerting pressure on the pulp, centrifugal force and diffusion (Mushtaq 2018).

The conventional approach for separating juice from fruit has involved pressing the pulped or mashed fruit to expel the liquid through a screen or cloth. 'Separators' can operate in both batch and continuous processes (Taylor 2016). Pressers are part of the widely used category of juice separators. The design and operation of pressers also vary greatly, and the porosity and type of pulp, juice viscosity, and amount of pressure applied affect the effectiveness of the process. A few commonly used pressers are explained in the below section (Mushtaq 2018).

9.2.2.3.4.1 Rack and Cloth Press

This is a prevalent technique for extracting fruit juice from the pulp. It comprises rectangular frames or beds made of hardwood or stainless steel materials that are encrusted with nylon or loose cloth. Several layers of frames are stacked one above the other, and they are pressed to extract the fruit juices. It consists of fluted rods to facilitate the stream of juice (Fig. 9.5). This technique has certain disadvantages including labour-intensive and lack of automation (Mushtaq 2018).

9.2.2.3.4.2 Hydraulic Presses

They are most used for grapes, berries and other similar fruits. It consists of a loading unit (called as trituration) where fruit mash is supplied from a continuous milling unit, an adjustable piston for applying pressure to the fruit mash, approximately 6000 psi, leading the cell vacuoles of fruits to burst and thereby releasing natural components (enzymes, bioactive, and minerals). A fine wire screen constructed of

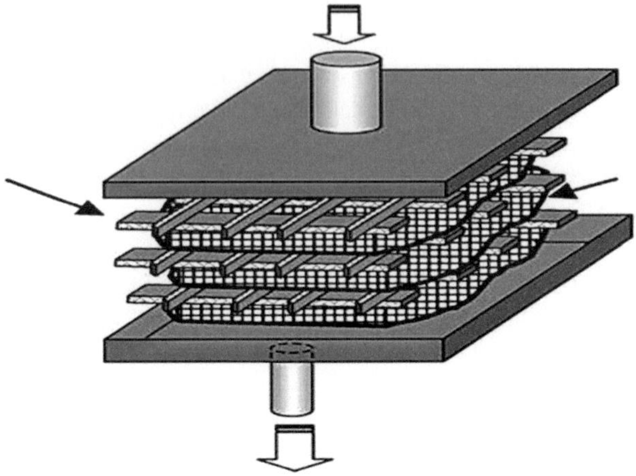

Fig. 9.5 Illustration of a pack press. (Adapted from Mushtaq 2018)

corrosion-resistant materials or muslin bags is subsequently used to filter the extracted juice. A Moyno-type moving cavity food pump is typically used for pulp pumping.

9.2.2.3.4.3 Screw Type Presses

They are one of the most widely used and conventional juice extraction devices. A screw press is employed for extracting oil from oilseeds (rape, flax, sunflower, etc.). Screw presses are currently becoming more prevalent, particularly for home use, on small farms, and among local business owners. They are made up of a cylindrical vessel containing a screw type press fixed above a perforated screening assembly. Several factors including the structure of the cell, assembly of the pressure plate and viscosity of pulp affect the efficiency of these extractors. Screw presses are classified as single- and double-screw type and mixing. The primary benefit of using a screw press is that it produces juice with large soluble solids and bioactive substances (Wilczyński et al. 2019; Mushtaq 2018).

The working mechanism of screw juicer consists of a screw rod rotating continuously and constantly chopping the materials which will be guided by the screw to the end of the extrusion cylinder. The material is continuously drawn together and compressed between the screw and the cylinder wall, creating the squeezing effect, raising the juice pressure in fruit and vegetable cells, breaking down the cell wall, and separating the juice from the fruit residue. There are three sections to the screw.

The first stage, called feed screw 1, is mainly used for material transportation, preliminary cutting, and extrusion to increase the juice production. The second stage, called the press screw, is to increase the pressing force applied to the materials, extrude them entirely, and push water out of the material to increase the yield of juice. In order to assist the separation of juice and fruit residue in the following process, the third section of the constant pressure screw is largely employed to maintain a constant pressure environment and to uphold the residual water rate of the treated material consistently (Fig. 9.6) (Wu et al. 2021).

9.2.2.4 Clarification and Stabilization

Fruit extracts typically have a turbid consistency because of the presence of colloid macromolecules (pectin, proteins, soluble-starch fractions, specific polyphenols, and their condensed or oxidized derivatives) and insoluble plant particles (fibres, cellulose, hemicellulose, protopectin, starch, and lipids). These materials need to be removed partially or completely, depending on the final product, to prevent turbidity and precipitation and to enhance sensory qualities (taste, flavour, and colour). Mechanical processes, physical–chemical methods, and their combined methods are common clarification approaches.

9.2.2.4.1 Physical–Chemical Clarification

This approach is used to completely eliminate the substances causing turbidity. Clarifying agents are used to form precipitates from macromolecules and insoluble plant particles, which are later separated by mechanical means. The first step involves removal of protective colloids (pectin, starch, and hemicellulose) as their gel-forming affinity inhibits the concentrations, and they hinder the settling of floating substances, along with subsequent turbidity in case of concentrates during storage.

Two different clarifying agents that are utilized in the fruit juice industry are solid silicic acid and bentonite. Bentonite is a member of the montmorillonite group and is derived from volcanic activity. It has a large surface area and good thickening

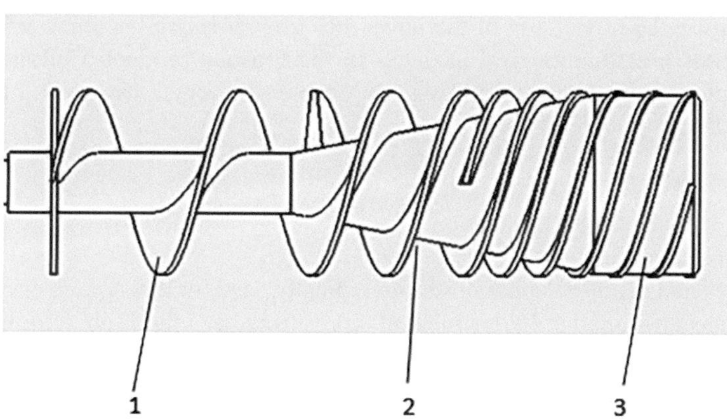

Fig. 9.6 Overall structure of Screw juicer (1. Feed Screw, 2. Press screw, 3. Constant press screw). (Adapted Wu et al. 2021)

capabilities; positively charged proteins are strongly adsorbed to its negatively charged particles. The stabilizing and clarifying qualities of bentonite are influenced by its thickening characteristics. The aqueous colloid solution of polymerized silica anhydride with negatively charged particles is known as silica sol and is used for clarifying.

Usually, it is combined with other clarifying agents or enzyme treatments. Aside from the brief clarification interval, it offers good clarification efficacy. Gelatin is a clarifying agent derived from proteins that precipitates negatively charged particles, such as degraded pectin and polyphenols. Tannin often finishes this process by interacting with the molecules of proteins. Juice flocculation is triggered by an electric discharge. As a result, turbidity-causing coarse particles precipitate and form flakes. On their surface, these precipitates absorb additional particles, resulting in cloudiness. The resulting precipitation settles and can be removed mechanically. Bentonite and silica sol are added to the juice after the enzymatic activity is finished, protecting the enzyme proteins in the event of enzymatic pectin breakdown. Active carbon is another clarifying agent used in the fruit industry, which is added in a lower concentration (100–200 mg per 1000 L of juice) (Taylor 2016).

Enzymes are used during this procedure to help break down the pectin molecules in the juice (Horvath-Kerkai 2012). Generally when using enzymes for soft fruits, the required dosage is added (e.g., 0.1%) to the well-mixed, pre-warmed mash and allowed to stand at a consistent temperature for the recommended amount of time (Taylor 2016).

Pectinase or poly-(1,4-alpha-D-galacturonide) glycanohydrolase is produced from fungal sources (i.e. Aspergillus sp., Rhizopus sp.), possessing optimum working conditions of pH between 2.5 and 6.0, temperatures between 30 and 60 °C. Other enzymes include esterase (polymethylgalacturonase esterase) used to de-esterify pectin, with the removal of methoxy groups, to form pectic acid. Amylases are employed to break down any residual starch in juice that results from using unripe fruit that has been picked before it reaches maturity and then matured in a cool store under regulated atmospheric conditions. This causes the unripe fruit to retain its starch. During the processing of the juice, this leftover starch becomes gelatinized, which causes precipitation and haziness in the finished product; Cellulases have been employed in fruit processing to enable the quick removal of colour (Taylor 2016).

9.2.2.4.2 Mechanical Clarification

Centrifuges and filtration apparatuses are typically used for this type of operation to remove suspended matter. In this method, fibres from hazy fluids are removed using decanters while the first filtration stage is carried out in settling centrifuges (Taylor 2016).

9.2.2.4.2.1 Decanter-Type Centrifugation Machine

A decanter-type centrifuge is commonly used in fruit juice processing industries as it effectively separates suspended solid components from the liquid/slurry fraction. It works on the principle of buoyancy effects, which states that when the mixture of juice containing suspended solids is rotated, denser components sink to the bottom of the liquid mixture first, followed by the lighter components. A typical decanter is made up of a conical solid wall bowl enclosing a cylindrical centrifuge with a reducing diameter covered with a permeable screen. This type of centrifuge has certain drawbacks including high-energy consumption, high equipment and maintenance expenses, along with the inability to remove biological solids with minute density differences (cells and viruses). New substitutes for the existing disc centrifuge have been generated, for example, self-cleaning disc stack centrifuges, which operate on the same buoyancy principle with added efficiency, automation and environmentally friendly nature advantages. The surface area of the vessel increases with the incorporation of large number of horizontal baffle plates (separation channels), which results in an increased surface area and hence facilitates a higher sedimentation rate (Mushtaq 2018).

9.2.2.4.2.2 Membrane-Based Filtration Technologies

Fruit juice clarity uses two different kinds of membrane materials: polymeric and ceramic. Because they are inexpensive and simple to make, polymeric membranes like polyvinylidene difluoride (PVDF) and polysulfone (PSF) are frequently employed in microfiltration (MF) and ultrafiltration (UF) procedures. Ceramic membranes are more durable and impervious to chemical and fouling attacks, but their greater expense restricts their use to specialized uses. Numerous investigations looked into a hybrid kind of membrane called a composite membrane, which is a necessary membrane made of multiple layers made of various materials (Sarbatly et al. 2023).

9.2.2.5 Drying/Evaporation and Concentration

To improve shelf life and transportation and storage qualities, concentration aims to raise the dry matter content of juices while lowering their water content. Minimal vital ingredient loss and minimal damage to sensory traits are key features to be taken care of during the concentration process. The techniques of freeze concentration, evaporation, and reverse osmosis are used in juice concentration.

9.2.2.5.1 Evaporation

Evaporation is the most frequently used concentration method which means evaporation of water by means of boiling, the energy required for boiling is provided by the steam; evaporators are heat exchangers where steam condenses on one end and the solution boils to remove some of the water content; vapours are then extracted from the vapour space followed by condensation. Generally, for retention of heat-sensitive components, low temperature and short time or the process is performed in a vacuum to lower the boiling point; multi-effect evaporators are used to minimize the energy costs. Tube, film, centrifugal and plate-based ones are the most widely used devices in fruit juice industry. Aroma recovery units are typically paired with evaporator systems; thus, condensed aromas are often remixed into the concentrate to retain their original flavour (Horvath-Kerkai 2012).

9.2.2.5.2 Concentration by Freezing

In comparison to high-temperature techniques, it is a gentle method for the concentration of essential, heat-sensitive fruit juices with minimal impact on their sensory and nutritional qualities. In this approach, the water content of the juice is converted into solid ice crystals, resulting in loss of solvent, which leads to concentration of the remaining solution. This two-phase system can then be easily separated by mechanical means. Higher energy usage and reduced concentration effectiveness are certain drawbacks compared with the high-temperature approaches (Horvath-Kerkai 2012).

9.2.2.5.3 Reverse Osmosis

It is a membrane concentration process used as an alternative technique to thermal concentration for concentrating fruit juices at room temperature, hence minimizing the heat damage; along with this, there is no physical transformation in the structure of water. The semipermeable membrane permits only water to pass through while preventing all substances dissolved in water. Therefore, it “filters out” some of the water from the solution. It is a cost and energy-efficient process with maximum retention in the nutritional and sensorial properties (Couto et al. 2011; Horvath-Kerkai 2012).

9.2.2.6 Preservation

Several resources like air, dust, handling, microorganisms, insects, pests, and animals such as birds, field and processing practices and conditions adopted throughout determine the quality of fruit and fruit juice.

9.2.2.6.1 Pasteurization

Pasteurization is a mild thermal treatment employed for inactivating heat resistant microorganisms (yeast and moulds, bacteria, etc.) and enzymes (polyphenol oxidase (PPO), lipoxygenase (LOX), etc.) responsible for food spoilage/ poisoning. It helps in extending the shelf life of the treated commodity up to several months while maintaining its safety and quality parameters. Heat processing, however, also causes flavour loss and other modifications affecting the fresh juice's inherent qualitative attributes. Different types of pasteurization can be employed for enhancing the shelf life of the commodity. Juice processing generally involves the use of HTLT (High Temperature Long Time) pasteurization, which involves temperatures between 80 °C and 100 °C for longer than 30 seconds. It is generally employed for low-acid juices with pH >4.5, which necessitates the need for stronger treatments to protect the food quality. This treatment increases the shelf life by destroying spoilage microorganisms and reduces the activity of some enzymes, but due to its severity, several health-benefiting bioactive compounds are affected negatively.

In order to minimize the intensity of heat treatment and maintain high product quality, high temperature short time (HTST) pasteurization was developed. It reduces the activity of deteriorative enzymes (PME, PPO, etc.). Food is subjected to a higher temperature above 80 °C for less than 30 seconds. There have been reports of increased viscosity, total phenolic, nutritional value, and colour hue of various juices or nectars under HTST treatment.

Mild temperature long time pasteurization, often known as MTLT, is the process of pasteurizing food at a temperature lower than 80 °C for a longer period than 30 seconds. It enhances the shelf life by exhibiting minimal quality deterioration.

Mild temperature short time (MTST) pasteurization uses temperatures below 80 °C for 30 seconds, making it a lighter process compared to other pasteurization methods. Nonetheless, MTST treatment can alter the sensory, physicochemical, and functional properties of juices (Ağçam et al. 2018; Rabie et al. 2014).

PEF is an emerging technology in the field of food preservation. With an effect comparable to traditional thermal pasteurization, PEF has the potential to nonthermally pasteurize several foods via exposure to high-voltage short pulses. In this, food items positioned between the two electrodes receive a high voltage (typically 50 kV/cm) short (μs to ms) pulse. The basic mechanism of inactivation of microorganisms is the modification of the cell wall because of the transmembrane potential—the potential difference between the two sides of the membrane caused by a high electric field.

Irreversible pore formation in the cell membrane (electroporation) takes place when this transmembrane potential reaches a particular critical value, causing cellular integrity loss and increased permeability and eventually destroying the affected cells (Koubaa et al. 2018; Góngora-Nieto et al. 2002).

High hydrostatic pressure (HHP) process is another nonthermal approach for food preservation techniques based on the isostatic and the Le Chatelier principle. It is used to preserve solid, fluid, and particulate food, as well. The isostatic principle assumes that the pressure is uniformly applied in all directions, leading to the homogeneous and instantaneous distribution of pressure, regardless of the size, shape, or composition of food. This principle ensures that high pressure processing (HPP) does not lead to any pressure gradient and thus preserves the original shape of the nonporous. Le Chatelier principle emphasizes that pressurization favours volume reduction, which occurs during several phenomenon including molecular configuration, phase transition, and chemical or biochemical reaction. Pre-packaged food is placed inside the chamber filled with pressurization fluid, where pressure is applied to the food via a piston directly or indirectly by increasing the pressure inside the chamber using pressurized fluid. A pressure of 600 MPa is reached after about two minutes of pressurization, at a rate of about 57 MPa/s. Equipment is depressurized after the residence time, and food is sent for dispatch. Due to the adiabatic heating caused by the pressurization, the temperature of the product rises. The temperature within the chamber tends to equilibrate with the chamber jacket, which is adjusted to the desired process temperature during the process. Additionally, the depressurization that occurs after the process causes an adiabatic cooling. By inhibiting the production of proteins and enzymes, altering the shape of cells, and changing the microorganisms' reproductive and survival strategies, this procedure serves to inactivate germs and prolong the shelf life of the product (Augusto et al. 2018).

9.2.3 *Other Fruit Beverages*

Fellows and Hampton classified fruit beverages as follows:

1. Juices, containing fruit juice without additives.
2. Nectars, containing 25% to 50% of fruit solid content, which must be consumed immediately after opening.
3. Syrups, are clear, concentrated pasteurized juices with high-sugar content.
4. Squash, containing 25% of fruit pulp with added sugar syrup and preservative, it requires dilution prior to serving.
5. Cordials generally referred to the squash with crystal clear appearance (Mani-López et al. 2019).

9.2.3.1 Nectars

It is defined as 20% fruit juice, 15% sugar, and 0.3% acidity by the Food Safety and Standards Authority of India (FSSAI) or Fruit Products Order (FPO). A laboratory pulper machine is a common type of extractor used to extract pulp from sound, healthy, mature, and ripe, firm fruits. The extracted pulp will be strained through a stainless-steel sieve or muslin cloth to obtain clear juice. The prepared nectar will be pasteurized for 15 minutes at 85 °C, along with class II type preservative (potassium metabisulphite, 350–600 ppm) to enhance the shelf life. The finished product will then be poured into a plain bottle or pouch that has been previously rinsed and sterilized. The filled bottle will be promptly sealed with a crown cork, will be boiled for 15 minutes, cooled and stored at room temperature (Ahmad et al. 2016; Srivastava and Kumar 2002).

9.2.3.2 Squash

It is a homogeneous sweetened beverage prepared from a single variety of fruit juice or mixture of fruit juices, either of single strength and/or concentrated, and preserved or otherwise mixed with sugar, potable water and various other ingredients in such proportions that will meet the requirements given in this standard. As per the specifications of Fruit Product Order, squash contains at least 25% fruit juice or pulp, 40–50% total soluble solids on a weight basis, about 1.0% acid and 350-ppm sulphur dioxide or 600 ppm sodium benzoate as preservative.

Squash is prepared from fruit juice or pulp (extracted using a suitable extractor). Ingredients such as sugar, citric acid, flavourings, colouring, and preservatives will be added in the required amount. The mixture will be poured into bottles (previously cleaned and sterilized), leaving a headspace of 1.2–2.5 cm. The bottles will be pasteurized for 10–15 minutes at 75–80 °C, chilled to 10 °C. Without significantly changing in colour, flavour, or aroma, the product lasts for more than a year (Saleem et al. 2011).

9.2.3.3 Cordials

This is fully pulp- and other suspended-material-free sparkling, transparent sweetened fruit juice, containing at least 25% juice, 30% total soluble solids, 1.5% acid and 350 parts per million of sulphur dioxide. Barrels lined with microcrystalline wax are used for storing juice. Potassium meta-bisulfite is used during the storage period for preservation. Clear juice remains at the top of the sediment-filled compacted layer that accumulates at the bottom during storage. Two to three months are needed for the clarification process, after which the clear juice is siphoned off. Gelatin and tannin can be added to speed up the slow clarification process. Sugar, water, colouring, and preservatives are added to the clear juice, which is then filtered using a filter press and then bottled (Srivastava and Kumar 2002).

9.2.4 Fruit Juice Concentrate

Fruit juice concentrate is a condensed form of fruit juice where water content has been removed, typically through processes like evaporation or freezing (Adnan et al. 2018). This process yields a thicker liquid, making the juice more compact and easier to transport, store, and reconstitute by adding water. Retaining most of the flavour and nutritional characteristics of the original juice, fruit concentrate is a key ingredient in various food and beverage products such as juices, soft drinks, jams, jellies, snacks, and desserts. Figure 9.7 illustrates the general processing flowchart for fruit juice production. Juice concentration is essential for several reasons: it reduces water activity, thereby extending shelf life; it minimizes packaging, storage, and transport costs; and it stabilizes and simplifies the handling of the final product (Buerman et al. 2020). Despite advances in separation science and automation, selecting the appropriate concentration method remains crucial. Improper methods can adversely affect sensory and nutritional profiles and increase production costs. Various concentration methods are practised, including thermal evaporation, freeze-drying, membrane filtration, and osmosis. Recent advancements in material science, engineering, and process control have improved efficiency and product consistency. Fruit juice concentrates must comply with specific regulations to ensure quality and safety. They are available in liquid, syrup, or frozen forms for reconstitution (Manjunatha and Raju 2015).

Quality assurance is critical, with concentrates retaining similar colour, flavour, and nutritional value to the original juice. Concentration processes are often carried out under low temperatures to preserve these characteristics (Salehi 2020).

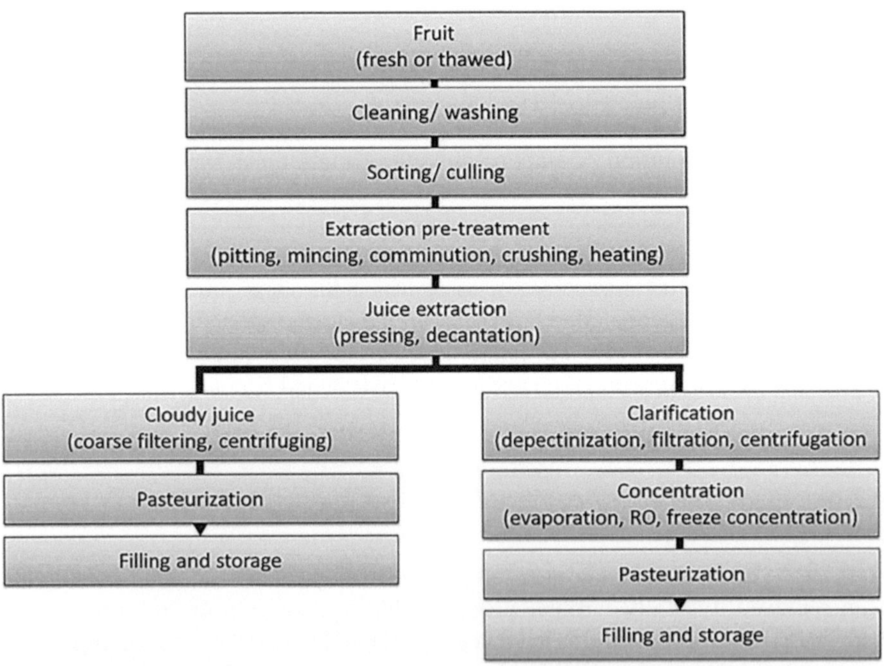

Fig. 9.7 General fruit juice processing flowchart. (Adapted from García 2018)

9.2.5 *Types of Concentrates*

9.2.5.1 Single-Strength Juice

Single-strength juice is the undiluted, unprocessed juice extracted directly from fruits. It is not concentrated and retains the original water content of the fruit. Single-strength juice is typically used for immediate consumption or as a base for making other juice products (Ramaswamy and Marcotte 2005).

9.2.5.2 Clarified Juice

Clarified juice is obtained by removing insoluble solids, pulp, and particulates from single-strength juice. This process involves filtration or centrifugation to produce a clear, transparent juice with reduced turbidity. Clarified juice is often used as a base for making transparent beverages or fruit-flavoured drinks (Manjunatha et al. 2012; De Oliveira et al. 2012).

9.2.5.3 Concentrated Juice

Concentrated juice is produced by removing a significant portion of the water content from single-strength juice. This is typically done through processes like evaporation, freeze concentration, or membrane filtration. Concentrated juice has a higher soluble solids content, making it more compact and easier to store and transport. It is commonly used in the food and beverage industry for manufacturing juices, nectars, and other fruit-based products (Trishitman et al. 2023).

9.2.5.4 Frozen Concentrate

Frozen concentrate is a type of concentrated juice that is further processed by freezing. After concentrating the juice, it is frozen to form a solid block or frozen cubes. Frozen concentrate is convenient for storage and transportation, and it can be reconstituted by adding water before consumption. Frozen concentrate is often used in households for making juice drinks and cocktails (Vásquez-Castillo et al. 2023).

9.2.5.5 Dehydrated Concentrate

Dehydrated concentrate, also known as powdered concentrate, is produced by further removing water from concentrated juice until it becomes a dry powder. This is typically done through spray drying or drum drying processes. Dehydrated concentrate has an even longer shelf life and is highly concentrated, requiring only minimal storage space. It can be reconstituted with water to produce juice or used as an ingredient in various food products (Kowalska et al. 2023).

9.2.5.6 Puree Concentrate

Puree concentrate is made by concentrating fruit puree, which includes both the juice and pulp of the fruit. The puree is concentrated to reduce its water content, resulting in a thick, viscous concentrate with high fruit solids content. Puree concentrate is commonly used in the production of fruit-based products like jams, jellies, and fruit snacks (Zhumaliyeva et al. 2023; Kiyak and Gunes 2024). Each type of concentrate offers unique advantages and is used in different applications based on factors such as processing requirements, shelf life, and end-product characteristics.

9.2.6 Method of Extraction of Concentrates

9.2.6.1 Cryoconcentration Techniques

Freeze concentration is a technology that concentrates aqueous food solutions by partially freezing them and separating the ice fraction from the concentrated residual solution (Aider and Ounis 2012). Unlike evaporation or membrane filtration, freeze concentration preserves the quality of the final product by minimizing the loss of volatile compounds and protecting heat-labile components (Morison and Hartel 2018). Recent innovations have focused on one-step systems like block freeze concentration or progressive freeze concentration, which offer simpler separation steps and equipment operation (Miyawaki and Inakuma 2021). Figure 9.8 illustrates the freeze concentration system employed for concentrated juice production, while Fig. 9.9 presents the typical process layout of the freeze concentration technique.

9.2.6.1.1 Suspension Crystallization

The cryoconcentration technology based on suspension crystallization involves the limited size of ice crystals, requiring a complex system composed of a surface-scraper heat exchanger for ice generation, a re-crystallization vessel for ice growth, and a washing tower for ice crystal separation (Miyawaki et al. 2005). This complexity renders the cryoconcentration process the most expensive among all liquid food concentration methods, limiting its practical application in the food industry. In the conventional suspension crystallization method, separation of ice crystals from the concentrated solution is crucial, often achieved through the Ostwald ripening mechanism (Samsuri et al. 2016). This method has been used to concentrate

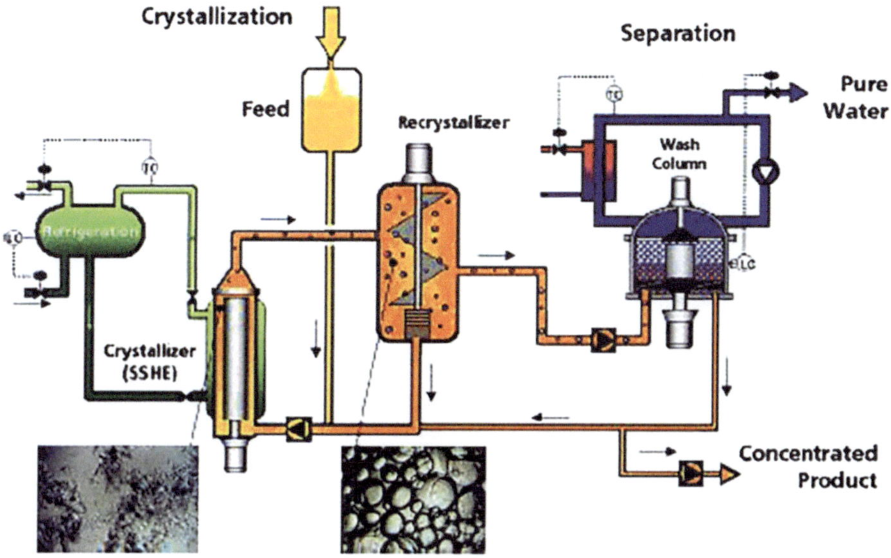

Fig. 9.8 Freeze concentration system for concentrated juice production. (Adapted from http://www.niro-pt.nl/)

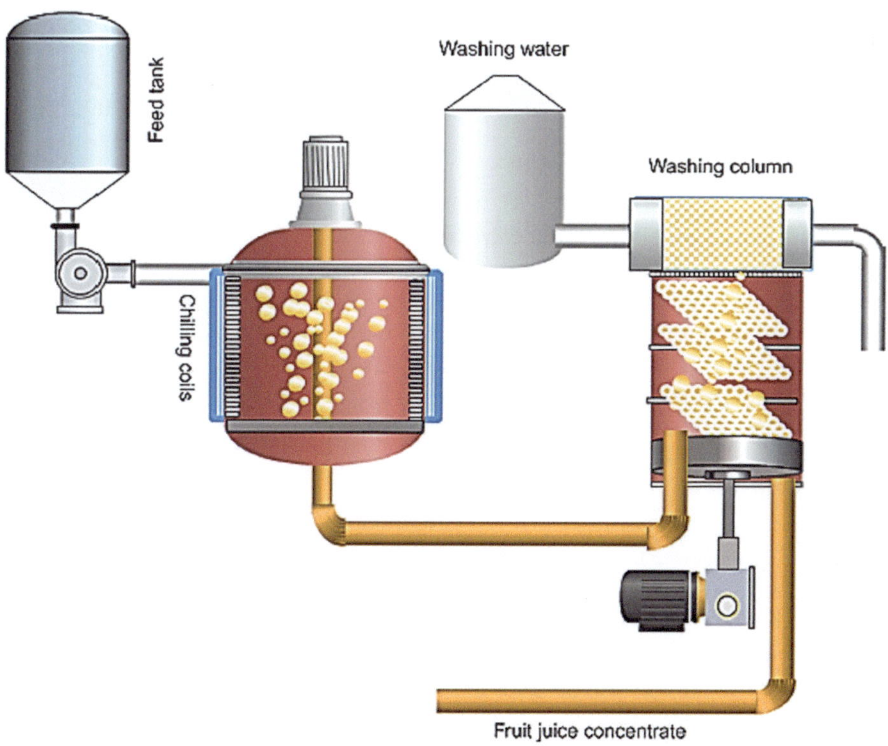

Fig. 9.9 A typical process layout of freeze concentration. (Adapted from Adnan et al. (2018))

coffee extract, fruit juice, milk, and wastewater treatment, including hazardous waste (Deshpande et al. 1984).

9.2.6.1.2 Progressive Cryoconcentration

Progressive cryoconcentration technology employs a different approach to crystallization compared to suspension crystallization, where a large single ice crystal forms and grows on a cooling surface in a crystallization vessel, simplifying the separation from the mother solution (Miyawaki et al. 2005). This method, a form of melt crystallization, has been historically used for the purification of organic materials (Samsuri et al. 2016), showing effectiveness in concentrating impurities from organic solvents. Although this method offers simplicity and reduced costs, its concentration and separation efficiency mechanisms require further investigation.

9.2.6.1.3 Eutectic Cryoconcentration

Eutectic cryoconcentration technology involves the simultaneous crystallization of solutes and ice, separating inorganic salts from aqueous solutions by cooling below the eutectic temperature. This technique operates near the eutectic point, where both ice and salt crystallize simultaneously. The separation of salt and ice crystals relies on their density differences and gravitational forces. Despite past investigations, renewed interest has focused on its low energy consumption and high purity of products, aiming for industrial applications (Aider and De Halleux 2009; Mazli et al. 2020).

9.2.6.1.4 Partial Block Cryoconcentration

Partial block cryoconcentration technology introduces a cooling agent into a crystallizer chamber, partially freezing the liquid solution from the centre and increasing its concentration. This approach reduces energy expenses and dry substance losses during separation, achieving efficient concentration of milk whey, fruit juices, coffee extract, and aromatic compounds from grape seeds (Moreno et al. 2013; Aider and De Halleux 2009).

9.2.6.1.5 Complete Block Cryoconcentration

Complete block cryoconcentration technology involves freezing the entire liquid solution and then thawing it, separating the concentrated fraction from the ice fraction. This method achieves high process efficiency by controlling the thawing temperature, allowing for minimal solute entrainment in ice crystals (Aider and De Halleux 2009).

9.2.6.2 Evaporation Technology

Drying, a traditional preservation method, reduces the water content and water activity (a_w) of fruits, vegetables, and their products. Various drying methods are employed, including natural or forced convective drying, sun drying, microwave drying, and notably, spray drying. Spray drying, a widely used technique, transforms liquid feedstocks into powders or concentrated products through evaporation.

9.2.6.2.1 Multi-effect Evaporators

Multi-effect evaporators utilize vapours from previous stages to evaporate water from juice at lower temperatures; maximizing thermal energy efficiency (Fig. 9.10). The process involves sequential boiling stages, each at a lower pressure or with increased surface area. By using vapours from one stage to evaporate the next, only the first stage requires external heating (Ruan et al. 2015). Multi-effect evaporators are designed to minimize heat exposure to preserve delicate nutrients in fruit juices (Sorour 2015).

9.2.6.2.2 Thermal Vapour Recompression

Thermal vapour recompression evaporators recycle vapours from the first effect for heating, reducing energy consumption. While mostly used for water desalination, they can enhance the efficiency of fruit juice concentration plants when integrated with multiple-effect evaporators (Sorour et al. 2013).

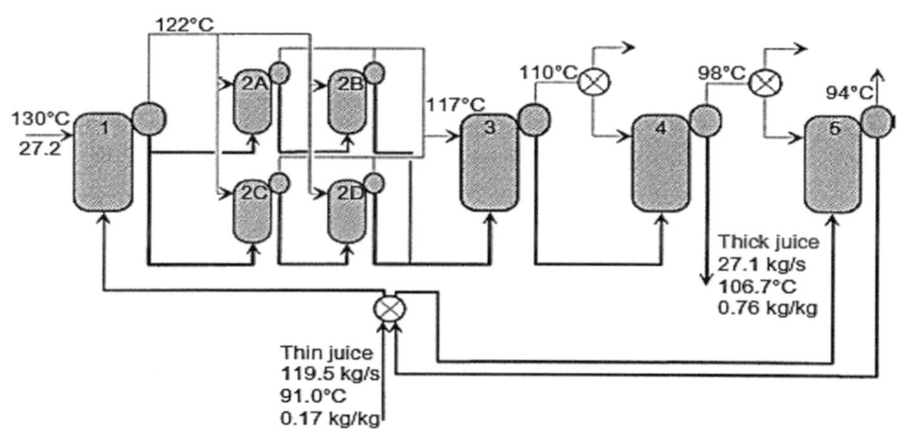

Fig. 9.10 Layout of fruit juice concentration plants based on multi-effect evaporation. (Adapted from Adnan et al. 2018)

9.2.6.2.3 Unstirred Open Pan or Kettle

Simple and cost-effective, these evaporators consist of heated pans for the evaporation of juices, sauces, jams, and confections. They are suitable for thermally stable products but unsuitable for heat-sensitive ones (Adnan et al. 2018).

9.2.6.2.4 Vacuum Pan Evaporators

Vacuum pan evaporators concentrate juices by boiling under vacuum, connected to high-power vacuum pumps. They can concentrate up to 200 kg of fruit juice per hour and operate at lower temperatures than atmospheric conditions (Adnan et al. 2018; Belbase 2022).

9.2.6.2.5 Rising/Falling Film Evaporators

These evaporators pass thin films of juice over rising or falling plates, allowing quicker evaporation at lower temperatures. They are used for heat-sensitive juices and offer modifications for specific products, such as plate-type and scraped-surface evaporators (Saravacos et al. 1970; Berk and Gökmen 2024).

9.2.6.2.6 Thermally Accelerated Short-Time Evaporator (TASTE)

TASTE evaporators operate on high-temperature, short-time exposure principles, using steam efficiently. They are suitable for low-pulp or clear fruit juices and distribute the product uniformly for treatment (Lin et al. 2002).

9.2.6.2.7 Spray Drying

Spray drying is a rapid, single-stage drying method used to produce fruit powders or concentrated juice products. Atomization creates spherical droplets for drying, but it is unsuitable for thermosensitive nutrients and high-sugar juices without a carrier agent. Modifications, such as using drying agents like maltodextrin, improve efficiency and reduce residue formation during drying (Goula and Adamopoulos 2010; Shishir and Chen 2017). Figure 9.11 presents the operational schematic of spray drying evaporators.

9.2.6.3 Membrane Technology

Membrane technology plays a crucial role in the concentration and purification of fruit juices, offering various methods tailored to specific needs. It is a promising approach for juice concentration and clarification, offering advantages like high selectivity, moderate operating temperatures, no need for chemical additives, easy

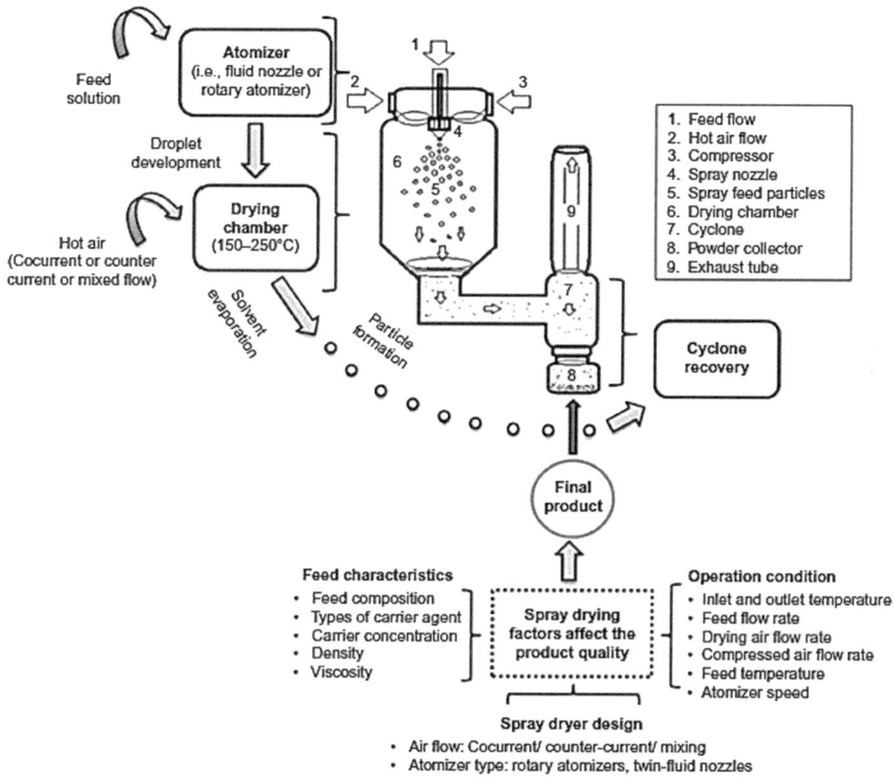

Fig. 9.11 Working layout of spray drying evaporators. (Adapted from Iwuozor et al. 2023)

scale-up, and low energy consumption (Sarbatly et al. 2023). Pressure-driven membrane operations such as microfiltration (MF), ultrafiltration (UF), nanofiltration (NF), and reverse osmosis (RO) are state-of-the-art technologies for juice clarification, fractionation, and concentration (Echavarría et al. 2011).

9.2.6.3.1 Osmotic Distillation, Membrane Distillation, and Pervaporation

Recent years have seen the application of osmotic distillation (OD), membrane distillation (MD), and pervaporation (PV) for juice concentration and recovery of aroma compounds. These processes align with sustainable food processing strategies due to their low energy consumption and mild operating conditions, preserving product quality (Conidi et al. 2020).

9.2.6.3.2 Membrane Distillation (MD)

MD, a relatively new process, utilizes hydrophobic membranes to separate aqueous solutions at different temperatures, achieving water flux at atmospheric pressure and low temperatures. MD is suitable for concentrating juices sensitive to high temperatures and osmotic pressure (Onsekizoglu Bagci 2015).

9.2.6.3.3 Direct Osmosis Concentration (DOC)

DOC establishes an osmotic pressure gradient between fruit juice and an osmotic agent across a semipermeable membrane, allowing water to flow from the juice to the agent, thereby concentrating the juice without altering its flavour, colour, or odour (Petrotos et al. 1999).

9.2.6.3.4 Reverse Osmosis (RO)

RO concentrates juices by applying pressure to move water through a semipermeable membrane from low to high solute concentration areas. It retains juice flavour and nutrients and can increase total solid concentration without compromising juice quality (Gurak et al. 2010).

9.2.6.3.5 Osmotic Evaporation

Osmotic evaporation, or osmotic distillation, employs a hydrophobic microporous membrane to separate two liquid phases differing in solute concentration. This process maintains the nutritional and sensory qualities of juices while concentrating them (Hongvaleerat et al. 2008).

9.2.6.3.6 Hydrate Separation Technology

Hydrate separation technology, initially developed for CO_2 conversion into hydrates, is used to concentrate fruit juices. This method forms solid clathrate hydrates that are mechanically separated. However, it may lead to declines in sensory and nutritional attributes of the final juice concentrate (Li et al. 2017).

9.2.7 *Equipment's Used in the Preparation of Fruit Juice Concentrate*

The preparation of fruit juice concentrate involves the utilization of various equipment to carry out different stages of the concentration process, ensuring efficiency and quality of the final product.

9.2.7.1 Evaporators

Evaporators are the core equipment used for juice concentration, operating on the principle of heating the juice to remove water content. Common types of evaporators include falling film, rising film, and forced circulation evaporators (Ramaswamy and Marcotte 2005). These evaporators facilitate the removal of water from the juice, resulting in a thick, concentrated liquid.

9.2.7.2 Membrane Systems

Membrane systems, such as microfiltration, ultrafiltration, nanofiltration, and reverse osmosis, are used for clarification, fractionation, and concentration purposes (Bhattacharjee et al. 2017). These systems utilize semipermeable membranes to separate components based on their molecular size, allowing for precise control over the concentration process.

9.2.7.3 Vacuum Systems

Vacuum systems are often employed to lower the boiling point of the juice, reducing the risk of heat damage, and preserving the quality of the final product. Vacuum evaporators, combined with other concentration methods like evaporation and osmotic distillation, enhance the efficiency of water removal while minimizing heat exposure (Orellana-Palma et al. 2018).

9.2.7.4 Centrifuges

Centrifuges are utilized for separation and clarification purposes, particularly in removing suspended solids and pulp from the juice. These machines operate by spinning the juice at high speeds, causing denser particles to settle at the bottom, allowing for the extraction of clarified juice (Beveridge et al. 1988).

9.2.7.5 Heat Exchangers

Heat exchangers are used to transfer heat between different streams of juice and heating or cooling media. Plate heat exchangers and tubular heat exchangers are commonly used in juice processing plants to optimize energy efficiency during heating and cooling stages (Córcoles et al. 2020).

9.2.7.6 Pumps

Pumps are essential for transferring juice between different stages of the concentration process, maintaining a consistent flow rate, and ensuring proper mixing (Ramaswamy and Marcotte 2005). Positive displacement pumps, centrifugal pumps, and peristaltic pumps are commonly used depending on the specific requirements of the process.

9.2.7.7 Spray Dryers

Spray dryers are used to convert concentrated juice into a powdered form, facilitating easier storage and transportation (Ramaswamy and Marcotte 2005). The concentrated juice is atomized into fine droplets and dried using hot air, resulting in a powdered product.

9.2.7.8 Homogenizers

Homogenizers are used to break down fat globules and ensure uniform distribution of particles in the juice (Ramaswamy and Marcotte 2005). This equipment enhances the stability and consistency of the final product, particularly in juices with high pulp content.

These equipment's, along with proper control systems and monitoring devices, play crucial roles in the production of high-quality fruit juice concentrates, meeting both industry standards and consumer expectations.

9.2.8 Preservation Technologies for Juice Concentrate

Preserving the quality and shelf life of juice concentrate is essential to maintain its freshness, flavour, and nutritional value. Various preservation technologies are employed to achieve this, ensuring that the concentrate remains safe and stable for consumption.

9.2.8.1 Aseptic Processing

Aseptic processing involves sterilizing the juice concentrate and packaging it under sterile conditions to prevent microbial contamination. The juice is heated to a high temperature for a short duration to kill pathogens and spoilage microorganisms, and then rapidly cooled before packaging in sterile containers. Aseptic processing preserves the quality of the concentrate while extending its shelf life without the need for refrigeration (Graumlich et al. 1986; Nanjegowda et al. 2022).

9.2.8.2 Thermal Processing

Thermal processing, including pasteurization and sterilization, is commonly used to eliminate microorganisms and enzymes that can cause spoilage (Heldman et al. 2018). Pasteurization involves heating the juice at a moderate temperature for a specific time to reduce microbial load, while sterilization involves heating the juice to higher temperatures to achieve complete microbial destruction. Thermal processing ensures microbial safety while maintaining the sensory and nutritional qualities of the concentrate (Farnworth et al. 2001).

9.2.8.3 High Pressure Processing (HPP)

HPP involves subjecting the juice concentrate to high levels of hydrostatic pressure to inactivate microorganisms and enzymes. This nonthermal preservation method effectively extends shelf life while preserving the fresh taste and nutritional content of the concentrate. HPP also helps in retaining the natural colour and aroma of the juice (Buerman et al.2021).

9.2.8.4 Chemical Preservation

Chemical preservatives, such as antioxidants, antimicrobials, and chelating agents, are used to inhibit microbial growth and prevent oxidation. Common additives include ascorbic acid (vitamin C), citric acid, and potassium sorbate. These preservatives maintain the quality and stability of the concentrate, particularly during storage and transportation (Rani et al. 2021; Zhang et al. 2021).

9.2.8.5 Vacuum Packaging

Vacuum packaging involves removing air from the packaging container to create a vacuum, which helps in preventing oxidation and microbial growth. Vacuum-sealed packages provide an oxygen-free environment, protecting the juice concentrate from spoilage and maintaining its freshness for an extended period (Adnan et al. 2018; Moradinezhad et al. 2019).

9.2.8.6 Controlled Atmosphere Storage (CAS)

CAS involves modifying the composition of the atmosphere surrounding the packaged juice concentrate to slow down enzymatic and microbial activity. By adjusting oxygen and carbon dioxide levels, CAS extends the shelf life of the concentrate while preserving its colour, flavour, and nutritional value (Shaw et al. 1990).

9.2.8.7 Refrigeration and Freezing

Refrigeration and freezing are traditional methods of preservation that slow down microbial growth and enzymatic activity. Cold storage at temperatures above freezing point and freezing at temperatures below freezing point inhibit spoilage and maintain the quality of the concentrate, respectively (Orellana-Palma et al. 2020; Rehman et al. 2014).

These preservation technologies, either individually or in combination, ensure the safety, quality, and shelf stability of fruit juice concentrate, meeting consumer demands for convenient and nutritious products.

9.3 Conclusion

In conclusion, fruit beverages and concentrates represent significant components of the global food industry, offering consumers a diverse range of flavourful and nutrient-rich options. These products are classified into distinct categories, including fruit juice and fruit drinks, each with specific compositions and regulatory definitions. Fruit juices, defined as 100% juice, and fruit drinks, containing a lower percentage of juice along with other additives like sugar, are derived from raw materials such as filtered-clarified juices, purees, and concentrates. The Fruit Juice Directive outlines six main classifications, including direct juice, juice from concentrate, concentrated juice, water-extracted juice, dehydrated/powdered juice, and fruit nectar, each reflecting different processing methods and compositions.

Technological advancements play a critical role in the production and preservation of fruit beverages and concentrates. Various methods such as evaporation, membrane filtration, freeze concentration, and hydrate separation are employed for juice concentration. Evaporation, a conventional method, involves water removal through heat application, while membrane filtration techniques such as microfiltration, ultrafiltration, and reverse osmosis offer advantages like low-temperature processing and minimal loss of volatile compounds. Freeze concentration, an innovative technique, utilizes partial freezing to separate ice from the concentrated solution, preserving heat-sensitive components effectively. Hydrate separation technology, although less common, provides an alternative method for water removal.

Preservation techniques are essential to maintain the quality and shelf life of fruit beverages and concentrates. Processes such as pasteurization, sterilization, and aseptic packaging are commonly employed to ensure microbial stability and extend product lifespan. Preservation methods are selected based on factors such as product type, processing method, and intended use. For instance, pasteurization is suitable for freshly squeezed juices, while aseptic packaging is preferred for concentrates to maintain freshness without refrigeration. Overall, the combination of proper classification, innovative technology, and effective preservation techniques ensures that consumers can access a wide variety of fruit beverages and concentrates that are both palatable and nutritious.

References

Adnan A, Mushtaq M, Islam T ul (2018) Fruit juice concentrates. In: Fruit juices: extraction, composition, quality and analysis. Academic Press, London, pp 217–240

Ağçam E, Akyildiz A, Dündar B (2018) Thermal pasteurization and microbial inactivation of fruit juices. In: Fruit juices: extraction, composition, quality and analysis. Academic Press, London, pp 309–339

Ahmad T, Raj D, Senapati A, Tandel Y, Tak M, Kumar V (2016) Guava nectar as a refreshing beverage an overall review. J Plant Dev Sci 8(7):321–328

Aider M, De Halleux D (2009) Cryoconcentration technology in the bio-food industry: principles and applications. LWT-Food Sci Technol 42(3):679–685

Aider M, Ounis WB (2012) Skim milk cryoconcentration as affected by the thawing mode: gravitational vs. microwave-assisted. Int J Food Sci Technol 47(1):195–202

Anwar R, Mattoo AK, Handa AK (2019) Ripening and senescence of fleshy fruits. In: Postharvest biology and nanotechnology of fruits, vegetables and flowers. Wiley, New Jersey, pp 15–51

Artés F, Allende A (2014) Minimal processing of fresh fruit, vegetables, and juices. In: Emerging Technologies for Food Processing, 2nd edn. Academic Press, London, pp 583–597

Augusto PED, Tribst AAL, Cristianini M (2018) High hydrostatic pressure and high-pressure homogenization processing of fruit juices. In: Fruit juices: extraction, composition, quality and analysis. Academic Press, London, pp 393–421

Aykas DP, Rodriguez-Saona L (2024) What's in your fruit juice?—rapid quality screening based on infrared (FT-IR) spectroscopy. Appl Sci 14(4). https://doi.org/10.3390/app14041654

Barreiro JA, Sandoval AJ, Rivas D, Rinaldi R (2006) Application of a mathematical model for chemical peeling of peaches (Prunus persica l.) variety Amarillo Jarillo. LWT - Food Sci Technol 40(4):574–578. https://doi.org/10.1016/j.lwt.2006.03.016

Başlar M, Biranger Yildirim H, Tekin ZH, Ertugay MF (2016) Ultrasonic applications for juice making. In: Handbook of ultrasonics and sonochemistry. Springer Singapore, Singapore, pp 1225–1246

Belbase C (2022) Effect of evaporation temperature on bioactive component and sensory quality of lemon juice concentrate. Doctoral dissertation, Dept of Food Technol, Central Campus of Technol, Inst of Sci and Technol, Tribhuvan Univ, Nepal

Berk E, Gökmen V (2024) Evaporation in the fruit juice industry. In: Heldman DR, Lund DB, Sabliov C (eds) Evaporation Technology in Food Processing. Woodhead Publishing, Cambridge, pp 105–136

Beveridge T, Harrison JE, McKenzie DL (1988) Juice extraction with the decanter centrifuge—a review. Can Inst Food Sci Technol J 21(1):43–49

Bhattacharjee C, Saxena VK, Dutta S (2017) Fruit juice processing using membrane technology: a review. Innov Food Sci Emerg Technol 43:136–153

Buerman EC, Worobo RW, Padilla-Zakour OI (2020) High pressure processing of spoilage fungi as affected by water activity in a diluted apple juice concentrate. Food Control 107:106779

Buerman EC, Worobo RW, Padilla-Zakour OI (2021) High pressure processing of heat and pressure resistant fungi as affected by pH, water activity, sulfites, and dimethyl dicarbonate in a diluted apple juice concentrate. Food Control 120:107551

Caceres LG, Andrade JS, da Silva Filho DF (2012) Effects of peeling methods on the quality of cubiu fruits. Food Sci Technol 32(2):255–260. https://doi.org/10.1590/s0101-20612012005000039

Caswell H (2009) The role of fruit juice in the diet: An overview. Nutr Bull 34(3):273–288. https://doi.org/10.1111/j.1467-3010.2009.01760.x

Chandel N, Kurrey VK, Minz RR, Thakur O (2018) Guava nectar as a refreshing beverage an overall review. Plant Archives 18(1): 1163–1169.

Conidi C, Castro-Muñoz R, Cassano A (2020) Membrane-based operations in the fruit juice processing industry: a review. Beverages 6(1):18

Córcoles JI, Marín-Alarcón E, Almendros-Ibáñez JA (2020) Heat transfer performance of fruit juice in a heat exchanger tube using numerical simulations. Appl Sci 10(2):648

Couto DS, Cabral LMC, Matta VM da, Deliza R, Freitas D de GC (2011) Concentration of pineapple juice by reverse osmosis: physicochemical characteristics and consumer acceptance. Food Sci Technol 31(4):905–910. https://doi.org/10.1590/s0101-20612011000400012

Danalache F, Mata P, Alves VD, Moldão-Martins M (2018) Enzyme-assisted extraction of fruit juices. In: Fruit juices: extraction, composition, quality and analysis. Academic Press, London, pp 183–200

Daud SMM, Sukri NM, Johari MH, Gnanou J, Manaf FA (2023) Pure juice supplementation: its effect on muscle recovery and sports performance. Malaysian J Med Sci 30(1):31–48. https://doi.org/10.21315/mjms2023.30.1.4

De Oliveira RC, Docê RC, de Barros STD (2012) Clarification of passion fruit juice by microfiltration: analyses of operating parameters, study of membrane fouling, and juice quality. J Food Eng 111(2):432–439

Deshpande SS, Cheryan M, Sathe SK, Salunkhe DK, Luh BS (1984) Freeze concentration of fruit juices. Crit Rev Food Sci Nutr 20(3):173–248

Echavarría AP, Torras C, Pagan J, Ibarz A (2011) Fruit juice processing and membrane technology application. Food Eng Rev 3:136–158

Farnworth ER, Lagace M, Couture R, Yaylayan V, Stewart B (2001) Thermal processing, storage conditions, and the composition and physical properties of orange juice. Food Res Int 34(1):25–30

Álvarez García C (2018) Application of Enzymes for Fruit Juice Processing. In: Fruit juices: extraction, composition, quality and analysis. Academic Press, London, pp 201–216

Góngora-Nieto MM, Sepúlveda DR, Pedrow P, Barbosa-Cánovas GV, Swanson BG (2002) Food processing by pulsed electric fields: treatment delivery, inactivation level, and regulatory aspects. Lwt 35(5):375–388. https://doi.org/10.1006/fstl.2001.0880

Goula AM, Adamopoulos KG (2010) A new technique for spray drying orange juice concentrate. Innov Food Sci Emerg Technol 11(2):342–351

Graumlich TR, Marcy JE, Adams JP (1986) Aseptically packaged orange juice and concentrate: a review of the influence of processing and packaging conditions on quality. J Agric Food Chem 34(3):402–405

Guldas M (2003) Peeling and the physical and chemical properties of Kiwi fruit. J Food Process Preserv 27:271–284

Gurak PD, Cabral LM, Rocha-Leão MHM, Matta VM, Freitas SP (2010) Quality evaluation of grape juice concentrated by reverse osmosis. J Food Eng 96(3):421–426

Heldman DR, Lund DB, Sabliov C (eds) (2018) Handbook of food engineering. CRC Press, USA

Hongvaleerat C, Cabral LM, Dornier M, Reynes M, Ningsanond S (2008) Concentration of pineapple juice by osmotic evaporation. J Food Eng 88(4):548–552

Horvath-Kerkai E (2012) Manufacturing fruit beverages. In: Handbook of fruits and fruit processing, 2nd edn. Wiley, Oxford, pp 213–228

Islam M, Bijjahalli S, Fahey T, Gardi A, Sabatini R, Lamb DW (2024) Destructive and non-destructive measurement approaches and the application of AI models in precision agriculture: a review. Precis Agric 25:1127. https://doi.org/10.1007/s11119-024-10112-5

Iwuozor KO, Adeniyi AG, Emenike EC et al (2023) Sugarcane juice powder produced from spray drying technology: a review of properties and operating parameters. Sugar Tech 25:497–507. https://doi.org/10.1007/s12355-022-01211-6

Kalaiselvan RR, Sugumar A, Radhakrishnan M (2018) Gamma irradiation usage in fruit juice extraction. In: Fruit juices: extraction, composition, quality and analysis. Academic Press, London, pp 423–435

Kiyak B, Gunes NC (2024) Life cycle assessment of apricot puree concentrate production. J Food Eng 368:111919

Koubaa M, Barba FJ, Bursać Kovačević D, Putnik P, Santos MD, Queirós RP, Moreira SA, Inácio RS, Fidalgo LG, Saraiva JA (2018) Pulsed electric field processing of fruit juices. In: Fruit juices: extraction, composition, quality and analysis. Academic Press, London, pp 437–449

Kowalska H, Trusinska M, Rybak K, Wiktor A, Witrowa-Rajchert D, Nowacka M (2023) Shaping the properties of osmo-dehydrated strawberries in fruit juice concentrates. Appl Sci 13(4):2728

Kumari P, Ahmad MF, Mir H (2018) Non-destructive quality evaluation by sensing. J Postharvest Technol 06(August):84–89

Li S, Qi F, Du K, Shen Y, Liu D, Fan L (2017) An energy-efficient juice concentration technology by ethylene hydrate formation. Sep Purif Technol 173:80–85

Lin J, Rouseff RL, Barros S, Naim M (2002) Aroma composition changes in early season grapefruit juice produced from thermal concentration. J Agric Food Chem 50(4):813–819

Mani-López E, Ríos-Corripio MA, Lorenzo-Leal AC, Palou E, López-Malo A (2019) 2 - characteristics and uses of novel and conventional preservatives for fruit drinks and beverages. In: Grumezescu AM, Holban AM (eds) Preservatives and preservation approaches in beverages. Academic Press, Cambridge, pp 31–67

Manjunatha SS, Raju PS (2015) Rheological characteristics of reconstituted spray dried beetroot (*Beta vulgaris* L.) juice powder at different solid content, temperatures, and carrier materials. Int Food Res J 22(6): 2333-2345

Manjunatha SS, Raju PS, Bawa AS (2012) Modelling the rheological behaviour of enzyme-clarified lime (*Citrus aurantifolia* L.) juice concentrate. Czech. J Food Sci 30(5):456

Martínez-Gonzáles NE, Martínez-Chávez L, Martínez-Cárdenas C, Castillo A (2011) Comparison of different washing treatments for reducing pathogens on orange surfaces and for preventing the transfer of bacterial pathogens to fresh-squeezed orange juice. J Food Prot 74(10):1684–1691. https://doi.org/10.4315/0362-028X.JFP-10-357

Mazli WNA, Samsuri S, Amran NA (2020) Study of progressive freeze concentration and eutectic freeze crystallization technique for salt recovery. IOP Conf Ser Mater Sci Eng 778(1):012167. IOP Publishing

Medvedkov Y, Nazymbekova A, Tlevlessova D, Shaprov M, Kairbayeva A (2021) Development of the Juice Extraction Equipment: Physico-Mathematical Model of the Processes. Eastern-European Journal of Enterprise Technologies 1(11–109):14–24. https://doi.org/10.15587/1729-4061.2021.224986

Mihalev K, Dinkova R, Shikov V, Mollov P (2018) Classification of fruit juices. In: Fruit juices: extraction, composition, quality and analysis. Academic Press, London, pp 33–44

Miyawaki O, Inakuma T (2021) Development of progressive freeze concentration and its application: a review. Food Bioprocess Technol 14(1):39–51

Miyawaki O, Liu L, Shirai Y, Sakashita S, Kagitani K (2005) Tubular ice system for scale-up of progressive freeze concentration. J Food Eng 69(1):107–113

Moradinezhad F, Khayyat M, Ranjbari F, Maraki Z (2019) Vacuum packaging optimizes quality and reduces postharvest losses of pomegranate fruits. J Hortic Postharvest Res 2(Special Issue-Postharvest Losses):15–26

Moreno FL, Robles CM, Sarmiento Z, Ruiz Y, Pardo JM (2013) Effect of separation and thawing mode on block freeze-concentration of coffee brews. Food Bioprod Process 91(4):396–402

Morison KR, Hartel RW (2018) Evaporation and freeze concentration. In: Heldman DR, Lund DB, Sabliov C (eds) Handbook of food engineering. CRC Press, Florida, pp 705–764
Mushtaq M (2018) Extraction of fruit juice: An overview. In: Fruit juices: extraction, composition, quality and analysis. Academic Press, London, pp 131–159
Mushtaq M, Akram S, Adnan A (2018) Novel extraction technologies. In: Fruit juices: extraction, composition, quality and analysis. Academic Press, London, pp 161–181
Nanjegowda M, Jani B, Devani B (2022) Aseptic processing. In: Thermal food engineering operations. CRC Press, Florida, pp 117–139
Onsekizoglu Bagci P (2015) Potential of membrane distillation for production of high-quality fruit juice concentrate. Crit Rev Food Sci Nutr 55(8):1098–1113
Orellana-Palma P, Petzold G, Torres N, Aguilera M (2018) Elaboration of orange juice concentrate by vacuum-assisted block freeze concentration. J Food Process Preserv 42(2):e13438
Orellana-Palma P, Tobar-Bolaños G, Casas-Forero N, Zúñiga RN, Petzold G (2020) Quality attributes of cryoconcentratedcalafate (*Berberis microphylla*) juice during refrigerated storage. Food Secur 9(9):1314
Oyeleke FI, Olaniyan AM (2007) Extraction of juice from some tropical fruits using a small scale multi-fruit juice extractor. Afr Crop Sci Conf Proc 8:1803–1808
Oyeleke FI, Olaniyan A, Sunmonu MO, Oyeniyi SK (2014) Development of a fruit washing machine. J Agric Eng Technol 22(1):29–36
Petrotos KB, Quantick PC, Petropakis H (1999) Direct osmotic concentration of tomato juice in tubular membrane–module configuration. II. The effect of using clarified tomato juice on the process performance. J Membr Sci 160(2):171–177
Prasad K, Jacob S, Siddiqui MW (2018) Chapter 2 - fruit maturity, harvesting, and quality standards. In: Siddiqui MW (ed) Preharvest modulation of postharvest fruit and vegetable quality. Academic Press, Cambridge, pp 41–69
Rabie MA, Soliman AZ, Diaconeasa ZS, Constantin B (2014) Effect of pasteurization and shelf life on the physicochemical properties of Physalis (Physalis peruvianaL.) juice. J Food Process Preserv. https://doi.org/10.1111/jfpp.12320
Ramadan MF (2019) Enzymes in fruit juice processing. In: Enzymes in food biotechnology: production, applications, and future prospects. Elsevier, Amsterdam, pp 45–59
Ramaswamy HS, Marcotte M (2005) Food processing: principles and applications. CRC Press, Boca Raton
Rani J, Shelke A, Bansal B, Gautam V (2021) Concentrate development from litchi juice and quality evaluation during storage. J Phys Conf Ser 2070(1):012072. IOP Publishing
Rehman MA, Khan MR, Sharif MK, Shabbir Ahmad SA, Faiz-ul-Hassan Shah FUHS (2014) Study on the storage stability of fruit juice concentrates
Ruan Q, Jiang H, Nian M, Yan Z (2015) Mathematical modeling and simulation of countercurrent multiple effect evaporation for fruit juice concentration. J Food Eng 146:243–251
Saleem N, Kamran M, Shaikh SA, Tarar OM, Jamil K (2011) Studies on processing and preparation of peach squash. J Biochem Mol Biol 44(1):12–17
Salehi F (2020) Physicochemical characteristics and rheological behaviour of some fruit juices and their concentrates. J Food Meas Charact 14(5):2472–2488
Samsuri S, Amran NA, Yahya N, Jusoh M (2016) Review on progressive freeze concentration designs. Chem Eng Commun 203(3):345–363
Saravacos G, Moyer J, Wooster G (1970) Concentration of liquid foods in a pilot-scale falling film evaporator
Sarbatly R, Sariau J, Krishnaiah D (2023) Recent developments of membrane technology in the clarification and concentration of fruit juices. Food Eng Rev 15:420. https://doi.org/10.1007/s12393-023-09346-2
Shaw PE, Carter RD, Moshonas MG, Sadler G (1990) Controlled atmosphere storage of oranges to enhance aqueous essence and essence oil. J Food Sci 55(6):1617–1619
Shishir MRI, Chen W (2017) Trends of spray drying: a critical review on drying of fruit and vegetable juices. Trends Food Sci Technol 65:49–67

Slavin JUM, Lloyd Beate PNGR (2012) Health benefits of Cassava-Karrapendalam. www.yadtek.com›Heal›. Diet Nutr 3(4):506–516. https://doi.org/10.3945/an.112.002154.506
Srivastava RP, Kumar S (2002) Fruit and Vegetable Preservation: Principles and Practices. International Book Distributing Company.
Sorour MA (2015) Optimization of multiple effect evaporators designed for fruit juice concentrate. American Journal of Energy Engineering 3(2): 6.
Sorour MA, Mostafa SR, Bo Samri SM (2013) A case study of orange juice concentration by recompression evaporation techniques. J Food Process Eng 36(3):337–342
Taylor B (2016) Fruit and juice processing. In: Chemistry and technology of soft drinks and fruit juices, 3rd edn. Wiley, London, pp 31–64
Trishitman D, Negi PS, Rastogi NK (2023) Concentration of pomegranate juice by forward osmosis or thermal evaporation and its shelf-life kinetic studies. Food Chem 399:133972
Varnam AH, Sutherland JP (1994) Fruit Juices. In: Beverages: technology, chemistry and microbiology. Springer US, Boston, pp 26–72
Vásquez-Castillo FM, Hernández E, Ureña-Peralta MO, Achaerandio I (2023) Comparative study of block cryoconcentration in pomegranate juice: centrifugation versus vacuum. J Food Process Eng 46(11):e14435
Vivek K, Mishra S, Pradhan RC (2018) Optimization of ultrasound-assisted enzymatic extraction of Sohiong (Prunus nepalensis) juice. J Food Process Eng 42(1):1–9. https://doi.org/10.1111/jfpe.12948
Wilczyński K, Kobus Z, Dziki D (2019) Effect of press construction on yield and quality of apple juice. Sustain For 11(13). https://doi.org/10.3390/su11133630
Wu C, Ge F, Shang G, Wang G, Zhao M, Wu L, Guo H (2021) Design of Screw Type Automatic Apple Juicer. J Phys Conf Ser 1750(1):012042. https://doi.org/10.1088/1742-6596/1750/1/012042
Zhang K, An X, Bai Y, Shen C, Jiang Y, Hu Y (2021) Exploration of food preservatives as draw solutes in the forward osmosis process for juice concentration. J Membr Sci 635:119495
Zhumaliyeva G, Chomanov U, Aktokalova G, Zhumaliyeva G (2023) Development of concentrate technology for puree. J Hyg Eng Des 44:35

Chapter 10
Fermented Fruits and Vegetables

Sangeeta Yadav, Pinki Saini, Unaiza Iqbal, and Mazia Ahmed

10.1 Introduction

Fruits and vegetables constitute essential sources of vitamins, including vitamins B and C as well as provitamin A, dietary fibers, phytosterols, phytochemicals, and minerals (Gebbers 2007), contributing significantly to the nutritional profile of the human diet. Research has emphasized the importance of incorporating fruits and vegetables into the regular diet as a preventive measure against chronic pathologies. For instance, Dauchet et al. (2007) have presented a positive relation between the intake of these plant-based foods and reduced cases of hypertension. Additionally, there is evidence supporting their role in lowering the possibility of stroke and heart diseases (He et al. 2007). The wealth of bioactive compounds found in vegetables and fruits, such as antioxidants and anti-inflammatory agents, contributes to their protective effects on cardiovascular health.

While each food processing method has its merits, fermentation stands out for its ability to preserve and enhance the nutritional benefits of fruits and vegetables in a natural and sustainable manner, with fewer drawbacks in certain aspects compared to alternative processing methods. Moreover, consumer preferences have shifted toward foods and beverages that align with a health-conscious lifestyle. There is a growing inclination toward fresh, minimally processed, and highly nutritious options. Consumers are seeking products that not only fulfill their nutritional requirements but also promote health and well-being. The current consumer preference toward processed foods and beverages that offer a fresh experience, high nutritive value, therapeutic attributes, and robust flavors is on the rise (Endrizzi et al. 2006).

S. Yadav · P. Saini (✉) · U. Iqbal · M. Ahmed
Centre of Food Technology, University of Allahabad, Prayagraj, India

OP Chauhan (ed.), *Fruits and Vegetables Technologies*,
https://doi.org/10.1007/978-981-96-8433-5_10

Fermentation, an ancient preservation technique and culinary art, has played a pivotal role in transforming raw fruits and vegetables into a plethora of flavorful and nutritionally enriched products (Katz 2012). This age-old practice involves the metabolic action of microbes, like yeast and bacteria, which convert sugars and other compounds present in fruits and vegetables into various beneficial byproducts. The result is a diverse array of fermented delights, from sauerkraut and kimchi to pickles and fruit preserves, each bearing a distinctive taste profile shaped by the microbial alchemy within (Steinkraus 2002). As we delve into the world of fermented fruits and fermented vegetables, it is essential to explore the historical roots and cultural significance of this transformative process. Fermented foods exhibit a diverse array of traditional and cultural choices across various topographical regions, reflecting the heterogeneous nature of their production. Throughout history, fermentation has played a crucial role in sustaining communities by helping people in temperate and cooler climates endure harsh winters and fostering resilience during droughts in tropical locations. In this natural preservation process, carbohydrates are gradually broken down by microbes and enzymes, resulting in the production of organic acids and alcohols (FAO 1998). The historical development of fermentation-based processes stems from the necessity of preserving fresh fruits and vegetables to be utilized during scarcity. Leveraging alcohols and organic acids during fermentation not only serves as a method of food preservation but also imparts desirable flavors and textures to the products. Additionally, the fermentation process has been instrumental in reducing the toxicity of certain foods and decreasing cooking time, contributing to both the safety and palatability of the preserved items (Rolle and Satin 2002). This multifaceted role of fermentation underscores its significance in the culinary heritage of diverse cultures and its adaptive utility across different climatic and environmental conditions.

This age-old method of preparation and preservation offers numerous benefits, such as improved digestibility, strengthened immunity, reduced serum cholesterol levels, and inhibition of carcinogenic compounds within the gastrointestinal tract by diminishing enzyme activity of fecal bacteria (Montet et al. 2014). Fermented fruits and vegetables stand out as essential dietary choices, not only for their rich nutritional and functional properties facilitated by the action of fermenting microbes but also for the exploration of new bioactive components. Ongoing research in this emerging field holds significant promise, unraveling the potential health benefits and nutritional innovations associated with fermented foods. Lactic acid fermentation emerges as a simple yet valuable biotechnological approach for preserving and enhancing the safety, nutritional content (Karovicová and Kohajdová 2003), sensory qualities, and reducing perishability of fruits and vegetables (Demir et al. 2006). Recent literature underscores the potential of combining this age-old preservation method with modern biotechnological techniques. This collaboration enables precise fermentation through deliberate selection of starter cultures, promoting greater intake of fresh-like fruits and vegetables (McFeeters 2004). The integration container of traditional and contemporary approaches holds promise for meeting the evolving demands of health-conscious and flavor-seeking consumers. The present chapter highlights the fermentation techniques and various conventional fermented foods developed from fruits and vegetables. An overview of the functional benefits of fermentation-based foods is also given. The chapter also delves into challenges, contamination issues, regulations, and safety aspects of fermented fruit and vegetable products.

10.2 Microorganisms Involved in Fermentation Process

Fruit and vegetable fermentation is a complex biological process facilitated by a diverse array of microorganisms, each contributing to the unique qualities of the final product. The population of microbes inhabiting raw fruits and vegetables is typically from 5.0 to 7.0 log cfu/g (Spurr 1994). Usually, fruits tend to be primarily colonized by yeasts and fungi, with dominant species of *Candida, Cryptococcus, Hansenula, Saccharomyces, Pichia, Rhodotorula* and *Debaromyces* genera (Nielsen 2007; Di Cagno 2010). Additionally, staphylococcus sp. and other fecal microbes may be present in uncooked vegetables, but their impact on human health is considered minimal due to their lower density, suppression due to adverse conditions, and microbial contamination. In raw fruits and vegetables, lactic acid bacteria make up a smaller percentage of the microbiota (2.0–4.0 log cfu/g). The identified heterofermentative and homofermentative species, including *Lactobacillus, Leuconostoc, Enterococcus*, *Weissella,* and *Pediococcus* genera, may vary depending on the vegetable (Spurr 1994). These microorganisms contribute to the natural microbial ecology of raw vegetables and fruits, influencing both their quality and safety.

The epiphytic microbial community of plants is greatly influenced by shifts in both physical and nutritional conditions (Lindow and Brandl 2003). Each plant species offers a unique ecological niche defined by its chemical composition, buffering capacity, competing microorganisms, and natural antimicrobial compounds (Buckenhüskes et al. 1997). Furthermore, fluctuations in temperature and harvesting practices play a key role in shaping plant microbiota dynamics. From existing research, it is evident that each plant species hosts a consistent and dominant microbiota (Yang et al. 2001). This emphasizes the specificity and stability of the microbial communities associated with different plants, highlighting the influence of environmental factors on their composition.

Fruits and vegetables undergo distinct processing pathways in fermentation due to their varied nutrient compositions. Fruits, abundant in sugars, vitamins, and minerals, are usually processed for alcohol based fermentation, producing beverages like beer and wine. The slightly acidic nature of fruit juices makes them conducive to the growth of yeasts, facilitating the rapid conversion of fruit sugars into alcohol. Vegetables are lower in sugar content but richer in vitamins and minerals with a neutral pH, which makes them an ideal medium for lactic acid bacteria (LAB) fermentation. Pickles and sauerkraut are examples of fermented vegetable products made possible by the microbial activity in vegetables. Thus, the eligibility of fruits and vegetables for particular fermentation methods is determined by their distinct nutrient profiles, resulting in a wide range of fermented foods and beverages. Raw vegetables and fruits undergo natural lactic fermentation when favorable anaerobic conditions, salt concentration, water activity, and temperature are present. In the primary stage of lactic acid fermentation, Gram-negative bacteria are typically inhibited. The fermentation process involves a succession of hetero- and homo-fermentative lactic acid bacteria, often accompanied by yeast. This microbial interplay, occurring naturally in the environment or on the surfaces of the raw produce, causes spontaneous fermentation of vegetables and fruits. The intricate balance of these microorganisms contributes to the unique flavors, textures, and preservation of the fermented products.

Lactic acid bacteria are commonly associated with the fermentation of fruits and vegetables, bringing about desirable changes in taste, texture, and shelf life. Strains of Lactobacillus, such as *L. brevis, L. plantarum,* and *L. acidophilus*, are frequently involved in this process, generating lactic acid as a byproduct, which contributes to the distinctive tanginess of fermented food in addition to serving as a preservative (Hutkins 2006). Yeasts, especially *Saccharomyces cerevisiae*, also play a pivotal role, particularly in the fermentation of fruits to produce alcoholic beverages and in some vegetable fermentations. These yeasts convert sugars into alcohol and carbon dioxide, imparting distinct flavors and aromas to the final product. Additionally, molds, such as *Penicillium* and *Aspergillus*, contribute to the fermentation of certain fruits and vegetables, including soybeans in the production of miso and tempeh (Tamang and Fleet 2009). The interaction between these microorganisms is dynamic, influenced by factors like pH, temperature, and the specific conditions of the fermentation environment. The selection of appropriate strains and the management of these microbial communities are critical for achieving the desired sensory and nutritional attributes in the fermented fruits and vegetables (Hwanhlem and Tanasupawat 2015).

10.3 Fermentation Techniques for Fruits and Vegetables

Fermentation techniques for fruits and vegetables encompass a variety of methods aimed at harnessing the transformative power of microorganisms to enhance flavor, texture, and nutritional value. Here are some common fermentation techniques used for fruits and vegetables:

10.3.1 Spontaneous or Natural Fermentation

Facilitating the proliferation of indigenous fermenting microorganisms present on vegetables is a practice that encourages the natural sequential development of fermentation. While this method may be somewhat unpredictable, it has the potential to yield the most intricate and well-developed flavors due to the diverse range of fermenting agents involved. The microbiota inherent in vegetables and fruits primarily comprises microbial entities responsible for natural lactic fermentation. The natural surface microflora, such as *Leuconostoc*, *Lactobacillus*, and *Pediococcus*, can "spontaneously" ferment fruits and vegetables (Kim and Chun 2005).

10.3.2 Back-Slopping

The incorporation of live bacteria from a pre-existing fermented product to initiate subsequent batches, a technique known as back-slopping, serves to expedite the initial fermentation phase by capitalizing on the presence of already active fermenting microorganisms. This approach mitigates the risk of fermentation failure. However, it comes with trade-offs as the natural progression of fermentation is disrupted, potentially leading to less complex flavor development. Moreover, the acidity of the back-slop can influence the success of fermentation initiation, where elevated acidity may result in suboptimal fermentation or yield a product with a softer texture. This method finds traditional application in sauerkraut fermentation, specifically employed to enhance the prevalence of well-adapted strains, thereby reducing fermentation duration and minimizing the chances of failure (Leroy and De Vuyst 2004).

10.3.3 Culture Inoculation

The initiation of fermentation process through the addition of a specific bacterial culture is a practice more commonly observed in industrial settings to ensure consistent and reliable fermentation outcomes. In contrast, home-based fermentations tend to employ this technique less frequently. To guarantee the presence and growth of only the starter culture bacteria, vegetables are typically heat-treated before inoculation. It is crucial to follow the accompanying directions if using a culture in home fermentations, as culture inoculation can lead to a less complex flavored product, especially when the culture comprises only a limited number of fermenters. However, the use of a starter culture containing probiotic strains such as *Lactobacillus plantarum, L. gasseri, L. rhamnosus,* and *L. acidophilus* can provide consistent and reliable results (Di Cagno et al. 2013). Despite the dependable value of natural fermentation in stabilizing and preserving raw vegetables and fruits, various factors support the preference of selected starter cultures. These factors encompass the potential for fermentation failure, insufficient suppression of spoilage and pathogenic microbes, and inconsistencies in sensory, nutritional, and textural attributes. Unlike traditional fermented products like cheeses, sausages, and leavened baked goods, the incorporation of starter cultures in vegetable and fruit fermentation has gained significant traction only in recent years (Molin 2001; Montet et al. 2006).

There are two main methods used in the regulated lactic fermentation of fruits and vegetables using autochthonous or allochthonous starters (Di Cagno et al. 2010,

2011). Autochthonous starter cultures are the ones isolated and reused on the same raw material, irrespective of geographic origin. On the other hand, allochthonous starter cultures are isolated from specific raw materials but employed for fermentation of different products. Commercial starters, commonly utilized for the fermentation of different vegetables and fruits, typically align with the category of allochthonous strains. This distinction between starters holds particular importance for plant species, as they represent diverse and distinct niches within the fermentation process.

The choice of fermentation technique depends on the desired end product, cultural traditions, and the specific characteristics of the fruits and vegetables being processed. Experimentation and a good understanding of the microbial ecology involved in each method are crucial for successful and safe fermentation.

10.4 Fermented Fruit and Vegetable Products

Before being recognized as probiotics, fermented foods had been a longstanding part of human diets for many years, serving as minimally processed food. However, the microbial processes involved were not well understood. Despite the numerous benefits, the incorporation of fermented foods into universal dietary guidelines has not been widespread. The World Health Organization (WHO) and Food and Agriculture Organization (FAO) have recommended a daily intake of fruits and vegetables to prevent chronic conditions, but fermented foods are not universally emphasized. Consumers generally prefer convenient, ready-to-eat or drink options that are fresh and nutritious. Fruits are frequently fermented with lactic acid to maintain and improve their nutritional value and flavor. Many lactic acid bacteria that are frequently found in fermented foods from Asia, including *Lactobacillus pentosus, Lactobacillus plantarum, Lactobacillus fermentum, Lactobacillus brevis, Lactobacillus casei, Lactobacillus kimchi, Leuconostoc mesenteroides, Lactobacillus fallax, Weissella koreenis, W. confusa,* and *Pediococcus pentosaceus*, are now known to be probiotics. The nutrients and acids in fruits and vegetables create a conducive environment for lactic acid bacteria fermentation (Anandharaj and Sivasankari 2013).

The fermentation of fruits exhibits a rich diversity shaped by various traditions and cultural preferences, influenced by the geographical regions of production. In temperate and cooler climates, fermentation played a crucial role in enabling ancestors to survive winter, while in tropical areas, it helped endure droughts. Fermentation is a slow process in which microorganisms or enzymes break down organic materials, turning carbohydrates into organic acids or alcohols. The methods for producing many traditional fermented foods have been passed down through generations as family customs. Ancient preservation techniques like drying and salting often accompany fermentation practices. Using organic acids and alcohols to preserve the food, improve texture and flavor, lower toxicity, and reduce cooking time, fermentation most likely developed as a way to preserve fruits during times of scarcity (Tamang et al. 2009).

Many globally processed vegetables, such as cucumbers, cabbage, olives, onions, and peppers, undergo commercial treatment with vinegar made from acetic acid (Roberts and Kidd 2005). In the Eastern Himalayan regions of India, various fermented vegetable products are created through bioprocessing to preserve and store perishable vegetables for later consumption. Examples of vegetables from Nepal, Sikkim, and Bhutan that undergo lactic acid fermentation include gundruk, sinki, and khalpi. The prevalent lactic acid bacteria (LAB) strains found in these traditional fermented vegetables include *L. plantarum, L. brevis, L. fallax, Pediococcus pentosaceus,* and *P. acidilactici.* Few of these LAB strains may possess functional and protective qualities, making them attractive candidates for use as starter cultures in the optimized and regulated production of fermented vegetable products (Tamang et al. 2009).

10.4.1 Gundruk

A naturally occurring acidic vegetable product from the Himalayas, is produced through a traditional fermentation process. During the gundruk fermentation process, the raw leaves of regional vegetables such as rayosag, mustard, cabbages and cauliflower wilt for 1–2 days. After wilting, the leaves are gently crushed, packed into an earthen pot, sealed, and kept for natural fermentation for a period ranging from 15 to 22 days. Following a satisfactory fermentation, the products are taken out and allowed to air dry for 2–4 days. Gundruk, similar to other fermented vegetable based products like Korean kimchi, German sauerkraut, and Japanese sunki, is consumed as a pickle or in soup preparations. The primary microbial flora of gundruk includes a variety of LABs such as *L. plantarum*, *L. fermentum,* and *L. casei* (Yan et al. 2008).

10.4.2 Sinki

It is a traditional delicacy crafted from fermented radish tap roots in the Sikkim Himalayas, is produced through a unique method known as pit fermentation, a special form of lactic fermentation for biopreserving food. The process involves digging a two to three-foot-diameter trench, burning, cleaning, and plastering it with mud in a dry location to create a pit. After removing the ashes, the pit is covered with bamboo leaves and paddy straw. Radish roots are allowed to wilt for 2–3 days, then flattened, compressed, soaked in warm water, and firmly packed. The mixture is covered with dried leaves and supported by large stones or planks. After applying mud to seal the pit, it is kept for fermentation for 2–3 weeks. The freshly fermented sink is taken out, chopped into small bits, sun dried for about 3 days, and stored for further use at room temperature (Yan et al. 2008). The pit fermentation technique, utilized for sinki, has also been employed to preserve other foods like breadfruit, taro, bananas, and cassava in regions such as Ethiopia and the South Pacific. The fermentation process for sinki involves LABs such as *Leuconostoc fallax, L. plantarum, L. brevis*, and *L. casei* (Tamang and Sarka 1996).

10.4.3 Khalpi

Cucumbers known as khalpi, sourced from Sikkim and Darjeeling, hold significance among Nepali Brahmins of the Bahun and Chettri castes who commonly prepare it at home for consumption. The fermentation of khalpi involves microbes identified as *Lactobacillus brevis, L. plantarum,* and *Leuconostoc fallax*. To prepare khalpi, full-grown and ripened cucumbers are precisely chopped and sun-dried. These dried cucumbers are then sealed within bamboo containers called dhungroo. The fermentation process occurs at room temperature, lasting for 4–7 days. The final product, when stored in an airtight container, can be preserved for approximately a week. Khalpi is typically consumed like a pickle after being combined with salt, chilies, and mustard oil (Tamang and Tamang 2009).

10.4.4 Sang

Similar to gundruk, Indian sang is kept fresh up to a year, even at room temperature in a tight container after being sun-dried for 4–5 days. In Nagaland, people typically consume Indian sang with steamed rice. This fermentation process is facilitated by a combination of lactic acid bacteria (LAB), including *Lactobacillus plantarum, Lactobacillus brevis*, and Pedipedococcus (Tamang et al. 2005).

10.4.5 Inziangsang

It is an indigenous fermented leafy vegetable preparation made from mustard leaves, which is a culinary tradition similar to gundruk and is enjoyed by people in Northeast India, particularly in Nagaland and Manipur (Yan et al. 2008). The method of preparation for Inziangsang is similar to that of gundruk. Locally known as hangam (*Brassica juncea L. Czern*), mustard leaves are collected, crushed, and immersed in hot water. After removing excess water by squeezing, the leaves are tightly packed into a container, ensuring an anaerobic environment, and sealed. After that, the container is kept for fermentation at room temperature (20–30 °C) for 7–10 days. Freshly made Inziangsang can be stored at ambient temperature in a closed container for up to a year after being sun-dried for 4–5 days. In Nagaland, Inziangsang is commonly consumed with steamed rice during the soup season (Tamang et al. 2009).

10.4.6 Soidon

A widely consumed fermented product in Manipur, is crafted from the tips of full-grown bamboo shoots, specifically focusing on the tips of bamboo varieties like *Dendrocalamus giganteus*, *Bambusa tulda,* and *Melocanna bambusoides*. The process involves removing the outer cover and lower portions of the bamboo shoots, submerging the whole ends in water within an earthenware pot. To initiate fermentation, a sour liquid, i.e., soijim, is added from a previous batch as a starter culture in a dilution ratio of 1:1. Fermentation occurs over 3–7 days at ambient temperature. Additionally, leaves of *Garcinia pedunculata* Roxb. (family: Guttiferae), known locally as heibungin in the Manipuri language, may be added during the process to increase the flavor of soidon. At the end of fermentation, soidon is taken out from the pot and stored in a closed container at ambient temperature for up to a year. The fermentation process is actively supported by lactic acid bacteria such as *Leuconostoc fallax, Lactococcus lactis*, and *L. brevis* (Tamang et al. 2008).

10.4.7 Goyang

A significant conventional fermented vegetable in Nepal and in Sikkim, prepared using the maganesaag leaves (*Cardamine macrophylla* Willd.), a member of the Brassicaceae family. The process involves collecting, washing, and cutting the leaves into pieces, followed by squeezing to remove excess water. Two to three layers of fig plant leaves are placed into bamboo baskets, and the prepared leaves are firmly packed inside. Additional fig plant leaves are kept on top of the baskets, and the combination is placed to naturally ferment for 25–30 days at temperature of 15–25 °C. During the fermentation process, the Goyang is actively separated from the yeasts of *Candida* spp., LAB as well as the microorganisms *Lactobacillus plantarum, L. brevis, Lactococcus Lactis, Enterococcus Faecium*, and *Pediococcus pentosaceus* (Tamang and Tamang 2007).

10.4.8 Kimchi

Traditionally prepared in homes and served as an accompaniment to meals, kimchi is a traditional fermented vegetable of Korea. Chinese cabbage, green onion, radish, red pepper powder, garlic, ginger, and fermented shrimp are commonly used in its preparation (Kim and Chun 2005). In Korea, various locally fermented vegetables fall under the umbrella term "kimchi" (Lee et al. 2005). The preparation involves pre-brining, followed by salting the primary ingredients (oriental cabbage or radish), incorporating diverse minor components (seasonings, cereals, fruits, and vegetables), and then fermenting the mixture at a cool temperature (ranging between 2 and 5 °C). Temperature affects how long kimchi ferments; ripening takes about a week at 15 °C and 3 days at 25 °C. However, to prevent excessive acidity and over-ripening and to maintain optimal flavor, it is recommended to maintain a low temperature during the kimchi fermentation (Nam et al. 2009). According to the classic recognition of bacterial isolates in kimchi, the predominant species are *Lactobacillus plantarum* and *Leuconostoc mesenteroides* (Cho et al. 2006). Other LAB sp involved in kimchi fermentation include *Lactobacillus brevis, L. plantarum, L. curvatus, L. lactis, L. sakei, P. pentosaceus, W. koreensis,* and *W. confusa*, as indicated by various studies. As the pH gradually decreases to 4.0, *Leuconostoc mesenteroides, L. pseudomesenteroides*, and *Lactococcus lactis* are considered significant species responsible for kimchi fermentation.

Kimchi contains beneficial components such as dietary fiber, vitamin C, chlorophyll, and *β*-carotene. Its antioxidative, antimutagenic, and angiotensin-converting enzyme inhibitory properties are believed to contribute to illness prevention. Bacteria isolated from kimchi, including *Leuconostoc mesenteroides*, produce beneficial enzymes like alcohol/acetaldehyde dehydrogenase and dextransucrase (Eom et al. 2007). The ideal pH and acidity levels for kimchi are approximately 0.5 and 0.6, respectively, which correspond to the point of maximal vitamin C concentration (Hong et al. 1999).

10.4.9 Sauerkraut

The sauerkraut fermentation process involves shredding of fresh cabbage and adding about 2.5–3.0% salt, letting it undergo natural fermentation. The creation of sauerkraut depends on an orderly microbial process, with homofermentative and heterofermentative lactic acid bacteria (LAB) playing distinct roles. *Leuconostoc* spp. is usually engaged in the first stage, while *Pediococcus* spp. and *Lactobacillus* spp. participate in the later stages of fermentation. The final sauerkraut product has a pH ranging from 3.5 to 3.8, contributing to its extended freshness. As a byproduct of cabbage fermentation, sauerkraut brine has practical applications. It can be utilized in various industrial processes, for instance, facilitating *Rhodotorula rubra*'s production of carotenoids or allowing *Candida wickerhamii* to manufacture β-glucosidase for commercial use (Yoon et al. 2006).

10.4.10 Paocai

It is a renowned traditional Chinese dish that stands out as a salty, fermented vegetable rich in lactic acid. In certain Chinese regions, surplus vegetables like radish, cucumber, celery, and cabbage were preserved through the creation of paocai during times of abundance. Typically served as a side dish to complement main courses, paocai also doubles as a snack on occasion. This pickle varies in preparation and taste across different regions. In Taiwan, paocai is crafted through anaerobic fermentation in a specialized container, utilizing an assortment of vegetables, spices, and additional ingredients. The end result boasts a crisp texture and a tangy flavor. A variety of microorganisms found in the basic components start the fermentation process of paocai, which eventually shifts to lactic bacteria (LAB) as the main culture species, giving it its unique properties (Chen et al. 2010).

10.4.11 Yan-Dong-Gua

A traditional Taiwanese fermented food crafted from wax gourd holds popularity as a distinctive culinary delight. Harvested wax gourd is first carefully cleaned, then chopped into little pieces and left to dry in the sun. After that, these dried bits are mixed with fermented soybeans, sugar, and salt to form a concoction that is heaped in a bucket. In the initial stages of fermentation, a small quantity of Mijiu (Taiwanese rice wine) is commonly added, and the bucket is securely sealed. The fermentation process typically spans around 1 month, although it can extend to over 2 months. The microbial agents responsible for the fermentation include *Weissella paramesenteroides* and *W. cibaria*, contributing to the unique and appetizing qualities of this traditional Taiwanese fermented food (Lan et al. 2009).

10.4.12 Sayur Asin

Sayur asin is an Indonesian culinary creation made by fermenting mustard cabbage leaves. A similar product (hum choy), is cultivated in Southeast Asian nations. The preparation involves crushing, rubbing, or wilting the leaves of mustard cabbage (*Brassica juncea var. rugosa*) in a solution containing 2.5–5% salt. The liquid from the boiled rice is added to supply fermentable carbs and to guarantee adequate acid production during fermentation. A series of lactic acid bacteria, such as *Lactobacillus confuses, L. curvatus, L. plantarum, Leuconostoc mesenteroides,* and *Pediococcus pentosaceus*, flourish throughout the fermentation. On the first day of fermentation, starch-breaking bacteria such as Bacillus, Staphylococcus, and Corynebacterium species only show minimal development. The fermentation process is facilitated by yeasts such as *Candida guilliermondii* and *Candida sake*, contributing to the distinctive qualities of sayur asin (Puspito and Fleet 1985).

10.4.13 Yan-Tsai-Shin

This particular variety of fermented broccoli, commonly used in Taiwan, is produced using *Brassica oleracea*. The process begins with harvested broccoli stems being cleaned, peeled, and chopped, then mixed with salt and kept in a bucket for about 6 hours. After draining off the released water, Sesame oil, soy sauce, and sugar are mixed with the fermented broccoli. For extra taste, some makers often include sliced hot pepper or rice wine. The mixture is then allowed to ferment for a day at a lower temperature, i.e., 6–10 °C. During fermentation, various bacterial species thrive, including *Weissella cibaria, W. paramesenteroides, W. minor, Leuconostoc mesenteroides, Lactobacillus plantarum*, and *Enterococcus sulfurous*, contributing to the unique taste and texture of yan-tsai-shin (Chen et al. 2013).

10.4.14 Jiang-Gua

In Taiwan, jiang-gua, a fermented cucumber, is used as a side dish and as a condiment. Harvested cucumbers (*Cucumis sativus* L.) are cleaned and chopped before being mixed with salt and piled in a bucket, usually with big stones on top. This phase typically takes 4–5 hours to complete, though processing times can vary. The cucumbers are combined with vinegar and sugar after the water that has been produced is drained off. The addition of soy sauce is optional and depends on the recipe. Usually, the fermentation process lasts for at least a day at a temperature of 6 to 10 °C. Numerous bacterial species, such as *Weissella hellenica, W. citbaria, Leuconostoc lactis*, *L. plantarum,* and *Enterococcus casseliflavus*, aid in the fermentation process. These microorganisms contribute to the distinctive flavors and characteristics of jiang-gua, making it a popular choice among Taiwanese people (Chen et al. 2012).

10.5 Fermented Pickles

A diverse array of fermented pickles, rooted in the cultural heritage of rural communities, remains minimally documented. Indigenous biopreservation methods, including pickling, are employed to ferment seasonal and perishable leafy vegetables, cucumbers, radishes, and young tender bamboo shoots (Tamang and Tamang 2009). In Turkey, Tursu is a traditional pickle crafted from a wide variety of vegetables and fruits including cabbages, cucumbers, green tomatoes, carrots, green peppers, green beans, eggplants, red beets, melon, and even sweet potatoes (FAO 1998). Pickled garlic has gained popularity among consumers due to its distinctive sensory properties. Spontaneous fermentation has been demonstrated to enhance health benefiting features of pickled garlic, particularly the polyphenol concentration and antioxidant activity (Montaño et al. 2004). Pak-Gard-Dong is fermented mustard leaf (*Brassica juncea*) traditionally made in Thailand, adding to the rich tapestry of fermented pickle varieties found across different cultures and regions. Nozawana-zuke is a type of low-salt pickle that is prominently consumed by the Japanese and is crafted using field mustard, locally known as Nozawana (*Brassica campestris var. rapa*), which is a leafy turnip plant. *Lactobacillus* and *Leuconostoc* are two of the plant-derived genera of lactic acid bacteria (LAB) that are essential to the fermentation process that produces Nozawana-zuke. By generating organic acids that add to Nozawana-zuke's distinctive flavors and prevent the growth of unwanted bacteria, these LAB are essential in forming the food's sensory qualities and protecting it from possible contamination.

Aadu pickle is a delightful and popular dish found in the Kullu Valley and other regions of Himachal Pradesh. It is made from the native peach, scientifically known as *Prunus mira*, also referred to as smooth stone peach or smooth-pit peach. This wild peach is abundantly found in the Himalayan foothills. *Aadoo* pickles are specifically crafted from the wild peach, *Prunus persica (L.) Batsch*, an indigenous fruit prevalent in the Kullu district of Himachal Pradesh. Rural women in the Kullu district use this wild peach exclusively to create pickles or local achar. The process of making Aadu pickle involves cleaning the fruit with water, followed by boiling and then draining and drying it in the sun using a sheeng, a traditional sieve. Various spices, including black mustard, fenugreek, and thyme seeds, are roasted and ground using a mortar and pestle. Aaroo is then mixed with turmeric powder, red chili powder, salt, mustard oil, and black mustard. The entire blend is placed in a sizable earthen pitcher for fermentation. Before adding the mixture to the pitcher, the smoke of Heeng (asafetida) is introduced to impart a pleasant aroma to the pickle (Monika et al. 2016).

Beedana (*Cydonia oblonga* Mill.) is a pear-shaped edible fruit primarily cultivated in the Kullu district of Himachal Pradesh. Consuming Beedana is believed to offer various health benefits and may aid in treating ailments such as tuberculosis, liver dysfunction, diarrhea, dysentery, gastric ulcers, liver diseases, and eye disorders. To prepare Beedana pickle, the fruits are first washed, chopped into pieces, and boiled for 23 minutes before being sun-dried. The dried pieces are then combined with red chili powder, salt, roasted black mustard powder, roasted fenugreek powder, turmeric powder, and mustard oil in a large pan. This mixture is transferred to a pitcher for fermentation, with the inner surface of the pitcher coated multiple times with mustard oil to enhance the process. Sealed and exposed to sunlight, the pickle develops a unique acidic taste due to the black mustard powder and can be stored for over 2 years without spoiling (Monika et al. 2016).

Galgal is a citrus fruit variety available in winter. *Galgal* pickles are prepared similarly to other fermented vegetable pickles. Fully ripened *Galgal* fruits are washed, immersed in boiling water for about 2 minutes, and sun-dried. The fruits are then chopped into small pieces. Mustard oil is heated, and fenugreek seeds are added and browned. The *Galgal* fruit pieces are mixed with salt, sugar, red chili powder, turmeric powder, mustard oil, thyme seeds, and fennel seeds in the heated pan and cooked for 5–10 minutes. After cooling, the mixture is kept for fermentation for 15–25 days. The fermentation process is done near traditional fireplaces called Tandoor, providing the required ambient temperature during winter. *Enterococcus faecalis* and *Leuconostoc mesenteroides* are isolated during the fermentation process. *Galgal* pickles have a distinct aroma due to the fennel seeds, which also give them a slightly sweet and sour taste (Monika et al. 2016). *Lasura* (*Cordia myna* L.) is a green fruit with a gum-like sticky substance, consumed by rural residents in Himachal Pradesh's Mandi, Bilaspur, Kangara, Una, and Hamirpur districts. It is eaten as pickles and vegetable curries and is considered to have anti-diabetic properties and the ability to lower blood pressure. The traditional preparation of *Lasura* pickle results in a product that can be stored for 1–2 years, exhibiting a distinct spicy flavor (Monika et al. 2016).

The Uchoi tribe has the unique tradition of creating *Amlai Ntoi*, a fermented product made exclusively from raw amla fruit (*Phyllanthus emblica* L.), commonly known as Indian gooseberry. The preparation of *Amlai Ntoi* involves boiling amla in water and then placing it in an earthen pitcher, layered with molasses at the top and bottom. The earthen pitcher is then buried to ferment for a duration of 5–6 months. Raw amla fruit is rich in nutrients and has a high vitamin C content. Its therapeutic properties are believed to contribute to better digestion, reduced blood pressure, and lowered cholesterol levels. The traditional fermentation process used by the Uchoi tribe adds to the unique characteristics and potential health benefits of *Amlai Ntoi* (Uchoi et al. 2015).

10.6 Traditional Fermented Beverages

Various traditional low-alcohol fermented beverages derived from fruits or vegetables are documented globally. The nutritional benefits associated with fruits, vegetables, and lactic-fermented products have led to the recent surge in the market for non-dairy lactic-fermented juices. African traditional fermented beverages predominantly utilize cereals such as cassava, maize, sorghum, and millet. The fermentation of these products involves lactic acid bacteria (LAB) like *Lactobacillus* sp., *Bacillus* sp., *Pediococcus* sp., *Weissella* sp., and yeasts (di Cagno et al. 2013). Additionally, certain regions, like Turkey (Altay et al. 2013), Albania, and Kosovo (Soukand et al. 2015), have a tradition of producing lactic-fermented beverages from fruits or vegetables. Most fruit-based non-alcoholic or low-alcoholic traditional beverages rely on complex microbiota, including yeasts, LAB, and acetic acid bacteria.

Kombucha, a fermented black tea beverage with added sugar, undergoes fermentation through a complex ecosystem of yeasts, acetic acid bacteria, and LAB. Unlike water kefir, kombucha's consortium is embedded in a cellulosic mat, and yeast domination occurs, especially by Zygosaccharomyces or Candida and Lachancea, depending on the kombucha origin. Acetic acid bacteria can represent 60% to over 90% of bacteria, with Lactobacillus, Leuconostoc, and Lactococcus appearing in early fermentation stages, and Lactobacillus species increasing by the end of the fermentation process (Marsh et al. 2014).

Water kefir, created from water, sugar, and dried fruit, undergoes fermentation with kefir grains that contain a stable microbial consortium surrounded by a polysaccharidic matrix named kefiran. Commercial water kefir grains often feature *Zymomonas mobilis* as the predominant bacteria, reaching up to 87% (Marsh et al. 2013). While research on water kefir microbiota is relatively recent compared to milk kefir, the microbial diversity is more influenced by grain origin than changes during fermentation and storage (Laureys and de Vuyst 2014).In Turkey, traditional fermented beverages from fruit and vegetables include *Shalgam* juice and *Hardaliye* (Altay et al. 2013). Hardaliye, prepared from grapes, mustard seeds, and cherry leaves, is primarily fermented by Lactobacillus spp. Additionally, Gilaburu juice in Turkey, made from water and European cranberry bush (*Viburnum opulus* L.), features identified LAB including *Lactobacillus* spp. and *Leuconostoc* spp. (Baschali et al. 2017). Shalgam juice, made from black or purple carrot slices, turnip slices, and wheat flour in salted water, results in a red-colored cloudy beverage after fermentation and filtration. Rich in minerals, vitamins, and polyphenols, its microbiota mainly consists of yeasts and *Lactobacillus* (89.63%), *Leuconostoc* (9.63%), and *Pediococcus* (0.74%).

Toddy, an alcoholic beverage with a sweet taste and distinctive flavor, is created by fermenting the sap of the coconut palm (*Cocus nucifera* L.) or date tree (*Phoenix sylvestris* L.). To harvest the sap, the peduncles of these trees are cut at the tip, and a small pot is placed beneath to collect the seeping sap. Fermentation starts immediately, especially if some toddy is already present in the pot. Proper fermentation takes 6–8 hours, and due to its short shelf life, toddy is usually sold promptly. The fermented sap is stored in glass or plastic bottles. Neera, also known as palm nectar, is a beverage made from sap extracted from the flowers of various toddy palms. Neera extraction is typically done in the morning, resulting in a transparent and pleasant-colored drink. Also referred to as palm wine, Neera is prone to spontaneous fermentation at room temperature within a few hours after extraction, transforming into toddy. Neera is widely consumed in India, Sri Lanka, Africa, Malaysia, Indonesia, Thailand, and Myanmar (Kumari et al. 2016). The juice from palm fruit is not termed Neera; instead, it is extracted by splitting the spathes of coconut, sago, and palmyra palms, as well as the sugar palm (*Arenga pinnata* L.). Neera's nutritional components promote the growth of lactic acid bacteria (LAB) and other microbes. Feni is a famous alcoholic beverage prepared from ripe cashew apples in the Indian state of Goa. The traditional process involves harvesting cashew apples, extracting juice by stomping with feet, allowing natural fermentation in earthen pots, and finally distilling and condensing the juice over a low wood fire in clay stills. After distillation, Feni is matured for several months before being bottled. Over time, some modifications have been made to the standard production process due to economic feasibility concerns, but Feni retains its unique natural flavor, scent, and high alcohol content (Deka 2024).

10.7 Health Benefits of Fermented Fruits and Vegetables

10.7.1 Probiotic Content and Gut Health

The recommended daily consumption of vegetables and fruits, advised by the World Health Organization (WHO) and the Food and Agriculture Organization (FAO), aims to prevent chronic conditions like hypertension, coronary heart disease, and strokes (Endrizzi et al. 2009). Consumers favor fresh, nutritionally rich, health-promoting foods that are convenient for immediate consumption. Fermented fruits and vegetables meet these preferences and additionally provide probiotics, which aid in digestion and nutrient absorption. Plant proteins generally have lower digestibility compared to animal-derived proteins, affecting the gastrointestinal environment by increasing protein excretion in feces. Fermentation, particularly lactic acid (LA) fermentation of vegetables and fruits, is widely used to preserve and enhance the nutritional and sensory qualities of food products. Lactic acid bacteria (LAB) are integral to this process and are crucial microorganisms involved.

LAB, a significant component of the human gut flora, functions as probiotics, benefiting the host by enhancing intestinal microbial balance (Cagno et al. 2013). These bacteria compete for adhesion sites with pathogens, antagonize them, and modulate the immune response, thereby improving nutrition, alleviating intestinal disorders, boosting immunity, optimizing the gut ecology, thereby promoting overall health (Cagno et al. 2013). Consumption of lactic acid-fermented fruits and vegetables enhances human nutrition by providing balanced nutrients such as essential vitamins, minerals, and carbohydrates. Moreover, the probiotic characteristics of these fermented foods aid in preventing diseases like diarrhea and liver cirrhosis (Yamano et al. 2006). In recently developed lactic-fermented fruit or vegetable juices, the population of lactic acid bacteria (LAB) exceeds the recommended limit of 6–7 log CFU/mL, as specified for probiotic products by literature and Food and Drug Administration (FDA)/WHO guidelines. This concentration is consistently maintained after fermentation and during storage of these beverages. LAB demonstrates resilience in food products, remaining viable throughout shelf life without being pathogenic or toxic. They survive transit through the gastrointestinal tract, attach to intestinal epithelial cells, colonize the intestinal lumen, produce antimicrobial substances against pathogens, stabilize the intestinal microflora, and contribute to health advantages. Research indicates that maintaining a healthy gut microflora protects against gastrointestinal disorders, including inflammatory bowel diseases and other infections. Probiotics found in fermented fruits and vegetables can substitute antibiotics in treating enteric infections or alleviating symptoms of antibiotic-associated diarrhea. In addition to enhancing intestinal microbial composition, fermented foods help prevent irritable bowel syndrome and combat internal and external pathogens. Their richness in probiotic LAB also aids immunomodulation, improving resistance to allergies and preventing respiratory ailments. Metabolically, LAB reduce toxigenic and mutagenic reactions in the gut. These fermented products enhance digestive functions, boost immune response, reduce colorectal cancer risk, regulate serum cholesterol levels, aid lactose digestion, prevent rotaviral diarrhea, address skin conditions, urinogenital infections, and remove undesirable antinutritional compounds from food materials.

Korean scholars have documented the benefits of kimchi, a fermented vegetable product, in aiding weight loss, reducing tumors, and lowering cholesterol and blood fat levels. Limited research exists on the health impacts of self-made probiotic fermented juices. However, *L. plantarum* NCU116, derived from traditional Chinese sauerkraut and sequenced using third-generation SMRT technology, has shown promise. It has demonstrated survival under gastrointestinal conditions in vitro and has exhibited effects such as lowering blood sugar, lipids, and cholesterol in vivo, along with assured laxative impacts in constipation models. This probiotic strain is noted for its strong ability to ferment fruits and vegetables, making it suitable for use as a starter culture, particularly with plant-based substrates (Guan et al. 2021–49). Research outlining the antimicrobial, functional, survival, and safety attributes of fermented fruits and vegetables is depicted in Fig. 10.1. The specific LAB species used exclusively for fermenting fruits and vegetables remain unclear. While strains like *L. plantarum* (Yang et al. 2018) and *L. acidophilus* (Battistini et al. 2018) derived from plants have been gradually adopted for fermenting fruit-vegetable juices, no single species is universally applicable due to the diversity of raw materials. Therefore, extensive screening is required to identify optimal strains for fermenting various fruit and vegetable substrates. Fermented foods are increasingly recognized not only for their preservation and sensory qualities but also for their broader functional properties. There is a critical need for further fundamental research to compare different fermentation-associated strains in terms of their core characteristics for food transformation, product formation, and interactions in the gastrointestinal tract. Additionally, randomized controlled clinical trials are necessary to quantify and validate the effects of various fermented foods on human health.

Fermented foods are widely acknowledged as functional probiotic products; however, in vivo studies often yield conflicting results, complicated by small study sizes that can lead to statistical inaccuracies. For example, research involving eight healthy volunteers who consumed 50 g of natto daily for 2 weeks showed increased

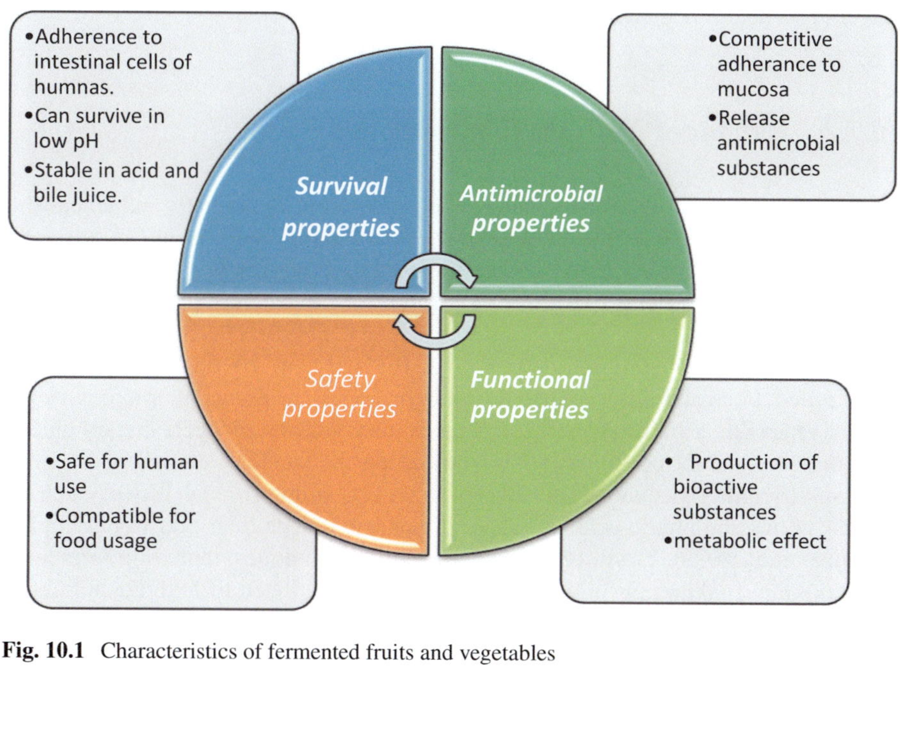

Fig. 10.1 Characteristics of fermented fruits and vegetables

levels of Bifidobacterium and Bacilli in their stool (Fujisawa et al. 2009). Another study examined the impact of daily intake of 60 g of kimchi over 4 weeks in six subjects with *Helicobacter pylori* infection, revealing higher levels of *Lactobacillus* and *Leuconostoc* bacteria in their feces, along with a decrease in Enterobacteriaceae (Kil 2004; Feng et al. 2015).

Furthermore, oral administration of sauerkraut juice to rats demonstrated probiotic effects, with microorganisms exhibiting high tolerance to low pH conditions. Additionally, an antimicrobial effect was observed due to the presence of bacteriocin paracin 1.7 (Yu et al. 2013). Sauerkraut was shown to prevent adhesion and invasion of *Salmonella enteritidis* in Caco-2 cells, linked to the induction of TNF-a and IL-12 expression by *Lactiplantibacillus plantarum* P2. However, a randomized double-blind controlled trial involving 58 individuals with irritable bowel syndrome found no significant alteration in gut microbiota composition after daily consumption of 75 g of sauerkraut for 6 weeks, despite improvements in scores on the irritable bowel syndrome-symptom severity scale (IBS-SSS) (Nielsen et al. 2018). The survival rate of probiotic microorganisms in the gastrointestinal tract, especially in specialized capsules resistant to hydrochloric acid, may not reach 100%. Various factors, both food-related (such as storage conditions, pH levels, and packaging) and physiological (including digestive pH and bile acids), influence the nutritional value and microbial diversity of the final fermented product. Notably, bacteria tend to lose viability at temperatures above 65 °C (Venema et al. 2020). Despite efforts to enhance probiotic survival in foods, particularly those not directly linked to vegetables, this remains an expansive and evolving area of research. Nevertheless, the consumption of fermented foods, including vegetables, holds probiotic potential and confers health benefits to the consumer.

10.7.2 Enhanced Nutrient Bioavailability

The activity of microorganisms during food processing significantly influences the composition and availability of nutrients in vegetables, such as macronutrients, polyphenols, and vitamins. While beverages and foods may contain sufficient amounts of macronutrients like zinc (Zn), iron (Fe), and calcium (Ca) to meet daily nutritional needs, their low bioavailability can lead to deficiencies in micronutrients, especially in developing regions (Skibsted 2016; Nair et al. 2016). Poor diet quality and limited bioavailability contribute to deficiencies in essential micronutrients such as zinc (Zn), vitamin A, and iron (Fe). Some countries advocate dietary guidelines that promote the consumption of whole grains as excellent sources of essential nutrients (Schaffer-Lequart et al. 2017). However, antinutritional factors such as polyphenolics and phytic acids in whole grains can reduce their nutritional impact (Ferruzzi et al. 2020). Traditional food processing techniques, including fermentation, cooking, soaking, puffing, and germination, are utilized to mitigate antinutritional factors and improve protein digestibility (Kumari and Platel 2020). This underscores the importance of fermented foods in human nutrition, particularly in developing countries where they hold significant cultural and nutritional value.

Carbohydrates from legumes and cereal starches are major contributors to caloric intake due to their high-carbohydrate content. Fermentation plays a crucial role by activating enzymes such as maltase and α-amylase, which break down starch into simpler sugars and maltodextrins. Early fermentation increases glucose levels through activated α-amylase and maltase, leading to enhanced production of ethanol and carbon dioxide, and decreased starch content in various millet varieties. Additionally, it stimulates phytase enzyme production and lowers pH (Castro-Alba et al. 2019). High-carbohydrate foods, especially starch, raise blood glucose levels, and their glycemic index (GI) reflects the speed of this increase. Fermentation reduces GI by converting glucose into short-chain organic acids such as propionic, acetic, and lactic acids. Lactic acid also lowers pH, reducing the activity of starch-hydrolyzing enzymes (Nilsson et al. 2002). When carbohydrate-rich vegetables like legumes undergo fermentation, they develop increased amounts of resistant starch. Furthermore, fermentation converts monosaccharides into polysaccharides. For example, fermented carrot juices showed a 52% increase in polysaccharides and a 27% decrease in monosaccharides, likely due to enzymatic conversion of simple sugars by fructosyltransferases produced by starter microorganisms like *Lactobacillus gasseri* DSM 20604 and DSM 20077 (Xu et al. 2020).

Several studies indicate that fermentation can alter both amino acid and protein contents in foods. Plant-based products generally have a less optimal amino acid profile compared to animal products due to lower levels of lysine and sulfur-containing amino acids like methionine and cysteine. However, a well-balanced vegan diet can meet amino acid requirements. Besides individual amino acid composition, protein quality is also influenced by digestibility. Fermentation has the ability to reduce the content of various plant compounds such as phytic acid, oxalate, tannins, carbohydrates, and complex proteins, while also promoting the production of microbial digestive enzymes. An in vitro study by Pranoto et al. (2013) examined the impact of 36 hours of natural fermentation and *Lactobacillus plantarum* on the digestibility of sorghum flour proteins. The study found that protein digestibility increased by 92% with *L. plantarum* and by 47% with natural fermentation. These results suggest that *L. plantarum* significantly enhanced proteolytic enzymes compared to natural fermentation, leading to better breakdown of protein-tannin complexes and other protein structures, releasing amino acids and peptides.

Fermentation not only influences proteins but also reduces lipid content through various mechanisms such as microbial metabolism and enzymatic breakdown by lipases, which utilize lipids as an energy source. However, research on the specific impact of fermentation on lipids in fresh food samples is limited. While most fruits and vegetables, except for soybeans, have low lipid content, further studies are needed to explore fermentation's effects on lipids across different types of food, considering all macronutrients. Mineral bioavailability in plant-based products is often hindered by complexes formed with non-digestible substances like phytates and myo-inositol, known as antinutritional compounds. Fermentation has been shown to effectively reduce these substances, thereby enhancing mineral bioavailability. By breaking down these complexes, fermentation makes minerals more accessible and usable by the body. This is particularly important for addressing mineral deficiencies, especially in vulnerable populations in developing countries where unprocessed pulses and cereals are staple foods. The impact of fermentation on minerals is influenced by food composition, with factors like fiber potentially affecting the availability of specific minerals. The acidic environment created during fermentation promotes the absorption of iron by converting ferric iron into ferrous iron, which is more readily absorbed. Furthermore, the optimal pH conditions resulting from fermentation are crucial for the activity of phytase, an enzyme that facilitates the breakdown of phytate.

A study highlighted the potential of *Leuconostoc mesenteroides* FSC2 and *Lactobacillus pentosus* FSC1 in enhancing mineral bioavailability. Fermentation of carrot juice with these microorganisms increased the solubility of minerals such as Cu, Zn, Fe, and Mn by 1, 1.2, 1.5–1.7, and 2 times, respectively (Bergqvist et al. 2006). Sauerkraut, a fermented food known for its health benefits, including strong anti-inflammatory effects, is rich in vitamins C and E (Palani et al. 2016). However, the microbial processing during fermentation can lead to a reduction in ascorbic acid compared to the fresh product. The vitamin C content varies depending on factors such as the brine composition and the specific microbial strain used. For example, the lowest ascorbic acid concentration (156.27 mg/100 g) was found in naturally fermented sauerkraut with 0.5% NaCl, while the highest level (266.25 mg/100 g) was achieved with *Lactiplantibacillus plantarum* fermentation using 1.5% NaCl (Penas et al. 2010). There is a noted relationship between NaCl concentration and vitamin C content, with higher levels observed at moderate NaCl concentrations and lower levels at lower concentrations. Although fermentation generally decreases vitamin C content, careful adjustment of fermentation conditions can help mitigate this loss. In addition to vitamin C, other vitamins like K, thiamine, and cobalamin are found in higher quantities in fermented vegetables compared to their fresh counterparts. However, there is limited information available on the fermentation of root vegetables and tomatoes.

10.7.3 Impact of Fermentation on Antioxidant Content

Vegetables are well-known for their high levels of antioxidants. Sawicki et al. (2018) conducted a study on the impact of boiling and spontaneous fermentation on the betalains in red beetroot. Their findings revealed that microbial activity and matrix softening during fermentation significantly increased the release of betalains. Initially, betalain content increased by 99% in the first 7 days of fermentation, followed by a subsequent decrease of 17% from days 8 to 12. Degradation processes were observed in the inner tissue of red beet slices by the 13th day of fermentation, with fermented red beets exhibiting the lowest antioxidant activity by the 14th day. This contrasts with earlier research suggesting that storage affects the antioxidant activity of fermented products. In a separate study, Wiczkowski et al.

(2015) examined the antioxidant capacity of red cabbage during fermentation, noting a lower antioxidant capacity in fermented red cabbage compared to fresh red cabbage due to reduced anthocyanin content.

In contrast, Olennikov et al. (2021) signified the antioxidant capabilities of a newly discovered melanoidin in fermented willowherb leaves, noting its concentration at 145.31 ± 4.35 mg per gram of dry plant weight and its potential applications as a dietary and medicinal antioxidant. Sauerkraut, a widely consumed fermented vegetable, contains essential antioxidants like isothiocyanates, ascorbates (ascorbigen), and indole-3-carbinol, which are formed through the breakdown of glucosinolates. These compounds possess antioxidant, anticancer, and anti-inflammatory properties, with their levels increasing during microbial fermentation. The concentration of isothiocyanates varies with fermentation duration and storage time, peaking on the 4th–5th day of the fermentation process. Selenium-enriched sauerkraut is recognized as a functional food, demonstrating elevated levels of indole-3-carbinol, indole-3-acetonitrile, and increased antioxidant activity compared to non-enriched varieties. In the case of fermented bamboo shoots, the content of flavonoids and total phenolic compounds (TPC) rises during fermentation due to enzymatic hydrolysis of phenolic glycosides by fermenting microorganisms, resulting in the release or formation of bioactive compounds such as protocatechuic acid, p-hydroxybenzoic acid, and syringic acid (Singhal et al. 2021).

The use of the *Lactococcus lactis* strain resulted in a notable increase of 113% in phenols and 147% in flavonoids. This heightened total phenolic content (TPC) in the fermented product is likely attributed to the enzymatic activity of β-glucosidase, which facilitates the release of phenols. This effect was observed in soy beverage analysis (Chien et al. 2006). However, it's important to note that not all strains of lactic acid bacteria (LAB) and types of lactic acid fermentation contribute equally to phenol content enhancement. For example, fermented tomatoes, classified as pickled vegetables, exhibit lower TPC compared to fresh samples (Ricci et al. 2020). Fresh tomato extracts measured using the Folin–Ciocalteu method showed a TPC of 2.4 mg GAE/mL, whereas fermented tomato extracts measured 1.6 mg GAE/mL (Ricci et al. 2020). Fermented vegetables are rich in bioactive compounds, vitamins, and minerals, which can potentially support the treatment of various diseases. The abundant polyphenol content in these foods plays a crucial role in reducing oxidative stress biomarkers and the risks associated with cancer and cardiovascular diseases.

In the realm of fruits, they are widely acknowledged as significant sources of natural antioxidants in human diets. A study by Dulf et al. (2016) explored the impact of solid-state fermentation (SSF) by *Rhizopus oligosporus* and *Aspergillus niger* on the antioxidant properties of plum pomaces. The research revealed heightened free-radical scavenging activities, leading to increased antioxidant potential in plum pomaces. Another investigation by Dachery et al. (2019) examined the influence of *Aspergillus carbonarius* on antioxidant activity in *Cabernet Sauvignon* (red) grapes and *Moscato Italico* (white) grapes. Their findings indicated an increase in antioxidant activity in red grapes but a decrease in white grapes. Conversely, Jiménez-López et al. (2018) observed only a slight rise in antioxidant activity during the fermentation of caper berries (*Capparis spinosa* L.). The breakdown of glucosinolate glucocapparin during fermentation resulted in methyl isothiocyanate as the primary degradation product. While most phenolic compounds remained unchanged during caper berry fermentation, there was hydrolysis of quercetin glycosides and degradation of glucosinolates, likely due to pH variations. Additionally, a reduction in epicatechin concentration was noted.

10.8 Changes of Antioxidant Constituents in Plant-Based Food Material During Fermentation

10.8.1 Phenolic Compounds

Phenolic compounds, found abundantly in plants, are considered vital phytochemicals due to their potent antioxidant properties, stemming from the reactivity of hydroxyl substitutions and their ability to scavenge free radicals. Ricci et al. (2019) noted that fermentation can enhance flavanol glycosides and anthocyanins, produce phenyllactic acids, and alter hydroxycinnamic acids. Their study also documented the generation of catechol and dihydrocaffeic acid during the fermentation of elderberry juice, alongside the consumption of protocatechuic acid and caffeic acid by lactic acid bacteria. Fermented elderberry juice exhibited characteristics such as flavonols, flavonol glycosides, anthocyanins, hydroxybenzoic acid, phenyllactic acids, and hydroxycinnamic acids.

10.8.2 Carotenoids

Carotenoids are essential antioxidants, and fermentation has been demonstrated to promote the release of these compounds. According to Oloo et al. (2014), lactic acid fermentation is particularly effective in preserving β-carotene compared to alternative methods such as deep-frying, blanching, and drying (Mapelli-Brahm et al. 2020).

10.8.3 Phytosterols

Phytosterols, natural antioxidants found in soy germ, contribute significantly to its antioxidant capacity. Examples of these phytosterols include stigmasterol, D-7-avenasterol, campesterol, β-sitosterol, and 5-avenasterol (Adeyeye 2013). Adeyeye's research, comparing changes in phytosterol content during the fermentation of African locust bean seeds, showed higher values in the fermented samples compared to the unfermented ones. However, Hubert et al. (2008) found a decrease in phytosterol content in soy germ fermented with lactic acid bacteria after 48 hours. This reduction might be due to dehydration or oxidation in an oxygen-rich environment when exposed to heat or ultraviolet light. Additionally, the study noted that phytosterols in soy germ exhibited limited efficacy in scavenging the superoxide anion radical.

10.8.4 Saponins

Saponins are chemical compounds consisting of a triterpenoid or steroid sapogenin nucleus with one or more carbohydrate branches. These natural glycosides contribute to the antioxidant capacity of soy germ, as noted by Hubert et al. (2008). However, soyasaponins were found to have limited effectiveness in scavenging the superoxide anion radical (O−2). The study documented significant alterations in the conjugation profile of soyasaponins during the fermentation of soy germ by lactic acid bacteria. Specifically, there was a notable decrease in 2,3-dihydro-2,5-dihydroxy-6-methyl-4H-pyran-4-one (DDMP)-conjugated saponin after 10 hours of fermentation, attributed to enzymatic hydrolysis leading to the loss of terminal sugars and conversion into other unidentified structures. Hubert et al. (2008) discussed the transformation of DDMP-conjugated soyasaponins into DDMP-moieties with the release of maltol molecules during fermentation, while soyasaponins B remained unchanged from 10 to 48 hours of incubation.

10.8.5 Antioxidant Peptides

Antioxidant peptides, known for their ability to inhibit peroxidation and scavenge free radicals, exhibit activity influenced by factors such as molecular weight, amino acid sequence, and side chain groups of amino acids. Muhialdin et al. (2020) highlighted rapeseed and bitter bean peptides, which demonstrated potent free-radical scavenging activities but low ferrous ion-chelating activity. Peptides possessing metal ion-chelating properties can act as indirect antioxidants. During fermentation, Muhialdin et al. (2020) observed that fermenting bitter beans increased peptide content with lower molecular weights than boiling processes. The enhanced antioxidant activity observed in *Lactobacillus fermentum* ATCC9338 fermented bitter beans was attributed to the higher levels of low-molecular-weight peptides. Furthermore, fermentation processes can reduce glucosinolate content, thereby enhancing the nutritional value of hydrolyzed peptide products. Various food items undergo peptide increase through protein hydrolysis facilitated by microbial proteolytic enzymes (He et al. 2012).

10.8.6 Antioxidant Polysaccharides

Polysaccharides play a vital role in various biological activities, particularly in demonstrating antioxidant properties, which are closely linked to their structural composition and molecular weight. According to Miao et al. (2020), the fermentation process significantly enhances the antioxidant activities of *Auricularia auricula* polysaccharides. During fermentation, the pyran-type polysaccharide transforms into a furan-type polysaccharide, resulting in improved antioxidant activity. Similarly, Liu et al. (2019) investigated polysaccharides from *Inonotus hispidus* obtained through fermentation. They identified two polysaccharide fractions with higher antioxidant activity and highlighted the advantages of fungal polysaccharides produced via solid-state fermentation, including shorter production cycles and enhanced bioavailability. The study illustrated that fermented polysaccharides exhibited superior antioxidant activity compared to polysaccharides derived from the fruiting body of *Inonotus hispidus* without fermentation. Furthermore, these fermented polysaccharides demonstrated a protective effect by reducing cellular damage induced by H_2O_2 and enhancing cell survival rates.

10.9 Challenges and Solutions in Fermenting Fruits and Vegetables

10.9.1 Microbial Spoilage of Fermented Fruits and Vegetables

As previously noted, microorganisms play a critical role in vegetable fermentation, influencing both the quality and safety of the final product. In the context of table olives, certain microbes such as *Enterobacter* sp., *Citrobacter* sp., *and Saccharomyces kluyveri* have been identified as potential contributors to gas pocket formation, which can lead to defects and significant economic losses (Garrido-Ferna'ndez et al. 1997). This issue is also observed in cucumber pickles, resulting in the development of superficial or internal gas pockets (alambrado) (Garrido-Fernandez et al. 1997; Fleming et al. 1975). Factors contributing to gas pocket formation include CO_2 production by yeasts, heterofermentative lactic acid bacteria (LAB), homofermentative LAB capable of metabolizing malic acid, respiration of fresh fruits prior to brining, storage without refrigeration, and field growing conditions (McFeeters et al. 1984). Effective strategies to mitigate these defects include the use of nitrogen purging systems, pH adjustment at the beginning of fermentation, and the application of preservatives targeting yeasts (Asehraou et al. 2002).

During lactic acid vegetable fermentations, the initial abundance of nutrients and a nearly neutral pH can contribute to the occurrence of putrid and butyric fermentations (Gililland and Vaughn 1943). Putrid fermentation, characterized by a decomposing organic matter odor, is typically associated with Clostridium spp., anaerobic bacteria that form spores and thrive in the lower regions of tall fermentation vessels under anaerobic conditions. Specifically, *Clostridium butyricum* is known to cause butyric fermentations, which are characterized by a smell resembling rancid butter (Gililland and Vaughn 1943). Putrid fermentations may also involve *Desulfovibrio* spp., which produce sulfhydryl compounds (Levin and Vaughn 1966). In lactic acid vegetable fermentations, poor sanitation practices for fermentation vessels or the use of contaminated water are significant contributors to spoilage. After completing lactic acid vegetable fermentation, the physicochemical conditions—high acidity and low pH—play a crucial role in preserving the product. However, undesired secondary fermentations or spoilage can arise, often initiated by propionic acid bacteria, notably Propionibacterium spp. These bacteria metabolize sugars or lactic acid during primary fermentation, producing propionic acid, acetic acid, and CO_2, which can elevate pH and volatile acidity (Gonzalez-Cancho et al. 1980). Such conditions create favorable environments for Clostridium species and Propionibacterium, promoting spoilage known as zapateria and causing abnormal odors and tastes in table olives (Plastourgos and Vaughn 1957). The resulting higher pH also encourages the growth of other spoilage microbes like spore-forming *Clostridium botulinum*, which produces the lethal botulinum toxin (Medina-Pradas and Arroyo-Lo'pez 2015). Controlling salt concentration and pH in brine is crucial to inhibit the occurrence of these spoilage microorganisms and their associated off-odor fermentations, particularly in warmer temperatures during summers (Gonzalez-Cancho et al. 1970). Despite the depletion of lactic acid, propionic acid generated by propionic acid bacteria can offer beneficial attributes in other fermented foods, such as sauerkraut (Babuchowski et al. 1999). It acts as a potent inhibitor of mold, inhibiting the occurrence of pathogenic fungi capable of releasing hazardous substances and off-flavors.

Gram-negative bacteria from genera like Enterobacter, Citrobacter, and Escherichia are commonly present in the initial phases of vegetable fermentations. These bacteria metabolize sugars, producing CO_2, which can affect the quality and safety of the preserved product (Garrido-Ferna'ndez et al. 1997). After lactic acid vegetable fermentation, product preservation relies on physicochemical factors such as high acidity and low pH. However, there is a risk of undesired secondary fermentation or spoilage, often initiated by propionic acid bacteria, particularly *Propionibacterium* spp. These bacteria metabolize sugars or lactic acid, generating propionic acid, acetic acid, and CO_2, which can raise pH and increase volatile acidity (Gonzalez-Cancho et al. 1980). Acetic acid bacteria (AAB) and *Lactobacillus* spp. are also implicated in the onset of spoilage in fermented cucumbers (Medina et al. 2016). AAB can convert lactic acid into acetic acid, while *Lactobacillus buchneri* produces acetic acid and 1,2-propanediol during the initial stages of undesired secondary fermentation (Johanningsmeier and McFeeters 2015).

During vegetable fermentation, yeast populations coexist alongside lactic acid bacteria (LAB). Yeasts primarily contribute to taste and aroma development, particularly in fermentation methods like direct brining of olives (Arroyo-Lo'pez et al. 2012). However, certain yeasts such as *Pichia kudriavzevii* can cause kimchi spoilage due to film formation, production of off-odors, and undesirable flavors (Moon et al. 2014). Overgrowth of fermentative yeasts like *Saccharomyces cerevisiae* and

Wickerhamomyces anomalus can lead to excessive CO_2 production, potentially damaging vegetables by creating gas pockets or causing "alambrado" (Garrido-Fernandez et al. 1997). Moreover, food spoilage can occur when oxidative yeasts consume lactic acids produced during olive fermentation under aerobic conditions (Ruiz-Cruz and Gonzalez-Cancho 1969). Yeasts such as *Pichia manshurica* and *Issatchenkia occidentalis* use lactic and acetic acids through aerobic metabolism, raising pH levels and initiating spoilage in fermented cucumbers (Franco et al. 2012). Certain yeast strains involved in vegetable fermentation also possess enzymatic properties (proteases, xylanases, and pectinases), which can contribute to fruit softening. Pectinolytic Gram-negative bacteria including *Aerobacter*, *Aeromonas*, *Achromobacter*, *Escherichia*, and *Paracolobactrum* are known to cause softening and skin sloughing in oxidized black olives, particularly under high-temperature conditions (Garrido-Fernandez et al. 1997). Pink sauerkraut, a common production defect, is caused by naturally occurring yeasts in cabbage, which thrive under conditions such as high sugar concentration, sodium chloride levels exceeding 3%, elevated acidity, and external factors like temperature and oxygen availability (Fred and Peterson 1922). *Lactobacillus brevis* has been associated with the development of a water-soluble red pigment in sterile cabbage juice, a phenomenon observed within pH ranges of 5.2–6.3 and suppressed under anaerobic conditions (Stamer et al. 1973). In sporadic instances within the US cucumber pickling industry, red-colored spoilage of fermented cucumbers has been linked to *Lactobacillus casei* and *L. paracasei*, attributed to their capacity to break down the azo dye tartrazine (FD&C yellow No. 5), which is commonly used as a yellow coloring agent in covering brines (D'ıaz-Mun˜iz et al. 2007).

10.9.2 Prevention of Spoilage

In numerous industries, vegetable fermentation often occurs naturally, which can make the process unpredictable and occasionally result in spoilage. Changes in vegetable packaging may also happen when the final product contains residual sugars and microorganisms. To address these challenges, production methods can be managed using either physicochemical or microbiological techniques.

10.9.2.1 Salting and Acidification

They are key factors in many vegetable fermentations, creating conditions unfavorable for spoilage and harmful microbes. Salt, particularly sodium chloride, influences the flavor of final products, lowers water activity, and affects microbial metabolism, thereby preventing vegetable tissue softening and promoting the diffusion of fruit components into brines. For instance, table olives require salt levels exceeding 8% (wt/vol.) for prolonged bulk storage to inhibit spoilage by *Propionibacteria* species (Garrido-Ferna'ndez et al. 1997). The pH and acidity levels dictate processing criteria and compliance with regulations. Proper acidification through fermentation or organic acids is crucial for preservation; inadequate acidification above pH 4.6 may lead to spoilage, posing health risks due to potential toxin production by *Clostridium* species (Garrido-Ferna'ndez et al. 1997; Pe'rez-D'ıaz et al. 2015). Recommended pH and salt concentrations vary significantly among different vegetables.

10.9.2.2 Hazard Analysis and Critical Control Point (HACCP) System

Fermented vegetables must be prepared and handled following the relevant sections of the General Principles of Food Hygiene (CAC/RCP1-1969, Codex Alimentarius Commission 2003), the Code of Hygienic Practice for Low-Acid and Acidified Low-Acid Canned Foods (CAC/RCP23-1979, Codex Alimentarius Commission 1993), and the Code of Hygienic Practice for Canned Fruit and Vegetable Products (CAC/RCP 2-1969). Additionally, the product must meet microbiological criteria established in accordance with the Principles for the Establishment and Application of Microbiological Criteria for Foods (CAC/GL21-1997, Codex Alimentarius Commission 1997).

10.9.2.3 Preservatives and Thermal Treatment

To extend the shelf life and prevent spoilage of fermented vegetables, the final products may undergo packaging involving pasteurization or sterilization, depending on their specific characteristics. Thermal treatment significantly reduces microbial levels in the products, thereby prolonging shelf life and ensuring stability during storage. However, thermal processing can affect texture and cause losses in the fresh appearance of preserved vegetables (Garrido-Ferna'ndez et al. 1997). Furthermore, processed vegetables treated thermally may require less acidic pH levels and lower salt concentrations for stability. In cases where processing constraints like the use of plastic bags or heat-sensitive seasonings exist, preservatives can serve as an alternative to thermal treatment. This practice is notably relevant in the United States, primarily for stabilizing products during transportation. Common preservatives used in fermented vegetables include sorbic and benzoic acids, along with their salts. Potassium sorbate, for instance, effectively inhibits yeast and mold growth in fermentation brines and certain processed products. Sodium benzoate is another key preservative used in specific fermented and fresh-pack products to prevent bacterial growth, especially that of lactic acid bacteria (LAB) (Pe'rez-D'ıaz et al. 2015).

While pathogenic microorganisms are a concern, European Union countries have also reported unauthorized or excessive levels of various substances in fermented vegetables to the Rapid Alert System for Food and Feed. These include unauthorized colors, excessive erucic acid, sulfites, and benzoic acid, among others, as common non-conformities. However, to date, no illnesses associated with consuming fermented vegetables due to these substances have been documented.

10.9.3 Substrate Cost

A significant challenge in lactic acid (LA) production is the substantial cost associated with raw substrate materials, which can account for 40–70% of the total production costs (Tejayadi and Cheryan 1995). Current LA production heavily depends on expensive food-related carbohydrates like refined sugars (e.g., glucose) or starches (Abdel-Rahman et al. 2013). To address this issue, there is growing interest in utilizing cost-effective non-food substrates such as lignocellulosic biomass, food waste, or algal substrates. Microalgae, for example, offer a promising approach to reducing production costs by utilizing carbon dioxide as a carbon source, thus eliminating the need for additional substrates. Lignocellulosic waste, including materials like corn stover, oil palm empty fruit bunches, rice straw, hardwood, and corncob residue, has been investigated as a viable feedstock for LA production due to its abundance, renewability, and cost-effectiveness (Bai et al. 2016). Additionally, food waste, which is rich in carbohydrates, presents an advantageous resource for LA production and contributes to effective environmental waste management. For instance, kitchen waste has been successfully used to produce 34.5 g/l of L-LA with high selectivity and optical purity using a bacterial consortium derived from marine animal-based compost (Tashiro et al. 2013).

One potential strategy to reduce substrate costs involves utilizing microalgae, which offer several advantages over lignocellulosic biomass. Microalgae thrive in diverse environments with short harvesting times, lack lignin content, which simplifies saccharification, and contain high levels of fermentable sugars (Nguyen et al. 2012a, b). Research using the green microalga *Hydrodictyon reticulatum* demonstrated its effectiveness as a substrate for producing L-Lactic Acid (L-LA) with Lactobacillus paracasei LA104. This study achieved a production of 36.1 g/l of L-LA, with a yield of 0.46 g/g-algae and a productivity of 1.03 g/l/h (Nguyen et al. 2012a). Furthermore, microalgae and cyanobacteria can act as LA producers without external feedstock costs, using light energy to fix CO_2. Some microalgae, containing starch, can undergo "self-fermentation" to convert starch into valuable products like lactic acid (LA) under anaerobic and dark conditions (Hirayama and Ueda 2004). For instance, *Nannochlorum* sp. 26A4 achieved a 70% conversion rate of starch to D-LA under such conditions.

10.9.4 Indirect Utilization of Polymeric Sugars (Cellulose and Xylan)

In the realm of LA production, lignocellulosic biomass is a favored substrate but faces challenges due to the indirect utilization of its polymeric sugars (cellulose and xylan) by most LA producers. Traditional pretreatment methods (physical, chemical, enzymatic) are essential for sugar hydrolysis prior to fermentation (Abdel-Rahman et al. 2011). However, these methods complicate the process, involve additional costs for enzymes and chemicals, and require detoxification due to inhibitory compounds from chemical hydrolysis. Direct conversion of polymeric sugars is thus critical. Novel approaches include exploring LA producers that utilize polymers directly, breeding strains to co-express cellulase/xylanase genes, or introducing the LA pathway into cellulase/xylanase-producing strains. For instance, genetically modified *Lactobacillus brevis* containing the xylanase gene (xynR8) from a rumen fungi metagenomic sample achieved direct conversion of xylan to LA (1.7 g/l) (Hu et al. 2016). Similarly, engineered *Thermoanaerobacterium aotearoense* LA1002 directly consumed 10 g/l of xylan, yielding 3.20 g/l of LA (Yang et al. 2013). Additionally, genome shuffling techniques have been employed to develop strains capable of directly utilizing starchy waste to produce LA (John et al. 2008).

10.9.5 Sensitivity to Inhibitory Compounds

Effective preparation and breakdown of complex lignocellulosic biomass structures are crucial for achieving high production of lactic acid (LA) (Abdel-Rahman et al. 2011). However, most pretreatment methods result in the formation of inhibitory byproducts, such as phenolic compounds from lignin breakdown, furan compounds like furfural and 5-hydroxymethylfurfural (HMF) from sugar degradation, as well as aliphatic acids (e.g., acetic acid, formic acid, and levulinic acid), inorganic ions, and bio-alcohols (Zhang et al. 2016). These byproducts significantly hinder the fermentation process by adversely affecting cell growth, enzyme activity, and LA production, posing challenges for the efficient utilization of lignocellulosic biomass. Recent efforts have identified inhibitor-tolerant strains suitable for LA production from natural sources. For example, *Pediococcus acidilactici* DQ2 has demonstrated high tolerance to inhibitors such as 3.0 g/l of furfural, 3.0 g/l of HMF, 3.6 g/l of acetic acid, and 3.2 g/l of levulinic acid (Zhao et al. 2013). Another isolated strain, Bacillus sp. P38 exhibits tolerance to 10 g/l of 2-furfural and 6 g/l of vanillin or acetic acid (Peng et al. 2013). Additionally, Zhang et al. (2014) isolated *Bacillus coagulans* IPE22, which shows resilience to various inhibitors (furfural, formic acid, acetic acid, vanillin, and sulfuric acid) found in lignocellulosic hydrolysates.

10.9.6 Contamination Challenges

The majority of Bacillus species identified for lactic acid (LA) production are thermophilic, thriving optimally at 50 °C. In contrast, industrially significant lactic acid bacteria (LAB) are mesophilic, with an optimal temperature of ≤37 °C, which increases the risk of contamination and may reduce fermentation efficiency (Abdel-Rahman et al. 2013). Utilizing thermophilic strains for LA production offers several advantages, including decreased contamination risk, the potential for non-sterilized fermentation (thus saving energy required for sterilizing and cooling media), and facilitating simultaneous saccharification and fermentation processes. This is particularly beneficial as many saccharification enzymes such as cellulases and xylanases are naturally thermophilic (Abdel-Rahman et al. 2015).

To tackle this issue, there is a growing interest in identifying new thermotolerant or thermophilic lactic acid (LA) producers, as well as developing strategies to enhance the temperature resilience of existing strains. Conducting lactic acid fermentations under extreme pH conditions (acidic or alkaline) can also help mitigate contamination risks. Several lactic acid bacteria (LAB) capable of producing lactic acid at elevated temperatures (>40 °C) have been isolated from natural sources. For example, *Lactobacillus delbrueckii* subsp. *lactis* QU 41 and *Enterococcus mundtii QU 25* have demonstrated optimal fermentation temperatures at 43 °C (Tashiro et al. 2011). Recently, *Enterococcus faecium* QU 50 was identified as the first thermophilic LAB strain capable of efficient homofermentation, yielding 1.0 g of L-LA per gram of consumed sugar from various lignocellulose-derived sugars at 50 °C (Abdel-Rahman et al. 2013). Ge et al. (2011) observed that *Lactobacillus casei* exhibited reduced growth and L-LA production with increased glucose concentration and temperature. They improved the thermo- and osmotic stress tolerance of an *L. casei* mutant by gradually increasing temperature by 2 °C and glucose concentration by 20 g/l, achieving levels of 45 °C and 360 g/l, respectively. This led to a 115% increase in L-LA yield and a 97.8% increase in productivity compared to the parent strain producing LA at 34 °C from 210 g/l glucose (Ge et al. 2011).

10.10 Safety Considerations in Fermentation

There is growing interest in traditional fermented foods worldwide, particularly for their potential health benefits and the presence of beneficial bacteria (Mukherjee et al. 2022; Campbell et al. 2022; Walsh et al. 2017). This interest has spurred the development of various new fermented food products in the market, many of which are associated with health claims. Fermented foods that offer health benefits are termed functional fermented foods, aligning with the definition of functional foods, which states that these foods positively impact one or more bodily functions beyond basic nutrition, contributing to improved health and well-being or reduced disease risk (Contor 2001). The growing interest in producing and marketing fermented goods has also emphasized the need for stricter regulations and increased scientific data to substantiate health claims and comply with regulatory requirements. Furthermore, this has prompted a renewed focus on ensuring safety, transparency, and quality standards. Regulatory frameworks governing fermented foods vary in their motivations, provisions, and enforcement levels, often influenced by regional, political, and cultural considerations. These regulations typically represent a collaborative effort involving scientific and political stakeholders to integrate data on foods, dietary practices, and health into culturally sensitive and actionable public health policies. However, international legislation concerning fermented foods often lacks cohesion and coherence, primarily concentrating on food safety issues (Laulund et al. 2017).

10.11 Regulatory Guidelines for Fermented Products

Most worldwide regulatory frameworks governing foods are designed in general accord with recommendations of the Codex Alimentarius Commission, hence simplifying harmonization among governments/nations. However, fermented foods are not substantially covered in the Codex Alimentarius. Fermented vegetables belong to the food category, and as such, their production, distribution, storage, and consumption are subject to regulations governing food, health, and safety. The safety of acidified vegetable products has been a worry owing to outbreaks of infections induced by vegetative cells of acid-resistant food pathogens, particularly acid foods. Both the Food and Drug Administration (FDA) and the industries responsible for acidified vegetables have taken steps to address these concerns. In 1979, the FDA implemented regulations (21 CFR part 114) to ensure the safety measures for manufacturing acidified vegetables. The primary goal was to avoid the occurrence of botulism resulting from inappropriate acidification. Acidified foods must maintain a final pH value at or below 4.6 for all ingredients, as studies have demonstrated that maintaining a pH below 4.6 prevents the production of toxins and spore outgrowth by *C. botulinum* (Breidt and Caldwell 2011). Freshly harvested cucumbers along with other vegetables that are being employed for commercial fermentation have been reported to harbor bacterial pathogens. For *E. coli* O157:H7 and other pathogens in acidified products, the FDA has set a 5-log reduction threshold. While no specific microbiological limits are defined for fermented fruits and vegetables, they are considered ready-to-eat (RTE) foods. The European Union is subject to European Commission Regulation No. 2073/2005 on microbiological standards for foodstuffs, which mandates that *L. monocytogenes* be monitored for the duration of the product's shelf life at a limit of 100 CFU/g. Based on pH and water activity levels, the regulation separates goods that can or cannot sustain *L. monocytogenes* development.

Fermented fruits and vegetables must meet maximum residual levels of pesticides or other pollutants in addition to microbiological requirements. For example, the CODEX Standard for Pickled Fruits and Vegetables deals with pesticide residues, whereas the CODEX Standard for Pickled Cucumbers specifies limitations for tin and lead. Achieving zero risk of foodborne disease and harm should be the goal of processors. Only if processors adhere to procedures that guarantee vegetables chosen for processing are grown using good agricultural practices (GAP), processed using the principles of good manufacturing practices (GMP), and produced at facilities equipped with staff capable of adhering to good hygiene practices (GHP) will this be possible. The hazard analysis and critical control point (HACCP) system and other tools to avoid deliberate adulterations, such as VACCP (vulnerability analysis and critical control points), TACCP (threat assessment and critical control points), and CARVER (criticality, accessibility, recuperability, vulnerability, effect, and recognizability), which are required in some countries, should be regarded as part of food safety management systems.

The 2001 publication of the Codex Alimentarius Standards for kimchi (CXS 223-2001) was followed by a recent revision (CXS 223-2001; FAO & WHO 2017). According to the Codex Standard, kimchi is made in three steps, with fermentation, ripening, and preservation making up the last stage. Additionally, it lists salt, spice mixes, and Chinese cabbage as necessary components. Important organoleptic characteristics include "a reasonably firm, crisp, and chewy" texture, a hot, salty, and/or sour flavor, and "red color originating from red pepper." Apart from a brief discussion on microbiological safety and hygiene, the text does not go into depth about the bacteria that ought to be present. Like the Codex Standard for fermented milks, CXS 223-2001 also outlines the acceptable components, quality standards, analytical techniques, weights and measures, pollutants, and labeling and hygiene guidelines. In addition to the Codex Alimentarius-approved Codex Standards for fermented foods, the FAO/WHO Regional Coordinating Committees have created a number of regional standards. The majority of these regional standards pertain to regionally specific traditional fermented foods and drinks that are only moderately eaten elsewhere in the globe. Fermented cooked cassava-based products (CXS 334R-2020, FAO & WHO, 2020) from Africa, doogh (CXS 332R-2018), fermented soybean (CXS 298R-2009, FAO & WHO, 2009a), gochujang (CXS 294R-2009, FAO & WHO, 2009b), and tempeh (CXS 313R-2013, FAO & WHO, 2013) from Asia are among the fermented foods covered by the Regional Codex Standards now in effect.

According to the Codex Standard, gochujang, a fermented dark red pasty snack that is popular in Korea, is made by fermenting saccharified material with strains of *Aspergillus* sp. that do not produce toxins or damage (CXS 294R-2009; FAO & WHO 2009b). The standard lists fermented soybean paste, fermented rice, fermented wheat protein, and powdered meju—a microbially fermented soybean product—as optional components for gochujang. The food is defined as fermented by either autochthonous, naturally occurring, or cultured microbes (non-pathogenic, non-toxic strains of *Bacillus* sp. and/or *Aspergillus* sp.), with soybean being an essential component (CXS 298R-2009; FAO & WHO 2009a). This definition was last updated in 2020. Seaweed, spices, herbs, yeasts, yeast extracts, and strains of *Lactobacillus* and/or *Lactococcus* are included as optional components. "A small, white, cake-form product, manufactured using dehulled boiled soy via solid-state fermentation with *Rhizopus* spp." is how the Codex Standards define tempe (CXS 313R-2013). Additionally, it states that the key ingredients of tempe are soybeans and a "mold of *Rhizopus* spp. (*R. oryzae, R. stolonifer,* and *R. oligosporus*) mixed with cooked rice powder, rice bran powder, and/or wheat bran powder as an inocula." According to CXS 313R-2013, tempeh cannot include any additives and must have a texture that is "compact and not readily fragmented upon cutting with knife." According to CXS 334R-2020 (FAO & WHO, 2020), the Codex Standard for fermented cooked cassava-based products, the dish is made from fresh cassava roots (*Manihot esculenta* Crantz) and then fermented in water, pressed, dried, and cooked; microbiological data is not available.

Additional information about additive standards, such as taste enhancers and preservatives, pollutants, weights and measures, analytical techniques, and labeling and hygiene guidelines, is provided by the regional standards for each of the fermented foods. According to data on the FAO website as of February 14, 2022, the Coordinating Committees of Europe, North America, and South America have not released any Regional Codex Standards pertaining to fermented foods. There isn't yet a Codex Standard for kombucha. At the World Trade Organization (WTO), however, a Ugandan-prepared Standards and Specification document for kombucha has been debated. In addition to other specifics, the proposed standard outlines quality standards for chemical (heavy metals, pesticides) and microbiological pollutants as well as analytical metrics such as acidity from acetic acid, alcohol content, total sugar, and acidity in lactic acid (De Miranda et al. 2022).

According to Mukherjee et al. (2022), the Food Safety and Standards Act (No. 34 of 2006; amended in No. 13 of 2008) established the Food Safety and Standards Authority of India (FSSAI), which is in charge of food surveillance and the act's implementation. It also established science-based guidelines for food production, storage, distribution, export, and import, among other things. The Food Safety and Standards (Food Products Standards and Food Additives) Regulations, 2011, deal with the regulation of fermented milks. Curd, an Indian fermented milk that resembles yogurt, is one of the milk and dairy products for which the rule specifies acceptable compositional (mostly nutritional) requirements. If appropriate statements are made on the label, *Bifidobacterium bifidum, Lactobacillus acidophilus*, or "other harmless lactic acid bacteria" may be added to yogurt, which is specifically defined as fermented milk made by lactic acid fermentation through *L. delbrueckii* subsp. *bulgaricus* and *S. thermophilus*. The microorganisms in the finished product must be "viable and abundant." The rule also outlines the different yogurt additions that are permitted. The paper also describes compositional requirements for different kinds of cheeses. The Food Safety and Standards Regulations, 2016, had another law that applies to fermented foods. These regulations went into effect on January 1, 2018. Although the law does not specifically address fermented foods, two appropriate classifications for fermented products with health claims are "food with added probiotic ingredients" and "novel foods." According to Indian regulations, "probiotic" labeling must be linked to a health claim and supported by substantial scientific proof that is assessed by the FSSAI. The legislation also establishes the spectrum of health claims that are permitted. Probiotic-containing foods must contain at least 10^8 CFU/g of live microorganisms. Additionally, the rule specifies certain labeling specifications for these goods. Schedule VII of the rule contains the list of authorized probiotic organisms, which is updated on a regular basis. Novel foods are generally understood to be those that have never been consumed by humans or that are created using cutting-edge technology to, for instance, change their nutritional content or lower their amounts of unwanted elements. Importing them requires FSSAI permission. The rule also applies to health claims made for new foods.

10.12 Potential Risks and Precautions for Consumers

Fermented meals have the same danger of pollutants that are harmful to human health as fresh or differently processed foods. Chemical hazards of microbiological origin and (micro) biological risks are the two categories of dangers. Numerous fermented foods, including cheese, sausages, and fermented cereals, have been linked to cases of microbial infections (Capozzi et al. 2017; Nout 1994). However, it is also necessary to take into account harmful byproducts of microbiological origin, such as mycotoxins, ethyl carbamate, and biogenic amines. As biological risk factors for this significant class of foods, a number of relevant pathogens have been previously identified, including *Escherichia coli* O157:H7, *Salmonella* sp., *Klebsiella, Staphylococcus aureus, Aeromonas, Listeria monocytogenes, Vibrio cholerae, Campylobacter, Shigella* sp., *Bacillus cereus,* and several molecules from the class of biogenic amines (Capozzi et al. 2017; Ostry et al. 2013, Gadaga et al. 2004).

Bacteriocins and organic acids that can stop the development of harmful microbes are produced by the LAB and its metabolites. *Lactobacillus plantarum* (prevalence of 67.6%), *Pediococcus pentosaceus* (21.6%), *Lactobacillus brevis* (9.5%), and *Lactobacillus farciminis* (1.3%) were found in the LAB of 30 *nem chua* samples (Tran et al. 2011). According to reports, *L. plantarum* is the most common in fermented products made from meat and legumes (La Anh 2015; Nguyen et al. 2013). Microbiological contamination can also originate from vegetables (Ha et al. 2019). Cucumbers, mustard greens, cabbages, immature melons, Chinese cabbages, bamboo shoots, papayas, bean sprouts with Asian spider flowers, and other vegetables are among the various types of fermented vegetables. They are often made on a modest scale with back-slopping fermentation sometimes, and largely spontaneous fermentation. In less developed nations, back-slopping and spontaneous fermentation are two dependable and affordable ways to produce fermented foods and drinks (Owens 2014). These fermented veggies are often sold in open containers at local markets, and the majority of vendors are farmers. A study that examined the microbiological quality of fermented vegetables sold in the markets of Phnom Penh, the capital of Cambodia, revealed a link between microbial contamination and hygienic practices (such as using the same rice paddle for all products, using gloves, using high-quality containers, and keeping containers open). The authors hypothesized that poor cultural norms and a lack of hygiene measures may extend to vegetable farming (García et al. 2019).

10.12.1 Toxin/Toxic Chemicals Produced by Microbes

10.12.1.1 Mycotoxin

The five filamentous fungal genera *Penicillium, Fusarium, Cladosporium, Aspergillus*, and *Alternaria* are the primary producers of mycotoxins, which are hazardous secondary fungal metabolites. Some of the mycotoxins generated by these molds are exclusive to a single species, but the majority may be produced by a variety of fungi (Bräse et al. 2009). Cereals and legumes, including wheat, rye, barley, rice, oats, maize, alfalfa, peanuts, beans, clover, peas, lentils, chickpeas, mesquite, lupine, soybeans, carob, tamarind, and some other cereal grains that are typically utilized as substrates for traditional fermented foods, are particularly susceptible to mycotoxin contamination. Mycotoxin contamination of those substrates can occur either prior to or during harvest (FAO 1991), which can lead to mycotoxin contamination of the final products. When it comes to processing and cooking temperatures, mycotoxin is comparatively constant. They cannot be safely removed after they have contaminated food and feed. This implies that if tainted foods are put on the market, human exposure is virtually certain. This issue is particularly severe on the Indian subcontinent because of the ideal humidity and temperature for the growth of mold. According to a number of studies, agricultural goods are heavily contaminated, and because mycotoxins are so stable, they continue to exist in agricultural soils (Tran-Dinh et al. 2009). Similar to unfermented items, mycotoxins might be anticipated in fermented ones (Sivamaruthi et al. 2018). Some mycotoxins have been found in soy fermented goods, despite the rarity of mycotoxin tests in fermented products (Petchkongkaew et al. 2008; Thanh and Viet Anh 2016). The presence of aflatoxin in 4 out of 14 brands of Tuong can be explained by the fact that *Aspergillus orizae*, which is extremely close to and thought to be the domesticated form of *A. flavus*, the aflatoxin producer, ferments the soy sauce, also known as Tuong. Mycotoxins are frequently found in this type of goods, according to a recent research that presented mycotoxins in fermented products worldwide (Sivamaruthi et al. 2018). The primary causes of mycotoxins are mold growth circumstances, which may be reduced by adhering to best practices, and the extremely stable nature of the relevant molecules, which can persist in the environment for a considerable amount of time after mold creation. However, there are other ways to reduce the amount of contamination, such as by employing adsorbent materials like microbial catalysts that can break down the toxins or bentionite mineral clays for swine husbandry (Thieu et al. 2008; Petchkongkaew et al. 2008).

10.12.1.2 Biogenic Amines

Low-molecular-weight nitrogen-based substances called "biogenic amines" are found naturally in microbes, plants, and mammals. They serve a variety of purposes in these species, such as regulating gene expression and promoting cell development and differentiation (Suzzi and Torriani 2015). Histamine, tyramine, cadaverine, putrescine, tryptamine, 2-phenylethylamine, spermine, and spermidine are among the biogenic amines that have been extensively documented to exist in fermented foods such as fish, pork, cheese, vegetables, and wines (Spano et al. 2010). Unfortunately, because biogenic amines can have harmful consequences, consuming foods or beverages that contain excessive levels of them poses a danger to consumer health (Park et al. 2019). Microbial degradation of amino acids is the primary source of the biogenic amines found in fermented foods (Mah et al. 2019). Many

Gram-positive and Gram-negative bacteria, such as *Staphylococci, Bacillus* species, and lactic acid bacteria (LAB), as well as *Enterobacteria* and *Pseudomonas,* are the primary microbial groups linked to the buildup of biogenic amines. Numerous biogenic amines are present in a number of traditional fermented foods, such as fermented soybean products and fish sauce (Prester 2011; Zaman et al. 2009; Park et al. 2019).

10.12.2 Chemical Contamination

Pesticides found in raw materials are the primary cause for worry when it comes to fermented food items. According to a survey conducted in the Red River delta, agricultural pesticides were often found in biota, resulting in repeated analyses that exceeded the permissible daily intakes in vegetables and fish (Hoai et al. 2011). In these settings, even prohibited organochlorine herbicides continue to exist. While some fruits (watermelon and durian) were deemed safe despite the presence of chemicals, other vegetables, such as Chinese cabbage and certain herbs, are cited as having reached the residue limits set by the European Union (Sapbamrer and Hongsibsong 2014; Wanwimolruk et al. 2015). In contrast to non-fermented products, fermented items are characterized by the presence of microorganisms. The presence of pesticides and their degradation products tends to limit the activity of microorganisms and modify the sensory properties of the product, but in the interim, microorganisms can degrade these products and help decrease the contamination of fermented products (Regueiro et al. 2015). The fate of pesticides in food products depends on the specific physicochemical properties of the compounds as well as the conditions of food preparation (pH, presence of water, temperature, etc.). Research on pesticides in food is still in its early stages, and more research is needed to fully understand how they degrade, the risks associated with the products of that breakdown, and if microbes can catalyze the process.

Pesticide-induced food contamination is a global issue, but when it comes to particular contaminations, agricultural practices are clearly a means to improve outcomes. For example, 97% of mangosteen fruits sold locally exceed MRLs, while no issues are found with GAP production of fruits meant for the EU (Wanwimolruk et al. 2015). This issue might also be improved by organic farming, although some measures are still needed to ensure production, particularly in areas with less control. Although it is possible to utilize untreated irrigation or washing water or improper organic fertilizers (such as compost or manure), the safety of organic foods is yet unknown (Taban et al. 2011; Nguyen et al. 2016). There are notable quantitative and qualitative variations between conventional and organic goods, according to data on the microbial populations connected to the surface of fresh fruits and vegetables (Leff and Fierer 2013; Bigot et al. 2015). Various studies have been conducted to assess the degree of contamination of organic vs conventional vegetables, particularly leafy ones, using E. coli, Salmonella, and Listeria as indicators. These bacteria are the main contaminating pathogens for leafy vegetables

(Tango et al. 2014; Karp et al. 2016). The results show that there is neither a significant trend nor a minor influence on vegetables cultivated organically (Leff and Fierer 2013; Tango et al. 2014). Research on the presence of mycotoxin of agricultural products has mostly focused on fruits or cereals, with leafy greens receiving less attention (van der Walt et al. 2006; Sanzani et al. 2016). Although the levels of mycotoxin contamination in organic cereal crops were inconsistent and unclear, they occasionally exceeded those seen in traditional crops (Baert et al. 2006; Magkos et al. 2006; Piqué et al. 2013). More research is required in this area before specific suggestions can be made. In every circumstance, the situation would be substantially improved by teaching farmers, merchants, and transformators.

10.13 Technological Advancements in Fermentation

Over the past few years, there has been a growing interest in new and developing fermentation-based food processing technology. They fall into two general categories: thermal and nonthermal processes. Ohmic heating (OH), radio frequency (RF), and microwave heating (MH) are examples of thermal processes; high-pressure processing (HPP), ultrasound (US) irradiation [gamma irradiation (ʤ-irradiation), microwave irradiation (MI), and pulsed electric field (PEF)] are examples of novel nonthermal processes. The latter may be used to increase shelf life, inactivate harmful and pathogenic microorganisms, enhance metabolic activities and enzyme production, and shorten the fermentation process. The former may be used to monitor fermentation and pasteurization, as well as to speed up chemical reactions (oxidation, polymerization, condensation, and esterification). Recent years have seen a thorough description and sufficient documentation of these methods (Garde-Cerdán et al. 2016; George and Rastogi 2016; Koubaa et al. 2016; Ojha et al. 2017).

Food goods that are already packaged and cannot be heated are typically subjected to HPP as a last mitigating measure (Bajovic et al. 2012). Despite their inconsistent outcomes, they have drawn a lot of attention as a method for getting rid of microorganisms in fermented foods. While some research has shown that HPP might not be ideal (Marcos et al. 2013; Omer et al. 2015), other research has supported the possibility of using HPP in fermented meals. There have been reports of a significant decrease and removal in the microbial loads of fermented foods (Omer et al. 2010; Gill and Ramaswamy 2008; Avila et al. 2016), while other research has shown that HPP improves composition and shortens the time that wine ages (Oey et al. 2008; Tchabo et al. 2017). Similar to this, microwave heating and irradiation have garnered a lot of attention due to their applications in material treatment, sterilization, and processing time reduction (Rasmussen et al. 2001; Hoai et al. 2011). Additionally, using US increases the activity of microorganisms and/or enzymes, guarantees the safety and quality of the product (Alouache et al. 2015), and encourages esterification, oxidation, and condensation reactions that produce more esters and acids in milk and aged processed wine (Tchabo et al. 2017; Nguyen et al. 2009).

Electric current has a wide range of present applications and potential future uses in fermentation and the creation of value-added products (Cho et al. 2016). Microorganisms have been effectively electroporated using OH (Sastry 2004; Loghavi et al. 2009). Cho et al. (2016) showed that using OH reduced the lag fermentation phase when compared to traditional heating, indicating that it is a superior method for pasteurizing and sterilizing thick foods. Numerous PEF uses in fermentation-related processes have been documented, including improvements in yeast metabolism, fermentation time reduction, phenolic substance and anthocyanin production, and browning (Delsart et al. 2015; Mattar et al. 2015). There aren't much research on the application of RF in fermentation-related processes, but one study found that it improved homogeneity, retained essential bacteria, and had no negative effects on the yogurt's storage stability (Siefarth et al. 2014). These innovative technologies may have drawbacks, such as high investment costs, additional process variables, and process standardization and optimization to comply with legal requirements. To properly comprehend these applications and make their eventual implementation easier, it is necessary to simulate them in industrial circumstances, as the majority of reported applications are also conducted in laboratories.

10.14 Market Trends and Consumer Preferences

The market for fermented foods and drinks is currently very large and includes a variety of food items, such as fermented fruits and vegetables, probiotics, and alcoholic beverages. The global market for fermented foods is expected to increase at a compound annual growth rate (CAGR) of 6% between 2011 and 2015. Given the aforementioned information on the growing need for fermented foods, new developments are crucial for the food and beverage sectors to meet this demand. In order to reduce the danger of sickness, illness, and death, the potential advantages of eating fermented foods can only be guaranteed in a setting free from contamination. Innovations should concentrate on the actions required to rule out such possibilities (Sahu and Panda 2018).

One of the primary goals of the food industry's technical advancements has been to satisfy customer acceptance as well as their desires for new and healthy foods. While certain technological breakthroughs, such as food irradiation and genetically modified foods (GMFs) in Europe, have been readily embraced, others have been readily rejected (Ronteltap et al. 2007). Therefore, customer acceptability and perception play a major role in a new food product's long-term success. Sensorial analysis and market research are often employed methods. In essence, innovation is the process of turning a realistic, financially viable idea into a successful product. This entails a wide range of activities, including technology, marketing, collaboration, and commercialization. Once more, consumers' responsibilities for innovative methods are crucial due to their increased awareness of and preferences for healthy meals. Consumers' acceptance of fermented food is influenced by a variety of elements, including their engagement in the product's production; cultural, social, and lifestyle considerations; the long-term consequences of innovative technology, including potential dangers and health issues; etc. In order to create successful food marketing strategies, it is crucial to evaluate how customers see the creation and promotion of goods (da Silva et al. 2014). Customers are continuously evaluating the possible health advantages of fermented foods and nutritional supplements as a result of growing consumer awareness brought about by technological advancements. Consumers are skeptical of the long-term health concerns associated with these fermented foods, despite scientific research demonstrating their safety. In light of these worries, a sizable portion of the public is looking for clean, green, minimally processed food items, such as organic and artisan foods.

10.15 Sustainable Practices in Fermented Food Production

One of the biggest worldwide issues facing food and agriculture today is sustainability (FAO 2017). According to SUSFOOD (2013), food sustainability is defined as "food system(s) that support food security, make optimal use of natural and human resources, respect biodiversity and ecosystems for present and future generations, are culturally acceptable and accessible, environmentally sound, economically fair and viable, and provide the consumer with nutritionally adequate, safe, healthy, and affordable food." As a result, both major, industrialized enterprises in the West and tiny, family-run, artisanal, and other small-scale firms in developing nations need to practice sustainable food production and processing.

Fermented foods were linked to many of the essential components of sustainability in agriculture even before it was acknowledged as a problem. Traditionally, the raw materials needed to manufacture fermented dairy, vegetable, or fruit products were sourced locally, giving customers access to wholesome, reasonably priced, and safe meals. Generally carried out under moderate settings, fermentation uses less energy than other food processing methods including retorting (such as canning), spray-drying, and freeze-drying. A small amount of waste or by-products are produced during the manufacturing of several fermented foods and drinks, such as yogurt, fish sauces, and fermented cereals. On the other hand, a lot of by-products are produced when various fermented foods and drinks are made. Grape sediments from winemaking, sour whey from cheesemaking, strained yogurt, and wasted grains and yeast lees from beermaking are a few examples. If untreated, these byproducts might cause environmental issues. Furthermore, a significant amount of by-products, including soy pulp (okara) and soy whey, are produced during the production of non-fermented meals and drinks, such as tofu (soybean curd). Food waste biovalorization may be possible by fermentation, which turns these byproducts into consumer goods and/or components with additional value (Vong and Liu 2016; Chua and Liu 2019).

Ecosystems that are social, cultural, and economic have always included fermented foods. The production of fermented foods in developing nations generates demand for raw materials, jobs, and food security, particularly for underprivileged groups (Adesulu and Awojobi 2014). Thus, many of the sustainability concerns shared by industrialized and developing nations are addressed by the fermentation of foods and drinks. Sustainability goals align with waste reduction, agricultural input minimization, productivity enhancement, and efficiency improvement (Capone et al. 2014). Additionally, it is equally important for small and big food producers to build sustainable food supply chains for fermented foods, use by-products, and conserve energy (Waché et al. 2018). Similarly, while approaches may differ, safety is a global sustainability problem as well. For non-pasteurized items, for instance, using protective microorganisms or adding herbs that both provide biopreservation may be useful in ensuring food safety and shelf life (Mahgoub 2019).

10.16 Conclusion

Understanding the vital link between nutrition, beneficial microorganisms, and individual health is crucial for improving food quality and preventing various diseases. To achieve large-scale production, it is recommended to replace mixed fermentation with high variability with pure cultures. Fermentation, particularly lactic acid fermentation, emerges as a practical and cost-effective method for increasing the daily consumption of fresh produce. The transition from mixed fermentation with high variability to pure culture fermentation is recommended for achieving significant scale production. The cultural and traditional roots of fermentation processes are highlighted, emphasizing the need to adapt new molecular approaches for a deeper understanding of microbiota and the selection of autochthonous starters. These approaches, successfully implemented in other fermented foods, can provide valuable insights for controlled fermentation processes in fruits and vegetables. Customizing lactic acid bacteria to diverse environmental conditions unlocks the full potential of vegetables and fruits, enhancing their hygiene, sensory attributes, nutritional content, and shelf life. Overall, the exploration of fermented fruits and vegetables opens avenues for innovation, offering a bridge between traditional practices and modern scientific advancements in the realm of food preservation and human well-being.

References

Abdel-Rahman MA, Tashiro Y, Sonomoto K (2011) Lactic acid production from lignocellulose-derived sugars using lactic acid bacteria: overview and limits. J Biotechnol 156(4):286–301

Abdel-Rahman MA, Tashiro Y, Sonomoto K (2013) Recent advances in lactic acid production by microbial fermentation processes. Biotechnol Adv 31(6):877–902

Adesulu AT, Awojobi KO (2014) Enhancing sustainable development through indigenous fermented food products in Nigeria. Afr J Microbiol Res 8(12):1338–1343

Adeyeye E (2013) The effect of fermentation on the dietary quality of lipids from African locust bean (*Parkia biglobosa*) seeds. Elixir Food Sci 58:14912–14922
Alouache B, Touat A, Boutkedjirt T, Bennamane A (2015) Monitoring of lactic fermentation process by ultrasonic technique. Phys Procedia 70:1057–1060
Altay F, Karbancıoglu-Güler F, Daskaya-Dikmen C (2013) A review on traditional Turkish fermented non-alcoholic beverages: microbiota, fermentation process and quality characteristics. Int J Food Microbiol 167:44–56
Anandharaj M, Sivasankari B (2013) Isolation of potential probiotic Lactobacillus strains from human milk. Int J Res Pharm Life Sci 1(1):26–29
Arroyo-Lopez FN, Romero-Gil V, Bautista-Gallego J, Rodriguez-Gomez F, Jimenez-Dıaz R, Garcıa-Garcıa P et al (2012) Yeasts in table olive processing: desirable or spoilage microorganisms. Int J Food Microbiol 160:42–49
Asehraou A, Peres C, Faid M, Brito D (2002) Reducing the bloater spoilage incidence in fermented green olives during storage. Grasas Aceites 53:330–334
Ávila M, Gómez-Torres N, Delgado D, Gaya P, Garde S (2016) Application of high-pressure processing for controlling *Clostridium tyrobutyricum* and late blowing defect on semi-hard cheese. Food Microbiol 60:165–173
Babuchowski A, Laniewska-Moroz L, Warminska-Radyko I (1999) Propionibacteria in fermented vegetables. Lait 79:113–124
Baert K, De Meulenaer B, Kamala A, Kasase C, Devlieghere F (2006) Occurrence 521 of patulin in organic, conventional, and handcrafted apple juices marketed in Belgium. 522. J Food Protect 69:1371–1378
Bai Z, Gao Z, Sun J, Wu B, He B (2016) D-Lactic acid production by *Sporolactobacillus inulinus* YBS1-5 with simultaneous utilization of cottonseed meal and corncob residue. Bioresour Technol 207:346–352
Bajovic B, Bolumar T, Heinz V (2012) Quality considerations with high-pressure processing of fresh and value-added meat products. Meat Sci 92(3):280–289
Baschali A, Tsakalidou E, Kyriacou A, Karavasiloglou N, Matalas AL (2017) Traditional low-alcoholic and non-alcoholic fermented beverages consumed in European countries: a neglected food group. Nutr Res Rev 30:1–24
Battistini C, Gullón B, Ichimura ES, Gomes AMP, Ribeiro EP, Kunigk L et al (2018) Development and characterization of an innovative synbiotic fermented beverage based on vegetable soybean. Braz J Microbiol 49(2):303–309
Bergqvist SW, Andlid T, Sandberg AS (2006) Lactic acid fermentation stimulated iron absorption by Caco-2 cells is associated with increased soluble iron content in carrot juice. Br J Nutr 96(4):705–711
Bigot C, Meile JC, Kapitan A, Montet D (2015) Discriminating organic and conventional foods by analysis of their microbial ecology: an application to fruits. Food Control 48:123–129
Bräse S, Encinas A, Keck J, Nising CF (2009) Chemistry and biology of mycotoxins and related fungal metabolites. Chem Rev 109(9):3903–3990
Breidt F Jr, Caldwell JM (2011) Survival of Escherichia coli O157: H7 in cucumber fermentation brines. J Food Sci 76(3):M198–M203
Buckenhüskes H et al (1997) Fermented vegetables. ASM Press, Washington DC, pp 595–609
Campbell R, Hauptmann A, Campbell K, Fox S, Marco ML (2022) Better understanding of food and human microbiomes through collaborative research on Inuit fermented foods. Microbiome Res Rep 1:5
Capone R, El Bilali H, Debs P, Cardone G, Driouech N (2014) Food system sustainability and food security: connecting the dots. J Food Secur 2:13–22
Capozzi V, Fragasso M, Romaniello R, Berbegal C, Russo P, Spano G (2017) Spontaneous food fermentations and potential risks for human health. Fermentation 3(4):49
Castro-Alba V, Lazarte CE, Perez-Rea D, Carlsson NG, Almgren A, Bergenståhl B et al (2019) Fermentation of pseudocereals quinoa, canihua, and amaranth to improve mineral accessibility through degradation of phytate. J Sci Food Agric 99(11):5239–5248

Chen YS, Miyashita M, Suzuki KI, Sato H, Hsu JS, Yanagida F (2010) *Lactobacillus pobuzihii* sp. nov., isolated from pobuzihi (fermented cummingcordia). Int J Syst Evol Microbiol 60(8):1914–1917

Chen YS, Wu HC, Lo HY et al (2012) Isolation and characterisation of lactic acid bacteria from jiang-gua (fermented cucumbers), a traditional fermented food in Taiwan. J Sci Food Agric 92(10):2069–2075

Chen YS, Liou MS, Ji SH, Yu CR, Pan SF, Yanagida F (2013) Isolation and characterization of lactic acid bacteria from Yantsai-shin (fermented broccoli stems), a traditional fermented food in Taiwan. J Appl Microbiol 115(1):125–132

Chien HL, Huang HY, Chou CC (2006) Transformation of Isoflavone phytoestrogens during the fermentation of soymilk with lactic acid bacteria and Bifidobacteria. Food Microbiol 23(8):772–778

Cho J, Lee D, Yang C, Jeon J, Kim J, Han H (2006) Microbial population dynamics of kimchi, a fermented cabbage product. FEMS Microbiol Lett 257(2):262–267

Cho WI, Yi JY, Chung MS (2016) Pasteurization of fermented red pepper paste by ohmic heating. Innov Food Sci Emerg Technol 34:180–186

Chua JY, Liu SQ (2019) Soy whey: more than just wastewater from tofu and soy protein isolate industry. Trends Food Sci Technol 91:24–32

Codex Alimentarius Commission (1993). Code of hygienic practice for low-acid and acidified low-acid canned foods (CAC/RCP 23-1979, Rev. 1-1993) Rome: Food and Agriculture Organization of the United Nations/World Health Organization

Codex Alimentarius Commission (CAC) (1997) Principles for the establishment and application of microbiological criteria for foods. CAC/GL 21–1997

Codex Alimentarius Commission (CAC) (2003) Recommended international code of practice general principles of food hygiene. CAC/RCP 1–1969, Rev.4–2003

Contor L (2001) Functional food science in Europe. Nutr Metab Cardiovasc Dis 11(4 Suppl):20–23

da Silva VM, Minim VPR, Ferreira MAM, de Paula Souza PH, da Silva Moraes LE, Minim LA (2014) Study of the perception of consumers in relation to different ice cream concepts. Food Qual Prefer 36:161–168

Dachery B, Hernandes KC, Veras FF, Schmidt L, Augusti PR, Manfroi V et al (2019) Effect of *Aspergillus carbonarius* on ochratoxin a levels, volatile profile and antioxidant activity of the grapes and respective wines. Food Res Int 126:108687

Dauchet L, Kesse-Guyot E, Czernichow S, Bertrais S, Estaquio C, Peneau S et al (2007) Dietary patterns and blood pressure change over 5-y follow-up in the SU.VI.MAX cohort. American. J Clin Nutr 85(6):1650–1656

de Miranda JF, Ruiz LF, Silva CB, Uekane TM, Silva KA, Gonzalez AGM et al (2022) Kombucha: a review of substrates, regulations, composition, and biological properties. J Food Sci 87(2):503–527

Deka DC (2024) Traditional rice-based alcoholic beverages in India. In: Mérillon JM, Riviere C, Lefèvre G (eds) Natural products in beverages. Reference series in Phytochemistry. Springer, Cham. https://doi.org/10.1007/978-3-031-04195-2_140-1

Delsart C, Grimi N, Boussetta N, Sertier CM, Ghidossi R, Peuchot MM, Vorobiev E (2015) Comparison of the effect of pulsed electric field or high voltage electrical discharge for the control of sweet white must fermentation process with the conventional addition of sulfur dioxide. Food Res Int 77:718–724

Demir N, Bachçeci K, Acar J (2006) The effects of different initial *Lactobacillus plantarum* concentrations on some properties of fermented carrot juice. J Food Proc Preserv 30:352–363

Di Cagno R et al (2010) Comparison of phenotypic (Biolog System) and genotypic (random amplified polymorphic DNA-polymerase chain reaction, RAPD-PCR, and amplified fragment length polymorphism, AFLP) methods for typing *Lactobacillus plantarum* isolates from raw vegetables and fruits. Int J Food Microbiol 143(3):246–253

Di Cagno R et al (2011) Exploitation of sweet cherry (*Prunus avium* L.) puree added of stem infusion through fermentation by selected autochthonous lactic acid bacteria. Food Microbiol 28(5):900–909

Di Cagno R, Coda R, De Angelis M, Gobbetti M (2013) Exploitation of vegetables and fruits through lactic acid fermentation. Food Microbiol 33:1–10

Díaz-Muniz I, Kelling R, Hale S, Breidt F, McFeeters RF (2007) Lactobacilli and tartrazine as causative agents of a red colored spoilage in cucumber pickle products. J Food Sci 72:M240–M245

Dulf FV, Vodnar DC, Socaciu C (2016) Effects of solid-state fermentation with two filamentous fungi on the total phenolic contents, flavonoids, antioxidant activities and lipid fractions of plum fruit (*Prunus domestica* L.). Food Chem 209:27–36

Endrizzi I, Framondino V, Ciaghi R, Gasperi F (2006) Towards a new fruit juice line with a high healthy power. Ingredienti Alimentari V:1–7

Endrizzi I, Pirretti G, Calò DG, Gasperi F (2009) A consumer study of fresh juices containing berry fruits. J Sci Food Agric 89(7):1227–1235

Eom HJ, Seo DM, Han NS (2007) Selection of psychotropic Leuconostoc spp. producing highly active dextransucrase from lactate fermented vegetables. Int J Food Microbiol 117(1):61–67

FAO (1998) Fermented fruits and vegetables-a global perspective, vol 134. FAO Agricultural Services Bulletin, Rome, Italy

Feng J, Liu P, Yang X, Zhao X (2015) Screening of immunomodulatory and adhesive Lactobacillus with antagonistic activities against Salmonella from fermented vegetables. World J Microbiol Biotechnol 31(12):1947–1954

Ferruzzi MG, Kruger J, Mohamedshah Z, Debelo H, Taylor JR (2020) Insights from in vitro exploration of factors influencing iron, zinc and provitamin A carotenoid bioaccessibility and intestinal absorption from cereals. J Cereal Sci 103126:103126

Fleming HP, Etchells JL, Thompson RL, Bell TA (1975) Purging of CO2 from cucumber brines to reduce bloater damage. J Food Sci 40:1304–1310

FAO & WHO (2009a) Food and Agriculture Organization of the United Nations. Codex Alimentarius. Regional Standard for Fermented Soybean Paste (CXS 298R-2009)

FAO & WHO (2009b) Food and Agriculture Organization of the United Nations. Codex Alimentarius. Regional Standard for Gochujang (CXS 294R-2009)

FAO & WHO (2013) Food and Agriculture Organization of the United Nations. Codex Alimentarius. Regional Standard for Tempeh (CXS 313R-2013)

FAO & WHO (2017) Food and Agriculture Organization of the United Nations. Codex Alimentarius. Standard for Kimchi (CXS 223–2001)

FAO & WHO (2020) Food and Agriculture Organization of the United Nations. Codex Alimentarius. Regional Standard for Fermented Cassava-based Products (CXS 334R-2020).

FAO (2017) The future of food and agriculture. Trends and challenges. Author, Rome. Retrieved from http://www.fao.org/3/a-i6583e.pdf

Food Nutrition and Agriculture (FAO) (1991) Food for the Future. FAO 1

Franco W, Perez-Diaz IM, Johanningsmeier SD, McFeeters RF (2012) Characteristics of spoilage-associated secondary cucumber fermentation. Appl Environ Microbiol 78:1273–1284

Fred EB, Peterson WH (1922) The production of pink sauerkraut by yeasts. J Bacteriol 7:257–269

Fujisawa T, Shinohara K, Kishimoto Y, Terada A (2009) Effect of Miso soup containing Natto on the composition and metabolic activity of the human Faecal Flora. Microb Ecol Health Dis 18:79–84

Gadaga TH, Nyanga LK, Mutukumira AN (2004) The occurrence, growth and control of pathogens in African fermented foods. Afr J Food Agric Nutr Dev 4(1)

García MG, Fernández-López C, Polesel F, Trapp S (2019) Predicting the uptake of emerging organic contaminants in vegetables irrigated with treated wastewater–implications for food safety assessment. Environ Res 172:175–181

Garde-Cerdán T, Arias M, Martín-Belloso O, Ancín-Azpilicueta C (2016) Pulsed electric field and fermentation. In: Novel food fermentation technologies, pp 85–123

Garrido-Fernandez A, Fernandez-Dıez MJ, Adams RM (1997) Table Olives production and processing. Chapman Hall, London
Ge XY, Yuan J, Qin H, Zhang WG (2011) Improvement of L-lactic acid production by osmotic-tolerant mutant of Lactobacillus casei at high temperature. Appl Microbial Biotechnol 89:73–78
Gebbers JO (2007) Atherosclerosis, cholesterol, nutrition, and statins – a critical review. Ger Med Sci 5:1–11
George JM, Rastogi NK (2016) High-pressure processing for food fermentation. In: Novel food fermentation technologies, pp 57–83
Gililland JR, Vaughn RH (1943) Characteristics of butyric acid bacteria from olives. J Bacteriol 46:315–322
Gill AO, Ramaswamy HS (2008) Application of high pressure processing to kill Escherichia coli O157 in ready-to-eat meats. J Food Prot 71(11):2182–2189
Gonzalez-Cancho F, Nosti-Vega M, Fernandez-Diez MJ, Buzcu N (1970) Propionibacterium spp. associated with olive spoilage. Factors influencing their growth. Microbiologia Espanola 23:233–252
Gonzalez-Cancho F, Rejano-Navarro L, Borbolla y Alcala J. (1980) Formation of propionic acid during the conservation of green table olives. III. Responsible microorganism. Grasas Aceites 31:245–250
Guan Q, Xiong T, Xie M (2021) Influence of probiotic fermented fruit and vegetables on human health and the related industrial development trend. Engineering 7(2):212–218
Ha TM, Shakur S, Do KHP (2019) Consumer concern about food safety in Hanoi, Vietnam. Food Control 98:238–244
He FJ, Nowson CA, Lucas M, MacGregor GA (2007) Increased consumption of fruit and vegetables is related to a reduced risk of coronary heart disease: meta-analysis of cohort studies. J Hum Hypertens 21(9):717–728
He R, Ju X, Yuan J, Wang L, Girgih AT, Aluko RE (2012) Antioxidant activities of rapeseed peptides produced by solid state fermentation. Food Res Int 49:432–438
Hirayama S, Ueda R (2004) Production of optically pure D-lactic acid by *Nannochlorum* sp. 26A4. Appl Biochem Biotechnol 119:71–77
Hoai PM, Sebesvari Z, Minh TB, Viet PH, Renaud FG (2011) Pesticide pollution in agricultural areas of Northern Vietnam: case study in Hoang Liet and Minh Dai communes. Environ Pollut 159(12):3344–3350
Hong SI, Kim YJ, Pyun YR (1999) Acid tolerance of *Lactobacillus plantarum* from Kimchi. Food Sci Technol 32(3):142–148
Hu J, Lin Y, Zhang Z, Xiang T, Mei Y, Zhao S, Liang Y, Peng N (2016) High-titer lactic acid production by *Lactobacillus pentosus* FL0421 from corn stover using fed-batch simultaneous saccharification and fermentation. Bioresour Technol 214:74–80
Hubert J, Berger M, Nepveu F, Paul F, Daydé J (2008) Effects of fermentation on the phytochemical composition and antioxidant properties of soy germ. Food Chem 109:709–721
Hutkins RW (2006) Microbiology and Technology of Fermented Foods. Blackwell Publishing
Hwanhlem N, Tanasupawat S (2015) Identification and application of lactic acid bacteria isolated from fermented fruits and vegetables. Proc Int Conf Food Sci Nutr 61
Jiménez-López J, Ruiz-Medina A, Ortega-Barrales P, Llorent-Martínez EJ (2018) Phytochemical profile and antioxidant activity of caper berries (*Capparis spinosa* L.): evaluation of the influence of the fermentation process. Food Chem 250:54–59
Johanningsmeier SD, McFeeters RF (2015) Metabolic footprinting of *Lactobacillus buchneri* strain LA1147 during anaerobic spoilage of fermented cucumbers. Int J Food Microbiol 215:40–48
John RP, Gangadharan D, Nampoothiri KM (2008) Genome shuffling of *Lactobacillus delbrueckii* mutant and *Bacillus amyloliquefaciens* through protoplasmic fusion for L-lactic acid production from starchy wastes. Bioresour Technol 99(17):8008–8015
Karovičová J, Kohajdová Z (2003) Lactic acid fermented vegetable juices. Hortic Sci (Prague) 30:152–158

Karp DS, Moses R, Gennet S, Jones MS, Joseph S, M'Gonigle LK, Kremen C (2016) Agricultural practices for food safety threaten pest control services for fresh produce. J Appl Ecol 53(5):1402–1412

Katz SE (2012) The art of fermentation: an in-depth exploration of essential concepts and processes from around the world. Chelsea Green Publishing

Kil JH, Jung KO, Lee HS, Hwang IK, Kim YJ, Park KY (2004) Effects of Kimchi on stomach and colon health of *Helicobacter pylori*-infected volunteers. Prev Nutr Food Sci 9(3):161–166

Kim M, Chun J (2005) Bacterial community structure in Kimchi, a Korean fermented vegetable food, as revealed by 16S rRNA gene analysis. Int J Food Microbiol 103(1):91–96

Koubaa M, Barba-Orellana S, Roselló-Soto E, Barba FJ (2016) Gamma irradiation and fermentation. In: Novel food fermentation technologies, pp 143–153

Kumari M, Platel K (2020) Impact of soaking, germination, fermentation, and thermal processing on the bioaccessibility of trace minerals from food grains. J Food Process Preserv 44:e14752

Kumari AN, Pandey AN, Ann A, Raj A, Gupta AN, Chauhan AR, Sharma A, Das AJ, Kumar A, Attri BL, Neopany B (2016) Indigenous alcoholic beverages of South Asia. CRC Press, New York, pp 501–566

La Anh N (2015) Health-promoting microbes in traditional Vietnamese fermented foods: a review. Food Sci Human Wellness 4(4):147–161

Lan WT, Chen YS, Yanagida F (2009) Isolation and characterization of lactic acid bacteria from Yan-dong-gua (fermented wax gourd), a traditional fermented food in Taiwan. J Biosci Bioeng 108(6):484–487

Laulund S, Wind A, Derkx PM, Zuliani V (2017) Regulatory and safety requirements for food cultures. Microorganisms 5(2):28

Laureys D, de Vuyst L (2014) Microbial species diversity, community dynamics, and metabolite kinetics of water kefir fermentation. Appl Environ Microbiol 80:2564–2572

Lee JS, Heo GY, Jun WL et al (2005) Analysis of Kimchi microflora using denaturing gradient gel electrophoresis. Int J Food Microbiol 102(2):143–150

Leff JW, Fierer N (2013) Bacterial communities associated with the surfaces of fresh fruits and vegetables. PLoS One 8(3):e59310

Leroy F, De Vuyst L (2004) Lactic acid bacteria as functional starter cultures for the food fermentation industry. Trends Food Sci Technol 15(2):67–78

Levin RE, Vaughn RH (1966) *Desulfovibrio aestuarii*, the causative agent of hydrogen sulfide spoilage of fermenting olive brines. J Food Sci 31:768–772

Lindow SE, Brandl MT (2003) Microbiology of the phyllosphere. Appl Environ Microbiol 69(4):1875–1883

Liu X, Hou R, Xu K, Chen L, Wu X, Lin W, Zheng M, Fu J (2019) Extraction, characterization and antioxidant activity analysis of the polysaccharide from the solid-state fermentation substrate of *Inonotus hispidus*. Int J Biol Macromol 123:468–476

Loghavi L, Sastry SK, Yousef AE (2009) Effect of moderate electric field frequency and growth stage on the cell membrane permeability of *Lactobacillus acidophilus*. Biotechnol Prog 25(1):85–94

Magkos F, Arvaniti F, Zampelas A (2006) Organic food: buying more safety or just peace of mind? A critical review of the literature. Crit Rev Food Sci Nutr 46:23–56

Mah JH, Park YK, Jin YH, Lee JH, Hwang HJ (2019) Bacterial production and control of biogenic amines in Asian fermented soybean foods. Food Secur 8(2):85

Mahgoub SA (2019) Fermented food in Egypt: a sustainable bio-preservation to improve the safety of food. In: Sustainability of agricultural environment in Egypt: part I: soil-water-food Nexus, pp 231–251

Mapelli-Brahm P, Barba FJ, Remize F, Garcia C, Fessard A, Mousavi Khaneghah A, Sant'Ana AS, Lorenzo JM, Montesano D, Meléndez-Martínez AJ (2020) The impact of fermentation processes on the production, retention and bioavailability of carotenoids: an overview. Trends Food Sci Technol 99:389–401

Marcos B, Aymerich T, Garriga M, Arnau J (2013) Active packaging containing nisin and high pressure processing as post-processing listericidal treatments for convenience fermented sausages. Food Control 30(1):325–330

Marsh AJ, O'Sullivan O, Hill C, Ross RP, Cotter PD (2013) Sequence-based analysis of the microbial composition of water kefir from multiple sources. FEMS Microbiol Lett 348:79–85

Marsh AJ, O'Sullivan O, Hill C, Ross RP, Cotter PD (2014) Sequence-based analysis of the bacterial and fungal compositions of multiple kombucha (tea fungus) samples. Food Microbiol 38:171–178

Mattar JR, Turk MF, Nonus M, Lebovka NI, El Zakhem H, Vorobiev E (2015) *Saccharomyces cerevisiae* fermentation activity after moderate pulsed electric field pre-treatments. Bioelectrochemistry 103:92–97

McFeeters RF (2004) Fermentation microorganisms and flavor changes in fermented food. J Food Sci 69(1):35–37

McFeeters RF, Fleming HP, Daeschel MA (1984) Malic acid degradation and brined cucumber bloating. J Food Sci 49:999–1002

Medina E, Perez-Diaz IM, Breidt F, Hayes J, Franco W, Butz N et al (2016) Bacterial ecology of fermented cucumber rising pH spoilage as determined by nonculture-based methods. J Food Sci 81:121–129

Medina-Pradas E, Arroyo-Lopez FN (2015) Presence of toxic microbial metabolites in table olives. Front Microbiol 16:Art 873

Miao J, Shi W, Zhang J, Zhang X, Zhang H, Wang Z, Qiu J (2020) Response surface methodology for the fermentation of polysaccharides from *Auricularia auricula* using *Trichoderma viride* and their antioxidant activities. Int J Biol Macromol 155:393–402

Molin G (2001) Probiotics in foods not containing milk or milk constituents, with special reference to *Lactobacillus plantarum* 299v. Am J Clin Nutr 73:380–385

Monika S, Kumari A, Angmo K, Bhalla TC (2016) Traditional pickles of Himachal Pradesh. Indian J Tradit Knowl 15(2):330–336

Montaño A, Casado FJ, de Castro A, Sánchez AH, Rejano L (2004) Vitamin content and amino acid composition of pickled garlic processed with or without fermentation. J Agric Food Chem 52:7324–7330

Montet D, Loiseau G, Kakhia-Rozis N (2006) Microbial technology of fermented vegetables. In: Ray RC, Ward OP (eds) Microbial biotechnology in horticulture, vol 1. Science Publishers, Inc., New Hampshire, pp 309–343

Montet D, Ray RC, Zakhia-Rozis N (2014) Lactic acid fermentation of vegetables and fruits. In: Ray RC, Montet D (eds) Microorganisms and fermentation of traditional foods. CRC Press, Boca Raton, pp 108–140

Moon SH, Chang M, Kim HY, Chang HC (2014) *Pichia kudriavzevii* is the major yeast involved in film-formation, off-odor production, and texture-softening in over-ripened Kimchi. Food Sci Biotechnol 23:489–497

Muhialdin BJ, Abdul Rani NF, Meor Hussin AS (2020) Identification of antioxidant and antibacterial activities for the bioactive peptides generated from bitter beans (*Parkia speciosa*) via boiling and fermentation processes. LWT-Food Sci Technol 131:109776

Mukherjee A, Gómez-Sala B, O'Connor EM, Kenny JG, Cotter PD (2022) Global regulatory frameworks for fermented foods: a review. Front Nutr 9:902642

Nair MK, Augustine LF, Konapur A (2016) Food-based interventions to modify diet quality and diversity to address multiple micronutrient deficiency. Front Public Health 3:277

Nam YD, Chang HW, Kim KH, Roh SW, Bae JW (2009) Metatranscriptome analysis of lactic acid bacteria during Kimchi fermentation with genome-probing microarrays. Int J Food Microbiol 130(2):140–146

Nguyen TMP, Lee YK, Zhou W (2009) Stimulating fermentative activities of bifidobacteria in milk by high-intensity ultrasound. Int Dairy J 19(6–7):410–416

Nguyen CM, Kim JS, Hwang HJ, Park MS, Choi GJ, Choi YH, Jang KS, Kim JC (2012a) Production of L-lactic acid from a green microalga, *Hydrodictyon reticulum*, by *Lactobacillus paracasei* LA104 isolated from the traditional Korean food, Makgeolli. Bioresour Technol 110:552–559

Nguyen CM, Kim JS, Song JK, Choi GJ, Choi YH, Jang KS, Kim JC (2012b) D-lactic acid production from dry biomass of *Hydrodictyon reticulatum* by simultaneous saccharification and co-fermentation using *Lactobacillus coryniformis* subsp. *torquens*. Biotechnol Lett 34:2235–2240

Nguyen DTL, Van Hoorde K, Cnockaert M, De Brandt E, De Bruyne K, Le BT, Vandamme P (2013) A culture-dependent and -independent approach for the identification of lactic acid bacteria associated with the production of *nem chua*, a Vietnamese fermented meat product. Food Res Int 50(1):232–240

Nguyen-The C, Bardin M, Berard A, Berge O, Brillard J, Broussolle V, Morris CE (2016) Agrifood systems and the microbial safety of fresh produce: trade-offs in the wake of increased sustainability. Sci Total Environ 562:751–759

Nielsen DS et al (2007) The microbiology of Ghanaian cocoa fermentations analysed using culture-dependent and culture-independent methods. Int J Food Microbiol 114(2):168–186

Nielsen ES, Garnås E, Jensen KJ, Hansen LH, Olsen PS, Ritz C, Krych L, Nielsen DS (2018) Lacto-fermented sauerkraut improves symptoms in IBS patients independent of product pasteurisation—a pilot study. Food Funct 9(10):5323–5335

Nilsson M, Liljeberg HGM, Êhl E, Molin G, Björck IME (2002) On the effect of lactic acid on blood glucose and insulin responses to cereal products: mechanistic studies in healthy subjects and *in vitro*. J Cereal Sci 36:339–346

Nout MJR (1994) Fermented foods and food safety. Food Res Int 27(3):291–298

Oey I, Lille M, Van Loey A, Hendrickx M (2008) Effect of high-pressure processing on colour, texture and flavour of fruit- and vegetable-based food products: a review. Trends Food Sci Technol 19(6):320–328

Ojha KS, Mason TJ, O'Donnell CP, Kerry JP, Tiwari BK (2017) Ultrasound technology for food fermentation applications. Ultrason Sonochem 34:410–417

Olennikov DN, Kirillina CS, Chirikova NK (2021) Water-soluble melanoidin pigment as a new antioxidant component of fermented willowherb leaves (*Epilobium angustifolium*). Antioxidants 10(8):1300

Oloo BO, Shitandi AA, Mahungu S, Malinga JB, Ogata RB (2014) Effects of Lactic Acid Fermentation on the Retention of B-Carotene Content in Orange Fleshed Sweet Potatoes. International Journal of Food Studies, 3(1):13–33. https://doi.org/10.7455/ijfs.v3i1.187

Omer MK, Alvseike O, Holck A, Axelsson L, Prieto M, Skjerve E, Heir E (2010) Application of high-pressure processing to reduce verotoxigenic *E. coli* in two types of dry-fermented sausage. Meat Sci 86(4):1005–1009

Omer MK, Prieto B, Rendueles E, Alvarez-Ordoñez A, Lunde K, Alvseike O, Prieto M (2015) Microbiological, physicochemical and sensory parameters of dry fermented sausages manufactured with high hydrostatic pressure processed raw meat. Meat Sci 108:115–119

Ostry V, Malir F, Ruprich J (2013) Producers and important dietary sources of ochratoxin A and citrinin. Toxins (Basel) 5(9):1574–1586

Owens JD (ed) (2014) Indigenous fermented foods of Southeast Asia. CRC Press

Palani K, Harbaum-Piayda B, Meske D, Keppler JK, Bockelmann W, Heller KJ, Schwarz K (2016) Influence of fermentation on glucosinolates and glucobrassicin degradation products in sauerkraut. Food Chem 190:755–762

Park YK, Lee JH, Mah JH (2019) Occurrence and reduction of biogenic amines in traditional Asian fermented soybean foods: a review. Food Chem 278:1–9

Peñas E, Frias J, Sidro B, Vidal-Valverde C (2010) Chemical evaluation and sensory quality of sauerkrauts obtained by natural and induced fermentations at different NaCl levels from *Brassica oleracea* var. *capitata* cv. Bronco grown in Eastern Spain. Effect of storage. J Agric Food Chem 58(6):3549–3557

Peng L, Wang L, Che C, Yang G, Yu B, Ma Y (2013) Bacillus sp. strain P38: An efficient producer of L-lactate from cellulosic hydrolysate, with high tolerance for 2-furfural. Bioresource Technology, 149:169–176. https://doi.org/10.1016/j.biortech.2013.09.047

Pérez-Díaz IM, Breidt F, Buescher RW, Arroyo-Lopez FN, Jiménez-Díaz R, Garrido-Fernández A et al (2015) Fermented and acidified vegetables. In: Downes P, Ito K (eds) Compendium of methods for the microbiological examination of foods, 4th edn. American Public Health Association, Washington, DC, pp 521–532

Petchkongkaew A, Taillandier P, Gasaluck P, Lebrihi A (2008) Isolation of *Bacillus* spp. from Thai fermented soybean (*Thua-nao*): screening for aflatoxin B1 and ochratoxin A detoxification. J Appl Microbiol 104(5):1495–1502

Piqué E, Vargas-Murga L, Gómez-Catalán J, de Lapuente J, Llobet JM (2013) Occurrence of patulin in organic and conventional apple-based food marketed in Catalonia and exposure assessment. Food Chem Toxicol 60:199–204

Plastourgos S, Vaughn RH (1957) Species of *Propionibacterium* associated with zapatera spoilage of olives. Appl Microbiol 5:267–271

Pranoto Y, Anggrahini S, Efendi Z (2013) Effect of natural and *Lactobacillus plantarum* fermentation on *in vitro* protein and starch digestibilities of sorghum flour. Food Biosci 2:46–52

Prester L (2011) Biogenic amines in fish, fish products and shellfish: a review. Food Addit Contam Part A 28(11):1547–1560

Puspito H, Fleet GH (1985) Microbiology of *sayur asin* fermentation. Appl Microbiol Biotechnol 22(6):442–445

Rasmussen MJ, Rea RF, Tri JL, Larson TR, Hayes DL (2001, June) Use of a transurethral microwave thermotherapeutic device with permanent pacemakers and implantable defibrillators. In: Mayo Clin Proc, vol 76, No. 6. Elsevier, pp 601–603

Regueiro J, Lopez-Fernandez O, Rial-Otero R, Cancho-Grande B, Simal-Gandara J (2015) A review on the fermentation of foods and the residues of pesticides—biotransformation of pesticides and effects on fermentation and food quality. Crit Rev Food Sci Nutr 55(6):839–863

Ricci A, Cirlini M, Calani L, Bernini V, Neviani E, Del Rio D, Galaverna G, Lazzi C (2019) *In vitro* metabolism of elderberry juice polyphenols by lactic acid bacteria. Food Chem 276:692–699

Ricci A, Marrella M, Hadj Saadoun J, Bernini V, Godani F, Dameno F et al (2020) Development of lactic acid-fermented tomato products. Microorganisms 8(8):1192

Roberts J, Kidd D (2005) Lactic acid fermentation of onions. LWT-Food Sci Technol 38(2):185–190

Rolle R, Satin M (2002) Basic requirements for the transfer of fermentation technologies to developing countries. Int J Food Microbiol 75(3):181–187

Ronteltap A, Van Trijp JCM, Renes RJ, Frewer LJ (2007) Consumer acceptance of technology-based food innovations: lessons for the future of nutrigenomics. Appetite 49(1):1–17

Ruiz-Cruz J, Gonzalez-Cancho F (1969) The metabolism of yeasts isolated from the brine of pickled Spanish-type green olives. I. The assimilation of lactic, acetic, and citric acids. Grasas Aceites 20:611

Sahu L, Panda SK (2018) Innovative technologies and implications in fermented food and beverage industries: an overview. In: Innovations in Technologies for Fermented Food and Beverage Industries, pp 1–23

Sanzani SM, Reverberi M, Geisen R (2016) Mycotoxins in harvested fruits and vegetables: insights into producing fungi, biological role, conducive conditions, and tools to manage postharvest contamination. Postharvest Biol Technol 122:95–105

Sapbamrer R, Hongsibsong S (2014) Organophosphorus pesticide residues in vegetables from farms, markets, and a supermarket around Kwan Phayao Lake of Northern Thailand. Arch Environ Contam Toxicol 67:60–67

Sastry SK (2004) Advances in ohmic heating and moderate electric field (MEF) processing. In: Novel food processing technologies. CRC Press, pp 513–522

Sawicki T, Wiczkowski W (2018) The effects of boiling and fermentation on betalain profiles and antioxidant capacities of red beetroot products. Food Chem 259:292–303

Schaffer-Lequart C, Lehmann U, Ross AB, Roger O, Eldridge AL, Ananta E et al (2017) Whole grain in manufactured foods: current use, challenges and the way forward. Crit Rev Food Sci Nutr 57(7):1562–1568

Siefarth C, Tran TBT, Mittermaier P, Pfeiffer T, Buettner A (2014) Effect of radio frequency heating on yoghurt, I: technological applicability, shelf-life and sensorial quality. Food Secur 3(2):318–335

Singhal P, Satya S, Naik SN (2021) Fermented bamboo shoots: a complete nutritional, anti-nutritional, and antioxidant profile of the sustainable and functional food to food security. Food Chem 3:100041

Sivamaruthi BS, Kesika P, Chaiyasut C (2018) Toxins in fermented foods: prevalence and preventions—a mini review. Toxins (Basel) 11(1):4

Skibsted LH (2016) Mineral nutrient interaction: improving bioavailability of calcium and iron. Food Sci Biotechnol 25(5):1233–1241

Sõukand R, Pieroni A, Biró M, Dénes A, Dogan Y, Hajdari A, Kalle R, Reade B, Mustafa B, Nedelcheva A et al (2015) An ethnobotanical perspective on traditional fermented plant foods and beverages in Eastern Europe. J Ethnopharmacol 170:284–296

Spano G, Russo P, Lonvaud-Funel A, Lucas P, Alexandre H, Grandvalet C et al (2010) Biogenic amines in fermented foods. Eur J Clin Nutr 64(3):S95–S100

Spurr H Jr (1994) The microbial ecology of fruit and vegetable surfaces: its relationship to postharvest biocontrol. In: Biological control of postharvest diseases: theory and practice

Stamer JR, Hrazdina G, Stoyla BO (1973) Induction of red color formation in cabbage juice by *Lactobacillus brevis* and its relationship to pink sauerkraut. Appl Microbiol 26:161–166

Steinkraus KH (2002) Fermentation in world food processing. Compr Rev Food Sci Food Saf 1(1):23–32

SUSFOOD (2013) Definitions. Sustainable Food Production and Consumption (SUSFOOD). Retrieved from https://www.susfoodera.net/index.php?index=30

Suzzi G, Torriani S (2015) Biogenic amines in foods. Front Microbiol 6:472

Taban BM, Halkman AK (2011) Do leafy green vegetables and their ready-to-eat (RTE) salads carry a risk of foodborne pathogens? Anaerobe 17(6):286–287

Tamang JP, Fleet GH (2009) Yeasts diversity in fermented foods and beverages. In: Yeasts in food and beverages. Springer, pp 169–197

Tamang JP, Sarkar PK (1996) Microbiology of *mesu*, a traditionally fermented bamboo shoot product. Int J Food Microbiol 29:49–58. https://doi.org/10.1016/0168-1605(95)00021-6

Tamang B, Tamang JP (2007) Role of lactic acid bacteria and their functional properties in goyang, a fermented leafy vegetable product of the Sherpas. J Hill Res 20(20):53–61

Tamang B, Tamang JP (2009) Traditional knowledge of biopreservation of perishable vegetables and bamboo shoots in Northeast India as food resources. Indian J Tradit Knowl 8(1):89–95

Tamang JP, Tamang B, Schillinger U, Franz CMAP, Gores M, Holzapfel WH (2005) Identification of predominant lactic acid bacteria isolated from traditionally fermented vegetable products of the Eastern Himalayas. Int J Food Microbiol 105(3):347–356

Tamang B, Tamang JP, Schillinger U, Franz CMAP, Gores M, Holzapfel WH (2008) Phenotypic and genotypic identification of lactic acid bacteria isolated from ethnic fermented bamboo tender shoots of North East India. Int J Food Microbiol 121(1):35–40

Tamang JP, Tamang B, Schillinger U, Guigas C, Holzapfel WH (2009) Functional properties of lactic acid bacteria isolated from ethnic fermented vegetables of the Himalayas. Int J Food Microbiol 135(1):28–33

Tango CN, Choi NJ, Chung MS, Oh DH (2014) Bacteriological quality of vegetables from organic and conventional production in different areas of Korea. J Food Prot 77(8):1411–1417

Tashiro Y, Kaneko W, Sun Y, Shibata K, Inokuma K, Zendo T, Sonomoto K et al (2011) Continuous D-lactic acid production by a novel thermotolerant *Lactobacillus delbrueckii* subsp. *lactis* QU 41. Appl Microbiol Biotechnol 89:1741–1750

Tashiro Y, Matsumoto H, Miyamoto H, Okugawa Y, Pramod P, Miyamoto H, Sakai K (2013) A novel production process for optically pure L-lactic acid from kitchen refuse using a bacterial consortium at high temperatures. Bioresour Technol 146:672–681

Tchabo W, Ma Y, Kwaw E, Zhang H, Xiao L, Tahir HE (2017) Aroma profile and sensory characteristics of a sulfur dioxide-free mulberry (*Morus nigra*) wine subjected to non-thermal accelerating aging techniques. Food Chem 232:89–97

Tejayadi S, Cheryan M (1995) Lactic acid from cheese whey permeate: productivity and economics of a continuous membrane bioreactor. Appl Microbiol Biotechnol 43(2):242–248

Thanh VN, Viet Anh NT (2016) Ethnic fermented foods and beverages of Vietnam. In: Ethnic fermented foods and alcoholic beverages of Asia, pp 383–409

The Food Safety and Standards Act (2006) *Gazette of India*

Thieu NQ, Ogle B, Pettersson H (2008) Efficacy of bentonite clay in ameliorating aflatoxicosis in piglets fed aflatoxin-contaminated diets. Trop Anim Health Prod 40:649–656

Tran KT, May BK, Smooker PM, Van TT, Coloe PJ (2011) Distribution and genetic diversity of lactic acid bacteria from traditional fermented sausage. Food Res Int 44(1):338–344

Tran-Dinh N, Kennedy I, Bui T, Carter D (2009) Survey of Vietnamese peanuts, corn and soil for the presence of *Aspergillus flavus* and *Aspergillus parasiticus*. Mycopathologia 168:257–268

Uchoi D, Roy D, Majumdar RK, Debbarma P (2015) Diversified traditional cured food products of certain indigenous tribes of Tripura, India

Untersmayr E, Jensen-Jarolim E (2008) The role of protein digestibility and antacids on food allergy outcomes. J Allergy Clin Immunol 121(5):1301–1308

Van der Walt AM, Van der Linde E, Alberts M, Modjadji P, Jivan SD, Bezuidenhout CC (2006) Fumonisin-producing *Fusarium* strains and fumonisins in traditional African vegetables (*morogo*). S Afr J Sci 102(3):151–155

Venema K, Verhoeven J, Beckman C, Keller D (2020) Survival of a probiotic-containing product using capsule-within-capsule technology in an *in vitro* model of the stomach and small intestine (TIM-1). Benef Microbes 11(5):403–409

Vong WC, Liu SQ (2016) Biovalorisation of okara (soybean residue) for food and nutrition. Trends Food Sci Technol 52:139–147

Waché Y, Do TL, Thi-Bao-Hoa D, Do TY, Haure M, Ho PH, Chu-Ky S (2018) Prospects for food fermentation in South-East Asia, topics from the tropical fermentation and biotechnology network at the end of the AsiFood Erasmus+ Project. Front Microbiol 9:2278

Walsh AM, Crispie F, Daari K, O'Sullivan O, Martin JC, Arthur CT, Cotter PD (2017) Strain-level metagenomic analysis of the fermented dairy beverage *nunu* highlights potential food safety risks. Appl Environ Microbiol 83(16):e01144–e01117

Wanwimolruk S, Kanchanamayoon O, Phopin K, Prachayasittikul V (2015) Food safety in Thailand 2: pesticide residues found in Chinese kale (*Brassica oleracea*), a commonly consumed vegetable in Asian countries. Sci Total Environ 532:447–455

Wiczkowski W, Szawara-Nowak D, Topolska J (2015) Changes in the content and composition of anthocyanins in red cabbage and its antioxidant capacity during fermentation, storage, and stewing. Food Chem 167:115–123

Xu Y, Hlaing MM, Glagovskaia O, Augustin MA, Terefe NS (2020) Fermentation by probiotic *Lactobacillus gasseri* strains enhances the carotenoid and fiber contents of carrot juice. Food Secur 9(12):1803

Yamano T, Iino H, Takada M, Blum S, Rochat F, Fukushima Y (2006) Improvement of the human intestinal flora by ingestion of the probiotic strain *Lactobacillus johnsonii* La1. Br J Nutr 95(2):303–312

Yan PM, Xue WT, Tan SS, Zhang H, Chang XH (2008) Effect of inoculating lactic acid bacteria starter cultures on the nitrite concentration of fermenting Chinese *paocai*. Food Control 19(1):50–55

Yang CH et al (2001) Microbial phyllosphere populations are more complex than previously realized. Proc Natl Acad Sci USA 98(7):3889–3894

Yang X, Lai Z, Lai C, Zhu M, Li S, Wang J, Wang X (2013) Efficient production of L-lactic acid by an engineered *Thermoanaerobacterium aotearoense* with broad substrate specificity. Biotechnol Biofuels 6:124

Yang XX, Zhou JC, Fan LQ, Qin Z, Chen QM, Zhao LM (2018) Antioxidant properties of a vegetable–fruit beverage fermented with two *Lactobacillus plantarum* strains. Food Sci Biotechnol 27(6):1719–1726

Yoon KY, Woodams EE, Hang YD (2006) Production of probiotic cabbage juice by lactic acid bacteria. Bioresour Technol 97(12):1427–1430

Yu Z, Zhang X, Li S, Li C, Li D, Yang Z (2013) Evaluation of probiotic properties of *Lactobacillus plantarum* strains isolated from Chinese sauerkraut. World J Microbiol Biotechnol 29(3):489–498

Zaman MZ, Abdulamir AS, Bakar FA, Selamat J, Bakar J (2009) A review: microbiological, physicochemical and health impact of high levels of biogenic amines in fish sauce. Am J Appl Sci 6(6):1199

Zhang Y, Chen X, Qi B, Luo J, Shen F, Su Y, Khan R, Wan Y (2014) Improving lactic acid productivity from wheat straw hydrolysates by membrane integrated repeated batch fermentation under non-sterilized conditions. Bioresour Technol 163:160–166

Zhang L, Li X, Yong Q, Yang ST, Ouyang J, Yu S (2016) Impacts of lignocellulose-derived inhibitors on L-lactic acid fermentation by *Rhizopus oryzae*. Bioresour Technol 203:173–180

Zhao K, Qiao Q, Chu D, Gu H, Dao TH, Zhang J, Bao J (2013) Simultaneous saccharification and high titer lactic acid fermentation of corn stover using a newly isolated lactic acid bacterium *Pediococcus acidilactici* DQ2. Bioresour Technol 135:481–489

Chapter 11
Jams, Jellies, Marmalades, Preserves, and Candied Products

Swati Sethi, Poonam Choudhary, and Pankaj Kumar

11.1 Introduction

Fruits play an important role in the diet as they are known to be protective foods owing to their high vitamin content. However, due to high moisture content, fruits are perishable in nature and are prone to spoilage throughout the food chain right from harvesting. Although India stands at the top in fruit and vegetable production, a substantial amount of the produce gets spoiled (72%) because of poor storage and transportation conditions. Therefore, shelf-life extension via preservation is necessary to moderate the wastage. Preservation of fruits in the form of jams, jellies, marmalades, and conserves with sugar extends the shelf life of fruits as well as diversify the product portfolio with different textures and flavors. Jams, jellies, marmalades, and preserves are categorized as intermediate moisture food products that are prepared by cooking fruits with sugar, pectin, and acid to increase the total soluble solids (TSS) content to >65%. Preservation of fruits with sugars is popular at domestic preservation as well as at the industrial level due to simple processing steps. The high levels of sugar lower moisture availability and create high osmotic pressure that prevents spoilage due to microbes (bacteria, yeasts, and molds). A minimum sugar concentration of 65% is necessary to attain the preservative effect.

S. Sethi (✉)
Food Grains & Oilseeds Processing Division, ICAR-Central Institute of Post-Harvest Engineering and Technology, Ludhiana, Punjab, India
e-mail: swati.fst@gmail.com

P. Choudhary
Agricultural Structures and Environmental Control Division, ICAR-Central Institute of Post-Harvest Engineering and Technology, Ludhiana, Punjab, India

P. Kumar
Transfer of Technology Division, ICAR-Central Institute of Post-Harvest Engineering and Technology, Ludhiana, Punjab, India

OP Chauhan (ed.), *Fruits and Vegetables Technologies*,
https://doi.org/10.1007/978-981-96-8433-5_11

Overripe fruits are preferably used for the preparation of jams and jellies due to high sugar content and optimal acid content which can further reduce the losses. Jams, jellies, marmalades, and preserves are high-sugar products that limit their application; however, many researchers have attempted to develop low-calorie jams and jellies with different sweeteners to partially replace the sugar. Additionally, food scientists have investigated newer processing methods like vacuum processing, high-pressure processing, ultrasonication, ultraviolet radiation, etc. to preserve such products.

11.2 Jam

A jam is a sugar preserved food product which is either made from pulp or the juice from fruits. As per Food Standards Authority of India, a jam means is a product prepared from sound, ripe, fresh, dehydrated, frozen or previously packed fruits including fruit juices, fruit pulp, fruit juice concentrate or dry fruit by boiling its pieces or pulp or puree with nutritive sweeteners namely sugar, dextrose, invert sugar or liquid glucose to a suitable consistency. It may also contain fruit pieces and any other ingredients suitable for the products. It may be prepared from any of the suitable fruits, singly or in combination. It shall have the flavor of the original fruit(s) and shall be free from burned or objectionable flavors and crystallization. The texture of the fruit jam should be consistent enough to spread and not to flow like a liquid. The specific requirements of jam include a minimum of 65% total soluble solids, and there should be a minimum of 45.0% of original fruit content, exclusive of any added sugar or other ingredients in the finished product, except for Strawberry or Raspberry fruit, where it shall contain minimum 25.0% fruit.

The fruit jams are high in calories and provide a suitable amount of carbohydrates and energy. An array of fruit jams is available in the market, including apple jam, strawberry jam, plum jam, pineapple jam, apricot jam, raspberry jam, mango jam, mixed fruit jam, etc. These jams are used as spread on different types of bread and can be used as toppings on cakes, ice creams, milkshakes, etc.

11.2.1 Role of Ingredients in the Manufacture of Jams

The main structural components to achieve an optimal texture of jam are sugar, pectin, and acid. The type and concentration of these components govern the texture and setting properties of the jam.

11.2.1.1 Gelling Agents

Gelling agents facilitate the gel formation and should not affect the aroma, flavor, or taste of the original product. The formation of gels is the result of the establishment of a continuous three-dimensional arrangement of cross-linked polymer molecules; an aqueous gel involves three elements (Jarvis 1984):

1. Junction zones where polymer molecules are joined together.
2. Moderately mobile interjunction segments of polymers.
3. Water entrapped in the polymer network.

Since gels are formed in an aqueous environment, the polymer (protein and polysaccharide) interactions with water are important factors in gel formation. Both protein and polysaccharides hydrate well in aqueous solution, and the loss of fluidity and increase in the elasticity of the network is the critical stage of gelation (Shomer 1991). Pectin is the predominant polymer responsible for the gel formation in jams and jellies. It is naturally present in fruits and can be substituted externally where insufficient quantities are present. Other commonly used gelling agents are gelatin, agar/alginates, gums, and modified starches.

Pectin consists of the partial methyl esters of polygalacturonic acid and their sodium, potassium, calcium, and ammonia salts. It has a molecular weight of up to 1,50,000 Daltons and is used for its gel-forming, thickening, and stabilizing properties (Lau et al. 1985). Pectin serves as a cementing and hydrating agent for the cellulose network in higher plants. Heat treatment with acidified water releases the protopectin from plant materials rich in pectin, which is then hydrolyzed into a water-soluble form. It occurs in plant tissues when fruits ripen, assisted by the enzyme protopectinase. An increasing amount of insoluble protopectin is transformed into soluble pectin as the fruit ripens (Woodmansee et al. 1959). It plays an important role as it forms the network with sugar and acid to provide a typical setting point for jam. Pectin is extensively used as a texturizer, stabilizer, and emulsifier in a variety of foods and other applications. It imparts firmness to the jam due to the formation of a long pectin chain at elevated temperatures (100–105 °C). Pectin is extracted via aqueous extraction, specifically from peels of citrus fruits and apple pomace, with subsequent alcohol or salt precipitation. The functional behavior of pectin is governed by the degree of methylation and molecular size (Doublier et al. 1992). Different grades of pectin varying from 100–500 are commercially available for varied applications. The desired texture, pH, processing temperature, ion content, protein content, and expected shelf life of the product are the variables that affect the selection of type of pectin for a certain application (Hoefler 1991). Based on degree of methoxylation (percentage of esterified galacturonic acid units to total galacturonic acid units), pectin is classified as high methoxyl and low methoxylpectins. Pectin is pre-mixed with sugar in the minimum ratio of 1:3 and dispersed in hot water at 85–90 °C, at <20% soluble solids, employing a high-speed stirrer. Pectin is recommended to be stored in a cool and dry condition and forms gel at a pH range of 2.8 and 4.7. Pectin is usually added at a level of 0.5 to 1.5% by weight, subject to the nature of the pectin used.

High methoxyl pectins are produced by conventional aqueous extraction and have 50% of methoxyl groups. This type of pectin forms gel at low pH values (2.0–3.8) and high soluble solids (>65%). HM pectin creates gel by cross-linking the polymer at junction zones where the regions for sugar and water entrapment are created by hydrogen bonds and the methyl groups' repulsion for the formation of network. Also, the disassociation of the carboxylic groups and the low pH value lessen the negative charges that are generated along the pectin chain. Consequently, electrical repulsion prevents pectin-pectin bond formation, resulting in gel formation. A low degree of methylation results in a longer setup period and lower setting temperature. The amount of hydrogen bonding decreases with increased total soluble solids, leading to stronger gels.

Low-methoxyl pectin is extracted via a modified extraction method with continued acid treatment. This type of pectin has less than 50% methylation and has a unique ability to form gels in the presence of bivalent Ca++ ions salts. Low-methoxyl pectin forms gel over a wide range of pH (2.6–7.0) and low total soluble solids. Intermolecular chelate binding of Ca^{++} with pectin leads to the development of macromolecular aggregates, also known as "egg-box" cavity. The higher the solids level and the lower the pH, the pectin is more reactive. The higher the calcium content and the lower the pH, the higher the setting temperature.

Different fruits have varied content of pectin, and fruits are classified into three categories based on the pectin content (Fig. 11.1):

(a) High pectin content: citrus fruits (lime, lemon, mandarin, grapefruit, orange, gooseberry) and apple. In citrus fruits, the peels are used for commercial extraction of pectin as the pectin is confined to the inner white layer known as albedo.

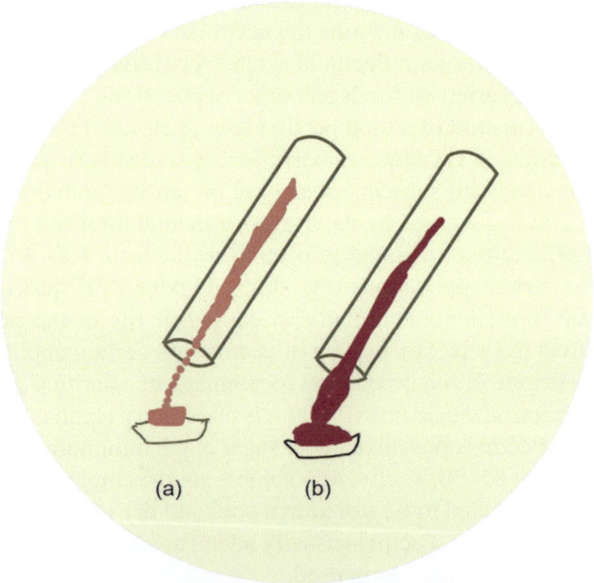

Fig. 11.1 Pectin strength test for gel-forming ability (**a**) low pectin and (**b**) high pectin

(b) Average pectin content: apricot, grapes and blackberry
(c) Low pectin content: pear, cherry, guava, peach, fig, litchi, pineapple, etc (Table 11.1).

Other gelling agents used in similar products are gelatin, agar/alginates, gums, and modified starches. Gelatin is a protein obtained from the degradation of collagen from animal skin and bones and is soluble in water. Gelatin results in soft or rubbery gels; therefore, it is generally added in combination with another gelling agent. Generally, thin boiling starches are used with gelatin for gel formation. Agar/alginates are obtained from algae, and these gels are very soft. Alginates with low levels of poly-d-mannuronic acid and poly-l-guluronic acid form gels that are transparent, stiff and brittle while high levels result in soft gels (Smidsred 1974). Gum Arabic is a water soluble, colorless, odorless, and tasteless gum and its use results in hard jellies. Modified starches improve the gelling properties of native starches. The amylose component of starch is involved in the formation of a gel. Gelling agent combinations are frequently utilized due to their ability to form the desired texture. Mixed systems are, therefore, extremely important from a technological standpoint (Christensen and Trudsoe 1980).

11.2.1.2 Sweetening Agents

The function of sweeteners is to impart sweetness, balance flavor, impart body, and control viscosity that governs the texture and prevents spoilage. Sweeteners bind free water and make it unavailable for the growth of microorganisms. Optimal amount of sugar is required in the formulation for the stated effects. Sucrose or table sugar or refined sugar, is the standard compared to which all sweeteners are measured in terms of quality and degree of sweetness. Inversion of sugar is essential to prevent the crystallization of sugar in jellies and jams during storage. The ideal proportion of inverted sugar is between 35–40% of the total sugar in the jam. The degree of inversion is governed by the pH of the product, boiling time, and temperature. Boiling, acid, or enzymatic treatment results in the inversion of sucrose, which is converted into dextrose and fructose in equal parts. Invert sugar has a higher molecular weight than sucrose due to the additional weight of water molecules added during the inversion process.

Sugar: Sucrose or table sugar is the most used sweetener in the preparation of jam. Apart from its preservative action and contribution to gel formation with pectin, it is essential for the desired flavor, taste, yield, and appearance of the product.

Table 11.1 Conditions for gel formation of the high-methoxyl and low-methoxyl pectin

Parameters	HM pectin	LM pectin
Degree of methoxylation (DM)	>50%	<50%
Soluble solid content (%)	≥65%	10–70%
pH	2.0–3.8	2.6–7.0
Bivalent ions, Ca^{++} (mg/g)	–	≥15

A general recommendation of 1:1 fruit: sugar is used for the manufacture of different types of jam. Optimal content of sugar is required according to the type of pectin for the gel formation. The level of acidity in the fruit also governs the amount of sugar required in the formulation. It maintains the required gel strength and acidity of the jam. The total sugar content in the fruit jam should range between 65 and 69%. The addition of excessive sugar results in crystallization, while a low amount of sugar creates poor gels. Sugar acts as a preservative in jams by reducing the available water and hence decreases water activity, creating unfavorable conditions for the growth of microorganisms.

In place of sucrose, other sugars including high fructose corn syrups, glucose, sugar alcohols, and maltose have been explored for partial replacement; however, replacing sucrose with these alternatives affects the textural and rheological properties of jam. Other alternatives such as jaggery, brown sugar, maple syrup, and molasses can alter the color and flavor of the jam, which may suppress the natural fruit flavor; therefore, these are not preferred for use in jams. Honey is not recommended to be heated therefore, it is not preferred to be used as a sweetener in jams; also, it has high moisture content and hence requires more boiling and creates form during boiling. Artificial sweeteners such as aspartame and saccharin cannot form a gel with pectin and are not suitable for the manufacture of jams.

Different types of sweeteners are used to partially substitute table sugar or sucrose, which include:

Starch hydrolysates: Under this category, glucose, fructose, and dried starch syrup are commonly used sweeteners. These syrups are derived from the saccharification of starch via acidic or enzymatic hydrolysis under controlled conditions to produce starch hydrolysates with distinctive characteristics for expanded uses.

High fructose corn syrup: HFCS is obtained after enzymatic liquefaction of starch, followed by saccharification and isomerization. The dextrose syrup thus obtained is then concentrated to 40–50%. The resultant HFCS has times more sweetness than glucose.

Anhydrous glucose: Starch from different sources is enzymatically saccharified to obtain a product with 95% glucose, which is then purified and evaporated to crystallize glucose as anhydrous α-D-glucose monohydrate.

Sugar alcohols: Polyols or sugar alcohols are industrially manufactured by the transformation of sugars. They are known for their laxative effects. Isomalt is the commonly used sugar alcohol in confectionery products. These impart slightly lower calories upon absorption when compared with sucrose. Other commonly used sugar alcohols are sorbitol, mannitol, xylitol, maltitol, and lactitol.

11.2.1.3 Acid

Optimal level of acid is essential for the setting of gel in the fruit jams. At pH higher than 3.3, there is ionization of the acetic acid component in the pectin, which results in negative charges on the molecules, leading to repulsive forces that prevent pectin chains from gelling. pH range of 2.8–3.3 prevents the ionization of the acetic acid groups and hence lowers the degree of the repulsive forces.

11.2.1.4 Chemical Preservatives

Chemical preservatives to further extend the shelf-life of jams, chemical preservatives are added for increased storage stability. Sodium salts of benzoic acids are more commonly used to preserve fruit jam. Other chemical preservatives used in jams are sorbic acid, potassium metabisulphite, and calcium propionate. These preservatives are required to be added in the jam confining to the legal limits or standards according to the country.

11.2.2 Preparation of Jam

Jam can be prepared from a single type of fruit or can be prepared using multiple fruits. Assorted combinations of different fruits are used to make jams where low pectin fruits can be combined with high pectin to have desired gelling properties. Similarly, the desired acidity can be achieved by combining different nature of fruits. The steps for the preparation of jam are:

Pulping: The ripe to overripe fruits of sound quality are washed either manually or in mechanical washers attached to brushes. These are then peeled, cored, or cut depending on the type of fruit. In case of apricots, steaming is done while scalding is done in plums to remove the peels. The method of pulp preparation varied with the nature of the fruit.

Addition of sugar: Sucrose or cane sugar or table sugar of good quality is used for the preparation of jam. A maximum of 25% replacement of sucrose with corn syrups are generally used for jam preparation. The level of sugar addition depends on the type and characteristics of fruit, degree of ripening, and type of pectin used for gel formation. A general ratio of 1:1 fruit to sugar is widely in practice.

Addition of acid: Fruits contain natural acids, the predominant acid in fruits varies with the type of fruit. Citric, ascorbic acid, malic, or tartaric acids are among the natural acids present in different fruits. Additionally, these can be added to maintain the desired level of acidity required in the formulation. The acidity of the jam ranged between 0.5 and 0.7%, with a corresponding pH of 3–3.2, depending on the nature of the jam. Acid is added at the end of the process; otherwise, it could lead to a weak gel due to the inversion of sugar in initial stages.

Processing/cooking: Fruit pulp is boiled with the optimal amount of sugar and pectin until a total Soluble Solids (TSS) of 68% is attained. Acids, permitted flavorings, colors, and preservatives are added at the end of cooking to attain the desired quality product. Cooking/boiling extracts the color and flavor from the fruits, partially inverting the sugar and inactivating the microorganisms responsible for spoilage. After this, the finished product is filled hot in sterilized jars or containers, followed by sealing.

Filling: There are three primary phases involved in the finishing steps in the manufacture of jams: (1) precooling before filling, (2) filling, and (3) cooling after filling. Extended heating has an impact on both the finished product quality and stability; therefore, better control with effective cooling can prevent deleterious effects.

11.2.3 Assessing the End Point

An optimum end point leads to better texture and prevents overcooking, resulting in economic losses, undercooking, and spoilage. Different types of refractometers are available to assess the endpoint at the industrial scale, and the endpoint is achieved when the refractometer shows 69 brix. However, at a small domestic level, the endpoint can be assessed by different methods.

Drop test: This is the easiest method, where a small amount of jam is taken in a teaspoon and allowed to cool, and then a drop of this is poured into water. If the drop settles down without dissolution, it indicates the endpoint.

Sheet test: Here, a small amount of jam is taken with a large spoon, which is then partially cooled and let to drop while the spoon is kept horizontally inclined. If the jam falls like syrup, more cooking is required. If it forms flakes or sheets, the endpoint is achieved (Fig. 11.2).

Boiling point test: Jam with a TSS of 69% exhibits a boiling point of 106 °C at sea level. Therefore, judging the temperature using a thermometer at the later stages can help in determining the endpoint.

Weighing method: When the weight of the finished reaches one and a half times of the weight of added sugar, it indicates the endpoint.

11.2.4 Defects in Jam

(a) Crystallization: Crystals are formed when the amount of sugar exceeds, and the product has a TSS of 70 degrees Brix. To prevent crystallization, there should be a 30–40% inversion of sugar during the processing. If the degree of inversion is less (<30), crystallization may occur upon storage, and if it is more (>50), instead of gel, a honey-like mass will be formed. Sugar can be replaced with corn syrup or glucose up to 25% to evade crystallization.

(b) Sticky or gummy jam: High levels of total soluble solids result in gummy or sticky jam. This can be rectified by the addition of pectin and/or citric acid.

Fig. 11.2 Sheet test to determine endpoint (**a**) unfinished and (**b**) finished

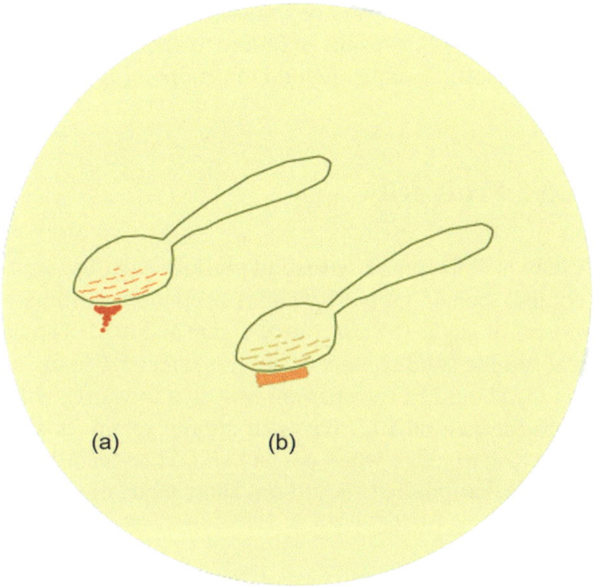

(c) Premature setting: Low TSS and high pectin may result in premature setting of jam. It can be avoided by adding a small amount of sodium bicarbonate, which will lessen the acidity and stop pre-coagulation.
(d) Surface graining and shrinkage: Evaporation of moisture during storage may result in shrinkage and can be prevented by suitably storing under cool conditions.
(e) Microbial spoilage: Microbial spoilage can be prevented by attaining the 68-degree brix and adding preservatives to a permissible limit. Storage in a cool and dry place, preferably below 80% RH, may prevent the chances of microbial growth.

11.2.5 Failure of Fruit Jam to Set

(i) Lack/excess of acid or pectin: Insufficient pectin content in the fruit jam results in a poor pectin network that is not adequately thick and firm enough to hold the sugar syrup. Excess acid causes hydrolysis of pectin, leading to the breakdown of the jam structure.
(ii) Prolonged Boiling: Prolonged cooking can also cause pectin hydrolysis, which ultimately prevents the setting of gel and results in syrupy jam. Optimal cooking to gel point 69 degrees Brix avoids this problem.
(iii) Low sugar or excess sugar content: High amount of sugar results in a syrupy or soft jelly, whereas low levels of soluble solids and sugar enable the pectin network to hold more liquid.

(iv) Cooking below and beyond the endpoint: Finishing below 65 percent of total soluble solids results in failure to set, and the texture of the jam remains syrupy and soft. Cooking beyond the endpoint results in hard or rigid gel.

11.3 Fruit Jelly

A jelly is defined as a semisolid product manufactured by boiling clear fruit extract (not less than 45 parts by weight) containing pectin after the addition of an optimal amount of sugar (55 parts by weight) and acid. The contents are heated to attain a total soluble solid content of not less than 65% in the finished product. A good quality jelly should be transparent, firm (not too stiff), with a pleasant color, and should retain the natural fruit flavor. It should retain its shape, while demolding should retain a sharp edge when cut and should be tender enough to quiver when pressed. The finished product should not show signs of syneresis, and it should not be rubbery, gummy, or sticky.

11.3.1 Theories of Jelly Formation

Different theories known to be associated with the formation of gel in jelly manufacture are:

Fibril theory: The addition of sugar to pectin solution causes the pectin-water equilibrium to become unstable, causing the pectin to gather and create an arrangement of fibrils throughout the jelly. The inter-fibrillar spaces of this network hold the sugar solution, leading to firmness of the gel, while the acid provides toughness to the fibrils (Fig. 11.3). The strength of the network is governed by the amount of sugar and acidity. Too low acid results in weak fibrillar structure, while too high levels result in loss of fibrillar elasticity due to pectin hydrolysis, which ultimately leads to runny or syrupy jelly.

Spencer's theory: Sugar precipitates the pectin in the optimal concentration of pectin in the solution.

Olsen's theory: The dehydrating effect of sugar destabilizes the equilibrium between water and pectin. Hydrogen ion concentration helps to lower the negative charge on pectin. With adequate levels of sugar, pectin precipitates, and aggregates to form a network of insoluble fibers.

Hinton's theory: It states that pectin is a complex mixture of variable composition. The coagulation of non-ionized type pectin forms the continuous network.

Fig. 11.3 Pectin gel network in jams and jellies

11.3.2 Role of Ingredients in Fruit Jelly Manufacture

Like jam, the optimal combination of sugar, pectin, and acid is required for the structure formation in jellies. However, the setting of strong gels that can retain their shape during storage is required in jellies.

Sugars: Successful gel formation in jelly requires sugar in the range of 60–65%. The proportion of sugar governs the pH optimum and maximum gel strength. The pectin-sugar gel arrangement happens due to the partial precipitation of pectin, leading to the formation of high binding forces at the surface. These bonds hold the solution with adequate strength to impart the desired rigidity. Sugar also improves the color, flavor, and yield of the finished product. The amount of sugar should be proportionate to pectin and acid levels.

Pectin: Gels are made up of polymeric molecules that have been cross-linked to create a complex web of interconnecting molecules submerged in a liquid media (Oakenfull 1991). As a solvent, water affects the kind and strength of intermolecular forces that keep the polymer network intact; in an acidic medium, the polymer network retains the water and keeps it from evaporating; the interaction of pectin with sugar modifies the pectin-water equilibrium and creates a web of fibers throughout the jelly which holds the liquids (Rees 1969). In general, 0.5–1.0% of pectin is adequate to form gel in jellies. Higher amount results in hard gel, while lower amount lead to weal gels or prevents gel setting.

Acid: The total acid content of the finished product varies between 0.5 and 1%. Citric, tartaric, and malic acids are generally added to maintain the required acid content in the fruit extract. Tartaric acid is more effective than citric and malic acid for gel formation.

11.3.3 Jelly Preparation

Selection of fruits: High-pectin fruits are most suitably used for the preparation of jelly; however, low-pectin fruits can also be converted to jelly by adding enough pectin in the formulation. Fruits are categorized into different groups based on their suitability for jelly preparation:

- High pectin and acid: lemon, oranges, sour apple, grape, jamun, and plum.
- High pectin but low acid: apple, ripe guava, pear, unripe banana, etc.
- Low pectin but high acid: apricot, peach, cherry, pineapple, and strawberry.
- Low pectin and acid: ripe apricot, pomegranate, and any other overripe fruit.

Extraction of pectin: Fruits are cooked to extract the juice and pectin. It converts protopectin into pectin and softens fruit tissues. Citrus fruits do not require the addition of water; they are crushed and boiled. However, pulpy fruits are crushed and boiled with water. The amount of water required for boiling as well as cooking duration varies with the type of fruit, e.g., citrus fruits require almost two to three times of water while apple requires 1–1.5 times of water for the extraction of pectin.

Filtration and clarification: The boiled fruit extract is filtered through cheesecloth or linen bags or a mechanical filter press to extract pectin. The extract is then analyzed for pH, acidity, soluble solids, and pectin content.

Addition of sugar and pectin: According to the level of pectin in fruit extract, sugar is added. In general, with high pectin extract, a 1:1 ratio of extract to sugar is added, while for moderate pectin extract, the addition of sugar is in the range of 650–750 to each kg of extract. Low pectin extracts are added with pectin in a range of 1–2% to obtain the gel.

Addition of acid: Acid enhances the strength of the gel under optimal levels of pH (3.2) and TSS (65%). In low-acid extracts, acid is added to improve the gel strength. Jellying strength depends on the quantity of pectin and the acid present in the original fruit extract.

Processing/boiling: The fruit extract is cooked to concentrate the extract till half the original weight. At this stage, the calculated amount of sugar is added gradually, and the contents are further cooked to achieve a TSS of 65%. It is recommended to remove foam/scum formed during boiling. Thereafter, the acid is added to maintain the pH of 3.2. Other additives like color, flour, and preservatives are added at the end to prevent the losses. Jellies are preferably prepared in smaller lots as longer cooking hours result in the degradation of pectin.

11.3.4 Judging the Endpoint

Finishing the product at the endpoint is important as prolonged heating may lead to the degradation of pectin and inversion of sugar. Like jam, the endpoint in jelly can be assessed by refractometer, sheet test, drop test, thermometer, and weight method. In jellies, the endpoint is achieved when the TSS reaches 65° Brix and the temperature of the finished product is around 105 °C.

11.3.5 Defects in Jelly Manufacturing

(a) Failure to set: Addition of excessive sugar and lack of acid or pectin may inhibit the setting of gel in jellies. Overcooking and undercooking also prevent the setting of gel.
(b) Gummy jelly: Excessive inversion of sugar due to extended heating can result in gummy jelly.
(c) Stiff jelly: Prolonged cooking and excessive use of pectin produce hard gels that lack spreadability.
(d) Cloudy or foggy jelly: The use of immature fruits, improper clarification of fruit extract, prolonged cooking, non-removal of scum, over-cooling, incorrect pouring, and premature gelation can result in cloudy or foggy jelly.
(e) Formation of crystals: Addition of excessive sugar, low levels of acid, and cooking beyond endpoint can result in crystallization in jelly.
(f) Syneresis or weeping of jelly: Syneresis is caused by the exudation of fluid from the jelly. It is caused by low levels of sugar in the formulation, insufficient pectin, excess of acid, and storage at a warm place.
(g) Change in color: The fading or lightening of color is caused due to storage at high temperatures or exposure to bright light. Prolonged cooking at elevated temperatures also results in a change of color. Darkening at the top of the jars can be caused due to faulty sealing and storage under warm conditions.
(h) Mold growth: insufficient levels of sugar in the formulation and faulty sealing can result in mold growth during storage.

11.4 Effect of Constituents and Processing on Bioactive Compounds

Fruits are high in bioactive substances, such as vitamins, carotenoids, flavonoids, and phenolics. These substances are heat-labile and are prone to deterioration during processing and further storage. The fruit and its cultivar, pH, type of pectin and concentration, sugar content, and other product characteristics can affect the bioactivity during processing. These changes may persist during storage, depending on the type of packaging, the temperature and duration of storage, and other factors.

Water-soluble anthocyanins are present in most fruits. Electron-deficient flavylium nucleus is responsible for the instability of the anthocyanins (Francis 1985). Pectin and sugar impart stability by absorbing flavylium cation (Fernandes et al. 2014). Sugars bind with water making it unavailable for the nucleophilic attack at the C-2 position. Conversely, at lower sugar levels, furfurals (sugar degradation products) quicken the degradation of anthocyanin. Similar findings were reported in blueberry jam, where lower sugar content showed higher total monomeric anthocyanin degradation (Scibisz and Mitek 2009).

The concentration and type of pectin also affect the retention of anthocyanins in jams. Pectin with a low degree of methylation and further amidation was reported to benefit the anthocyanin retention. This effect was attributed to the formation of additional hydrogen bonds between the amide groups of pectin and the hydroxyl groups of anthocyanins, preventing nucleophilic attack by water or condensation reactions between anthocyanins and procyanidins. An increase in pectin content from 0.3 to 1% expressed 13–17% higher total monomeric anthocyanins. This could be attributed to stronger electrostatic contacts between the dissociated carboxyl groups of pectin chains and the flavylium cations (Poiana et al. 2013). Total monomeric anthocyanins with 31–35% higher levels were reported with low-methoxyl pectin amidated when compared with high methoxyl pectin in blackberry jam (Poiana et al. 2013). Low degree of esterification and more amidated groups reduced the loss of total phenols during the cooking of jam. Strawberry jam with LMPA reported 20% higher total phenol content than HM (Kopjar et al. 2009). Similarly, reduced-calorie jam reported 5–14% higher TPC in LMPA than in LMP. The concentration of pectin in the formulation also changes the TPC in the jam. The jam with higher (1%) pectin reported 5–8 and 12–15% higher TPC than the jam with 0.7 and 0.3% pectin, respectively, depending on the type of pectin (Poiana et al. 2012).

Bioactive compounds are prone to thermal degradation and the effect varies with temperature and duration of the treatment. Due to the complex formulation, the degree of degradation varies with the nature of the fruit and the concentration of added constituents. Cooking under vacuum at 78 °C/15 min and pasteurization at 92 °C resulted in a 17–41% loss in total selected anthocyanins with different cultivars in raspberry jam. Among the selected anthocyanins, cyanidin-3 glucoside reported a maximum loss of 20–40% in different cultivars Garcia-Viguera et al. (1999). In another study, cooking at 80–105 °C resulted in significant degradation of 60–97% total monomeric anthocyanins in strawberry, cherry, apricot, fig, and orange jams, with maximum and minimum loss in strawberry and fig jam, respectively Rababah et al. (2011). This effect might be linked to the thermal hydrolysis of the glycoside linkage in anthocyanins and the resultant formation of degradation products like chalcones. Likewise, a 90% reduction in total monomeric anthocyanins was reported in reduced-calorie strawberry jam under vacuum cooking at 80 °C/45–60 min and pasteurization at 80 °C/10 min (Kovacevic et al. 2009). While only 7–10% reduction in total phenol content was observed in blueberry (Scibisz and Mitek 2009). A significant loss (25–51%) of TPC was reported in reduced-calorie sour cherry, sweet cherry, strawberry, and bilberry jams (Poiana et al. 2011, 2012). Also, thermal processing degrades the ascorbic acid or vitamin C via oxidation to dehydroascorbic acid and then to 2,3-diketogulonic acid and other products (Davey et al. 2000). Reduced-calorie jams from different fruits expressed degradation of L-ascorbic acid up to 78% due to thermal processing, and the loss increased with a decrease in pectin content (Poiana et al. 2012).

11.5 Encapsulation to Enhance the Stability of Bioactive Compounds in Fruit Jam and Jelly

Encapsulation of bioactive compounds in food products reduces the degradation of bioactive during processing and storage. Encapsulation entraps bioactive components (phenols, anthocyanins, vitamins, etc.) within a wall material to protect them from harsh processing conditions and stabilize them. The different wall materials used for encapsulation include starch, gums, maltodextrin, gelatin, etc. These components have highly branched structures that establish a hydrodynamic volume, thus improving the intermolecular interactions and resulting in effective entrapment of the bioactive compound.

The jelly featured 7% barberry anthocyanin, which was comparable to the control sample in relation to color, flavor, and texture. The wall material maltodextrin and gum Arabic increased the half-life of barberry's anthocyanin to 7 and 5 months in comparison to control samples with only 3.6- and 2.5-months half-life at 4 and 25 °C, respectively (Mahdavi et al. 2016). Tomato jam (*Solanum betaceum*) with aloe vera (*Aloe Barbadensis* Miller), stevia (*Stevia rebaudiana*) as a sweetener, and encapsulated *Lactobacillus bulgaricus* as a probiotic was developed by González-Cuello et al. (2018).

11.6 Novel Food Processing Technologies for Jams and Jellies

Different emerging technologies are used to minimally process jams and jelly products, including high-pressure processing, ultraviolet light, membrane filtration, ultrasonication, osmotic dehydration, and pulsed electric field.

These technologies better preserve the bioactive components in the food matrix and prevent microbial growth (Chakraborty et al. 2016). These processing techniques have displayed favorable results in jams. Strawberry jam processed at 200–800 MPa displayed better retention of anthocyanins in comparison to thermally processed jam (Gimenez et al. 2001). High-pressure processing has also improved the stability of anthocyanins and vitamins during storage for up to 60 days compared to thermally processed strawberry jam (Kimura et al. 1994). Peel of high pressure-treated yuzu marmalade at 500 MPa better retained the natural color (Kuwada et al. 2010). The effect of HPP on artificially contaminated (with *Listeria*) apple and plum jam baby food was evaluated. No viable cells were found after high-pressure treatment of 350 MPa for 30 min at 5 °C also, low temperature helps to maintain the sensory properties of the product (Prestamo et al. 1999). Thermal-assisted HPP (550–700 MPa/45–75 °C/5 min) in combination with osmodehydration (40 °C for 3 h) has also been applied for the processing of grapefruit jam to inactivate pectin methyl esterase and peroxidase. The aimed high-pressure preservation improved the enzymatic stability of jam with osmotic dehydration without altering the bioactive content (Igual et al. 2013).

The influence of ultraviolet light and ultrasound irradiation on the fiber-enriched pear fruit jam's physicochemical properties, sensory profile, and microbiological characteristics was evaluated by Manev and Petkova (2021). The sensory parameters such as graininess, elasticity, and juiciness presented better approval for jam processed with ultrasound irradiation. Ultraviolet light-treated jam showed the lowest microbial growth. Ultraviolet light and ultrasound irradiation caused insignificant differences in pH, titrable acidity, dry matter, and sucrose content. Low-calorie blackberry (*Rubus irasuensis* Liebm.) jelly was manufactured with clarified juice, attained from a crossflow microfiltration technique. The statistical model was used to optimize the factor levels for maximum acceptability to obtain a jelly with less than eight calories per serving (Acosta et al. 2006).

Thermal processing in combination with vacuum better preserved the bioactive (total phenolics, vitamin C) and antioxidant activity (DPPH) compounds in low-calorie mixed Brazilian cerrado fruit jam (Brandão et al. 2021). High-pressure processing at 600 MPa reported a decline in microorganisms, greater clarity, less browning, and better sensory acceptance in olive jam than thermally pasteurized olive jam (Delgado-Adamez et al. 2013). Likewise, strawberry jam processed at 600 MPa expressed better retention of ascorbic acid, anthocyanins, phenolics, flavonoids and antioxidant activities when compared with thermally processed jam (Sravani et al. 2017). Electron beam irradiation (3, 5, and 7 kGy), ultrahigh-pressure (400, 500, and 600 MPa for 6 min), and conventional thermal pasteurization treatments (80 °C for 30 min) were compared for their effects on the physicochemical properties, active components, and flavor volatiles of jujube jam (Chen et al. 2023). Nonthermal technologies outperformed the thermally pasteurized jam in terms of color, retention of bioactive compounds (organic acids, cyclic adenosine monophosphate, and phenolic components), and aroma components. Electron beam irradiation increased total polyphenols by 15.4–26.3% and ABTS+ scavenging activity by 9.5–36.89%. High-pressure processed (400 MPa at 27 ± 1.5 °C for 10 min) sapodilla jam was developed with the aim of reducing the calories by Shinwari and Rao (2020).

11.7 Reduced-Calories Jams and Jellies

Sour cherry jam with 36.5% sugar, 0.277% pectin, and stevia 0.30% with encouraging physicochemical, textural, and rheological properties was developed (Nourmohammadi et al. 2021). Sucrose was partially and fully replaced by stevioside and fructose in tomato pomace jam. These jam formulations were characterized by a lower total carbohydrate, lower energy value, and high fiber when compared to commercial products (Belović et al. 2017). Response surface methodology with mixture design was applied to estimate the optimum ratio of pineapple, papaya, and carambola in the reduced-calorie tropical mixed fruit jam. The optimum mixed fruit jam formulation contained one-third (106 kcal/100 g) of the caloric value of a normal jam (Abdullah and Cheng 2001). Orange-based low-calorie jam with pumpkin and papaya and alternative sweeteners like fructose, stevioside, and sucralose were developed by Abolila et al. (2015). Addition of stevioside and sucralose in low-calorie jam augmented total phenol and antioxidant capacity and enhanced color and taste. Response surface methodology was used to optimize grape juice jellies with low-methoxyl pectin. Based on overall balance, a jelly prepared with 38° Brix of total sugar, 1.2% of low-methoxyl pectin, and a cooking temperature of 69 °C expressed better acceptability (Sousa et al. 1997). Tropical mixed fruit (pineapple, banana, and passion fruit) jelly with sorbitol, low-methoxyl pectin, and calcium chloride. Sodium benzoate and potassium sorbate were optimized based on overall acceptability (Acosta et al. 2008). In another study, Acosta et al. (2006) optimized the sorbitol, low-methoxyl pectin, and calcium chloride for the preparation of low-calorie blackberry jam. Development of sugar-free grape jelly with sucralose, low-methoxyl pectin, maltodextrin, and xanthan gum and/or locust bean gum was attempted by Khouryieh et al. (2005). Compared to their use separately in the formulations, the combination of xanthan gum and locust bean gum significantly reduced syneresis and enhanced firmness and springiness in the jelly.

11.8 Marmalade

In the category of preserves, marmalade holds a unique position due to its bright colors and tangy taste. These products are generally made from citrus fruits that contain shredded peel slices as a suspended ingredient. It is a type of fruit jelly that contains slices of fruit and uniformly suspended fine shreds of peel in the gel throughout. The history of marmalade dates to the ancient Greeks and Romans, who boiled quince fruits with honey to produce a thick paste. Mostly lime, grapefruit, lemon, and orange fruits are used for marmalade production. According to the FSSAI, marmalade is defined as a product that is made by boiling wholesome fruits, including their peels, pulps, and juices, either with or without water, adding nutritive sweeteners, and concentrating to the point where the product cools and gelatinizes. It should be clear and transparent and not syrupy, sticky, or gummy.

The specific requirements for marmalade preparation as per FSSAI are as follows:

1. Total soluble solids—65%
2. Fruit content except peel—45%
3. Citric acid—2 g
4. Preservative—40 ppm of SO_2
5. Shreds—62 g/l

11.8.1 Basic Ingredients and Their Role

Fruits

Marmalades are made from ripe and under-ripe fruit, rich in pectin and organic acids. Mostly citrus fruits alone or in combination with other fruits are mainly used for manufacturing marmalades. The most popular combinations of fruits for marmalade preparation are orange-peach, orange-pear, pear-pineapple, and grape. Marmalades made from sour oranges (*Citrus aurantium* L.) are becoming more popular because they contain a variety of bioactive chemicals, including flavonoids, flavones, and furocoumarins (Barreca et al. 2011). Fruits provide the sugars, flavor, coloring agent, vitamins, inorganic compounds, and organic acid that are used for balancing the sweetness of the marmalade, and the pectin plays an important role as a natural thickening agent which helps in achieving the gel-like consistency. Over-ripened fruits are not suitable for preparing marmalade due to the degradation of pectin and aroma compounds during the overripening process (Varzakas and Özer 2012).

Sugars

Sugar is a key ingredient in marmalade preparation, which contributes flavor and texture to the final product. During the cooking process, sugars react with organic acids and pectin to form a thick, spreadable consistency. Sugar functions as a preservative, inhibiting microbiological growth and extending the shelf life of the marmalade. It also improves the fruit flavors by counteracting the inherent acidity of citrus fruits. Sugar syrup or sweeteners can be used instead of sugars to prevent the marmalade from crystallizing during storage (Varzakas and Özer 2012).

Acids

Low acidity fruits need the addition of a source of acids to correct the acidity. Lemon juice or citric acids are sometimes added to marmalade to ensure acceptable gelling. Acids enhance taste, aroma, balance the sweetness, and contribute to gel structure by activating pectin molecules. The pulp or fruit juice pH should range between 3.0 and 3.3.

Pectin

Pectin is a polysaccharide consisting of small molecules of galacturonic acid, and pectin concentration differs in fruit. Although the peel and seeds of citrus fruits are rich in pectin, some fruits have a lower pectin concentration; hence, commercial pectin is usually added to ensure a firm set. In marmalade, pectin and acid content are kept higher than in jelly preparation.

11.8.2 Types of Marmalades

There are two types of marmalades that are commercially prepared: Jelly marmalade and jam marmalade.

11.8.2.1 Jelly Marmalade

The jelly marmalades are prepared from a combination of sweet orange and sour orange in a ratio of 2:1, and the shredding of sweet orange peels is used as suspended ingredients.

Procedure of Jelly Marmalade
Selection and Preparation of Fruits

Mature, ripened fruits are selected and avoid use of over-ripened fruits. Selected fruits are washed with water and dried at room temperature. The method of fruit preparation depends upon the type of fruits chosen for marmalade preparation. The colored part or peel of the fruit is removed with the help of peeler, and then peeled fruits are either used for juice extraction or cut into slices. The seeds are removed and tied in cheesecloth or muslin cloth. This will be helpful in pectin extraction.

Boiling of Peel
The outer layer (yellow colored) of citrus fruits is peeled, and the stripped-off peel is cut into slices about 20–25 mm long and 1–1.2 mm thick. These shreds are cooked in water with 0.25% sodium bicarbonate or ammonia solution (0.1%). The water is boiled over a medium-high heat. After boiling, reduce the heat gently for 60–90 minutes or until the shreds become softer and translucent by occasional mixing to avoid sticking and maintain even cooking. The shreds may be placed into the heavy syrup for some time to increase their bulk density and prevent them from floating on the surface when it is mixed with jelly.

Prepare the Fruit Pulp and Boiling
Chopped small pieces of fruits or fruit slices are placed in a food processor to prepare pulp or puree, depending on the desired consistency of marmalade. Once the peel is soft and translucent, chopped fruit pulp and any collected juices are added to the pot with the peel. Heating sliced fruits in boiling water also aids in the conversion of protopectin to pectin and softening fruit tissues. To extract pectin, boil the fruit and pass it through linen bags or cheesecloth. For large-scale manufacture, fruit extract is filtered via presses for clarity. If the consistency of fruit extract is not

achieved as per the requirement, then commercial pectin should be added to obtain the desirable consistency.

Boil and Test for Setting Point
The fruit extract is boiled at 100–103 °C to bring the mixture to a rapid boil. The mixture is mixed occasionally to avoid scorching. After boiling the fruit extract, sugar is added. To achieve a bright and sparkling marmalade, boiling should not last longer than 20 minutes after the sugar is added. Continue the boiling until the consistency of 65° Brix and the temperature reaches about 104–105 °C (219–221 °F) in the thermometer, which is the setting point for marmalade.

Packaging and Storage
The foam from the upper surface of the marmalade must be removed using a spoon. The hot marmalade (at a temperature of around 85 °C) is filled into sterilized cans or jars. About 1 cm of headspace should be kept in the jar. The jars are cleaned with cloth and sealed tightly with sterilized lids. Then, jars are allowed to cool completely at room temperature. Label the jars with the date and store them in a cool (temperature 10–12 °C and relative humidity at about 75%), well-ventilated, dark place. These steps are important because marmalade is a hygroscopic product that absorbs water, creating favorable conditions for mold development. Properly sealed jars have a shelf-life of 12 months.

11.8.2.2 Jam Marmalade

Jam marmalade is manufactured using the same procedure as jelly marmalade but without clarifying the pectin extract. After removing the albedo, slice the orange peel into 3 mm thick pieces and prepare as jelly marmalade. To soften sliced orange, lemon, or grapefruit, remove the peel and combine with a small amount of water before boiling. The boiling liquid is squeezed through a coarse pulper to remove seeds and produce a thick pulp. Cook the pulp and sugar until it reaches 65° Brix, like jam consistency. Mix the treated shreds into the jam when it has cooled slightly. Filling and packaging are done in a similar way as adopted for the packaging of jelly and jelly marmalade.

11.8.3 Recent Advances in Marmalade Preparation

The development of new confectionery items with the lowest sugar content is the current trend in creating a healthy diet, moving them from the "risk" group to a category of healthy food products. Perfilova et al. (2018) developed marmalade with improved nutritional value and lower sugar content with agar-based apple jam and carrot scraps by replacing the white crystalline sugar. Savaş et al. (2020) used the wild fruit named Cornelian cherry to prepare a marmalade rich in antioxidant substances due to their higher phenolic compounds. The prepared marmalade showed excellent antimicrobial activities against *Listeria monocytogenes, Staphylococcus aureus, Pseudomonas fluorescens, and Yersinia enterocolitica.* The organoleptic and antioxidant properties of the marmalades can be improved by the addition of plant cryogenic additives during storage (Shmatchenko et al. 2018). The authors used plant-based cryogenic pastes from quince, apple, carrot, pumpkin, grapes, and cryopowders from rose hips, sea buckthorn, and grapes to improve the shelf-life and antioxidant power of the jelly marmalade. The results showed that the prepared marmalade met the physico-chemical, microbiological, and organoleptic properties for 90 days of storage period, and the antioxidant capacity was improved 2.3–8 folds as compared to the control. Özbek et al. (2019) prepared strawberry marmalade enriched with chia seeds which are rich in omega-3 fatty acids, dietary fiber, protein, and antioxidant compounds. They also replaced the refined sugar with sorbitol and found that the addition of sorbitol and chia seeds (5%) increased the phenolic content by 15% and dietary fiber by 168% and decreased the caloric value by 48% as compared to the control. A novel antioxidant-rich marmalade was prepared with black plum peel puree (50%), and pectin (0.4%) showed the highest acceptability (Estaji et al. 2020). Glycemic control is of great importance for people suffering from metabolic disorder type 2 diabetes. The agar-, gelatin-, and pectin-based marmalade were prepared with maltitol, dried fruits, and berries to control the glycemic index (Zhilinskaya et al. 2018). Similarly, marmalade was prepared by replacing the sucrose with *Stevia* extracts up to the level of 50%, giving equally acceptable results (Ruiz and Campos 2019). Gum Arabic (1–5%) from *Acacia senegal* var. *kerensis* was used to prepare jams and marmalades, and the developed products do not show any evidence of syneresis (Kavaya et al. 2019). The addition of peel also improves the dietary fiber content of the marmalade. The addition of mango peel (30%) in the marmalades showed 7.8% total dietary fiber as compared to the jam that was not added with mango peel (Iuit-González et al. 2019). A novel diabetic jelly marmalade was prepared by replacing sugars with a combination of sorbitol and glycosyl stevioside (Tabatorovich and Reznichenko 2019). The dry bracts of hibiscus were used to give its color and bioactive ingredients to the marmalade and low-calorie prebiotic polydextrose was used as a filler material. The marmalade was enriched with succinic acid, which maintains the acidity, increases insulin resistance, and lowers the risk of diabetic complications.

11.9 Preserves

Fruits are seasonal in nature and have a short shelf life. Therefore, seasonal fruits are conserved in an innovative way of preserving for consumption during the off-season or making the fruit of choice available for consumers to relish throughout the year. Fruit preserves are a sort of fruit-based food that is often boiled with sugar and pectin to get a thick, spreadable consistency. They are like jams and jellies but may include bigger pieces or whole fruits. Preserve is a mature fruit or vegetable or a portion of it that has been soaked or immersed with thick or heavy sugar syrup till it becomes smooth and transparent. Due to the osmosis process, water comes out of the fruit, and sugar enters the fruit until it reaches a state of equilibrium. Sugar can extend the shelf life of fruit products by exerting osmotic pressure on microorganisms. Preserves can be made from apples, cherries, figs, apricots, anola, pineapple, pear, peaches, plums, mango, grapes, papaya, strawberries, quinces and tomatoes, and other fruits.

11.9.1 Preparation Method

Preparation involves primary operations like the selection of fruits, peeling, puncturing (to promote sugar penetration), and blanching. Blanching may be done with or without additives to inactivate natural enzymes and to reduce the oxidative decolorization. The blanched fruits are then treated to firm the texture of the product. Now, sugar is added concentrated and packed after the addition of preservatives. Different processes employed for the preparation of preserves from fruits and vegetables at the commercial level are explained below.

Selection and Preparation of Fruits
For making the preserves, mature-ripe uniform sized fruits are selected. The over-ripened soft fruit is not suitable for making preserved products. The fruits are washed thoroughly, and stems, pits, or seeds are removed carefully. After that, depending on the fruit type, the peel of the fruit is removed. The selected fruits are diced into uniform size pieces that will help in uniform cooking. The fruit preparation and primary processing of fruits depends on the type of fruit used. When the whole fruits of apples and pears are used for making preserves, then fruit peels are removed, and fruits are pierced. In case half or quartered pieces are preserved, then the peels are removed, and fruits are sliced into half or quartered size, cored, and punctured. In the case of mangoes, the fruits are peeled first, then sliced and punctured. In case of apricots and cherries, the fruits are pitted and used for making preserves. Oranges, lemons, grapefruits, and other citrus fruits are halved, and the pulp is removed. Pumpkins are sliced, peeled, punctured, and immersed in weak lime water for a few hours to harden their texture. Strawberries and raspberries are utilized as such, with no prior treatment.

Cooking
After fruit preparation, the fruits are cooked in water to make them soft and more capable of absorbing sugar. Cooking fruits directly in syrup might be challenging since the syrup should be kept at a precise consistency to confirm that it permeates the fruit evenly without causing it to shrink or become tough. Primarily, the fruit should be boiled in water before being transferred to the syrup or cooked first in dilute syrup with a low sugar concentration. However, highly juicy fruits can be placed directly into thick syrup, as the excess juice in the fruit will help in diluting the syrup.

Cooking in Syrup

There are three ways of cooking the fruit in syrup. These are: (i) open kettle one period process, (ii) open kettle slow process, and (iii) vacuum cooking process.

11.9.1.1 Open Kettle One Period Process

Fruit and other items can be preserved by cooking them in an open pot or kettle, a technique known as the "open kettle" process. In this process, fruits are placed in a large open kettle or pot and cooked until they become soft and reach the desired consistency. During this stage, the fruits are typically cooked with sugar and possibly other ingredients, such as spices or lemon juice. When beginning to cook the fruit, the concentration of the sugar syrup should be minimal. When fruits are heated quickly, they will become rough, especially if it is done in large shallow pans with a tiny amount of syrup. Hard fruits like apples, pears, and peaches require prolonged heating to become soft, whereas soft fruits like strawberries and raspberries only need a little boiling to become soft. Sugar syrup should have a final concentration of at least 68° Brix, or 106 °C at the final stage. The fruits are packed, still hot, into sterilized jars as soon as it is cooked. This process is essential to preserving the quality and safety of the preserved products. Immediately after filling, the jars are covered with lids and let them cool. In order to preserve the fruit and keep it from spoiling, heat is needed to assist in producing a vacuum seal.

11.9.1.2 Open Kettle Slow Process

A traditional food preservation technique known as the "open kettle slow process" involves cooking food—typically fruit—at a low temperature in an open kettle or pan. Although this process is labor- and time-intensive, when it is done properly, it can yield preserves of excellent quality. In this process, fruits and vegetables are placed in an open kettle or pan. The fruits are cooked slowly at low heat to soften them and release the juices. This process enables the fruit to break down and cook evenly. As the fruit boils, slowly add sugar and any additional ingredients (such as lemon juice or spices). Then, fruits are cooked slowly and continuously and mixed occasionally until they reach the required consistency. Depending on the fruit and how much is being cooked, this procedure may take many hours. The objective is to get a thick, syrupy consistency that will work well to keep the fruit fresh. After initial boiling, the fruits are further boiled by placing them in a jar in alternating layers with sugar equal to half the weight of the fruit, and the mass is let to stand for 24 hours. The fruit releases extra water during this process, allowing the sugar to dissolve and form a syrup of about 38° Brix. More sugar is also added at later stages to make the syrup stronger to around 60° Brix. After that, the fruit is kept in the syrup for 3–4 days, and when the consistency reaches around 70° Brix, the preserves are packed into containers.

11.9.1.3 Vacuum Cooking

Cooking food in a vacuum-sealed bag at a precise and controlled temperature is the method known as sous-vide cooking or vacuum cooking. Vacuum-cooked preserves hold their color, texture, shelf-life, and flavor longer than those prepared in open kettles. In this process, cooking the fruits is carried to make them soft. After that, the cooked fruits are placed in the syrup solution having 30–35° Brix. Then, the syrup blends are placed in the vacuum pan and concentrated at a controlled temperature for a specified amount of time under low pressure until it reach 70° Brix. The cooking time varies according to the type of fruit and desired texture.

11.10 Candid Products

Candied products are fruits, vegetables, or other dried products such as dry fruits that have been preserved by coating them with sugar or syrup and drying them. Candied fruit is made by boiling the fruit in sugar syrup until it is completely saturated with sugar and then drying. The procedure is identical to that used to make fruit preserves, except that the fruit is impregnated with a higher concentration of sugar or glucose; its overall sugar content is approximately 75.0%. This approach yields a sweet, chewy dessert that may be consumed on its own or utilized in a variety of recipes. Cane sugar is replaced with a certain amount of inverted sugar or glucose. Fruits with a strong flavor, such as pineapple, peach, orange, lemon, grapefruit, citrus, and cherry, are appropriate for such preparations. The under-ripened fruits are more suitable for making candid products to prevent the formation of jam-like consistency during the cooking of fruits in syrup solutions. The process of candid fruits manufacturing includes blanching of the fruits to kill the mass to increase the permeation of sugar into and through the tissues, allowing the free osmosis process until the equilibration to take place between the surrounding syrup solution and the juice in the fruit tissues within as specified time. There are two main methods for producing candid products: (1) the traditional slow method and (2) the fast method (Dalgleish 1990). In both methods, preparations of fruit or peel are the same as the method of preserves. In the traditional slow method, the fruits are blanched properly in steam or very hot water. Before candying, the fruit pieces are pierced with needles for faster syrup penetration, depending on their size, shape, density, etc. The needles made of stainless steel or silver plated are suitable to prevent material discoloration, eliminate colored metal traces, and conceal the piercing. Traditionally, the fruits are immersed in 40° Brix sugar syrup at room temperature. After a 2-week period, the sugar concentration is gradually increased up to 75° Brix. Following the syrup treatment, the fruit is taken out of the syrup and allowed to drain for around 30 minutes. A moist sponge is used to wipe the fruit or peel. Occasionally, it is briefly submerged in boiling water to get rid of the extra concentrated syrup. The next step is slow drying in the shade or for 8 to 10 hours at 66 °C in a dryer. Drying of candid fruits also affects the texture of the final products.

The 9 h drying time of candid fruit made from dewandaru fruit was hard and showed shrinkage and dried appearance compared to the 8-h dried candid fruit (Asri et al. 2021). Likewise, drying plum fruit candid products using an infrared dryer showed a minor reduction in mineral and polyphenolic compounds and a greater loss of vitamins (Prichko and Droficheva 2021).

Sometimes, the fruit is dusted with crystalline sugar before drying. In the fast method, the process of candying is completed within 24 hours or less time. The peel or fruit pieces are rapidly blanched in hot water in a weak syrup solution and under vacuum and with continuously rising syrup concentration and temperature. The temperature rises from about 50 °C to 80 °C. The candid products covered or coated with a thin transparent layer of sugar to give it a shiny or glossy appearance are called glazed fruit, and the candid products that are coated with crystalline sugar are called crystallized products. Candied fruits are consumed with a coating or dusting of crystallized sugar after drying or from a container or jar containing a portion of the final 75% sugar syrup. A small amount of glucose syrup or inverted sugar can be added to regulate the texture, tenderness, and sweetness of candid products.

11.10.1 Spoilage of Glazed and Crystallized Fruit

Fermentation can cause spoiling, particularly in the early phases of making preserves and candies, when the sugar content in the syrup is low. This can be regulated by boiling the product at regular intervals. If candied and crystallized fruit are stored in humid environments, they lose some of their sugar due to moisture absorption from the air. If they are stored in moist rather than dry containers, they will eventually develop moldy. If the fruit is not thoroughly dried, mold infestation may occur. The ozone-microwave sterilization method for candid products inhibited the infestation of mold by destroying the cell membrane of molds cell (Huang et al. 2024).

11.11 Specifications/Standards for Jams, Jellies, and Marmalades

Jams, jellies, and marmalades are widely consumed. Institutional bulk packs, retail packs, and single-serving packs of these products are available in the market. In the market, jams, jellies, and marmalades prepared from conventional fruits are easily available throughout the globe. Also, exotic varieties of jams and jellies are available to cater to the specified consumer. The overall quality of these products has been greatly improved by advances in processing methods, a better comprehension of fruit qualities, and competitive pressures. The quality of jams and jellies is monitored using various global standards. The Codex Alimentarius guidelines for jams and jellies are intended to help standardize jams and jellies across the globe, which will facilitate regulatory harmonization and further the growth of the global food trade. FPO Act controls the production, storage, and sale of fruit and vegetable products including jams and jellies. FPO specifications for jam jelly, marmalade, and preserve are described in Table 11.2.

Many jams and jellies manufactured today do not fulfill the requirements of FDA standards for fruit (or juice) to sugar ratios or optimal ingredients. These products are marketed as "imitation" jams or jellies and are frequently manufactured with corn syrup and artificial flavors.

11.12 Conclusions

Fruits are being preserved in the form of jams, jellies, marmalades, candied and glazed food products since ages. A variety of conventional and non-conventional fruits are converted into such products to prevent food loss and wastage and to ensure year-round availability. The current market, however, requires newer products with fresher qualities and less calories. Some researchers have tried to develop such products using various nonthermal processing techniques and non-caloric sweeteners, but the comprehensive studies must be examined from the perspective of commercial production. Further research is required to determine whether bioactive chemicals are retained in these products and whether they may be fortified with improved delivery methods. To investigate the commercial possibilities of such products, it is also necessary to determine their cost economics. Expanding

Table 11.2 FSSAI specifications for jam jelly, marmalade, preserved, candied, and crystallized or glazed fruit and peel

Type of product	Specifications	
	TSS % in finished product (minimum)	% of fruit in finished product (minimum)
Fruit jam	65	45 (25% in strawberry and raspberry jam)
Fruit jelly	65	45
Marmalade	65	45
Fruit preserve	68	55
	Specifications	
	Total sugar %	*Reducing sugar as % of total sugar*
Candied and crystallized or glazed fruit and peel	Not less than 70	Not less than 25
Permitted additives		
Preservatives Sulfur dioxide 40 ppm Benzoic acid 200 ppm Permitted colors 200 ppm		

scientific inputs in these domains may help create newer products with improved nutritional profiles and minimum processing.

References

Abdullah A, Cheng T (2001) Optimization of reduced calorie tropical mixed fruits jam. Food Qual Prefer 12(1):63–68

Abolila RM, Barakat H, El-Tanahy HA, El-Mansy HA (2015) Chemical, nutritional and organoleptical characteristics of orange-based formulated low-calorie jams. Food Nutr Sci 6(13):1229–1244. https://doi.org/10.4236/fns.2015.613129

Acosta O, Viquez F, Cubero E, Morales I (2006) Ingredient levels optimization and nutritional evaluation of a low-calorie blackberry (*Rubus irasuensisLiebm.*) jelly. J Food Sci 71(5):S390–S394

Acosta O, Viquez F, Cubero E (2008) Optimisation of low calorie mixed fruit jelly by response surface methodology. Food Qual Prefer 19(1):79–85

Asri NKT, Devi M, Soekopitojo S (2021) Effect of drying on antioxidant capacity, sugar content, water content, physical and organoleptic properties of dried candied dewandaru fruit (*Eugeiauniflora L.*). J Phys Conf Ser 1882(1):012110. https://doi.org/10.1088/1742-6596/1882/1/012110

Barreca D, Bellocco E, Caristi C, Leuzzi U, Gattuso G (2011) Distribution of C- and O-glycosyl flavonoids, (3-hydroxy-3-methylglutaryl) glycosyl flavanones, and furocoumarins in *Citrus aurantium L.* juice. Food Chem 124(2):576–582

Belović M, Torbica A, Pajić-Lijaković I, Mastilović J (2017) Development of low calorie jams with increased content of natural dietary fibre made from tomato pomace. Food Chem 237:1226. https://doi.org/10.1016/j.foodchem.2017.06.045

Brandão TM, Carvalho EEN, Lima JP, Carmo EL, Elias HS, Martins GAS, Borges SV (2021) Effects of thermal process in bioactive compounds of mixed Brazilian cerrado fruit jam. Food Sci Technol (Campinas) 41(Suppl. 2):439–446. https://doi.org/10.1590/fst.36020

Chakraborty S, Rao PS, Mishra HN (2016) Changes in quality attributes during storage of high-pressure and thermally processed pineapple puree. Food Bioprocess Technol 9(5):768–791. https://doi.org/10.1007/s11947-015-1663-0

Chen M, Wang Z, Yu J, Wang J, Xu H, Yue X (2023) Effects of electron beam irradiation and ultrahigh-pressure treatments on the physicochemical properties, active components, and flavor volatiles of jujube jam. LWT Food Sci Technol 187:115292. https://doi.org/10.1016/j.lwt.2023.115292

Christensen O, Trudsoe J (1980) Effect of other hydrocolloids on the texture of kappa carrageenan gels. J Texture Stud 11(2):137–147

Dalgleish JM (1990) Fruit-based snacks including dried and candied fruit. In: Booth RG (ed) Snack food. Springer, Boston, pp 113–126. https://doi.org/10.1007/978-1-4613-1477-6_12

Davey MW, Montagu MV, Inze D, Sanmartin M, Kanellis A, Smirnoff N et al (2000) Plant L-ascorbic acid: chemistry, function, metabolism, bioavailability and effects of processing. J Sci Food Agric 80(7):825–860

Delgado-Adamez J, Franco MN, Sánchez J, De Miguel C, Ramírez MR, Martín-Vertedor D (2013) Comparative effect of high-pressure processing and traditional thermal treatment on the physicochemical, microbiology, and sensory analysis of olive jam. Grasas Aceites 64(4):432–441

Doublier JL, Launay B, Cavelier G (1992) Visco-elastic properties of food gels. In: Rao MA, Steffe JF (eds) Visco-elastic properties of food. Elsevier Applied Science, New York, pp 371–434

Estaji M, Mohammadi-Moghaddam T, Gholizade-Eshan L, Firoozzare A, Hooshmand-Dalir MAR (2020) Physicochemical characteristics, sensory attributes, and antioxidant activity of marmalade prepared from black plum peel. Int J Food Prop 23(1):1979–1992

Fernandes A, Brás NF, Mateus N, de Freitas V (2014) Understanding the molecular mechanism of anthocyanin binding to pectin. Langmuir 30(29):8516–8527

Francis FJ (1985) Pigments and other colorants. In: Fennema OR (ed) Food chemistry, 2nd edn. Marcel Dekker Inc., New York, pp 545–584

Garcia-Viguera C, Zafrilla P, Romero F, Abellan P, Artes F, Tomas-Barberan FA (1999) Colour stability of strawberry jam as affected by cultivar and storage temperature. J Food Sci 64(2):243–247

Gimenez J, Kajda P, Margomenou L, Piggott JR, Zabetakis I (2001) A study on the colour and sensory attributes of high-hydrostatic-pressure jams as compared with traditional jams. J Sci Food Agric 81(13):1228–1234. https://doi.org/10.1002/jsfa.935

González-Cuello RE, Pájaro K, Acevedo W, Ortega-Toro R (2018) Study of the shelf-life of a low-calorie jam added with microencapsulated probiotics. Contemp Eng Sci 11(25):1235–1244

Hoefler AC (1991) Other pectin food products. In: Walter RH (ed) The chemistry and technology of Pectin. Academic Press, New York, pp 51–56

Huang H, Ni Z, Xie J (2024) Good feasibility of ozone-microwave treatment as a sterilization technology to extend the edible life of candied fruit as a post-processed fresh fruit product. Food Bioproc Technol 17:3086. https://doi.org/10.1007/s11947-023-03315-7

Igual M, Sampedro F, Martínez-Navarrete N, Fan X (2013) Combined osmode hydration and high-pressure processing on the enzyme stability and antioxidant capacity of a grapefruit jam. J Food Eng 114(4):514–521. https://doi.org/10.1016/j.jfoodeng.2012.09.006

Iuit-González M, Betancur-Ancona D, Santos-Flores J, Cantón-Castillo CG (2019) Marmalade enriched with dietary fiber from mango (*Mangifera indica L.*) peel. Revista Tecnologíaen Marcha 32(1):193–201

Jarvis MC (1984) Structure and properties of pectin gels in plant cell walls. Plant Cell Environ 7(3):153–164

Kavaya RI, Omwamba MN, Chikamai BN, Mahungu SM (2019) Sensory evaluation of syneresis reduced jam and marmalade containing gum Arabic from *Acacia Senegal* var. *kerensis*. Food Nutr Sci 10(11):1334–1343

Khouryieh HA, Aramouni FM, Herald TJ (2005) Physical, chemical and sensory properties of sugar-free jelly. J Food Qual 28:179–190

Kimura K, Ida M, Yoshida Y, Ohki K, Fukumoto T, Norihiro S (1994) Comparison of keeping quality between pressure-processed jam and heat-processed jam: changes in flavor components, hue, and nutrients during storage. Biosci Biotechnol Biochem 58(8):1386–1391

Kopjar M, Pilizota V, Tiban NN, Subaric D, Babic J, Ackar D et al (2009) Strawberry jams: influence of different pectins on colour and textural properties. Czech J Food Sci 27:20–28

Kovacevic DB, Levaj B, Dragovic-Uzelac V (2009) Free radical scavenging activity and phenolic content in strawberry fruit and jam. Agric Conspec Sci 74:155–159

Kuwada H, Jibu Y, Teramoto A, Fuchigami M (2010) The quality of high pressure induced and heat-induced yuzu marmalade. High Pressure Res 30(4):547–554. https://doi.org/10.1080/08957959.2010.534090

Lau JM, McNeil M, Darvill AG, Albersheim P (1985) Structure of the backbone of rhamnogalacturonan I, a pectic polysaccharide in the primary cell walls of plants. Carbohydr Res 137(1):111–125

Mahdavi SK, Jafari SM, Assadpour E, Ghorbani EM (2016) Storage stability of encapsulated barberry's anthocyanin and its application in jelly formulation. J Food Eng 181:59–66

Manev Z, Petkova N (2021) Influence of ultraviolet light and ultrasound irradiation on the physicochemical, sensory, and microbiological characteristics of pear jam enriched with fibers. Biointerf Res Appl Chem 11(6):14304–14315

Nourmohammadi A, Ahmadi E, Heshmati A (2021) Optimization of physicochemical, textural, and rheological properties of sour cherry jam containing stevioside by using response surface methodology. Food Sci Nutr 9(5):2483–2496

Oakenfull DG (1991) The chemistry of high methoxylpectins. In: Walter RH (ed) The chemistry and technology of Pectin. Academic Press, New York, pp 87–108

Özbek T, Şahin-Yeşilçubuk N, Demirel B (2019) Quality and nutritional value of functional strawberry marmalade enriched with chia seed (*Salvia hispanica L.*). J Food Qual 2019:2391931. https://doi.org/10.1155/2019/2391931

Perfilova OV, Babushkin VA, Magomedov GO, Magomedov MG (2018) Quality of jelly marmalade from fruit and vegetable semi-finished products. Int J Pharm Res 10(4):721

Poiana M A, Moigradean D, Dogaru D, Mateescu C, Raba D, Gergen I (2011) Processing and storage impact on the antioxidant properties and color quality of some low sugar fruit jams. Rom. Biotechnol. Lett. 16(5)

Poiana MA, Alexa E, Mateescu C (2012) Tracking antioxidant properties and colour changes in low-sugar bilberry jam as effect of processing, storage and pectin concentration. Chem Cent J 6:1–11

Poiana M, Munteanu M, Bordean D, Gligor R, Alexa E (2013) Assessing the effects of different pectins addition on colour quality and antioxidant properties of blackberry jam. Chem Cent J 7:1–13

Prestamo G, Sanz PD, Fonberg-Broczek M, Arroyo G (1999) High pressure response of fruit jams contaminated with *Listeria monocytogenes*. Lett Appl Microbiol 28(4):313–316. https://doi.org/10.1046/j.1365-2672.1999.00531.x

Prichko T, Droficheva N (2021) The influence of varietal characteristics on the quality indicators of candied fruits from plum fruits. Slovak J Food Sci 15:18–28

Rababah TM, Al-Mahasneh MA, Kilani I, Yang W, Alhamad MN, Ereifej K (2011) Effect of jam processing and storage on total phenolics, antioxidant activity, and anthocyanins of different fruits. J Sci Food Agric 91(6):1096–1102

Rees D (1969) Structure conformation and mechanism in the formation of polysaccharide gels and networks. Adv Carbohydr Chem Biochem 24:267–332

Ruiz JC, Segura Campos MR (2019) Development of nopal-pineapple marmalade formulated with stevia aqueous extract: effect on physicochemical properties, inhibition of α-amylase, and glycemic response. Nutr Hos 36(5):1081–1086

Savaş E, Tavşanlı H (2020) The antimicrobial and antioxidant properties of garagurt: traditional Cornelian cherry (*Cornus mas*) marmalade. Qual Assur Safe Crops Foods 12(2):12–23

Scibisz I, Mitek M (2009) Effect of processing and storage conditions on phenolic compounds and antioxidant capacity of highbush blueberry jams. Polish J Food Nutr Sci 59(1):45–52

Shinwari KJ, Rao PS (2020) Development of a reduced-calorie high pressure processed sapodilla (*Manilkara zapota L.*) jam based on rheological, textural, and sensory properties. J Food Sci 85(9):2699–2710. https://doi.org/10.1111/1750-3841.15364

Shmatchenko N, Artamonova M, Aksonova O, Oliinyk S (2018) Investigation of the properties of marmalade with plant cryoadditives during storage. Food Sci Technol 11(4):82–89. https://doi.org/10.15673/fst.v12i1.843

Shomer J (1991) Protein coagulation cloud in citrus fruit extract. I. Formation of coagulates and their bound pectin and neutral sugars. J Agric Food Chem 39(12):2263–2266

Smidsrød O (1974) Molecular basis for some properties in the gel state. Faraday Discuss Chem Soc 57:263–274

Sousa IMN, Matias EC, Laureano O (1997) The texture of low calorie grape juice jelly. Z Lebensm Unters Forsch 205:140–142

Sravani VJ, Ravi N, Roopa N, Kumar S, Pandey AK, Chauhan OP (2017) Use of high pressure technology for the development of novel jam and its quality evaluation during storage. J Food Sci Technol 54(11):3562–3568. https://doi.org/10.1007/s13197-017-2814-2

Tabatorovich AN, Reznichenko IY (2019) Formulation and quality assessment of diabetic jelly marmalade 'Karkade' fortified with succinic acid. Food Proc Tech Technol 49(2):320–329

Varzakas T, Özer B (2012) Application of sweeteners in food and drinks (bakery, confectionery, dairy products, puddings, fruit products, vegetables, beverages, sports drinks, hard candies, loukoumia, marmalades, jams, jellies, baked goods, sorbet). In: Varzakas T, Özer B (eds) Sweeteners: nutritional aspects, applications, and production technology. Springer, p 209

Woodmansee CW, McClendon JH, Somers GF (1959) Chemical changes associated with the ripening of apples and tomatoes. Food Res 24(5):503–514

Zhilinskaya NV, Sarkisyan VA, Vorobieva VM, Vorobieva IS, Kochetkova AA, Smirnova EA, Glazkova IV (2018) Development of a marmalade for patients with type 2 diabetes: sensory characteristics and acceptability. Food Sci Technol Int 24(7):617–626

Chapter 12
Sauces and Chutneys

Neha Pandey, Prerna Nath, S. J. Kale, Nita Khandekar, Shivangi Mishra, and OP Chauhan

12.1 Introduction

Processing of fruits and vegetables is crucial in developing countries where a significant amount is wasted due to negligence and lack of processing facilities (Nilar 2012). The aim of fruit and vegetable processing and value addition is to supply consumers with safe, nutritious, sensory acceptable, wholesome food throughout the year. Due to the change in lifestyle of the people living in the cities, the consumption of processed, ready-made, and fast foods has gradually increased. Among these, the production of sauces and chutneys is a very common and popular way of preserving fruits and vegetables. Sauces and chutneys are microbiologically safe products with a long shelf life mainly due to the hurdle effect of the heat, acidity, and low water activity altogether. They can be homemade, as well as manufactured on an industrial level. A variety of fruits and vegetables can be utilized to

N. Pandey
ICAR-National Soybean Research Institute, Indore, India

P. Nath (✉)
ICAR-Research Complex for Eastern Region, RS, FSRCHPR, Ranchi, Jharkhand, India

S. J. Kale
ICAR-National Institute of Secondary Agriculture, Ranchi, Jharkhand, India

N. Khandekar
ICAR-Central Institute of Agricultural Engineering, Bhopal, Madhya Pradesh, India

S. Mishra
National Institute of Food Technology Entrepreneurship and Management, Kundli, Sonipat, Haryana, India

OP Chauhan
Defence Food Research Laboratory, Mysore, Karnataka, India

OP Chauhan (ed.), *Fruits and Vegetables Technologies*,
https://doi.org/10.1007/978-981-96-8433-5_12

make sauces and chutneys such as chilies, tomatoes, carrots, mushrooms, onion, apples, and mangoes.

As per FSSAI, Sauces refer to "culinary preparations used as an adjunct to food, prepared from the edible portion of any suitable fruit/vegetable including, roots, tubers & rhizomes, their pulps/purees, dried fruits, singly or in combination by blending with nutritive sweeteners, salt, spices and condiments, and other ingredients appropriately to the product." Sauces include ready-to-eat gravies, dressings, as well as dry mixes that can be reconstituted before consumption (García-Casal et al. 2016). These are either liquid consistency using stock or semi-solid consistency using thickening agents, such as flour, cornstarch, and seasonings (to enhance flavors). Sauces can be used to moisten food and add flavors (Marcus 2019). The basic steps of sauce making are mentioned in Fig. 12.1.

They can be simple, made with a few key ingredients, or complex, involving elaborate cooking techniques and a multitude of flavorings. Sauces are divided into various categories based on the method of preparation, end-use, the region where they originated, and complexity.

1. *Emulsified Sauces and Dips:* These include rich, creamy sauces like mayonnaise, various salad dressings (French, Italian, ranch), and fatty dips (cheese, onion).
2. *Nonemulsified Sauces:* This category encompasses a wide range of sauces, from simple condiments like ketchup to complex preparations like cheese, cream, and brown gravy. It also encompasses water, and milk-based sauces, dressings, and gravies such as BBQ sauce, cheese sauce, tomato ketchup, white sauce, Worcestershire sauce, to name a few.
3. *Sauce Mixes:* These are concentrated powdered mixes that can be easily prepared into various sauces and gravies by adding water, milk, or other liquids. Popular mixes include cheese sauce, bearnaise, hollandaise, tomato, and pasta sauce mix (arrabbiata, alfredo, white, and Parma rosa).
4. *Clear Sauces:* These are thin sauces that are nonemulsified and often used as condiments or ingredients rather than as main courses. Examples include fish sauce, oyster sauce, and Thai fish sauce.

Chutneys are semi-solid or liquid condiments prepared primarily from fruits, vegetables, herbs, and spices. They are typically ground or pounded into a paste, creating a range of textures from chunky to smooth. Chutneys are known for their intense flavor profiles, often incorporating a combination of sweet, sour, spicy, and salty notes. They can be categorized into fresh, cooked, and fried based on the preparation process.

Sauces and chutneys are one type of food product that becomes a favorite among consumers and the Indian market. Sauces and chutneys have a rich history of adding

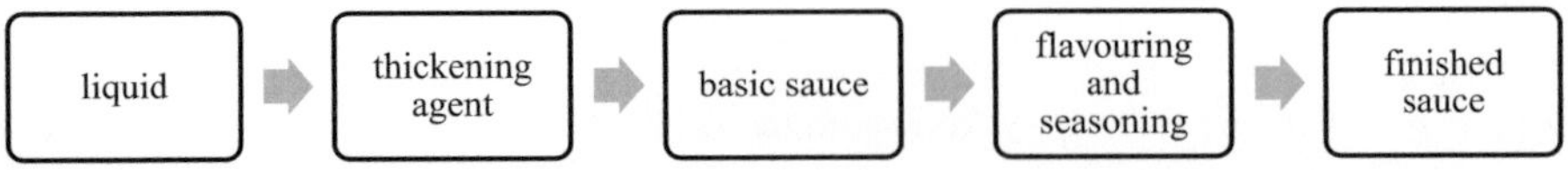

Fig. 12.1 Basic steps of sauce preparation

flavor to diverse cuisines, from main courses to snacks. This vibrant market is constantly evolving, driven by innovation and the global exchange of culinary traditions. Brands in this competitive landscape must stay ahead by identifying and catering to consumer preferences, while also addressing the rising interest in health and wellness. This chapter elaborates origin, significance, types, and preparation methods of sauces and chutneys. This chapter also sheds light on market share, global scenario, research trends, packaging requirements, and microbial quality of sauces and chutneys.

12.2 Market Trend

The sauces (mayonnaise, dressings, mixes, emulsified and nonemulsified sauces) and chutneys sector was valued at $176.4 billion in 2023 and estimated to grow at a CAGR of more than 10% during 2023–2028 (Global data 2023). The graph of chutney and sauce consumption has gone up due to the rising concern about health and wellness among consumers. Especially, the demand for sauces and chutneys with claims such as natural, organic, preservative-free, and low in sugar/fat has escalated. Moreover, busy lifestyles, rapid urbanization, rising disposable incomes, and the availability of innovative flavors have also spurred the demand for convenient seasonings, dressings, and sauces. In 2023, the condiment sauces category accounted for the highest value sales, followed by dressings (Table 12.1) (Global data 2023).

Sauces are widely consumed across various cultures. In China, approximately 70% of the population incorporates soy sauce into their daily diet (Huo 2020), while in Vietnam, the population consumes fish sauce daily (Thuy et al. 2003). Due to their frequent and substantial consumption, sauces can provide essential nutrients and antioxidants, such as vitamins, minerals, and antioxidants (Avallone et al. 2008). Moreover, sauces can be utilized as effective vehicles for fortification programs, particularly for iron, as evidenced by the fortification of fish and soy sauces

Table 12.1 Global growth of sauces and chutneys

Market size (2023)	$176.4 Billion
CAGR (2023–2028)	11.8%
Key regions (market share)	Asia-Pacific (APAC) (20%) Middle East and Africa (15%) The Americas (30%) Europe (25%) Latin America (10%)
Innovation in packaging types	Flexible packaging Paper & board Rigid plastics Squeezy bottles Aluminum cans Glass

Source: Global data (2023), Cavil (2024)

(Huo 2020; Whitefield et al. 2016). The taste and aroma of the sauces and chutneys are linked to the presence of high contents of sapid substances (salt, pepper, etc.) but also to their richness in volatile compounds, as well as sauces and chutneys containing spices, pepper, or mustard (Avallone et al. 2008).

12.3 Significance

Sauces and chutneys are more than just condiments; they are cultural ambassadors. They represent the ingenuity and resourcefulness of communities in utilizing local ingredients to create flavor and preserve food (Fig. 12.2). They are also a testament to the interconnectedness of cultures, with influences and adaptations traveling along trade routes, enriching culinary traditions worldwide. APAC was the highest revenue generator for the seasonings, dressings, chutney and sauce market in the world in 2023 (Global data 2023). The APAC market is driven by rising disposable income levels and urbanization in major APAC countries such as China, Japan, and India, prompting the growing working class to spend more on seasonings, dressings, and sauces due to the convenience these foods provide. In 2023, there is potential for growth in the sauces and chutneys market share in countries like Malaysia, Japan, the US, Canada, the UAE, Saudi Arabia, Romania, Poland, the UK, and Spain (Global data 2023).

Contrary to popular belief, sauces can be a healthy addition to your diet. Many options exist that are low in calories, fat, and sodium, while still providing significant flavor and moisture. For example, chutney, guacamole, harissa, hummus, pesto, tahini, and tzatziki can elevate lean proteins like fish, poultry, and seafood. These

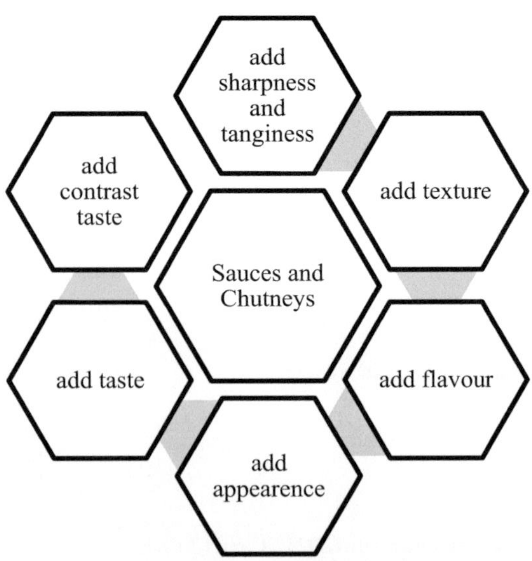

Fig. 12.2 Significance of sauces and chutneys in food

sauces often draw inspiration from various ethnic cuisines, making them appealing to a wider ethnicity (Marcus 2019). Some sauces and chutneys help in digestion, for example, mint chutney. Some chutneys, like coriander chutney, are packed with vitamins and antioxidants and can be good for people with diabetes. They are also used to moisten foods like white sauce and act as gravy in firm and dry food. They also contribute color to vegetables like hollandaise sauce. Sometimes they are used to give a contrasting taste to food, like cranberry sauce with roast pork. Fruit and vegetable sauces even enhance the nutritional value of the dish. These sauces dress and complement the food and sometimes make the food more palatable; for example, high-energy density sauce was prepared for old age patients to increase the acceptability of hospital food (Tsikritzi et al. 2011). It gives tartness to balance bland food; for example, chutneys are served with samosas and other Indian snacks. A drizzle of tangy vinaigrette elevates a salad, while a spoonful of rich gravy adds comfort to a roasted chicken. Chutneys, similar to sauces, are used to enhance the flavor of many dishes, from fruits to vegetables. In conclusion, sauces and chutneys are more than just ingredients; they are threads that weave together history, culture, and taste.

12.4 Origin and History

The art of sauce-making boasts a rich history dating back centuries. Ancient civilizations in Mesopotamia and Egypt used simple sauces made with herbs, spices, and oils. In ancient Rome, sauces were served as a means of flavor enhancement, as well as to mask imperfections in stale meat. Fermented fish sauce in Rome, also known as garum, was used to be prepared using the viscera and blood of mackerel (Badham 1854). The principle involved blood coagulation under high salt concentration and 9-month-long fermentation by halotolerant enzymes from mackerel viscera to obtain brown clear liquid garum (Beddows et al. 1976, 1979; Beddows 1985). Another fish sauce known as garos was produced in Greece, from the liver of the fish Scomber colias (Kelaiditis 1949). The production of this fish sauce was faster due to the presence of a high quantity of proteolytic enzymes in the liver. Aimeteon was also a Greek fish sauce made from the blood and viscera of tuna fish (Lopetcharat et al. 2001). While the French word “sauce” literally translates to “relish,” and the French are credited with developing many popular varieties, sauces are truly a global culinary treasure. The sixteenth, seventeenth, and eighteenth centuries marked a significant period in the culinary world, with the invention of several world-famous sauces and chutneys. Pesto sauce, a beloved Italian staple, originated in sixteenth-century Italy. The seventeenth century witnessed the introduction of béchamel, chasseur, and chutney, while the mid-eighteenth century brought the elegant hollandaise sauce to the table. While tartar sauce-like concoctions existed in the Middle Ages, the modern version appeared in the 1800s. Worcestershire sauce, a unique blend of soy sauce, garlic, onion, molasses, and anchovies, was invented in 1835 by John Lea and William Perrins, initially as an attempt to replicate a favorite sauce for Lord Marcus Sandys (Bottone 2014). BBQ sauce, a versatile condiment ranging from thin to thick, typically combines ketchup, vinegar, sugar, garlic, onion, and mustard. Its origins are debated, with some attributing its invention to seventeenth-century American settlers seeking to mask unfamiliar flavors, while others trace it back to the fifteenth century when Christopher Columbus introduced a Spanish sauce (Goldwyn 2014). The origins of mayonnaise can be traced back to 1756 during the French naval attack on Minorca. Some accounts suggest it was created by Duke de Richelieu's chef to commemorate the victory, while others claim it was a target of the attack itself. The first commercial mayonnaise appeared in 1905 at Richard Hellman's New York deli, and it was mass-marketed as Hellman's Blue-Ribbon Mayonnaise in 1912 (Bellis 2014; Fillers 2008). Regardless of their origin story, most sauces share a common foundation: a liquid base thickened with various agents like flour, fat, eggs, or starches (cornflour). Chutneys date back over 2000 years as part of Indian cuisine. They may have originated as a way to preserve seasonal produce before it went bad. Each state in India has its unique chutneys, reflecting the country's diverse culinary landscape. Indian chutneys have gained international acclaim, with some being included in lists of the world's best dips.

12.5 Components and Functionality

Sauces, chutneys, and dressings are essential components of the semi-solid food market (Sheldrake 2003). This chapter will explore the importance of raw material selection and interaction in developing and controlling the texture of chutneys and sauces. A thorough understanding of the components' role is essential for developing and controlling the sensory properties of these products.

A. Liquid: Liquid makes up the body of the sauce and chutneys. Different sauces use different liquids, for example, veloute—white stock, white sauce—milk, tomato sauce—vegetable stock or tomato puree, brown sauce—brown stock, mayonnaise—salad/olive/vegetable oil, hollandaise sauce—butter, chutneys—fruit or vegetable pulp with water.
B. Acidulants: *Vinegar* and *citric acid* add a tangy flavor and act as preservatives.
C. *Salt (NaCl):* Salt enhances the overall flavor of the sauce.
D. Sweeteners: *Sugar* is another key ingredient that affects the physicochemical and organoleptic properties of the sauces. It contributes to the sweetness, color, flavor, and viscosity of the sauce.
E. Spices, Herbs, and Aromatics: Spices like pepper, garlic, and onion give heat to the sauce, and herbs like oregano, thyme, and basil enhance the flavor component of the sauce.
F. Thickeners: These are used to thicken sauces, the most common being corn starch (approximately 10%). Modified starch, corn starch, arrowroot, potato starch, and tapioca thicken the sauce. Some thickening agents bind with water molecules that have the effect of reducing the fluidity and increasing the

viscosity of the sauce (Gibson and Newsham 2018). A paste should be made of cold liquid and starch, and then stirred into boiling liquid and allowed to boil, till the starch is gelatinized. Other thickeners include xanthan gum, pectin, sodium carboxymethyl cellulose, carrageenan, and gums. They prevent layer separation, which can shorten the shelf life of the sauce. Chutneys generally use groundnut and gram as thickeners.

G. Fats and Oils: Fats have a pivotal role in carrying flavors, creating mouthfeel, and influencing heat transfer during cooking. Manie butter is chiefly used for fish sauces. Equal quantities of flour and butter are kneaded together, and very little quantity is added at a time to the boiling liquid and stirred well to form a smooth consistency batter, named as roux. As per the degree of cooking, the color and characteristics of the sauce change (white, blond, brown). An equal proportion of flour, butter, or margarine is taken to prepare the roux of different colors based on the degree of cooking of the flour. While preparing sauce, cold liquid should be added to a hot roux, as adding hot liquid may cause lumpiness.

H. Cooked chutneys use oils but fresh chutneys do not require oil, except fried chutneys.

I. Emulsifiers: Egg yolks and cream function as a liaison, finishing agent added at the end of cooking. The product is never boiled after the addition of liaison, as it may cause curdling. The liaison's purpose is to thicken soups. Egg yolks are used as emulsifiers in the preparation of mayonnaise. Other than egg yolks, nowadays, various food-grade natural and synthetic emulsifiers are used. Although consumers have started preferring additives of natural origin and fewer chemicals. Also, some consumers prefer a replacement for egg yolk in mayonnaise and salad to address allergenic, vegan, or dietary requirements. Some of these alternatives include polysorbate 60, a variety of modified plant-sourced proteins and starches (Campbell 2019).

12.6 Process Details

12.6.1 Preprocessing

Chutney and sauce preparation starts with a selection of fruits and vegetables, as well as spices of good quality. Manufacturing generally includes preprocessing (Fig. 12.3), pulp extraction, mixing ingredients, processing for preservation, the addition of preservatives, packaging, quality check, and storage (Fig. 12.4). Process the fruit through a pulper or pulp finisher to separate seeds and skin from the pulp. Pass the remaining seeds and skin through the pulper a second time to maximize juice extraction. The pulp's acidity should be 4.0 or lower. For new recipes, measure the pH using a pH meter or pH paper. If it is too high, add lemon juice to reduce the acidity. You can customize your chutneys and sauces with various spices. For

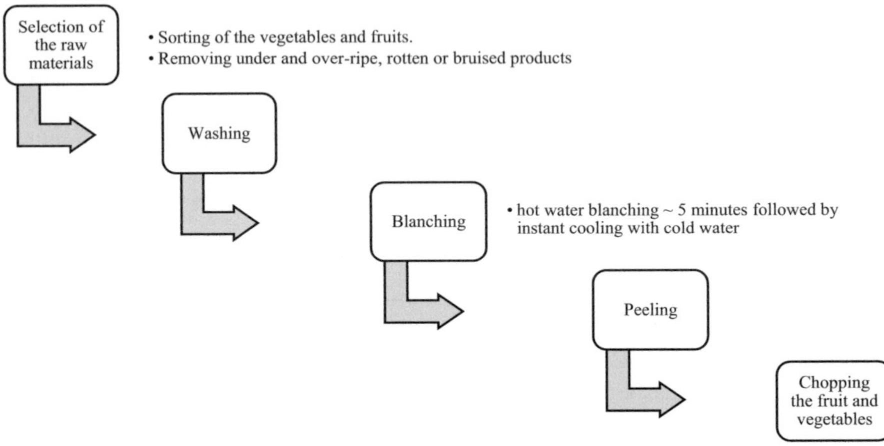

Fig. 12.3 Preprocessing of fruits and vegetables for chutneys and sauce manufacturing

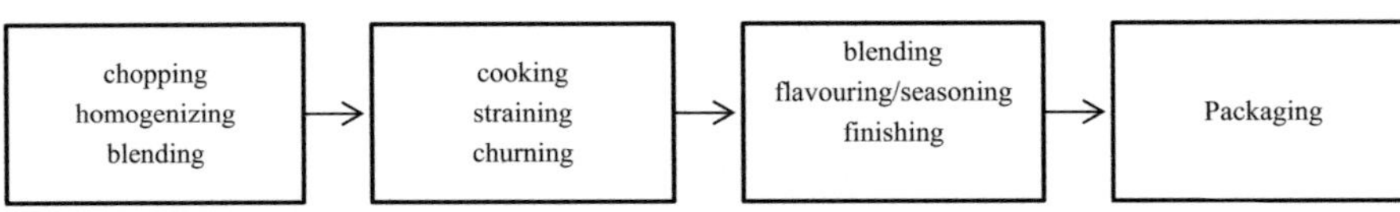

Fig. 12.4 Process flow of sauce and chutney manufacturing (Dauthy 1995)

consistent product quality, especially when selling, maintain a standardized recipe and use the same amount of spices in each batch.

12.6.2 Thermal Treatment

Heating is a way of releasing flavors in the prepared sauce. It also acts as a method of preservation by killing microorganisms, depending on the extent of cooking. The sauce mixture of pulp and added ingredients is heated in a large steam-jacketed kettle to avoid scorching. Once the fruit or vegetables are softened, sugar is added and the mixture is heated slowly to dissolve it, before increasing the heat to boil the mixture. Stirring is done continuously with a wooden spoon during heating to prevent burning. The bright red color of tomato sauces can be preserved by very slow heating.

12.6.3 Finishing

Some finishing techniques that are used to develop variety of sauces from basic mother sauce are described further in this chapter.

Reduction: Reduction involves simmering sauce at a low flame for a long time to evaporate some water after cooking, which results in more flavorful sauce. Some reduction takes place in nearly all sauces depending on how long they are simmered (Chawan 2022). The reduction can be used to adjust textures, as concentration also thickens the sauce when water evaporates. If a sauce is too thin, it may be simmered until it reaches the desired thickness. It may also add new flavors to sauces, especially red and white wines, or herbs and spices are used a great deal in this way. Reduction is also used to enhance flavors in cooked chutneys.

Straining: Straining is done to attain a velvety smooth texture in the sauce. Straining is usually done before the final seasoning. A colloidal mill is also used for making homogeneous sauce. On the other hand, chutneys have a chunky pulpy texture.

Deglazing: Deglazing is an important finishing technique that involves swirling liquid in a pan to dissolve the remaining food particles on the bottom of the pan. A liquid like wine, meat or vegetable stock is used to deglaze the pan which are reduced to half its volume by heating. This reduction is then added to the sauce and chutneys before serving.

Enriching: It is done by adding a liaison of egg yolks and cream as a thickening agent to give extra richness and creaminess to the finished sauce. Enriching is generally accomplished by addition of heavy cream and butter which given rich flavor, shine, and smoothness to sauces and fried chutneys.

Adjusting Seasoning: Before final packaging or serving, sauces and chutneys are tasted to be adjusted for seasoning. Salt and vinegar are the most important seasoning for sauces. Cayenne and white pepper are perhaps third and fourth in importance. Wines like Sherry and Madeira are frequently used as final flavorings at the end of cooking because their flavors are easily evaporated by heat.

Chemical Preservatives: Salts of sorbates and benzoates are sometimes added to sauces and chutneys to preserve the products for longer times. All countries have maximum permitted levels for preservatives. The recommended level as per FSSAI for additives is given in Table 12.2.

12.6.4 Filling and Packaging

Hot fill the sauce or chutney into hot, clean jars or bottles. If the glass jars are cold, there is a risk of breaking when the hot liquid is added. It is preferable to use glass jars but plastic jars are also used (Fig. 12.4).

12.6.5 Pasteurization and Storage

Pasteurization is a method of preservation by killing pathogens using heat. It is optional for certain sauces with high sugar (low water activity) and acid content (low pH). Sauces and chutneys are packed in glass jars or bottles before

Table 12.2 Additives permitted by FSSAI for use in sauces and chutneys

Additive	Permitted limit
Sodium benzoate	750 ppm
Modified starch	0.5%
Xanthan gum	0.5%
Calcium or sodium propionate	Maximum of 5000 ppm
Sorbic acid or its salts	Maximum of 1000 ppm
Acid calcium phosphate	Maximum of 10,000 ppm
Sodium diacetate	Maximum of 4000 ppm
Acid sodium pyrophosphate	Maximum of 5000 ppm
Ponceau 4R	Maximum of 100 ppm
Aspartame	Maximum of 2200 ppm
Acesulfame K	Maximum of 1000 ppm
Sucralose	Maximum of 750 ppm
Ammonium carbonate	Maximum of 5000 ppm

Source: Appendix A: List of food additives, FSSAI (2011)

pasteurization. Both the time and temperature of pasteurization are critical to achieve the correct shelf life and retain the color and flavor of the sauce. Cool the bottles immediately to room temperature by immersing them in clean, cold water. The shelf life of sauces and chutneys is generally up to 12 months due to high acidity and preservative content. The products should be stored away from direct sunlight in a cool and dry place to avoid color changes.

12.6.6 Packaging Requirements and Quality Control

Packaging Material: Glass jars are washed by putting them in water in a large saucepan and boiling water for 5 minutes. Dry the jars properly in trays drier. If recycled plastic jars are to be used, clean them with a solution of chlorinated water (100 ppm). Turn it upside down so all the water can drain out. The glass was the most used pack material in the chutneys, seasonings, dressings, and sauces industry in 2023. Additionally, the jar pack type was the most used in the seasonings, dressings, and sauces industry in the same year. Besides glass bottles, retort pouches with convenient spout openings are also in trend (Global data 2023). For example, barbecue sauce, a popular condiment for meat preparations, can be stored in glass bottles and retort pouches at room temperature (15–30 °C) and low temperature (5–7 °C) for three months. The pH, TSS, viscosity, and spreadability of the sauce are the parameters used to check the quality of sauces and chutneys. Storage for a longer time increases the viscosity of sauces and chutneys due to the formation of polymers or aggregates, thus decreasing the spreadability.

When preparing a chutney or sauce, the correct levels of acidity and TSS must be achieved, to ensure that the product does not spoil during storage. FSSAI has given specific standards to be met in the production of chutneys such as mango chutney, and sauces like tomato sauce, soy sauce, vegetable purees, and soup powders. In this chapter, we have summarized the quality standard requirements of important sauces and chutneys.

Tomato Sauce: "Tomato Sauce means the product prepared by blending tomato juice/Puree/Paste of appropriate concentration with nutritive sweeteners, salt, vinegar, spices, condiments, and any other ingredients suitable to the product and heating to the required consistency. The product may contain food additives permitted in these regulations, including Appendix A. The product shall conform to the microbiological requirements given in Appendix B." It shall meet the following requirements:

(i) Total Soluble solids not less than (NLT) 25.0% salt-free basis
(ii) Acidity as acetic acid NLT 0.2%

Fruits and Vegetable Sauces: "Fruits and Vegetable Sauces are a culinary preparation used as an adjunct to food, prepared from edible portions of any suitable fruit/vegetable including, roots, tubers & rhizomes, their pulps/purees, dried fruits, singly or in combination by blending with nutritive sweeteners, salt, spices and condiments, and other ingredients appropriate to the product. The product may contain food additives permitted in these regulations including Appendix A. It may contain caramel but shall not contain any other added color, whether natural or synthetic. The product shall conform to the microbiological requirements given in Appendix B." It shall meet the following requirements:

1. Chili Sauce NLT 8.0% TSS and 1.0% acidity (as acetic acid)
2. Fruits / Vegetable Sauces NLT 15.0% TSS, and 1.2% acidity (as acetic acid)
3. Culinary Paste/ Sauce NLT 8.0% TSS, and 1.0% acidity (as acetic acid)

Soybean Sauce: "Soybean Sauce is the product obtained from wholesome soybeans, by fermenting the soybean paste in which trypsin inhibitors have been completely inactivated & blending with salt, and nutritive sweeteners. It may contain spices, condiments, and other ingredients appropriate to the product preserved by using permitted preservatives" (FSSAI 2011). The product may contain food additives permitted in FSSAI regulations for sauces and chutneys, like acidifying agents (mainly acetic acid), acidity regulators, salt, sugar, and preservatives (Table 12.2).

It shall meet the following requirements, namely:

1. Total Soluble Solids NLT 15.0% salt-free basis
2. Acidity as acetic acid NLT 0.6%
3. Total nitrogen NLT 1.0%

12.7 Fruits and Vegetable Chutney

Fruits and Vegetable Chutney means the product prepared from washed, clean, sound raw fruit(s) and vegetable(s) of any suitable variety, which have been peeled, chopped, sliced, shredded, or minced and cooked with a nutritive sweetener. It may contain salt, spices, and condiments. It may also contain any other ingredients, seasonal and suitable to the product, and preserved by thermal processing or hurdle technology, or other means. The product may contain food additives permitted and included in Appendix A of the regulation, and shall conform to the microbiological requirements included in Appendix B.

(i) Total soluble solids (m/m): (a) Fruit Chutney NLT 50.0%, (b) Vegetable Chutney NLT 25.0%, (c) Hot and Sour Chutney (Spicy) NLT 25.0%.
(ii) Fruits and Vegetable content NLT 40.0%.
(iii) pH 4.6 (max.).
(iv) Total ash 5.0% (max.).
(v) Ash insoluble in hydrochloric acid 0.5% (max.).
(vi) The container shall occupy not less than 90.0% of the container.

Beyond palatability and nutrition, ensuring sauce safety and stability is crucial for both consumers and manufacturers. Several strategies can achieve the absence of harmful microorganisms and maintain shelf life stability: (i) lowering pH with organic acids, (ii) reducing water activity (aw) with sugars or salt, (iii) adding chemical or natural preservatives, and (iv) applying thermal or nonthermal preservation techniques (Institute of Food Technologists 2003).

Microbial safety of food products is the most important criterion for the development of any kind of food for human consumption. It is generally determined through challenge tests, traditional or accelerated shelf life studies (IFT 2003). Challenge tests validate food manufacturing processes through testing for the elimination of target pathogens, considering factors like inoculation levels, pathogen selection, and analytical methods (National Advisory Committee on Microbiological Criteria for Foods 2010). In case of sauces and chutneys, preservation is usually ensured by vinegar incorporation, with a target pH of 4.4, which is below the reported inhibitory pH of 4.5 for foodborne pathogens. With pH this low, the commercial sauces and chutneys usually have a shelf life of 9–12 months (Smittle and Cirigliano 1992). When addressing pathogens like *E. coli 0157:*H7, *Salmonella, L. monocytogenes, S. aureus,* and *Y. enterocolitica*, the pH played the most important role in the destruction of these pathogenic bacteria in mayonnaise and other dressings (Smittle 2000). For salad dressing, pH 3.5 to 3.6 was considered safe from spoilage occurring due to yeasts and *L. fructivorans* (Meyer et al. 1989). Additives like potassium sorbate, 1% sucrose monoesters of lauric, myristic, or palmitic acid or 0.1% sodium benzoate or ethylene diamine tetraacetic acid, and ascorbic acid are also used to inhibit the growth of *Z. bailii* and *L. fructivorans* in salad dressings (Yang et al. 2003; Castro et al. 2002). Shelf life studies determine the rate of microbial changes during storage, often employing accelerated conditions like high temperatures to obtain faster data for shelf life estimation (IFT 2003). Generally, sauces and chutneys, if processed following good manufacturing practices, have a long shelf life. Researchers contribute to the synergistic effect of hurdle technology like mild pasteurization, lower pH (around 4.6), and lower water activity (below 0.92) (Maoloni et al. 2022). The product shall conform to the microbiological requirements given in FSSAI 2011, Appendix B (Table 12.3).

Table 12.3 Microbiological requirements for sauces and chutneys as per FSSAI guidelines

Products	Parameters	Limits
Tomato ketchup, tomato sauce, chutneys, and sauces	(a) Mould count (b) Yeast and spores (c) Total plate count	Positive in not more than 40.0% of the field examined Not more than 125 per 1/ 60 cm^2 Not more than 10,000 /ml

12.8 Common Sauces and Chutneys

Sauces and chutneys are a fundamental element in global cuisine, adding layers of flavor, texture, and moisture to our dishes. From fiery hot flavor to cool and refreshing accompaniments, cooked staples to versatile dips, the world of sauces and chutneys offers diverse culinary possibilities (Figs. 12.4 and 12.5). This section delves into the key types of sauces and chutneys, exploring their ingredients, variations, and applications.

12.8.1 Béchamel Sauce or White Sauce

Béchamel sauce, commonly known as white sauce, is named after Louis de Bechamel. This sauce is made with butter, flour, milk, and flavored with salt, onions, and garlic (Tahmaz et al. 2023) (Fig. 12.6). This smooth velvety white sauce adds a creamy touch to chicken, soy chaps, vegetables, and eggs.

Ingredients: Flour, milk, butter, onion, garlic (optional), clove, bay leaf.

12.8.2 Brown Sauce or Espagnole Sauce

Espagnole sauce is a rich, complex sauce made by combining a dark brown roux with meatstock, meat, bones, vegetables, and seasonings (Fig. 12.7). The mixture is then simmered, skimmed from time to time, and reduced to concentrate its flavors. Once reduced, tomato sauce is added, and the mixture is further reduced. This time-consuming process can take hours, or even days, to complete. Due to its intense flavor, Espagnole sauce is rarely served directly on food. Instead, it serves as a base for other sauces, such as demi-glace. Demi-glace is a reduced version of brown stock and is widely used in various culinary applications. It is important to avoid over-browning the roux and stock, as this can impart a bitter taste to the sauce.

Ingredients: Flour, fat, tomato puree, brown roux, carrots, onions, Bacon trimmings & pork rind, celery, parsley, bay leaf, and Mirepoix.

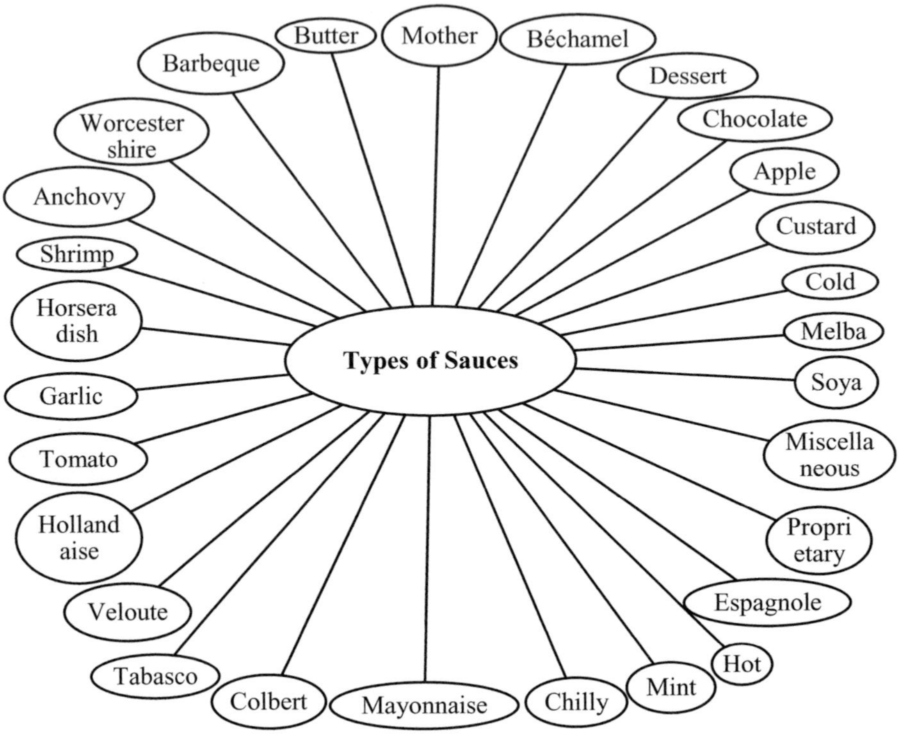

Fig. 12.5 Classification of different global sauces

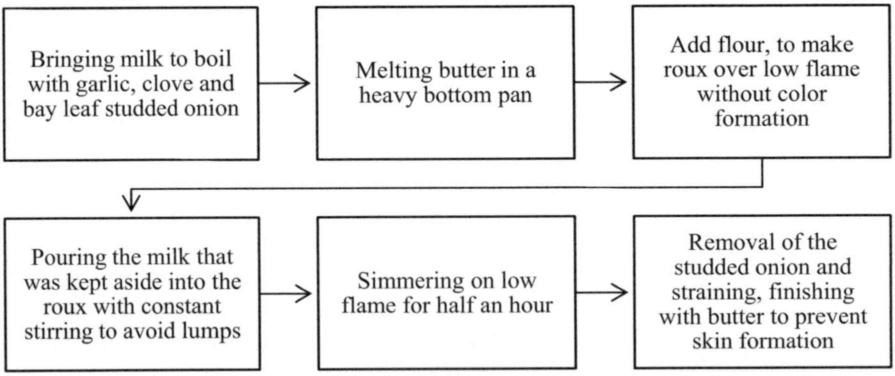

Fig. 12.6 Béchamel sauce or white sauce manufacturing process

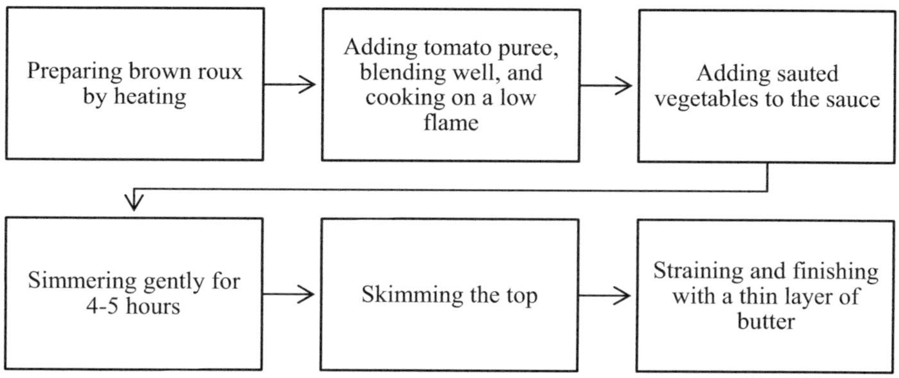

Fig. 12.7 Espagnole sauce or brown sauce manufacturing process

12.8.3 Veloute Sauce or Blond Sauce

Veloute sauce is a fundamental sauce made by combining a light brown (blond) roux with a clear stock (Fig. 12.8). The type of stock used determines the specific type of veloute, such as fish veloute, which is made with fish stock and a blond roux. To create this smooth, ivory-colored sauce, a pale roux is added to a white stock and cooked until fully incorporated. The sauce should have a luxurious, glossy appearance and be thick enough to coat the back of a spoon. When making fish veloute, it is important to cook the sauce for about 20 minutes. Wine should be added for finishing to enhance the flavor without overpowering the delicate nature of the fish.

Ingredients: flour, Butter, Flour, Stock, and Mushroom trimmings.

12.8.4 Tomato Sauce

India faces significant postharvest losses of tomatoes, especially during peak production when supply is abundant and prices are low (Fig. 12.9). To address this issue, effective preservation technologies are needed. It is therefore processed into various products such as sauce, paste, puree, juice, and ketchup (Acedo et al. 2008). Tomato sauce should have a deep brown color, and be translucent and glossy when correctly reduced.

Ingredients: Tomatoes, Sugar, Onions, Salt, Vinegar, Mace, Cinnamon, Cardamom, Cumin, Ground black pepper, Ground white pepper, Ground ginger.

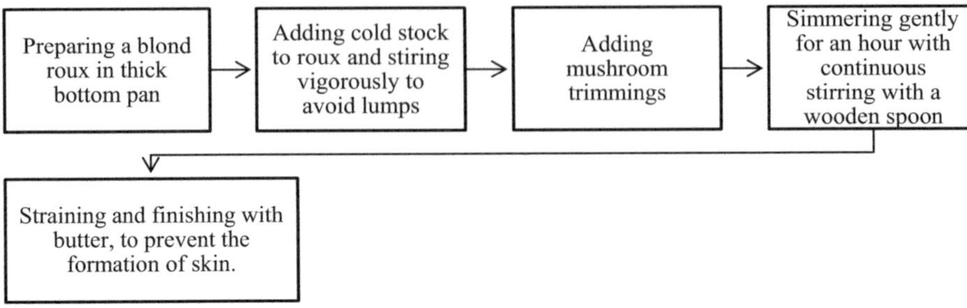

Fig. 12.8 Veloute sauce or blond sauce manufacturing process

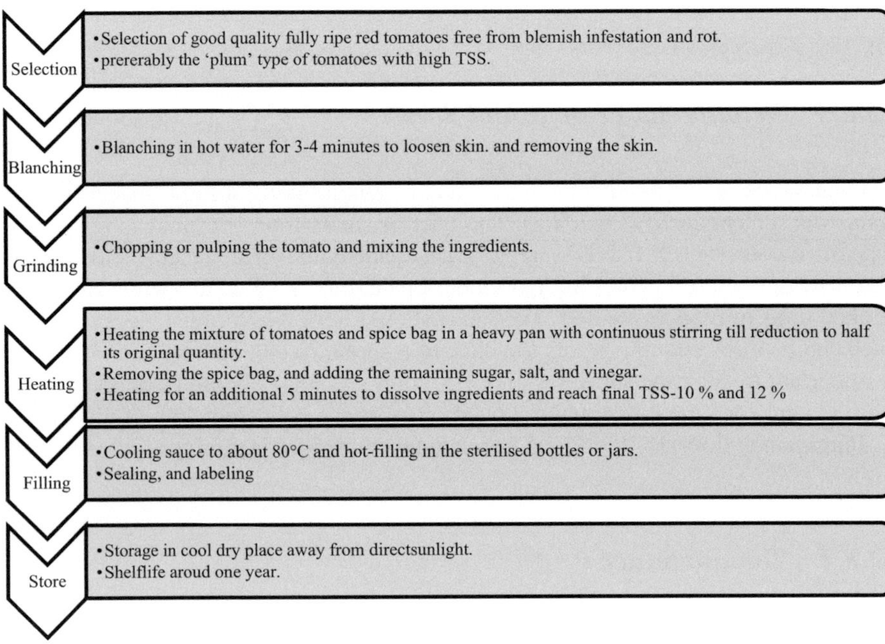

Fig. 12.9 Tomato sauce manufacturing process

12.8.5 Hollandaise Sauce

France imported butter from Holland during World War I. During this time, the Isigny sauce became Hollandaise sauce. It is an emulsion of egg yolks, butter, and lemon juice (Fig. 12.10). It is a rich and warm yellow sauce. Hollandaise is served over grilled or baked vegetables and fish. Making Hollandaise sauce right requires practice. It contains a high amount of egg yolk and butter.

Ingredients: Egg yolks, Butter, Vinegar, peppercorns, Lemon juice.

If hollandaise sauce is too thick or curdles, the addition of a little hot water at a time saves the sauce. On the other hand, if it is too thin, add a little lemon juice to the sauce in a warm bowl and whisk until it is thick by adding sauce little by little.

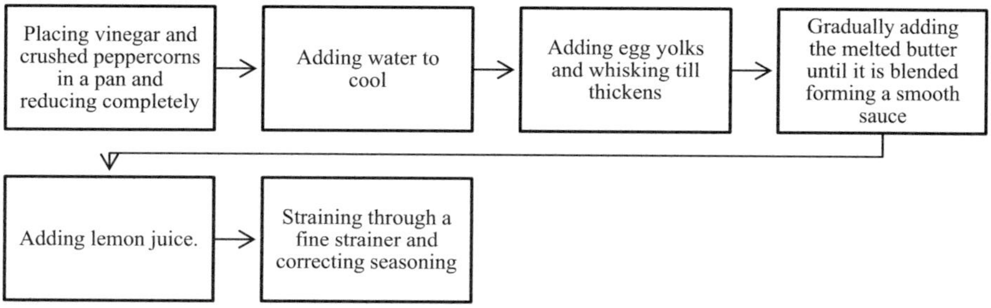

Fig. 12.10 Hollandaise sauce manufacturing process

Hollandaise can also be rectified by whisking egg yolk and hot vinegar while adding the curdled hollandaise.

12.8.6 Mayonnaise

Mayonnaise is gaining popularity in Indian cuisine. It is important to know that the ingredients must be at room temperature to make good mayonnaise. Another thing to remember is to add oil very slowly, at the beginning, until the sauce begins to emulsify (Fig. 12.11). If the sauce does separate, whisk in a teaspoon of mustard in a warm, dry bowl until creamy (mustard helps to emulsify the sauce). Then, gradually whisk in the remaining mayonnaise. Mayonnaise is a basic cold sauce. It is used as a salad dressing and as an accompaniment. It has various uses, particularly in hors d'oeuvres and salads. It is a rich sauce, as it is thickened with egg yolk and has a high percentage of fat. Various reduced-fat and eggless versions are available in markets as per consumer preference.

Ingredients: Egg yolks, Salad or olive oil, French mustard, Vinegar, Lemon, Seasoning.

Special points to be noted	Mayonnaise should not be exposed to too low a temperature during storage
	The oil should be added slowly in starting
	Oil temperature should not be too low
	Use hot water if necessary to act as stabilizer

If the mayonnaise is curdled, add boiling water or vinegar in a clean bowl and add curdled mayonnaise gradually with constant whisking. Add egg yolk as an emulsifier and whisk with curdled mayonnaise. Sometimes, mustard powder can also rectify small quantities of curdled sauce.

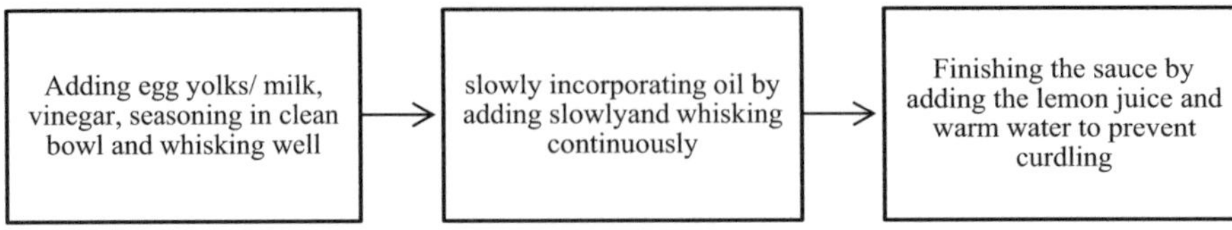

Fig. 12.11 Mayonnaise manufacturing process

12.8.7 Soy Sauce

Soy sauce originated in China around 2500 years ago and is a traditional liquid condiment that is still a key condiment in Asian countries (Zhu and Tramper 2013). It is popular due to its unique umami flavor, as well as the presence of phytochemicals with well-proven health benefits, such as the cholesterol-lowering (Larkin et al. 2009), alleviation of menopausal symptoms (Kurzer 2008), antidiabetic (Kwon et al. 2010), and anticancerous (Song et al. 2018). Owing to its taste and potential health benefits, its demand has spread internationally, increasing the global annual production to more than 10 million metric tons (Zhang et al. 2020). The nonsterile processing conditions and lengthy fermentation process during soy sauce production foster a complex interdependence among diverse microbial species, such as bacteria, yeast, and fungi (Detudom et al. 2019; Devanthi and Gkatzionis 2019). Production of traditional soy sauce involves solid fermentation (koji) followed by liquid state fermentation (moromi) (Yee et al. 2021). The microorganisms involved in the koji fermentation, A. *Oryzae* or *A. Sojae*, produce enzymes to break down soybean and wheat into simple compounds used by halophilic bacteria, including yeasts, lactic acid bacteria (LAB), and Bacillus species at the time of the moromi stage in 18%–20% salt concentration. This mash is then fermented for a few months to years of maturation to produce soy sauce (Det-udom et al. 2019; Harada et al. 2017) (Fig. 12.12). Commercial manufacturers also produce soy sauce via acid hydrolyzation, which significantly shortens production time to days. This method is faster and cheaper, but the quality of soy sauce is unsatisfactory due to the absence of aroma and flavors, as well as the presence of carcinogens such as 3-chloropropane-1,2-diol and 1,3-dichloropropanol (Lee et al. 2006; Liu 2012). Recently, due to health awareness in consumers and health claims associated with soy sauce, a revolution in the field of soy sauce manufacturing has taken place for target-specific functionality improvement.

12.8.8 Fish Sauce

Fermented fishery products are one of the oldest types of sauces. In APAC, and especially in Thailand, fish sauce production has extended into international markets. In Malaysia, fish sauce is called nampla in Thailand, budu in the Philippines,

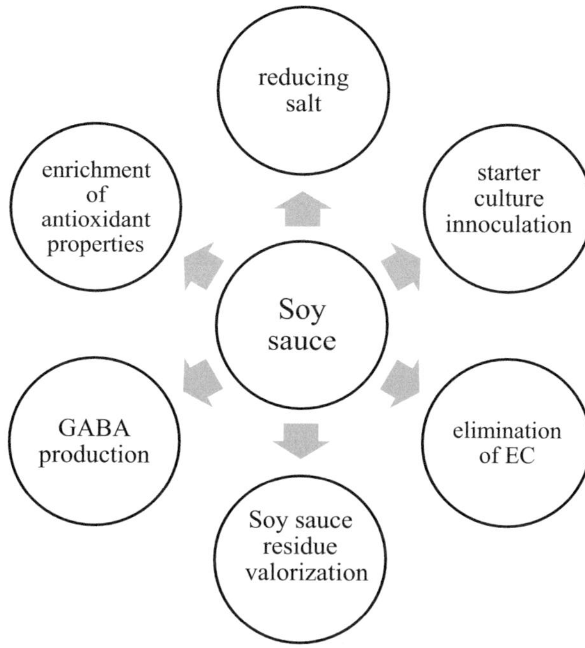

Fig. 12.12 New age soy sauce manufacturing: scope for research and innovation

ketjap-ikan in Burma, patis in Indonesia, nouc-nam in Vietnam,ngapi in Cambodia, ishiru in Japan (Itoh et al. 1993), Colombo-cure in India and Pakistan, yeesu in China, and aekjeot in Korea (Beddows 1985; Saisithi et al. 1966). Fish sauce is classified into three types as per production method in Thailand, viz., pure fish sauce, hydrolyzed fish sauce, and diluted fish sauce (Wilaipan 1990). Pure fish sauce is derived from fresh fish obtained from fish fermented with brine or salt. Hydrolyzed sauce is prepared from the hydrolysates of fish or other kinds of animals, using hydrochloric acid or other hydrolyzing processes. Diluted fish sauce is obtained by diluting pure fish sauce or hydrolyzed fish sauce.

12.8.9 Tomato Ketchup

Tomato is a versatile, most commonly grown and used vegetable in the whole world (Kumar et al. 2015). It is rich in antioxidants like lycopene, which has antiproliferative, anticancer, hypocholesterolemic, cognition enhancement, and neuroprotective properties (cited by Kumar et al. 2015). Tomato ketchup is a sauce generally made from tomato paste by diluting it to 15% TSS content. It is used as a condiment. Ingredients like sugar, vinegar, salt, and spices (red pepper, onion, garlic, extracts of spices and herbs) are added to tomato paste to give flavor to ketchup (Drdak 1989).

12.8.10 Hot Sauces

Hot sauces ignite our taste buds with the fiery kick of chili peppers. These globally popular condiments come in a vast array of styles and flavors with long history of consumption and commerciallisation (Pribil et al. 2025). From the iconic Sriracha (Southeast Asia) to the vinegary bite of Tabasco (USA), hot sauces have transcended geographical boundaries. Different regions offer unique hot sauce experiences—the smoky heat of Harissa (North Africa) or the intense chili paste known as Sambal Oelek (Indonesia). The key ingredient in hot sauces is chili peppers, with capsaicin content dictating the heat level. Different chili varieties, from the fiery habanero to the milder jalapeño, contribute distinct levels of spiciness. Hot sauces offer more than just heat. Preparation methods like fermentation (e.g., Sriracha) or pickling (e.g., Tabasco) add depth and complexity. Vinegar is a common ingredient, providing a balancing acidity. Garlic, ginger, and other spices further enrich the flavor profile.

12.8.11 Cold Sauces

Cold sauces provide a cool and refreshing counterpoint to rich or spicy dishes. They are often milk (mayonnaise) or yogurt-based (raita), offering a soothing texture and a tangy flavor (Khatko and Tamakhina 2021). A Touch of Coolness: Raita (India), a yogurt-based sauce with cucumber and mint, and Tsatziki (Greece), with yogurt, cucumber, garlic, and dill, are popular examples. These sauces alleviate the heat of curries or grilled meats. Cold sauces showcase regional preferences. Guacamole (Mexico), made with mashed avocado, lime juice, onions, and cilantro, offers a creamy richness. Chimichurri (Argentina), a blend of parsley, garlic, olive oil, and vinegar, adds a bright garlicky flavor to grilled meats. Common ingredients in cold sauces include yogurt, cucumbers, herbs like mint and cilantro, and sometimes vegetables like onions or tomatoes. The variations are endless, allowing chefs to customize them based on regional ingredients and palates.

12.8.12 Chili Sauce

Chili sauce is prepared from the edible portion of fresh, clean, good quality, sound chilies (Capsicum spp.) or processed chilies. It may also contain other fruits and vegetables such as mango, tamarind, tomato, onion, garlic, carrot, other spices and herbs, and other edible ingredients (CODEX STAN 306R-2011).

12.8.13 Cooked Sauces

Cooked sauces are culinary workhorses, offering versatility and forming the base for countless dishes. They are prepared through simmering, reducing, or thickening various ingredients. Tomato sauce, a global staple, provides the base for countless pasta dishes, pizzas, and stews. Bechamel, a classic French sauce made with butter, flour, and milk, adds creaminess to gratins and casseroles. Cooked sauces often utilize thickening agents to achieve the desired consistency. Starches like flour or cornstarch create a smooth viscosity, while roux (a paste of butter and flour) adds richness and body. With their complex blend of spices and aromatics, curry pastes from Thailand form the foundation for fragrant curries. Demi-glace, a rich brown sauce obtained by reducing stock, is a cornerstone of French cuisine, adding depth and complexity to sauces and braises.

12.8.14 Dipping Sauces

Soy sauce, chimichurri, and salsa are just a few examples of dipping sauces, meant to enhance the flavor of appetizers or finger foods. Barbecue sauce is a versatile liquid condiment with a watery and thick consistency (Nerurkar et al. 1999). It transcends specific dishes, commonly used to enhance the flavor of meats after cooking (grilling, barbecuing, baking), as a dipping sauce for fries, and even as a substitute for tomato sauce on barbecue pizzas. The first commercially produced barbecue sauce is credited to Louis Maullco in 1962 (reference not provided in original text). Across the globe, regional preferences for barbecue sauce styles and variations exist. For instance, the southern United States often favors vinegar and mustard-based sauces, while Asian countries tend to lean toward ketchup and corn syrup varieties (Nerurkar et al. 1999). This diversity showcases a spectrum of flavors, ranging from tangy and hot to sweet and spicy. Tomato, typically in the form of paste or puree, reigns supreme as the most common ingredient in barbecue sauce (Nerurkar et al. 1999). Vinegar plays a crucial role as well, acting as an acidic flavor enhancer and promoting meat penetration (Nerurkar et al. 1999).

12.8.15 Cheese Sauce

Cheese sauce is a very popular type of dipping sauce (Salek et al. 2019). Recently, there has been an increase in demand for cheese sauces in the Indian market due to changes in food preferences of consumers (Szafrańska and Sołowiej 2020). One of the reports on the cheese sauce market stated that it will grow exponentially; therefore, the number of research and solutions dedicated to the production of cheese sauces is growing constantly (Orbis Research 2017). Cheese sauces' popularity and demand are increasing due to their versatility, for example, for topping, grilling, or dipping. Therefore, manufacturers are trying to come up with innovations to even further enhance functionality and production technologies. Some examples of recent research interest include bioactive cheese sauces like cheese sauces with essential oils (Shalaby et al. 2017) or cheese sauces with *Lactobacillus salivarius* strain LS-33, acting preventively against myocardial infarction (Stenman et al. 2018).

12.8.16 Marinades

Marinades infuse meats, fish, or vegetables with flavor and moisture before cooking. They often contain acidic elements like vinegar or yogurt, alongside spices and herbs.

12.8.17 Salad Dressings

These sauces bring life to salads, offering a balance of sweet, sour, salty, and sometimes fatty components. Popular examples include vinaigrettes (oil and vinegar-based) and creamy dressings like ranch.

12.8.18 Chutneys

Chutneys are a diverse family of condiments originating in India, known for their vibrant flavors, textures, and colors. They play a crucial role in Indian cuisine, adding a burst of taste and refreshment to various dishes. This section explores the different types of chutneys, their characteristics, and popular examples (Fig. 12.13).

12.8.19 Fresh Chutneys

Fresh chutneys, also known as "pachadi" in South India, are vibrant and full of life. These uncooked or minimally cooked chutneys showcase the fresh flavors of their ingredients. Fresh chutneys often boast bright green, red, or orange hues due to the use of fresh herbs, vegetables, and fruits. These chutneys are typically coarsely ground, retaining some texture from the chopped or grated ingredients. Fresh herbs like coriander and mint, vegetables like onions and tomatoes, and grated coconut are prominent features of fresh chutneys. Due to their minimal processing, fresh chutneys have a shorter shelf life and are best consumed soon after preparation.

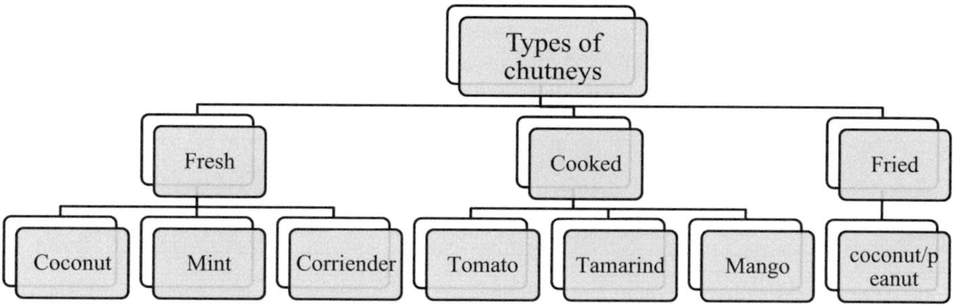

Fig. 12.13 Common chutney types

Examples

Mint Chutney: A vibrant green chutney made with fresh mint leaves, cilantro, chilies, and lemon juice. It offers a refreshing and cooling accompaniment to savory dishes.

Coconut Chutney: This South Indian classic features are grated coconut, green chilies, and fresh coriander leaves. It has a balanced flavor profile with a hint of sweetness from the coconut.

Tomato Onion Chutney: Chopped tomatoes, onions, chilies, and spices are blended to create this tangy and slightly spicy chutney, often served with samosas or pakoras.

12.8.20 Cooked Chutneys

Cooked chutneys, known as "chaatni" in North India, undergo a cooking process that intensifies flavors, concentrates sugars, and extends shelf life. Cooking helps break down ingredients and release their oils, resulting in a thicker and smoother texture compared to fresh chutneys. The cooking process and the addition of preservatives (like sugar or vinegar) enhance the shelf life of cooked chutneys. Some ingredients, like onions or tomatoes, may be caramelized during cooking, adding sweetness and depth of flavor. Cooked chutneys often incorporate a wider range of spices for more complex flavor profiles.

Examples

Tamarind Chutney

Tamarind chutney is a tangy, spicy chutney that is eaten as an accompaniment to Indian snacks like samosas, dahi vada, potato tikki, and other main meals. It is prepared using tamarind, salt, sugar, coriander, black cumin, cloves, pepper, caraway, cardamom, cinnamon, red chilies, and oil (Fig. 12.14).

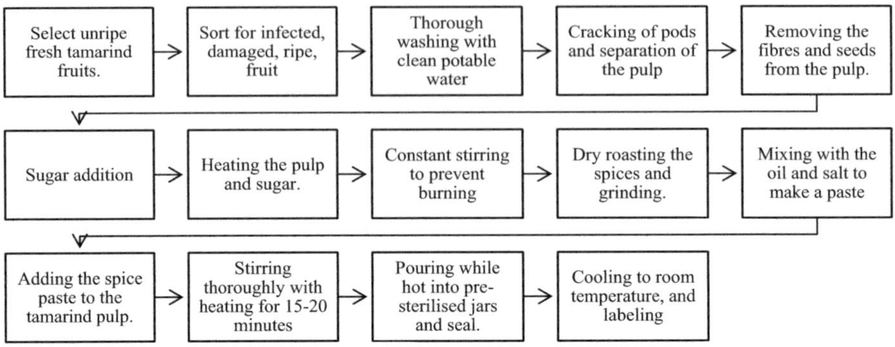

Fig. 12.14 Tamarind chutney manufacturing process

12.8.21 Vegetable Chutney

Ingredients: onions, carrots, tomato, capsicum, garlic, chili and ginger powder, salt, oil, vinegar (Fig. 12.15).

Mango Chutney: Ripe mangoes are cooked with sugar, vinegar, and spices like fenugreek and ginger to create this sweet and tangy chutney. It pairs well with curries and grilled meats.

Tomato Chutney: This chutney features cooked tomatoes, spices like cumin and coriander, and sometimes jaggery or raisins for a sweet and spicy flavor profile. It is a versatile condiment for various Indian dishes.

Peanut Chutney: Roasted peanuts are ground with spices, chilies, and sometimes coconut to create a rich and flavorful chutney, often served with dosas or idlis.

12.9 Global Influence of Sauces and Chutneys

Sauces and chutneys have always played a significant role in global cultures, and their growing use in daily life underscores their increasing importance from both nutritional and economic perspectives (Sikora et al. 2008; García-Casal et al. 2016). Hundreds of commercially available sauces exist worldwide, encompassing popular condiments like ketchup, mustard, and mayonnaise (Sikora et al. 2008). Classification of sauces and chutneys can be based on factors like origin (Italian, Indian, etc.), temperature (warm or cold), making process (fresh or cooked), flavor (mild or hot), acidity, sweetness, and color (Sikora et al. 2008).

Sauces and chutneys have transcended their regional origins to become staples in kitchens worldwide. This global influence can be attributed to several factors like trade and exploration, by facilitating the exchange of ingredients and culinary techniques. For example, the introduction of chilies from the Americas to Asia led to the creation of new hot sauces like Sriracha in Southeast Asia. Migration of people

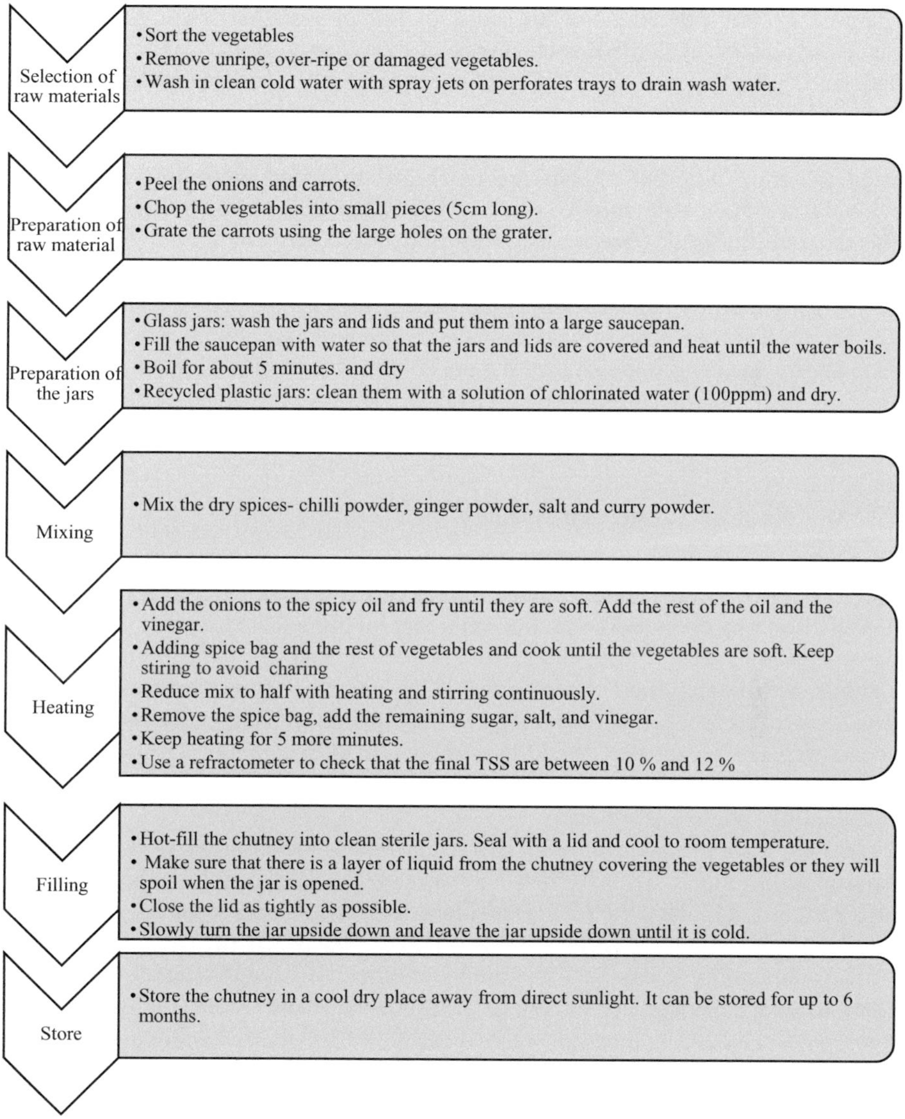

Fig. 12.15 Vegetable chutney manufacturing process

across continents also plays a crucial role in globalization and adaptation of traditional sauces and chutneys. This is evident in the popularity of Indian chutneys in Caribbean cuisine, where they are often infused with local fruits and spices. Another factor to be considered is the rise of international food chains, and the global food market has made a wider variety of sauces and chutneys accessible to a broader audience. This has led to increased awareness and appreciation for these condiments. For example, tomato sauce in Italian variations includes marinara sauce and arrabbiata, while French cuisine utilizes tomato-based sauces like béchamel. Similarly, Mexican salsa and southwest sauce have become quite popular in food chains and restaurants.

Sauces and chutneys play a significant role in fusion cuisine, where culinary traditions from different regions are combined to create innovative dishes. For example, a fusion dish might utilize a Thai peanut sauce with Caribbean jerk chicken, or a European cheese sauce with Indian samosas. The versatility of sauces and chutneys allows chefs to experiment with flavor profiles and create unique culinary experiences. By understanding the global influence of sauces and chutneys, we appreciate their role in bridging cultures and enriching our culinary tapestry. Their adaptability and ability to enhance flavors continue to inspire culinary innovation, ensuring their continued presence on tables worldwide.

12.10 Recent Trends in Sauces

Earlier in the early twentieth century, sauces and chutneys were prepared fresh in households or by chefs and local food suppliers for immediate use. Due to increased consumption of these commodities, the processing industry introduced sauce and chutney with longer shelf life products. This involved innovation, research, and development of processing technology in bulk quantities using automated machinery, as well as the inclusion of additives such as acidity regulators, emulsifiers, and preservatives to extend shelf life in terms of microbiological safety of the product and stability (phase separation).

Modern life's fast pace in developed countries has led to busier work schedules, a rise in subhealthy individuals, and a growing emphasis on health. As health awareness increases and the demand for nutritious foods rises, chia seeds have become a hot topic in food research. Tomato sauce enjoys global popularity due to its versatility and adaptability across diverse culinary cultures. Ripening tomatoes on the vine could be an excellent strategy to positively influence the nutritional quality of tomatoes (Ferro et al. 2021). Functional tomato sauce with enhanced antioxidant activity or lycopene content using mild processing conditions might be the direction of future research (Özer 2021). Another area of research can be adding tomato byproducts or antioxidants-rich ingredients like carrots, chia seeds, pumpkin, cherries, and purple cabbage (У and Мельник 2022; Giménez et al. 2022; Tagliamonte et al. 2023) to retain more nutrients and add functionality to cooked sauces. Energy-dense sauces could also mitigate the problem of malnutrition due to palate appeal among consumers (Tsikritzi et al. 2011). Therefore, targeting specific research while developing new types of sauces and chutneys could help prevent various health problems.

The rising income levels in India have fueled a growing trend toward fast food consumption (Martinez et al. 2006). This includes an increased popularity of items like chicken sausages, burgers, patties, and pizzas among Indian consumers. Recognizing this shift and the expanding availability of processed food products due to organized retail growth (Martinez et al. 2006; Cullen et al. 2003), the researchers now aim to develop world-famous sauces tailored to the Indian palate (Cullen et al. 2003). By delving into the diverse world of sauces and chutneys, we unlock a universe of flavor possibilities. From the fiery heat of hot sauces to the cool comfort of cold sauces, and from the versatility of cooked sauces to the variety of dips and dressings, these culinary creations elevate our dishes and transform the way we experience food.

Food waste poses a significant global challenge, contributing to environmental damage and food insecurity. Upcycling, the process of transforming food byproducts into new products, offers a sustainable solution (UPCYCLED FOOD 2021). By utilizing these otherwise discarded ingredients in sauce formulations, we can reduce waste, create high-value products, and minimize environmental impact (Bennet et al. 2022).

Prepackaged sauces are a double-edged sword in modern cuisine. While convenient, their frequent consumption has been linked to chronic health issues like diabetes, obesity, and heart disease due to high sugar, salt, and unhealthy fat content. Current food labels often lack a nuanced picture of a product's nutritional value, making it difficult for health-conscious consumers to make informed choices. Artificial intelligence (AI) can be explored as a promising solution through the development of nutritional scoring systems. AI algorithms trained on extensive food composition and health data can analyze prepackaged sauce ingredients and generate a score encompassing factors like total calories, sugar content (differentiating added sugars from natural ones), sodium content, unhealthy fat profile, and presence of beneficial nutrients (Kusuma and Vijaykumar 2024). This information, displayed clearly on labels, empowers consumers to make informed choices based on their dietary needs. Furthermore, AI can guide food product development by allowing manufacturers to compare the nutritional profiles of their existing sauces and identify areas for improvement, like reducing sugar, salt, and unhealthy fats. This paves the way for healthier prepackaged sauce options, ultimately improving public health outcomes. Future research can explore integrating AI-based scoring systems into food labeling regulations and investigate consumer behavior changes based on such information to optimize public health initiatives.

References

Avallone S, Estelle Tiemtore T-W, Mouquet-Rivier C, Trèche S (2008) Nutritional value of six multi-ingredient sauces from Burkina Faso, J Food Compos Anal, 21(7): 553-558, ISSN 0889–1575, https://doi.org/10.1016/j.jfca.2008.04.012

Acedo Jr A, Thanh CD, Van LTS, Linh NTT, Weinberger K (2008) Tomato precoolingusing simple hydrocooler and fruit quality changes during ambient and cold storage. In Asia Pacific Symposium on Assuring Quality and Safety of Agri-Foods 837; pp. 135–140

Badham CD (1854) Prose Halieutics (ancient and modern fish tattle). Parker and Sons, London

Beddows CG, Ismail M, Steinkraus KHJ (1976) Food Technol 11:379

Bennett A, Dubey S, Lee WTK, Damen B, Bucatariu C (2022) FAO Regional Strategy on Food Loss and Waste Reduction in Asia and the Pacific. Bangkok, FAO. https://doi.org/10.4060/cb8959en.

Beddows CG, Ardeshir AG, Daud WJ (1979) J Sci Food Agric 30:1097

Beddows CG (1985) In: Wood BJB (ed) Microbiology of fermented foods, vol 2. Elsevier, London, p 3

Bellis M (2014) History of mayonnaise and Miracle Whip http://inventors.about.com/od/foodrelatedinventions/a/mayonnaise.htm

Bottone E (2014) Worcester sauce Teaching dictionary. http://www.edwardbottone.com/index.php?s=TEACHING

Campbell B (2019) Current emulsifier trends in dressings and sauces. In: Hasenhuettl G, Hartel R (eds) Food emulsifiers and their applications. Springer, Cham. https://doi.org/10.1007/978-3-030-29187-7_9

Castro M, Garro O, Campos CA, Gerschenson LN (2002) Interactions between additives: its effects on sorbate stability and Z. bailii minimum inhibitory concentration in model aqueous systems resembling salad dressings. Food Sci Tech Int 8(1):33–39

Cavil (2024) Emerging Trends in the Vegetable Chutney Market: Detailed Overview of Applications (Household, Food Services, Food Process, Others) and Types (2024–2032)

Chavan A (2022) Basic continental food production in chapter 3 sauces, basic continental food production Sem II

Cullen PJ, O' Donnel CP, Houska M (2003) Rotational rheometry using complex geometries— A Rev J Texture Stud, 34(3): 1–20.

CODEXSTAN192-1995 (2013) CodexAlimentarius.Codex general standard for food additives. Adopted in 1995. Revision 1997, 1999, 2001, 2003, 2004, 2005, 2006, 2007, 2008, 2009, 2010, 2011, 2012, 2013. FAO/WHO. http://www.fao.org/ag/agn/jecfa-additives/search.html?lang=en

Dauthy ME (1995) Fruit and vegetable processing. FAO Agricultural Services Bulletin No.119 Food and Agriculture Organization of the United Nations, Rome

Detudom R, Prakitchaiwattana C, Mahawanich T (2019) Autochthonous microbes and their key properties in browning reduction during soy sauce fermentation. LWT – Food Sci and Tech 111:378–386. https://doi.org/10.1016/j.lwt.2019.05.042

Devanthi PVP, Gkatzionis K (2019) Soy sauce fermentation: microorganisms, aroma formation, and process modification. Food Res Int 120:364–374. https://doi.org/10.1016/j.foodres.2019.03.010

Drdák M (1989) Technológiarastlinnýchneúdržnýchpotravín. ALFA, Bratislava. Ferrer E, AlegríaA, Farré R, Abellán P, Romero F (2002). J Chromatogr A, 947: 85-95

Ferro Y, Mazza E, Angotti E et al (2021) Effect of a novel functional tomato sauce (OsteoCol) from vine-ripened tomatoes on serum lipids in individuals with common hypercholesterolemia: tomato sauce and hypercholesterolemia. J Transl Med 19:19. https://doi.org/10.1186/s12967-020-02676-3

Fillers L (2008) History of mayonnaise Culinary Arts 360: spices, herbs, and condiments http://www.culinaryarts360.com/index.php/history-of mayonnaise-3-30189/

FSSAI (2011) Fruit & vegetable products in chapter 2 food product standards https://fssai.gov.in/upload/uploadfiles/files/FSSAI-regulations.pdf

García-Casal MN, Peña-Rosas JP, Malavé HG (2016) Sauces, spices, and condiments: definitions, potential benefits, consumption patterns, and global markets. Ann N Y Acad Sci 1379:3–16. https://doi.org/10.1111/nyas.13045

Gibson M, Newsham P (2018) Sauces. In: Food science and the culinary arts. Elsevier, pp 303–321

Giménez CG, Traffano-Schiffo MV, Sgroppo SC, Sosa CA (2022) Development of a bioactive sauce: effect of the packaging and storage conditions. Chem Eng 6:34. https://doi.org/10.3390/chemengineering6030034

Khatko ZN, Tamakhina MA (2021) Development of white and red low-calorie functional purpose sauces using vegetable powders. New Technol 17:67–76. https://doi.org/10.47370/2072-0920-2021-17-2-67-76

Goldwyn M (2014) The story of barbecue. http://amazingribs.com/BBQ_articles/barbecue_history.html
Harada R, Yuzuki M, Ito K, Shiga K, Bamba T, Fukusaki E (2017) Influence of yeast and lactic acid bacterium on the constituent profile of soy sauce during fermentation. J Biosci Bioeng 123(2):203–208. https://doi.org/10.1016/j.jbiosc.2016.08.010
https://www.globaldata.com/store/report/seasonings-dressings-and-sauces-market-growth-analysis/
Huo J, Jing S, Jiyong Y, Huang J, Chen J (2020) The contribution of iron fortified soy sauce in the decrease of Anemia prevalence in China. Biomed J Sci Tech Res 24(4):BJSTR. MS.ID.004089
Institute of Food Technologists (IFT) (2003) Microbiological safety of food. Blackwell Publishing Professional
Itoh H, Tachi H, Kikuchi S (1993) In: Lee CH, Steinkraus KH, Alan Reilly PJ (eds) Fish fermentation technology. United Nations University Press, Tokyo, p 177
Kelaiditis GC (1949) Praktika of the Hellonic Hydrobiological Institute 3,5
Kumar V, Kumar L, Kumar K, Goyal SK, Kumar A, Jain G (2015) Physico-chemical and quality evaluation of tomato ketchup during storage. South Asian J Food Tech& Environ ISSN 2394–5168 Society for World Environ, Food & Technol:250–255
Kusuma T, Vijayakumar V (2024) Analyzing the health effects of packaged food sauces through AI-powered nutritional scoring. Int J Innov Eng Man Res 13(5) Available at SSRN: https://ssrn.com/abstract=4857030
Kurzer MS (2008) Soy consumption for reduction of menopausal symptoms. Inflammopharmacology 16(5):227–229. https://doi.org/10.1007/s10787-008-8021
Kwon DY, Daily JW, Kim HJ, Park S (2010) Antidiabetic effects of fermented soybean products on type 2 diabetes. Nutr Res 30(1):1–13. https://doi.org/10.1016/j.nutres.2009.11.004
Larkin TA, Astheimer LB, Price WE (2009) Dietary combination of soy with a probiotic or prebiotic food significantly reduces total and LDL cholesterol in mildly hypercholesterolaemic subjects. Eur J Clin Nutr 63(2):238–245. https://doi.org/10.1038/sj.ejcn.1602910
Lee SM, Seo BC, Kim Y-S (2006) Volatile compounds in fer mented and acid- hydrolyzed soy sauces. J Food Sci 71(3):C146–C156. https://doi.org/10.1111/j.1365-2621.2006.tb15610.x
Liu KS (2012) Soybeans: chemistry, technology, and utilization. Springer. ISBN 10: 146135711X
Lopetcharat K, Choi YJ, Park JW, Daeschel MA (2001) Fish sauce products and manufacturing: a review. Food Rev Intl 17(1):65–88. https://doi.org/10.1081/FRI-100000515
Marcus JB (2019) Culinary considerations for the aging. In: Aging, nutrition and taste. Elsevier, pp 297–337. https://doi.org/10.1016/B978-0-12-813527-3.00009-0
Maoloni A, Cardinali F, Milanović V, Garofalo C, Osimani A, ozzon M, Aquilanti L. (2022) Microbiological safety and stability of novel green sauces made with sea fennel (Crithmum maritimum L.), Food Research International, vol 157: 111463, ISSN 0963-9969, https://doi.org/10.1016/j.foodres.2022.111463
Meyer RS, Grant MA, Luedecke LO, Leung HK (1989) Effectsof pH and water activity in microbiological stability of salad dressing. J. FoodProt., 52(7):477–479
National Advisory Committee on Microbiological Criteria for Foods [NACMCF] (2010)
Nilar (2012) A Study on the Factors Influence the Processing of Some Fruit Sources. Univ Res J. University of Mandalay Open Access Repository
Nerurkar PV, Marchand LL, Cooney RV (1999) Effects of Marinating With Asian Marinades or Western Barbecue Sauce on PhIP and MeIQx Formation in Barbecued Beef,Nutrition and Cancer, 34(2): 147–152.
Orbis Research (2017) 2017–2022 Global Top Countries Cheese Sauce Market Report. Dallas, TX
Özer Ç. (2021) Effects of ingredients and cooking time on total phenolic content and antioxidant activity of different homemade tomato sauces. Food and Health, 7(2): 84-90. https://doi.org/10.3153/FH21010
Saisithi P, Kasemsarn B, Liston J, Dollar AM (1966) J Food Sci 31:105
Salek RN, Vašina M, Lapcˇík L, Černíková M, Lorencová E, Li P, Buňka F (2019) Evaluation of various emulsifying salts addition on selected properties of processed cheese sauce with the

use of mechanical vibration damping and rheological methods. LWT— Food Sci Technol, 107: 178–184. https://doi.org/10.1016/j.lwt.2019.03.022

Shalaby SM, Mohamed AG, Bayoumi HM (2017) Preparation of a novel processed cheese sauce flavored with essential oils. Int J Dairy Sci 12:161–169. https://doi.org/10.3923/ijds.2017.161.169

Sheldrake P (2003) Controlling textures in soups, sauces and dressings. In B. M. McKenna (Ed.), Texture in food, vol 1: Semi-solid foods. Boca Raton, FL: CRS Press and Cambridge, England: Woodhead Publishing

Smittle RB (2000) Microbiological safety of mayonnaise, salad dressings andsauces produced in the United States: A review. J. Food Prot., 63(8):1144–1153

Smittle RB, Cirigliano, MC (1992) Salad dressings. In: compendium of methods for the microbiological examination of foods. eds. Vanderzant C, Splittstoesser DF pp. 975–983. American Public Health Association.Washington, D.C. US.

Sikora M, Badrie N, Deisingh AK, Kowalski S (2008) Sauces and dressings: a review of properties and applications. Crit Rev Food Sci Nutr 48:50–77. https://doi.org/10.1080/10408390601079934

Stenman L, Lahtinen S, Konhilas J (2018) Lactobacilli for treating cardiac dysfunction.EP 3 322 480 B1. European Patent Office. https://patentimages.storage.googleapis.com/73/25/11/ad9773c013e6ca/EP3322480B1.pdf

Song J, Wei Q, Wang X, Li D, Liu C, Zhang M, Meng L (2018) Degradation of carotenoids in dehydrated pumpkins as affected by different storage conditions. Food Res. Int., 107, 130–136. https://doi.org/10.1016/j.foodres.2018.02.024

Szafrańska JO, Sołowiej BG (2020) Cheese sauces: characteristics of ingredients, manufacturing methods, microbiological and sensory aspects. J Food Process Eng 43:e13364. https://doi.org/10.1111/jfpe.13364

Tagliamonte S, Romano R, Chiacchio MF et al (2023) Enrichment of tomato sauce and chopped tomatoes with tomato by-products increases antioxidant activity upon in vitro digestion. LWT 184:115002. https://doi.org/10.1016/j.lwt.2023.115002

Tahmaz J, Biščević S, Sarić Z, Oras A, Operta S(2023) Physical properties and quality of bechamel sauce prepared with different proportions of cow's milk to soy and rice drinks. Poster Presentation. In: 44th International Dairy Expert Symposium. Rovinj, Croatia; October 12–15

Van Thuy P, Berger J, Davidsson L, Khan NC, Lam NT, Cook JD, Hurrell RF, Khoi HH (2003) Regular consumption of NaFeEDTAfortified fish sauce improves iron status and reduces the prevalence of anemia in anemic Vietnamese women2. Am J Clin Nutr, 78(2): 284–290, ISSN 0002-9165, https://doi.org/10.1093/ajcn/78.2.284.

Tsikritzi R, Wang J, Wong M et al (2011) Development of sauces with high energy density, flavour and taste impact for older people, aiming to increase food acceptability and consumption. Proc Nutr Soc 70:E176. https://doi.org/10.1017/S0029665111002278

У Ч, Мельник ОЮ (2022) УДОСКОНАЛЕННЯ ТЕХНОЛОГІЇ СОУСІВ З ВИКОРИСТАННЯ ОВОЧЕВО-ЯГІДНОЇ СИРОВИНИ. Bull Sumy Natl Agrar Univ Ser Mech Autom Prod Process 50:3–7. https://doi.org/10.32782/msnau.2022.4.1

UPCYCLED FOOD (2021) Moshtaghian H, Bolton K, Rousta K. Challenges for upcycled foods:definition, inclusion in the food waste management hierarchy and publicacceptability, 10:2874, https://doi.org/10.3390/foods10112874

Wilaipan PMS (1990) Halophilic Bacteria Producing Lipase in Fish Sauce. [MSc thesis]. Bangkok, Thailand: Chulalongkorn University

Yang CM, Ludecke LO, Swanson BG, Davidson PM (2003) Inhibition of microorganisms in salad dressing by sucrose and methylglucose fatty acid monoesters. J Food Proc Pres 27(4):285–298

Yee CS, Sohedein MNA, Suan OP, Loen AWW, Abd Rahim MH, Soumaya S, Ilham Z, Wan WAAQIJ (2021) The production of functional γ-aminobutyric acid Malaysian soy sauce koji and moromi using the trio of Aspergillus oryzae NSK, Bacillus cereus KBC, and the newly identified Tetragenococcus halophilus KBC in liquid-state fermentation Future Foods, 4:100055, https://doi.org/10.1016/j.fufo.2021.100055

Zhang LJ, Zhang L, Xu Y (2020) Effects of Tetragenococcus halo philus and Candida versatilis on the production of aroma- active and umami- taste compounds during soy sauce fermentation. J Sci Food Agric 100(6):2782–2790. https://doi.org/10.1002/jsfa.10310

Zhu Y, Tramper J (2013) Koji – where east meets west in fermentation. Biotech Adv 31(8):1448–1457. https://doi.org/10.1016/j.biotechadv.2013.07.001

Pribil E, Payne J, McGlynn W, Jadeja, R (2025) Food safety considerations in hot sauce production: hazards, regulations, and industry challenges. MOJ Food Process Technol 13(1):8–15. DOI: 10.15406/mojfpt.2025.13.00318

Chapter 13
Dehydration of Fruits and Vegetables

S. Ganga Kishore, Singamayum Firdosh Nesha, C. P. Rakshith Patel, and Madhuresh Dwivedi

13.1 Introduction

Fruits and vegetables are essential elements of the human diet with significant health benefits as it is valuable sources of nutrients, including fibres, sugars, minerals, vitamins and a variety of bioactive compounds. There is a growing global demand for fruits and vegetables, and their consumption is rising worldwide due to globalisation and urbanisation as it provides energy to the human body and it contains anti-carcinogenic properties that prevent cardiovascular diseases (Srivastava et al. 2024). According to the Food and Agriculture Organisation of the United Nations (FAO) and the World Health Organisation (WHO), the human body requires at least 400 g of fruits and vegetables every day for nutrition to attain a balanced diet (Fanzo and Miachon 2023; Kumar and Agrawal 2023).

India is the second-largest producer of fruits and vegetables with the overall production of approximately 331 million tonnes which contributes to the global fruit and vegetable production of 11.38% and 11.78%, respectively (Neeraj et al. 2017; Sah et al. 2022). India exports fruits and vegetables with a total net worth of 1342.14 USD million, of which fruits account for 674.53 USD million and vegetables contribute to 667.61 USD million (Sharma 2021). India's inability to meet quality standards and high post-harvest losses results in a low export market share with an annual loss of about 33.1% of total production (Kumar and Agrawal 2023).

Fresh fruits and vegetables are highly perishable as they contain more than 80% moisture content which results in increased losses of up to 25–40% which lead to an increase in food waste and costs allied with inputs, labour and pre-harvest resources (Sharma et al. 2024). These losses are due to the improper transportation, inadequate cold storage facilities, insufficient packing materials and open-air markets (Spiker et al. 2023). Within the fruits and vegetables category, the percentage of

---

S. G. Kishore · S. F. Nesha · C. P. Rakshith Patel · M. Dwivedi (✉)
Department of Food Process Engineering, National Institute of Technology, Rourkela, India

OP Chauhan (ed.), *Fruits and Vegetables Technologies*,
https://doi.org/10.1007/978-981-96-8433-5_13

production losses was 12.44% for tomatoes, 15.88% for guava, 10.39% for apples, 9.56% for cauliflower, 9.69% for citrus and 9.37% for cabbage, etc. (Kumar and Agrawal 2023).

Dehydration is a widely employed preservation method that removes water from vegetables and fruits to increase their shelf life by preventing the growth of microorganisms and lowering enzymatic and chemical reactions (Pateiro et al. 2022). In countries like India, where distribution and facilities for handling cold temperatures are scarce, dehydration is a practical solution for post-harvest management. Dehydration enhances transportation infrastructure, lowers handling expenses and expands the availability of fruits and vegetables during off-peak seasons. It reduces weight and volume, packing, storage and shipping expenses. Some of the limitations of available dehydration methods include higher cost (Freeze drying), longer drying time (solar drying), quality degradation in heat-sensitive products (spray drying and drum drying) and difficulty in drying sugar-rich products. To overcome this, there is a need for the development of an efficient dehydration method that is cost-effective which results in a high-quality product in a short time (Sultana et al. 2024). This chapter reviews different existing traditional mechanised, advanced and combined dehydration methods that are used in the food industry.

13.2 Mechanism

Dehydration is the process of removing moisture from a hygroscopic material by moving it to the surface and evaporating it with heat. It is a complex process that depends on the rate of mass and heat transfer, and it relies on the water's chemical potential present in the product. Water travels from the material's interior (greater chemical potential) to its outer surface (lower potential), followed by vapourisation of moisture from the surface of the material to the surrounding (Ahmed 2018). Heat transfer mechanisms use convection, conduction, or radiation to move energy from a source of heat (such as a metal surface or hot gas) to food. However, the rate at which moisture diffuses (whether in vapour or liquid form) within the food matrix, as well as the rate of moisture evaporation from the product's surface to the drying medium, can be controlled by the mechanisms of mass transfer during the drying process (Sabarez 2018). Moisture can be transported throughout solids via several mass transfer methods which include:

1. Liquid Diffusion: It occurs when the wet solid is at a lower temperature than the liquid's boiling point.
2. Vapour Diffusion: It occurs when the liquid vapourises within the material.
3. Knudsen Diffusion: It occurs at extremely low pressure and temperature, such as in freeze drying.
4. Surface Diffusion: It occurs due to liquid movement due to moisture diffusion in pores.
5. Difference in Hydrostatic Pressure: It occurs when the internal rate of vapourisation exceeds the vapour transport through the material to the surroundings.
6. Capillary Flow: It occurs due to liquid movement by surface forces.

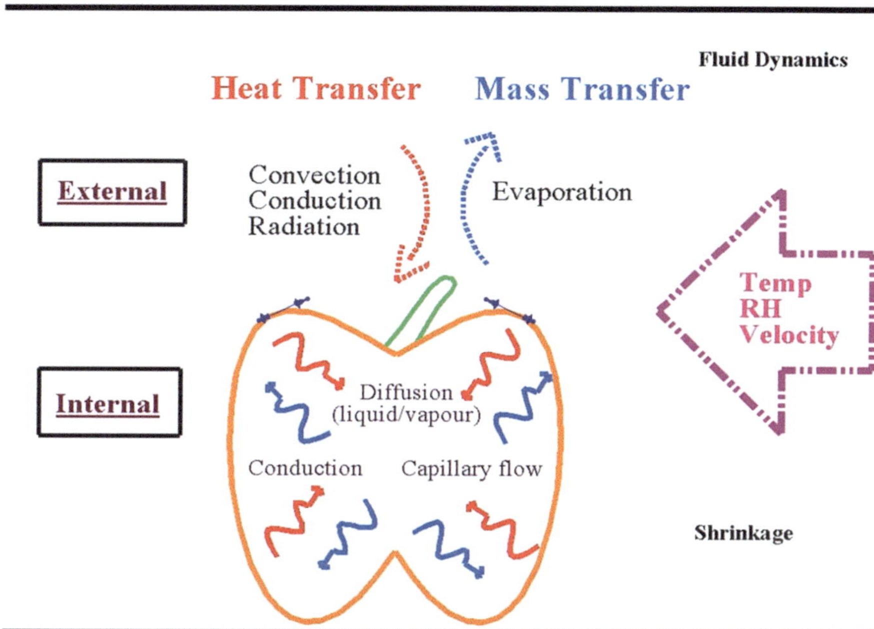

Fig. 13.1 Dehydration process of the product

Different dryers use different methods of heating such as convection, conduction or radiation. Convective or air dryers are commonly used in the food sector. They can work in batches or continuously depending on the nature and composition of the product. The knowledge of the mechanism and kinetics of the food product to be dried are important in designing a dryer. This includes establishing quantitative correlations between time and drying conditions (Grau et al. 2014). During the dehydration process, a higher temperature shortens the process time. However, high temperatures also cause modification of sensory characteristics caused by chemical or enzymatic reactions, loss of thermo-sensitive nutrients and undesirable changes in product quality attributes (González-Pérez et al. 2021). Figure 13.1 shows transport processes that occur throughout the dehydration process of the product (Sabarez 2018).

13.3 Types of Dehydration Methods

Dehydration methods are classified into traditional, mechanised, advanced and combined methods based on source of energy used, heat and mass transfer. The advantages and disadvantages of different dehydration methods are given in Table 13.1.

Table 13.1 Advantages and disadvantages of traditional, mechanised, advanced and combined drying methods

Method/ applications	Advantages	Disadvantages
Traditional methods		
Sun drying: Tomatoes Raisins Apricots Figs	Reduce energy consumption Simple construction Inexpensive labour Minimal loss of colour and vitamins	Some food exhibits a loss of volatile and sensitive to light compounds Labour-extensive, with a possibility of livestock and weather-related harm
Shade drying: Beans Green and red chillies Okra	Preventing damage from direct sunlight to maintain nutrients Low cost	Failure to manage large quantities Failure to attain consistent quality standards
Air drying: Mango Passion fruit Pineapple Mushrooms Blackberry	Simplicity Low cost	Energy-intensive, with less effective drying Inaccurate control
Oven drying: Broccoli Cauliflower Banana chips Mushrooms	Flexibility during operation Low labour cost	Longer drying period An explosion might happen because of induced evaporation Case hardening of the product may occur due to excessive heat
Smoke drying: Berries Garlic Figs	Smoke prevents bacteria-carrying insects throughout the drying process	Smoke can alter the flavour of the product
Mechanised methods		
Solar drying: Apple Tomatoes Cabbage Mango Banana slices	Requires less drying space There is less contamination as it is in a storage unit	It cannot be used on cloudy days More expensive instead of sunlight drying
Cabinet dryer: Fruits (grapes, dates and apples) Vegetables (onion and cabbage)	Convenient, cost-effective and time-saving method	High initial cost for setup
Tray drying: Plum Apricots Cantaloupe Ginger Green beans	Uniform drying Simple design Cost-effective Capacity to handle a wide range of products	Consume more energy They are made for specific applications and might have little flexibility in regards to accommodating different types of products or adjusting drying conditions Complex operation

(continued)

Table 13.1 (continued)

Method/ applications	Advantages	Disadvantages
Tunnel drying: Strawberry Blueberry Peach Paprika Carrot	Uniform heat distribution Effective drying process Appropriate for a large quantity of product	Possibility of excessive drying Higher energy consumption
Drum drying: Pumpkin Corn Sweet potatoes Banana Guava Papaya Raisin	Possess a rapid rate of drying Excellent energy efficiency Efficient in drying slurries	Product might have a charred flavour Solubility of the product is significantly reduced (85%) due to protein denaturation Changes in the feed's thickness and moisture content
Spray drying: Tamarind powder Tomato powder Fruit powders like apple, amla, banana and papaya	High production rate Gentle drying Short drying period	Loss of heat-sensitive compounds Broad distribution of particle size
Fluidised bed dryer: Broccoli florets Corn Pears Tomatoes	Preserves the colour, flavour and nutritional value of fruits and vegetables while decreasing drying time	Limited application for sticky materials and particle attrition
Hot air drying: Garlic slices Sweet potatoes Chinese wolfberry Carrot	Preserving bioactive compounds and inactivating pathogens in fruits like mulberry	Fruits that are hot-air dried at higher temperatures lose important nutrients like phenolics and ascorbic acid, which affects the antioxidant activity of the fruits
Advanced methods		
Freeze drying: Asparagus Date Pumpkin Grapes Orange juice Guava	Oxidisable materials have excellent protection in vacuum conditions Little contamination due to the aseptic process Minimal loss of heat-sensitive nutrients, volatile chemicals and fragrant components	Expensive operation unit High vacuum can eliminate volatile compounds Lengthy processing time It can be only used for high-value products
Vacuum drying: Potatoes Corn Cherries Raisins Cranberries	Pressure at a lower temperature causes water to boil when in a vacuum, which shortens the drying time Enhanced drying rates Appropriate for solvent recovery from solid products that contain solvents	Excessive pressure can result in product darkening Working with minimal pressure only

(continued)

Table 13.1 (continued)

Method/ applications	Advantages	Disadvantages
Superheated steam drying: Potato slices Apples Carrot Cabbage Mangosteen	Free of pollutants since no hazardous gases or particulates are released into the atmosphere by a closed system Lower energy consumption and faster drying times There is the possibility of in situ particle crystallisation	Lack of evidence supporting its suitability for fruits and vegetables that are heat-sensitive
Osmotic-dehydration: Orange Pineapple Kiwi Apple Cherry Papaya Strawberry	Reduced shrinkage and better quality of the product Increased activity of antioxidants Fruits and vegetables retain their aroma, colour and nutritional value	Extreme browning and potential changes to phytochemical content are caused by high temperatures Takes a longer time to process than alternative techniques Decrease in nutrition and requirement for extra processing phases Health risks arise from using sugar as an osmotic agent
Heat pump dryer: Apple Guava Potato Ginger	Effective polyphenol preservation during herb drying Efficient drying process	Heat pump performance degrades as the air bypass ratio increases
Explosion puff drying: Blueberries Apple Carrot Sweet potatoes Onions	Short running time Excellent rehydration Honeycomb structure	As moisture content falls, there is an increase in water precipitation Procedure requiring a lot of energy High puffing temperature may cause harm to the product
Low-pressure superheated steam drying: Apple Cabbage Carrot Mangosteen Potato slices Banana	In comparison to other drying techniques, this method saves energy	Longer drying times at high vacuum because of inefficient heat transfer
Electrohydrodynamic drying: Wolfberry Spinach Cucumber Tomato slices Okara Grape pomace Quince	Better food appearance, flavour, less hardness, less shrinkage (between 25% and 32%) Higher capacity for rehydration	High operating cost For industrial applications, its drying kinetics ought to be enhanced

(continued)

Table 13.1 (continued)

Method/ applications	Advantages	Disadvantages
Electromagnetic radiation techniques: Banana Carrot Grapes	Improving food safety Extending shelf life	Some loss of nutrients in fruits and vegetables, particularly sensitive vitamins like thiamine, ascorbic acid and vitamins A and E High cost of infrastructure
Refractance window drying: Lucuma fruit Papaya puree Mango powder Asparagus Sweet corn Tomatoes	Reduced reaction to browning in pomegranate leather Asparagus, sweet corn and tomatoes showed better antioxidant activity and polyphenols retention	Processing limitations High maintenance requirements
Microwave drying: Moringa Apple Banana Carrot Potato	Less drying time Improve product quality and flavour	There may be overheating High energy efficiency There may be a decline in quality
Infrared heating: Butter mushroom Eggplant Green peas Goji berries Lemon Longan	Provides a high rate of drying without burning the material by applying direct, fast heat concentration Low cost of operation	Needs electricity to function Possesses a rising radiation intensity Repairing a dryer can be challenging due to its advanced design
Radio frequency drying: Carrot Banana Mango	Higher energy efficiency Uniform drying	Huge size Higher operating costs
Combined methods		
Hybrid microwave convective drying: Apple Carrot Grapes	Shorten the drying time of fruits and vegetables by up to 90%	Equipment complexity
Convective drying followed by vacuum-microwave drying: Cranberries Tomatoes Kiwifruit Pear	It enhances food quality Nutritional quality retention	Less flexibility Equipment complexity High maintenance cost
Fluidised bed drying—Assisted by far infrared rays, microwaves and ultrasounds: Carrot Strawberry Tomatoes	Reduce negative effects like shrinking, discolouration, breaking and nutritional changes Easy to use	Equipment complexity High capital cost

(continued)

Table 13.1 (continued)

Method/ applications	Advantages	Disadvantages
Intermittent drying assisted by pressure, temperature, convection, radiation humidity and microwave: Kiwi Papaya Banana Guava	Reduce browning effects and chemical reactions Flexibility in process control Preserve the quality attributes of food products	Process optimisation Equipment complexity Potential variability in results

13.3.1 Traditional Methods

13.3.1.1 Sun Drying

Sun drying is the process of dehydrating foods by exposing them directly to solar radiation (Ekechukwu and Norton 1999). This method has a lower limit for moisture content of about 15%. Sun drying is an intermittent drying process which requires low-humidity areas. This method is used to dry large quantities of fruits and vegetables. Sun drying is commonly used for drying tomatoes, raisins, apricots and figs globally.

13.3.1.2 Smoke Drying

This method combines drying and smoky flavouring. It is traditionally used for meats but can also be applied to some vegetables and fruits. The smoke drying exposes vegetables and fruits to smoke created by burning wood, leaves and crop waste. This slow process of drying enables the produce to absorb smoke composition, potentially adding additional flavours by means of substance interaction (Qin et al. 2022).

13.3.1.3 Oven Drying

The technique of oven drying is applied for drying a small quantity of fruits and vegetables (Swanson 2003). A conventional oven is appropriate for drying fruits and vegetables with a thermostatic setting of 60 °C. An oven thermometer is used to test the oven temperature.

13.3.1.4 Shade Drying

This technique is applicable for products that lose their colour when subjected to direct radiation from sun. Usually beans, green and red chillies, okra and herbs are dried in a good air circulation, shaded area.

13.3.1.5 Air Drying

Air drying is a substitute for sun drying which involves fruits and vegetables drying on a small scale. It relies on natural air movement and ventilation to dry produce.

13.3.2 Mechanised Methods

13.3.2.1 Solar Drying

The term 'solar dryer' refers to a wide range of convective dryers in which products are dried indirectly by air heated by solar energy rather than being exposed to the sun. Figure 13.2a, b represents the direct and indirect solar dryer. Solar drying employs specially designed structures to collect and maximise solar radiation. It produces high air temperatures and low humidity, resulting in faster drying (Fellows 2022). Solar dryers are classified into two types:

1. Absorption or Hot Box Dryers: Product is heated due to the radiation from the sun.
2. Indirect or Convection Dryers: The hot air generated due to the irradiation of the sun is used to dry the product.

13.3.2.2 Spray Dryer

Spray drying is the process of converting feed from a liquid to a dried particle form by spraying it into a hot drying medium. Spray drying is one of the most effective techniques for drying fruits and vegetable juices. This method is used for fluids that are high in moisture and slightly paste-like consistency. Some of the common uses of spray drying include fruit juice drying and herbal extracts (Jangam and Law 2010). In a spray dryer, fluid is pumped to a nozzle or rotary valve disc atomiser, which sprays fine droplets of feed into the drying chamber. A stream of hot air moves against or alongside the falling droplets. As a result, the fluid droplets are dried, leaving the dry matter as powder particles that fall to the bottom of the chamber and are constantly removed.

13.3.2.3 Drum Drying

This technique is an indirect drying method that keeps the fruit and vegetable slices to be dried in a steam-heated rotating drum. The diagrammatic representation of drum drying is represented in Fig. 13.3. The slices are spread on the dryers outside surface. As the hot drum rotates, the clinging and drying continue. The product reaches a 'doctor blade' that scrapes the dried slices from the drum after completing

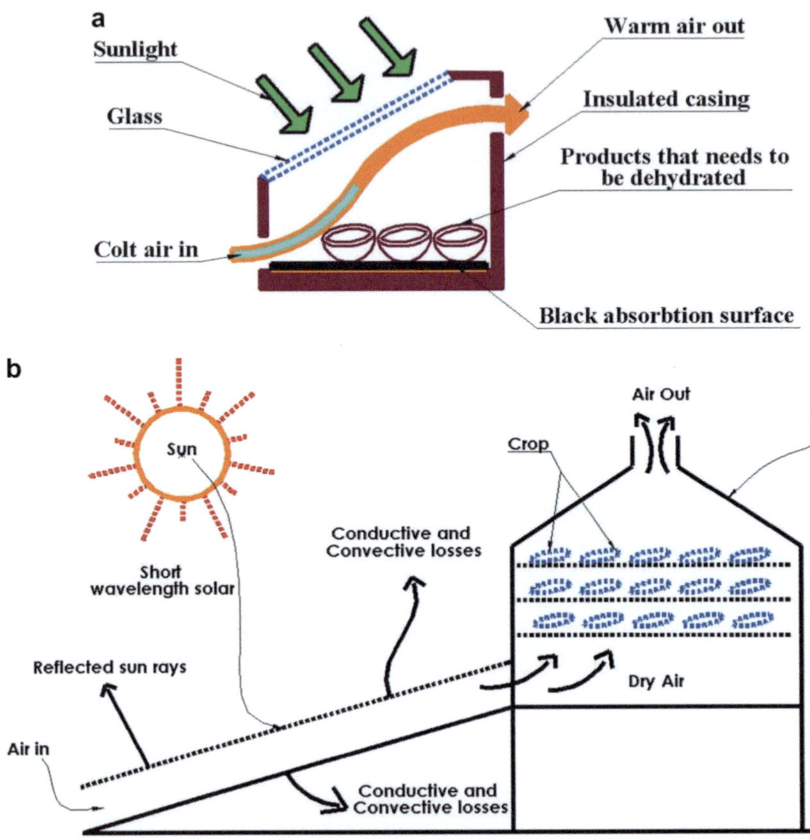

Fig. 13.2 (**a**) Direct method of solar drying (**b**) Indirect method of solar drying

about three-quarters of a full rotation on the drum surface. The method is also referred to as roller drying or film drying. The proportion of consumption of steam compared to the evaporation of water ratio ranges between 1.2 and 1.6:1.

13.3.2.4 Cabinet Drying

Cabinet drying is a convenient, cost-effective and time-saving method that preserves the product's nutrients (Precoppe et al. 2015). Cabinet dryers feature a closed compartment where trays containing the product to be dried are placed. The trays are placed on the shelves with enough space between them. Heated dry air flows between the shelves. Tray bottoms are often slotted or perforated to allow air to flow through the trays. Air inlet temperatures typically range from 60 to 80 °C. Air velocity is modified based on the density, shape and size of the food particles to prevent dry matter from being entrained by the wind. A batch's duration typically ranges

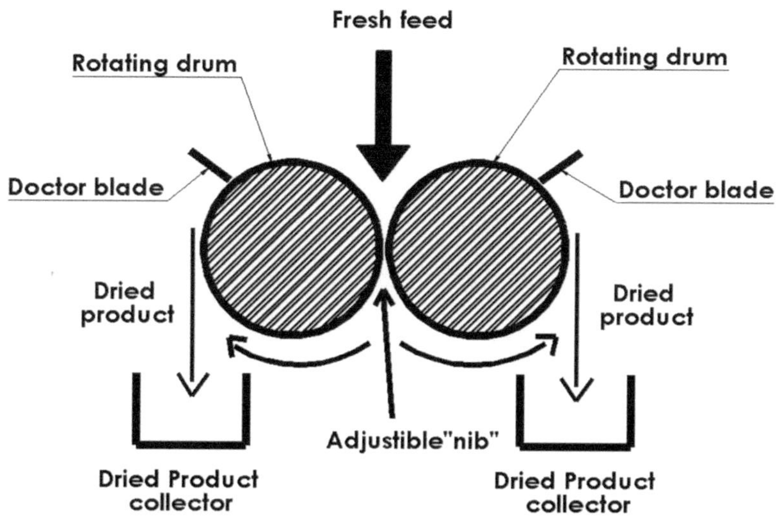

Fig. 13.3 Various parts of drum drying

from 2 to 10 h, according to the product and conditions. Figure 13.4 represents the components of cabinet dryer. Cabinet dryers are used for drying vegetables (cabbage and onion), fruits (dates, grapes and apples) and herbs (basil and parsley).

13.3.2.5 Tray Drying

Tray drying is an approach of drying in which foods are spread on trays and exposed to hot air to remove moisture. The product to be dried is distributed in a uniform layer on trays. The trays are subsequently arranged inside a drying chamber. Hot air is circulated through the drying chamber via forced convection with a fan. The hot air removes moisture from the product on the trays. Effective air flow distribution is essential in ensuring that the product is dried uniformly across all the trays. This aids in maintaining the drying rates and quality throughout the batch. The working principle of tray drying is given in Fig. 13.5. Tray drying is one of the popular drying methods in the food industry because of its simplicity, affordability and capacity to dry various products (Misha et al. 2013).

13.3.2.6 Tunnel Drying

It is made up of narrow chambers and several parallel sloping that are placed above the furnace room. A conveyor belt is used to transport trays to the upper end and is removed at the lower end. The fan circulates air during the entire drying process. Humidity is maintained by allowing air to enter on occasion. Heating can be done directly or through radiation. In direct heating, sufficient heat is passed over the

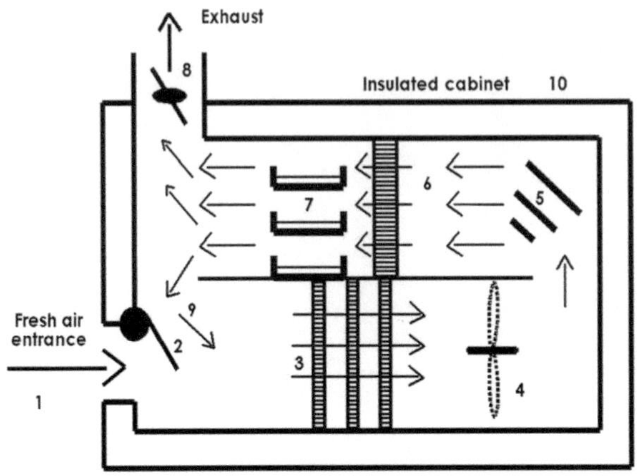

Fig. 13.4 Components of cabinet dryer

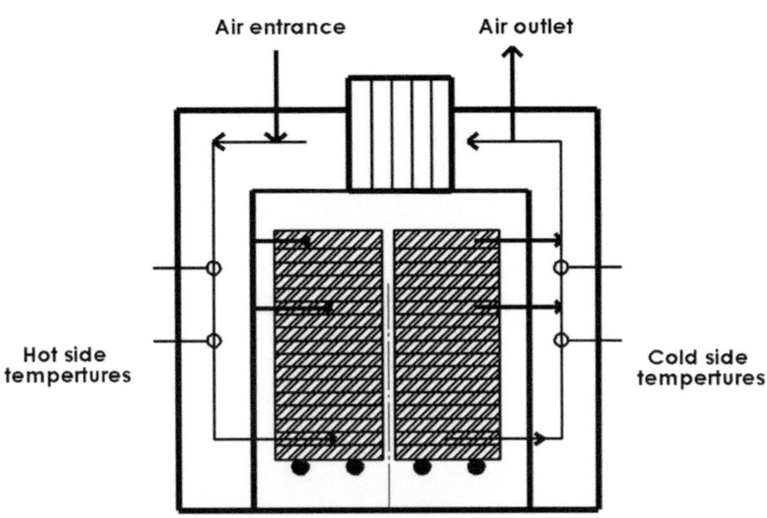

Fig. 13.5 Working principle of tray drying

material, and there is an overheating risk. Whereas, air traveling over pipes carrying hot gases from a furnace is heated by radiation.

13.3.2.7 Fluidised Bed Dryers

Fluidised bed drying involves placing the product on a porous medium and blowing hot air through it to suspend the product in a gentle boiling motion. The air velocity is carefully controlled so that it exceeds the product settling rate. As a result, the product is suspended and behaves like a fluid, hence the term ‘fluidised bed’. The fluidised state of the product enables efficient heat and mass transfer. The high surface area contact between the product and the drying medium (air or gas) allows moisture to quickly evaporate from the product. The fluidised bed promotes even drying throughout the product. This uniform drying reduces the formation of hot spots and ensures consistent quality in the finished product. Fluidised bed drying is widely used for drying a variety of fruits and vegetables such as berries, mangoes, apples and tomatoes. This method preserves the nutritional value, flavour and colour of vegetables and fruits while decreasing drying time (Senapati et al. 2021).

13.3.2.8 Hot Air Oven Drying

Currently, most of the dehydrated fruits and vegetables are produced by the technique of hot air drying, which is the simplest and most economical among the various methods. This technique allows mass and heat transfer by convection by exposing the wet product to heated air. The movement of water to the product’s surface and the removal of water vapour from the surface are two vital aspects of mass and heat transfer. Raw material, quality attributes and economy are the factors to be considered during the choice of drying methods. Vegetables and fruits are typically dried using belt-trough conveyor, kiln, cabinet, tunnel, pneumatic and bin dryers. Among this, the tunnel dryer is semi-continuous, the belt-trough dryer is continuous, followed by the cabinet, kiln and bin dryers are batch operated (Mujumdar 2006).

13.3.3 Advanced Methods

13.3.3.1 Freeze Drying

Freeze drying, or lyophilisation, is the process of preserving food by freezing it and then evaporating the water (in the form of ice) into vapour via sublimation. This drying method is used for products with sensitive heat compounds that are prone to oxidation due to their low temperature and high vacuum operation. The initial phase of freeze drying involves freezing the product with conventional freezing equipment. The following step is to eliminate water through subsequent drying. The process of freeze drying involves subjecting the product to a pressure that is lower than the triple point (610 Pa at 0 °C) and heating the frozen material which causes ice sublimation to vapour. This technique is commonly used for high-quality dehydrated products with heat-sensitive ingredients like antibiotics, microbial culture and vitamins. Accelerated freeze drying method is used for drying products without altering their texture and shape. The dried product will have good taste, reconstitution property and flavour.

13.3.3.2 Vacuum Drying

Vacuum drying is the technique for removing moisture from the product under vacuum conditions. The boiling point of moisture or solvent is reduced using a vacuum. The low pressure causes molecules to diffuse to the surface and evaporate at the same time. Heaters are frequently used to increase drying efficiency. It is carried out at a pressure of above 0.6 kPa and below 101 kPa, with heat transfer typically occurring via conduction. Drying time is shortened in this process as water boils under vacuum due to pressure at a low temperature. This drying improves water circulation of the product resulting in better mass transfer. In contrast, high pressure could result in product darkening (Chibuzo et al. 2021). A vacuum drying chamber with a temperature control system, a desiccator and a vacuum pump are some of the common parts of laboratory-scale vacuum drying equipment (Parikh 2015).

13.3.3.3 Superheated Steam Drying

This drying technique is widely accepted as a non-polluting, safe drying method that requires little energy consumption (Mujumdar 2006). The method operates with superheated steam and recovers heat by means of a vapour recompression cycle. In the heat treatment chamber, superheated steam acts as the drying medium which removes moisture from the product and condensing the evaporated water using a heat exchanger. In the treatment chamber, heat is transferred to the recycled steam by compressing and diverting a portion of the saturated steam from the main recycling line. Heating can be accomplished with the warm condensate.

13.3.3.4 Heat Pump Drying

This technique of drying retrieves latent heat through the condensation of water using the refrigerator. The process results in the product being exposed to dry heated air, which releases humid air. The drying air gets warmed by the latent heat of vapourisation when the air is condensed in the heat pump evaporator. Figure 13.6 represents the components of heat pump dryer. Heat pump dryers use less fossil fuels and are more energy-efficient (Fayose and Huan 2016). This method of drying depends on the drying media, heat pump dryer operation, supplementary heating, HPD cycle, drying temperature, number of heat pump stages and functioning mode (Jangam and Law 2010; Mujumdar 2006). A well-designed heat pump can be

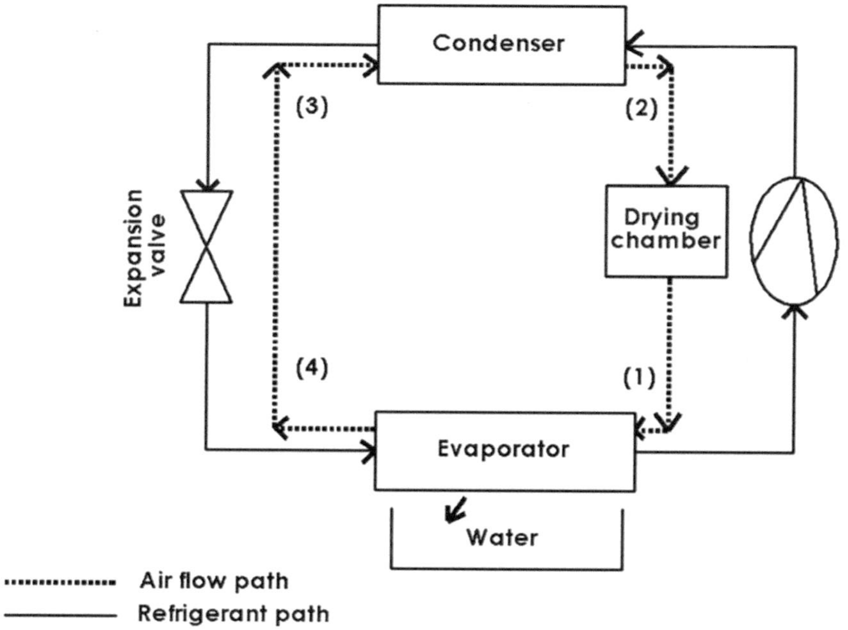

Fig. 13.6 Components of heat pump dryer

installed in any dryer that uses convection as its main heat source. On the other hand, HP is not compatible with dryers that need a large quantity of drying air, like spray or flash dryers. Some of the shortcomings of heat pump drying can be addressed by combining it with other drying techniques that result in high thermal efficiency, lower energy consumption, high coefficient of performance and improved product quality. Heat pump-assisted drying methods include infrared drying, chemical heat pump-assisted drying, microwave drying, atmospheric freeze drying, solar drying, radiofrequency drying and fluidised bed drying. This dryer applies to heat-sensitive materials that require only low temperature such as fruits and vegetables. For instance, combining solar drying and HPD improves drying and lowers cost. A heat pump is appealing as it provides more heat energy with less electrical energy consumption (Fayose and Huan 2016).

13.3.3.5 Explosion Puff Drying

Explosion puff drying is an approach in which food products such as fruits and vegetables are exposed to a sudden and intense release of pressure usually using steam to cause a rapid increase in the product's internal moisture content. This unexpected increase leads to the product puffing up, resulting in a porous structure that allows hot air drying. The equipment internal pressure is raised to 0.2 MPa by the air compressor as fruit or vegetable samples are heated with steam to 95 °C and held for 5 min. Puff samples are vacuum-dried when the decompression valve is opened, which reduces the pressure. The schematic diagram of explosion puff drying is given in Fig. 13.7. The resulting explosion-puffed products are porous, easily rehydrated and have appealing sensory characteristics. Explosion puffing allows for efficient food drying while maintaining quality and reducing drying time when compared to traditional methods. One disadvantage of explosion puff drying is that it is not appropriate for all product types (Feng et al. 2021).

13.3.3.6 Low-Pressure Superheated Steam Drying

Low-pressure superheated steam drying is a technique for drying products using superheated steam at low pressure levels and the setup is given in Fig. 13.8. This method of dehydration occurs in a closed drying chamber with a vacuum pump that maintains low pressure. To prevent additional condensation of steam, the reservoir is equipped with a steam trap that removes the drying agent from the boiler. During the initial startup cycle, the initial condensation of steam is greatly reduced by using a heater equipped with a temperature control system (Chibuzo et al. 2021). In comparison to traditional drying methods, this method saves energy and improves the retention of bioactive components in dried products. This method is used effectively for drying onions (Sehrawat and Nema 2018).

13.3.3.7 Electrohydrodynamic Drying

Electrohydrodynamic (EHD) drying is the direct use of electricity to remove moisture from wet products like fruits and vegetables. EHD drying produces a large difference in voltage between the collector and emitter electrodes in the range of kilovolts. Because of its large curvature, this voltage causes an electrical discharge,

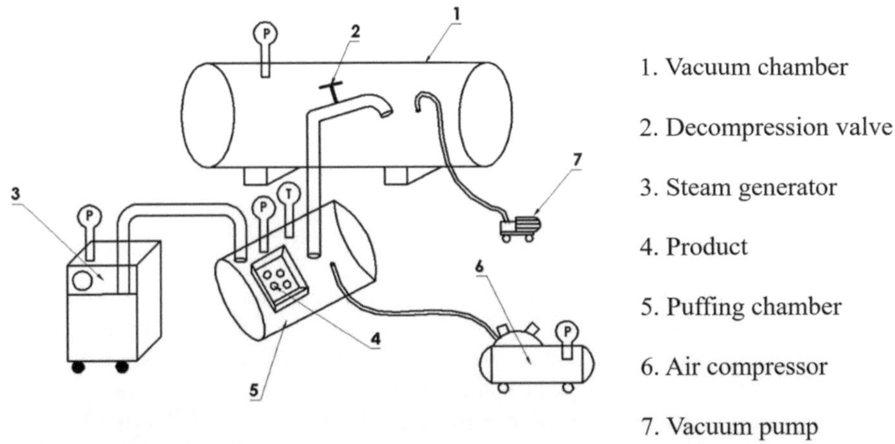

Fig. 13.7 Schematic diagram of explosion puff drying

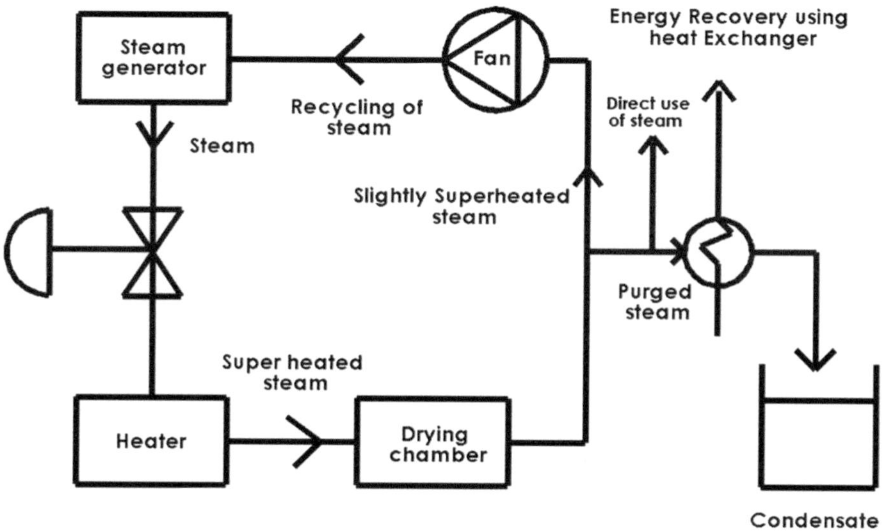

Fig. 13.8 Low-pressure superheated steam drying setup

resulting in air ionisation at the emitter electrode. The Coulomb force attracts the ionised air towards the collector electrode, resulting in an ionic wind. When positively ionised air collides with neutral particles in the air during movement, a secondary airflow is formed. This ionic wind promotes the convective removal of moisture from the product that needs to be dried. In general, EHD drying offers advantages over traditional convective air-drying methods, especially for heat-sensitive foods, by helping preserve the sensory qualities and nutritional content of the produce while reducing energy consumption and negative environmental effects. Combining EHD drying with other commercial drying methods like hot air-drying needs careful optimisation to accomplish desired results regarding energy efficiency, drying time and product quality (Onwude et al. 2021).

13.3.3.8 Electromagnetic Radiation Techniques

The electromagnetic wavelength spectrum is used as an energy source in a variety of food applications. The product is exposed to a specific wavelength of electromagnetic wave, which produces heat and accelerates the drying process (Khan et al. 2020). This drying method heats the product using electromagnetic radiation (Marra et al. 2009). The following are the drying methods which employ this concept.

13.3.3.8.1 Refractance Window Technology

Refractance Window Drying (RWD) uses a thin transparent infrared film, usually made of Mylar, as a 'window' to dry heat-sensitive products like fruit juices and purees. The method operates by permitting radiant energy to travel through the film to dry the product by maintaining it at a low temperature. The working principle of refractance window technology is given in Fig. 13.9. This approach assures quick drying by utilising every type of heat transfer including evaporative cooling and convective heat transfer into the surrounding air. In an RWD system, the product is distributed as a slurry over a transparent film, allowing infrared energy to travel at the speed of light. When transparent plastic and moist food come into contact, an infrared window is established, enabling the immediate transfer of infrared energy to the material. The hot water underneath the film, typically kept below the boiling point (around 99 °C), aids in the drying process. Even though the water temperature is high, the product temperature is within a range of 60–70 °C due to evaporative cooling and convective heat transfer to the surrounding air. Refractance Window Drying has many benefits, such as product quality retention and energy efficiency. It is critical to consider the potential drawbacks of equipment complexity, maintenance and processing limitations when determining its suitability for specific applications (Mahanti et al. 2021).

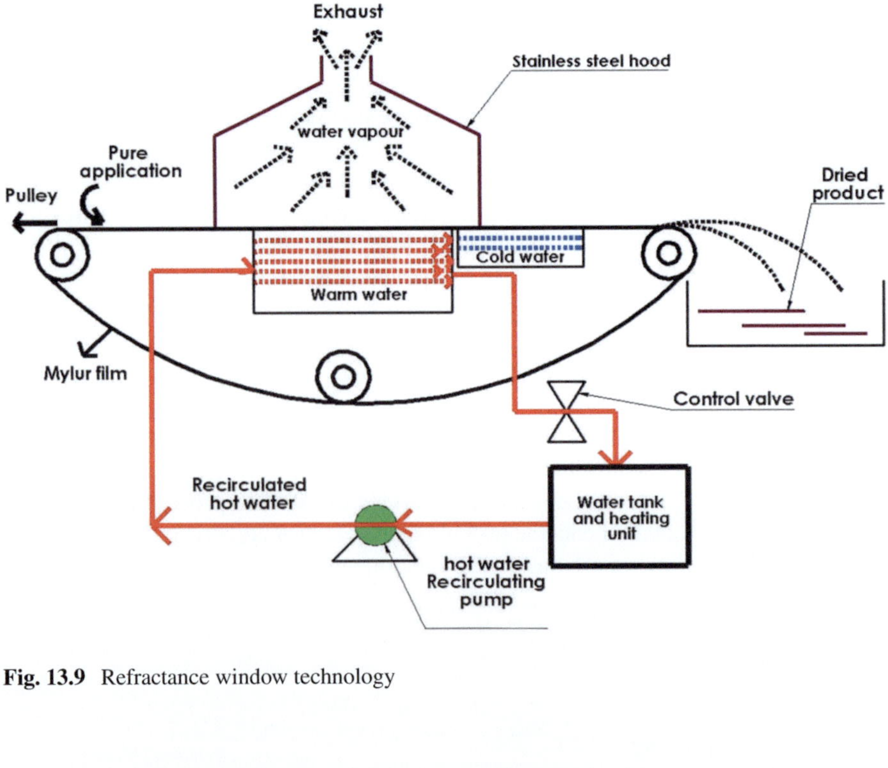

Fig. 13.9 Refractance window technology

13.3.3.8.2 Infrared Heating

In infrared drying, fruits or vegetables are exposed to electromagnetic radiation with a specific range of 0.8–1000 μm wavelength, resulting in drying (Askari et al. 2013). The product surface acquires infrared energy from the heating component without affecting the ambient air temperature. Exposure to infrared radiation results in the accumulation of charge at the atomic and molecular levels. As a result, the temperature of the product rises resulting in drying. Using infrared drying, agricultural commodities like tomatoes, sweet potatoes and carrots have been dried. Because of its ease of use and energy saving, this method is ideal for integrating with traditional drying methods.

13.3.3.8.3 Radio Frequency Drying

The interaction of a product's molecular species with an electromagnetic field generated by a radio frequency generator is known as radio frequency heating (Calín-Sánchez et al. 2020). The product is placed between two parallel electrodes that are exposed to an electric field that changes 40,000,000 times per second, and the frequencies range between 1 and 60 MHz. The electric field causes friction and heating throughout the product. The water heats and evaporates due to its bipolar nature (Babu and Gupta 2008). High operating costs and huge size are major disadvantages. Radio frequency drying is a substitute for the traditional hot air-drying of apple slices (Marra et al. 2009).

13.3.3.8.4 Microwave Drying

In microwave drying, the product is exposed to very high-frequency electromagnetic waves between 0.3 and 3 GHz. Like radiant heat transfer, these waves are also transferred to the product. Water molecules are polarised and tend to shift in orientation due to high-frequency waves. Sufficient heat is produced during orientation to drive out moisture from the product. The heat produced by the system decreases as the frequency increases. The outside surface remains wet until the drying cycle is finished because high-frequency heating evaporates water inside the material and creates convective cooling on the surface. As a result, cracking is avoided. Because of the quick temperature rise, microwave drying also has the benefit of increasing the rate at which bacteria are killed. Citrus fruit concentrates can be dried using microwave drying under vacuum (Mujumdar 2006).

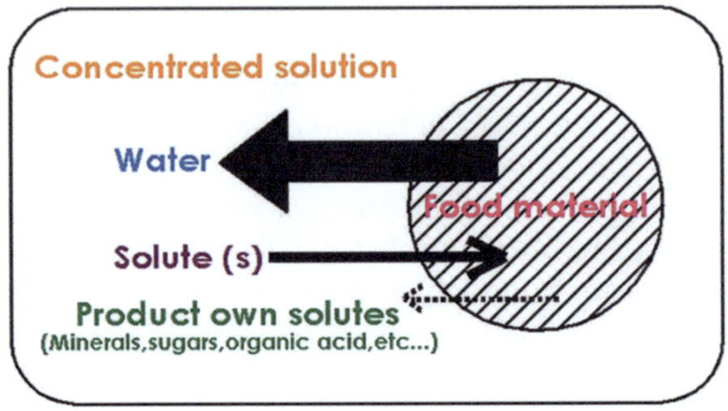

Fig. 13.10 Principle of osmotic dehydration

13.3.3.9 Osmotic Dehydration

This technique is a water extraction process in which the product, typically fruits and vegetables, is soaked in a solution of hypertonic sugar, salt or a combination of both to decrease water content and increase soluble solid content. Figure 13.10 represents the principle of osmotic dehydration. Osmotic dehydration is a minimal processing technique carried out under vacuum or at atmospheric pressure which is used to lower the product's water activity (Lewicki and Lenart 2020). The raw material is soaked in concentrated soluble solid solutions that have lower water activity and a higher osmotic pressure. Compared to dissolved substances, water flows through the selectively permeable membrane much more easily. Sodium chloride (NaCl) is an excellent osmotic agent for vegetables, and the use of other products derived from animals in fruits is limited due to taste alteration and colour bleaching. The driving force for drying was increased when NaCl was added to the osmotic solution (Rajanya and Singh 2021). Using blends with two or more solutes appears to be a desirable substitute. A gentle heat treatment results in less heat damage to the product's colour and flavour while maintaining superior sensory attributes. Osmotic dehydration prevents enzymatic and oxidative browning in fruits and vegetables, resulting in good colour retention with little or no use of sulphur dioxide. One of the drawbacks of this method is that a decrease in acidity level reduces the distinct flavour of some products. This can be overcome by including fruit acid in the solution.

13.3.4 Combined Dehydration Methods

13.3.4.1 Microwave-Aided Convective Drying

Ongoing microwave drying may result in overheating and deterioration of the dried material. This can be overcome with intermittent microwave-convection drying (İzli and Polat 2018). In microwave-assisted convective drying, heated air minimises free moisture present on the product surface, whereas the microwave energy removes bound moisture from the interior surface of the product using volumetric heating. The application of microwave energy can reduce the drying time of fruits and vegetables by up to 90% (Musielak et al. 2024).

13.3.4.2 Convective Drying Assisted by Vacuum-Microwave Drying

Convective pre-drying removes free moisture from fruits and vegetables without affecting bioactive components, while moisture content is minimised to the desired level by vacuum-microwave drying. The hybrid convective drying followed by vacuum-microwave drying is more efficient than either method used alone. This method increased the quality of beetroot, blackcurrant, plum, pumpkin, jujube, orange peel and sour cherries (Calín-Sánchez et al. 2020).

13.3.4.3 Fluidised Bed Drying Followed by Ultrasounds, Far Infrared Rays and Microwaves

Microwave energy can be used in fluidised bed drying like convective drying. This method requires more research and multiple drying phases to find its application to various fruits and vegetables. The researchers were able to monitor the influence of allicin using far-infrared rays-assisted fluidised bed drying of garlic (Calín-Sánchez et al. 2020). Fluidised bed drying assisted by ultrasound showed reduced negative effects like discolouration, breaking, nutritional changes and shrinking. It is recognised as an energy-efficient and low-cost system with an easy user experience (Singh and Kingsly 2008). To generate a high-intensity ultrasonic field inside the fluidised bed drying chamber, an aluminium cylinder and piezoelectric transducers are used (Jafari and Zare 2017).

13.3.4.4 Intermittent Drying Assisted by Temperature, Pressure, Radiation, Humidity, Microwave and Convection

Intermittent drying occurs when the drying conditions change over time. This drying method can be made feasible by changing the drying air temperature, airflow rate, humidity, thermal energy source and pressure, which can increase the energy efficiency and product quality (Kumar and Karim 2019). Intermittent drying can reduce chemical reactions and browning effects that help to improve product quality and preserve the bioactive ingredients present in the product (Pham et al. 2020).

13.4 Effect of Dehydration on Quality of Fruits and Vegetables

Nutrient retention, water activity, visual appeal, flavour, colour, texture, shape, microbial load and rehydration properties are some of the major quality parameters associated with fruits and vegetables dehydration. The primary structure and shape of the products are protected during freeze drying due to the solid state of water, which keeps the cells almost intact and produces high porosity end products. In contrast, hot air drying typically destroys the cell structure. It is recommended to use a convective dryer along with ultrasounds to uphold the internal assembly of the dehydrated product. Dehydration increases the bulk density, whereas shrinkage is caused due to decrease in the porosity of the product. Freeze drying and vacuum-microwave drying are the most effective techniques for reducing bulk density to about 50% when compared to convective drying. Combined vacuum-microwave drying with convective drying prevents the product from overheating and can lower the bulk density of the dehydrated products by about 55% in comparison to the dehydrated products obtained from a convective dryer. Freeze-dried fruits and vegetables typically exhibited lower shrinkage when compared to vacuum-microwave and hot air dried products (Téllez-Pérez et al. 2020).

Preserving the colour of dried products is crucial for marketability. An extended drying time and high temperature cause colour degradation. Most colour changes that occur during dehydration are associated with browning reactions due to both enzymatic and non-enzymatic processes. Fruits and vegetables undergo enzymatic browning because of phenolic compounds oxidation by polyphenol oxidase, which initiates the production of o-quinones (brown pigments). Non-enzymatic browning is caused due to caramelisation, ascorbic acid oxidation and the Maillard reaction, which are impacted by temperature, pH, water activity and product composition. Conventional dehydration techniques, such as convective drying, resulted in significant colour losses. Advanced dehydration technologies, such as combined dehydration methods, limited the colour degradation by using pH adjustment and high temperature combined with short times or by minimising exposure time to heat (Calín-Sánchez et al. 2020).

Factors such as duration of exposure, temperature, light levels and oxygen concentrations can influence the changes in phytochemical content. Dried goods produced at low temperature had a higher phytochemical content than dried goods produced at high temperature. Ascorbic acid content decreases with longer drying time, and carotenoids were less sensitive to drying time than temperature. Dried products using combined drying methods of microwave-assisted and refractance window drying showed higher polyphenol, anthocyanin and ascorbic acid content. Higher lycopene content was observed in osmotic-vacuum dried produce than in produce dried solely by vacuum or convective method (Sablani 2006). Freeze-dried fruits and vegetables significantly retained more vitamin C than those that are oven or sun-dried. Oxidation-susceptible components can be preserved by dehydrating fruits and vegetables in an oxygen-free environment. Similarly, refractance window, vacuum, low-pressure-assisted microwave and superheated steam dehydration methods prevent vitamin loss due to low oxygen levels (Sagar and Suresh Kumar 2010).

13.5 Conclusion

Dehydration is one of the best methods to preserve fruits and vegetables for a longer time. The development and enhancement of innovative dehydration methods are directly related to the quality of dehydrated fruits and vegetables. Colour, shrinkage, porosity, bulk density, phytochemicals, antioxidant capacity, volatile compounds, proteins, sugars and sensory characteristics are among the quality parameters that are to be maintained during the dehydration process. Freeze drying is one of the best techniques for maintaining functional quality. Moreover, dehydrated products obtained from freeze drying can lose elasticity and become more viscous when rehydrated. This problem can be solved by combining freeze drying with osmotic pre-treatment. The combination of dehydration techniques ensures high-quality dehydrated produce. Nonetheless, further research is still needed to optimise novel dehydration techniques and their combinations with other methods to enhance the quality parameters of the dehydrated fruits and vegetables. It is necessary to properly evaluate the dehydration protocols to recommend specific conditions, as the ideal dehydration technique varies depending on the produce being dehydrated.

References

Ahmed J (2018) Handbook of vegetables and vegetable processing. Wiley, New York, pp 381–405. https://doi.org/10.1002/9781119098935.ch16

Askari G, Emam-Djomeh Z, Mousavi S (2013) Heat and mass transfer in apple cubes in a microwave-assisted fluidized bed drier. Food Bioprod Process 91(3):207–215. https://doi.org/10.1016/j.fbp.2012.09.007

Babu B, Gupta S (2008) Adsorption of Cr (VI) using activated neem leaves: kinetic studies. Adsorption 14(1):85–92. https://doi.org/10.1007/s10450-007-9057-x

Calín-Sánchez Á, Lipan L, Cano-Lamadrid M et al (2020) Comparison of traditional and novel drying techniques and its effect on quality of fruits, vegetables and aromatic herbs. Food Secur 9(9):1261. https://doi.org/10.3390/foods9091261

Chibuzo NS, Osinachi UF, James MT et al (2021) Technological advancements in the drying of fruits and vegetables: a review. Afr J Food Sci 15(12):367–379. https://doi.org/10.5897/AJFS2021.2113

Ekechukwu OV, Norton B (1999) Review of solar-energy drying systems II: an overview of solar drying technology. Energy Convers Manag 40(6):615–655. https://doi.org/10.1016/S0196-8904(98)00093-4

Fanzo J, Miachon L (2023) Harnessing the connectivity of climate change, food systems and diets: taking action to improve human and planetary health. Anthropocene 42:100381. https://doi.org/10.1016/j.ancene.2023.100381

Fayose F, Huan Z (2016) Heat pump drying of fruits and vegetables: principles and potentials for Sub-Saharan Africa. Int J Food Sci 2016(1):9673029. https://doi.org/10.1155/2016/9673029

Fellows PJ (2022) Food processing technology: principles and practice, 5th edn. Woodhead Publishing, Cambridge

Feng L, Xu Y, Xiao Y et al (2021) Effects of pre-drying treatments combined with explosion puffing drying on the physicochemical properties, antioxidant activities and flavor characteristics of apples. Food Chem 338:128015. https://doi.org/10.1016/j.foodchem.2020.128015

González-Pérez JE, Ramírez-Corona N, López-Malo A (2021) Mass transfer during osmotic dehydration of fruits and vegetables: process factors and non-thermal methods. Food Eng Rev 13:344–374. https://doi.org/10.1007/s12393-020-09276-3

Grau R, Andres A, Barat JM (2014) Principles of drying. Handbook of fermented meat and poultry. Wiley, New York, pp 31–38. https://doi.org/10.1002/9781118522653.ch5

İzli N, Polat A (2018) Intermittent microwave drying of apple slices: drying kinetics, modeling, rehydration ratio and effective moisture diffusivity. J Agric Sci 26(1):32–41. https://doi.org/10.15832/ankutbd.464715

Jafari A, Zare D (2017) Ultrasound-assisted fluidized bed drying of paddy: energy consumption and rice quality aspects. Dry Technol 35(7):893–902. https://doi.org/10.1080/07373937.2016.1230742

Jangam SV, Law AS (2010) Drying of foods, vegetables and fruits

Khan MIH, Rahman M, Karim M (2020) Recent advances in micro-level experimental investigation in food drying technology. Dry Technol 38(5–6):557–576. https://doi.org/10.1080/07373937.2019.1657145

Kumar A, Agrawal S (2023) Challenges and opportunities for agri-fresh food supply chain management in India. Comput Electron Agric 212:108161. https://doi.org/10.1016/j.compag.2023.108161

Kumar C, Karim M (2019) Microwave-convective drying of food materials: a critical review. Crit Rev Food Sci Nutr 59(3):379–394. https://doi.org/10.1080/10408398.2017.1373269

Lewicki PP, Lenart A (2020) Osmotic dehydration of fruits and vegetables. In: Handbook of industrial drying. CRC Press, pp 691–713. https://doi.org/10.1201/9780429289774

Mahanti NK, Chakraborty SK, Sudhakar A et al (2021) Refractance WindowTM-drying vs. other drying methods and effect of different process parameters on quality of foods: a comprehensive review of trends and technological developments. Future Foods 3:100024. https://doi.org/10.1016/j.fufo.2021.100024

Marra F, Zhang L, Lyng JG (2009) Radio frequency treatment of foods: review of recent advances. J Food Eng 91(4):497–508. https://doi.org/10.1016/j.jfoodeng.2008.10.015

Misha S, Mat S, Ruslan M et al (2013) Review on the application of a tray dryer system for agricultural products. World Appl Sci J 22(3):424–433. https://doi.org/10.5829/idosi.wasj.2013.22.03.343

Mujumdar AS (2006) Handbook of industrial drying, 3rd edn. CRC Press, Boca Raton, p 1312. https://doi.org/10.1201/9781420017618

Musielak G, Mieszczakowska-Frąc M, Mierzwa D (2024) Convective drying of apple enhanced with microwaves and ultrasound- process kinetics, energy consumption, and product quality approach. Appl Sci 14(3):994. https://doi.org/10.3390/app14030994

Neeraj AC, Bisht V, Johar V (2017) Marketing and production of fruits and vegetables in India. Int J Curr Microbiol App Sci 6(9):2896–2907

Onwude DI, Iranshahi K, Martynenko A, Defraeye T (2021) Electrohydrodynamic drying: can we scale-up the technology to make dried fruits and vegetables more nutritious and appealing? Compr Rev Food Sci Food Saf 20(5):5283–5313. https://doi.org/10.1111/1541-4337.12799

Parikh DM (2015) Vacuum drying: basics and application. Chem Eng 122(4):48–54

Pateiro M, Vargas-Ramella M, Franco D et al (2022) The role of emerging technologies in the dehydration of berries: quality, bioactive compounds, and shelf life. Food Chem X 16:100465. https://doi.org/10.1016/j.fochx.2022.100465

Pham ND, Khan M, Karim M (2020) A mathematical model for predicting the transport process and quality changes during intermittent microwave convective drying. Food Chem 325:126932. https://doi.org/10.1016/j.foodchem.2020.126932

Precoppe M, Janjai S, Mahayothee B et al (2015) Batch uniformity and energy efficiency improvements on a cabinet dryer suitable for smallholder farmers. J Food Sci Technol 52:4819–4829. https://doi.org/10.1007/s13197-014-1544-y

Qin H, Yang T, Yang S et al (2022) Effects of different pre-drying and drying methods on volatile compounds in the Pericarp and Kernel of Amomum tsao-ko. Front Plant Sci 13:803776. https://doi.org/10.3389/fpls.2022.803776

Rajanya DR, Singh G (2021) Recent trends in osmotic dehydration of fruits: a review. Plant Arch 21(1):09725210. https://doi.org/10.51470/PLANTARCHIVES.2021.v21.no1.013
Sabarez HT (2018) Thermal drying of foods. In: Rosenthal A, Deliza R, Welti-Chanes J, Barbosa-Cánovas G (eds) Fruit preservation, Food engineering series. Springer, New York, pp 181–210. https://doi.org/10.1007/978-1-4939-3311-2_7
Sablani SS (2006) Drying of fruits and vegetables: retention of nutritional/functional quality. Dry Technol 24(2):123–135. https://doi.org/10.1080/07373930600558904
Sagar V, Suresh Kumar P (2010) Recent advances in drying and dehydration of fruits and vegetables: a review. J Food Sci Technol 47:15–26. https://doi.org/10.1007/s13197-010-0010-8
Sah S, Johar V, Karthi J (2022) Status and marketing of fruits and vegetables in India: a review. Asian J Agric Ext Econ Sociol 40(7):1–11. https://doi.org/10.9734/ajaees/2022/v40i730911
Sehrawat R, Nema PK (2018) Low pressure superheated steam drying of onion slices: kinetics and quality comparison with vacuum and hot air drying in an advanced drying unit. J Food Sci Technol 55:4311–4320. https://doi.org/10.1007/s13197-018-3379-4
Senapati A, Varshney A, Sharma V et al (2021) Fluidized bed drying of food: a review. e-planet 19(1):25–30
Sharma S (2021) Value chain approaches for effective fruits and vegetables marketing in India. Value addition opportunities in fruits 93
Sharma S, Barman K, Krishna H et al (2024) Advances in freeze drying to improve efficiency and maintain quality of dehydrated fruit and vegetable products. In: Freeze drying of food products: fundamentals, processes and applications. Wiley, pp 229–253. https://doi.org/10.1002/9781119982098.ch9
Singh D, Kingsly A (2008) Effect of convective drying on quality of anardana. Indian J Hortic 65(4):413–416
Spiker ML, Welling J, Hertenstein D et al (2023) When increasing vegetable production may worsen food availability gaps: a simulation model in India. Food Policy 116:102416. https://doi.org/10.1016/j.foodpol.2023.102416
Srivastava P, Ramakanth D, Hakim L et al (2024) Package designing aspects for marketing of fruits and vegetables. In: Novel packaging systems for fruits and vegetables. Apple Academic Press, pp 265–287. https://doi.org/10.1201/9781003415701
Sultana A, Aghajanzadeh S, Thibault B et al (2024) Exploring conventional and emerging dehydration technologies for slurry/liquid food matrices and their impact on porosity of powders: a comprehensive review. Compr Rev Food Sci Food Saf 23(3):e13347. https://doi.org/10.1111/1541-4337.13347
Swanson MA (2003) Drying fruits and vegetables. Washington State University Extension, Washington, DC. https://hdl.handle.net/2376/7147
Téllez-Pérez C, Cardador-Martínez A, Tejada-Ortigoza V (2020) Antioxidant content of frozen, convective air-dried, freeze-dried, and swell-dried chokecherries (Prunus virginiana L.). Molecules 25(5):1190. https://doi.org/10.3390/molecules25051190

Chapter 14
Plantation Crops and Products

Aashika Ashok, Rudragoud Policegoudra, and OP Chauhan

14.1 Introduction

Plantation crops are valuable commercial crops that are crucial in their agricultural economy and export trade. With the World Trade Organization (WTO) placing increased focus on agriculture, the commercial cultivation of these crops is becoming increasingly important from an economic perspective. In nations like Kenya and Cameroon in Africa, crops such as tea, coffee, and cocoa are vital to their economies. Meanwhile, oil palm is a key economic driver in Malaysia and Indonesia among Asia. In Latin America, rubber serves as a significant source of foreign exchange. The primary crops grown extensively on plantations include areca nut (betel nut), cardamom, cashew, cocoa, coconut, coffee, oil palm, rubber, and tea. The cultivation of these crops is labor-intensive as they are perennial in nature. Growing crops creates a significant number of job opportunities, both through direct work on the farm and through processing activities off the farm. A key characteristic of plantations is the presence of monocultures, where large, contiguous areas are dedicated to the cultivation of the same crop, usually consisting of plants of similar age. The exceptions to this pattern are coconut, areca nut, cocoa, and coffee, which may be grown either alone or in combination with other crops. These crops are mainly found in tropical regions between 20 degrees north and 20 degrees south of the equator. Tea, cardamom, and coffee, on the other hand, thrive in subtropical climates and tend to flourish in shaded conditions (Chopra et al. 2016).

Plantation refers to a property or extensive piece of land. Plantation crops are crops that are grown commercially in extensive land areas and overseen by an individual or corporation. The perennial crops hold significant economic value and are

A. Ashok (✉) · R. Policegoudra · OP Chauhan
Defence Food Research Laboratory, DRDO, Mysore, India
e-mail: aashika.dfrl@gmail.com

OP Chauhan (ed.), *Fruits and Vegetables Technologies*,
https://doi.org/10.1007/978-981-96-8433-5_14

crucial for our country's economy. Tea, coffee, rubber, cocoa, coconut, betel nut, oil palm, palmyra palm, betel leaf, bamboo, etc., are among the crops grown (Alam 2020).

14.2 Definition of Plantation Crops

"A group of perennial agricultural plants grown in tropical and subtropical regions that require year-round labor and are typically consumed after being processed." Plantation crops are crops that are utilized or whose product is utilized only after undergoing processing. They can make use of waste land in rain fed dry land, hilly, arid, and coastal areas to guarantee nutritional security, generate foreign exchange, and secure livelihoods. India ranks as one of the top producers of plantation crops globally.

Plantation crops are crops cultivated on a large area owned and managed by an individual or company. These cash crops are valuable commercial crops with significant economic importance and are essential to the Indian economy. In the traditional sense, plantation crops are large-scale cultivated crops such as tea, coffee, and rubber. In this context, plantation and estate are used interchangeably. Estate or plantation is typically a large agricultural area focused on cultivating a single crop.

14.3 Scope and Economic Importance of Plantation Crops

14.3.1 Expansion in Non-traditional Areas

Due to the limited geographical distribution of population and spice crops, expansion opportunities within traditional cultivation areas are increasingly constrained. However, significant potential exists for expanding cultivation into non-traditional regions, such as the North Eastern States of India, where there is untapped irrigation potential. Furthermore, advancements in technologies like drip irrigation have facilitated the expansion of plantation crops into new and non-traditional areas, increasing the scope for agricultural development in these regions.

14.3.2 Export Potential

Plantation crops contribute significantly to foreign exchange earnings, particularly through the export of products like coir-based goods and coffee. Both the primary products and their by-products offer strong export prospects while also meeting substantial domestic demand across several ancillary industries. Earnings from the export of plantation crops account for approximately 27% of the total agricultural commodity exports and 4.8% of the nation's overall exports, underscoring their economic importance.

14.3.3 Employment Generation

The cultivation of plantation crops generates consistent, year-round employment opportunities, both on farms and in processing factories. For instance, the cultivation of coconut alone provides 78 man-days of employment per hectare annually, translating to 70 million man-days per year in Kerala. This illustrates the substantial role plantation crops play in rural employment generation.

14.3.4 Crop Diversification

Plantation crops provide significant opportunities for agricultural diversification, which enhances the sustainability and resilience of farming systems. By incorporating these crops, farmers can reduce the risks linked to monoculture and improve the overall economic viability of agriculture.

14.3.5 Availability of Technology and Yield Gap

Institutions such as the Central Plantation Crops Research Institute (CPCRI) in Kerala, India, the National Research Centre on Cashew in Puttur, Karnataka, India, and various agricultural universities and research stations, have developed technologies aimed at improving yield and production efficiency in plantation crops. Despite the availability of modern technologies, yield gaps persist in many areas, indicating further potential for productivity improvements through the adoption of these advancements.

14.4 Economic Importance

Tea, rubber, and coconut are contributing a lot to the country through its exports. India exports a wide variety of plantation crops and is the leading producer on the global share. In addition to this, the plantation industry provides a livelihood for millions through both direct and indirect employment. For instance, in India, the tea industry alone is directly employing one million people and indirectly providing employment to another million. Around 300,000 people work in cashew processing factories and around 200,000 farmers are involved in growing the plant. This way, other than generating employment in itself, the plantation sector spawns a multitude of by-product and rural industries spurring overall regional economic development. In addition, plantation crops are important in conservation. Tea plantations on mountain slopes and cashew plantations in arid, barren areas help mitigate the impact of heavy rainfall and soil conservation caused by intense rains or violent storms, thus playing a role in protecting soil health and supporting ecosystems.

14.5 Importance of Plantation Crops

Plantation crops are of significant commercial value and play a critical role in the global economy, particularly in developing countries. They generate substantial revenue and often serve as a primary source of foreign exchange earnings. In addition, they provide employment to millions of people, not only in cultivation but also in associated sectors such as processing, transportation, and value-added industries. The following are some of the key contributions of plantation crops:

14.5.1 Economic Impact

Plantation crops are vital to the economies of many countries, generating income through both domestic and international trade. They frequently represent a primary source of foreign exchange earnings, thus supporting economic growth and enhancing financial stability.

14.5.2 Employment Generation

Due to the labor-intensive nature of plantation agriculture, these crops provide extensive employment opportunities globally. This workforce extends beyond farming to include positions in processing, transportation, and related industries, offering livelihood to millions of individuals.

14.5.3 Supply of Industrial Products

Plantation crops serve as essential raw materials for various industries. For example, rubber is indispensable to the manufacturing sector, particularly for tire production; cocoa is fundamental to the chocolate industry; and cotton is a critical input for the textile industry. These crops are integral to the functioning and growth of these industries.

14.5.4 Contribution to Food Security

Certain plantation crops, such as bananas and pineapples, contribute to global food security by providing essential nutrients to populations. Their role in food supply chains helps ensure a reliable source of nourishment, especially in tropical regions.

14.5.5 Socio-economic Development

Revenues generated from plantation crops often lead to improvements in infrastructure, healthcare, education, and other critical social services. These economic benefits foster broader socio-economic development, particularly in rural areas, contributing to improved living standards.

14.5.6 Environmental Benefits

When cultivated sustainably, plantation crops can have positive environmental impacts, such as aiding in carbon sequestration, maintaining soil health, and supporting biodiversity. Crops grown on slopes or hillsides, such as tea and cashew, help prevent soil erosion by stabilizing the land and serving as cover crops that reduce runoff during heavy rains.

14.5.7 Cultural Significance

Many plantation crops, such as tea and coffee, hold deep cultural importance in numerous societies. These crops often play an integral role in social customs, traditions, and rituals, making them culturally significant in addition to their economic value.

14.5.8 Potential for Value-Added Products

Plantation crops offer considerable scope for developing high-value goods, which enhance their economic impact. For instance, sugarcane can be processed into sugar, biofuel, and molasses, while coconuts can yield copra, coconut oil, and coir. The creation of such products increases the value derived from these crops, benefiting both local economies and industries.

14.6 Basic Characteristics of Plantation Crops

Plantation crops are mainly grown for commercial purposes and often need specific climate conditions to thrive. The following characteristics differentiate plantation crops from other agricultural systems:

14.6.1 Monoculture

Plantation crops are predominantly cultivated as monocultures, meaning that a single crop is grown over a large area. While this uniformity can simplify management practices and increase yields, it also increases vulnerability to diseases, pests, and environmental stressors, given the lack of biodiversity.

14.6.2 High-Value Cash Crops

These crops are typically high-value cash crops grown primarily for commercial sale rather than local consumption. Plantation agriculture is aimed at generating profits through large-scale market production.

14.6.3 Specific Climatic Requirements

Plantation crops generally thrive under specific climatic conditions, most often in tropical or subtropical regions. These crops require a warm, humid environment for optimal growth, making them particularly suited to these geographic zones.

14.6.4 Labor-Intensive

The cultivation and harvesting of plantation crops are highly labor-intensive processes. Large amounts of manual labor are required for tasks such as planting, crop care, and harvesting. This labor demand creates significant employment opportunities, particularly in developing regions where plantation agriculture is prominent.

14.6.5 Capital-Intensive

The large-scale cultivation of plantation crops requires significant capital investment. Extensive land areas, high-quality seeds, fertilizers, pesticides, and often advanced machinery for planting and harvesting are necessary, making this form of agriculture financially demanding.

14.6.6 Long Gestation Period

Many plantation crops have long gestation periods before they reach maturity. For instance, rubber trees may take up to 7 years from planting to produce latex. This long maturation time necessitates patience and long-term planning from farmers and investors.

14.6.7 Processing Requirements

Most plantation crops require substantial processing before they are market-ready or consumable. For example, coffee beans must undergo drying, roasting, and grinding before being used to make coffee, and rubber must be processed to be used in manufacturing industries.

14.6.8 Large-Scale Production

As the term "plantation" implies, these crops are usually cultivated on a large scale. The extensive scale of production is primarily driven by the commercial objectives of plantation agriculture, with a focus on maximizing output and profit.

14.6.9 Export-Oriented

Plantation crops are often grown for export, as they are high-value commodities in the global market. As a result, they play a crucial role in the export earnings of many countries, particularly in the Global South, contributing significantly to national economies.

14.6.10 Sensitivity to Market Fluctuations

The prices of plantation crops on the international market can be highly volatile. Fluctuations in global demand and supply, trade policies, and other factors can directly impact the income of farmers and countries that rely heavily on the export of these crops. This economic sensitivity poses challenges for both producers and national economies.

14.7 Status of Plantation Crops

From 2007 to 2016, India had the largest area of cropland, covering nearly 170 million hectares (Mha), followed by the United States at 158 Mha, and both the Russian Federation and China with approximately 120 Mha each. About 10% of the total cropland is allocated to permanent crops, such as oil palm, fruit trees, and cocoa plantations (FAO 2020) (Tables 14.1 and 14.2).

Table 14.1 Global trends in the area and production of key plantation crops

Plantation crop	Leading producer of the crop (country)	2000–2010		2011–2020		2020–2024 (estimated production)	
		Area (mha)	Production (mmt)	Area (mha)	Production (mmt)	Area (mha)	Production (mmt)
Coconut	Indonesia	3.565	1.62	4.272	2.79	4.200	16.44
Tea	China	1.518	1.85	2.573	2.75	3.170	3.24
Coffee	Brazil	2.580	38.6	2.890	41.7	2.506	62.08
Cocoa	Ivory coast	1.890	1.89	2.843	1.79	2.140	2.06
Cashew nut	Ivory coast	0.575	0.435	0.775	0.895	0.700	1.25

Source: Directorate of Economics and Statistics, Ministry of Agriculture, GOI and FAOSTAT

Table 14.2 Recent developments in the area, production, and productivity of key plantation crops in India during 2023–24

Plantation crop	Area (mha)	Production (mmt)	Productivity (Kg/ha)	India's share in global production (%)	Leading state
Coconut	2.4	19.7	8709.50	31.45%	Karnataka
Tea	0.6	0.0013	1600	22.68%	Assam
Coffee	0.48	0.37	Arabica—360 Robusta—1200	3.14%	Karnataka
Cocoa	0.018	0.015	570	0.65%	Kerala
Cashew nut	1.18	0.75	760	17.80%	Maharashtra

14.8 Categories of Plantation Crops

14.8.1 Coconut

14.8.1.1 Introduction

The coconut palm, *Cocos nucifera L.*, is a species with diverse ecological and economic uses that provides several key natural products, generating numerous economic opportunities and socio-environmental benefits to humanity. It is also important in terms of the economy and society for more than eighty tropical countries. The coconut is often called Kalpavriksha, or the "tree of heaven that provides everything." It is also known as the "King of Tropical Palms" due to its thriving growth in warmer climates.

14.8.1.2 Origin

There are many theories about the origin of the coconut. One theory suggests it comes from Tropical America and was expanded further to the Pacific. The likeliest theory, however, is that coconuts from Mesoamerica were brought to the Pacific Islands by equatorial sea currents. Other theories suggest the coconut palm first originated in South Asia or the Pacific and later spread naturally to the Americas. References to coconuts in India can be traced back to the time of Kalidasa (*Raghuvamsa*) and Sangam literature. The coconut is also rumored to symbolize the wealth of Kerala, often termed "the country of coconuts." However, the exact origin of the coconut in India remains debated.

14.8.1.3 Distribution

Coconuts are utilized in both tender and mature forms. Tender coconuts are primarily consumed as a natural mineral-rich drink, while mature coconuts serve multiple purposes, including direct consumption, and the production of copra, oil, and various other products. Copra, in particular, holds significant promise as a source of edible oil, contributing approximately 6% to India's edible oil supply. Around 38% of the total coconuts produced in India are used for copra production.

14.8.1.4 Varieties

The coconut palm exhibits a wide range of varieties and forms, which have been classified over time by various researchers. Coconut palms are generally classified into two main types—talls and dwarfs—based on their plant structure and morphological characteristics. Tall coconut palms are typically cross-pollinated and, as a result, exhibit high heterozygosity. Tall coconut palms generally produce medium to large fruits that yield high-quality copra with a substantial oil content. In India, notable indigenous tall cultivars include West Coast Tall, East Coast Tall, Benaulim Tall, Tiptur Tall, Andaman Ordinary Tall, and Laccadive Ordinary Tall, each found in different regions. Prominent exotic tall cultivars globally include Fiji, Rennell, Philippines Ordinary, Sri Lankan, Malayan, Jamaican, and San Ramon Tall. In contrast, dwarf coconut palms are commercially valued for their early fruiting, shorter height, tender nuts of high quality, and resistance to certain diseases. Dwarf palms tend to be more homozygous due to increased self-pollination and generally produce smaller to medium-sized nuts. In India, popular dwarf cultivars include Chowghat Green, Chowghat Orange, Kenthali Orange, and Gangabondam Green Dwarf. Exotic dwarf varieties such as Malayan Yellow, Malayan Orange, and Malayan Green Dwarf are also widely cultivated in many coconut-producing countries. Furthermore, natural selection and hybridization have resulted in intermediate types, combining the early flowering trait of dwarf palms with the hardiness and high yield of tall varieties, thereby, exploiting hybrid vigor (Narayana and John 1949; Liyanage 1958; Swaminathan and Nambiar 1961; Purseglove 1968).

14.8.1.5 Establishment of Plantation

Coconuts are propagated solely from seeds and can grow successfully in various soil types, as long as the soils are well-drained, aerated, and have a pH range between 4.3 and 8.0. Effective disease control in coconut cultivation relies on good sanitation practices and the use of appropriate systemic treatments for pest management (Owoyemi et al. 2022). According to John (1970), the steps involved in establishing a coconut plantation are as follows:

1. Selection of mother plant.
2. Selection of seed nuts.
3. Harvest and preservation of seed nuts.
4. Preparation of seedlings for nursery planting.
5. Planting of seedlings in the nursery.
6. Maintenance of seedlings in the nursery.
7. Selection of best seedlings.
8. Preparation of land for planting.
9. Planting in the plantation.
10. Care of young plantation.
11. Maintenance of mature plantation.
12. Harvesting.

14.8.1.6 Harvesting

Coconuts serve as an essential source of vegetable oil, utilized for edible and industrial applications. Approximately 40% of the total coconut production is used for culinary and religious purposes, while 35% is processed into copra in India. Additionally, 6% is dedicated to value-added products, including 1% for virgin coconut oil (VCO), 1% for coconut milk or cream, and 4% for desiccated coconut powder. Around 17% of coconuts are harvested as tender nuts, and 2% are set aside for seed production. Depending on the region, coconut harvesting occurs between 6 and 12 times per year. Well-maintained coconut plantations often have monthly harvests, whereas neglected plantations, where coconut bunches do not develop consistently, may yield no more than six harvests annually. In most coconut-growing countries, fully mature nuts, typically around 12 months old, are harvested every 2 months.

14.8.1.7 Harvesting Methods

Harvesting coconuts is a specialized task that requires skill and is generally performed by experienced climbers. Traditionally, climbers ascend to the top of the coconut tree using ladders approximately 3–4 meters long. Once they reach the top of the ladder, they use a rope ring around their feet or ankles to continue the climb. In some regions, climbers also use a second rope ring for their hands. Another method involves cutting notches into the tree trunk at intervals to facilitate climbing, though this is not recommended as it can damage the tree. In some places, tied-on husks are used as makeshift steps for climbing. Additionally, some climbers employ a rope loop that encircles both the trunk and their body, aiding their climb. In recent years, various mechanical climbing devices have been developed to make the process of harvesting coconuts less labor-intensive and to reduce the physical strain involved in climbing (Thamban et al. 2011). These innovations are aimed at improving efficiency and safety in coconut harvesting practices.

14.8.1.8 Processing

A range of coconut processing devices have been designed by Indian Council of Agricultural Research-Central Plantation Crops Research Institute (ICAR-CPCRI) and other institutions to enhance efficiency, minimize processing time, and reduce the physical effort required in coconut.

14.8.1.8.1 Dehusking of Coconut

Dehusking is the initial step in processing of coconut, traditionally done manually using a spike. To streamline this process, a power-driven semi-automatic coconut dehusking machine has been designed. It removes the husk by guiding coconuts between two spiked rollers that rotate in opposite directions at varying speeds. A pneumatic cylinder presses the coconut toward the rollers, and the rotating spikes remove the husk. This machine can process up to 200 coconuts per hour. Another dehusker, *Keramitra*, crafted by Kerala Agricultural University (KAU), widely used in households due to its simplicity, although it has a relatively low output (Varghese and Jacob 2014). Other dehusking tools, including pedal-operated dehuskers, are also available.

14.8.1.8.2 Coconut Splitting

To address the safety concern of splitting coconuts by hand, a manually operated coconut splitting device has been created. The machine features a cutting blade made of spring steel, positioned at a 25-degree bevel angle to reduce the frequency of sharpening. The nut is manually split using the blade's impact force, while the coconut water is collected in a container placed at the bottom.

14.8.1.8.3 Copra Drying

Copra, the dried form of the coconut kernel, is usually obtained by drying fresh coconut kernel by sun on sand or cement for a period of 7–9 days. This method can lead to contamination from dirt and prolonged drying periods, especially during the monsoon season, can cause microbial infections. To address these issues, energy-efficient copra dryers have been developed, ensuring clean, microbial-free, and dust-free copra in a shorter time frame (Vidhan Singh et al. 1999). Different types of copra dryers include smallholder dryers, shell-fired dryers, forced circulation dryers, and solar tunnel dryers.

14.8.1.8.4 Deshelling of Coconut

A power-driven, batch-type coconut deshelling machine has been designed to remove the shell from copra after partial drying. This machine has the capacity to process 400 half coconut shells per batch. For optimal deshelling efficiency (92.16%), the moisture content should be around 35% (dry basis). The machine operates at an optimal speed of 10 RPM and can deshell a batch in approximately 4 min (Singh and Udhayakumar 2006).

14.8.1.8.5 Copra Moisture Meter

Producing premium coconut oil with a low level of free fatty acids—ideally less than 1%—requires high-quality copra grinding. One important factor in determining the quality of copra is its moisture level; high-quality copra has less than 6% moisture, which lowers the possibility of damage from insects, molds, or microbes. Using the electrical conductivity of the coconut kernel, an electronic moisture meter was created to determine the moisture level of copra. This device is capable of measuring moisture concentrations between 5% and 40% (Madhavan 1987).

14.8.1.8.6 Tender Coconut Punch and Cutter

These simple tools are designed to easily puncture and open tender coconuts. A punch creates a hole, sufficient to introduce a straw, allowing convenient consumption of fresh coconut water.

14.8.1.8.7 Snowball Tender Nut Machine

The snowball tender coconut is a round-shaped coconut filled with water, harvested when the nut is approximately 7–8 months old. Using a specialized machine, the shell is cut, and the tender coconut ball is scooped out. After cleaning, the snowball coconut can be packaged hygienically for immediate consumption or refrigerated for later use.

14.8.1.8.8 Coconut Shell Separating Machine

In coconut processing, removal of shell is the following step after dehusking, particularly when fresh coconut kernel is used as the primary commodity, the shell is commonly removed manually using a knife. However, modern shell removal machines have been developed to mechanize and expedite this process, improving efficiency in coconut processing industries.

14.8.1.8.9 Coconut Slicing Machine

The process of slicing coconut kernels into uniformly thick chips is a crucial operation in the production of coconut chips. Traditionally, this task is performed manually, which is both labor-intensive and tedious. The consistency in thickness and overall quality of the chips are mainly determined by the operator's expertise.

14.8.1.8.10 Coconut Chips Dryers

An electrically operated dryer has been developed to optimize the drying process for coconut chips. The dryer is equipped with ten trays that have wire-mesh screens for placing coconut slices. An automated sensor and electronic control unit manage the temperature, maintaining consistency in the drying process. While the standard dryer is designed to handle 50 coconuts at a time, its capacity can be scaled up as needed.

14.8.1.8.11 Coconut Grating Machines

To improve the efficiency of coconut grating, two types of machines have been designed. The first machine is a single-user model, while the second is a multi-user variant with four scraping blades. Both machines are equipped with stainless steel blades that finely grate the coconut kernel. The unique-user machine can process up to 60 coconuts per hour, while the multi-user machine operates at four times that capacity.

14.8.1.8.12 Coconut Pulverizer

The coconut pulverizer is designed to produce fine coconut powder through a motorized rotating blade system. The machine processes coconut kernel bits manually placed in the cutter. The rotary blade's exert influence, combined with rubbing against a static blade, transforms the coconut into fine powder. The pulverizer can handle up to 250 coconuts per hour.

14.8.1.8.13 Coconut Milk Extractors

In order to increase the efficacy of coconut milk extraction, manually operated machines resembling hand-operated vertical screw presses have been developed. Among the different types of coconut milk extractors, screw-type expellers exhibit the optimal extraction rate. The single-screw expeller can process 300 coconuts per hour, while the double-screw expeller is able to accommodate 1000 coconuts per hour.

14.8.1.8.14 Virgin Coconut Oil (VCO) Cookers

The production of virgin coconut oil (VCO) through thermal methods has been standardized, and commercial guidelines have been established. Traditionally, VCO is extracted by slowly heating coconut milk in an open vessel with constant manual stirring. This traditional process is labor-intensive, and improper stirring can cause the milk to burn, affecting the quality of the oil. To address these challenges, two types of VCO cookers have been developed, automating the stirring process to prevent charring and optimize fuel usage (Mathew et al. 2014).

14.8.1.9 Storage of Coconuts

Storing harvested coconuts in shaded heaps for several days, a process called seasoning, is a common practice before further processing. This method offers several benefits:

1. It is easier to husk.
2. Shelling is more efficient, with shells becoming dry and hard, which improves their utility as fuel by allowing continuous burning.
3. As the moisture content of the coconut meat reduces, its thickness increases, leading to higher yields of copra and oil.
4. Copra made from matured nuts is of higher quality compared to that produced from freshly harvested produce.

Research has revealed that storage is advantageous primarily for fully ripe coconuts, as they yield high-quality copra even when processed immediately after harvest. While storing, immature nuts can increase the copra yield per nut, these nuts are prone to spoilage during storage.

14.8.1.10 Food Products

An extensive variety of food products are derived from the coconut palm, either consumed in their natural state or processed into various forms. These products cater to diverse culinary and nutritional needs, utilizing different parts of the coconut, especially the kernel.

14.8.1.10.1 Products from Wet Kernel

The soft texture and high protein content of the coconut kernel, especially from nuts that are 7–8 months old, make it a highly prized food to eat with the sweet, tender nut water. The sugar and protein content of the kernel is at its peak at this point and progressively drops as the nut ages. The kernel is regarded as fully mature when the nut is 11–12 months after pollination. 90.50% moisture, 0.80% protein content, 1.30% lipids, 6.80% carbohydrates, and 0.60% ash make up the unripe kernel's composition at this stage (Banzon et al. 1990). The nut's moisture, protein, and ash content all decline with maturity, while fat content increases. The coconut kernel is nutrient-rich, containing essential vitamins and minerals. Coconut fat comprises a balanced mix of saturated and unsaturated fatty acids, making it highly digestible compared to other fats. Coconut protein has a digestion rate of 86–94% and a biological value of 71–77%. A variety of carbohydrates, such as sucrose, glucose, fructose, starch, and cellulose, are also found in coconut. Although not a significant source of minerals, fresh coconut kernels contain about 7% sugar, along with small amounts of thiamine, vitamin C, and tocopherol.

14.8.1.10.2 Sweet Coconut Chips

Sweet coconut chips are a popular dehydrated snack made from coconut kernels with intermediate moisture content (35–40% wet basis). The preparation involves partially dehydrating mature coconut kernels through osmotic dehydration using sugar syrup. The resulting product, known as sweet coconut chips, is freeze-dried for extended shelf life, although this process is costly. With a six-month shelf life, these crunchy chips are packed in laminated aluminum bags.

14.8.1.10.3 Coconut Milk and Its Derivatives

Coconut milk is a water-based combination of coconut oil enhanced with soluble ingredients from the meat of the coconut. In order to obtain it from mature, fresh coconuts, the white kernel is ground into a slurry and then separated from the milk through pressing. In large-scale production, a screw press is used for continuous extraction. For the production of milk concentrate, coconut cream, and coconut milk powder, the extracted milk is processed subsequently (Monera and Del Rosario 1982). Beyond its food applications, the milk from coconuts can be used to replace dairy cream in a variety of food products, such as yogurt, soft cheeses, drinks, and evaporated milk (Banzon 1978; Sanchez and Rasco 1983; Sanchez and Rasco 1984). Although there may be variances in content, the Central Food Technological Research Institute (CFTRI) in Mysore, India, states that the approximate composition of coconut milk is 41% water content, 5.8% proteins, 38–40% lipids, 6.2% mineral substances, and 9–11% carbohydrates. Yoghurt and curd-type products have also been prepared from coconut milk and tender coconut water (Chauhan et al. 2009; Manoj Kumar et al. 2017).

14.8.1.10.4 Preserved Coconut Milk Products

Coconut milk and cream items that have been preserved, such as canned milk and cream, as well as dehydrated coconut milk, are becoming a growing trend. The preservation process involves heating pared coconut kernels in hot water (75–80 °C) for 10 min to control microbial growth. A hammer mill is then used to grind the kernels or disintegrator, with a 3 mm perforation to achieve the optimal texture. For maximum milk yield, coconuts aged 11–12 months are preferred, and the kernel gratings are maintained at 40–45% moisture content. A screw press is used to extract milk, and the resulting milk is filtered to remove any remaining grated material, which should not exceed 0.5%. To increase the milk's shelf life, preservatives such sodium metabisulphite or the antibiotic nisin are added after it has been pasteurized for 10 min at 75–80 °C. Stabilizers, such as carboxymethyl cellulose (CMC) and guar gum, and emulsifiers, like Tween 80, are incorporated to prevent separation and ensure smooth emulsification. After pasteurization, the milk undergoes pressure homogenization at 4000 psi, which improves the product's stability and texture.

14.8.1.10.5 Coconut Cream

Made from fresh coconut kernels, coconut cream is a concentrated form of coconut milk. Approximately 2500 kg of cream are produced from 10,000 ripe coconuts with 500 kg of byproduct residue. This versatile product can be applied straight or diluted with water for a number of culinary applications such as curries, desserts, puddings, and sweets. Additionally, coconut cream is valuable in the manufacturing of baked goods and for food flavoring. Through preliminary testing, the technology for storing condensed coconut milk has been refined and shown to be economically viable. In domestic areas, the product has also received favorable consumer reception (Ghosh 2015).

14.8.1.10.6 Dehydrated Coconut Milk

Dehydrated coconut milk is produced by blending fresh coconut milk with trace amounts of casein or maltodextrin added, then spray drying. The resultant powder can be reconstituted by dissolving it in water to create a milky liquid that preserves the taste and consistency of fresh coconut milk. It is usually sold in laminated foil packets.

14.8.1.10.7 Coconut Jam

The watery part of coconut milk is used to make coconut jam. The procedure, initially originated in the Philippines, begins with the extraction of coconut milk by mixing grated coconut with equal amounts of water. After the milk has been extracted, brown sugar and glucose (10.25% and 5.5% of the milk's weight, respectively) are added, and the mixture is heated slowly for about 20 min while being constantly stirred. After straining to get rid of any suspended solids, the mixture is heated over higher heat. Before thickening, 0.25% citric acid (by weight of the milk) is added, and the mixture is cooked until it reaches the desired consistency. The jam is hot-filled into sterilized containers and sealed. The flavor of this coconut jam is rich and creamy (Muralidharan and Jayashree 2011). Tender coconut pulp has also been utilized to make a jam-like product (Chauhan et al. 2013a).

14.8.1.10.8 Coconut Syrup

Coconut syrup is produced by extracting milk from freshly grated coconut meat. After homogenization, adding equal amounts of sugar and 0.05% citric acid (or 0.25% sodium phosphate), the mixture is steam-cooked until 65–68% of the total soluble solids are present. After pouring the hot syrup into lacquered tin cans, they are sealed and allowed to cool under running water. The syrup can be used as a bread spread or in confections because it has 63.47% total soluble solids, 3.11% protein, and 27.18% moisture. When mixed with water, it produces a milk-white, flavorful drink (Sangamithra et al. 2013).

14.8.1.10.9 Coconut Honey

Coconut honey is thick, a great alternative to real honey is a golden paste produced from coconut milk. Grated, unpared coconut meat is combined with an equal amount of water to extract the milk. The mixture is combined with brown sugar (60% by weight) and glucose (30% by weight), then cooked until thick in containers heated with steam. The resultant coconut honey is sealed after being hot-filled into lacquered bottles or tins. This nutty-flavored produce can be used as a spread or a base for soft drinks (Muralidharan and Jayashree 2011).

14.8.1.10.10 Coconut Cheese

Coconut cheese is made by grating fresh coconut meat and extracting its milk. The cream rises to the top and is delicately scraped off after the milk has stood for around 8 h. After the proteins are coagulated by heating the leftover skim milk with vinegar, they are combined with the cream and kneaded with salt. Adoption of this procedure as a small-scale domestic industry is feasible due to its low complexity.

14.8.1.10.11 Fermented Beverage

Lactobacillus bulgaricus serves as the starter culture and skim milk serves as the basis for this fermented beverage. The total solids content of the milk base must be 10%. Non-fat dry milk (NFDM) is almost entirely solid, whereas coconut skim milk (CSM) is normally 14.5% sugar and 85.5% water. Different CSM and NFDM combinations are employed to maintain the necessary 10% solids. Three percent of the *Lactobacillus bulgaricus* culture is added to the mixture after it has been pasteurized for 30 min at 90 °C and cooled to 40 °C. After that, it is incubated for 24 h at 37–38 °C. The milk is heated to 60 °C and homogenized for 5 min after curdling. The liquid is heated to 80 °C, cooled to 60 °C, and flavored with 0.5% extract after adding sugar in a 1:1 ratio. After being bottled, the final product is pasteurized for 30–60 s at 70 °C. Diluted at a 1:3 ratio, the resulting product has 1.01% proteins,

0.74% lipids, 18.7% sugar, and 79.25% moisture. It can be stored at room temperature or in a refrigerator for up to 2 months.

14.8.1.10.12 Sweetened Condensed Milk

The foundation for making sweetened condensed milk is skim milk. Pasteurized milk is combined with sugar, corn oil, and coconut cream before being run through a colloid mill. After that, the milk is boiled in a steam-jacketed kettle while being constantly stirred until 68% of the milk is made up of soluble solids. The finished product is promptly chilled after being hot-packed into sterile tin cans. The finished condensed milk contains 5.53% protein, 60.68% carbohydrates, 10.35% fat, and significant quantities of calcium (112.62 mg/100 g), magnesium, iron, zinc, and manganese (Banzon 1978).

14.8.1.10.13 Coconut Flour

Coconut flour is produced as a byproduct of the manufacturing of coconut cream and associated goods. After expelling the milk, the fibrous residue is dried and ground into a powder. Usually, this flour has 17% oil, 3–5% moisture, and 7–8% protein. Its high fiber content makes it a valuable ingredient in dietary foods.

14.8.1.10.14 Edible Coconut Oil

Because it contains a lot of saturated fatty acids, coconut oil is very resistant to oxidative rancidity, which also impart a pleasant flavor. For this reason, it is preferred as a fat source for making instant milk powder and filled milk, as well as for use as a confectionary fat in goods like whipped cream, ice cream, and imitation cream. Hardened coconut oil and sugar in a 40:60 ratio are frequently used as fillings in wealthy nations in biscuits and wafers. Due to its ease of digestion, coconut oil is also an important ingredient in many ghee substitutes (Rethinam and Bosco 2006).

14.8.1.10.15 Desiccated Coconut

Desiccated coconut consists of finely grated white coconut kernel, dried till less than 3% moisture remains. The common grade is typically smaller than 5 mm in particle size. Its composition includes 67.5% fat, 9.3% protein, 5.9% carbohydrates, and 8.9% pentosan. Vacuum drying at 100 °C with 200 mm Hg of pressure produces high-quality desiccated coconut. Dessicated coconut flakes have been used for making instant coconut chutney (Chauhan et al. 2013b).

14.8.1.10.16 Tender Coconut Water

Made from 7–8 month-old coconuts, tender coconut water is a refreshing and nourishing beverage that is especially well-liked in the summer. It contains minerals like sodium, calcium, magnesium, potassium, and iron, as well as B-group vitamins like riboflavin, ascorbic acid, pantothenic acid, nicotinic acid, and biotin. Tender coconut water has a pH of 4.8–5.3 and an ascorbic acid level of 2.2–3.7 mg per 100 ml, which steadily declines as the coconut matures and the kernel hardens. Several consumables can be derived from tender coconut water, including snowball tender coconut (soft kernel), bottled coconut water, frozen coconut water, coconut water concentrate, coconut vinegar, and coconut honey. The water obtained from tender and mature coconuts have been utilized to make refreshing beverages by blending with fruit juices (Chauhan et al. 2008, 2014).

14.8.1.10.17 Toddy

It is collected by tapping the coconut palm's closed spadix. Freshly tapped toddy is mildly acidic and contains rich carbohydrates, proteins, minerals, and vitamins, which makes it a great fermentation medium and a nutrient-dense beverage. It contains 10–15% sugar, and if not collected under sterile conditions, it ferments rapidly, producing 5–8% alcohol. This fermented toddy, when distilled, produces arrack, a potent alcoholic beverage. If fermentation continues, the toddy undergoes acetic fermentation, resulting in 4–7% acetic acid in coconut vinegar. Furthermore, baker's yeast can be naturally found in fresh toddy.

14.8.1.10.17.1 Products Derived from Toddy

- *Jaggery*: Unfermented toddy solidifies into a substance called coconut jaggery or gur when it is cooked to 118–120 °C. Scum and other impurities are skimmed out during manufacture, and the concentrated toddy is then put into molds to solidify. With a calorific value of 321.6 kcal per 100 g, coconut jaggery is composed of 10.92% moisture, 6.58% reducing sugars, 2.17% inorganic matter, and 1.64% protein.
- *Refined Sugar*: Toddy is treated with 2% lime to coagulate its albuminous impurities in order to extract raw sugar. After that, the mixture is carbonated twice while being filtered to get rid of extra lime. The concentrated sugar content of the clarified juice is 75% followed by crystallization in vacuum pans. The crystals are separated through centrifugation. Though nutritionally valuable for its low glycemic index, this process is not economically viable.
- *Vinegar*: Coconut toddy vinegar is produced when toddy ferments for over 24 h, converting the sugars into acetic acid. A standardized process for large-scale production of coconut vinegar was developed. It was found that producing vinegar using submerged fermentation with strains of *Acetobacter* or jaggery isolates worked well.

- *Treacle*: Treacle is a golden syrup produced by boiling toddy until concentrated. About 16% of the toddy used is recovered. Many places view treacle as a delicacy, particularly in the Lakshadweep Islands, where its production is common.
- *Arrack*: Arrack is a distilled alcoholic beverage derived from fermented toddy. Arrack can recover between 17.5% and 25% of the initial toddy volume. Esters (164–258), total acids (116–158), stable acids (6–13), volatile acid compounds (105–152), and furfural (0.45–1.32) are among the normal analytical values for arrack. Coconut feni, a double-distilled, premium form of arrack, is offered.

14.8.1.10.18 Other Food Products

- *Haustorium*: The haustorium is a spongy, ball-like structure that develops inside coconuts when they start to germinate. A variety of culinary products are produced in this haustorium, such as jams and marmalades. It can also be cut into slices and kept in sugar syrup to be used as a component in fruit salads. The composition of haustorium includes 15.23% moisture, 9.39% ash, 10.60% protein, 15.68% fat, 38.42% carbohydrates, and 10.59% fiber. Another notable product from the coconut palm is the terminal bud, also known as "cabbage." This part is considered a delicacy and can be consumed raw, used in salads, or preserved through pickling, canning, or bottling.
- *Copra*: Copra, along with coconut oil, is one of the primary products of the coconut palm. It refers to the dried coconut endosperm, which is highly valued for its rich oil content, ranging from 65% to 70%. Although there can be variations in the oil content of copra depending on the region, the Indian Standards classify copra into two main types: edible and milling copra. Edible copra is frequently used as an ingredient in a variety of processed foods and in sweetmeat dishes. Usually, it comes in two varieties: cup copra and ball copra.

14.8.2 Tea

14.8.2.1 Introduction

Tea ranks as the world's second most popular drink, surpassed only by water. Based on recent figures, worldwide tea production reaches over USD 17 billion each year, whereas the international tea market is estimated to be worth around USD 9.5 billion. Consequently, tea cultivation and consumption are prevalent, especially in nations like China and India. However, driven by globalization and consumer demand, nations like the United States and Europe also have significant markets and suitable growing conditions for tea (FAO 2022; Pan et al. 2022a, b; Clarke et al. 2023). Tea is primarily categorized into five main types: white, green, oolong, black, and pu-erh (in Chinese tea culture). Currently, black tea (fermented tea) and green tea (non-fermented tea) are the most widely consumed varieties. In Western countries, black tea dominates, accounting for over 90% of total tea sales. Some of the most renowned black teas in the world include Qi Men black tea from China, Darjeeling and Assam black teas from India, and Uva black tea from Sri Lanka. On the other hand, all ten of the most well-known green teas are produced in China, with Xi Hu Long Jing being the most recognized green tea globally. A significant amount of scientific research has been directed towards green tea and its primary key components, particularly flavan-3-ols, also known as catechins (Pan et al. 2022a, b; Da Silva Pinto 2013).

14.8.2.2 Origin

In 1915, the Dutch botanist Combertus Pieter identified the Eastern Himalayan region as the origin of tea, referring to the species as *Camellia theifera* (Van Driem 2020). Modern studies have used biochemical analysis to trace tea's origins and explore factors that influence its chemical composition. China is often acknowledged for making tea popular, but the tea plant, Camellia sinensis, actually originates from Southern China, North India, Myanmar, and Cambodia. One well-known variety of this species is Assam tea (*C. Assamica* Kitamura) from India. The traditional method involves using the dried leaves and buds of the plant to make tea, which are steeped in boiling water to create the beverage (Hicks 2001). In modern times, tea cultivation extends as far north as 43°N (Georgia, USSR) and as far south as 27°S (Corrientes, Argentina). However, most tea-growing regions, such as China, Japan, Formosa, Java, Sumatra, Sri Lanka, and India, fall within a narrower band from 6°S in Java to 35°N in Japan and between 80°E and 140°E longitude.

14.8.2.3 Distribution

Tea continues to be a vital cash crop worldwide, playing a crucial role in rural development, reducing poverty, and enhancing food security in numerous nations. This crop is cultivated in 58 nations spanning five continents, with a significant proportion of its production concentrated in Asia and Africa. In 2015, the amount of land used for growing tea was 4.35 million hectares, resulting in a yearly production of 5.32 million tons (Indian Tea Club 2016).

14.8.2.4 Production

Global tea cultivation has increased by about 36% in the last 10 years. This remarkable increase was largely driven by China, which saw an 83.96% rise in tea production in 2020 compared to 2011. Bangladesh also experienced significant growth, with tea production increasing by 31.08% during the same period.

14.8.2.5 Varieties of Tea Plants

Tea comes in many varieties, each distinguished by its unique aroma, taste, color, and brewing qualities. In India, there are numerous varieties and cultivars of tea, but three major types stand out: *Camellia sinensis* var. *sinensis* (small-leaf), *Camellia sinensis* var. *cambodi* (medium-leaf), and *Camellia sinensis* var. *assamica* (large-leaf). The *sinensis* variety, whose Latin name means "Chinese," is the small-leaf cultivar commonly grown in China, Taiwan, and Japan. This variety is used to produce green teas, oolong teas, and certain black teas, and the plants grow as shrubs. The tea made from *sinensis* leaves is known for its delicate, aromatic flavors. The *cambodi* variety (*C. sinensis* var. *parvifolia*) features a small tree with several stems, producing leaves which fall in-between the size of Assam and Chinese tea leaves. Often described as a cross between the Assam and China tea plants, this variety is not widely used. The most widely consumed tea variety is *Assamica*, which is primarily grown in India's Assam state. This tea, native to Northeast India, Myanmar, Vietnam, and southern China, has a strong malty flavor and a bright color. Assam tea is slightly stimulating due to its caffeine content, which ranges from 60 to 112 mg per 240 ml serving (Hrishikesh Lanjekar 2022).

14.8.2.6 Characteristics of Tea

The freshness of tea leaves and its polyphenolic compounds plays a significant role in determining the strength of tea. Several factors contribute to tea quality, including standard plucking techniques, the appropriate management of fresh foliage, and the production methodology. The International Organization for Standardization (ISO) delineates two methodologies for assessing the quality of tea: **B**iochemical testing, which analyzes the chemical composition, and organoleptic testing, which assesses taste, aroma, and other sensory characteristics (Table 14.3).

Table 14.3 ISO requirements for black tea

Characteristics	ISO requirements
Water extract (%) minimum	32.0
Total Ash (%)	4.0–8.0
Water-soluble ash as percentage of total ash (minimum)	45.0
Alkalinity of water-soluble ash (as KOH) %	1.0–3.0
Acid insoluble ash (%)	1.0
Crude Fiber (%)	16.5
Total Polyphenols (%)	9

14.8.2.7 Composition

Freshly prepared tea contains an appropriate quantity of caffeine, along with essential oils, tannins, and a set of water-soluble vitamins. Aroma of tea mostly comes from its essential oils, whereas its bitterness and color are due to tannins. Tea is a low-energy beverage, with a cup containing as few as four calories. However, when sugar and milk are added, this rises to approximately 40 cal. per cup. Consuming six cups of tea with sugar and milk provides about 240 calories and contributes 10% of the Recommended Daily Allowance (RDA) for B-complex vitamins (Hicks 2001).

14.8.2.8 Processing

Tea harvesting occurs in cycles, with plucking typically done every 10 days at elevations of 0–1200 meters, and every 3 weeks at altitudes above 1200 meters. Tea can be "fine plucked," which involves selecting only the flush (the top two leaves and the bud), or "coarse plucked," where a sprig with more than two leaves is collected. Each picker gathers between 20 and 30 kilograms of foliage, which are then weighed and deposited in wicker baskets before being sent to the factory.

The typical procedures involved in tea production are as outlined below.

- *Withering*: Freshly harvested foliage are laid out in order to lower their moisture levels. The foliage becomes softer as a result of this process, increasing their adaptability for subsequent processing steps.
- *Rolling*: The leaves are crushed by using a roller to rupture their cell walls, which releases vital oils and initiates oxidation process. Rolling also shapes the leaves and helps release flavor compounds.
- *Roll-breaking (Green Leaf Sifting)*: The rolled leaves are broken up and sifted to separate smaller particles from whole leaves, ensuring uniform oxidation and processing.
- *Fermentation (Oxidation)*: During this stage, the leaves are exposed to air, allowing them to oxidize. The extent of oxidation determines the type of tea produced. For example, black tea undergoes full oxidation, whereas green tea is minimally oxidized.
- *Drying (Firing)*: In this stage, leaves are subjected to heat treatment in order to inhibit oxidation process, remove any remaining moisture, and preserve the tea. This process assures that the tea persists stable and contributes to the development of the intended flavor profile.
- *Sorting (Grading)*: The processed tea leaves are sorted and graded based on their size and quality. Different grades influence both the flavor and price of the tea.
- *Packing for Sale*: Finally, the processed tea is packaged and labeled according to its quality and destination, preparing it for sale or shipment (Fig. 14.1).

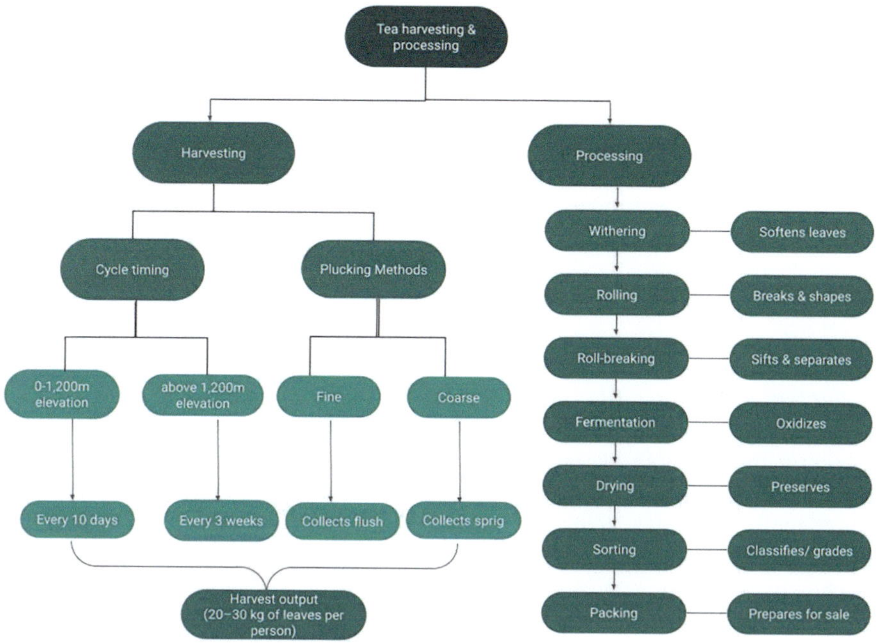

Fig. 14.1 Harvesting and processing of Tea

14.8.2.9 Types of Tea and its Processing Methods

14.8.2.9.1 White Tea

Production and Processing White tea undergoes minimal processing compared to other types of tea. In the production of tea, it is customary to select primarily the buds or the initial leaves of the plant. After being picked, these buds and leaves are primarily subjected to a prolonged withering process lasting between 36 and 48 h, followed by immediate drying without any heat fixation. This extended withering allows for slight fermentation, as oxidation begins during this stage (Tan et al. 2017; Chen et al. 2020).

Chemical Changes The withering process is the primary driving force behind chemical transformation. During withering, cell membranes break down gradually, enabling interactions between catechins and oxidative enzymes like polyphenol oxidase (PPO) and peroxidase. As a result, catechins might undergo oxidation and create dimeric compounds. Although drying through heat application may affect some compounds, including catechins, its impact on the phytochemical profile is expected to be minimal compared to the changes that occur during withering (Tan et al. 2017; Chen et al. 2020).

Characteristics White tea is named for the delicate white hairs that adorn the buds during the harvesting process. Its infusion produces a light yellow color, reminiscent to specific types of green teas. The taste is characterized by mild umami and slight sweetness, accompanied by a fresh, green aroma. The rich taste of these foods is due to the abundance of peptides and soluble sugars, which enhance both savory and sweet flavors. Additionally, the reduction in catechins and polyphenols during processing diminishes the astringency and bitterness typically associated with tea (Hilal 2017; Chen et al. 2020).

14.8.2.9.2 Green Tea

Production and Processing Green tea processing involves heat treatment to deactivate oxidative enzymes, including polyphenol oxidase and peroxidase, that are inherently found in the tea leaves. Once the leaves are heated, they are rolled or shaped, releasing plant juices that subsequently cover the surface of the leaves. Subsequently, the leaves undergo a drying process to decrease their moisture content, resulting in the final dried tea product. Japanese green tea, exemplified by Sencha, is typically subjected to steaming at approximately 110 °C, whereas Chinese green tea undergoes a pan-firing process at temperatures between 300 °C and 350 °C. Additionally, pan-firing is sometimes used for premium Japanese teas like Kamari-cha. The rolling techniques also differ: Chinese green teas are typically rolled in one direction, while Japanese teas are rolled bidirectionally. In some cases, the withering duration for Chinese-style green tea may vary depending on processing facility capacities, especially when handling large quantities of fresh leaves (Hilal 2017; Ye et al. 2018).

Chemical Changes The catechin content in green tea remains relatively stable during processing because the oxidative enzymes responsible for catechin oxidation are inactivated, preventing fermentation. Consequently, catechins may constitute as much as 30% of the dry weight of green tea, which is comparable to the composition of fresh leaves. However, catechin levels may decrease if withering occurs and the PPO enzyme is not yet deactivated, allowing oxidation to begin. Additionally, high-temperature pan-firing and drying can further reduce catechin content. Green tea typically contains eight identifiable catechins, with epigallocatechin gallate (EGCG) being the most abundant. Other catechins, such as gallocatechin gallate (GCG) and catechin gallate (CG), are more commonly found in green tea due to epimerization of EGCG and ECG during pan-firing and drying (Donlao and Ogawa 2019; Svoboda et al. 2015; Turkmen et al. 2009).

Characteristics Infusions of green tea typically exhibit a pale yellow color and are frequently characterized by their astringent and bitter flavor profiles. However, the flavor profile also consists of umami and a mild sweetness, which is mainly because of the amino acid l-theanine. Astringency is directly related to catechin contents, particularly the gallated catechins. A decrease in catechin levels could result in a diminished characteristic flavor of green tea. As a result, green tea is produced in a manner that maintains the levels of catechins present in fresh leaves (Ye et al. 2018; Zhang et al. 2020).

14.8.2.9.3 Oolong Tea

Production and Processing Oolong tea undergoes a unique process that involves sunning (withering), fermenting (also called shaking and setting), panning, rolling, and firing. The leaves are initially laid out in the sun to decrease the amount of fluids in them, which enables catechins to shift within the leaves. Next, during the shaking and setting stage, the leaves are subjected to shaking multiple times periodically in order to induce bruising along their edges. This crushing process disintegrates the cells, blending catechins with oxidases, which initiates fermentation. Oolong tea undergoes partial fermentation, with the level of fermentation varying from 10% to 80% based on the type of tea. The tea master monitors changes in the aroma of the leaves during shaking to determine when to stop the process. The fermentation is then halted through panning, a heat-fixation step that also removes any grassy odors. Subsequently, the leaves undergo a process of rolling and firing, resulting in the production of the final dried tea (Ho et al. 2018; Lin et al. 2016).

Chemical Changes Throughout the fermentation process, the catechins found in oolong tea undergo oxidation, transforming into theaflavins and theasinensins (colorless catechin dimers). However, since oolong is only semi-fermented, the degree of oxidation is much lower than in fully fermented black tea. Lightly fermented oolong teas tend to have more theasinensins, while theaflavins are more common in heavily fermented varieties. Oolong tea also produces unique compounds called oolongtheanins, which are characteristic dimers found in this tea type (Wang et al. 2019; Lin et al. 2016).

Characteristics The coloring pigments in tea leaves may vary from light to dark green or even brown. When brewed, the tea infusion is typically golden, which results from a combination of catechins, theaflavins, thearubigins, and flavones. The flavor profile of this product is characterized by its floral scent and well-developed notes of ripe fruits, sweetness, and a velvety smooth finish (Xu et al. 2018; Chen et al. 2011).

14.8.2.9.4 Black Tea

Production and Processing To minimize moisture content, black tea leaves are first withered at room temperature before processing, which makes them more flexible and easier to roll. During rolling or maceration, the leaves' cells are broken down, allowing catechins to mix with enzymes like polyphenol oxidase (PPO) and peroxidase, starting the fermentation process. This fermentation is critical for black tea, as it fully oxidizes the tea leaves. The rolling method can vary: Orthodox rolling, a more delicate technique, is commonly utilized in regions such as Sri Lanka, while more aggressive methods like the Lawrie Tea Processor (LTP) and crush-tear-curl (CTC) machines are used in regions such as Africa and India. Following fermentation, the leaves undergo a drying process to halt enzymatic activity, ultimately leading to the production of dried black tea (Caffin et al. 2004; Das et al. 2019; Turkmen et al. 2009).

Chemical Changes The primary change that occurs at the time of black tea processing is the alteration of catechins through oxidation, forming theaflavins and thearubigins. Catechin oxidation starts during withering and continues through rolling, where the tea's cell structures are disrupted, allowing catechins and enzymes to interact. Theaflavins are primarily formed during fermentation, creating oxidized catechin dimers like theaflavin 3-gallate and theaflavin 3,3′-gallate. During the continued process of fermentation, the compounds undergo further oxidation, resulting in the formation of thearubigins, which contribute to the tea's color and flavor. Black tea also sees an increase in gallic acid content, produced through enzymatic and microbial processes during fermentation (Techome 2019; Lee et al. 2019; Wang et al. 2019).

Characteristics Black tea leaves turn from green to deep brown during processing, and when brewed, they produce a reddish-brown infusion. Theaflavins give the tea its brisk, astringent taste and golden color, while thearubigins contribute to a rich mouthfeel and a darker, brownish-red color (Hilal 2017; Ho et al. 2018).

14.8.2.9.5 Pu-erh Tea (Dark Tea)

Production and Processing Pu-erh tea, a type of dark tea, is distinct due to its additional microbial fermentation step. The production process occurs in two stages. First, sun-dried green tea is produced from the larger-leafed variety of the tea plant, *C. sinensis var. assamica*. The leaves are withered, pan-fried to reduce moisture, and then sun-dried. The second stage involves microbial fermentation, which can be done naturally or through inoculation with specific microorganisms like *Aspergillus* species to speed up the process. Modern methods favor inoculation, reducing fermentation time to just days or hours. The fermented tea is then dried to produce the final loose or compressed form of pu-erh tea (Ho et al. 2018; Lv et al. 2013; Qin et al. 2012).

Chemical Changes Microbial fermentation causes significant changes to the chemical content of pu-erh tea. Catechins degrade and oxidize into theabrownins, dark brown pigments considered the primary bioactive components of the tea. Other notable compounds, such as puerins and lovastatin, have also been detected, though it remains unclear if these form during post-fermentation (Zhang et al. 2019a).

Characteristics Pu-erh tea has a reddish to brownish-red color, thanks to its high theabrownin content. The flavor is full-bodied, complex, and mellow, with a stale taste and silky-smooth mouthfeel. These characteristics are likely due to the high theabrownin levels and the reduced polyphenol content (Lee and Foo 2013).

14.8.2.9.6 Yellow Tea

Production and Processing Yellow tea, also known as huángchá, is a delicately oxidized tea that is distinctively Chinese in origin. Its production process is similar to that of green tea, with steps including fixing, rolling, yellowing, and drying. The yellowing process is what sets this tea apart, giving it its distinctive color and flavor (Yuming et al. 2021).

Characteristics Yellow tea is known for its high levels of theanine, glutamic and aspartic acids, and flavonoids. Compared to other teas like green and white, yellow tea contains the highest flavonoid content, along with phenolic compounds, amino acids, soluble sugars, and vitamins, making it both flavorful and nutritious (Jiao et al. 2016; Zhang et al. 2019b) (Fig. 14.2).

14.8.2.10 Tea and Its Products

14.8.2.10.1 Dark Tea

Dark tea distinguishes itself compared to other tea varieties because of its unique processing methods, which involves solid-state fermentation facilitated by microorganisms. These microorganisms transform the tea leaves by performing a range of chemical reactions, including degradation, oxidation, condensation, methylation, and glycosylation. As a result, the tea's phytochemicals are significantly modified. Throughout history, dark tea has been commonly employed for addressing digestive problems (Li et al. 2018; Zhu et al. 2020).

14.8.2.10.2 Loose Tea

Loose tea refers to tea leaves that are sold without being prepackaged in teabags. Unlike bagged tea, loose leaf tea preserves its complete flavor profile, aromatic qualities, and health advantages because the leaves are not crushed or confined to a tea bag. Loose tea is typically packaged in canisters, paper bags, or containers like tea chests. This type of tea is available worldwide and loose leaf tea is considered to be of higher taste and quality than bagged tea. Additionally, it is also more environmentally sustainable, as it can be composted easily, and does not require the use of teabags (Gebely 2016).

14.8.2.10.3 Tea Bags

A tea bag is a compact, permeable pouch that is tightly sealed and filled with either tea leaves or additional spices. The tea bag is placed into water to steep and create a flavorful infusion. Tea bags are typically crafted from filter paper or food-safe

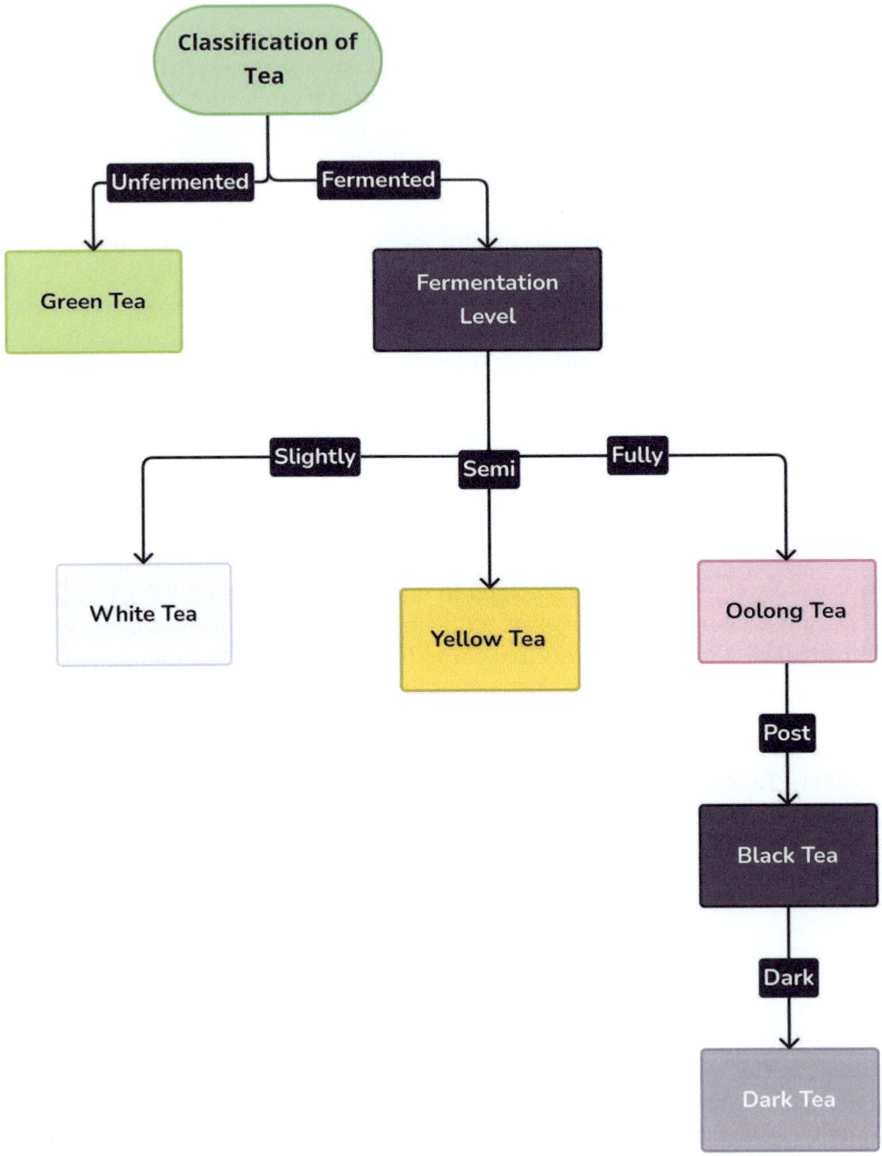

Fig. 14.2 General classification of Tea

plastic. They serve a similar function as tea infusers. Tea bags can be utilized for multiple brews before their flavor diminishes. A typical feature of tea bags is an attached string with a paper label, facilitating easy removal from the cup and providing branding information (Hernandez et al. 2019).

14.8.2.10.4 Handmade Tea

Handmade or handcrafted tea refers to tea that is produced in smaller quantities with meticulous attention to detail, often at the garden level, and marketed exclusively. Sometimes referred to as "artisanal tea," handmade tea is not mass-produced in factories using large machinery. Instead, it is harvested by hand or machine, with most or all processing steps done manually, usually in small batches. Handmade teas can be fully handcrafted (where all steps are done by hand) or partially handcrafted (where most steps are done by hand). Popular handmade teas include varieties from China, Japan, and India, such as Silver Needle white tea, blooming teas, premium oolongs like Wuyi yancha, Assam orthodox black and green teas, and Singpho or Darjeeling designer teas shaped into flowers, balls, or other forms (Baruah et al. 2005).

14.8.2.10.5 Tea Snacks and Confectionery

Flour-based confectionery products have long been popular with consumers, and recent innovations have incorporated tea extracts or tea powders to enhance these products with plant-based components rich in micronutrients. These ingredients are not only easily digestible but also add health benefits. For example, research has shown that adding green tea infusion to bread recipes can slow the staling process (Edygova 2018). When green tea powder is added to ice cream, it enhances both the flavor and nutritional profile of the dessert (Swarnathilake and Don 2020). Moreover, the antioxidant properties of cakes made by embedding green tea have been acknowledged as superior when compared to regular cakes (Lu et al. 2010).

14.8.2.10.6 Tea Staple Foods

Tea leaves or extracts serve as the primary component in a variety of common foods. Popular dishes include green tea noodles (Xu et al. 2019), green tea rice, and tea dumplings. One popular dish often cited is Ochazuke, a classic Japanese meal consisting of steamed rice paired with flavorful green tea. "It is a perfect choice for warm summer days due to its refreshing yet filling nature." One more example is Cha Soba, a Japanese noodle concoction blended with green tea, crafted from a combination of buckwheat flour and wheat flour.

14.8.2.10.7 Canned Tea

Canned tea represents a modern approach to tea marketing, in contrast to the traditional loose leaf and bagged varieties. This method involves canning tea as a ready-to-drink product, which offers the convenience of minimal preparation. The initial canned tea product, an unsweetened oolong tea, was launched in Japan by the beverage corporation Ito En. Following its success, a wide range of canned tea products became available in the market.

14.8.2.10.8 Bottled Tea

Bottled tea is now easily accessible in nations including Japan, China, India, Korea, and Sri Lanka. An example of this would be bottled drinks containing catechins, which are infused with additional beneficial compounds like β-carotene and vitamins. These beverages have shown significant potential for neutralizing harmful free radicals due to the strong antioxidant properties of catechins (Kuroda and Hara 2004).

14.8.2.10.9 Instant Tea

Instant tea powder quickly reconstitutes into a cup of tea when mixed with water. The United kingdom was the birthplace of instant tea in 1885. The tea paste was highly concentrated and transformed into a refreshing beverage with the addition of hot water. India and Sri Lanka stood out as the top producers of instant tea in the year 2020, with exports totaling over 2893 million kilograms (ITC 2021).

14.8.2.10.10 Carbonated Tea

Carbonated tea, while available in some foreign markets, has a limited demand. The potential market for bottled carbonated tea in Sri Lanka is limited due to the elevated costs associated with bottling, transportation, and distribution, ultimately deterring consumers from purchasing this product.

14.8.2.10.11 Tea Wine

Tea wine is an innovative beverage developed by the Institute of Himalayan Bioresource Technology (IHBT). It is produced through a conventional winemaking process that involves fermenting secondary-grade black tea dust with sugar using the yeast *Saccharomyces cerevisiae*. After fermenting, the tea wine is matured for one year and typically contains 12–15% alcohol. Two main varieties have been developed: the "pure tea wine" and "Palam Belle," the latter being a mild, sweet fruit-infused tea wine.

14.8.2.10.12 Tea Cider

Tea cider is an alcoholic beverage produced through the fermentation of sugar with a blend of microorganisms. The growth of these organisms is fueled by the amino acids and vitamins derived from tea liquor. The resulting tea cider is comparable to traditional apple cider and champagne perry, offering a refreshing alternative.

14.8.2.10.13 Blended Tea

Tea blends contain the combinations of various tea leaves to create a final product, a common practice primarily with black teas, which are often blended for use in tea bags. Blending may also take place with various tea varieties, where leaves from various regions are combined prior to compression. Popular blends include breakfast teas, afternoon teas, and Russian Caravan tea (Smith 2016).

14.8.2.10.14 Altered Tea

"Modified teas go through supplementary processing either during or after their initial processing prior to being made available for sale." These modifications can include flavoring, scenting, blending, smoking, aging, decaffeinating, or grinding. The majority of teas available for sale in the United States undergo modifications to improve or change their flavor and characteristics.

14.8.2.10.15 Roasted Tea

Tea roasting is a prevalent practice in the production of oolongs from China and Taiwan, as well as in the creation of Japanese Hojicha. While modern roasting typically uses electric ovens, traditional roasting methods involve bamboo baskets placed over charcoal. During roasting, the amino acid in the tea leaves undergo the Maillard reaction, which produces aromatic compounds, giving the tea a toasty, nutty flavor. Roasting plays a crucial role in the aging of oolongs by aiding in the reduction of moisture content acquired during the aging period.

14.8.2.10.16 Decaffeinated Tea

All teas derived from *Camellia sinensis* naturally contain caffeine, although decaffeinated teas retain a small residual amount. The decaffeination process uses solvents to extract caffeine from tea leaves. Historically, methylene chloride was the primary solvent, but due to health concerns, its use has been banned in the United States. Today, common solvents for decaffeination include ethyl acetate and supercritical carbon dioxide (Chin et al. 2008).

14.8.2.10.17 Flowering Tea

Flowering tea is a visually appealing variety in which dried tea leaves are carefully wrapped around one or more dried flowers, forming a compact bundle. Once steeped, the bundle unfurls, mimicking the blooming of a flower with the flowers serving as the centerpiece. Popular flowers used in flowering teas include globe amaranth, chrysanthemum, jasmine, lily, hibiscus, and osmanthus (Richardson 2009).

14.8.2.10.18 Bubble Tea

Bubble tea, also known as pearl milk tea, boba tea, or tapioca milk tea, originated in Taiwan in the early 1980s. This popular tea-based beverage typically includes tea with chewy tapioca pearls or other toppings. Over time, bubble tea has gained international popularity, with variations including different tea bases, flavors, and toppings (Wu 2020).

14.8.2.10.19 Flavored and Scented Teas

Flavored and scented teas are often infused with flowers, herbs, spices, or smoke to create distinctive aromas and flavors. More specialized flavored teas may involve the use of added flavorings or perfumes, especially for blends with pronounced fruit or floral notes. Certain companies, like Mariage Frères and Kusmi Tea, are well known for their perfumed teas. Jasmine is commonly used to scent delicate white and green teas, while bergamot oil is the hallmark of Earl Grey tea.

14.8.3 Coffee

14.8.3.1 Introduction

The coffee plant belongs to the genus *Coffea* and the family *Rubiaceae*, is a woody perennial tree that thrives at higher altitudes. Coffee has been consumed for over a millennium and remains the most popular beverage globally, with more than 400,000,000 glasses being used each year. The spread of coffee culture is largely attributed to the Arabian Peninsula. Caffeine, the most noticeable of coffee's several phytoconstituents, is the main reason it is widely used for its stimulating qualities. As one of the highest concentrations of caffeine, with 240 ml of instant coffee containing roughly 100 milligrams of caffeine. While coffee's historical roots extend far back, its introduction as a cash crop dates to the fifteenth century (Belay et al. 2008). Debate continues on whether coffee benefits or harms human health. On the other hand, consuming it has been associated with a significant decrease in chronic illnesses like diabetes, Parkinson's disease, and some types of cancer.

14.8.3.2 Origin

The earliest documented references to coffee culture date back to 575 AD in Yemen. However, it was not until the sixteenth century in Persia that coffee beans were roasted and brewed into the beverage we recognize today. Coffee was first introduced to Europe in 1615 by travelers, leading to its rapid spread. Europeans, including Germans, French, and Italians, aimed to cultivate coffee within their territories, but the Dutch were the first to be successful. The initial seedlings were grown in the Amsterdam botanical garden's stoves, which contributed to coffee becoming one of the most consumed beverages in Europe. The French later received coffee plants from the mayor of Amsterdam, leading to cultivation in the Sandwich Islands and Bourbon Island. These Dutch and French efforts laid the foundation for coffee cultivation in other European colonies. European demand for coffee fueled the expansion of coffee plantations in Africa, and European colonists later introduced coffee to regions such as Puerto Rico, Suriname, São Domingos, Cuba, and the Guianas. From the Guianas, coffee eventually reached northern Brazil, spreading the once-guarded secret of coffee cultivation worldwide.

14.8.3.3 Distribution

The majority of coffee plants are grown in tropical and subtropical areas with temperate and humid temperatures, including as Africa, Central and South America, and Southeast Asia provide optimal growing conditions (Schenker 2000). The global consumption of coffee continues to rise, with increasing sales particularly evident in developing countries like India and Pakistan. With a combined yearly production of almost 2.5 million tons, Brazil continues to be the world's top producer and green coffee bean exporter, with Vietnam, Colombia, Indonesia, Ethiopia, and India (Franca and Oliveira 2009).

14.8.3.4 Production

Coffee is produced extensively in about 60 tropical and subtropical countries, where it serves as a primary agricultural export for many. India ranks as the seventh-largest coffee producer globally, with most production occurring in its southern states, notably the Monsooned Malabar variety. Coffee is a key plantation crop in India, covering 415,341 hectares, with Karnataka accounting for 55.5% of production, followed by Kerala (20.6%), Tamil Nadu (7.6%), and smaller regions such as coastal Andhra Pradesh, Orissa, and the northeastern states (14.9%). India exports the majority of its coffee, accounting for 4.55% of the worldwide coffee market. Indian coffee exports reach over 40 countries, with the largest markets being the Russian Federation, and European Union nations such as Italy, Germany, Belgium, Spain, Finland, and Greece, as well as the United States, which together account for 70% of India's coffee exports. Specialty coffee varieties exported from India include Malabar Monsoon Coffee, Mysore Nuggets Extra Bold, and Robusta Kaapi Royale. The major export ports are Cochin, Chennai, and Tuticorin. In the past few years, India has brought in limited quantities of coffee from Indonesia and Vietnam, while domestic consumption is largely concentrated in Tamil Nadu and Karnataka (Table 14.4).

Table 14.4 Country-wise global production of Arabica and Robusta coffee in the year 2022

Country	Arabica coffee production[a]	Robusta coffee production[a]
Brazil	43,406	21,382
Vietnam	1461	27,760
Colombia	12,384	–
Ethiopia	7932	–
Honduras	6047	–
Peru	3993	–
Mexico	3962	209
Guatemala	3720	–
Nicaragua	2895	–
Indonesia	2517	9467
India	1940	4617

[a]*Are in thousand bags where 1 bag = 60Kg*

14.8.3.5 Varieties

Coffea L., a plant that contains more than 70 species, is the source of coffee beans. Globally, only two types of coffee plants are grown for commercial purposes which are *Coffea arabica*, considered the best coffee type and making up 75% of worldwide production, and *Coffea Robusta,* which is characterized by high acidity and increased resistance to pests and accounts for a quarter of global production (Belitz et al. 2009; Etienne 2005).

14.8.3.6 Harvesting of Coffee Cherries

Coffee cherries are harvested when they exhibit a bright red, glossy appearance and a firm texture. The ways in which crops are collected can involve carefully choosing individual fruits by hand, removing entire branches without discrimination, or using machines to gather the produce. Although the hand-picking method requires a lot of labor and takes considerable time, it results in a superior quality product. Mature coffee cherries usually appear 9–11 months after the flowering stage. The coffee fruit known as cherry has an outer skin called the pericarp that is smooth and durable. When the fruit ripens, the pericarp changes from its typical green color to a deep red. Beneath the pericarp is a yellowish fibrous flesh called the outer mesocarp, which has a sweet taste. As the fruit ripens, a thin, hydrated, slimy layer known as mucilage (or the pectin layer) forms. This is followed by a thin, yellowish endocarp, commonly referred to as parchment. There is a thin membrane called the silver skin that surrounds the coffee beans and serves as a protective seed coat. Normally, every cherry contains two seeds. However, if there is only one seed, it will have a rounder shape and is called a peaberry (Belitz et al. 2009) (Fig. 14.3).

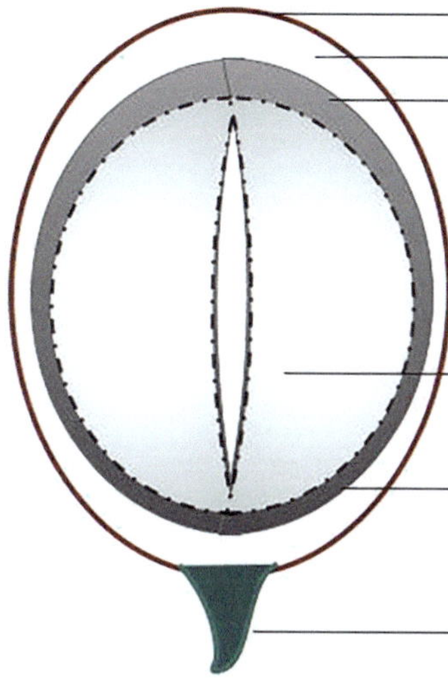

Fig. 14.3 Cross-section of a coffee cherry

14.8.3.7 Chemical Composition and Nutritional Value of Coffee

Green coffee beans mainly consist of cellulose and hemicellulose, making up around 50% of their total weight. They also contain various soluble carbohydrates, including the monosaccharides fructose, glucose, galactose, and arabinose, as well as oligosaccharides like sucrose, raffinose, and stachyose, and polymers made up of galactose, mannose, arabinose, and glucose. Soluble sugars are important for aroma retention, foam stabilization, sedimentation improvement, and viscosity enhancement in coffee extract. Additionally, both non-volatile aliphatic acids (including citric, malic, and quinic acids) and volatile acids (such as acetic, propanoic, butanoic, isovaleric, and hexanoic acids) are present.

Oils and waxes are a significant element of coffee, comprising approximately 10–20% of the dry weight, in addition to proteins and amino acids, which comprise 9–12% of the weight, and minerals, accounting for 3–5%. Coffee beans also contain other significant constituents, comprising cellulose, minerals, sugars, fats, and polyphenolic compounds. The mineral content includes potassium, magnesium, calcium, sodium, iron, manganese, rubidium, zinc, copper, strontium, chromium, vanadium, barium, nickel, cobalt, lead, molybdenum, titanium, and cadmium. Among the sugars present are sucrose, glucose, fructose, arabinose, galactose, and mannose. Coffee beans are rich in several amino acids, including alanine, arginine, asparagine, cysteine, glutamic acid, glycine, histidine, isoleucine, leucine, lysine, methionine, phenylalanine, proline, serine, threonine, tyrosine, and valine. Furthermore, they contain B-complex vitamins, particularly niacin (vitamin B3 or PP), and chlorogenic acid, with proportions ranging from 7% to 12%, which is three to five times higher than that of caffeine (Arya and Rao 2007; Belitz et al. 2009; Grembecka et al. 2007) (Table 14.5).

Table 14.5 Basic characteristics and chemical composition of Arabica and Robusta

Characteristics/ component	Arabica	Robusta
Date species described	1753	1895
Time from flower to ripe cherry	9 months	10–11 months
Ripe cherries	Fall	Stay
Yield (kg beans/ha)	1500–2000	2300–4000
Optimum temperature (yearly average)	15–24 °C	24–30 °C
Optimal rainfall	1500–2000 mm	2000–3000 mm
Optimum altitude	1000–2000 m	0–700 m
Caffeine content of beans	0.8–1.4%	1.7–4%
Shape of bean	Flat	Oval
Polysaccharides[a]	49.8	54.4
Sucrose[a]	8.0	4.0
Reducing sugars[a]	0.1	0.4
Lipids[a]	16.2	10.0
Oil	7.7–17.7	7.7–17.7
Wax	0.2–0.3	0.2–0.3
Lignin	1–3	1–3
Non-volatile aliphatic acids	2–2.9	1.3–2.2
Proteins[a]	9.8	9.5
Amino acids[a]	0.5	0.8
N compounds	11–15	11–15
Aliphatic acids[a]	1.1	1.2
Quinic acids[a]	0.4	0.4
Chlorogenic acids[a]	6.5	10.0
Caffeine[a]	1.2	2.2
Trigonelline[a]	1.0	0.7
Minerals (as oxide ash)[a]	4.2	4.4
Volatile aroma[a]	traces	traces
Water[a]	8–12	8–12

[a]Chemical composition of green arabica and green robusta coffee beans are expressed in (g/100 g)

14.8.3.8 Processing of Green Coffee

Green coffee can be produced using two methods, dry or wet, both with the goal of extracting the coffee beans from the fruit pulp around them. The dry process, being simpler and more cost-effective, involves drying whole coffee cherries in the sun. The green coffee beans are revealed by removing the outer layers, which include the pulp and parchment, after drying. On the other hand, the wet processing method is more intricate and necessitates a larger financial investment. However, it yields higher quality coffee. Here, the outer pulp, composed of the exocarp and mesocarp, is mechanically separated, while the parchment layer remains intact on the beans. The beans are subsequently subjected to a drying process, which may occur either

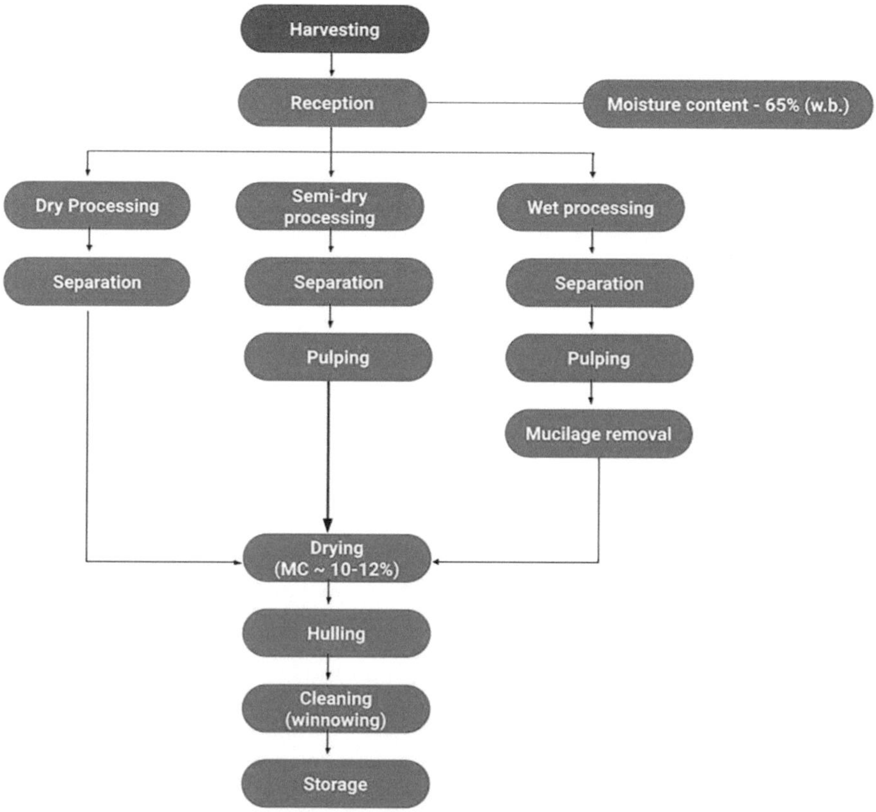

Fig. 14.4 General steps in coffee processing

through natural sunlight exposure or via the use of mechanical drying equipment. The beans are subsequently subjected to a drying process, which may occur either through natural sunlight exposure or via the use of mechanical drying equipment (Illy and Viani 1995) (Fig. 14.4).

14.8.3.9 Dry Processing

Dry processing is a commonly used, simplified technique for Robusta production in Brazil and for Arabica production in Sri Lanka. The process involves drying coffee cherries in open yards until their moisture content reaches around 12% (wet basis). A critical aspect of this process is "racking," which involves frequently turning the cherries to ensure uniform drying and enhance overall quality of the end product.

14.8.3.9.1 Drying

Immediately following the harvest, the coffee cherries are usually dried by exposure to sunlight on clean, dry surfaces or mats. It is important that the depth of the cherries does not exceed 40 mm, and it is essential to rake them regularly to prevent fermentation or discoloration.

However, this method presents several challenges. A significant concern is the contamination from dust and dirt. Additionally, sudden rain storms can occur, even during the dry season, soaking the cherries rapidly. Moreover, employing labor is necessary to mitigate risks of damage or theft. Due to these issues, avoiding sun drying is typically advised.

14.8.3.9.2 Hulling

Once the cherries have been properly dried, they undergo hulling to eliminate pericarp. This can be accomplished either manually, using a pestle and mortar, or through mechanical hullers. Steel screw hullers usually have a design where the pitch of the screw gradually increases towards the outlet, effectively eliminating the pericarp.

14.8.3.9.3 Cleaning

After removing the outer hull, the coffee goes through a cleaning process called winnowing, where any defective beans are removed. Sorting will be done in producing as well as manufacturing nations to assure product standards. This can be achieved mechanically or optically. During the mechanical process, faulty beans are carefully selected by hand and then put into air classifiers (catadors) to undergo separation through an adjustable upward air current.

14.8.3.10 Wet Processing

In the wet processing method, the cherries are crushed using a pulping machine or pestle and mortar, removing the outer fleshy layers (mesocarp and exocarp) and leaving the beans coated in mucilage. The mucilage undergoes fermentation and is subsequently washed and dried.

14.8.3.10.1 Pulping

Pulping involves removing the outer red skin (exocarp) and the white fleshy pulp (mesocarp) while separating the pulp from the beans. Immature cherries, which are hard and green, are particularly difficult to pulp, making correct harvesting crucial for wet processing. For small-scale operations, cherries can be pulped using a pestle and mortar, although this method is labor-intensive. The most common pulpers suited for small-scale units are drum and disc pulpers.

- *Drum Pulpers*: This system employs a rotating drum with a punched sheet surface and an adjustable breastplate, allowing for the pulping of coffee cherries while separating the pulp from the beans. The spacing between the drum and breastplate must be carefully adjusted to remove the pulp without damaging the beans. Drum pulpers can be manually operated or powered by treadles or bicycles, with motorized options available for larger-scale operations.
- *Disc Pulpers*: Similar to drum pulpers, disc pulpers use a disc with a roughened surface instead of a breastplate. The principle of operation remains the same: the cherries are pressed between the disc and a surface to facilitate pulping.

14.8.3.10.2 Mucilage Removal

The fermentation of the mucilage occurs in large tanks over 24–40 h. The two-stage "dry" fermentation process is particularly effective for achieving high-quality results while maintaining an efficient factory routine. In the first stage, mucilage degradation occurs, followed by soaking in water for 24–48 h in the second stage. The mucilage surrounding the beans consists of hemicelluloses, pectic substances, and sugars, which are insoluble in water. While chemical methods, warm water, or an "aqua pulper" can be employed for mucilage removal, fermentation remains the most practical approach for small-scale units. This involves placing the beans in plastic buckets or tanks until the mucilage breaks down. Natural enzymes and bacteria in the environment facilitate this process. The beans should be stirred occasionally, and a sample should be washed to assess readiness. When the mucilage can be easily washed off and the beans feel gritty rather than slippery, they are ready for the next stage. Timely washing is crucial to prevent the development of undesirable flavors. After fermentation, the coffee is referred to as "parchment coffee," as the seed retains its endocarp layer. It must be dried to approximately 10–12% (wet basis) moisture content to ensure stability.

14.8.3.10.3 Drying (Wet Process)

To prevent cracking, the coffee beans should be dried gradually to reach 10% moisture content (wet basis). This drying should occur immediately after processing to avoid off-flavors. Similar drying methods can be employed as those used for dry-processed coffee.

14.8.3.10.4 Hulling (Wet Process)

Following drying, the coffee should rest for 8 h in a well-ventilated area. The thin parchment surrounding the beans can then be removed manually, using a pestle and mortar, or in a small huller.

14.8.3.10.5 Cleaning (Wet Process)

The hulled coffee undergoes a final cleaning through winnowing.

14.8.3.11 Drying Process in Coffee Production

Drying is a critical step in coffee processing that directly influences the quality of the final product. Traditionally, natural sun drying on terraces is widely used by coffee producers. While effective, this method is labor-intensive, time-consuming, and dependent on favorable weather conditions. As coffee production increases, reliance on sun drying alone presents challenges, prompting the need for mechanical drying. Mechanical drying can streamline the harvesting process, improve both the quality and quantity of coffee, and free up usable space for other activities. The primary goal of drying is to reduce the moisture content of the coffee beans to approximately 12% (wet basis). Proper drying helps slow down the respiration rate of the beans, extending their shelf life while minimizing quality loss. However, over-drying can cause the beans to become brittle, while inadequate drying can result in mold growth. Achieving even and consistent drying is crucial for maintaining the beans' color, size, and pest resistance during storage, which can last up to 8 months. Mathematical simulations based on thin-layer drying principles have been employed to design and optimize drying systems for commercial use (Berbert et al. 1995).

- *Sun Drying (Natural Drying)*: Sun drying remains a popular method, especially in areas where mechanical drying may not be feasible. Usually, the coffee cherries are arranged in a layer that is 4–6 centimeters thick, requiring approximately 17,000 kJ/kg of energy. Removing the fruit tissue during the pulping step in wet processing can significantly reduce this energy requirement. However, choosing the right drying method is a complex decision that varies by region and goes beyond just energy efficiency. Although sun-dried coffee was once considered superior, studies now show that both sun-dried and mechanically dried coffee can achieve comparable quality. Patios with a small slope for drainage that are composed of concrete, asphalt, or tiles are commonly used for sun drying. Depending on weather conditions, drying parchment coffee on patios can take 7–15 days, while cherries may take 12–21 days. Parchment coffee requires more gentle handling to prevent cracking and physical damage. To avoid scorching in tropical countries, parchment is frequently covered during the warmest hours of the day, which can damage the beans. Drying racks, which provide better air circulation and prevent contamination from the ground, can also shorten drying time by allowing wind exposure.

- *Mechanical Drying*: Mechanical drying involves heating the coffee beans using hot air, which simultaneously removes moisture. Careful temperature control is essential to avoid overheating, which can occur in both natural and mechanical drying methods. The optimal drying temperature is typically below 40 °C for parchment and 45 °C for cherries. Various types of mechanical dryers are used, including column dryers, rotary dryers, cascade dryers, vertical dryers, and tray dryers with stirrers. These dryers are powered by fuels such as wood, coffee husks, fuel oils, diesel, and gasses. Depending on the type of dryer, the drying process can take anywhere from 20 to 60 h.
- More recent innovations include solar-assisted dryers, which provide an environmentally friendly alternative. Using a vibratory bed drier, Finzer et al. (2007) investigated the drying of coffee berries in Brazil, determining dispersion coefficients that ranged from 2.32×10^{-4} to 76.81×10^{-4} m^2/s. The kind of drying surface (soil, concrete, or asphalt) and the thickness of the drying layers were found to affect coffee quality in a study by Reinato et al. (2012). The best beverage quality was obtained from washed parchment coffee that was thinly layered and dried on concrete and asphalt surfaces.

14.8.3.12 Physicochemical Changes During Drying

Drying plays a crucial role in decreasing the moisture content of coffee beans from 45–50% (wet basis) to 10–12% (wet basis), ensuring the beans are suitable for long-term storage. Moisture content significantly affects the size and dimensions of the beans, leading to size expansion, which is likely due to moisture absorption. For example, lentil seed thickness rises linearly with moisture content, a pattern also observed in pistachio nuts and caper seeds (Dursun and Dursun 2005). In the case of Liberica beans, even as moisture content decreases during storage over an 8-month period, the size of the beans continues to expand, though the exact cause of this expansion is unknown. The beans' brightness is also significantly influenced by their moisture content, with higher moisture levels producing greater brightness and lower moisture levels producing less brightness (Isamil et al. 2012). The retention of volatile flavor compounds is better when beans are dried using a combination of hot air and microwave drying compared to conventional hot air drying alone. In a study conducted in 2012 by Nilnont et al. investigated coffee drying at 40 °C, 50 °C, and 60 °C. They discovered that beans dried at 50 °C and 60 °C had higher overall sensory acceptability. Microwave drying was also found to produce coffee with a good cup quality, characterized by a mild flavor and slight astringency. Sensory analysis indicated that interruptions in the drying process due to moisture content did not significantly affect coffee quality, and overall beverage quality improved as the rest period increased.

14.8.3.13 Grading

After curing, coffee is graded based on various factors, including color, odor, size, shape, and density.

- *Color grading*: A color sorter is used to reject beans that don't match color requirements, such as white, black, opalescent, grayish, glassy, foxy red, or blotchy/spotted beans.
- *Aroma-based grading*: Beans with off-putting odors, such as "stinkers" that emit a putrid smell, rancid or acidic beans, musty beans, and those with a distinct "rio" flavor (an undesirable off-flavor), are removed, as they can result in undrinkable coffee.
- *Other defects*: Other abnormal beans, such as droughted, broken, crushed, pitted, and elephant beans, as well as debris and foreign materials, are also removed throughout the grading procedure.

Coffee is graded into the following categories:

1. *Pea berry*: Oval-shaped beans.
2. *A grade*: Large, flat, bold beans those are dense and firm.
3. *B grade*: Beans of B grade are marginally smaller than those in A grade.
4. *C grade*: Beans smaller than B-grade.
5. *Triage grade*: Beans that are pale, discolored, or have black spots.

14.8.3.14 Commercially Processed Coffee Products

Instant coffee, decaffeinated coffee, and coffee powder are examples of commercially processed coffee products.

14.8.3.14.1 Coffee Powder

Roasting One of the most important processes in deciding the final coffee flavor is roasting. The process of roasting is largely influenced by temperature and time, and it is usually evaluated visually by looking for the browning of a thin white line on either side of the bean. Traditionally, coffee beans are roasted at 200 °C in a pan with sand while being constantly stirred; however, large-scale production uses automated roasters. Roasting conditions significantly influence the final flavor profile of the coffee. The most commonly used roasters in both home and commercial coffee production are fluidized/spouted bed roasters, which expose beans to hot air or gases, and drum roasters, which expose beans to fire or heated surfaces. Many companies like the latter because it can achieve color uniformity and speed up the roasting process by precisely controlling the temperature and velocity of the air/gas within the roaster. Throughout the process, this control makes it possible to modify the heat input and regulate the bean temperature. Industrial fluidized bed roasters usually employ maximum roasting temperatures of 210–230 °C.

The beans lose both free and bound water at the start of the roasting process, and as the temperature rises to around 130 °C, browning and sugar caramelization start, which causes the beans to swell. At this point, chemical reactions are quite small in comparison to those that take place later. As soon as the beans reach 160 °C, several exothermic and endothermic reactions occur, causing the beans to expand significantly and take on a light brown color, while aroma development begins. When the beans reach about 190 °C, the Maillard and Strecker reactions begin, which are the main processes that give roasted coffee its distinctive flavor and aroma. Various low and high molecular weight molecules, including melanoidins, are broken down and formed as a result of these processes, which involve the interaction of proteins, carbohydrates, and other substances. The beans' color changes from pale brown to almost black as the procedure goes on.

When the beans reach the appropriate stage—typically determined by their color or a certain time—the roasting process is stopped. In order to avoid over-roasting, the beans are quickly chilled, usually with cool air. In order to prevent the beans from becoming too wet, which might encourage the growth of microbes, air cooling is favored over water chilling. The beans are ground after roasting and either used to make instant coffee or sold as ground roast coffee. Because different countries have varied tastes, ground coffee comes in a range of roasting degrees, from very light to very black. As an illustration, light to medium roasts are common in the United States and the United Kingdom (Clarke 2003), However, in other parts of Europe, darker roasts are preferred. There are noticeable flavor variations between beans roasted for longer periods of time at lower temperatures and beans roasted for shorter periods of time at higher temperatures because the former contain a different volatile component profile, more soluble solids, and more acidity.

Grinding Two primary types of grinders are used for producing coffee powder: manual grinders and motorized mechanical grinders.

- *Manual Grinders*: Traditional manual grinders, once used for small-scale coffee grinding, have evolved into more efficient systems. For instance, A person may grind up to 30 kg of coffee a day using a two-wheeled treadle mechanism or a cycle connected to a grinding wheel. Multiple grinders can be used together to grind up to 100 kg per day.
- *Motorized Grinders*: For large-scale production, motorized grinding mills are used. These mills may feature horizontal or vertical plates or hammer mills, with adjustable openings to achieve the desired fineness of the ground coffee. To maintain its quality, the ground coffee is then packed and kept in a cold, dry location.

  An investigation of how coffee's levels of anti-nutritional diterpenes are affected by roasting and brewing techniques, specifically cafestol and kahweol, used High-Performance Liquid Chromatography (HPLC) for analysis. The results showed that the profiles of these diterpenes in roasted beans are strongly influenced by the temperature and roasting time.

14.8.3.14.2 Instant Coffee

In order to make instant coffee, roasted and ground coffee must first have its volatile components and soluble coffee solids extracted. The extract is then dried to create powder or granules.

Pre-stripping High-pressure steam is used to remove more than 700 volatile components from a layer of ground coffee prior to extraction. The resulting condensate, containing these volatile compounds, is collected by a condenser. This step ensures the preservation of key aromatic components.

Coffee Solids Extraction The solvent used in this procedure is water, conducted at a temperature of 175 °C. The goal is to extract the coffee solids from the ground beans. There are three commonly practiced methods of extraction:

- *Percolation Calandrias*: The most popular technique for removing coffee solids is percolation calandrias. A continuous sequence of five to eight containers holds ground and roasted coffee, and each vessel is fed hot water that is between 170 °C and 175 °C. As the water moves through, it extracts the soluble coffee solids. Once extraction is complete in one vessel, it is replaced with a new one, and the process continues until the solution reaches a concentration of 15–25% w/w soluble coffee solids.
- *Countercurrent System*: In this method, two helical screws rotate at a rate of 10–22 revolutions per hour, moving the coffee up and down as it is continually delivered into the bottom of an inclined cylindrical vessel. The extracted coffee solids are gathered from the bottom of the vessel while hot water is added at the top. This system operates under vessel pressure at 180 °C.
- *Slurry Extraction*: In this process, coffee is mixed with water and agitated in a tank. The mixture is then separated using centrifugation to isolate the extracted solids.

Drying Methods The extracted coffee slurry is dried using one of three methods: spray drying, drum drying, or freeze drying.

- *Spray Drying*: Spray drying involves atomizing the concentrated coffee solution into droplets (10–200 μm in diameter) using a centrifugal atomizer. The liquid is sprayed into a large drying chamber where hot air (150–300 °C) dries the droplets rapidly, within approximately 10 s. A screw conveyor is used to collect the dry powder. While effective, spray drying is an expensive method due to its high energy costs.

- *Freeze Drying*: The sublimation concept is the basis of freeze drying. In this process, after the coffee is frozen, the ice is directly converted to vapor by heating it under a vacuum pressure of 610 Pa., which causes it to sublimate—or turn straight into vapor. A vacuum pump extracts the vapor, which is then condensed on refrigeration coils. Evaporative drying at extremely low pressure is the method used for final drying. Coffee is frozen as foam by adding gas to the liquid to make the procedure easier. Because of its intricate procedure, this approach is the most expensive despite being the most successful.
- *Drum Drying*: Drum drying, a less common method implies directly exposing coffee extracts to a heated cylindrical surface in order to dry them. While effective, it is not as widely used as other drying methods.

14.8.3.14.3 Decaffeinated Coffee

While coffee is often valued for its stimulating effects, a portion of consumers prefer decaffeinated options. Decaffeination, commonly referred to as "decaf," is the process of removing caffeine from coffee beans. Despite this removal, decaffeinated coffee still retains about 1–20% of its original caffeine content. The process typically involves exposing unroasted beans to steam, followed by immersing them in a solvent that effectively extracts caffeine while leaving most other components of the coffee intact. This extraction process is repeated several times until approximately 97% of the caffeine is removed, in line with International Standards. Given that coffee contains over 400 volatile compounds, achieving decaffeination without impacting flavor can be challenging.

Packaging To preserve the aromatic components of decaffeinated coffee, packaging is typically done in polypropylene packages. This prevents the loss of volatile compounds, ensuring the product maintains its intended flavor profile.

Storage For optimal storage, modern techniques such as Modified Atmospheric Storage (MAS) are recommended. Generally, coffee should be stored in cool, low-humidity environments that are clean and free from contaminants. Shady and dry areas are ideal for maintaining the quality of coffee over time.

14.8.3.14.4 Steam-Treated Coffee

Coffee can be made "less irritating" to the gut by steam-treating it, making it more suitable for individuals with digestive sensitivities. This process, which occurs before roasting, reduces the irritant effect of coffee, because chlorogenic acids were eliminated (Maier 1994). As a result, steam-treated coffee is often referred to as "stomach-friendly" coffee (Steinhart and Luger 1997).

14.8.3.14.5 Monsooned Coffee

A distinctive specialty coffee from India, monsooned coffee is made using a natural curing method. The wet monsoon winds of India's Malabar coast expose dry green Arabica and Robusta beans, particularly in regions like Mangalore and Tellicherry. Historically, the flavor of monsooned coffee originated during the era of sailing ships, when beans transported to Europe by sea underwent a transformation due to prolonged exposure to the high humidity and monsoon winds during the six-month voyage. The beans swelled, turned pale yellow, and developed a mild yet potent flavor as a result of these circumstances. With the advent of faster steamships and the opening of the Suez Canal, European importers noticed that the coffee no longer had its distinctive flavor, prompting Indian exporters to replicate the aging effect by developing the monsooning process. Monsooned coffee, known for its full body, reduced acidity, and distinct aroma, is now exported worldwide, including to Europe, Asia, Africa, and North America. For this procedure, only dry-processed Arabica and Robusta beans are typically utilized, and the resulting coffee is often blended with other varieties to enhance richness and mellow acidic profiles (Coffee Board of India 2007).

14.8.4 Cashew

14.8.4.1 Introduction

The cashew tree (*Anacardium occidentale*) is a tropical species renowned for producing cashew nuts and cashew apples. Originally native to Brazil, the tree is now extensively cultivated in various tropical regions worldwide, particularly in West Africa, India, and Vietnam. Cashew nuts represent a crucial agricultural commodity with considerable economic significance on a global scale. The cultivation and processing of cashews involve diverse practices aimed at ensuring high-quality yields while addressing challenges associated with market dynamics and labor conditions. Economically, the cashew industry holds great importance, particularly in West Africa, which accounts for nearly 50% of the world's cashew production. In 2023, global production of raw cashew nuts reached approximately 3.9 million tons, making cashews one of the most commercially viable nut crops, following almonds and walnuts (Babatunde et al. 2023). Various cultivation methods are employed in cashew plantations, with an increasing emphasis on sustainable practices. These include organic farming, which avoids synthetic pesticides and fertilizers, and agroforestry, where cashew trees are integrated into existing ecosystems to enhance biodiversity and resilience. Organic farming methods for cashew production are environmentally friendly and promote sustainability through a series of ecological practices aimed at improving soil health and biodiversity. Key practices, such as cover cropping and crop rotation, enhance soil fertility and reduce erosion.

The advantages of organic cashew farming are noteworthy. It encourages environmental stewardship by eliminating synthetic pesticides and fertilizers, thereby protecting water sources and local ecosystems. Furthermore, organic farming can yield higher economic returns for farmers, as there is a growing demand for organic cashew nuts due to their perceived health benefits and environmental advantages. Nonetheless, challenges persist in the organic cashew sector. Farmers often grapple with price volatility, inadequate infrastructure, and the complexities of complying with international trade regulations, which can complicate the export process. Overall, the transition to organic cashew farming is viewed as a crucial step towards achieving food sovereignty and empowering local communities. Agroforestry systems that incorporate cashew cultivation involve integrating trees and shrubs with crop and livestock farming, offering numerous benefits. These systems enhance soil fertility, mitigate erosion, and provide habitats that promote biodiversity, ultimately leading to improved cashew yields. For example, farmers practicing agroforestry in Nigeria and Vietnam have reported yield increases of 30% and a 15% reduction in soil erosion, respectively (Fahad et al. 2024). The integration methods often entail combining cashew orchards with native vegetation, fostering biodiversity and resilience within agricultural landscapes. Moreover, sustainable practices such as organic farming and integrated pest management (IPM) are increasingly being adopted, demonstrating reductions in pesticide use and improvements in water efficiency (Baker et al. 2020). Recent developments as of 2024 indicate a rising trend towards organic cashew nut production, recognized as vital for local empowerment and food sovereignty.

14.8.4.2 Climatic Requirements

Cashew trees flourish in tropical climates, with specific climatic conditions critical for their successful growth and fruit production. The ideal climate for cashew plantations includes temperatures ranging from 24 to 28 °C, although the trees can endure brief periods of higher temperatures, exceeding 36 °C. Conversely, low temperatures below 10 °C can harm the tree, particularly during its early growth stages. Adequate rainfall is essential, with optimal distribution particularly important during the critical growing period from September to October. Cashew trees require annual rainfall of 1000–2000 mm (40–80 inches) and thrive in regions with a distinct dry season. While rainfall during the growing season is crucial for initial growth, a dry period during flowering and fruiting is beneficial for nut production. Cashew trees also require moderate humidity levels, ideally between 50% and 80%, for effective growth. High humidity combined with elevated temperatures can increase the risk of fungal diseases, negatively impacting crop yields. These trees grow best in low-altitude areas, although they can tolerate elevations up to 1000 meters (3280 feet) above sea level. Higher altitudes may lead to cooler temperatures, which can hinder tree growth. In terms of soil conditions, cashew trees thrive in red sandy loam but are adaptable to various soil types. A soil pH ranging from 4.5 to 6.5 is considered optimal for cashew cultivation.

14.8.4.3 Varieties of Cashew Nuts

Cashew nuts come in various varieties, each differing in size, quality, flavor, and the regions where they are cultivated. These varieties are typically classified based on characteristics such as nut size, shape, quality, and adaptability to diverse climatic and growing conditions.

14.8.4.4 Varieties by Region

14.8.4.4.1 Indian Varieties

India ranks among the largest producers of cashews and has developed numerous high-yielding and disease-resistant varieties:

- *Vengurla 4 (V-4)*: This variety is particularly well suited to India's coastal regions and exhibits resistance to pests, such as the tea mosquito bug. It is known for its favorable kernel recovery rate and high yields.
- *BPP-8*: Characterized by a shorter gestation period, this high-yielding variety begins nut production within 3–4 years. It is ideal for regions with moderate rainfall.
- *VRI-1*: Recognized for its large nut size and high-quality kernels, VRI-1 is cultivated in Tamil Nadu. It demonstrates drought tolerance and produces high yields.
- *Dhana*: Popular in Goa and Maharashtra, this variety is noted for its high productivity and adaptability to varying climatic conditions.
- *Ullal-1*: Primarily grown in Karnataka, Ullal-1 is acknowledged for its moderate growth rate and high yield, producing large nuts (Babatunde et al. 2023).

14.8.4.4.2 African Varieties

Several African nations, including Ivory Coast, Nigeria, Tanzania, and Mozambique, are significant producers of cashew nuts. Improved varieties cultivated in Africa include:

- *K-578 (Tanzanian Variety)*: Known for its larger kernel size and high yields, K-578 is commonly grown in Tanzania and other parts of East Africa.
- *Kola*: This variety is cultivated in West Africa and is favored for its adaptability to various climatic conditions and high productivity.

14.8.4.4.3 Brazilian Varieties

As the native region of the cashew tree, Brazil has developed several high-yielding and large-nut varieties, including:

- *Embrapa 51*: This high-yielding variety is known for its resistance to pests and adaptability to a range of growing conditions.
- *CCP 76*: This variety features large kernels and is highly productive, making it suitable for both smallholder and commercial farming systems.

14.8.4.5 Grading of Cashew Kernels

Cashew kernels are graded according to size and quality, adhering to international standards that help determine their commercial value. The grading system typically indicates the number of whole kernels per pound, categorized as follows:

- *W180*: Known as the "King of Cashews," this grade consists of the largest and most premium cashew nuts, with approximately 180 nuts per pound. They are regarded as the highest quality.
- *W210*: Slightly smaller than W180, this grade includes around 210 nuts per pound and maintains a high-quality standard.
- *W240*: A popular variety for export, W240 contains around 240 nuts per pound and strikes a balance between size and price.
- *W320*: The most widely sold grade, W320 contains about 320 nuts per pound and offers good quality at a more accessible price point.
- *W450*: This grade features smaller kernels, with around 450 nuts per pound. These cashews are often used in snacks or mixed nut assortments.
- *Broken Grades (B Grade)*: This category includes broken pieces of cashews, such as splits, large pieces, small pieces, or bits. These are primarily used in cooking, baking, or confectionery applications and are less expensive than whole nuts while retaining the same flavor (Vidyarthi et al. 2020) (Fig. 14.5).

14.8.4.6 Post-harvest Processing

The post-harvest processing of cashew nuts encompasses several steps to transform raw cashew seeds into edible kernels ready for consumption or sale. Proper processing ensures the production of high-quality nuts, extends storage life, and enhances market value. The key steps in cashew nut processing include:

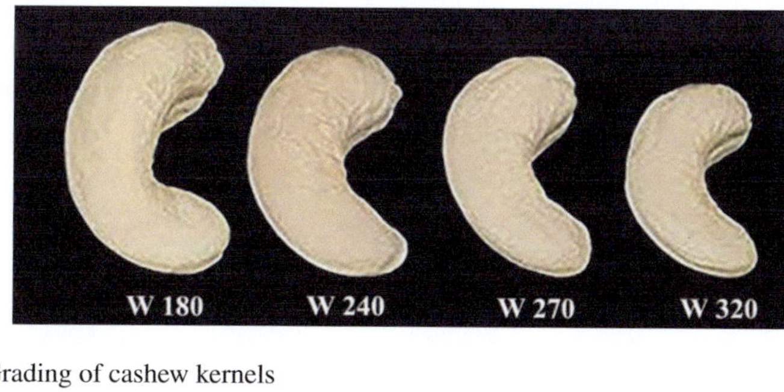

Fig. 14.5 Grading of cashew kernels

14.8.4.6.1 Collection and Sorting

- *Collection*: After harvesting, cashew nuts are collected manually from the ground as the cashew apples fall naturally. Care is taken to avoid immature or damaged nuts.
- *Sorting*: The collected nuts are sorted to eliminate debris, damaged, or rotten nuts. High-quality nuts are set aside for further processing, while lower-quality nuts may be discarded or repurposed for animal feed or oil extraction.

14.8.4.6.2 Drying of Raw Cashew Nuts

- *Sun Drying*: Following sorting, the raw cashew nuts are sun-dried to reduce moisture content, a crucial step for preserving the nuts and preventing mold or fungal growth. The nuts are spread out in a single layer and turned periodically to ensure even drying. The ideal moisture content after drying should be between 8% and 10%.
- *Mechanical Drying*: In regions with limited sunlight or during rainy seasons, mechanical drying methods may be employed. These systems utilize heat to achieve the desired moisture level.

14.8.4.6.3 Roasting (Shelling Preparation)

Roasting is essential for making the outer shell brittle and easier to remove. Two common methods are used:

- *Drum roasting (traditional method)*: In this method, the nuts are roasted in large drums at high temperatures (around 200–250 °C) for 15–20 min, causing the shell to become brittle and releasing the cashew nut shell liquid (CNSL) contained within.
- *Steam roasting (modern method)*: Instead of roasting, nuts are steamed. This technique helps preserve the quality of the kernel and reduces contamination from the toxic CNSL.

14.8.4.6.4 Cooling

After roasting, the nuts are cooled, either by spreading them in a cool area or using cooling systems, to facilitate easier handling during shelling.

14.8.4.6.5 Shelling (Removal of Outer Shell)

- *Manual shelling*: In traditional processing methods, cashew shells are cracked manually with a hammer or shelling tool. Workers must handle the shells with care due to the presence of toxic CNSL, which can cause skin burns.
- *Mechanical shelling*: Modern processing facilities utilize machines to crack the shells, making the process faster, safer, and reducing the likelihood of kernel damage compared to manual methods.

14.8.4.6.6 Removal of Cashew Nut Shell Liquid (CNSL)

CNSL is a toxic, corrosive oil found in the outer shell of cashew nuts. After roasting or shelling, CNSL is collected as a byproduct and can be utilized in various industrial applications, including paints, varnishes, lubricants, and insecticides (Scavo et al. 2022).

14.8.4.6.7 Drying of Kernels (Second Drying)

Following shelling, the cashew kernels still retain a thin outer skin known as the testa. To facilitate the removal of this skin, the kernels are dried again to further reduce moisture content.

- *Sun Drying*: The shelled kernels are sun-dried for a few hours.
- *Hot Air Drying*: Alternatively, hot air systems can be employed to dry the kernels, aiming to reduce moisture content to 5% or lower.

14.8.4.6.8 Peeling (Removal of Testa)

- *Manual peeling*: After the second drying, the testa is removed manually by rubbing the kernels. This process requires careful handling to avoid breaking the delicate kernels.
- *Mechanical peeling*: In large-scale operations, peeling machines are utilized to gently remove the testa without damaging the cashew kernels.

14.8.4.6.9 Quality Control (Inspection)

Quality control is critical throughout the processing of cashew nuts:

- *Visual inspection*: Graded nuts undergo visual inspections to ensure uniformity in size, color, and the absence of defects. Defective or off-color nuts are separated from the batch.

- *Weight and moisture check*: Kernels are assessed for weight consistency and moisture content, with the final moisture level ideally maintained at around 3% to 5% to ensure longevity and prevent mold growth during storage.

14.8.4.6.10 Packaging and Storage

Effective packaging and storage are essential for maintaining the freshness of cashew kernels:

- *Vacuum packaging*: To preserve the freshness of cashew kernels and prevent oxidation, they are often vacuum-packed or stored in nitrogen-flushed containers.
- *Storage conditions*: Packaged cashews should be stored in cool, dry, and well-ventilated environments to prevent moisture accumulation, which can lead to mold growth or spoilage. The optimal storage temperature is below 25 °C (77 °F) with low humidity levels (Fig. 14.6).

14.8.4.7 Products

Cashew nuts are not only popular as a snack but also serve as the foundation for a diverse range of products derived from both the nuts and the cashew apple. Their versatility allows for widespread use in food, health, and industrial sectors.

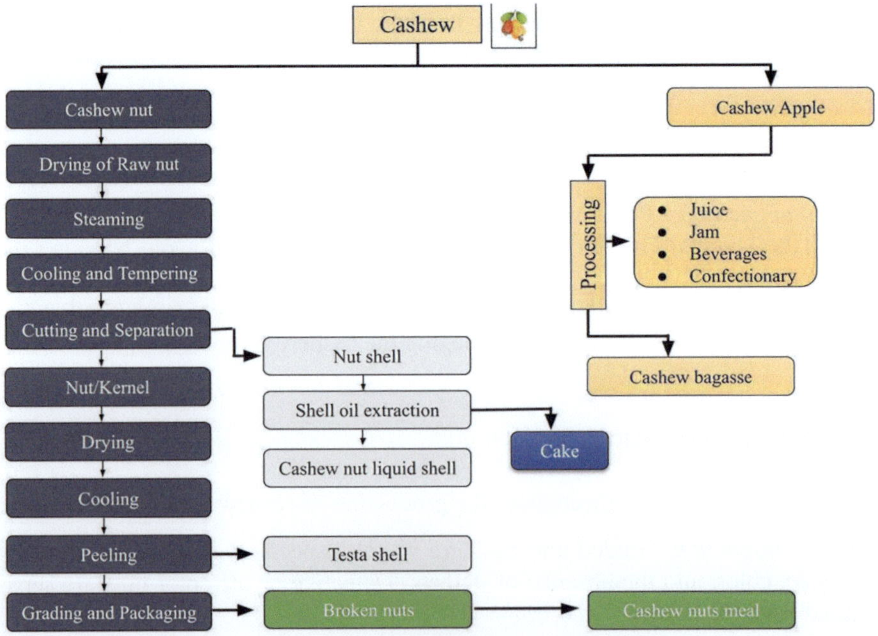

Fig. 14.6 Schematic diagram of cashew nut and cashew apple processing

14.8.4.7.1 Whole Cashew Kernels

- *Whole cashew kernels*: The most common product, whole cashew nuts are enjoyed globally as snacks or incorporated into various culinary applications.
- *Raw cashew nuts*: These are available in their raw form and are commonly used in cooking, baking, and as ingredients in diverse dishes.
- *Roasted cashew nuts*: Cashews can be dry-roasted or oil-roasted, sometimes with added salt, resulting in a crunchy snack.
- *Flavored cashew nuts*: Seasoned cashews are available with a variety of spices, herbs, or sweeteners, including honey-roasted, chili, or garlic-flavored options.

14.8.4.7.2 Cashew Nut Butters

- *Cashew nut butter*: A creamy spread created by grinding cashew nuts into a paste, it is used in various culinary applications.
- *Flavored cashew butter*: This variant can be enhanced with flavors such as honey, cocoa, or cinnamon to provide additional variety.

14.8.4.7.3 Cashew Milk and Cream

- *Cashew milk*: A popular plant-based milk alternative, cashew milk is made by blending cashews with water and straining out the solids. It has a creamy texture and can be sweetened or flavored as desired.
- *Cashew cream*: A thicker version of cashew milk, this product is often utilized in cooking or as a dairy substitute in sauces and desserts.

14.8.4.7.4 Cashew Flour and Meal

- *Cashew flour*: This gluten-free alternative to wheat flour is produced by grinding cashews and is commonly used in baking for cakes, cookies, or bread, as well as serving as a thickening agent in sauces.
- *Cashew meal*: A coarser version of cashew flour, cashew meal is frequently used in coatings for fried foods or blended with other flours.

14.8.4.7.5 Broken Cashew Kernels

- Cashews are also classified based on size and shape, with broken kernels being more economical than whole nuts. Despite this, they retain similar nutritional value and are employed in various products.
- *Split cashews*: These consist of halves or larger pieces of cashew nuts, often used in confectionery, cooking, and baking.

- *Cashew pieces*: Small pieces of cashew nuts are commonly found in trail mixes, desserts, granola, or as a garnish.

14.8.4.7.6 Cashew Nut Shell Liquid (CNSL)

- *CNSL*: The toxic liquid found in the cashew nut shell has valuable industrial applications. It is utilized in the production of resins, coatings, paints, lubricants, and brake linings, owing to its heat-resistant and anti-corrosive properties.
- *Anacardic acid*: This chemical compound present in CNSL possesses medicinal properties and finds applications in pharmaceuticals and cosmetics.

14.8.4.8 Products from Cashew Apples

The cashew apple, often overlooked, can be processed into a variety of valuable products:

- *Cashew apple juice*: A sweet and tangy juice rich in vitamin C, it is commonly consumed in regions where cashews are cultivated.
- *Cashew apple jam*: Made from the pulp of the cashew apple, this jam is often sweetened and cooked to create preserves.
- *Cashew apple cider or vinegar*: The juice of the cashew apple can undergo fermentation to produce cider or be converted into vinegar.
- *Cashew apple wine*: In some cultures, the juice is fermented to create alcoholic beverages, such as feni, a traditional drink from Goa, India.

14.8.4.8.1 Cashew Oil

- *Culinary cashew oil*: Extracted from the cashew kernel, this oil is used in cooking, salads, and as a flavoring oil. It has a mild, nutty flavor, similar to other nut oils.
- *Cosmetic cashew oil*: Known for its moisturizing properties, cashew oil is rich in fatty acids and vitamin E, making it beneficial for skincare and haircare products.

14.8.4.8.2 Cashew-Based Confectionery

Cashew nuts are a popular ingredient in various desserts and sweets, particularly in Asian and Middle Eastern cuisines:

- *Cashew brittle (Kaju chikki)*: This hard candy is made from cashew nuts and caramelized sugar or jaggery, akin to peanut brittle.
- *Cashew halwa*: A traditional Indian dessert prepared with ground cashews, sugar, and ghee.
- *Cashew fudge (Kaju katli)*: A favored Indian sweet made from cashew powder and sugar, often flavored with cardamom or saffron.

14.8.4.8.3 Cashew-Based Snacks

Cashews can be transformed into a variety of snack products, either on their own or mixed with other ingredients:

- *Trail mix*: Cashews are frequently combined with other nuts, dried fruits, and chocolate to create a healthy, energy-rich snack.
- *Cashew bars*: Snack bars made with cashew nuts, often mixed with dates, seeds, or other nuts, provide a convenient and nutritious option.
- *Cashew paste*: Made by grinding cashews into a thick, smooth texture, this paste is used in various culinary applications.
  - *Culinary cashew paste*: Often employed in sauces, gravies, and curries, particularly in Indian and Southeast Asian cuisines, cashew paste adds a creamy texture and mild flavor to savory dishes.
  - *Dessert cashew paste*: This paste is used in the preparation of cashew-based desserts, pastries, and ice creams.

14.8.4.8.4 Cashew Alcoholic Products

In some regions, cashew apples are fermented to produce alcoholic beverages:

- *Cashew wine*: This beverage is created from fermented cashew apple juice, resulting in a mildly alcoholic drink.
- *Feni*: A popular distilled spirit produced primarily in Goa, India, feni is made from fermented cashew apple juice.

14.8.4.8.5 Cashew Protein and Supplements

Cashews are rich in protein and can serve as a base for nutritional supplements:

- *Cashew protein powder*: This plant-based protein powder is made from cashews and is commonly used in smoothies, protein bars, and dietary supplements, providing a nutritious option for health-conscious consumers.

14.9 Conclusions

Plantation crops play a pivotal role in shaping both the agricultural landscape and global economies. Defined by large-scale cultivation and management, these crops are characterized by extensive acreages, specialized farming techniques, and significant economic impact. Beyond their contribution to national and international trade, plantation crops are essential for providing raw materials for various industries, sustaining livelihoods, and influencing environmental dynamics. From the sprawling tea estates of Assam to the vast rubber plantations in Southeast Asia, plantation crops have become synonymous with specific regions, each contributing its unique flavor to the global agricultural. The cultivation and management of plantation crops require specialized knowledge, incorporating a blend of traditional wisdom and modern agricultural practices. Plantation crops, including coconut, coffee, tea, and cashew, present substantial opportunities for advancing agricultural and economic development through diversification and value-added processing. These crops not only enhance nutrition and support livelihoods but also play a pivotal role in global trade and promoting sustainable practices. Advancements in cultivation techniques, processing technologies, and market strategies can enhance their contribution to meeting the growing demand for high-quality, eco-friendly products. Maximizing their potential can drive rural economic growth and pave the way for a sustainable and resilient future in the plantation crop sector.

References

Alam M (2020) Role of plantation crops to improve socio-economic condition in Bangladesh

Arya M, Rao LJM (2007) An impression of coffee carbohydrates. Crit Rev Food Sci Nutr 47:51–67

Babatunde OP, Adeigbe O, Sobowale O et al (2023) Cashew production and breeding in 5 West African countries. J Sci Res Rep 29:28–39

Baker BP, Green TA, Loker AJ (2020) Biological control and integrated pest management in organic and conventional systems. Biol Control 140:104095

Banzon JA (1978) Reconstitution of milk using coconut milk and nonfat dry milk. Philippine J Cocon Stud 3:1–8

Banzon JA, Gonzalez ON, De Leon SY, Sanchez PC (1990) Coconut as food. Philippines Coconut Research Development Foundation, Pasig

Baruah S, Tamuly P, Bordoloi P et al (2005) Technology for product diversification and value added items of tea. In: Proceedings of 34th Tocklai conference, tea research association, Tocklai Experimental Station, Jorhat, Assam, pp 63–69

Belay A, Ture K, Redi M et al (2008) Measurement of caffeine in coffee beans with UV/Vis spectrometer. Food Chem 108:310–315

Belitz HD, Grosch W, Schieberle P (2009) Coffee, tea, cocoa. In: Belitz HD, Grosch W, Schieberle P (eds) Food Chem, 4th edn. Springer, Leipzig, pp 938–951

Berbert PA, Queiroz DM, Silva JS et al (1995) Simulation of coffee drying in a fixed bed with periodic airflow reversal. J Agric Eng Res 60:167–173
Caffin N, D'Arcy B, Yao L et al (2004) Developing an index of quality for Australian tea. RIRDC Publication (Issue 04/033). https://www.agrifutures.com.au/wp-content/uploads/publications/04-033.pdf
Chauhan OP, Singh A, Raju PS et al (2008) Development of pomegranate juice blended tender coconut water beverage. Indian Coconut J 24:46–60
Chauhan OP, Singh A, Raju PS et al (2009) Development of tender coconut water blended yoghurt. J Food Sci Technol 46:427–430
Chauhan OP, Archana BS, Singh A, Raju PS, Bawa AS (2013a) Utilization of tender coconut pulp for jam making and its quality evaluation during storage. Food Bioproc Technol 6:1444–1449
Chauhan OP, Archana BS, Roopa N et al (2013b) Stabilization of ready-to-eat coconut chutney using combination processing technology. Indian Food Packer 67:77–83
Chauhan OP, Archana BS, Singh A et al (2014) A refreshing beverage from mature coconut water blended with lemon juice. J Food Sci Tech 51:3355–3361
Chen YL, Duan J, Jiang YM et al (2011) Production, quality, and biological effects of oolong tea (*Camellia sinensis*). Food Rev Int 27:1–15. https://doi.org/10.1080/87559129.2010.518294
Chen Q, Shi J, Mu B et al (2020) Metabolomics combined with proteomics provides a novel interpretation of the changes in nonvolatile compounds during white tea processing. Food Chem 332:127412. https://doi.org/10.1016/j.foodchem.2020.127412
Chin JM, Michele M, Goldberger BA et al (2008) Caffeine content of brewed teas. J Anal Toxicol 32:702–704. https://doi.org/10.1093/jat/32.8.702
Chopra VL, Krishnakumar V, Peter KV (2016) Plantation crops and plantations. https://doi.org/10.1016/B978-0-12-394807-6.00174-X
Clarke RJ (2003) Coffee: green coffee/roast and ground. In: Caballero B, Trugo LC, Finglas P (eds) Encyclopedia of food science and nutrition, 2nd edn. Academic Press, Oxford
Clarke C, Richter BS, Rathinasabapathi B (2023) Genetic and morphological characterization of United States tea (*Camellia sinensis*): insights into crop history, breeding strategies, and regional adaptability. Front Plant Sci 14:1149682
Coffee Board of India (2007). http://www.indiacoffee.org/default.html
Da Silva Pinto M (2013) Tea: a new perspective on health benefits. Food Res Int 53:558–567
Das S, Samanta T, Datta AK (2019) Analysis and modeling of major polyphenols during oxidation in production of black tea. J Food Proc Preserv 43:1–10. https://doi.org/10.1111/jfpp.14283
Donlao N, Ogawa Y (2019) The influence of processing conditions on catechin, caffeine and chlorophyll contents of green tea (*Camelia sinensis*) leaves and infusions. LWT Food Sci Tech 116:108567. https://doi.org/10.1016/j.lwt.2019.108567
Dursun E, Dursun I (2005) Some physical properties of caper seed. Biosyst Eng 92:237–245
Edygova SN (2018) Wheat bread development using green tea infusion. In: The collection: improving the quality and safety of foodstuff materials of the VIII All-Russian Scientific and Practical Conference, pp 151–152
Etienne H (2005) Somatic embryogenesis protocol: coffee *(Coffea arabica L. and C. canephora P.)*. In: Jain SM, Gupta PK (eds) Protocol for somatic embryogenesis in woody plants. Springer, Dordrecht, pp 167–168
Fahad S, Saud S, Nawaz T et al (2024) The Green revolution: promoting environmental stewardship and plant growth. In: Environment, climate, plant and vegetation growth. Springer, pp 425–469
FAO (2020) Plant production and protection. https://openknowledge.fao.org/server/api/core/bitstreams/9f3800f4-0577-472e-89bb-a5f6c22f2582/content. Accessed on 6 Aug 2024
Finzer JRD, Sfredo MA, Sousa GDB et al (2007) Dispersion coefficient of coffee berries in vibrated bed dryer. J Food Eng 79:905–912

Food and Agriculture Organization of the United Nations (2022) International tea market: market situation, prospects and emerging issues. Food and Agriculture Organization of the United Nations, Rome

Franca AS, Oliveira LS (2009) Alternative uses for coffee husks – a solid waste from green coffee production. In: Chemical, biological and environmental engineering. Word Scientific Publishing Co. Pte. Ltd, Singapore, pp 21–24

Gebely T (2016) Tea User Guide Print. Eggs and Toast Media, LLC. ISBN: 9780998103006 0998103004

Ghosh DK (2015) Postharvest, product diversification and value addition in coconut. In: Value addition of horticultural crops: recent trends and future directions. Springer, New Delhi, pp 125–165

Grembecka M, Malinowska E, Szefer P (2007) Differentiation of market coffee and its infusions in view of their mineral composition. Sci Total Environ 383:59–69

Hernandez LM, Xu EG, Larsson HCE et al (2019) Plastic tea bags release billions of microparticles and nanoparticles into tea. ACS Publications. Journal contribution 53:12300–12310. https://doi.org/10.1021/acs.est.9b02540.s001

Hicks A (2001) Review of global tea production and the impact on industry of the Asian economic situation. AU JT 5:1–8

Hilal Y (2017) Morphology, manufacturing, types, composition and medicinal properties of tea (*Camellia sinensis*). J Basic Appl Plant Sci 1:107

Ho KKHY, Haufe TC, Ferruzzi MG et al (2018) Production and polyphenolic composition of tea. Nutr Today 53:268–278. https://doi.org/10.1097/NT.0000000000000304

Illy A, Viani R (1995) Espresso coffee: the chemistry of quality Londres, UK. Academic

Isamil I, Anuar MS, Shamusdin R (2012) Effect on the physicochemical properties of liberica green coffee beans under ambient storage. Int Food Res J 20:255–264

ITC: International Tea Committee (2016) Annual bulletin of statistics. International Tea Committee LTD, London, p 52

Jiao W, Xiao Y, Qian X et al (2016) Optimized combination of dilution and refined QuEChERS to overcome matrix effects of six types of tea for determination eight neonicotinoid insecticides by ultra-performance liquid chromatography-electrospray tandem mass spectrometry. Food Chem 210:26–34. https://doi.org/10.1016/j.foodchem.2016.04.097

John CM (1970) Coconut cultivation, 6th edn. Indian Council of Agricultural Research, New Delhi, pp 5–25

Kuroda Y, Hara Y (2004) Health effects of tea and its catechin. Kluwer Academic Plenum Publishers, New York, p 109. ISBN: 030648207X, 9780306482076. https://doi.org/10.1007/978-1-4757-5390-5

Lanjekar H (2022) Tea: varieties and types. Acta Sci Agric 68:8–9

Lee LK, Foo KY (2013) Recent advances on the beneficial use and health implications of Pu-Erh tea. Food Res Int 53:619–628. https://doi.org/10.1016/j.foodres.2013.02.036

Lee MK, Kim HW, Lee SH et al (2019) Characterization of catechins, theaflavins, and flavonols by leaf processing step in green and black teas (*Camellia sinensis*) using UPLC-DAD-QToF/MS. Eur Food Res Technol 245:997–1010. https://doi.org/10.1007/s00217-018-3201-6

Li Q, Chai S, Li YD et al (2018) Biochemical components associated with microbial community shift during the pile-fermentation of primary dark tea. Front Microbiol 9

Lin SY, Lo LC, Chen IZ et al (2016) Effect of shaking process on correlations between catechins and volatiles in oolong tea. J Food Drug Anal 24:500–507. https://doi.org/10.1016/j.jfda.2016.01.011

Liyanage DV (1958). Varieties and forms of the coconut palm grown in Ceylon

Lu TM, Lee CC, Mau JL et al (2010) Quality and antioxidant property of green tea sponge cake. Food Chem 119:1090–1095

Lv H, Zhang Y, Lin Z et al (2013) Processing and chemical constituents of Pu-erh tea: a review. Food Res Int 53:608–618. https://doi.org/10.1016/j.foodres.2013.02.043
Madhavan K (1987) Design and development of copra moisture meter. J Plantn Crops 16:113–116
Maier HG (1994) Status of research in the field of non-volatile coffee components. In: Proc 15th Coll Sci Int Café, Montpellier Paris, France, ASIC, pp 567–576
Manoj Kumar CT, Chauhan OP, Rajani CS et al (2017) Effect of coconut milk, skim milk powder, and banana pulp on sensory and functional properties of coconut curd and its applicability as a carrier for probiotic microorganisms. J Food Proc Preserv 42:13460. https://doi.org/10.1111/jfpp.13460
Mathew AC, Madhavan KA, Arumuganathan T (2014) Development and performance evaluation of virgin coconut oil cooker. Agric Mech Asia Africa Latin America 45(1):56–59
Monera OD, Del Rosario E (1982) Physico-chemical evaluation of the natural stability of coconut milk emulsion. Ann Trop Res: Tech J Visayas State College of Agriculture
Muralidharan K, Jayashree A (2011) Value addition, product diversification and by-product utilization in coconut. Indian Cocon J 7:4–10
Narayana GV, John CM (1949) Varieties and forms of coconut. Madras Agric J 36:349–366
Nilnont W, Thepa S, Janjai S et al (2012) Finite element simulation for coffee drying. Food Bioprod Process 90:341–350
Owoyemi JM, Akinwamide TO, Ibidokun AO (2022) Development of coconut plantation in Nigeria; problems and prospects. Int J Agric Plant Sci 4(1):58–66
Pan R, Zhao X, Du J et al (2022a) A brief analysis on tea import and export trade in China during 2021. China Tea 44:6
Pan SY, Nie Q, Tai HC et al (2022b) Tea and tea drinking: China's outstanding contributions to mankind. Chin Med:17–27
Purseglove JW (1968) The origin and distribution of the coconut. Trop Sci 10:190–199
Qin JH, Li N, Tu PF et al (2012) Change in tea polyphenol and purine alkaloid composition during solid-state fungal fermentation of post fermented tea. J Agric Food Chem 60:1213–1217. https://doi.org/10.1021/jf204844g
Reinato CHR, Borem FM, Cirillo MÂ et al (2012) Quality of the coffee dried on grounds with different surfaces and thickness layers. Coffee Sci 7:223–237
Rethinam P, Bosco SJB (2006) Coconut- harvest, processing and products. Asia & Pacific Coconut Community, Jakarta, p 36
Richardson LB (2009) Tea with a twist: entertaining and cooking with tea. Harvest House Publishers, Eugene, p 12. ISBN 978-0-7369-2579-2
Sanchez PC, Rasco PM (1983) Utilization of coconut in white soft cheese production. Res Los Banos 2(2):13–16
Sanchez PC, Rasco PM (1984) Coconut milk (Gata) in yoghurt manufacture. Philippine J Cocon Stud 8:42–50
Sangamithra A, Swamy GJ, Sorna PR, Chandrasekar V, Sasikala S, Hasker E (2013) Coconut: an extensive review on value added products. Indian Food Ind Mag 32(6):29–36
Scavo A, Fontanazza S, Restuccia A et al (2022) The role of cover crops in improving soil fertility and plant nutritional status in temperate climates. A review. Agron Sustain Dev 42:93
Schenker SR (2000) Investigations on the hot air roasting of coffee beans. Swiss Federal Institute of Technology
Singh TV, Udhayakumar R (2006) Development of a coconut de-shelling machine. Trop Agric Res 18:20
Smith K (2016) World atlas of tea. Mitchell Beazley, London, p 107. ISBN 978-1-78472-124-4
Steinhart H, Luger A (1997) An analytical distinction between untreated and steam-treated roasted coffee. In: Proc. of 17th Int. Sci. Coll. on Coffee (Nairobi). Paris, ASIC, pp 155–160

Svoboda P, Vlčková H, Nováková L (2015) Development and validation of UHPLC-MS/MS method for determination of eight naturally occurring catechin derivatives in various tea samples and the role of matrix effects. J Pharm Biomed Anal 114:62–70. https://doi.org/10.1016/j.jpba.2015.04.026

Swaminathan MS, Nambiar MC (1961) Cytology and origin of the dwarf coconut palm. Nature 192:85–86

Swarnathilake D, Don A (2020) Improving the quality and the taste of green tea ice cream by identifying suitable cultivar, agronomic practices and processing techniques of Assamica tea grown in Sri Lanka. World J Adv Res Rev 8:062–068. https://doi.org/10.30574/wjarr.2020.8.1.0354

Tan J, Engelhardt UH, Lin Z et al (2017) Flavonoids, phenolic acids, alkaloids and theanine in different types of authentic Chinese white tea samples. J Food Compos Anal 57:8–15. https://doi.org/10.1016/j.jfca.2016.12.011

Techome K (2019) Effect of tea processing methods on biochemical composition and sensory quality of black tea (*Camellia sinensis (L.) O. Kuntze*): a review. J Horticul For 11:84–95. https://doi.org/10.5897/JHF2019.0588

Thamban C, Mathew AC, Muralidharan K, Subramanian P, Singh VT, Madhavan K (2011) Coconut climbing methods and devices: a participatory analysis of constraints and strategies. J Plantn Crops 39:45–47

Turkmen N, Sarı F, Velİoglu YS (2009) Factors affecting polyphenol content and composition of fresh and processed tea leaves. Akademik Gida 7:29–40

Van Driem G (2020) The primordial origins of tea. In: The tale of tea. Brill, pp 1–38

Varghese A, Jacob J (2014) A study on the KAU coconut husking tool. In: Annual international conference on emerging research areas: magnetics, machines and drives. https://doi.org/10.1109/AICERA.2014.6908254

Vidhan Singh T, Swamy KGN, George SA (1999) Design and development of smoke free collapsible copra dryer for medium holdings. J Plantn Crops 27:112–120

Vidyarthi SK, Singh SK, Tiwari R et al (2020) Classification of first quality fancy cashew kernels using four deep convolutional neural network models. J Food Process Eng 43:e13552

Wang Y, Kan Z, Thompson HJ et al (2019) Impact of six typical processing methods on the chemical composition of tea leaves using a single *camellia sinensis* cultivar, Longjing. J Agric Food Chem 67:5423–5436. https://doi.org/10.1021/acs.jafc.8b05140

Wu J (2020) What makes bubble tea popular? Interaction between Chinese and British tea culture. Front Soc Sci Technol 16:97–102. https://doi.org/10.25236/FSST.2020.021614

Xu YQ, Liu PP, Shi J et al (2018) Quality development and main chemical components of Tieguanyin oolong teas processed from different parts of fresh shoots. Food Chem 249:176–183. https://doi.org/10.1016/j.foodchem.2018.01.019

Xu M, Wu Y, Gary G et al (2019) Evaluation of different tea extracts on dough, textural, and functional properties of dry Chinese white salted noodle. LWT 101:456–462. ISSN 0023-6438. https://doi.org/10.1016/j.lwt.2018.11.066

Ye Y, Yan J, Cui J et al (2018) Dynamic changes in amino acids, catechins, caffeine and gallic acid in green tea during withering. J Food Compos Anal 66:98–108. https://doi.org/10.1016/j.jfca.2017.12.008

Yuming W, Li T, Xu S et al (2021) The profile of dynamic changes in yellow tea quality and chemical composition during the yellowing process. LWT 139. https://doi.org/10.1016/j.lwt.2020.110792

Zhang L, Ho CT, Zhou J et al (2019a) Chemistry and biological activities of processed *Camellia sinensis* teas: a comprehensive review. Comp Rev Food Sci Food Safety 18:1474–1495. https://doi.org/10.1111/1541-4337.12479

Zhang Q, Wu S, Li Y et al (2019b) Characterization of three different classes of non-fermented teas using untargeted metabolomics. Food Res Int 121:697–704. https://doi.org/10.1016/j.foodres.2018.12.042
Zhang L, Cao QQ, Granato D et al (2020) Association between chemistry and taste of tea: a review. Trends Food Sci and Tech 101:139–149. https://doi.org/10.1016/j.tifs.2020.05.015
Zhu MZ, Li N, Zhou F et al (2020) Microbial bioconversion of the chemical components in dark tea. Food Chem 312:126043

Chapter 15
Fruit and Vegetable By-Products and Their Utilization

M. K. Mrithula Mahalakshmi, Thejesh Kathirvelu, Janifer Raj Xavier, and OP Chauhan

15.1 Introduction

With increased global population, there is a rise in demand for food with a huge generation of food waste from unutilized part of the plant causing environmental issues. Hence, efficient use of fruit and vegetable by-products could be a solution to sustain food demands and meet the goals of the United Nations Sustainable Development (SDG) programme. The agricultural by-products could be an alternate nutrient source through functional foods, aiding in overcoming the challenges of food insecurity. Fruit and vegetable by-products are incorporated into bakery and dairy products in the form of functional foods. Post-harvest treatment steps are highly crucial to ensure the food safety and quality of fresh produce while retaining its nutritional value until consumed by the end consumer (Carré et al. 2016). During vegetable processing, at least one-third of it gets discarded or wasted where some parts are deliberately wasted or discarded due to its unfavourable taste, odour or texture like hulls, bagasse and seeds. In the case of broccoli, cauliflower and pumpkin, the stems and leaves are not consumed, and therefore, they get wasted. Similarly, the outer layer of beans or seeds called seed coat is not consumed, and their only purpose is providing the protection to the bean. Peanut seed is an exception because of its thin texture, taste and digestibility (Vicentini et al. 2016).

Approximately 22% of oilseeds and pulses get wasted worldwide, with the highest contribution from North Africa, West Asia and Central Asian countries. The hard protection over seeds called hull is removed, while processing due to its hardness of texture with 30% reported loss of whole cereals during post-harvest activities which is further used as a building material, fuel, fertilizer and functional ingredients (Tlais et al. 2020). Vegetable peels as found in tubers, gourd and allium families are

M. K. Mrithula Mahalakshmi · T. Kathirvelu · J. R. Xavier (✉) · OP Chauhan
Defence Institute of Bio-defence Technologies, Defence Research and Development Organization, Mysore, Karnataka, India

OP Chauhan (ed.), *Fruits and Vegetables Technologies*,
https://doi.org/10.1007/978-981-96-8433-5_15

disregarded during production, contributing up to 5,814,000 tonnes of by-product production in North America and Oceanic regions. Unlike vegetable by-products, fruit by-products end up in municipal solid waste, making it impossible for further processing or converting it into a value-added product (Carré et al. 2016). The outer layer of fruits, the peel, is mostly disregarded by the end consumer during consumption, and a very minor portion is contributed by industrial processing, especially from the beverage manufacturing industry. Fruit peels are discarded due to its texture and bitter aftertaste. Smaller seeds present in fruits like watermelon and kiwi are consumed, while large-sized seeds are mostly discarded and end up as by-products. These seeds are rich in nutritional sources with numerical possibilities in the area of functional food development. Beverage or juice manufacturing industry produces fruit pomace that includes seeds, skin, pulp and stems, which can be further processed to utilize the nutritional benefits present in them. The vegetable and fruit by-products are produced during harvest and post-harvest or production and consumption stages, and most often used as burning material, construction material or animal feed. By the addition of agricultural by-products into food products, the nutritional value increases while lowering the production cost and reducing solid waste accumulation. With increased consumer demand for healthier food at economical value, the food industry is thriving to utilize the nutritional benefits from by-products to produce value-added products with high functionality or bioactivity (Lau et al. 2021). Even with the installation of post-harvest management techniques such as drying, cooling and blanching, the efficiency of preservation depends on the successful design of the techniques linking material science, fluid dynamics, mechanical deformation, food chemistry and process control (Verma and Joshi 2000).

During handling and marketing, the harvested fruits and vegetables undergo huge quantitative losses measured by consumer acceptability and qualitative losses like losing inedibility, nutritional quality and calorific value. In India, post-harvest losses amount to around 25–33%, majorly caused due to insufficient infrastructure for processing and packaging. As per All India Coordinated Research Project (AICRP) on post-harvest technology of the Indian Council of Agricultural Research (ICAR), the post-harvest losses from different handling techniques account for up to 13%, storage losses about 6% and at market about 12% loss adding up to one-third average loss of the harvested produce. Research scientists from esteemed ICAR institutes such as Central Institute of Post-Harvest Technology (CIPHET) and State Agricultural Universities conducted studies on post-harvest management to identify the gap/missing links in the current technologies (Adhikary and Kumar 2021). It is highly crucial to identify the state of maturity in each product to analyse if the produce is ready to be harvested and therefore less susceptible to physiological disorders and diseases. For universal understanding, maturity indices have been developed for fruits and vegetables such as mango, pomegranate, apple, grapes and mandarins. Agricultural waste will not only have an adverse economic effect on farmers but also has a high impact on consumers, especially in countries like China and the United States of America, where the cost of food waste is high. Around $940 billion worth of food is estimated to be lost annually. With simple control of

1% prevention of these losses World Bank could benefit with $40 million in economic gains, eventually benefitting the farmers (Lau et al. 2021).

15.2 Generation of By-Products: Scope, Significance and Recent Market Trends

Fresh fruits and vegetables are highly fragile and perishable, and therefore, poor storage conditions can accelerate the deterioration of produce. High care should be taken after harvest to maintain the freshness, nutrition and sensory attributes of the produce. Processing steps such as handling, sorting, grading, washing, storing, transportation, temperature-controlled storage and transportation are the crucial points in the post-harvest supply chain to implement treatments for retaining the quality and shelf life of the products. These treatments have been grouped into the following categories, namely: field operations, packinghouse operations and special treatments. In field operations, the fruits and vegetables are graded based on the size and colour of the produce which is a direct indicator of the quality and maturity of the product. The harvest index and matured harvest are the key indicators for adequate shelf life and ripening quality of the fruit. Though manual harvest is the most common practice worldwide, some cultivators also use pickup machines and conveyors for harvest where the immature produce gets damaged due to handling; therefore, special care must be taken while handling operations to prevent product damage and subsequent decay. The matured produce from the harvest is then hauled to packinghouses, storage, or loading facilities in bulk bins, gondolas or bulk field trailers or trucks (Zhao et al. 2017).

Bulk bins are used for the storage of hard rind squashes, capsicums and pink tomatoes; while mature green tomatoes and peppers are stored in gondolas and bulk field trailers are used for muskmelons and hard rind squashes. For sweetcorn, it is preferred to harvest at night when the products are coolest, as this technique helps in reduction of time and costs for cooling the produce, and the night atmosphere aids in achieving a uniform cooling while maintaining the product quality better. Agricultural practice shows an increased trend towards field packing of fruits and vegetables where processes such as grading, sorting, sizing, packing and palletizing are carried out in the field and central cooling facility is used for transportation and storage. In field pack operations, the produces entail less handling than packinghouses, for example, mobile packing facilities are commonly towed through the fields for rock melon, honeydew melon, eggplant, cucumber, summer squashes and capsicums. Through this method, we can achieve maximum reduction in product damage at the time of harvest and increase the product yield. Reduced rolling, dumping and dropping will prevent the loss of the seed cavity from the pericarp wall, thereby reducing the “shaker” problem and preventing water loss by reducing net scuffing of the product. Even though field pack operations incur reduced handling costs, it requires increased supervision to maintain consistent quality in the packed product, therefore making it difficult for the application of products requiring sorting according to colour and size, similar to tomatoes. In packinghouse operations, the loaded field transporting vehicles are allocated parking in shaded areas to prevent warming and sunburning of the products with varied unloading processes according to the product type. Some highly fragile products such as soft-rind squashes, eggplant, muskmelons, cucumbers and watermelon are preferred to be unloaded by hand. In another method of unloading called dry dumping, products are released onto sloping, padded ramps or onto moving conveyors. In wet dumping, products are placed into tanks of moving water to minimize physical damage such as bruising, scratching, abrading and splitting as observed in dry dumping. In the wet-dumping process, it is critical to maintain slightly warmer temperature in the tanks to prevent water absorption and microbial growth within the products by maintaining the chlorination in the tank via clean water separation or using spray to enhance overall handling sanitation. The most common preliminary operations are pre-sizing, sorting or selection, grading, waxing, sizing, packing, palletizing and cooling (Sudheer and Indira 2007).

The commodities that require to be within a certain size for commercial and sensory reasons such as undersized fruit are processed or diverted to a culling conveyor. Through the sorting process, the cull, overripe, misshapen or defective products are separated by colour, maturity and category of ripeness. Grading is carried out mostly for fruits, where they are sorted according to various standards of quality and stored in separated grades as per the international grading standards for respective commodities. Waxing is generally carried out in products to prevent further water loss and improves aesthetic appearance of the product as the natural wax gets removed during the cleaning operations. Wax might contain fungicides and is done either before or after sizing in commodities such as cucumber, eggplant, sweet peppers, rock melon and tomato during the packinghouse handling. Then the produce such as legumes, squashes, cucumber, eggplant, capsicum, okra, pumpkin, muskmelons, rock melons and watermelon are sorted for defects and colour differences into various categories by volumetric weights, diverging roll sizers, diverging bar sizers, by diameter with belt sizers or by weight (Zhao et al. 2017). Followed by packing these sorted products into their respective containers such as cartons, flats and barrels. Conventionally, mechanical harvesters are used to do the preliminary pre-sizing, sorting and packing operations, though these incur a high investment, they cut down on the operating costs and pay for themselves within a few years. Packed containers are then palletized for further shipment which is usually carried out at a centralized cooling facility. Palletizing is carried out after hydrocooling but before forced-air cooling. In field pack operations, this step is usually carried out at the fields. Different cooling methods are employed according to specific product types, where cooling parameters are tailored to the needs. Forced-air cooling is used for beans, rockmelons, cucumbers, muskmelons, peas, capsicum, soft-rind squashes and tomatoes. While chilling-sensitive commodities such as squashes, peppers, eggplants and cherry tomatoes are cooled using forced-air evaporative cooling. Cantaloupe, sweet corn and okra are cooled before grading using the hydrocooling technique. During hot weather, hydrocooling cycles might be insufficient; therefore after palletizing, the product is stored in cold rooms before being transported to markets. For sweet corn and rock melons, both package icing and liquid icing methods are utilized. For large-volume operations, products are placed in temporary cold storage rooms after cooling and before shipment.

In smaller farm operations, products are transported to central cooperatively owned or distributor-owned facilities for cooling. Refrigerated railcars are used to transport tomatoes, rock melons and muskmelons, but most commonly refrigerated trucks or vans are used for transportation. During transportation, it is essential to segregate ethylene-sensitive products to prevent ripening and spoilage from ethylene exposure. Ripening methods and modified atmospheres are some special treatment methods employed during the storage of products. Ethylene is applied onto products such as tomatoes, honeydew, casaba and crenshaw melons for ripening either while shipping or at the destination. To achieve the best results, it is recommended to apply ethylene soon after harvest at a desired temperature range of 12.5°–25 °C (55°–77 °F). At temperatures greater than 30 °C (86 °F), the development of red colour in tomatoes gets inhibited (Rizk et al. 2014). At an ethylene concentration of 100 ppm, ripening is carried out for 24–72 h. Treating the produce before packaging is advantageous in preventing the growth of pathogens under warmer conditions. Ethylene treatment before packing aids in eliminating infected produce, while ethylene treatment after packing facilitates uniform packed product.

With increased demand for ready-to-eat packed food due to rapid industrialization, population growth and changed lifestyle, packaging is highly crucial to ensure safe and efficient delivery of the commodity by providing protection, convenience, communication and containing the produce. With the emergence of microbial outbreaks, it has led to the demand for smart packaging with antimicrobial functions to preserve the quality and safety of the produce. The crucial function of packaging is to minimize physical damage to fresh produce and obtain better shelf life; therefore, it is critical to identify the appropriate type of packaging. Individual packaging such as moulded trays is suitable when required to ensure no physical contact with other products, ensuring improved texture and nature. Packaging films utilized at the source require high gas permeability and anti-fog properties. For a packaging to considered efficient, it must carry out the following functions—protection from physical damage, contamination, odour, toxins and the material must be non-toxic, usage friendly, act as a barrier for moisture and oxygen ingress, filter harmful UV lights, should be transparent, economical and preferably biodegradable. Polymeric films are commercially preferred for food packaging owing to the fact of high controllability in their movement and concentration of gases, thereby abridging the respiration rate and promoting shelf life of the produce. Temperature is a directly influential factor in MAP as it affects the respiration rate and has an effect on the product shelf life (Xie et al. 2020). The reduced O_2 concentration in MAP causes fermentation of tissues, contributing to the undesirable flavour and odours in the commodity. Modified atmosphere packaging is of two types—active and passive MAP. In active MAP, a vacuum is created using the desired gaseous concentration with absorbers controlling it. In passive MAP, the atmosphere within the product is achieved because of its respiration. Modified atmospheres are seldom preferred for commercial purposes to transport melons, tomatoes for long-distance markets with oxygen levels of 3–5% with carbon dioxide below 5% to avoid injury. Ripen fruits and vegetables are highly sensitive to low temperatures, causing a delay in colour changes and the development of sensory attributes. Around 30% of the harvest gets wasted due to poor establishment of post-harvest management methods. By-products are cheaper sources of functional ingredients such as peptides, carotenoids and phenolic compounds which are derived from primary agro-food production processes, thereby promoting a circular economy. Food functionalization is an interesting market where there is a constant requirement for bioactive ingredients to be used for the development of functional products (Chen and Tang 1998).

Studies project a consistent increase in demand for functionalized food in the twenty-first century to meet the nutritional requirements of an ever-changing world. According to the European Union (EU), the strategic approach of reduce, reuse, recover and recycle is the ideal action plan to achieve a circular economy. Reusing agricultural by-products can either represent a renewable source for existing food additives or originate a new value-added ingredient with functional compounds and properties benefitting the entire food system. By-products are rich in polysaccharides, organic acids, proteins and other bioactive compounds which at no additional production and at low industrial expense can be further processed to make a rich source of food additives.

15.3 Classification and Composition of By-Products

15.3.1 By-Products from Fruit Processing

Banana or *Musa* spp. the most consumed fruit worldwide with a production of over 120 million tonnes per year. Banana peels, rhizomes, leaves and pseudo-stems are the most utilized by-products of the banana plant which are highly beneficial macromolecules. The peel consists of dietary fibres like cellulose, lignin, resistant starch, pectin, hemicellulose, protein, amino acids, polyunsaturated fatty acids and potassium, antioxidants including hydroxycinnamic acids, flavonoids, phytosterols, carotenoids, anthocyanins and vitamins. Pomegranate juice produces peels, pulp and bagasse as by-products where significant presence of ellagitannins, flavonoids and anthocyanins with antimicrobial, anti-mutagenic, anti-carcinogenic, anti inflammatory and anti-diabetic properties have been identified. *Phoenix dactylifera L.* or date palms are extensively grown in arid regions around the world. Date palm consists of flesh (pulp) and seed (pits) with a rich source of carbohydrates, dietary fibres, vitamins, fatty acids and proteins. The seed is the only by-product of dates which is an excellent source of dietary fibres, proteins and lipids (Teshome et al. 2023). The anti-carcinogenic properties are contributed by the phenolic and antioxidant compounds present in the dates. Mango or *Mangifera indica L.* cultivated mostly in tropical countries is highly rich in essential nutrients with peels and seeds as major by-products. The peel accounting for 24%, of the total mango fruit weight, is highly rich in dietary fibres, carotenoids, vitamins, pectin, edible fats, proteins, fatty acids like stearic and oleic acid, polyphenols such as hydrolysable tannins, flavonoids, proanthocyanidins, alkylresorcinols, mangiferin, quercetin, kaempferol, ellagic, gallic, protocatechuic and ferulic acids. Polyphenols found in mango have been extensively studied for its anti-carcinogenic activity against breast and colon and antimicrobial activity against Gram-positive and Gram-negative bacteria. *Citrus* spp. are extensively used for juice production with huge by-products consisting of peel and seeds. Citrus fruits are in general made up of 35%, internal tissues, with 65% peel and 10% seeds. Ascorbic acid, flavonoids like hesperidin, naringin and narirutin, flavanols such as rutin and quercetin and flavones like diosmin and tangeretin have been found in citrus by-products. Citrus peel is also an excellent source of essential oils, with D-limonene, natural antioxidants, ethanol, organic acids, pectic oligosaccharides and pectin (Nour et al. 2015).

15.3.2 By-Products from Vegetable Processing

Brassica oleracea L.spp Botrytis or cauliflower has a very high waste index with tonnes of by-products from stems and leaves generated annually. These by-products are found to contain beneficial components such as dietary fibre, phenolic compounds, vitamin C, glucosinolates, leaf protein, carotenoids, flavonoids like kaempferol and quercetin and hydroxycinnamic acids such as caffeic and sinapic acids. The protein hydrolysates were studied for their glucose and glycogen regulation properties along with angiotensin I-converting enzyme inhibitory activities in cell-free systems and antioxidant properties. Carrot juice has a low yield of only 50%, where the raw material remains as pomace which is later used as manure. This pomace contains valuable compounds such as carotenoids, dietary fibre, uronic acids and neutral sugars. The pomace disposal is a threat to the environment in areas with congested space and inadequate water supply. Around 50% of the carotene present in carrots is lost with the pomace (Sharma et al. 2012). Fresh unblanched carrots contain 17% of total α- and β-carotenes, while the blanched carrots contain around 35% of α- and β-carotenes. The dietary fibre constituents of carrot pomace contain (%) 3.8 pectin, 12.3 hemicellulose, 51.6 cellulose and 32.1 lignin. Dehydrated carrot shreds are an alternative source for the availability of carrots all throughout the year. Fortification method helps in utilizing carrot pomace in baked products, dressings, pickles, fortified bread and biscuits and functional drinks, and however, this addition affects the sensory quality of these products, therefore raising the need to study consumer acceptability (Smith and Yu 2015). Carrot pomace consists (%) of 5.0 protein, 8.0–9.0 reducing sugar, 6.0 minerals and 48.0 total dietary fibre.

Tomato (*Solanum lycopersicum*) being one of the most consumed vegetables is an annual crop cultivated in temperate regions. Juice, sauces, purees, pastes and canned tomatoes are the most commercially manufactured tomato products with solid wastes such as vascular tissues, seeds and peels. Tomato by-products are rich in lycopene with high antioxidant properties that aid in the prevention of chronic diseases. Some phenolic compounds such as caffeic, ferulic, chlorogenic acids, quercetin and quercetin 3-β-*O*-glucoside are also important components present in tomato puree (Banga and Bawa 2002). *Daucus carota L.* (carrot) is a root vegetable that is highly rich in phytonutrients like carotenoids, polyphenols, vitamins, minerals, dietary fibres consisting of D-galacturonic acid, L-arabinose, L-rhamnose, D-galactose and some monosaccharides.

15.3.3 Chemical Composition and Nutritional Value

A huge source of bioactive compounds and nutrients such as pectin, proteins, fat, minerals, vitamins and fibres are found in fruit by-products like peel, seeds, skin, pomace and stones (Table 15.1). Banana peel comprises phytochemicals with high antioxidant capacity including phenolic compounds, dietary fibres and carotenoids, contributing to about 40% of the edible portion of the fruit. Other phytochemicals such as catecholamines, flavonoids, phenols, steroids, phenols, steroids, glycosides, phytosterols and terpenoids; and macro-minerals such as potassium, calcium, phosphorus, magnesium; and trace minerals—iron and zinc and also vitamins C and A are found abundant in the banana peel. The fibre content of banana peel is a rich source of pectin, therefore used as an additional product for emulsification in industrial processes. Banana peel is utilized in nutraceutical and pharmaceutical industries as a value addition to the products; in the food industry, banana peel is employed to promote health factors by contributing bioactive compounds from a cheap and high-rich source. Forty percent of fatty acids in banana peel consists of linoleic acid (Omega-6) and α-linoleic acid (Omega-3); these components act as anti-inflammatory properties on liver and obesity, respectively. The antihypertensive, anti-diabetic, anti-inflammatory and antibacterial properties of banana peel are contributed by the bioactive compounds—flavonoids, tannins, phlobatannins, alkaloids, glycosides, anthocyanins and terpenoids. Phenolic compounds such as gallic acid, catechin, gallocatechin and epicatechin are found in higher concentration in banana peels compared to other fruits, indicating the peel to be a great source of antioxidants. Other phenolic compounds such as flavonols, hydroxycinnamic acid, flavan-3-ols and catecholamines exhibit antioxidant activity by preventing reactive oxygen species (ROS) formation, direct ROS scavenging, and induction of antioxidant enzymes. The ferulic acid compound found in banana peel may act as an anti-melanogenesis factor by regulating vascular endothelial growth factor expression, initiating nitric oxide synthase and acting as tumour suppressor gene. The apple pomace is highly nutritious in fibres, vitamins, minerals, phenolic compounds and pigments, which play a crucial role by treating gastrointestinal disorders, reducing serum triglycerides and LDL cholesterol and regulating glycemia. Dihydrochalcones, procyanidins, flavan-3-ol monomers, flavonols, anthocyanidins and hydroxycinnamic acid are the most common phenolic compounds found in apple pomace. The

Table 15.1 Utilization of fruit and vegetable by-products

S. No.	By-product	Application	Reference
1	Raspberry, blueberry, blackcurrant, chokeberry, carrots, red beet and sour cherries in the form of juices	Fruit and vegetable flour for the production of bread, dough products, dairy and confectionery products	Ivanova et al. (2023)
2	Pineapple, acerola and passion fruit (pulp) residues	Production of fruit and vegetable flours	Martins et al. (2023)
3	Legume flour, starch flour or other dietary fibres extracted from by-products of fruits and vegetables	Manufacture of various gluten-free bread and pasta items	Rout et al. (2024)
4	Potato peel wastes and powders	Used as a supplement for making *pandesal*, a Filipino breakfast staple.	Lagmay (2021)
5	Fruit and vegetable waste-based green films (FVBGF)	Used as packaging material for livestock products such as meat, poultry and dairy	Thakur and Raposo (2023)
6	Fruits and vegetable waste-based starch and millet starch made from the waste of underused crops	Used as a biodegradable film in food packaging materials	Sarma et al. (2023)
7	Fruit and vegetable by-products	Used as an additive in meat products to improve their functional characteristics	Campos-González et al. (2023)
8	Peels of banana, papaya, orange, tomatoes and gourd	Green synthesis of metallic nanoparticles such as silver, gold, zinc oxide, iron, copper, palladium and titanium	Kaur et al. (2023)
9	Fruit and vegetable dietary waste	Production of enzymes such as phytase, pectinase, amylase, cellulase, pectinase, protease and organic acids like acetic acid, lactic acid	Zahid and Khedkar (2023)
10	By-products and wastes derived from fruits and vegetables, dairy products, bakery products, meat, fish and sawdust	Heterotrophic cultivation of *Chlorella vulgaris*	Marques (2023)
11	Citrus peel waste	Manufacture of animal feed Increases soil fertility Controls hypercholesterolemia and other health problems Exhibits inhibitory activity against pathogenic bacteria and fungi Used in biogas production Used in production of biodegradable polymers Used in production of cosmetics Used as preservatives, flavouring agents and food additives in food industries	Swamy et al. (2023)
12	Citrus pomace of orange, lemon, red grapefruit, mandarin, lime and pomelo	Used as a feed raw material in animal's rations	Neshovska (2023)

(continued)

Table 15.1 (continued)

S. No.	By-product	Application	Reference
13	Organic wastes from vegetables, fruits, empty fruit bunches (EFB), urban waste and cow manure	Used as biogas sludge to improve and maintain sandy soil chemistry	Siregar et al. (2023)
14	Underused wastes and biomass (seed, peel, rind and pomace) from carrot and turnip processing	Used to create biofuels such as bioethanol, biomethane, biohydrogen and biobutanol	Kaur and Ghoshal (2024)

antioxidant, anti-inflammatory and antimicrobial properties in the apple pomace are contributed by the phenolic compound, phlorizin due to its ability to change glucose absorption and excretion. Phlorizin also inhibits sodium/glucose cotransporters in the intestine and kidneys, thereby benefitting postprandial hyperglycemia therapy in diabetes and related illnesses such as obesity.

Most common by-products of mango are mango seed and mango peel, both of which are rich sources of vitamin C, beta-carotene, polyphenols and dietary fibre. Mango seed kernels are highly utilized for nutraceutical applications owing to the phytochemical composition of phenolic acid, flavonoids, catechins, hydrolysable tannins and xanthanoids. Mango seed kernels are a common raw material source in producing energy-rich functional foods due to their constituents of good amounts of fat, protein and carbohydrates (Melo et al. 2019). Mango peels are found to comprise 45–78% of dietary fibre; similarly, citrus peels are also a source of highly valuable phytochemicals with high antioxidant, anti-inflammatory and anticancerous properties. The most common by-product of grapes is from wine processing which can be utilized as a cheap source of dietary supplements as it is a good source of dietary fibres, lipids, proteins, natural antioxidants and phenolic compounds. Mango peel, accounting for almost 24% of the mango fruit, is highly rich in phytochemicals with broad bioactivities contributing to the functional and antioxidant properties in the food applications. Ethyl gallate and penta-O-galloyl-glucoside have strong scavenging activities for hydroxyl radicals (-OH), superoxide anion (O_2^-) and singlet oxygen (O_2) and also possess anti-tumour, antioxidant, anti-cardiovascular and hepatoprotective properties". (OECD, F 2012).

By-product of apple, the pomace is a rich source of bioactive compounds such as phenolic acids, flavonoids and dihydrochalcones apart from carbohydrates, proteins, vitamins and minerals. The pomace comprises mainly insoluble sugars like cellulose, hemicellulose, lignin and simple sugars such as glucose, fructose and galactose. Also, some minerals like phosphorus, calcium, magnesium, iron and fatty acids like linoleic acid (18:2 n-6) and oleic acid (18:1 n-9) have been found in apple pomace. Grape pomace contains dietary fibre in the highest concentration accounting for almost 70% of the total content followed by carbohydrates, minerals and vitamins. Where the insoluble dietary fibres such as cellulose and hemicellulose account for 78% and water soluble fibres including α-glucans, pectin and gums account for almost 11% of the pomace. Health benefits of pomace are linked to its fibre fermentation capability in the colon, producing short-chain fatty acids with probiotic potential. Phenolic compounds are one of the most important non-nutrient elements in the pomace (OECD, P 2015). The monomeric phenolic acids, oligomeric proanthocyanidins and glycosylated anthocyanins and catechins, epicatechin, dimers and trimers of procyanidins, resveratrol and anthocyanins found in the pomace have antioxidant, antimicrobial properties and anti-mutagenic properties. The grape pomace also contains high concentrations of tannic compounds and other common phenolic compounds such as hydroxybenzoic and hydroxycinnamic acids, flavonols, stilbenes and anthocyanins. Extensive studies have been carried out to understand the cardio-protective, anticancer, anti-inflammatory, anti-ageing and antimicrobial properties in grape pomace. Consuming grape pomace has been found to have a variety of health benefits such as glucose absorption regulation, obesity prevention, blood cholesterol reduction, reduced cardiovascular risk, regulated bowel functions and water retention.

Grape seeds are densely composed of stilbene, resveratrol, rutin, gallic acid and catechinalate compounds exhibiting cardiovascular, antimicrobial, antioxidant and anticancerous properties. Both avocado peel and seed are found to comprise good source of carbohydrates (70%), proteins (8%), lipids (4%) and ash (3%), with high in phytochemicals such as phenolic acids, condensed tannins and flavonoids exhibiting antioxidant and anti-inflammatory properties (Ajila et al. 2007). Pineapple by-products are a great source of dietary fibre, finding its application in the production of fibre-rich foods and also as a substrate for the production of organic acids with high demand for acidification and flavour enhancers of low-acid foods. The phytochemical components found in fruit by-products promote human health development by functioning as antioxidants by acting against free radicals and oxidative stress.

Citrus (*Citrus* sp.) by-products are rich in sugars, carbohydrates, proteins, lipids, essential oils, carotenoids, vitamins, polyphenols, flavonoids and phenolic acids. Commercial extraction of polyphenols from citrus by-products is highly advantageous due to their quantitative presence in fruit contributing to antioxidant, anti-inflammatory and anti-carcinogenic effects. Citrus by-products are rich in phenolic acids classified based on their free radical scavenging activity as hydroxybenzoic acid (gallic, vanillic, syringic acids) and hydroxycinnamic acids (caffeic, ferulic, p-coumaric and sinapic acids).

Avocado (*Prunus americana*) by-products and seeds are used to treat gastric disorders due to their anti-inflammatory, antioxidant and antimicrobial properties. Leaves and seeds were found to contribute anti-cholinesterase and antioxidant properties. Phenolic compounds such as gallic acid, 3,4-dihydroxyphenylacetic acid, 4-hydroxybenzoic acid, vanillic acid, p-coumaric acid, ferulic acid and quercetin found in avocado pulp contain antioxidant properties. Avocado peels were found to have higher concentrations of phenolic compounds, especially B-type procyanidins and epicatechin when compared with seeds. The peel extract was found to protect transgenic parkin Drosophila melanogaster fly against paraquat-induced oxidative stress, movement impairment and lipid peroxidation, as a model of Parkinson's disease. Anti-radical activity of avocado by-product is contributed by polyphenols (+), catechin (−), epicatechin, 3-O-caffeoylquinic acid (chlorogenic acid isomer), quercetin, naringenin and sakuranetin, where these compounds are linked to the reduction of a variety of deteriorative processes by reducing free radical formation and scavenging. Pineapple peel accounts for the highest contributor of food waste accounting for 57%, which is also high in nutrition, containing polyphenol compounds such as myricetin, salicylic acid, tannic acid, trans-cinnamic acid, p-coumaric acid, ferulic acid and syringic acid with antioxidant and antimicrobial activity (Carré et al. 2016).

15.4 Utilization of By-Products

Agricultural waste and by-products involving intensive agriculture practices originate from all activities related to crop cultivation, vegetable and fruit production and processing as well as from dairy, poultry, aquaculture and meat production. Substantial amounts (30% approx.) of world food production (1.3 billion tonnes) are wasted for one or more reasons, notably fruits and vegetables as opposed to cereals, root and tuber crops, dairy and beverages. Rapidly expanding global population demands increased food and fodder supply, resulting in increased agricultural waste biomass output. The resulting biomass when used effectively can produce value-added products, which will not only utilize this abundant biomass but also alleviate numerous environmental concerns while increasing farmer income. Rather than burning and releasing sequestered carbon into the atmosphere, the biomass can be converted into useful metabolites and products such as antioxidants, charcoal, polysaccharides, peptides, polyphenolics, fertilizer, compost, animal feed and biofuels, as well as various other innovative uses (Table 15.2). Effective use of existing resources can assist in solving numerous socioeconomic and environmental problems, as well as improve the standard of living in rural regions, particularly in developing nations (Arora et al. 2023).

15.4.1 Food Industries/Food Applications

Waste management processes have been devised by manufacturing organizations with the goal of transforming them into goods for other uses (such as fertilizers or animal feed) while preventing environmental pollution. The by-products are determined by the various raw materials and technical procedures utilized in the manufacturing of fruit and vegetable products (Ivanova et al. 2023).

Table 15.2 Fruit/vegetable residues and their potential benefits (Joshi 2020; Mohd Basri et al. 2021; Zhu et al. 2023)

Type of fruit/vegetable	Nature of residue	Approx. wastage (%)	By-products and its class of Residue	Potential benefits
Mango	Peel, stones	45	Pectin, cattle feed, alcohol, fat, tannins, syrup, bromelain	Antioxidant, colour, flavour and texture enhancer
Apple	Peel, pomace and seeds	30	Juice, pectin, lactic acid, citric acid, vinegar, wine; Flavanols—Catechins, Epicatechins, Procyanidins, Proanthocyanidins	Antioxidant, colour, flavour and texture enhancer
Citrus	Peel, pulp, rag and seed	50	Essential oil, pectin, cattle feed, candy; Beta-carotene, lutein, Zeaxanthin, flavanones	Antioxidant, antibacterial
Pineapple	Skin, Core	33	Cattle feed, biogas	Antioxidant, antibacterial
Grape	Stem, skin and seeds	20	Flavanols—Catechins, Epicatechins, Procyanidins, Proanthocyanidins	Antioxidant, antibacterial
Banana	Peel	35	Fibre	Antioxidant, colour/flavour enhancer
Tomato	Core, Peel and seeds	30	Animal feed, seed oil	Antioxidant, antibacterial
Potato	Peel	15	Animal feed, single cell protein	Texture enhancer

15.4.1.1 Manufacture of Flour

A technique was developed for manufacturing fruit and vegetable flour from by-products and waste generated by the industrial processing of fruit and vegetables. Processed by-products of fruits such as raspberry, blueberry, blackcurrant, chokeberry and those of vegetables like carrot, red beet and sour cherries were used in this study. The developed fruit and vegetable flours were a rich source of dietary fibres and biologically active substances, intended for their use as natural components in recipes for the production of bread, dough products, dairy and confectionery products (Ivanova et al. 2023).

Drying process and its effect on the total carotenoid content of pineapple, acerola and passion fruit residues were investigated. Drying was carried out using an air circulation oven at temperatures of 60 and 70 °C for a total time of 180 min. The drying kinetics of the fruit residues were similar, although the moisture ratios fell rapidly at a temperature of 70 °C. The total carotenoid content was found to be higher in the drying kinetics at 70 °C for the pineapple residues, while no significant difference was observed for the acerola and passion fruit residues at the two evaluated temperatures ($p \leq 0.05$). It was concluded that fruit pulp residues from agro-industries can be utilized as alternatives for producing fruit and vegetable flours. These residues maintain a stable chemical composition during drying at the temperatures studied, implying that they might be used as by-products in various food applications (Martins et al. 2023).

In response to rising customer demand for gluten-free food products, food technologists have analysed a range of gluten-free components derived from diverse sources in order to reproduce the network structure generated by gluten in a wheat-based dough. Due to its functional properties, such as solubility and water-binding capacity, which have been critical in the development and production of gluten-free meals, the utilization of legume flour, starch flour and other components such as dietary fibres derived from fruit and vegetable by-products has been in focus in recent years. Their nutritional composition may also compensate for the lack of nutrients typically highlighted in commercial gluten-free bread and pasta items, which has a positive influence on people's health (Rout et al. 2024).

Dietary fibres are plant-based carbohydrates made up mostly of cellulose, hemicelluloses, pectic polysaccharides and lignin, and their sources have become popular in the food and nutrition business due to their importance in the prevention and control of various diseases. According to research, fruit and vegetable by-products such as potato peel waste contain significant levels of dietary fibres. Lagmay (2021) conducted research to assess the effects of adding different amounts of potato peel powder (2 g, 5 g and 8 g) to pandesal, a Filipino breakfast staple, on total dietary fibre content and sensory acceptability. A sensory evaluation was conducted using the 7-point Hedonic Scale, with 70 adult respondents chosen at random using convenience sampling. Analysis of the gathered data revealed a significant difference ($p < 0.05$) for appearance, taste, odour, texture and overall acceptability between the pandesal samples. With increasing levels of potato peel powder supplementation, dietary fibre content increased as well. The findings showed that pandesal bread samples were acceptable, while potato peels were high in dietary fibre. It was concluded that potato peel waste would be a crucial element in boosting the nutritional content of traditional pandesal formulations (Lagmay 2021).

15.4.1.2 Food Packaging Material

Agricultural waste pollution is a major global concern for environmentalists. A solution lies in economically valorizing agricultural residues, turning by-products into raw materials for other industries. For example, green films made from fruit and vegetable waste can be used for food packaging. This area is well-researched, with many biomaterials explored. These biofilms not only provide barrier properties but also offer antioxidant and antimicrobial benefits from bioactive additives like essential oils. Advanced technologies such as encapsulation, nano-emulsions and radio-sensors further enhance their performance and sustainability. Perishable livestock products, like meat and dairy, rely on effective packaging to extend shelf life. Thakur et al. reviewed these aspects, presenting fruit and vegetable-based green films (FVBGFs) as a viable packaging material for livestock products, discussing their bio-additives, technological enhancements, properties and applications (Thakur and Raposo 2023).

Non-commercial starches from fruit and vegetable waste were explored as alternatives to commercial starches for biodegradable film production. This aligns with circular economy principles and sustainable manufacturing goals, emphasizing waste recycling and innovative concepts. Starches from neglected fruits and vegetables offer significant research potential for packaging technology. For example, millet starch derived from crop waste presents new opportunities for biodegradable film development. Incorporating these starches into circular manufacturing practices is essential to reduce carbohydrate waste. Additionally, optimizing biodegradable film composition to enhance barrier properties and shelf life is crucial. Further research may employ hybrid optimization, neural networks or deep learning models to refine film composition and monitor materials within packaging systems (Sarma et al. 2023).

15.4.1.3 Processing of Meat

Reformulating meat products with beneficial vegetable components is a key focus for food programmes and the meat industry. Using fruit and vegetable by-products helps to revalorize them into functional products and reduce pollution. However, proving the functional properties of these compounds requires in vitro, in vivo and in silico studies during development. Campos-González et al. showed how the compounds interact with meat matrices and benefit human health. Understanding their metabolism and bio-transformation in the body is crucial for consumer health benefits. They reviewed the studies validating the functionality of fruit by-products in meat. Overall, incorporating these by-products in meat products is a promising strategy to enhance functionality and introduce new market products (Campos-González et al. 2023).

15.4.2 Pharmaceutical and Nutraceutical Applications

15.4.2.1 Synthesis of Metallic Nanoparticles

Nanoparticle synthesis through chemical, physical and biosynthesis methods is essential in nanotechnology. Traditional methods are costly, energy-intensive, use hazardous chemicals and generate significant by-products. Consequently, using natural sources for nanoparticle synthesis is becoming popular due to its simplicity, non-toxicity, cost-effectiveness and sustainability. Fruit and vegetable peels, often discarded by food manufacturers, are rich in bioactive compounds like polyphenols, flavonoids, antioxidants and essential oils, which aid in reducing and stabilizing metal ions in green nanoparticle synthesis. Peels, from fruits such as banana, papaya, orange and tomato, and vegetables like gourd, have been used to synthesize metallic nanoparticles, including silver, gold, zinc oxide, iron, copper, palladium and titanium. Kaur et al. reviewed nanoparticles synthesized from these peels and their biological activities, including anticancer, antimicrobial, antioxidant and catalytic properties (Kaur et al. 2023).

15.4.2.2 Production of Industrial Enzymes

Various food and vegetable wastes can be used to produce a wide range of industrial enzymes, beneficial for food research, pharmaceuticals, cosmetics, organic acids and chemicals. Zahid et al. explored the sustainable use of fruit and vegetable waste to produce key food industry enzymes like phytase, pectinase, amylase, cellulase and protease, as well as organic acids such as acetic and lactic acid, using various microorganisms isolated from these wastes (Zahid and Khedkar 2023).

15.4.2.3 Cultivation of *Chlorella vulgaris*

Microalgae present a promising solution due to their adaptability and efficiency in converting food waste into valuable biomass. Marques evaluated the use of food waste by-products as substrates for the heterotrophic cultivation of microalgae. Various recipes from seven product categories (fruits, vegetables, dairy, bakery, meat, fish and sawdust) were developed and analysed. Different pre-treatment methods (autoclaving, conventional heating and ohmic heating) were tested for their effectiveness in controlling microbial contamination and their impact on reducing sugar content, which is crucial for the growth of *Chlorella vulgaris*. Characterization showed that these by-products contain essential nutrients for microalgae cultivation. Autoclaving was the most effective pre-treatment, eliminating microbial contamination and reducing sugar content. *C. vulgaris* successfully grew in the by-product medium, achieving a biomass density of 19–26 × $10^{\wedge 7}$ cells/mL with an initial reducing sugar concentration of 20 g/L. However, this was about 80% lower than the biomass density achieved with a commercial medium. Supplementing the by-product medium with vitamins and trace metals did not improve growth, indicating that the by-products already contained sufficient micronutrients. These results demonstrate that retail sector by-products can be used for the heterotrophic cultivation of *C. vulgaris* (Marques 2023).

15.4.3 Agricultural Applications

15.4.3.1 Use as Raw Material for Animal Feed

Citrus fruits, primarily from the *Rutaceae* family, are the most consumed fruits globally, grown in over 140 countries. They are rich in dietary fibre and sugars like sucrose, glucose and fructose, which help prevent gastrointestinal disorders and improve cholesterol circulation. Widely processed, citrus fruits generate large amounts of by-products high in cellulose, hemicellulose, soluble sugars, pectin and essential oils. Citrus peel waste is especially rich in ascorbic acid, essential oils, pectin, flavonoids, polyphenols, sugars, dietary fibres and biomass, containing trace elements. Citrus waste biomass is used as animal feed and enhances soil fertility. It contains diverse phytochemicals with antimicrobial, anti-inflammatory, anti-diabetic, antidepressant and anti-carcinogenic properties. Studies indicate that citrus waste has nutraceutical potential for managing hypercholesterolemia and other health issues and exhibits inhibitory activity against pathogenic bacteria and fungi due to compounds like alkaloids, flavonoids and terpenoids. Flavonoids also reduce viral infection risks. Additionally, citrus waste is used in biogas production, biodegradable polymer manufacturing, cosmetics and nutraceuticals. Certain compounds from citrus waste serve as preservatives, flavouring agents and food additives. Swami et al. reviewed the bioactive compounds in citrus waste and their applications, providing insights into their nutritional, nutraceutical, biological and industrial uses. They offered comprehensive knowledge that may help optimize and sustain the extraction of novel bioactive compounds from citrus waste.

Chemical and mineral composition of citrus wastes from orange (*Citrus sinensis*), lemon (*Citrus limon*), red grapefruit (*Citrus paradisi*), mandarin (*Citrus reticulata*), lime (*Citrus aurantifolia*) and pomelo (*Citrus maxima*) were investigated. These waste products were sourced from a juice production factory in Musachevo, Western Bulgaria. The study measured indicators such as dry matter, moisture, crude protein, crude fibre, crude ash, ether extract, manganese, zinc, magnesium, calcium, phosphorus, sodium and potassium. Mineral content was assessed using an atomic absorption spectrophotometer. The analyses revealed variations in nutrient and mineral distribution among the different citrus wastes. The findings indicate that citrus pulp, rich in essential nutrients, could effectively replace some primary raw materials in animal feed (Neshovska 2023).

15.4.3.2 Used to Improve Soil Chemistry

The abundance of organic waste can be managed without burning by converting it into biogas and producing by-products like sludge. Biogas sludge, derived from organic waste, can be used to improve soil fertility, especially in sandy soils. Siregar evaluated the impact of biogas sludge from various organic waste materials—such as vegetables, fruits, empty fruit bunches (EFB), urban waste and cow manure—on the chemical properties and nutrient content of corn plants grown in sandy soil. This study was conducted in Sei Semayang Village, Sunggal District, at an altitude of +31 metres above sea level, using sandy soil from Namo Mirik Village in the Kutalimbaru District. The researchers employed a factorial randomized block design with two factors: the source materials for the biogas sludge (vegetables, fruits, EFB, urban waste and cow manure) and the dosage levels (0 L ha^{-1}, 400 L ha^{-1}, 800 L ha^{-1}, 1200 L ha^{-1}). The application of biogas sludge showed varying effects on the chemical properties and nutrient content of the soil and corn plants. Different dosages of sludge improved chemical properties, total microbial content, nutrient levels and corn plant growth. The study recommends using biogas sludge in combination with chemical fertilizers, particularly from municipal waste, cow dung and vegetable waste, to optimize soil fertility and plant growth (Siregar et al. 2023).

15.4.3.3 Manufacture of Biofuels

Underused fruit and vegetable wastes (such as seeds, peels, rinds and pomace) are rich in beneficial bioactive substances, including functional nutrients, amylopectin, phytochemicals, vitamins, enzymes, dietary fibres and oils. These wastes are among the most common agro-industrial by-products. This discarded biomass can be utilized to produce biofuels like bioethanol, biomethane, biohydrogen and biobutanol. Kaur specifically highlighted the applications of carrot and turnip processing wastes in the creation of various value-added products (Kaur and Ghoshal 2024).

15.5 Conclusions and Future Prospects

The fruit and vegetable processing industry generates a substantial amount of by-products, which have traditionally been considered waste. However, the significant potential of these by-products as valuable resources is being explored for various food, pharmaceutical, nutraceutical and agricultural sectors. By-product utilization has been explored in the era of circular economy due to the nutritional value of the bioactive compounds in the residues generated. This would promote resource efficiency, reduce the environmental impact of wastes generated during food processing and create new revenue streams for the food processing industry. The effective utilization of by-products would lead to the development of functional foods, pharmaceuticals and biofertilizers, which not only add economic value but also contribute to environmental sustainability. Furthermore, the utilization of fruit and vegetable by-products would pave the way for sustainable and economically viable future food processing and agricultural practices.

Future trends and innovations in the field of utilization of fruits and vegetable processing residues may include efficient and cost-effective advanced extraction methods of bioactive compounds which will enhance their usability in food, pharmaceutical and nutraceutical industries. Development of clear guidelines by standardized regulatory frameworks would also ensure safe and effective use of the value-added produce across different industries as well as facilitate wider industry adoption and consumer acceptance. By addressing the technological and regulatory challenges, the widest potential of residue utilization could be unlocked, contributing to UN global efforts of Sustainable Development Goals 2030 in waste reduction and resource optimization.

References

Adhikary T, Kumar DH (2021) Advances in postharvest packaging systems of fruits and vegetables. In: Postharvest technology-recent advances, new perspectives and applications. IntechOpen, London, p 170

Ajila CM, Naidu KA, Bhat SG, Rao UP (2007) Bioactive compounds and antioxidant potential of mango peel extract. Food Chem 105(3):982–988

Arora J, Ramawat KG, Mérillon JM (2023) Disposal of agricultural waste and its effects on the environment, production of useful metabolites and energy: potential and challenges. In: Agricultural waste: environmental impact, useful metabolites and energy production. Springer Nature Singapore, Singapore, pp 3–20

Banga R, Bawa AS (2002) Studies on dehydration of grated carrots. J Food Sci Technol (Mysore) 39(3):268–271

Campos-González N, Gómez-Salazar JA, Cerón-García A, Lorenzo JM, Santos EM, Campagnol PC, Sosa-Morales ME (2023) Approaches to assess the functionality of meat products made from the incorporation of vegetable by-products: a review. Food Rev Intl:1–25

Carré P, Citeau M, Robin G, Estorges M (2016) Hull content and chemical composition of whole seeds, hulls and germs in cultivars of rapeseed (Brassica napus). OCL 23(3):A302

Chen BH, Tang YC (1998) Processing and stability of carotenoid powder from carrot pulp waste. J Agric Food Chem 46(6):2312–2318

Ivanova P, Georgiev D, Dobrev I (2023) Technology for processing fruit and vegetables flours. J Mt Agric Balkans 26(2):428–444

Joshi VK (2020) Fruit and vegetable processing waste management-an overview. Int J Food Ferment Technol 10(2):67–94

Kaur A, Ghoshal G (2024) Valorization of carrot and turnip processing wastes and by-products. In: Roots, tubers, and bulb crop wastes: management by biorefinery approaches. Springer Nature Singapore, Singapore, pp 183–214

Kaur S, Chopra HK, Panesar PS (2023) Fruit and vegetable peels for nanoparticles synthesis and applications. In: Nanomaterials from agricultural and horticultural products. Springer Nature Singapore, Singapore, pp 243–259

Lagmay EMC (2021) Dietary Fiber content and sensory acceptability of Pandesal supplemented with potato Peel powder

Lau KQ, Sabran MR, Shafie SR (2021) Utilization of vegetable and fruit by-products as functional ingredient and food. Front Nutr 8:661693. https://doi.org/10.3389/fnut.2021.661693

Marques FA (2023) Heterotrophic cultivation of Chlorella vulgaris using agri-food by-products (Doctoral dissertation)

Martins IR, Prudente EDS, Tiburco X, Bezerra FWF, Martins LHDS, Waughon T, Joele MRSP (2023) Drying kinetics and mathematical modeling of agroindustrial fruits residues from the Brazilian Amazon region

Melo PE, Silva APM, Marques FP, Ribeiro PR, Brito ES, Lima JR, Azeredo HM (2019) Antioxidant films from mango kernel components. Food Hydrocoll 95:487–495

Mohd Basri MS, Abdul Karim Shah NN, Sulaiman A, Mohamed Amin Tawakkal IS, Mohd Nor MZ, Ariffin SH et al (2021) Progress in the valorization of fruit and vegetable wastes: active packaging, biocomposites, by-products, and innovative technologies used for bioactive compound extraction. Polymers 13(20):3503

Neshovska H (2023) Determination of the chemical and mineral composition of citrus by-products in relation to its utilization as a feed raw material. Bulg J Animal Husb 60(4):42–48

Nour V, Ionica ME, Trandafir I (2015) Bread enriched in lycopene and other bioactive compounds by addition of dry tomato waste. J Food Sci Technol 52:8260–8267

OECD, F (2012) Agriculture Organization of the United Nations: "OECD-FAO agricultural outlook 2012

OECD, P (2015) OECD-FAO agricultural outlook 2015–2024

Rizk EM, El-Kady AT, El-Bialy AR (2014) Charactrization of carotenoids (lyco-red) extracted from tomato peels and its uses as natural colorants and antioxidants of ice cream. Ann Agric Sci 59(1):53–61

Rout S, Gupta RK, Jadhav HB, Srivastav PP, Annapure U (2024) Understanding the role of dietary fibers on gluten-free pasta's functional quality aspects. In: Development of gluten-free pasta. Academic, pp 157–178

Sarma M, Chakraborty S, Kesavan R, Dash KK, Nayak PK (2023) Starch-based biodegradable film from fruit and vegetable waste and its standardization modules based on neural networks and response surface methodology. Starch-Stärke 2300082

Sharma KD, Karki S, Thakur NS, Attri S (2012) Chemical composition, functional properties and processing of carrot—a review. J Food Sci Technol 49(1):22–32

Siregar N, Rauf A, Hanum H (2023) Improvement in sandy soil chemistry using biogas sludge based on organic waste. In: AIP conference proceedings (Vol. 2913, No. 1). AIP Publishing

Smith IN, Yu J (2015) Nutritional and sensory quality of bread containing different quantities of grape pomace from different grape cultivars. Ec Nutr 2(1):291–301

Sudheer KP, Indira V (2007) Post-harvest technology of horticultural crops, vol 7. New India Publishing

Swamy CT, Sivadurga K, Reddy MP, Marimuthu G, Prashantkumar CS, Premkumar C, Purewal SS (2023) Citrus waste: a treasure of promised phytochemicals and its nutritional-nutraceutical value in health promotion and industrial applications. In: Recent advances in citrus fruits. Springer International Publishing, Cham, pp 395–426

Teshome E, Teka TA, Nandasiri R, Rout JR, Harouna DV, Astatkie T, Urugo MM (2023) Fruit by-products and their industrial applications for nutritional benefits and health promotion: a comprehensive review. Sustain For 15(10):7840

Thakur N, Raposo A (2023) Development and application of fruit and vegetable based green films with natural bio-actives in meat and dairy products: a review. J Sci Food Agric 103:6167

Tlais AZA, Fiorino GM, Polo A, Filannino P, Di Cagno R (2020) High-value compounds in fruit, vegetable and cereal byproducts: an overview of potential sustainable reuse and exploitation. Molecules 25(13):2987

Verma LR, Joshi VK (2000) Post-harvest technology of fruits and vegetables, vol 1. Oxford Book Co, Jaipur, pp 1–76

Vicentini A, Liberatore L, Mastrocola D (2016) Functional foods: trends and development of the global market. Italian J Food Sci 28(2):338

Xie Y, Niu X, Yang J, Fan R, Shi J, Ullah N et al (2020) Active biodegradable films based on the whole potato peel incorporated with bacterial cellulose and curcumin. Int J Biol Macromol 150:480–491

Zahid A, Khedkar R (2023) Sustainability in production of enzymes from fruit and vegetable waste. In: Sustainable food systems (volume II) SFS: novel sustainable green technologies, circular strategies, food safety & diversity. Springer, Cham, pp 111–140

Zhao HM, Guo XN, Zhu KX (2017) Impact of solid state fermentation on nutritional, physical and flavor properties of wheat bran. Food Chem 217:28–36

Zhu Y, Luan Y, Zhao Y, Liu J, Duan Z, Ruan R (2023) Current technologies and uses for fruit and vegetable wastes in a sustainable system: a review. Food Secur 12(10):1949

Chapter 16
Post-harvest Processing, Packaging, and Commercialization of Underutilized Fruits of India

Vasundra Vijay Bhat, Manpreet Kaur, Akash Deep Shukla, and Yashi Srivastava

16.1 Introduction

India's varied soil and climate contribute significantly to the world's horticultural industry. According to Ravindran et al. (2007), it grows tropical, subtropical, and temperate fruits. Major fruit crops in India include citrus, guava, banana, mango, and apple, which together account for more than fruit crops cover 72% of the total land, although native fruit crops, which are rather productive, only make up 6.56% of the area (0.437 mha) (Meena et al. 2022).

India produced 107.24 million metric tons of fruits, according to the NHB's Database (third Advance Estimates) APEDA in 2021–2022, of which 5.53 million metric tons were underutilized. Often called lesser-known or underutilized fruits, minor fruits are less grown and less popular. These fruits are not widely grown using contemporary agricultural techniques and therefore are not as desirable to mainstream markets. They can, however, survive in hostile environments because they are resilient. When used appropriately, underutilized fruits can improve food security, nutrition, health, income creation, and the environment (Hossain et al. 2021). There have also been reports about their potential as food items and as commercial fruit substitutes. Although the popularity of fruit crops varies depending on the location and crop, exposure might increase awareness of these areas (Bachheti et al. 2023).

Although these minor fruit species are used locally or globally, they have the potential to produce even more food resources because they are frequently ancient and indigenous (Srivastava et al. 2017a, b). According to Meena et al. (2022), many minor fruit crops could be utilized as fresh fruit, for cooking, and for health, as they provide essential nutrients. Some of these crops also have ornamental value. Minor fruits, on the other hand, are nutritious and can lower the likelihood of diabetes,

---

V. V. Bhat · M. Kaur · A. D. Shukla · Y. Srivastava (✉)
Department of Applied Agriculture, Central University of Punjab, Bathinda, Punjab, India

OP Chauhan (ed.), *Fruits and Vegetables Technologies*,
https://doi.org/10.1007/978-981-96-8433-5_16

cancer, and heart problems by containing vitamins, minerals, fiber, and polyphenols (Narzary et al. 2013).

In addition to providing employment opportunities and a source of revenue, these fruits are vital to the nutrition and way of life of rural and tribal populations, and local communities are often aware of their therapeutic qualities. Furthermore, because of their distinct characteristics, small fruits can be useful for processing and product development (Srivastava et al. 2017a, b).

The GOI, through its MIDH scheme, made special efforts to establish 57 underutilized fruit species orchards in 2005–2006 (Meena et al. 2022). In addition to being one of the Indian Gene Center's hotspots for biodiversity, the northeastern hill region is renowned for its abundance of traditional culture and ethnic varieties (Hazarika et al. 2016). North East India, a region noted for its varied soil, climate, and geography, makes up 7.7% of India's total geographic region and is habitat to 50% of the country's biodiversity. It has over 3% of the nation's total fruit varieties, making it rich in diversity.

Investigating ways to harness the potential of minor fruit species is becoming more and more important in order to supply the growing population's needs. They are becoming more popular because they can adapt to local agricultural practices, agroecosystems, degraded and marginal soils, and low-input farming. They also give rural communities access to food and nutrition. In addition, minor fruit crops offer the potential for value addition, high genetic variety, low pest risk, and multiple uses (Phukan and Das 2017).

Meghalaya, Assam, Nagaland, Mizoram, Arunachal Pradesh, Manipur, Sikkim, and Tripura are the eight states that make up the northeastern area of India. The region's average temperature typically ranges from 18 to 29 °C, while the colder months see temperatures between 9 and 24 °C (Singh and Sarkar 2022). The area has an average of 2000 mm of rainfall per year, acidic to strongly acidic soil, and rich organic matter in the soil. The northeast region has a particularly distinct environment that provides ideal growing conditions for almost all fruit crops, making it extremely rich in genetic variety for all horticultural crops (Paul and Panda 2021) (Table 16.1).

The majority of the Northeast region's underutilized fruits are largely unknown outside of their local markets and are frequently found in local markets. Consumers today are much more cognizant of the nutritional requirements and overall health of the foods they purchase. There is a trend away from artificial sweeteners and processed foods and toward natural foods that are higher in nutrients. Most of the horticulture fruit crops that are not being fully utilized are cultivated in residential gardens and provide an additional source of income for the rural people and tribal groups in the region. These fruit crops require value-added techniques to become more widely recognized in the market.

The cultivation of a substantial number of these fruits is possible even in poor conditions. These fruits are also recognized for their medicinal and nutritional properties, and they are able to fulfill the requirements of consumers who are concerned

Table 16.1 Seasonal pattern of underutilized fruits of India

Fruit name	Common name	Flowering season	Fruiting time
Actinidia callosa	Himalayan kiwi vine	NA	October–December
Baccaurea sapida	Khattaphal/Latkan	Summer months	May–July
Castanopsis indica	Indian chestnut	April–May	August–September
Pyrus pashia	Wild Himalayas pear	May	November–December
Rubus niveus	Mysore raspberry	NA	NA
Citrus medica	Fingered citron	NA	NA
Livistona jenkinsiana	Assam fan palm	February–March	September–December
Viburnum foetidum	Himalayan Evergreen viburnum	June–July	September–November
Dillenia indica	Elephant apple	May–June	July–August
Malus baccata	Siberian crabapple	NA	September
Machilus edulis	Wild avocado	NA	November–March
Phyllanthus acidus	Malay gooseberry	January–May	NA
Mangifera sylvatica	Himalayan mango	NA	April–May
Docynia indica	Wild apple		November–December
Elaeagnus latifolia	Bastard oleaster	September–December	March–April
Citrus macroptera	Wild orange	NA	NA
Grewia hirsute	Kukurbicha/Veronica lolia	June–July	NA
Physalis minima	Little Gooseberry	NA	July–October
Rubus ellipticus	Yellow Himalayan raspberry	NA	March–June
Elaeocarpus floribundus	Indian olive	April–may	August–October
Ziziphus mauritiana	Indian plum	September–November	
Carissa carandus	Karonda	May–June	August–September
Feronia limonia	Wood Apple	February–may	July–December
Grewia subinaequalis	Phalsa	March–April	April–May
Anona reticulata	Ramphall	May–July	September–January
Averrhoa carambola	Carambola	July–August	October–January
Averrhoa bilimbi	Bilimbi	February–April	May–October
Artocarpus lakoocha	Monkey Jack/Barhal	February–may	July–August
Curcur bita foetidissima	Buffalo gourd	NA	NA
Cucumis melo	Long melon	August–September	NA
Capsicum chinense	Ney mulaku, karanampotti	NA	NA
Carya illinoinensis	Pecan nut	April–may	September–October
Alternanthera sessilis	Joyweed	NA	NA
Hibiscus sabdariffa	Roselle	August–September	NA

(continued)

Table 16.1 (continued)

Fruit name	Common name	Flowering season	Fruiting time
Phyllanthus acidus	Star aonla	NA	December–January
Hylocereus undatus	Dragon fruit	NA	June–December
Flacourtia indica	Governor's Plum	NA	NA
Flacourtia jangomas	Paniala	March–April	NA
Pandanus andamanensis	Kewra	NA	NA
Nephelium lappaceum Lamk	Tader	NA	NA
Mimusops elengi	Bakul	NA	NA
Citrus deumana	Rabab	Spring	NA
Ficus auriculata	Fig	NA	July–September
Antidesma bunius	Bigony	Septermber–October	NA
Passiflora foetida	Giant granadilla	NA	NA
Artocarpus altilis park	Bread fruit	December–January	NA
Prunus persica	Soh-setanghi	NA	March–June
Diospyros kaki thumb	Persimmon	NA	September–November
Iberian peninsula	Sambucus nigra L	NA	NA
Morinda citrifolia L	Indian Noni	All year	All year
Aegle marmelos	Indian Bael	April–May	March–June

about their health. However, the market now offers the fewest value-added products made from these commercially available crops. The lack of marketing support, infrastructure for transportation, storage, and processing facilities, proper seeds, planting material, awareness about nutrition and therapeutic significance, and low production are the major challenges.

16.2 Small/Underutilized Indian Northeastern Fruits and Their Nutritional Composition

The term "underutilized crops" is frequently used to characterize plant species with a lot of promise but are only seldom grown, consumed, or utilized locally. Several other names are regularly used to describe these crops, including underutilized, neglected, orphaned, minor, promising, traditional, and specialized. The characterization of these fruit crops is generally accepted by the scientific community as having the potential to contribute to food security, health, nutrition, income, and environmental benefits.

Numerous fruit species that are underused have excellent nutritional value and are well-suited for low-input farming. They are abundant in vitamins, minerals, and other healthy elements needed for human well-being. Minor fruits are important in the prevention of various diseases since they have antioxidant and free radical-scavenging properties (Hossain et al. 2021) (Table 16.2).

In times of food scarcity, minor fruits like wild, underutilized, and wasted fruits can be used as a substitute for larger meals and are regarded as the most important dietary supplements (Hossain et al. 2021). Micronutrients can be found in minor fruits and can be utilized as a substitute for foods high in fat, such as those lacking in vitamins A and C, which can cause concealed hunger. A lack of some micronutrients also hinders growth, reduces immunity to diseases, and even raises the chance of dying. Additionally, they demonstrate plant secondary metabolites with anticancer, antidiabetic, antihypertensive, anti-inflammatory, antibacterial, antioxidant, immunomodulatory, or neuroprotective properties (Tripathi 2021).

16.2.1 Actinidia callosa *(Common Name: Himalayan Kiwi Vine)*

They are rich in minerals, dietary fiber, vitamins E and vitamin C, antioxidants, phytonutrients, enzymes, and other bioactive ingredients. They begin to bear fruit in October and continue through December. According to Padhan et al. (2023), they possess antipyretic and anti-inflammatory properties. They are used to relieve and eliminate dampness and stomach ache (Wu et al. 2019).It contains 83.8% moisture and 5.135 g/100 g total carbohydrate and is low in reducing sugar and protein. It has 0.212 g GAE/100 g total phenol, 0.002 g QE/100 g total flavonoid, and 0.000421 g/ml IC50s antioxidant activity (Padhan et al. 2023). It is typically consumed in its ripe state and is eaten raw (Khruomo and Deb 2018) (Fig. 16.1a).

16.2.2 Baccaurea sapida *(Common Name: Khattaphal/Latkan)*

The harvesting period of fruits is from May to July. In addition to having vitamin C and antiviral, antioxidant, and diuretic properties (Mann et al. 2016), they also have anti-inflammatory properties that make them useful in the treatment of skin conditions, injuries, arthritis, and abscesses (Mann et al. 2015). The fruit has 81.17% moisture and 18.83% dry matter, 86.14% total carbohydrates, 2.39% reducing sugar, 5.43% protein, and 1.24% fat. It also contains 3.60% fiber and 3.59% ash. The fruit exhibits antioxidant activity with 273 mg/100 g and is rich in minerals like calcium, magnesium, potassium, iron, copper, manganese that are 23.77 ± 2.0 μg/L, 21.67 ± 1.2 μg/L, 375.37 ± 9.68 μg/L, 29.55 ± 2.6 μg/L, 14.33 ± 1.2 μg/L and 37.29 ± 3.8 μg/L, respectively, and its energy is 377.44 ± 3.26 Kcal/g (Pandey et al. 2018a, b; Banik et al. 2013). In India, raw bark is chewed, or juice is used internally for constipation. In Bangladesh, in addition to fresh food consumption, it is mostly

Table 16.2 Proximate composition, Phenolic content, flavonoid content, mineral content, and energy value of underutilized fruits of North-East India

S.N.	Fruits	Moisture (%)	DM (%)	Carbs (%)	RS (g/100g)	Protein (g/100g)	Fat (%)	Fiber (%)	AC (%)	TPC	TFC	AA	Ca(mg/100g)	Vit C (mg/100g)	Mg (mg/100g)	K (mg/100g)	P (mg/100g)	Mo (µg/L)	Na (mg/100g)	Zn (mg/100g)	Fe (mg/100g)	Cu (mg/100g)	Mn (mg/100g)	Na (mg/100g)	Energy value (Kcal/100g)	Reference
1	*Actinidia callosa*	83.8	–	5.135g/100g	0.097	0.026	–	–	–	0.212gGAE/100g	0.002gQE/100g	0.000421g/m1IC50s	–	–	–	–	–	–	–	–	–	–	–	–	–	Padhan et. al, 2023
2	*Baccaurea sapida*	81.17±0.65	18.83±0.65	86.14±1.49	2.39±0.62 %	5.43±0.87%	1.24±0.09	3.60±0.03	3.59±0.72	–	–	–	23.77 ± 2.0 µg/L	273	21.67 ± 1.2 µg/L	375.37 ± 9.68 µg/L	–	1.03 ± 0.12	7.99 ± 1.8µg/L	0.97 ± 0.18 µg/L	29.55 ± 2.6 µg/L	14.33 ± 1.2 µg/L	37.29 ± 3.8 µg/L	–	377.44±3.26 Kcal/g	Pandey et. al, 2018 Banik et. al, 2013 Pandey et. al, 2018
3	*Castanopsis indica*	18.90 ± 0.02	–	64.99 ±0.71	–	10.83 ± 0.76 %	0.30 ± 0.04	3.72 ± 0	1.26 ± 0	–	–	–	5.85	4.84		16.60	–	–	0.95	0.488	0.162	0.004	0.125	–	305.98	Prakash et. Al, 2020 Seal et. Al, 2011
4	*Pyrus pashia*	60.36 ± 0.25	–	28.38± 0.12	–	3.28 ± 0.04%	1.62 ± 0.20	5.26 ± 0.05	1.10 ± 0.05	–	–	–	0.75 ± 0.06	1.80± 0.05%	0.12± 0.02	3.21± 0.05	0.86 ± 0.04	–	–	–	0.008 ± 0.02	–	–	0.68± 0.06	141.12± 0.12	Saklani et. al, 2012 Prakash et. al, 2021
5	*Rubus niveus*	78.56 ± 0.87	–	85.35 ± 0.25g/100	–	3.28 ± 0.87	1.10 ± 0.07g/100	5.90 ± 0.25g/100	4.37 ± 0.09g/100	–	–	–	390 ± 0.06	10.23 ± 0.02 g/100	–	720 ± 0.59	1.48 ± 0.02	–	56.3 ± 0.05	8.13 ± 0.05	3.26 ± 0.52	1.070 ± 0.01	2.43 ± 0.04	500 ± 0.06	364.42 ± 0.56	Pancholi et. al, 2020
6	*Citrus medica*	81.78–86.03	–	9.19–16.60 g/100 g	–	0.80–2.99	0.39–0.56 g/100 g	–	0.44–1.23 g/100 g	–	–	–	107.39–195.91	–	5.86–16.29	126.04–263.27	–	–	6.74–27.92	0.24–0.51	0.82–2.92	0.061–0.45	0.052–0.266	–	–	Benedetto et. al, 2023

7	*Livistonia jenkinsiana*	63.09	–	–	1.76 mg/g	2.46	–	1.24	–	–	–	–	–	–	–	–	–	–	–	–	–	–	–	–	–	Yatung et. al, 2023 Bhatt et. al, 2010
8	*Viburnum foetidum*	78.1±1.3	–	8.84±0.22	–	5.45±0.05%	2.96±0.07	1.04±0.05	2.20±0.17	–	–	–	153.67±2.66		12.53±0.02	323.66±2.33	15.6±0.52g/100g			0.539±0.02	0.49±0.003	0.05±0.002	1.51±0.04		83.0±0.18	Seal et. al, 2020 Chyne at. al, 2019
9	*Dillenia indica*	4.98± 0.38	–	60.55± 0.35	2±0.01 %	9.36± 0.03%	1.26± 0.09%	18.41± 0.05	5.18± 0.04	–	–	45.5±0.45%	1.12 ± 0.03	0.004 ± 0.02	0.23 ± 0.02 mg		0.026 ± 0.07	–	–	–	–	–	–	–	58.85 ± 0.04	Meeprathom et. al, 2018 Rahman et. al, 2023 Basumatary et. al, 2016
10	*Malus baccata*	78.53	–	19.01	3.12%	0.43%	0.36%	1.26	0.41	–	–	–	–	17.13	–	–	–	–	–	–	–	–	–	–	–	Kumari et. al, 2017
11	*Machilus edulis*	71.97	–	51.50%	1.25± 0.010%	4.51%	25.50±1.24%	10.30±0.49	2.65	–	–	–	0.150 -+ 0.0002%	0.006±0.0016%	0.339_+ 0.067%	0.610 -+ 0.012%	0.118 +- 0.0001%	–	0.024_+ 0.002%	392.50 -+ 23.28µg/g	0.253_+ 0.001%	100.66 --- 7.91 µg/g	–	0.726_+ 0.056%	–	Sundriyal et. al, 2004
12	*Phyllanthus acidus*	86.7	–	4.8	0.04%	0.25%	–	–	–	–	–	–	163.22	36.7	–	223.44	–	–	17.5	1.63	2.43	0.2	1.31	–	–	Brooks et. al, 2020
13	*Mangifera sylvatica*	85	-	1.95g	–	5g	3.2g	2g	1.93g	–	–	–	–	–	–	–	–	–	–	–	–	–	–	–	–	Esparza et. al, 2017
14	*Docynia indica*	74.60 ± 1.81	13.91 ± 0.11	89.06 ± 0.52	–	2.61 ± 0.58%	0.87 ± 0.02%	4.57 ± 0.06	2.90 ± 0.49	29.13 ± 1.03mg GAE/g	2.10 ± 0.07mg QE/ g	–	1.98 ± 0.02µg/l	9.7 ± 0.59	6.88 ± 0.8µg/l	111.37 ± 4.38µg/l		0.79 ± 0.05	7.05 ± 1.4µg/l	0.16 ± 0.2µg/l	23.29 ± 2.1µg/l	4.16 ± 0.9µg/l	3.09 ± 0.18µg/l		393.14 ± 1.96	Biswas et. al, 2021

(continued)

Table 16.2 (continued)

15	*Elaegnus latifolia*	87.31	–	74.06g/100g	4.0%	7.19%	0.55	7.22%	4.41%	–		10.06 m mol trolox/g	1.47g/100g	16-19.2	–	487	95	–	–	4.90	0.18g/100g	1.90	4.50	1022	—	Lepcha et. al, 2018 Bachheti et. al, 2023 Patel el. al,2013
16	*Citrus macroptera*	83.80 ± 3.26	–		4.31 ± 1.03%	0.088 ± 0.03%	–	–	4.49 ± 1.26%	–	–	–	–	220.75 ± 12.23	–	–	–	–	–	–	–	–	–	–	29.01 ± 2.45	Biswas et. al, 2022
17	*Grewia hirsute (approx.)*	–	–	21.1	–	1.57	0.1	5.53	1.10	–	–	–	136	5.21	–	372	24.2	–	–	1.9	20.8	1.5	1.08	17.3	–	Qamar et al., 2021
18	*Physalis minima*	89.9	–	–	–	2.04g	0.056g	–	–	–	–	–	25	29	7.8	2.43	30	–	1.8	–	0.28	–	–	–	–	Musinguzi et. al, 2007
19	*Rubus ellipticus*	66.36 ± 0.58	–	86.4 ± 0.38g/100g	–	4.37 ± 0.52	2.73 ± 0.06g/100g	3.53 ± 0.17g/100g	2.97 ± 0.01g/100g	72.0 ± 1.25 GAE/100g	86.4 ± 2.04 QE/100g	196.4 ± 1.80 DPPH/µg/ml	450 ± 0.22	44.0 ± 4.95 AAE/100g	118.72 ± 0.48	680.16 ± 1.27	1.26 ± 0.001	–	–	12.77 ± 0.05	4.249 ± 0.15	0.020 ± 0.01	1.948 ± 0.03	700 ± 0.08	374.0 ± 1.56	Pandey et. al, 2016
20	*Elaecarpus floribundus*	77.63	–	19.65 %	–	0.72%	2.54	1.5%	2.75%	–	–	–	–	49	–	–	–	–	–	–	–	–	–	–	94.44 kcal/g	Dey et. al, 2022
21	*Ziziphus mauritiana*	81-83	–	17	22	0.8%	0.07	1.3g	–	–	–	–	0.03%	65-76	–	–	0.03%	–	–	–	0.76-1.8%	–	–	–	104	Muchuweti et al., 2005
22	*Carissa carandus*	91	–	2.9	–	1.1%	2.9	–	–	–	–	–	2.1	200-500	–	–	28	–	–	–	–	–	–	–	42	Tripathi et al.,2013
23	*Ferronia limmonia*	85	–	42.2	–	2.43	29	11.52	8.5	–	–	–			–	–	–	–	–	–		–	–	–	–	Oo & Khine, 2017

24	*Grewia subinaequalis*	76.80	–	21.1g/100g	1.24	1.57g/100g	<0.1g/100g	5.53g/100g	1.1g/100g	–	–	–	136	4.385	-	372	24.2	–	-	144	1.08	0.48	–	17.3	–	Sinha et al., 2015
25	*Anona reticulata (approx.)*	80	–	15.8g	-	1.4g	0.6g	2.5g	0.4g	–	–	–	17	43	32	250	–	–	–	0.3	–	–	–	4	310	Yassine, 2014
26	*Averrhoa carambola*	92.6g	–	9.38g/100g	–	0.38	–	0.8g/100g	–	–	–	–	6.37-6.40	25.8	11.85-12.05	167.13-168.0	17.87-17.88	–	–	0.29.0.51	0.34-0.45	0.19-0.45	0.04-0.52	3.8-3.85	35.7	Muthu et al., 2016 Dhyani & Mehta, 2023
27	*Averrhoa bilimbi*	16.84	–	32.02	–	11.50	3.38	15.31	4.62	–	–	–	–	–	–	–	–	–	–	–	–	–	–	–	–	Yan et al., 2013
28	*Artocarpus lakoocha*	89.6	–	13.9	–	1.6%	1.2%	2.8	–	–	–	–	0.067%	–	–	–	0.025%	–	–	–	–	–	–	–	–	Gopalan et al.,1993
29	*Curcurbita foetidissima*	–	95.6		–	6.6%	2.5%	48.4	7.0	–	–	–	–	–	–	–	–	–	–	–	–	–	–	–	–	Mejía-Morales, 2021
30	*Cucumis melo*	–	–		–	2.16-3.67%	0.482	4.12	3.059	29.5%	–	–	–	–	–	–	–	–	–	1.498	0.46	–	–	–	–	Panday et al., 2021
31	*Capsicum chinense*	84.56g/100g	–	6.89g/100g	–	1.45g/100g	0.1g/100g	1.67g/100g	1.46g/100g	–	–	–	–	118.45	–	–	–	–	–	–	–	–	–	–	42.68	Malakar et al., 2019
32	*Carya illinoinensis*	–	–	13.86g/100g	—	9.17g/100g	71.97g/100g	9.60g/100g	1.49g/100g	–	–	–	70	1.10	–	410	277	–	–	–	–	–	121	–	691	USDA-FAS

(continued)

Table 16.2 (continued)

33	*Alternanthera sessilis*	71.40	–	16.30g/100g	–	4.50g/100g	2.90g/100g	3.40g/100g	1.50g/100g	–	–	–	436.70	11.50	252.30	412.20	48.80	–	–	–	14.30	0.40	–	246.70	109.00	Hwong et al., 2022
34	*Hibiscus sabdariffa*	84.5	–	12.3	–	1.9%	0.1%	2.3	1.2	–	–	–	0.0017%	0.014%	–	–	0.057%	–	–	–	0.029%	–	–	–	–	Duke, 1983
35	*Phyllanthus acidus*	86.7	–	4.8	–	0.25%	–	–	–	–	–	–	–	–	–	–	–	–	–	–	–	–	–	–	–	Mahapatra et al., 2012
36	*Hylocereus undatus*	82.5-83	–	11.2g	–	0.15-0.22%	0.21-0.61	0.7-0.9%	–	–	–	–	2.3-10.2	8-9	31.3-38.9	272-328	27.5-36.1	–	–	–	0.6-3.4	–	–	7.3-8.9	283	Liaotrakoon, 2013
37	*Flacourtia indica*	–	–	24.2g/100g	–	0.5g/100g	0.6g/100g	1.2g/100g	–	–	–	–	33	5	–	171	17	–	–	–	0.7	–	–	–	94	Eramma, 2016
38	*Flacourtia jangomas*	76.80	–	89.73	9.82%	6.04	0.60	1.13	2.56	1.28 mg $100g^{-1}$	–	–	–	24	–	1554.19	133.39	–	–	–	–	–	–	73.66	388.43	Barua et al., 2022
39	*Pandanus andamanansium*	80	–	17g/100g		1.3g/100g	0.7mg/100g	3.5g/100g	–	–	–	–	134	5	–	–	108	–	–	–	–	-	–	–	–	Singh & Parle, 2015

40	*Nephelium lappaceum Lamk*	72.05	–	23.78g/100g		2.04g/100g	0.23g/100g	0.7g/100g	1.2g/100g	–	–	–	–	7.43	–	–	–	–	–	–	–	–	–	–	–	Mahmood et al., 2018
41	*Mimusops elengi*	79.27	–	–	8.9%	1.29%	2.76	1.13	–	–	–	–	–	3.27	–	98.54	–	–	–	–	0.59	–	–	5.16	–	Nazarudeen, 2010
42	*Citrus deumana (approx.)*	87.8g	–	10.9	–	1.5g	1g	1.3g	–	–	–	–	90	63	–	–	20	–	–	–	0.3	–	–	–	59	Abobatta, 2019
43	*Ficus auriculata*	46.64±0.15	–	27.09±0.11	–	5.32±0.04%	0.65±0.25	16.96±0.09	3.7±0.08	–	–	–	1.35±0.10	0.09±0.04	0.90±0.06	2.11±0.08	0.28±0.04	–	–	–	–	–	–	–	135.51±0.09	Saklani & Chandra, 2012
44	*Antidesma bunius*	4.530±0.351g/100g	95.466±0.351g/100g	92.743±0.428	–	1.231±0.050g/100g	0.970±0.026g/100g	9.433±0.305	64.46±0.251g	–	–	–	–	–	–		–	–	–	–	–	–	–	–	384.651±1.296	Islarya et al., 2017
45	*Passiflora foetida*	76±1	–	–	–	0.15±0.01	13±2g/100g		1.7±0.3	1.66mg/g	–	–	1.06%	–	6.71%	75.70%	14.44%	–	–	0.17%	0.15%	0.04%	0.05%	1.68%	–	Sripathi et al., 2023
46	*Artocarpus altilis park*	73.22	–	22.11	–	2.99%	0.46	0.14	1.09	–	–	–	–	–	–	–	–	–	–	–	–	–	–	–	–	Dada et al., 2023

(continued)

Table 16.2 (continued)

47	*Prunus persica*	89g/100g	–	9.9g/100g	–	0.9g/100g	0.3g/100g	1.5g/100g	0.8g/100g	–	–	–	1%	6.6	9	190	20	–	–	0.2	2%	0.1	–	–	163	Hussain et al., 2021
48	*Diospyros kaki thumb*	68.9±1.27	–	17.30g/100g	2.87±0.24	0.64	0.25	2.60g/100g	0.32±0.02	–	–	–	8	6.9±0.24	9.25	230	19.50	–	–	0.11	0.20	–	–	2.50	71.50	Nazir et al.,2013 Dierito et al., 2021
49	*Iberian peninsula (approx.)*	–	–	–	2.8-8.55%	2.7-2.9%	–	–	–	–	–	–	574–1528	13-35	396–739	2953–5494	735–1337	–	–	1.9-11.3	12.4–84.7	1.7-2.9	3.6-9.5	13-146	–	Mlynarczyk,2018
50	*Marinda citrifolia L*	54.21	–	35.14	–	2.18	3.25	4.49	0.73	–	–	–	–	134.10	–	–	–	–	–	–	–	–	–	–	–	Saah & Adu-Poku, 2020
51	*Aegle marmelos*	51.7±0.8	–	1.82±0.85	–	10.4±0.1	–	12.9±0.4	23.3±0.3	–	–	–	–	–	–	–	–	–	–	–	–	–	–	–	–	Emurotu et al., 2017

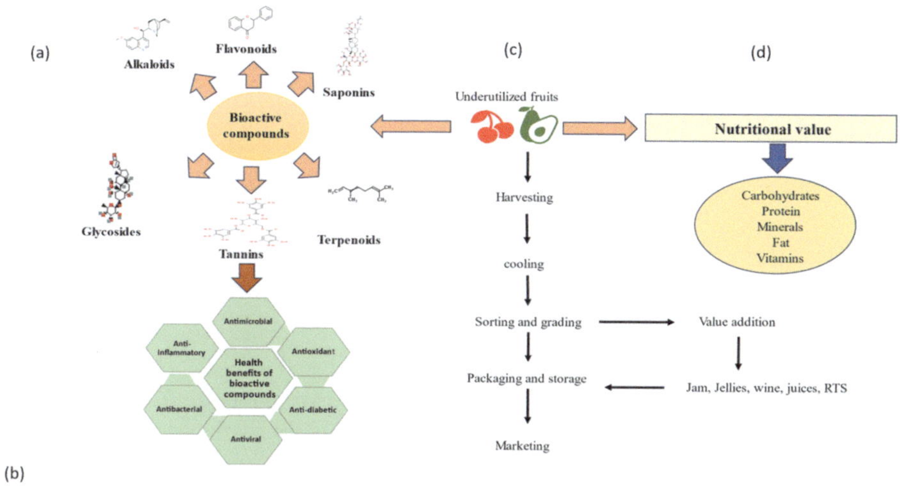

Fig. 16.1 (**a**) Bioactive compounds from underutilized fruits (**b**) Health benefits of bioactive compounds (**c**) Processing of underutilized fruits (**d**) Nutritional value. (Adapted from Tripathi 2021; Padhan et al. 2023; Chemical structure from PubChem and ChemSpider)

grown for the manufacturing of the important dye annatto seeds. 4.8–6% of the annatto dye is present in seeds. Adding value to this fruit by processing it into products like jams, juices, or dried fruits can increase their commercial value. These processed products may have a longer shelf life and can be sold at higher prices (Malik et al. 2024).

16.2.3 Castanopsis indica *(Common Name: Indian Chestnut)*

The flower blooms from April to May and fruits mature from August to September (Hasnat et al. 2019). They exhibit antioxidants and are used to treat diarrhea, headaches, indigestion, and chest pain (Prakash et al. 2020). Mature nuts can be eaten raw or can be roasted for consumption. They show anticancer and anticonvulsant action (Dolai et al. 2012). *C. indica* contains 18.90% moisture, 64.99 ± 0.71% carbohydrate, and 10.83 ± 0.76% protein, and is low in fat, fiber, and vitamin C content. The fruit is rich in potassium 16.60 mg/g and calcium 5.85 mg/g, and its energy value is 305.98 cal/100 g (Prakash et al. 2020; Seal 2011).

16.2.4 Pyrus pashia *(Common Name: Wild Himalayas Pear)*

They are seasonal, and fruit ripens from November to December after beginning to form in May (Prakash et al. 2021a, b). The fruits of this plant exhibit sedative, febrifuge, laxative, and antioxidant properties. Additionally, they help in the curing of epilepsy, hysteria, diaphoretic perspiration, fever, migraines, and gastrointestinal disorders (Siddiqui et al. 2015). They exhibit antimicrobial properties (Tsering et al. 2012). The fruit has 60.36 ± 0.25% moisture and 28.38 ± 0.12% total carbohydrate and provides 141.12 ± 0.12 Kcal energy (Prakash et al. 2021a, b). The study found that Himalayan pear fruit can be a great resource for jam production having retained great nutritional and sensory attributes.

16.2.5 Rubus niveus *(Common Name: Mysore Raspberry)*

The fruit can be used to treat diabetes, burns, wounds, and inflammation. According to Tolentino et al. (2015), they additionally show antigastropathic, anti-inflammatory, antioxidant, antibacterial, analgesic, and antinociceptive attributes. The moisture content in *R. niveus* is 78.56 ± 0.87 g/100, and the carbohydrate content recorded is 85.35 ± 0.25 g/100. It contains 10.23 ± 0.02 g/100 of vitamin C and is high in calcium, potassium, and nitrogen, 390 ± 0.06 mg/100 g, 720 ± 0.59 mg/100 g, and 500 ± 0.06 mg/100 g, respectively. Its energy value was calculated to be 364.42 ± 0.56 Kcal (Pancholi and Rana 2020). Fruits serve multiple purposes, such as fresh consumption, juice production, and the making of jams and jellies (Bhatt et al. 2023).

16.2.6 Citrus medica *(Common Name: Fingered Citron)*

A powerful antiscorbutic, stomachic, cardiac tonic, stimulant, sedative, and analgesic, ripe fruits are used to treat earaches, colds, fevers, palpitations, sore throats, coughs, asthma attacks, and vomiting. Antioxidant, antidiabetic, hypocholesterolemic effect reduction, hypolipidemic effect, anticancer, anticholinesterase, anthelmintic, anti-ulcer, and antibacterial properties have all been demonstrated (Panara et al. 2012). It contains 81.78–86.03 g/100 g of moisture and 9.19–16.60 g/100 g of total carbohydrate, protein, and fat, which are found in trace amounts. Fruit is a rich source of calcium, 107.39–195.91 mg/100 g, and potassium, 126.04–263.27 mg/100 g (Benedetto et al. 2023). Citron's versatile applications in carbonated drinks, alcoholic beverages, syrups, candied peels, jams, marmalades, and cordials highlight its potential as a raw material for healthy processed foods (Chhikara et al. 2018).

16.2.7 Livistona jenkinsiana *(Common Name: Assam fan Palm)*

This fruit ripens from September to December after flowering from February to March (Singh et al. 2010). Fruits have properties that are anticancer, antioxidant, antibacterial, antiviral, anti-inflammatory, anti-allergenic, and anti-proliferative and cardioprotective (Payum 2018). The fruit has 63.09% of moisture content and is low in protein and fiber (Yatung et al. 2023; Bhatt et al. 2010). Fruits are typically consumed raw to prevent stomach ailments (Hazarika and Pongener 2018).

16.2.8 Viburnum foetidum *(Common Name: Himalayan Evergreen Viburnum)*

The flowers appear in June and July, and the fruits ripen from September to November (Rymbai et al. 2016). Fruit is taken internally for postpartum hemorrhage and menorrhagia. It has been proven that Methanolic extract of Viburnum foetidum showed strong antioxidant, hypoglycemic, and hypolipidemic effects in fructose-induced diabetic mice. It reduced blood glucose, cholesterol, triglycerides, SGPT, SGOT, and improved antioxidant enzymes and liver-pancreas health, with the best effects at 400 mg/kg/day (Hasnin et al. 2023). The moisture content in *V. foetidum* is 78.1 ± 1.30 g/100 g and is rich in calcium, 153.67 ± 2.66 mg/100 g, potassium, 323.66 ± 2.33 mg/100 g, and gives an energy value. Carbohydrates, proteins, fat, and fiber are found in low amounts (Seal 2020; Chyne et al. 2019).

16.2.9 Dillenia indica *(Common Name: Elephant Apple)*

The fruits ripen in July and August, the flowers bloom in May and June, and the fruits fall in November and December. A decoction of the fruit is used to cure food poisoning, fever, stomach aches, and flatulence. Possess laxative, hepatoprotective, cytotoxic, anti-inflammatory, antioxidant, antiproliferative, antileukemic, and antifungal qualities (Gandhi and Mehta 2013). The fruit is low in moisture content but high in carbohydrates, 60.55 ± 0.35%, fiber, 18.41 ± 0.05%, and antioxidant content, 45.5 ± 0.45%. Its energy value is 58.85 ± 0.04 Kcal (Meeprathom et al. 2018; Rahman et al. 2023; Basumatary et al. 2016). The elephant apple is a wild fruit valued for its nutritional and medicinal benefits. With an acidic flavor, it can be enjoyed raw or cooked and is often used in various dishes like juice, pickles, curry, jams, chutney, and jellies. Its versatility also allows it to be processed into marketable value-added products such as squash and RTS beverages (Saikia et al. 2023) (Fig. 16.1b).

16.2.10 Malus baccata *(Common Name: Siberian Crabapple)*

Fruit that ripens in September prevents newborns from constipation, diarrhea, and dysentery and lowers their chance of developing certain malignancies, cardiovascular illnesses, asthma, and diabetes. Display antioxidant qualities (Dadwal et al. 2018). Moisture content in *M. baccata* is 78.53%, carbohydrate 19.01%, reducing sugar 3.12%, protein 0.43%, fat 0.36%, fiber 1.26%, ash content 0.41%, and vitamin C 17.13 mg/100 g. Crab apples could be used to make jam and fruit bars (Kumari and Dhaliwal 2017).

16.2.11 Machilus edulis *(Common Name: Wild Avocado)*

Fruits are available from November to March (Sundriyal and Sundriyal 2003). They exhibit antioxidant properties and are used to manage diabetes (Chhetri et al. 2020). The moisture in *M. edulis* is 71.97%, 51.50% carbohydrate, 25.50 ± 1.24% fat, 392.50 ± 23.28 μg/g, and 100.66 ± 7.91 μg/g of copper (Sundriyal and Sundriyal 2004).

16.2.12 Phyllanthus acidus *(Common Name: Malay Gooseberry)*

They produce flowers and fruits from January to May (Brooks et al. 2020). Numerous ailments, such as diabetes, rheumatism, bronchitis, asthma, respiratory problems, and inflammatory and hepatic diseases, can be cured with it. It demonstrates antioxidant, antibacterial, and anti-inflammatory qualities (Tan et al. 2020). It has 86.7% moisture, 36.7 mg/100 g vitamin C content, 163.22 mg/100 g calcium, 223.44 mg/100 g potassium, and 17.5 mg/100 g sodium content (Brooks et al. 2020). *P. acidus* leaves are cooked as vegetables in Indonesia, Thailand, and India. In Malaysia, the fruit is made into relish and syrup for stomach issues. The Philippines uses the fruit for vinegar and drinks, while its tart flesh serves as a flavoring agent in Indonesia (Tan et al. 2020).

16.2.13 Mangifera sylvatica *(Common Name: Himalayan Mango)*

The maturing period is from April to May (Akhter et al. 2022). Fruit has antiallergic, anticancer, antiviral, anti-inflammatory, antimicrobial, antidiabetic, and anti-ulcer properties. It also helps against heart disease and hepatic steatosis (Akhter et al. 2021). Fruit contains 85 g of moisture and provides 59 kcal of energy. It is low in carbohydrates, protein, fat, and fiber (Anaya-Esparza and Montalvo-González 2017). Wild mango kernels produce a butter that can replace cocoa butter, opening new market opportunities. The unripe fruit is also sold to the pickle industry, generating income during the fruiting season (Akhter 2016).

16.2.14 Docynia indica *(Common Name: Wild Apple)*

Fruits are available from November to December (Chhetri et al. 2020). Blood dysentery and diarrhea are treated with fruit decoction. Additionally, it has antioxidant and antimicrobial action, as well as hypoglycemic and hypolipidemic effects (Shende et al. 2016). *D. indica* has 74.60 ± 1.81% moisture, 13.91 ± 0.11% dry matter content, 89.06 ± 0.52% total carbohydrate, 29.13 ± 1.03 mgGAE/g ash content, and is rich in potassium, 111.37 ± 4.38 µg/l, and has a good amount of iron content, 23.29 ± 2.1 µg/l. It contains 393.14 ± 1.96 Kcal/100 g of energy. It is low in protein and fiber content (Biswas et al. 2021). It is traditionally used in the form of syrup, alcohol, wine, and vinegar (Khruomo and Deb 2018).

16.2.15 Elaeagnus latifolia *(Common Name: Bastard Oleaster)*

Flowers bloom from September to December, and fruits are collected from March to April. According to Dutta et al. (2018), it has antioxidant and antibacterial action. Research has indicated that it may be able to lower the incidence of cancer as well as stop or reverse the progression of tumors (Patel et al. 2008). Fruit contains 87.31% moisture, 74.06 g/100 g of total carbohydrate, and 10.06 m mol trolox/g of antioxidant content. It is also considered to be a good source of minerals like 487 mg/100 g of potassium, 95 mg/100 g of phosphorus, and 1022 mg/100 g of nitrogen (Lepcha et al. 2018; Bachheti et al. 2023; Patel et al. 2013).

16.2.16 Citrus macroptera *(Common Name: Wild Orange)*

It exhibits hepatoprotective, hypoglycemic, thrombolytic, antidepressant, cytotoxic, antibacterial, thrombolytic, and anxiolytic properties. It is used to treat several chronic illnesses, including diabetes, arthritic conditions, cancer, liver disorders, inflammation, and stroke. *C. macroptera* has 83.80 ± 3.26% moisture content and is a great source of vitamin C, 220.75 ± 12.23 mg/100 g. It has 29.01 ± 2.45 Kcal of energy value (Biswas et al. 2022). Wild orange has various uses: Its dried rind can flavor meat dishes, the juice treats stomach ailments and aids digestion, the pulp serves as a washing detergent, and essential oil can be extracted from the leaves, which would otherwise be wasted (Nongalleima et al. 2019).

16.2.17 Grewia hirsute *(Common Name: Kukurbicha/ Veronica Lolia)*

The blooming period is from June to July (Chikkamath et al. 2019). It is used to cure ulcers, cholera, rheumatism, headaches, sore throats, and diarrhea. According to Das et al. (2019), it exhibits analgesic, antimicrobial, anti-inflammatory, and antioxidant properties (Chikkamath et al. 2019).

16.2.18 Physalis minima *(Common Name: Little Gooseberry)*

Fruits are available from July–October. They are utilized to cure fever and toothaches (Chhetri et al. 2020). It has palatable, diuretic, and laxative properties and is beneficial for inflammations, antigonorrhea, spleen enlargement, and stomach issues. It is also used to treat bleeding disorders and skin conditions like leprosy. Display antibacterial, anti-ulcer, anti-inflammatory, antipyretic, and antimalarial properties (Chothani and Vaghasiya 2012). In *P. minima,* the moisture was calculated to be 89.9%, with 29 mg of vitamin C content, minerals like 25 mg of Ca, 7.8 mg of Mg, and 30 mg of P content (Musinguzi et al. 2007).

16.2.19 Rubus ellipticus *(Common Name: Yellow Himalayan Raspberry)*

Fruits are available from March to June. Vitamin C-rich fruits are prepared to treat diarrhea and cough (Chhetri et al. 2020), diabetes, diarrhea, gastritis, wound healing, and epilepsy, and it has antioxidant, analgesic, antibacterial, and antifertility qualities (Sharma and Kumar 2011). The moisture content in fruit is 66.36 ± 0.58 g/100 g and 86.4 ± 0.38 g/100 g of total carbohydrate. Its total phenol content was 72.0 ± 1.25 GAE/100 g, 86.4 ± 2.04 QE/ 100 g of total flavonoid, high content of antioxidant activity 196.4 ± 1.80 DPPH/μg/ml, and it was found to have vitamin C content 44.0 ± 4.95 AAE/100 g. It contains a high amount of calcium, 450 ± 0.22 mg/100 g, 118.72 ± 0.48 mg/100 g of magnesium, and 700 ± 0.08 mg/100 g of nitrogen. Its energy value calculated was 374.0 ± 1.56 Kcal/100 g (Pandey and Bhatt 2016). The delicious fruits, commonly consumed raw in the Himalayan region, are also used for juice production and in making jams, jellies, and syrups (Kewlani et al. 2023; Bhatt et al. 2023).

16.2.20 Elaeocarpus floribundus *(Common Name: Indian Olive)*

The flowering period is from April–May. Fruits mature from August–October and are ready for harvest (Bhowmick 2017). Its antibacterial, antioxidant, anti-aging, antitumor or anticancer, and antiseptic properties have all been demonstrated (Sircar and Mandal 2017). According to Saraswathy (2023), it is used to treat gum inflammation, diarrhea, ulcers, and insect bites. The moisture of *Elaeocarpus floribundus is 77.63%, 19.65% of total carbohydrate, its vitamin C content was 49 mg/100 g, and 94.44 kcal/g energy value was calculated* (Dey et al. 2022). The fruits are edible and are used to make pickles or chutney (Ogundele et al. 2021).

16.2.21 Ziziphus mauritiana *(Common Name: Indian Plum)*

It is also known as ber, desert apple, or Indian plum, and is rich in nutrients. It contains 81–83% moisture, 17.0% carbohydrates, 0.8% protein, and minimal fat (0.07%). It is a good source of iron (0.76–1.8%), along with trace amounts of calcium and phosphorus. The fruit also provides ascorbic acid (65–76 mg/100 g), sugars (22 g), and fiber (1.3 g) per 100 g and 104 calories (Morton 1987).

It contains sugar like galactose, fructose, and glucose, while significant phenolic compounds are p-hydroxybenzoic, caffeic, ferulic, and p-coumaric acids. Plum contains notable carotenes (4–6 mg/100 g dry weight) (Muchuweti et al. 2005). Ber can also be pickled by mixing it with salt, oil, spices, and vinegar (Hazarika and Pongener 2018).

16.2.22 Carissa carandus *(Common Name: Karonda)*

Christ thorn, or carandas plum, is native to India, Sri Lanka, Java, Malaysia, Myanmar, and Pakistan. Due to their high acidity and astringency, these berries are usually not eaten raw but are processed into products like jellies, pickles, and chutneys. Its berries are used as an anthelmintic, astringent, and appetite stimulant, and help treat fevers, biliary issues, and rheumatism (Kirtikar and Basu 2003). Extracts from the plant have been shown to exhibit cardiotonic, anticonvulsant (Hegde et al. 2009), antiviral (Rajasekaran et al. 1999), and nematicidal properties (Saleem et al. 2015). With their commercial potential and nutritive value, it is also seen as a viable crop for renewable energy and phytochemicals in the Western Ghats (Augustus et al. 2003). When ripe, the fruits can be eaten as a dessert or used to make candies and jellies (Srivastava et al. 2017a, b).

16.2.23 **Feronia limonia** ***(Common Name: Wood Apple)***

The wood apple is also called as kaitha, monkey fruit, elephant apple, kath bel, curd fruit, and "poor man's food" in India. It is the native fruit of dry plains in Ceylon and India and commonly utilized for the preparation of jam, chutneys, and jelly (Morton 1987). The whole part of the wood apple fruit, i.e., seed, pulp, and oil, has medicinal properties (Ramakrishna et al. 1979). It is mainly utilized as liver and cardiac tonic and an unripe one for halting diarrhea, dysentery, hiccough, sore throat, and gum treatment in India (Mondal et al. 2002). In Poultice, the pulp is used to help with bites and stings of venomous insects as well as having insecticidal and antifungal properties (Hiwale 2015). In India, the fruit is consumed as syrup, drink, murabba, chutney, and juices (Srivastava et al. 2020).

16.2.24 **Grewia subinaequalis** ***(Common Name: Phalsa)***

It is one of the hardiest, drought-resistant, and quick-growing fruit crops of India. It is an exotic bush plant generally consumed as fresh, desserts, and processed product form like soft drinks, syrup, jam, during hot summer months in India (Salunkhe and Desai 1984). It has short shelf life, so it is suitable for local marketing only (Anand 1960). The unripe fruits may alleviate inflammation but are useful in cardiac, respiratory, and blood disorders, also in fever reduction. Phalsa is an important commercial plant known for its high drought resistance and nutritional and medicinal value. The fruit can be consumed fresh or processed into juices, jams, and squashes (Hassan et al. 2022).

16.2.25 **Anona reticulata** ***(Common Name: Ramphall)***

The fruit known as bullock's heart or custard apple has a sweet flavor and offers various pharmacological benefits, including antioxidant, anticancer, analgesic, and antimalarial properties. Species such as *A. cherimola, A. chrysophylla,* and *A. squamosa* belong to the same genus (Kaladhar and Rayavarapu 2014). This fruit is indigenous to the West Indies and parts of India. It boasts medicinal qualities, including anti-inflammatory and wound healing effects, and contains phytochemicals like tannins, alkaloids, and flavonoids (Ngbolua et al. 2018). Additionally, it is used in making kulfi (Thorat et al. 2024).

16.2.26 Averrhoa carambola *(Common Name: Carambola)*

Carambola, or "Kamrakh" (star fruit), is a popular substitute for tamarind in South Indian cooking and has become a commercial crop in India. The ripe fruit can be processed into drinks, jams, and jellies. Its high acidity allows it to be enjoyed fresh or as a dessert, while the sweet variety is used to make wine in Surinam. The fruit contains sucrose as major carbohydrate while pectin and volatile components are also present (Dasgupta et al. 2013).

16.2.27 Averrhoa bilimbi *(Common Name: Bilimbi)*

Bilimbi, also known as bilimbino, biri-biri, caramboleiraamarela, or limão de caiena, is a tropical fruit believed to be native to India and introduced to Brazil centuries ago (Correa 1926). The cylindrical fruits, with five rounded lobes, mature in clusters, changing from green to light yellow (Mathew et al. 1993). Due to their sourness, they are used in vinegar, wine, pickles, jams, and jellies and are also applied in traditional remedies for coughs, mumps, rheumatism, pimples, and scurvy. The fruit's juice contains high levels of oxalic acid, allowing it to remove rust stains and shine brassware (Correa 1926; Joseph and Mendonca 1989; Lennox and Ragoonath 1990; Wong and Wong 1995).

16.2.28 Artocarpus lakoocha *(Common Name: Monkey Jack/Barhal)*

Lakoocha is a perennial tree found along India's west coast, from Kokan to Kerala and Tamil Nadu, as well as in West Bengal and Assam. Its fruit is harvested before ripening to maintain texture and serves as a liver tonic. Unripe fruits can cause appetite loss, while ripe fruits are sweet or sour and beneficial to health. The seeds act as a purgative for children and possess medicinal properties. The fruit contains 89.6% moisture and 13.9% carbohydrates. Raw fruits and male flower spikes are used in pickles. The timber is valuable for construction, and the fodder is nutritious for lactating animals, boosting milk production (Pai and Akhilraj 2021).

16.2.29 Curcur bita foetidissima *(Common Name: Buffalo Gourd)*

The plant has several common names, including buffalo gourd, Missouri gourd, chili coyote, calabazillaloca, and fetid gourd. Its range extends from the northern Great Plains to central Mexico. This perennial plant has the potential to be cultivated as a food crop, producing edible oil and protein-rich seeds. Additionally, it develops an extensive storage root system that is rich in starch (Neelamma et al. 2016). It can be used to make various valuable products, such as edible vegetables, fruits, seeds, seed oils, pickles, salads, dessert fruits, candied treats, and confectionery items (Salehi et al. 2019).

16.2.30 **Cucumis melo** *(Common Name: Long Melon)*

The fully ripe long melon is yellow, oval-shaped, and known for its delicious flavor. It is widely marketed from Asia to the USA and is rich in vitamins B1, B2, and folic acid. Different parts of the fruit have varying concentrations of sugars, carotene, and other nutrients (Gill et al. 2011). However, it has a short shelf life and can deteriorate quickly. Key compounds like (Z, Z)-3, 6-nonadien-1-ol and phenylethyl alcohol give it a fresh, sweet-floral scent (Vishwakarma et al. 2017). Long melons are enjoyed as desserts, snacks, or breakfast items, and recent innovations like pre-cut displays and salad bars target the single-serving market (Saltveit 2011).

16.2.31 **Capsicum chinense** *(Common Name: Ney Mulaku, Karanampotti, Malimulaku)*

It is a perennial plants that produce medium-sized, fleshy, and pungent fruits. It can be intercropped with coconut, rubber, pepper, and banana plantations. Rich in vitamin C and other bioactive compounds like capsaicin, these fruits are found in the Maldives, Midwest and Northeast USA, and the Amazon Basin. Peppers are vital in Brazil's fresh vegetable market and are essential globally for condiments and preserved foods (Segura-Campos et al. 2016).

16.2.32 **Carya illinoinensis** *(Common Name: Pecan Nut)*

It is native to North America, while it originated from Canada and slowly moved to south. It grows in Jammu, Himachal Pradesh, Uttarakhand, and northeastern hill region in India. It is underutilized nut apart from hazelnut and chestnut. It is a highly nutritious and energy-providing nut. It can be harvested from October to January upon cultivar, rootstock, and agro-climatic conditions (Atanasova et al. 2018). The nuts are used to add aroma, flavor, crispness to a large number of dishes, but the major use is in the baking industry dishes like fruit cakes, pecan pie, custard, ice cream, and cake filling (Peter 2008).

16.2.33 Alternanthera sessilis *(Common Name: Joyweed)*

Othman et al. (2016) studied the nutritional composition and antioxidant properties of Joyweed and identified catechin, ferulic acid, and vanillic acid as phenolic compounds and noted a positive relationship between these compounds and antioxidant capacity, which aids in free radical scavenging. Additionally, factors like geographical location, climate, ripeness, storage conditions, and fruit color affect the levels of bioactive compounds and antioxidant properties (Gonçalves et al. 2004).

Joyweed is widely used as a vegetable and herbal medicine in Asia. Its leaves and shoots can be eaten raw, cooked, or incorporated into various local dishes. In India, the young leaves are often consumed as a vegetable or mixed with meat and fish, while in Sri Lanka, it is typically chopped and served with grated coconut and onions as a salad alongside rice and curry (Hwong et al. 2022).

16.2.34 Hibiscus sabdariffa *(Common Name: Roselle)*

It is effective against hypertension, inflammation, and liver issues (Alarcon-Alonso et al. 2012). Its methanolic fruit extract inhibits the α-glucosidase enzyme and enhances wound healing. The fruit of roselle is used in the preparation of a refreshing, radiation-resistant tea drink which can nourish the liver, benefit wisdom, awaken the brain, and lower the glycaemic index, weight, and toxins. The *Hibiscus sabdariffa* fruit's fresh extract shows anticlastogenic effects in bone marrow cells of mice after 7 days of feeding (Adetutu et al. 2004). Sukkhaenga et al. (2018) analyzed the physico-chemical properties of the roselle calyx of 13 roselle (*Hibiscus sabdariffa L.*) genotypes. In Mexico and other regions, it is commonly used to make refreshing drinks, jams, jellies, liquors, and biscuit flour (Cid-Ortega and Guerrero-Beltrán 2015).

16.2.35 Phyllanthus acidus *(Common Name: Star Aonla)*

It is closely related to the Indian gooseberry found in Malaysia, Madagascar, the Malayan islands, Mizoram, Andhra Pradesh, Karnataka, and Maharashtra. The name of this fruit varies, as per region and acid taste, not by the varieties of star gooseberry. The tree bears flowers and fruits twice (maturing time 40–45 days) in a year with 30–60 green fruits in each inflorescence, which after maturation become pale yellow with 8–10 days shelf life at room temperature (Kishore et al. 2015). The fruits have antibilious, laxative, diaphoretic properties which are used to cure diarrhea, jaundice, and cough (Bose and Mithra 1990). Young leaves of *P. acidus are* cooked in Indonesia, Thailand, and India. In Malaysia, the fruit is prepared as relish and syrup, while the Philippines uses it for vinegar and beverages. Indonesia also utilizes the tart flesh as a flavoring (Tan et al. 2020).

16.2.36 Hylocereus undatus *(Common Name: Dragon Fruit)*

The dragon fruit (non-climacteric fruit) is a popular fruit crop of southern China, night night-blooming, hemi-epiphytic climbing cactus well recognized for its nutritive value and market potential. Its whole fruit juice, syrup can be enjoyed as a cool drink, color pastries and candies, while an unopened flower bud may be consumed as cooked vegetable. The dragon fruit pulp has kiwi fruit flavor which can be processed into jelly, jam, sharbat, syrup, preserve, candy, and ice cream. Its medicinal properties are due to antioxidants which prevent colon cancer, reduce blood pressure and diabetes, and combat anemia. The fruit is rich in vitamin C, phosphorus, and calcium which may help to make strong bones, teeth, and skin (de Freitas and Mitcham 2013).

16.2.37 Flacourtia indica *(Common Name: Governor's Plum)*

It is also known as ramontchi, or batoko plum, or baichi, or katai, Indian plum, which is native to Africa and Asia (tropical and temperate parts in Bangladesh and India). The fruit is rich in phytosterol (β, D-glucopyranoside, β-sitosterol), butyrolactone lignan disaccharide, ramontoside, flacourtin, coumarin (aesculetin and scoparone), flavonoids, and polyphenols (Nazneen et al. 2002). It is considered to be the most useful traditional and valuable source of medicinal plants in India for targeting various diseases and medicinal activities like antimicrobial, anti-asthmatic, anti-inflammatory, antioxidant, hepato-protective, antidiabetic, antimalarial, and antibacterial activity. Each part of the plant, leaves, bark, stem, fruits, root, and even the whole plant is useful for its health benefits. The fruits are used as a diuretic, appetizer, and digestive in jaundice and enlarged spleen (Ndhlala et al. 2007; Eramma and Devaraja 2013). It is typically consumed raw and processed into products like jams, jellies, pickles, and even wine (Vijay and Singh 2024).

16.2.38 Flacourtia jangomas *(Common Name: Paniala)*

This semi-cultivated fruit plant is traditionally found in the Brahmaputra Valley of Assam and surrounding regions in northeastern India, the Malay Peninsula, Bangladesh, and Myanmar. It has medicinal uses, including the treatment of asthma, bleeding gums, toothaches, diabetes, and digestive issues such as diarrhea and dysentery. The leaves are brewed for blood purification during prenatal and postnatal care. Additionally, its edible fruits can be eaten raw or used in jams and preserves, while its wood is also valuable (Sarkar et al. 2011).

16.2.39 Pandanus andamanensis, P. odoratissimus *(Common Name: Kewra)*

It occurs in coastal areas and forests of Southeast Asia, including parts of the Philippines, Indonesia, Papua New Guinea, and various Pacific islands. It is prevalent in India's coastal regions, particularly in Orissa, Andhra Pradesh, and Tamil Nadu. It is used in traditional Ayurvedic medicine to treat various ailments such as headaches, rheumatism, and infections. It contains beneficial phytochemicals, including lignans, isoflavones, and antioxidants, which contribute to health benefits. Additionally, it enhances overall well-being in tropical climates and can be chewed for fresh breath or used as a preservative in rice dishes (Adkar and Bhaskar 2014).

16.2.40 Nephelium lappaceum Lamk *(Common Name: Tader)*

Rambutan, known as the hairy fruit due to its spiny exterior, is a tropical fruit that originates from Malaysia and Indonesia but is now cultivated in various countries, including Mexico, Thailand, and the Philippines (Smith et al. 1992; Tindall 1994). Notable for its vibrant red color and sweet, juicy white aril, rambutan is enjoyed fresh or as processed products like squash (Nakasone and Paull 1998). In Mexico, it is primarily grown from seeds, although grafting could enhance production. This fruit is popular in many tropical Asian regions and is known for its nutritional benefits, including high vitamin C content (Wall 2006). Additionally, remnants such as peels and seeds are utilized for their antioxidant properties (Thitilertdecha et al. 2008; Saeed et al. 2012). Rambutan fruits are categorized based on quality into "Extra," "Category I," and "Category II" as per Codex standards.

16.2.41 Mimusops elengi *(Common Name: Bakul)*

Mimusops elengi, or Bakul/Spanish cherry, possesses fruit and seeds with a range of medicinal properties, including astringent, tonic, antibiotic, anti-inflammatory, analgesic, and effects on conditions like gingival bleeding, gastric ulcers, and hypotension. These benefits are due to the phytochemicals found in the plant.

16.2.42 Citrus deumana *(Common Name: Rabab)*

In India's Northeast region, particularly in Assam, Arunachal Pradesh, Meghalaya, Manipur, Mizoram, Tripura, Nagaland, and Sikkim, around 300 species of Rabab are found, covering 7.79% of the country's geographical area. Additionally, approximately 800 plant species serve as food for the tribal populations, with many lesser-known fruits being rich in minerals like Ca, Fe, P, and Mn. These fruits are primarily used to make a pinkish-white pulp for squash (Joseph and Murti 1981).

16.2.43 Ficus auriculata, F. retusa L, F. religiosa, F. sycomorus, F. carica *(Common Name: Fig)*

The common fig (*Ficus carica L.*) is valued for its commercially important fruit. In folk medicine, Egyptians use the fruits of both *F. carica L.* and *F. sycomorus L.* to treat respiratory disorders and skin diseases (Mousa et al. 1994). Figs are marketed through a careful process in which they are hand-picked from trees, sorted, and packed into small boxes to ensure their quality. They are then refrigerated for shipment to retail markets. Figs are processed into various products, including preserves, fig syrup, jam, jelly, spiced figs, pickles, fig bread, and fig meat (Peter 2008).

16.2.44 Antidesmabunius *(Common Name: Bigony)*

The fruits are oval in shape, fleshy, and red, offering a tangy flavor, and become wrinkled upon drying. They are a valuable source of calcium and iron, known for their effectiveness in treating dysentery, quenching thirst, and providing essential nutrients. The fruits are astringent in nature and utilized for the preparation of wine (Shariful et al. 2018).

16.2.45 Passiflora foetida, P. edulis, P. quadrangularis *(Common Name: Giant Granadilla)*

P. foetida L. is a fast-growing perennial vine native to the USA, India, Southeast Asia, and Hawaii. *P. amethystina* Mikan, found mainly in Brazil's rainforests, is another wild species. Both *P. foetida* and *P. incarnata* are indigenous to West Tropical Africa and India, while *P. quadrangularis*, also known as Giant Tumbo, originates from Tropical America. Globally, there are over 500 species of passion fruit, with 20 varieties being edible. The yellow passion fruit (*P. edulis flavicarpa*) is the most cultivated, making up 95% of orchards and thriving in moist tropical lowlands of Central and South America, as well as other regions like Hawaii, India, and West Africa. Passion fruit is economically significant for its edible fruits and medicinal properties. *P. edulis, P. quadrangularis,* and *P. ligularis* are used to make juices, sherbets, ice cream, and candies, with 65% of Colombian production targeting international markets. *P. incarnata* is known for its sedative effects, while

P. quadrangularis has applications in treating snake bites and may neutralize hemorrhaging in mice (Otero et al. 2000). Additionally, it exhibits chemopreventive properties against mammary cancer in rats (Crowell 1997). Biochemical compounds from these plants, such as schaftoside, hydrocyanic acid, and harman alkaloids, have notable medicinal benefits, including antimutagenic properties (Peyrt et al. 1992) and are used to treat anxiety, stress, insomnia, skin inflammation, coughs, and fever in humans.

16.2.46 Artocarpus altilis park, A. camansi, A. mariannensis, A. chaplasha, A. hirsutus, A. lakoocha *(Common Name: Breadfruit)*

Breadfruit is a multipurpose tree crop found in the Philippines, New Guinea, and possibly the Moluccas. This nutritious fruit is rich in carbohydrates, minerals, and vitamins and is known for being highly perishable and climacteric. It is utilized in breadfruit chips, bagi preparation, soup, fruit leather, and afara masi (fermented fruit in the form of a pasty mass) (Rouse-Miller and Duncan 2000).

16.2.47 Prunus persica, Prunus jenkinsii *(Common Name: Soh-Setanghi, Nectarine)*

Peaches are a vital part of cultivated fruit trees in Lebanon, valued for their nutrition and economic importance. Originating in China over 4000 years ago, they can be classified as melting or non-melting, hairy or smooth, and clingstone or freestone. Southern China is known for its melting peaches, prized for their soft texture, juicy flesh, and sweet flavor, making them competitive in the fresh fruit market. *P. jenkinsii* fruit is a drupe, acid in taste, ovoid to subglobose in shape. Its ripened fruits are dark purplish or black, stone pointed or rounded at the apex, and look like sohiong fruit (Rishi Kumar and Shukla 2018).

16.2.48 Diospyros kaki Thumb *(Common Name: Persimmon)*

The persimmon (Diospyros kaki), native to Asia and grown in temperate and subtropical regions, is rich in health benefits. It contains high levels of phenolic compounds, including polyphenols and carotenoids, along with antioxidants, fiber, vitamins, and minerals (Del Bubba et al. 2009; Veberic et al. 2010; Nugraheri and Rahmawati 2013; Jimenez-Sánchez et al. 2015). Persimmons can be processed into jam, wine, vinegar, ice cream, sweet jelly, and other treats.

16.2.49 Iberian peninsula *(Common Name:* Sambucus Nigra *L)*

Sambucus nigra is a shrub or small tree commonly found throughout Europe, thriving at sea level and at elevations exceeding 1600 meters. Known for its versatility and historical significance, this plant has been used extensively for various purposes, both popular and industrial. Its primary applications include medicinal uses for treating various ailments and culinary uses, such as in producing jams and beverages. Furthermore, *S. nigra* has many other applications across different fields. For example, its pith (the inner part of the stems and branches) is traditionally used as an embedding medium for preparing microscope slides (Peter 2008).

16.2.50 Morinda citrifolia L *(Common Name: Indian Noni)*

The Indian noni, also known as Indian mulberry, is native to Southeast Asia and Australia, grows in the wild in coastal areas and the Western Ghats of India, and is widespread throughout the Pacific region, valued for its medicinal properties among the people of the Pacific islands.

The suitable temperature range for noni is 20–38 °C. Flowering and fruiting are continuous throughout the year, fruits are greenish white to pale yellow in color, and fruit juice is a rich source of Carbohydrates, vitamins (A, B, C, E), minerals (Ca, Fe, Zn, Cu, Mg, Mn, Mo, Na, K), antioxidants, and nutraceuticals, and helps to strengthen the immune, digestive, metabolic, and nervous systems (Saleha et al. 2007). It is commonly used as a dietary supplement for cancer treatment. Key industrial products from this plant include juice beverages, dried fruit powders, seed oil, and leaf powders (Almeida et al. 2019).

16.2.51 Aegle marmelos *(Common Name: Indian Bael)*

Aegle marmelos belongs to the family Rutaceae and is native to central India and the Eastern Ghats. It has therapeutic properties and is effective for treating diabetes, cancer, and oral and gastric ulcers. Aegle marmelos contains a high moisture content (61%), minerals (Ca, Fe, P, and K), fat, fiber, carbohydrates (36.80–41.70%), protein (7.52–8.81%), vitamins (A, B, and C), amino acids, fatty acids (C16:0), bioactive compounds, and phytochemicals. It can be processed into jam, powder, wine, and syrup (Sharma et al. 2022).

16.3 Production and Trade

Only 6.56% of India's 0.437 mha of underutilized fruit crops are used despite their high production of 11.47 tons/ha (Meena et al. 2022). The northern Himalayan region (J & K, Himachal Pradesh, Ladakh, and Uttarakhand) is home to underutilized fruits, *Ficus auriculata, Rubus ellipticus, Myrica esculenta, Ficus palmata, Pyracantha crenulate, Prunus armeniaca, Berberis asiatica, Pyrus pashia, Rubus niveus, Carissa spinarum, Cordia dichotoma, Flacourtia indica, Malus sikkimensis, Physalis peruviana, Ziziphus mauritiana, Viburnum mullaha, Olea ferruginea, Elaeagnus latifolia, Corylus jacquemontii, and Pinus gerardiana* (Bachheti et al. 2023).

In the Northeast region of India, *Actinidia callosa, Baccaurea sapida, Musa velutina, Musa ornata, Castanopsis indica, Sterculia hamiltonii, Nephelium lappaceum, Lithocarpus spp, Mangifera sylvatica, Pyrus pashia, Rubus niveus, Citrus medica, Livistona jenkinsiana, Viburnum foetidum, Artocarpus chaplasha, Dillenia indica, Malus baccata, Machilus edulis, Phyllanthus acidus, Garcinia lanceifolia, Passiflora edulis, Musa rosea, and Mangifera sylvatica, Lemon citrus, Citrus maxima, Citrus megaloxycarpa, Citrus jambhiri, C. macroptera, C. assamensis, Garcinia lanceifolia, Myrica farquhariana, Artocarpus lakoocha, Dillenia indica, Averrhoa carambola, Phyllanthus acidus, Baccaurea saipda, Flacourtia indica, Elaeagnifolia, Passiflora edulis, C. Phyllanthus acidus, Citrus aurantium, ichangensis, Juglans regia, Malus baccata; Actinidia strigosa, Machilus edulis, Bassia butyracea, Musa sikkimensis, Spondias axillaris, Baccaurea sapida, Elaeagnus latifolia; Citrus macroptera, Baccaurea sapida, Averrhoa carambola, Zizyphus uniculisa, Antidesma bunius, A. ghaesembilla, A. Carissa canards, Rubus ellipticus, Elaeocarpus floribundus, Citrus maxima, Dillenia indica, Physalis minima, Grewia hirsuta, Grewia sapida, Physalis minima, Psidium guineense* are among the underutilized fruits found (Shankar et al. 2020; Deka et al. 2012).

India's arid and semi-arid zones include the states of Gujarat, Rajasthan, Haryana, and Punjab. The underused fruits found in the states of Karnataka, Tamil Nadu, Andhra Pradesh, and Maharashtra are *Salvadora oleoides, Capparis decidua, Grewia subinaequalis, and Cordia myxa. Persica, Carissa carandus, Sclerocarya birrea subsp., Manilkara hexandra, Buchnania lanzan, Opuntia ficus indica, Aegle marmelos, Feronia limonia, Syzygium cumini, and Morus* spp.; *Caffra, Ziziphus rotundifolia, Acacia Senegal, Madhuca indica, Pithecellobium dulce, Ceratonia siliqua, Cordeauxia edulis, Uapaca kirkiana, Dovyalis caffra, Lannea acida* (Krishna et al. 2019; Meghwal and Singh 2016). Eastern Central Indian states like Jharkhand have *Madhuca longifolia, Diospyros melanoxylon Roxb., Buchanania lanzan, Syzygium cumini (L.) Skeels, Syzygium fruticosum (Roxb.) DC., Ziziphus mauritiana Lam., Aegle marmelos (L.) Corr., Spondias dulcis L., Artocarpus lakoocha Roxb., Dillenia indica Linn., Dilleniapentagyna Roxb.* minor fruits (Sarkar et al. 2018).

FAO reports that underused fruits are grown worldwide but are used domestically instead of being exported. Its trading is limited to the local market and is used to increase food security and supply all of the vital nutrients needed for proper bodily function. However, some of the underutilized fruits are being traded outside of their home nation, where the demand for quality fruits has improved economic growth. The increased awareness of the nutritional advantages of minor fruits and the advancement of health knowledge among Western consumers are driving up demand for these fruits in industrialized nations. Retailers in the USA and EU have begun to carry an increasing variety of underutilized fruits on a regular basis, with the most significant being fruits that are comparable to lychee, guava, and passion fruit (Altendorf 2018).

16.4 Commercialization

According to market equilibrium customers may be able to obtain the goods, but their quality may be subpar when they go to buy it; they may even be willing to purchase the product, just not at the locations where it is marketed. These factors could explain the apparent lack of demand. The need might only be met by people who live nearby, work in rural regions, are elderly, have poor incomes, or are part of a group that uses products from minor plant species in a custom that is unknown to outsiders. Customers are constantly in need of fresh, delectable, healthy, and visually appealing food goods. Even though there is a high demand for products derived from minor plant species, efforts are constantly made to develop products from a variety of sources. In addition, inefficiencies may limit the supply or quality of available products, and an ineffective or incomplete marketing channel may increase transaction costs.

According to Chandra et al.'s research from 2020, an uncontrolled marketing route can significantly restrict the market for underused plant species. This can be the case whether the marketing channel is straightforward (collection and distribution) or more complicated (wholesale, processing, and retail distribution). Horticultural products that are seasonal and have a short shelf life are examples of underutilized fruits. These qualities are the main barriers to the industrial and commercial applications of these underutilized fruits (Bachheti et al. 2023). The only ways to solve the issue are to develop methods for value addition, create markets, and inform the indigenous population about the nutritious value of underutilized commodities (Fig. 16.1c, d).

It is feasible to increase the economic value of underutilized commodities by processing and adding value to tribal fruits and vegetables in order to create products that are easier to digest, tastier, more convenient, and more helpful. The nation has been successful in creating appropriate processing technologies for the purpose of enabling the commercial exploitation of these fruits and the addition of value to them. As a result products like juice from *Actinidia callosa* and *Rubus niveus*, wine from *Actinidia callosa, Rubus niveus, Baccaurea sapida* and *Docynia indica,* flavoring agent from *Baccaurea sapida, Dillenia indica, Citrus macroptera* and *Physalis minima*, pectin from *Baccaurea sapida*, chutney from *Baccaurea sapida,*

Dillenia indica, Livistona jenkinsiana, Elaeagnus latifolia and *Elaeocarpus floribundus*, tea from *Pyrus pashia*, purees from *Pyrus pashia* and *Machilus edulis* paste from *Pyrus pashia*, jam from *Pyrus pashia, Rubus niveus, Docynia indica, Dillenia indica, Rubus ellipticus, Elaeagnus latifolia* and *Malus baccata,* jelly from *Mangifera sylvatica, Malus baccata, Rubus niveus, Dillenia indica* and *Elaegnus latifolia*, beverage from *Phyllanthus acidus, Malus baccata, Dillenia indica, Elaegnus latifolia* and *Citrus medica*, pickle from *Phyllanthus acidus, Mangifera sylvatica, Dillenia indica, Citrus macroptera* and *Elaeocarpus floribundus*, mayonnaise from *Machilus edulis*, dried fruits from *Phyllanthus acidus*, vinegar from *Phyllanthus acidus* and *Docynia indica*, syrup from *Rubus niveus, Docynia indica* and *Phyllanthus acidus*, cocoa butter from *Mangifera sylvatica*, sherbets from *Grewia hirsute* and *Physalis minima*, and squash from *Dillenia indica* and *Rubus ellipticus* have gone through the process of development, and they have a promising commercial value (Brooks et al. 2020; Yatung et al. 2023; Benedetto et al. 2023; Pancholi and Rana 2020; Sharma et al. 2022; Prakash et al. 2021a, b; Padhan et al. 2023).

The techniques for the following value-added product preparations have been established by the ICAR Research Complex for the Northeast: Sohiong RTS and squash, Sohiong candy, Sohiong powder, Sohiong wine, Sohiong fruit powder, Sohshang leather, and Soshang jam are all products that are manufactured from Sohiong (Deka and Rymbai 2014).

16.5 Challenges in Post-harvest Processing of Underutilized Fruits

The primary barriers to the cultivation of neglected crops are plating supplies cost, and inadequate cultural and post-harvest management practices, because of their low nutritional content, consumption patterns, lack of study, and government agency development policies about their possible exploitation, these minor fruits are not particularly popular and are sold in local markets for extremely low costs (Meena et al. 2022). According to Patel et al. (2013), several factors, such as lack of research, technology, and access to high-quality planting material, ignorance of nutritional value, and economic benefits, contribute to the low recognition of certain crops in horticultural promotion programs.

Indigenous fruits are stigmatized excessively because of the belief that they are a poor man's diet and have poor nutritional value. With appropriate public sensitization, such stigmatization can be dismantled. Thus, more native fruits should be the subject of research, with the public being informed of the findings. The bulk of the developing world's impoverished and marginalized communities will benefit from increased food security and dietary diversity as a result (Musinguzi et al. 2007) (Fig. 16.2).

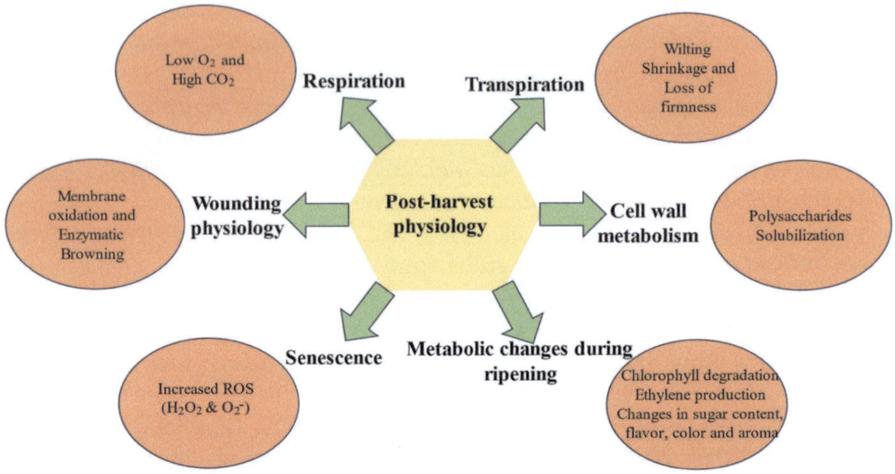

Fig. 16.2 Post-harvest physiology of underutilized fruits. (Adapted from Hossain et al. 2020)

16.6 The Future of Neglected Fruits

To boost fruit output, more care must be taken to provide the right planting materials, support farmers in developing their skills and capability, and encourage them to use more advanced production techniques for higher productivity. Vegetative propagation can shorten the gestation time; homestead cultivation can domesticate prospective wild fruit species; pruning and training can control the canopy; and superior genotypes can be chosen for a range of applications, including fruit juice, wine, and the extraction of natural color.

To enhance the cultivation of underused fruits, strategies include standardizing post-harvest management, adding value through processing procedures, multiplying high-quality planting materials, and documenting indigenous knowledge to increase value and multipurpose usage (Patel et al. 2013). Technology demos and end-user training can speed up the promotion of neglected fruit crops.

Planting nurseries creation, to ensure the availability of premium planting materials and processing facilities to increase the production of their value-added goods at a commercial scale through self-help organizations, FSSAI, and entrepreneurship, etc., are examples of developmental activities that require special attention. Planting these crops in public spaces, such as gardens, parks, and the grounds of places of worship, might encourage their acceptance. In addition to offering recreational opportunities and esthetic benefits, these plantings will provide people with nutrient-rich fruits.

To build commercial orchards of diverse fruit crops, especially underused fruits, the government provides high-quality planting materials through a number of programs, including MIDH, TSP, and SCSP. Diseases and insect pests are less likely to affect these crops. The use of pesticides is essentially nonexistent, and therefore, they fit very well with organic farming (Meena et al. 2022). Farmers need to be made aware of the nutritional value of underutilized fruits. Extension agents can do this by planning exhibitions, special awareness camps and campaigns, and other events at both the local and macro levels. Similar to this, raising awareness among farmers can be accomplished through the effective use of mass media, including radio, television, newspapers, and other written materials (Deka and Rymbai 2014).

16.7 Conclusion

This paper highlights the value of underutilized fruits in India, especially in the northeast. These minor fruits, which are high in nutrients and have therapeutic qualities, may benefit the environment, food security, health, and nutrition as well as generate income. The northeast region has a wide variety of underutilized fruit species that present special opportunities for cultivation and use, appealing to health-conscious consumers and bringing economic benefits to rural communities. The most crucial post-harvest physiological aspects that should be minimized in fruits for prolonged storage without compromising their original qualities are transpiration, ethylene production, and respiration.

To commercialize underused fruits, better packaging and storage conditions are needed to extend their shelf life. The fruits are processed into value-based food products like jam, jelly, and pickles to boost their market value and profitability for local producers. This increases the fruits' utility, digestibility, flavor, and convenience. The reason behind the low global popularity of these underutilized fruits is a lack of awareness about them. Schemes and management that are required should be implemented in order to improve its demand.

References

Abobatta WF (2019) Nutritional benefits of citrus fruits. Am J Biomed Sci Res 3:4. https://doi.org/10.34297/AJBSR.2019.03.000681

Adetutu A, Odunola OA, Queade OA, Adeleke OA, Amuda OS (2004) Anticlastogenic effects of Hibiscus sabdariffa fruits against sodium arsenite-induced micronuclei formation in erythrocytes in mouse bone marrow. Phytother Res 17

Adkar PP, Bhaskar VH (2014) Pandanus odoratissimus (Kewda): a review on ethnopharmacology, phytochemistry, and nutritional aspects. Adv Pharmacol Pharm Sci 1:120895

Akhter S (2016) Conservation of Mangifera sylvatica: a wild fruit species for health and livelihoods. Bangor University, Bangor

Akhter S, McDonald M, Marriott R, Jashimuddin M, Bashirul-Al-Mamun M (2021) Morphological, nutritional, and medicinal traits of wild mango (Mangifera sylvatica Roxb.): implications for increased use and options for cultivar development. Trees For People 6:100126

Akhter S, McDonald M, Jashimuddin M, Bashirul-Al-Mamun M, Sarker P (2022) Agroforestry potential of a wild mango species (Mangifera sylvatica Rexb.). Trees For People 7:100194

Alarcon-Alonso J, Zamilpa A, Aguilar FA, Herrera-Ruiz M, Tortoriello J, Jimenez-Ferrer E (2012) Pharmacological characterization of the diuretic effect of Hibiscus sabdariffa Linn (Malvaceae) extract. J Ethnopharmacol 139:751–756

Almeida ÉS, de Oliveira D, Hotza D (2019) Properties and applications of Morindacitrifolia (noni): a review. Compr Rev Food Sci Food Saf 18:883–909

Altendorf S (2018) Minor tropical fruits. Food Outlook 1:67–75

Anand JC (1960) Efficacy of sodium benzoate to control yeast fermentation in phalsa (Grewia asiatica L.) juice. Indian J Hortic 17:138–141

Anaya-Esparza LM, Montalvo-González E (2017) Nutritional quality of Mangifera species. In: Bioactive molecules in food, pp 1–19

Atanasova AG, Sabharanjakc SM, Zengin G, Mollica A, Szostaka A, Simirgiotis M, Huminieckia Ł, Horbanczukh OK, Nabavii SM, Mocan A (2018) Pecan nuts: a review of reported bioactivities and health effects. Trends Food Sci Technol 71:246–257

Augustus GDP, Jayabalan MJ, Seiler GJ, Seiler GJ (2003) Alternative energy sources from plants of Western Ghats (Tamil Nadu, India). Biomass Bioenergy 24(6):437–444

Bachheti A, Deepti BRK, Singh A, Zebeaman M, Hunde Y, Husen A (2023) Bioactive constituents and health promoting compounds of underutilized fruits of the northern Himalayas of India: a review. Food Prod Process Nutr 5(1):24

Banik BC, Ubale NB, Pain R, Bhowmick N (2013) Burmese grape (Baccaureasapida Muell. Arg.)-an underexploited fruit crop of nutritional and economic importance

Barua U, Das RP, Das P, Das K, Gogoi B (2022) Morpho-physiological, proximate composition and antioxidant activity of Flacourtiajangomas (Lour.) Raeus. Plant Physiol Rep. https://doi.org/10.1007/s40502-022-00644-1

Basumatary J, Khobragade CB, Paul SK, Kumar P (2016) Study of physical, optical and biochemical characteristics of elephant apple (Dillenia indica) for its utility potential in food processing industries. J Agric Eng Food Technol 3(1):54–57

Benedetto N, Carlucci V, Faraone I, Lela L, Ponticelli M, Russo D, Milella L (2023) An insight into Citrus medica Linn.: a systematic review on phytochemical profile and biological activities. Plants 12(12):2267

Bhatt BP, Sarangi SK, De LC (2010) Fuelwood characteristics of some firewood trees and shrubs of eastern Himalaya, India. Energy Sour Part A Recov Utilization Environ Eff 32(5):469–474

Bhatt SC, Naik B, Kumar V, Gupta AK, Kumar S, Preet MS, Rustagi S (2023) Untapped potential of non-conventional rubus species: bioactivity, nutrition, and livelihood opportunities. Plant Methods 19(1):114

Bhowmick N (2017) Fruiting characteristics of jalpai-the Indian olive (Elaeocarpus floribundus Blume.). J Pharmacogn Phytochem 6(5):608–609

Biswas J, Barman P, Rahman M, Ghosh RA (2021) Comprehensive review on the endemic plant Docynia indica: its ethnobotanical, phytochemical, and Ethnomedicinal perspective. Asian J Biol Life Sci 10:515

Biswas SC, Kumar P, Kumar R, Das S, Misra TK, Dey D (2022) Nutritional composition and antioxidant properties of the wild edible fruits of Tripura, Northeast India. Sustainability 14(19):12194

Bose TR, Mithra SK (1990) Fruit's tropical subtropical. Nayan Prokash, Bidhan Sarani, Kolkata, India, pp 12–13

Brooks R, Goldson-Barnaby A, Bailey D (2020) Nutritional and medicinal properties of Phyllanthus acidus L. (Jimbilin). Int J Fruit Sci 20(sup3):S1706–S1710

Chandra MS, Naresh RK, Thenua OVS, Singh R, Geethanjali D (2020) Improving resource conservation, productivity and profitability of neglected and underutilized crops in the breadbasket of India: a review. Breeding Sci 70(1):19–31

Chaudhary C, Khatak A, Devi R, Rai D, Yadav BS (2017) Study of fermentation variables for the preparation of wine from jamun fruit. J Pure Appl Microbiol 11(3):1623–1631

Chhetri G, Bhujel D, Rai YK (2020) Ethno-medicinally significant edible wild fruits of Sikkim Himalaya. J Adv Plant Sci 10(2):45–57

Chhikara N, Kour R, Jaglan S, Gupta P, Gat Y, Panghal A (2018) Citrus medica: nutritional, phytochemical composition and health benefits–a review. Food Funct 9(4):1978–1992
Chikkamath V, Kulkarni VH, Habbu PV, Kulkarni PV (2019) Grewia hirsuta Vahl chemical constituents and biological activities. RGUHS J Pharm Sci 9(3)
Chingtham C, Banik AK (2014) Studies on preparation of puree from ripe banana. J Environ Res Dev 9(1):224
Chothani DL, Vaghasiya HU (2012) A phyto-pharmacological overview on Physalis minima Linn
Choudhury A, Patel NA (2016) Pharmacognostical and physico-chemical studies of the bark of Baccaurearamiflora lour. Int J Pharmacogn Phytochem Res 8(12):1935–1939
Chyne DAL, Ananthan R, Longvah T (2019) Food compositional analysis of indigenous foods consumed by the Khasi of Meghalaya, north-East India. J Food Comp Anal 77:91–100
Cid-Ortega S, Guerrero-Beltrán JA (2015) Roselle calyces (Hibiscus sabdariffa), an alternative to the food and beverages industries: a review. J Food Sci Technol 52:6859–6869
Correa MP (1926) Diccionario das PlantasÚteis do Brasil e das ExoticasCultivadas. Imprensa Nacional, Rio de Janeiro, 307 p
Crowell PL (1997) Monoterpenes in breast cancer chemoprevention. Breast Cancer Res Treat 46(2–3):191–197
Dada SO, Ehie GC, Osukoya OA, Anadozie SO, Adewale OB, Kuku A (2023) In vitro antioxidant and anti-inflammatory properties of Artocarpus altilis (Parkinson) Fosberg (seedless breadfruit) fruit pulp protein hydrolysates. Sci Rep 13:Article 1493. https://doi.org/10.1038/s41598-023-1493-y
Dadwal V, Agrawal H, Sonkhla K, Joshi R, Gupta M (2018) Characterization of phenolics, amino acids, fatty acids and antioxidant activity in pulp and seeds of high altitude Himalayan crab apple fruits (Malus baccata). J Food Sci Technol 55:2160–2169
Daheriya M, Verma DK, Kuruba EK, Mahajan KC, Joshi NU (2023) Advances in multi-fruit and vegetable grading: a comprehensive review. Int J Environ Clim Change 13(9):1239–1245
Darkwa I, Boakye NAB (2016) The preparation of jam: using star fruit. Global J Educ Stud 2(2):2377–3936
Das M, Debnath D, Hoque MA, Rahman MS, Alam S, Islam A, Begum MM (2019) Preliminary phytochemical and biological investigations of ethanolic extract of Grewia hirsute Vahl. Orient Pharm Exp Med 19:145–156
Dasgupta P, Chakraborty P, Bala NN (2013) Averrhoa Carambola: an updated review. Int J Pharma Res Rev 2(7):54–63
de Freitas ST, Mitcham EJ (2013) Quality of pitaya fruit (Hylocereus undatus) as influenced by storage temperature and packaging. Sci Agric 70(4):257–262
De LC, Singh DR (2017) Natural resources in North East region of India. Int J Agric Sci Res 7(5):51–66
Deka BC, Rymbai H (2014) Status of underutilized fruit crops in Northeast India. In: Proc Natl seminar strat conserv improve Util underutilized fruits. Central Horticultural Experiment Station and Society for Promotion of Horticulture, Bangalore
Deka BC, Thirugnanavel A, Patel RK, Nath A, Deshmukh N (2012) Horticultural diversity in North-East India and its improvement for value addition. Indian J Genet Plant Breed 72(02):157–167
Del Bubba M, Giordani E, Pippucci L, Cincinelli A, Checchini L (2009) Changes in tannins, ascorbic acid and sugar content in astringent persimmons during on-tree growth and ripening and in response to different postharvest treatments. J Food Comp Anal 22:668–677
Dey K, Ghosh A, Bauri FK, Sahu PK (2022) Studies on fruit developmental stages and maturity of Indian olive (Elaeocarpus floribundus Bl.)
Dhyani S, Mehta H (2023) Challenges of star fruit Averrhoa carambola: a comprehensive overview. Int J Sci Res 12(8). https://www.ijsr.net
Direito R, Rocha J, Sepodes B, Eduardo-Figueira M (2021) From Diospyros kaki L. (persimmon) phytochemical profile and health impact to new product perspectives and waste valorization. Nutrients 13(9):3283. https://doi.org/10.3390/nu13093283

Dolai N, Karmakar I, Kumar RBS, Haldar PK (2012) CNS depressant activity of Castanopsis indica leaves. Orient Pharm Exp Med 12:135–140
Duke J (1983) Handbook of energy crops
Dutta IN, Gogoi BA, Sharma RU, Sharma HK (2018) In vitro evaluation of leaves and fruits of Elaeagnus latifolia L. for antioxidant and antimicrobial activities. Asian J Chem 30(11):2433–2436
Emurotu JE, Onojah PK, Musa AS (2017) Proximate and phytochemical screening of the fruit and leaves extract of Bael (Aegle marmelos). IOSR J Appl Chem 10(9 Ver. II):22–28. https://www.iosrjournals.org
Eramma NK (2016) A comprehensive review on pharmacology of Flacourtia Indica (BURM.F.) Merr. (Governor's Plum)
Eramma N, Devaraja G (2013) Antibacterial potential and phytochemical analysis of Flacourtia indica (Burm.f.) Merr. root extract against human pathogens. Indo Am J Pharm Res 3(5):3832–3846
Fraire V (2001) El rambutan: alternativa para la producciónfrutícola del tropicohúmedo de México. INIFAP, 41 p. (Folleto, 1).
Gami B, Pathak S, Parabia M (2012) Ethnobotanical, phytochemical and pharmacological review of Mimusopselengi Linn. Asian Pac J Trop Biomed 2(9):743–748
Gandhi D, Mehta P (2013) Dillenia indica Linn. and Dilleniapentagyna Roxb.: pharmacognostic, phytochemical and therapeutic aspects. J Appl Pharm Sci 3(11):134–142
Gill NS, Bajwa J, Sharma P, Dhiman K, Sood S, Sharma PD (2011) Evaluation of therapeutic potential of traditionally consumed Cucumis melo seeds. Asian J Plant Sci 10(1):86–91
Gonçalves B, Landbo AK, Knudsen D, Silva AP, Moutinho Pereira J, Rosa E, Meyer AS (2004) Effect of ripeness and postharvest storage on the phenolic profiles of cherries (Prunus avium L.). J Agric Food Chem 52(3):523–530
Hasnat GN, Hossain MA, Hossain MK, Uddin MM (2019) Effect of pre-sowing treatments on germination and initial seedling growth of Castanopsis Indica-an endangered tree species in Bangladesh. J For Environ Sci 35(4):223–231
Hasnin I, Akter S, Khatun R, Khatune NA, Islum MM, Rahman AA (2023) Hypoglycemic and hypolipidemic effects of Viburnum Foetidum (Linn.) methanolic extract in a mice model of fructose induced type ii diabetes. Available at SSRN 4132385
Hassan H, Amin M, Rajwana IA, Ullah S, Razzaq K, Faried HN, Sahar HW (2022) Nutritional functions and antioxidative enzymes in juice extract from two different maturity stages of low temperature stored phalsa (Grewia subinaequalis DC) fruit. LWT 153:112552
Hazarika TK, Pongener M (2018) Potential wild edible fruits of Nagaland, North-East India and its significance in the livelihood and nutritional security of rural, indigenous people. Genet Resour Crop Evol 65:199–215
Hazarika TK, Marak S, Mandal D, Upadhyaya K, Nautiyal BP, Shukla AC (2016) Underutilized and unexploited fruits of Indo-Burma hot spot, Meghalaya, North-East India: ethno-medicinal evaluation, socio-economic importance and conservation strategies. Genet Resour Crop Evol 63:289–304
Hegde K, Karunakar S, Shalin P, Thakker AB, Joshi AB, Joshi KS (2009) Anticonvulsant activity of Carissa carandas Linn. Root extract in experimental mice. Trop J Pharm Res 8(2):117–125
Hiwale S (2015) Wood apple (Feronia limonia Linn.). In: Sustainable horticulture in semiarid dry lands. Springer, New Delhi, pp 225–235
Hossain MA, Karim MM, Juthe SA (2020) Postharvest physiological and biochemical alterations in fruits: a review. Bangladesh Agricultural University
Hossain MM, Rahim MA, Haque MR (2021) Biochemical properties of some important underutilized minor fruits. J Agric Food Res 5:100148
Hussain SZ, Naseer B, Qadri T, Bhat TA (2021) Peach (Prunus persica)—morphology, taxonomy, composition and health benefits. In: Fruits grown in Highland regions of the Himalayas. Springer Nature Switzerland AG, pp 207–232. https://doi.org/10.1007/978-3-030-75502-7_16

Hwong CS, Leong KH, Aziz AA, Junit SM, Noor SM, Kong KW (2022) Alternanthera sessilis: uncovering the nutritional and medicinal values of an edible weed. J Ethnopharmacol 298:115608

Islary A, Sarmah J, Basumatary S (2017) Nutritional value, phytochemicals and antioxidant properties of two wild edible fruits (Eugenia operculata Roxb. and Antidesmabunius L.) from Assam, North-East India. Mediterr J Nutr Metab 10:29–40. https://doi.org/10.3233/MNM-16119

Jimenez-Sánchez C, Lozano-Sanchez J, Marti N, Saura D, Valero M (2015) Characterization of polyphenols, sugars, and other polar compounds in persimmon juices produced under different technologies and their assessment in terms of compositional variations. Food Chem 182:282–291

Joseph J, Mendonca G (1989) Oxalic acid content of carambola (Averrhoa carambola L.) and bilimbi (Averroha bilimbi L.). Proc Interam Soc Trop Hortic 33:117–120

Joseph J, Murti SK (1981) Plant resources of North East India. In: Seminar on Resource Potential of North-East India, April 29–May 1, 1981, Meghalaya Science Society, Shillong

Kader A (2009) Rambutan: recommendations for maintaining postharvest quality. Available at: http://postharvest.ucdavis.edu/Produce/ProduceFacts/Fruit/rambutan.html. Accessed 18 Feb 2009

Kaladhar DSVGK, Rayavarapu KA (2014) Phytochemical analysis, antioxidant and antimicrobial activities of Annona reticulata raw fruit peel extracts. World J Pharm Sci 3(11):1226–1234

Kalia A, Parshad VR (2015) Novel trends to revolutionize preservation and packaging of fruits/fruit products: microbiological and nanotechnological perspectives. Crit Rev Food Sci Nutr 55(2):159–182

Kewlani P, Tiwari D, Rawat S, Bhatt ID (2023) Pharmacological and phytochemical potential of Rubus ellipticus: a wild edible with multiple health benefits. J Pharm Pharmacol 75(2):143–161

Khruomo NK, Deb CR (2018) Indigenous wild edible fruits: sustainable resources for food, medicine, and income generation – a study from Nagaland, India

Kirtikar KR, Basu BD (2003) Indian medicinal plants, Lalit Mohan Basu, Allahabad, Vol. II, pp. 1546. Knowledge 9:398–402

Kishore K, Singh HS, Kurian RM (2015) Paclobutrazol use in perennial fruit crops and its residual effects: a review. Indian J Agric Sci 85(7):863–872

Krishna H, Saroj PL, Maheshwari SK, Singh RS, Meena RK, Chandra R, Parashar AV (2019) Underutilized fruits of arid & semi-arid regions for nutritional and livelihood security. Int J Minor Fruits Med Aromat Plants 5:1–14

Kumari A, Dhaliwal YS (2017) A study on nutritional composition and value addition of crab apple (Malus baccata). Am J Food Sci Technol 5(1):19–22

Lennox A, Ragoonath J (1990) Carambola and bilimbi. Fruits 45(5):497–501

Lepcha R, Upadhyay S, Muddarsu VR, Manivannan S, Bhutia RW (2018) Study on value addition and products diversification of Mallero (Elaeagnus latifolia)

Liaotrakoon W, Van Buggenhout S, Christiaens S, Houben K, De Clercq N, Dewettinck K, Hendrickx ME (2013) An explorative study on the cell wall polysaccharides in the pulp and peel of dragon fruits (Hylocereus spp.). Eur Food Res Technol 237:341–351. https://doi.org/10.1007/s00217-013-1997-7

Mahapatra AK, Mishra S, Basak UC, Panda PC (2012) Nutrient analysis of some selected wild edible fruits of deciduous forests of India: an explorative study towards non-conventional bio-nutrition. Adv J Food Sci Technol 1:15–21

Mahendran R, Jayashree GC, Alagusundaram K (2012) Application of computer vision technique on sorting and grading of fruits and vegetables. J Food Process Technol 10:2157–7110

Mahmood K, Kamilah H, Alias AK, Ariffin F (2018) Nutritional and therapeutic potentials of rambutan fruit (Nephelium lappaceum L.) and the by-products: a review. J Food Meas Charact. https://doi.org/10.1007/s11694-018-9771-y

Malakar S, Sarkar S, Kumar N (2019) King chilli (Capsicum chinense Jacq.), "the India's hottest chilli" – an overview. J Appl Hortic 21(1):53–56

Malik DSA, Paul R, Sarkar T (2024) Rambai (Baccaureasapida): a prospective lesser-known fruit with the potential to uplift the socio-economic conditions of rural communities in West Bengal

Mani A, Kuchi VS, Mitra S, Banik AK, Chakraborty I, Das S, Bauri FK (2020) Standardization of recipe of Barbados cherry pickle. Eur J Nutr Food Saf 12(4):43–50

Mann S (2016) Studies on some underutilized edible fruits in India

Mann S, Sharma A, Biswas S, Gupta RK (2015) Identification and molecular docking analysis of active ingredients with medicinal properties from edible Baccaureasapida. Bioinformation 11(9):437

Mann S, Satpathy G, Gupta RK (2016) Evaluation of nutritional and phytochemical profiling of Baccaurearamiflora Lour. syn. Baccaureasapida (Roxb.) Mull. Arg. fruits

Mathew L, George ST, Babylatha AK, Geetha CK (1993) Flowering and fruit development in bilimbi (Averrhoa bilimbi L.). South Indian Hort 41(1):41–42

Meena VS, Gora JS, Singh A, Ram C, Meena NK, Rouphael Y et al (2022) Underutilized fruit crops of Indian arid and semi-arid regions: importance, conservation and utilization strategies. Horticulturae 8(2):171

Meeprathom N, Nukit K, Jongrattanavit K, Kooprasertying P (2018) Proximate compositions, phenolic compounds, antioxidant capacity and antibacterial activity of Chulta (Dillenia indica Linn.) fruits: effects of maturity stage and extraction solvent. J Food Health Bioenv Sci 11(1):57–78

Meghwal PR, Singh AK (2016) Underutilized fruits research in arid regions: a review. Ann Arid Zone 56(1&2):23–36

Mejía-Morales C, Rodríguez-Macías R, Salcedo-Pérez E, Zamora-Natera JF, Rodríguez-Zaragoza FA, Molina-Torres J, Délano-Frier JP, Zañudo-Hernández J (2021) Contrasting metabolic fingerprints and seed protein profiles of Cucurbita foetidissima and C. radicans fruits from feral plants sampled in Central Mexico. Plan Theory 10:2451. https://doi.org/10.3390/plants10112451

Mlynarczyk K, Walkowiak-Tomczak D, Łysiak GP (2018) Bioactive properties of Sambucus nigra L. as a functional ingredient for food and pharmaceutical industry. J Funct Foods 40:377–390. https://doi.org/10.1016/j.jff.2017.11.025

Mondal MEA, Goswami JN, Deomurari MP, Sharma KK (2002) Ion microprobe 207Pb/206Pb ages of zircon from the Bundelkhand massif, northern India: Implication for crustal evolution of the Bundelkhand–Aravalli protocontinent. Precambrian Res 117:85–100

Morton JF (1987) Indian jujube. In: Morton JF (ed) Fruits of warm climates. Florida Flair Books, Miami, pp 272–275

Mousa O, Vuorela P, Kiviranta J, Wahab SA, Hiltohen R, Vuorela H (1994) Bioactivity of certain Egyptian Ficus species. J Ethnopharmacol 41:71–76

Muchuweti M, Zenda G, Ndlala AR, Kasiamhuru A (2005) Sugars, organic acids, and phenolic compounds of Ziziphus mauritiana fruit. Eur Food Res Technol 221:570–574

Musinguzi E, Kikafunda JK, Kiremire BT (2007) Promoting indigenous wild edible fruits to complement roots and tuber crops in alleviating vitamin A deficiencies in Uganda. In: Proc 13th ISTRC Symp, pp 763–769

Muthu N, Lee SY, Phua KK, Bhore SJ (2016) Nutritional, medicinal, and toxicological attributes of star-fruits (Averrhoa carambola L.): a review. Bioinformation 12(12):420–424. https://doi.org/10.6026/97320630012420

Nakasone H, Paull R (1998) Tropical fruits. CAB International, New York. 445 p

Narzary H, Brahma S, Basumatary S (2013) Wild edible vegetables consumed by Bodo tribe of Kokrajhar District (Assam), North East India. Arch Appl Sci Res 5(5):182–190

Nazarudeen A (2010) Nutritional composition of some lesser-known fruits used by the ethnic communities and local folks of Kerala. Indian J Tradit 5:50–60

Nazir A, Wani SM, Gani A, Masoodi FA, Haq E, Mir SA, Riyaz U (2013) Nutritional, antioxidant, and antiproliferative properties of persimmon (Diospyros kaki) – a minor fruit of J&K. India Int J Adv Res 1(7):545–554

Nazneen M, Mazid MA, Kundu JK, Bachar SC, Rashid MA, Datta BK (2002) Phytochemical and biological studies of F. indica. Dhaka Univ J Biol Sci 11(2):183–187

Ndhlala AR, Kasiyamhuru A, Mupure C, Chitindingu K, Benhura MA, Muchuweti M (2007) Phenolic composition of Flacourtia indica, Opuntia megacantha, and Sclerocaryabirrea. Food Chem 103:82–87

Neelamma G, Durai Swamy B, Dhamodaran P (2016) Phytochemical and pharmacological overview of Cucurbita maxima and future perspective as a potential phytotherapeutic agent. Eur J Pharm Med Res 3(8):277–287

Ngbolua KT, Inkoto CL, Bongo GN, Moke LE, Lufuluabo GL, Ashande CM, Tshibangu DST, Tshilanda DD, Mpiana PT (2018) Phytochemistry and bioactivity of Annona reticulata L. (Annonaceae): a mini-review. South Asian Res J Nat Prod 1(1):1–11

Nongalleima K, Ajungla T, Singh CB (2019) Morphological phytochemical and therapeutic efficacy of Citrus macroptera Montruz. Food Biores Ethnic Foods Manipur Northeast India, pp 59–70

Nugraheri M, Rahmawati F (2013) Potential of yospirus Khaki beverage as sources of natural antioxidant. Pak J Nutr 12:620–627

Ogundele AV, Yadav A, Haldar S, Das AM (2021) Antimicrobial activities of extract, fractions, and isolated compounds from the fruits of Elaeocarpus floribundus growing in North-East India. J Herbal Med 30:100511

Oo WM, Khine M (2017) Pharmacological properties of Feronia limonia fruit pulp – a review. https://doi.org/10.5281/zenodo.2526043

Otero R, Nunez V, Barona J, Fonnegra R, Jimenez SL, Osorio RG, Saldarriaga M, Diaz A (2000) Sankebites and ethnobotany in the northwest region of Columbia: part in neutralization of the haemorrhagic effect of Bothrops atrox venom. Ethnopharmacol 73(1–2):233–241

Othman A, Ismail A, Hassan FA, Khatib A (2016) Comparative evaluation of nutritional compositions, antioxidant capacities, and phenolic compounds of red and green sessile joyweed (Alternanthera sessilis). J Funct Foods 21:263–271

Padhan A, Kumar A, Pathirana R, Sharma DP, Thakur DS, Rana VS, Kumar P, Chauhan A (2023) Potential of wild, underutilized Actinidia callosa and Actinidia strigosa from Northeast India for kiwifruit breeding. Genet Res Crop Evol:1–14

Pai V, Akhilraj TM (2021) Domestication and value addition of Artocarpus lakoocha. Eco Env Cons 28

Panara K, Joshi K, Nishteswar K (2012) A review on phytochemical and pharmacological properties of Citrus medica Linn. Int J Pharm Biol Arch 3(6):1292–1297

Panchal JB, Gaikwad RS, Dhemre JK, Chavan UD (2018) Studies on preparation and storage of jelly from dragon fruit (Hylocereus undatus). J Pharm Phytochem 7(4):2648–2655

Pancholi B, Rana AC (2020) Traditional uses, phytochemistry, and pharmacological aspects of Rubus niveus thumb plant-a review. J Phytopharmacol 9(6):438–444

Pandey Y, Bhatt SS (2016) Overview of Himalayan yellow raspberry (Rubus ellipticus Smith.): a nutraceutical plant. J Appl Nat Sci 8(1):494–499

Pandey Y, Upadhyay S, Bhatt SS (2018a) Phyto-chemical constituent of some wild edible fruits of Sikkim Himalaya. J Pharm Phytochem 7(3):1045–1047

Pandey Y, Upadhyay S, Bhatt SS, Sharma L, Manivannan S, Chanbisana C (2018b) Nutritional compositions of Baccaureasapida and Eleaocarpussikkimnesis of Sikkim Himalaya. Int J Curr Microbiol Appl Sci 7(2):2101–2106

Pandey A, Ranjan P, Ahlawat SP, Bhardwaj R, Dhariwal OP, Singh PK, Malav PK, Harish GD, Prabhu P, Agrawal A (2021) Studies on fruit morphology, nutritional and floral diversity in less-known melons (Cucumis melo L.) of India. Genet Res Crop Evol 68(1):1453–1470. https://doi.org/10.1007/s10722-020-01075-3

Patel RK, Singh A, Deka BC (2008) Soh-shang (Elaeagnus latifolia): an under-utilized fruit of north east region needs domestication. ENVIS Bull Himalayan Ecol 16(2):1–2

Patel RK, Verma VK, Deshmukh NA, Jha AK, Nath A, Patel RS (2013) Underutilized fruits of Meghalaya: an option for livelihood and nutritional security. In: Horticultural for economic prosperity and nutritional security in 21st century, pp 256–263

Paul JR, Panda AK (2021) Traditional medicinal knowledge of underutilized minor fruits as medicine in Northeast India: a preview. J Hortic 8
Payum T (2018) Chemical composition and ethnobotany of Livistona jenkinsiana Griff: an endangered thatching palm tree of Eastern Himalayas. Pharm J 10(6)
Peter KV (ed) (2008) Underutilized and underexploited horticultural crops, vol 04. New India Publishing
Peyrt B, Szymczyk T, Lesca P (1992) Mechanism of antimutagenicity of wheat sprout extracts. Mutat Res 269-272:201–215
Phukan PT, Das RP (2017) Diversity of some minor fruits of North East India. Int J Agric Environ Res 3(1):2410–2416
Prakash S, Bindu M, Prasad AS, Timilsina H, Ram B (2020) Total phenolic and flavonoids contents, antioxidant activities, phytochemical and nutritional analysis of Castanopsis indica (Indian chestnut). Sustain Dev 10(2)
Prakash O, Selvi MK, Vijayaraj P, Kudachikar VB (2021a) Lipidome, nutraceuticals, and nutritional profiling of Pyrus pashia Buch.-ham ex D. Don (Kainth) seeds oil and its antioxidant potential. Food Chem 338:128067
Prakash OM, Chauhan AS, Kudachikar VB (2021b) Traditional uses, nutrition, phytochemistry and various pharmacological properties of Indian wild pear. Int J Funct Nutr 2(4):1–13
Qamar M, Akhtar S, Ismail T, Wahid M, Barnard RT, Esatbeyoglu T, Ziora ZM (2021) The chemical composition and health-promoting effects of the Grewia species—a systematic review and meta-analysis. Nutrients 13(10):4565. https://doi.org/10.3390/nu13124565
Raghavan M, Ramjan M (2018) Burmese grape (BaccaurearamifloraLour.): a promising fruit crop for future generations. J Med Plants Stud 6(3):50–52
Rahman SS, Reja MM, Islam MR, Islam MM, Rouf SM, Rahman MH (2023) Proximate nutrient analysis of elephant apple (Dillenia indica) fruit and its hypoglycemic, and hypolipidemic potentials in alloxan-induced diabetic rats. Food Hum 1:1355–1361
Rajasekaran A, Jeyasudha V, Kalpana B, Jayakar B (1999) Preliminary phytochemical and antipyretic evaluation of Carissa carandas. Indian J Nat Prod 15(1):27–29
Ramakrishna G, Azeemoddin G, Ramayya DA, Rao ST, Devi KS, Pantulu AJ (1979) Characteristics and composition of Indian wood apple seed and oil. J Am Oil Chem Soc 56(9):870–871
Ravindran C, Kohli A, Murthy BS, Srinivasa A (2007) Fruit production in India. Chronica Horticuturae 47:21–26
Rishi Kumar RK, Shukla A (2018) A review on peach (Prunus persica): an asset of medicinal phytochemicals. Int J Res Appl Sci Eng Technol 6:I
Rouse-Miller JR, Duncan JE (2000) In vitro propagation of Artocarpus altilis (park.) Fosberg (breadfruit) from mature plant material. In Vitro Cell Dev Biol Plant 36(2):115–117
Rymbai H, Roy AR, Deshmukh NA, Jha AK, Shimray W, War GF, Ngachan SV (2016) Analysis study on potential underutilized edible fruit genetic resources of the foothills track of Eastern Himalayas. India Genet Res Crop Evol 63(1):125–139
Saah SA, Adu-Poku D (2020) Phytochemical, proximate, and vitamin C content in Morindacitrifolia (Noni). J Trop Pharm Chem 5(3):274. https://doi.org/10.25026/jtpc.v5i3.274
Saeed N, Khan MR, Shabbir M (2012) Antioxidant activity, total phenolic and total flavonoid contents of whole plant extracts of Torilisleptophylla L. BMC Complement Altern Med 12:221–228
Saikia D, Kesavan R, Inbaraj BS, Dikkala PK, Nayak PK, Sridhar K (2023) Bioactive compounds and health-promoting properties of Elephant Apple (Dillenia indica L.): a comprehensive review. Food Secur 12(16):2993
Saklani S, Chandra S (2012a) In vitro antimicrobial activity, nutritional profile of medicinal plant of Garhwal, Himalaya. Int J Pharm Sci Res 3:268–272
Saklani S, Chandra S (2012b) In vitro antimicrobial activity, nutritional profile and phytochemical screening of wild edible fruit of Garhwal Himalaya (Ficus auriculata). Int J Pharm Sci Rev Res 12(2):Article-011

Saleem M, Wongsrisujarit N, Boonyarattanakalin S (2015) Removal of nickel (II) ion by adsorption on coconut copra meal biosorbent. Desalination Water Treat. https://doi.org/10.1080/19443994.2015.1005155

Saleha A, Kamath LM, Peter PI (2007) Indian Noni (Morindacitrifolia L.) – a unique fruit crop with high medicinal value. In: Peter KV (ed) Underutilized and underexploited horticultural crops, vol 1. New India Publishing Agency, pp 301–312

Salehi B, Sharifi-Rad J, Capanoglu E, Adrar N, Catalkaya G, Shaheen S, Cho WC (2019) Cucurbita plants: from farm to industry. Appl Sci 9(16):3387

Saltveit ME (2011) Melon (Cucumis melo L.). In: Postharvest biology and technology of tropical and subtropical fruits. Woodhead Publishing, pp 31–45

Salunkhe DK, Desai BB (1984) Postharvest biotechnology of vegetables, vol 2. CRC Press, Boca Raton, pp 70–75

Saraswathy SD (2023) An overview on the ethnopharmacological, nutritional, and phytochemical perspectives of Elaeocarpus floribundus Blume. Curr Pharmacol Rep 9(5):377–389

Sarkar CG, Zahan R, Alam BM, Islam S, Mosaddik AM, Haque KME (2011) Antibacterial activity of Flacourtiajangomas and Flacourtiasepiaria. Int J Pharm Life Sci 2(7):878–883

Sarkar PK, Bishnoi SK, Shinde R (2018) Under-exploited fruits of the forests: eastern Indian plateau is a potential hub. Agrobios (India), Jodhpur

Seal T (2011) Determination of nutritive value, mineral contents and antioxidant activity of some wild edible plants from Meghalaya state. India Asian J Appl Sci 4(3):238–246

Seal T (2020) Effect of boiling and microwave cooking on nutritional, anti-nutritional, and toxicity of wild edible plants of north-eastern region in India. Int J Pharm Pharm Sci 12(11):61–70

Segura-Campos MR, Ruiz Ruiz JC, Chel-Guerrero LA, Betancur D, Betancur D (2016) Capsicum chinense: composition and functional properties. In: Functional properties of traditional foods, pp. 20–30. DOI: https://doi.org/10.1007/978-1-4899-7662-8_20

Shamra S (2022) Value addition to underutilized wild Himalayan pear (Pyrus Pashia): development, composition, and sensory characteristics of jam. SciOpen Preprints

Shankar K, Haokip SW, Ramjan M, Sheikh KA, Ansari MT, Angami T, Kumar V (2020) Genetic diversity of fruits in north east region of India. J Pharmacogn Phytochem 9(1):207–209

Shariful I, Sabiha F, Koly S, Ferdowsy K (2018) A review on phytochemical and pharmacological potentials of Antidesmabunius. J Anal Pharm Res 7(5):602–604

Sharma US, Kumar A (2011) In vitro antioxidant activity of Rubus ellipticus fruits. J Adv Pharm Technol Res 2(1):47

Sharma N, Radha KM, Zhang B, Kumari N, Singh D, Chandran D, Sarkar T, Dhumal S, Sheri V, Dey A, Rajalingam S, Viswanathan S, Mohankumar P, Vishvanathan M, Sathyaseelan SK, Lorenzo JM (2022) Aegle marmelos (L.) Correa: an underutilized fruit with high nutraceutical values: a review. Int J Mol Sci 23(18):10889. https://doi.org/10.3390/ijms231810889

Shende KM, Singh NI, Negi PS (2016) Phytochemical characterization and biological activities of Docynia indica (Wall) fruit extracts. J Mol Genet Med 10(204):1–5

Siddiqui SZ, Ali S, Rubab K, Abbasi MA, Ajaib M, Rasool ZG (2015) Pyrus pashia: a persuasive source of natural antioxidants. Pak J Pharm Sci 28(5)

Siddiqui MW, Rahman MS, Wani AA (eds) (2018) Innovative packaging of fruits and vegetables: strategies for safety and quality maintenance. CRC Press

Singh M, Pandey A (2021) Phytochemicals, mineral contents and antioxidant property of wild edible fruits of Sikkim Himalaya

Singh G, Parle A (2015) Unique pandanus – flavour, food and medicine. J Pharmacogn Phytochem 5(3):8–14

Singh KR, Sarkar A (2022) A review: the significance and status of native fruits under NE region

Singh TM, Sharma A (2021) Constraints faced during the production and marketing of cabbage and potato crops: a comparative study of Manipur and Nagaland states of India. Plant Arch 21(2)

Singh RK, Srivastava RC, Mukherjee TK (2010) Toko-Patta (Livistona jenkinsiana Griff): Adi community and conservation of culturally important endangered tree species in eastern Himalaya

Sinha J, Purwar S, Chuhan SK, Rai G (2015) Nutritional and medicinal potential of Grewia subinaequalis DC. (syn. G. asiatica.) (Phalsa). J Med Plant Res. https://doi.org/10.5897/JMPR2015.5724

Sircar B, Mandal S (2017) Indian olive, Elaeocarpus floribundus fruit: perspective to the antioxidative capacity and antibacterial activity. EC Microbiol 12(6):273–282

Smith NJ, Williams JT, Pucknett DL, Talbot JP (1992) Tropical forests and their crops. Cornell University Press, Ithaca, p 568

Sripathi SK, Sruthi RD (2023) A comprehensive review of the medicinal plant Passiflora foetida Linn. Int J Pharm Sci Rev Res 14(6)

Srivastava A, Bishnoi SK, Sarkar PK (2017a) Value addition in minor fruits of eastern India: an opportunity to generate rural employment. In: Fruits for livelihood: production technology and management practices. Agrobios (India), Jodhpur, pp 395–417

Srivastava A, Sarkar PK, Bishnoi SK (2017b) Value addition in under-exploited fruits of Karonda (Carissa carandus L.): an earning opportunity for rural communities in India. Rashtriya Krishi 12(2):161–163

Srivastava R, Mishra N, Tripathi S, Mishra N (2020) Effect of solvents on antioxidant activities of Feronia limonia fruit. Int J Pharm Sci Res 11(7):3385–3391

Sukkhaenga S, Promdanga S, Doung-ngern U (2018) Fruit characters and physico-chemical properties of roselle (Hibiscus sabdariffa L.) in Thailand—a screening of 13 new genotypes. J Appl Res Med Aromatic Plants

Sundriyal M, Sundriyal RC (2003) Underutilized edible plants of the Sikkim Himalaya: need for domestication. Curr Sci:731–736

Sundriyal M, Sundriyal RC (2004) Wild edible plants of the Sikkim Himalaya: nutritive values of selected species. Econ Bot 58(2):286–299

Tan SP, Tan ENY, Lim QY, Nafiah MA (2020) Phyllanthus acidus (L.) Skeels: a review of its traditional uses, phytochemistry, and pharmacological properties. J Ethnopharmacol 253:112610

Thitilertdecha N, Teerawutgulrag A, Rakariyatham N (2008) Antioxidant and antibacterial activities of Nephelium lappaceum L. extracts. LWT Food Sci Technol 41:2029–2035

Thorat DD, Sonwane RS, Naik AP (2024) A review of the potential of incorporating local wild fruits into value-added dairy products. IJRAR Int J Res Anal Rev 11:203–211

Tindall HD (1994) Chapter 5: Sapindaceous fruits: botany and horticulture. In: Janick J (ed) Horticultural reviews, vol 16. Wiley, New York, pp 143–195

Tolentino F, de Araújo PA, de Souza ME, Petreanu M, de Andrade SF, Niero R, Maistro EL (2015) In vivo evaluation of the genetic toxicity of Rubus niveus Thunb. (Rosaceae) extract and initial screening of its potential chemoprevention against doxorubicin-induced DNA damage. J Ethnopharmacol 164:89–95

Tripathi PC (2021) Medicinal and therapeutic properties of minor fruits: a review. Int J Minor Fruits Med Aromat Plants 7:1–28

Tripathi PC, Karumakaran G, Sankar V, Senthilkumar R (2013) Karonda: a potential fresh fruit of future. Central Horticultural Experiment Station, Indian Institute of Horticultural Research, Chettalli, Kodagu

Tsering J, Gogoi BJ, Tag H (2012) Ethnobotany and phytochemical analysis of Pyrus pashia leaves. Int J Pharm Sci Res 3:2721

USDA-FAS (2004) Foreign Agricultural Service. U.S. Department of Agriculture. http://www.ars.usda.gov/Services/docs.htm?docid=15866

Veberic R, Jurhar M, Mikulic-Petkovsek F, Stampar F, Schmitzer V (2010) Comparative study of primary and secondary metabolites in 11 cultivars of persimmon fruit (Diospyros kaki L.). Food Chem 119:477–483

Vijay MK, Singh N (2024) Nutritional composition, photochemistry, and pharmacognostic activities of Flacourtia indica (Burm. F.) Merr.: an important wild edible fruit species of Central India. Int J Adv Biochem Res 8:765–771

Vino SA, Sinija VR (2016) Underutilized fruits in India. Indian Food Ind Mag 35:45–46

Vishwakarma VK, Gupta JK, Gupta PK, Upadhyay PK (2017) Pharmacological importance of Cucumis melo L.: an overview. Asian J Pharm Clin Res 10:8–12

Wall MM (2006) Ascorbic acid and mineral composition of longan (Dimocarpus longan), lychee (Litchi chinensis), and rambutan (Nephelium lappaceum) cultivars grown in Hawaii. J Food Compos Anal 19:655–663
Wills RB, Lee TH, Graham D, McGlasson WB, Hall E (1981) Postharvest: an introduction to the physiology and handling of fruits and vegetables. AVI Publishing Company, Oxford
Wong KC, Wong SN (1995) Volatile constituents of Averrhoa bilimbi L. fruit. J Essent Oil Res 7:691–693
Wu H, Li M, Wang D, Liu H, Xu X (2019) The complete chloroplast genome sequence of Actinidia callosa var. henryi. Mitochondrial DNA Part B 4:652–653
Yan SW, Ramasamy R, Alitheen NBM, Rahmat A (2013) A comparative assessment of nutritional composition, total phenolic, total flavonoid, antioxidant capacity, and antioxidant vitamins of two types of Malaysian underutilized fruits (Averrhoa bilimbi and Averrhoa carambola). Int J Food Prop 16:1231–1244
Yassine FA (2014) State of Annona cultivation in Lebanon. Master's Thesis, Lebanese University
Yatung T, Bhargav V, Shivashankara KS, Geetha GA, Lokesha AN (2023) Biochemical profiling of 'Toko' (Livistona jenkinsiana Griff.): an endangered underutilized fruit of North-East India. Biochem Syst Ecol 107:104610

MIX
Papier aus verantwortungsvollen Quellen
Paper from responsible sources
FSC® C105338

If you have any concerns about our products,
you can contact us on
ProductSafety@springernature.com

In case Publisher is established outside the EU,
the EU authorized representative is:
Springer Nature Customer Service Center GmbH
Europaplatz 3, 69115 Heidelberg, Germany

Printed by Libri Plureos GmbH
in Hamburg, Germany